Electronic and Electrical Systems

Fundamentals of Service

PUBLISHER

Fundamentals of Service (FOS) is a series of manuals created by Deere & Company. Each book in the series is conceived, researched, outlined, edited, and published by Deere & Company. Authors are selected on the basis of their expertise in each book's subject matter.

PUBLISHER: DEERE & COMPANY SERVICE PUBLICATIONS, Dept. FOS/FMO John Deere Road, Moline, Illinois 61265-8098; Dept. Manager: Alton E. Miller.

SERVICE PUBLICATIONS EDITORIAL STAFF

Managing Editor: Louis R. Hathaway
Editor: John E. Kuhar
Publisher: Lori J. Lees
Art Director: Bernard F. DeKazel
Promotions: Cindy S. Calloway

TO THE READER

THE SCOPE OF THIS BOOK

This text and its supporting materials are intended only as an educational media and should not be considered a substitute for service manuals for specific machines. Always refer to the pertinent service manual for specific repair procedures and safety precautions.

APPLICATION OF ELECTRICITY AND ELECTRONICS IN THIS MANUAL

"Electricity" and "Electronics" are very broad fields. But in this manual, the prime interest is in electricity and electronics as they are commonly used in mobile machines.

Persons not familiar with electronic components/electrical systems should start with Chapter 1 and study the chapters in sequence. Experienced persons can find what they need on the "Contents" page.

Answers to "Test Yourself" questions, at the end of each chapter, appear at the end of this manual on pages 22–25 following the index.

ACKNOWLEDGEMENTS

John Deere gratefully acknowledges help from the following groups: Delco-Remy and AC spark Plug, Divisions of General Motors Corporation; ITT Cannon Electric; Allen Electric and Equipment Company; AMP Inc.; Champion Spark Plug Company; Cole-Hersee Company; Marquette Manufacturing Company; Motorola Automotive Products Inc.; Prestolite Company; Mort Schultz-Plantation, FL; Sun Electric Corporation; Wico Corporation; and Triplett Corporation. **Thanks also to a host of John Deere people for their valuable suggestions and comments.**

AUTHOR: *Julius DeFauw* is a John Deere retiree with 29-1/2 years of experience in service publications, field service, and service training. One of his last active primary responsibilities was as a factory service supervisor.

COPY EDITOR: *Bruce Kinnaird* is a San Francisco-based writer and editor who has worked on numerous magazines, including Satellite Orbit and PC World. He was the editor of the first two editions of The World Satellite Almanac, a landmark guide to global telecommunications satellites.

CONSULTING EDITORS: *Donald V. "Deve" Detloff*, Instructor with the John Deere Training Center, has 20 years of experience in designing, developing and troubleshooting electrical and electronic systems. He has been extensively involved in field training and training material development.

Lon R. Shell, Ed.D., professor in Agricultural Systems Management and Agricultural Mechanics, Agriculture Department, Southwest Texas State Unviersity, San Marcos, Texas. Dr. Shell has authored numerous papers and instructional materials in agricultural mechanics and engineering in his 32 years working in agriculture education and research.

FOR MORE INFORMATION

This text is part of a series of texts and visuals on agricultural and industrial machinery entitled Fundamentals of Service (FOS). For information, request a free Catalog of Teaching Materials. Send your request to: John Deere Service Publications, Dept. FOS/FMO, John Deere Road, Moline, Illinois, 61265-8098.

Other manuals in the Fundamentals of Service (FOS) series:

- **Engines**
- **Hydraulics**
- **Power Trains**
- **Shop Tools**
- **Welding**
- **Hoses, Tubing and Connectors**
- **Identification of Parts Failures**
- **Air Conditioning**
- **Bearings and Seals**
- **Belts and Chains**
- **Fuels, Lubricants, and Coolants**
- **Mowers and Sprayers**
- **Tires and Tracks**
- **Fiber Glass**
- **Fasteners**
- **A Glossary of Technical Terms**

Each manual is backed up by a set of 35 mm color slides for classroom use. Transparency masters, instructor's guides, and student guides are also available for the first three subjects and this one.

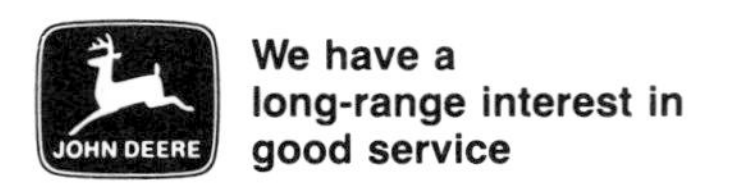

We have a long-range interest in good service

 ISBN 0-86691-182-0

FOS—20 Litho in U.S.A.

CONTENTS

FOS—20 Litho in U.S.A.

 Litho in U.S.A.

ELECTRICITY—How It Works / CHAPTER 1

INTRODUCTION

In this chapter we will cover the basics of electronics and electricity:

- **Electron Theory**
- **Putting Electrons to Work**
- **Electronics vs. Electricity**
- **Introduction to Current, Voltage, and Resistance**
- **Magnetism**
- **Electromagnetism**
- **Electromagnetic Induction**

Let's discuss each of these subjects in detail.

ELECTRON THEORY

Because all matter contains electrons, all matter has an essential ingredient called electricity.

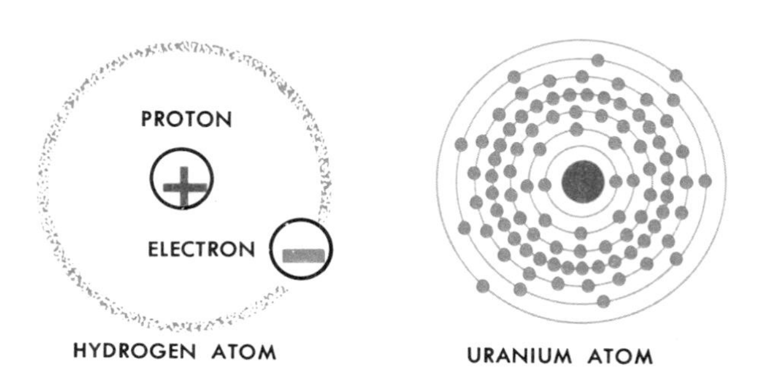

Fig. 1—All Matter is Made Up of Atoms

To understand this, let's look at the smallest unit of all matter—the atom (Fig. 1). All **atoms** have particles called **electrons** in orbit around a core of **protons.**

The simplest element is hydrogen. As shown, its atom has a single electron in orbit around a core of one proton.

One of the most complex elements is uranium. It has 92 electrons in orbit around a core of 92 protons.

Each element has its own atomic structure.

But each atom of an element has an *equal number of protons and electrons.*

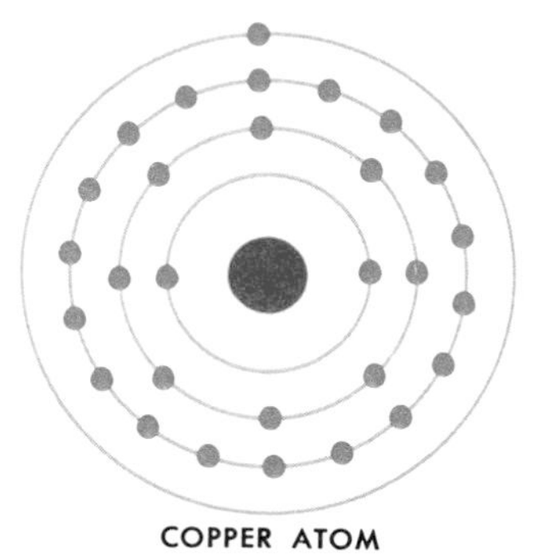

Fig. 2—Structure of a Copper Atom

The element copper is widely used in electrical systems because it is a good conductor of electricity.

The reasons for this can be seen in Fig. 2. The copper atom contains 29 protons and 29 orbiting electrons. The electrons are distributed in four separate orbits or rings. Those electrons beyond the second orbit are more or less free to move from one orbit to another, but still keeping the same amount of electrons per orbit.

But notice that the outer ring has only one electron. This is the secret of a good conductor of electricity.

Elements whose atoms have less than four electrons in their outer rings are generally good **conductors.**

Elements whose atoms have more than four electrons in their outer rings are poor conductors and are called **insulators.**

Litho in U.S.A.

The fewer electrons in the outer ring of conductors are more easily dislodged from their orbits by a low voltage to create a flow of current from atom to atom.

In summary:

- **Atoms have electrons in orbit around a core of protons.**
- **Each atom contains an equal number of electrons and protons.**
- **The electrons occupy shells or rings in which they orbit around the core.**
- **Atoms which have less than four electrons in their outer rings are good conductors of electricity, as with copper.**
- **Atoms which have more than four electrons in their outer rings are good insulators of electricity, as with plastic or rubber.**

PUTTING ELECTRONS TO WORK

We have seen that atoms contain particles called protons and electrons.

These particles have a potential force:

- **Protons = positive (+) charges**
- **Electrons = negative (–) charges**

The protons in the core attract the electrons and hold them in orbit. Since the positive charge of the protons is equal to the negative charge of the electrons, the atom is said to be *neutral.*

However, this neutral state can be altered. If the orbiting electrons can be forced away from the atom, the atom becomes positive (+) charged and the collection of the orbiting electrons taken away become negative (–) charged. Thus:

Positive charged atoms = too few electrons

Negative charged atoms = too many electrons

The atom does not give up its orbiting electrons except by force. This force must be used to take the electrons from their position around the neutral atom and induce them into another atom's orbit.

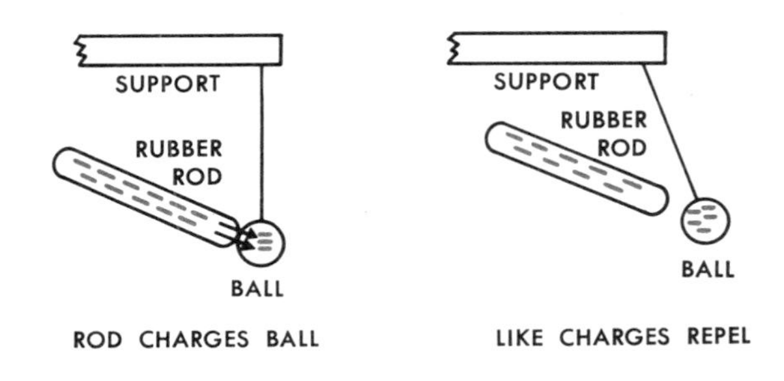

Fig. 3—Like Charges Repel

Let's show this transfer of electrons in an experiment.

When a rubber rod is rubbed with wool, orbiting electrons from the wool are removed and collected on the rod. The wool now has too few electrons and becomes positive (+) charged and the rod has too many electrons and becomes negative (–) charged.

Now let's touch the negative charged rubber rod to a hanging pith ball and remove the rod (Fig. 3). What happens is that some of the extra electrons on the rod move into the orbit of the atom of the ball. The ball then becomes negative (–) charged while the rod also retains part of its negative (–) charge.

When the rod is moved toward the ball again, the ball will swing away from the rod as shown.

In other words, *like charges repel.*

In the experiment, both charges were negative. If both charges were positive (+) the same thing would occur.

 Litho in U.S.A.

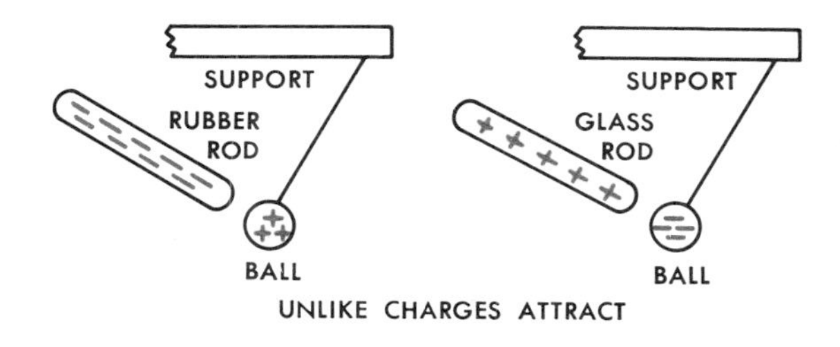

Fig. 4—Unlike Charges Attract

What happens if we move a positive (+) charged rod toward a negative (–) charged ball?

When a glass rod is rubbed with silk, the silk becomes negative (–) charged and the glass rod becomes positive (+) charged. Fig. 4 shows that a negative charged hanging pith ball will be attracted to the positive charged glass rod. (In the same way a positive charged pith ball will be attracted to a negative charged rubber rod.)

In other words, *unlike charges attract.*

The rubbing force that causes this electron movement is called *static electricity.*

In summary:

- **Electrons can be made to leave their atoms in some materials**
- **A force such as friction is needed to cause electrons to leave their atoms.**
- **Like charges repel and unlike charges attract.**

DYNAMIC ELECTRICITY

Now let's look at another type of electron flow. Fig. 5 shows what happens with a conductor such as copper wire when it has a negative charge on one end and a positive charge at the other end. This can be accomplished by connecting the ends of the copper wire to the positive and negative terminals of a dry cell battery.

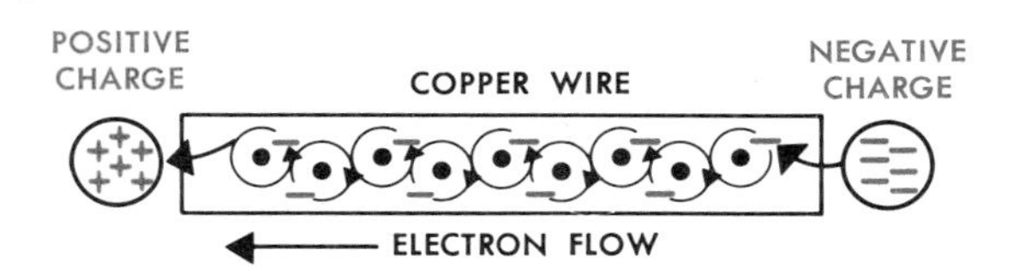

Fig. 5—Flow of Electrons in a Conductor

What happens is that an electron (–) of the copper atom is forced out of its orbit and attracted to the positive (+) end of the battery. This atom is now positive (+) charged because it now has too few electrons. It in turn attracts an electron from its neighbor. The neighbor in turn receives an electron from the next atom and so on until the last copper atom receives an electron from the negative (–) end of the battery.

The net result of this chain reaction movement of electrons is that the electrons move through the wire from the negative end of the battery to the positive end of the battery.

This flow or current of electrons will continue as long as the positive and negative charges from the battery are maintained at each end of the wire (unlike charges attracting each other).

The use of a battery to force electrons to flow through a conductor is known as *dynamic electricity.*

ELECTRONICS VS. ELECTRICITY

From our previous study of the flow of electrons we would conclude: **Electricity is the flow of electrons from atom to atom in a conductor.**

What then is electronics?

Litho in U.S.A.

Electronics is the control of electrons and the study of their behavior and effects.

This control of electrons is accomplished by devices that resist, carry, control, select, steer, switch, store, manipulate, and exploit the electrons.

For a better understanding let's review the previous experiment of electron flow (electricity). The pith ball would not have reacted to the electrons in the rubber and glass rods unless the rods were rubbed with wool and silk. Thus, the rubber and glass rods, along with the respective rubbing of the wool and silk, constitute electronic devices that control electron flow.

INTERESTING SIDELIGHTS

Fig 6—Lightning During an Electrical Storm

During warm weather clouds acquire a static charge that moves from cloud to cloud and cloud to earth. This static charge between clouds and earth may grow to more than one million volts before it is discharged as lightning.

Lightning with its accompanying thunder is the result of a great number of electrons forcing their way through the atmosphere, which is normally an insulator and does not allow electrons to flow through it. The release of this much energy is what can cause vast damage during a bad electrical storm.

INTRODUCTION TO CURRENT, VOLTAGE, AND RESISTANCE

When electricity goes to work, we are dealing with three basic factors:

- **Current**
- **Voltage**
- **Resistance**

These terms are basic to the understanding of electricity as we'll see now.

CURRENT

The flow of electrons through a conductor is called a current and is measured in amperes. It is represented by the letters A or I.

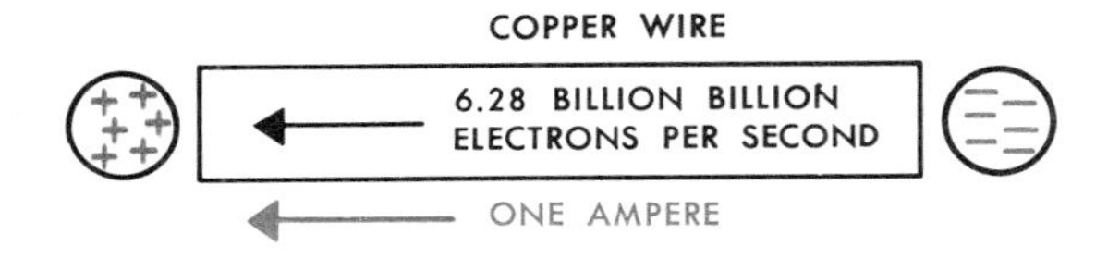

Fig. 7—How Current Is Measured

One ampere is an electric current of 6.28 billion *billion* electrons passing a certain point in the conductor in one second (Fig. 7).

Thus, current is the rate of electron flow and is measured in amperes or electrons per second. You can compare this with hydraulics where the flow of oil in a pipe is measured in gallons per minute.

There are two ways to describe current flow through a conductor (Fig. 8). In the electron flow theory, of which we are now familiar, the electron flow or current through a conductor is from a negative (–) power source to a positive (+) power source.

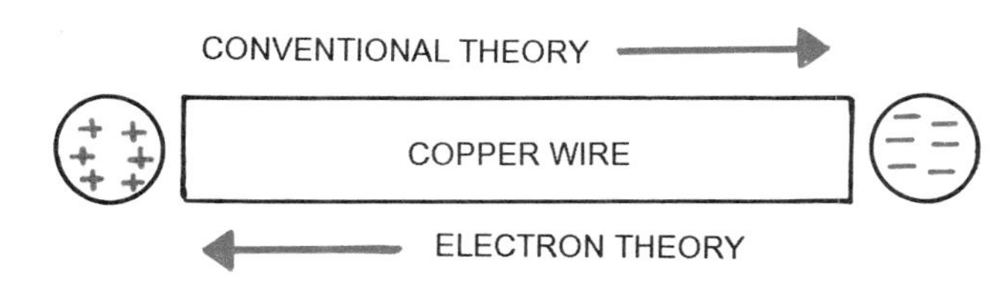

Fig. 8— Two Theories of Current Flow through a Conductor

The second theory is called the *conventional flow pressure theory,* where the current flow through a conductor is from a positive (+) power source to a negative (–) power source.

In the sixteenth century, long before the electron flow theory was discovered and when the basic laws of electricity were being developed it was believed that the current flow through a conductor was due to positive carriers. Even today, the majority of educational and industrial institutions teach the conventional flow theory.

Either theory can be used, but we will use the more popular conventional flow pressure theory (+ to –) in the remainder of this manual.

VOLTAGE

Voltage is the force that causes a flow of current in a conductor. Voltage is measured in **volts.** It is represented by the letters V or E (electromotive force). Voltage depends on the difference in the charges at each end of the conductor.

Voltage can be generated by a storage battery using chemicals, or by a generator using mechanical means. Voltage is a *potential* force and can exist even when there is no current flow in a circuit.

A storage battery, for example, may have a potential of 12 volts between its (+) and (–) terminal posts, and this potential exists even though no current-consuming devices are connected to the posts.

Thus, voltage can exist without current, but current cannot exist without the "push" of voltage.

Fig. 9—Voltage

Voltage is produced between two points when a positive charge exists at one point and a negative charge exists at the other point (Fig. 9).

The greater the charges at each point, the greater the voltage.

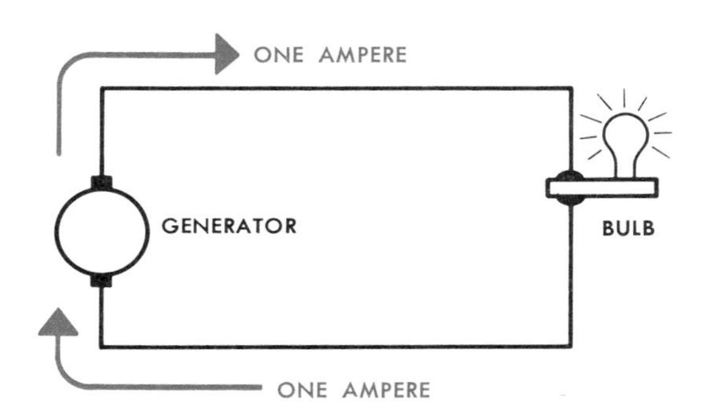

Fig. 10—The Generator as an Electron Pump

Look at a battery or generator as an electron pump (Fig. 10). The generator, for example, will supply a continuous flow of electrons (current) through the light bulb connected to it. The movement of electrons is continuous: In the generator, if one ampere of current is leaving, one ampere is entering to keep a constant flow of current.

RESISTANCE

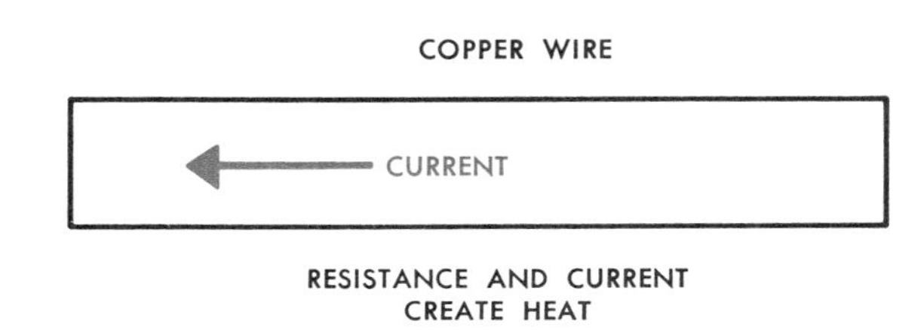

Fig. 11—Resistance to the Flow of Current in a Conductor

Resistance is the *opposition* of electron flow in a conductor. It is represented by the letter R and is measured in **ohms.** One ohm is the resistance that will allow one ampere to flow when the potential is one volt.

All conductors offer some **resistance** to the flow of current. Resistance is caused by:

1. Each atom resisting the removal of an electron due to attraction toward the core.

2. Collisions of countless electrons and atoms as the electrons move through the conductor.

The collisions create resistance and cause heat in the conductor.

This is an expression of Ohm's Law. (See later in chapter 2.) Resistance is often shown by the Greek symbol omega (Ω); thus 5Ω means five ohms.

MAGNETISM

Another form of force that causes electron flow or current is magnetism.

The effects of magnetism were first observed when fragments of iron ore called lodestone, found in nature, were seen to attract other pieces of iron (Fig. 12).

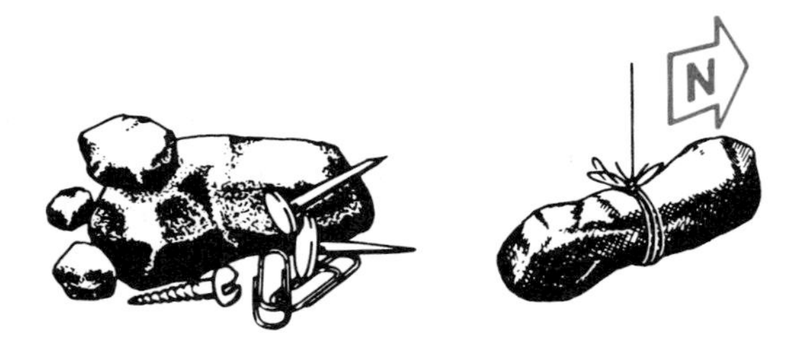

Fig. 12—Magnetism

It was further discovered that a long piece of this iron ore suspended in air would align itself so that one end always pointed toward the North Pole of the earth. This end of the iron bar was called the north pole, or **N pole,** and the other end the south or **S pole.** Such a piece of iron ore was called a **bar magnet.** This principle became the basis for the compass, which has been used as an aid in navigation for over 1000 years.

MAGNETIC FIELDS

Further study of the bar magnet revealed that an attractive force was exerted upon bits of iron or iron filings even though the iron filings were some distance away from the bar magnet. From this it was clear that a force existed in the space close to the bar magnet. This space around the magnet in which iron filings are attracted is called the field of force or **magnetic field.**

The magnetic field is described as invisible lines of force which come out of the N pole and enter the S pole.

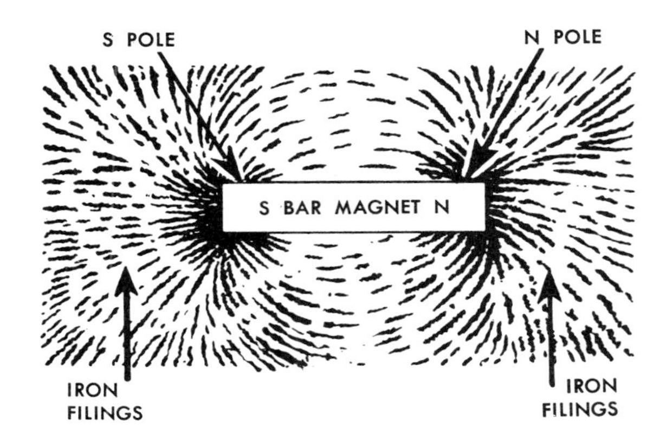

Fig. 13—Magnetic Field of a Bar Magnet

The theory of magnetic lines of force can be dramatically shown by sprinkling iron filings on a piece of paper resting on top of a bar magnet. When the paper is lightly tapped by hand, the iron filings line up to form a clear pattern around the bar magnet (Fig. 13).

The pattern shows that the lines of force are heavily concentrated at the N and S poles of the magnet, and then spread out into the surrounding air between the poles. The concentration or number of lines at each pole is equal, and the attractive force on the iron filings at each pole is equal. Notice that the force of attraction on bits of metal is greatest where the concentration of magnetic lines is greatest. For a bar magnet, this area is next to the two poles.

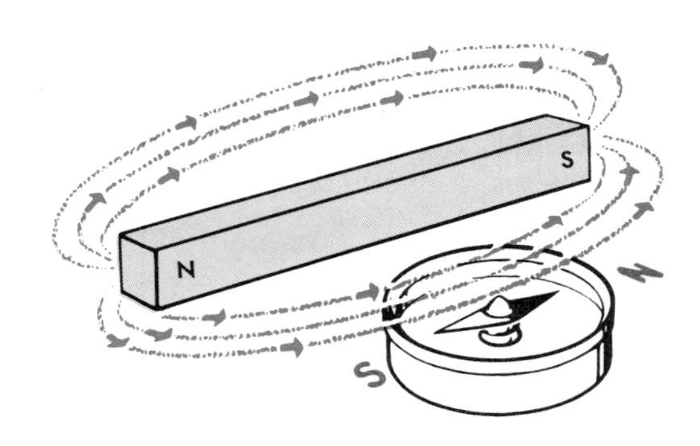

Fig. 14—Magnetic Lines of Force Come Out of N Pole and Enter S Pole

 Litho in U.S.A.

We have said that the lines of force always leave the N pole and enter the S pole of a magnet. When a small compass needle, which is a small bar magnet, is located in the magnetic field of a strong bar magnet, the compass needle will align itself so it is parallel with the lines of force of the bar magnet (Fig. 14).

This alignment takes place because the strong magnetic lines from the bar magnet must enter the S pole and leave the N pole of the compass needle.

We can also see that the unlike poles of the two magnets are attracted towards each other.

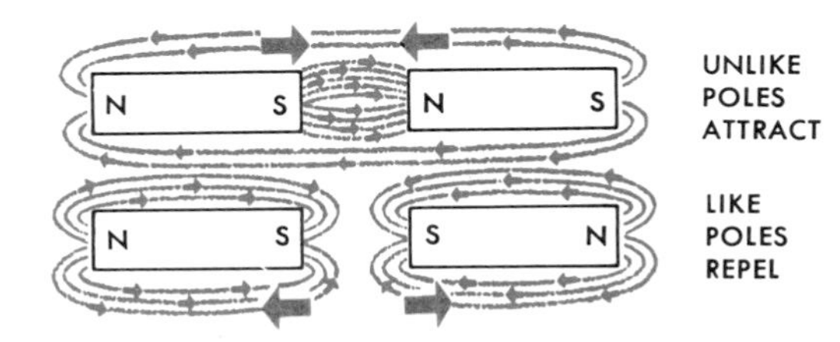

Fig. 15—Magnetic Forces between Poles of Bar Magnets

To demonstrate further the force of attraction between the unlike poles of two magnets, a force of attraction is seen to exist between two bar magnets lying end to end with an N and S pole facing each other (Fig. 15). The force of attraction increases as the two magnets are moved closer together.

If, on the other hand, the magnets are aligned so the N poles or the S poles face each other, a force of repulsion is seen to exist between the two magnets, and this repulsion increases as the two magnets are moved closer together.

From these experiments, a fundamental law of magnetism can be stated:

Unlike poles attract each other and like poles repel each other.

THEORIES OF MAGNETISM

Exactly what magnetism is, and how it exerts a field of force, can best be explained by either one of two theories.

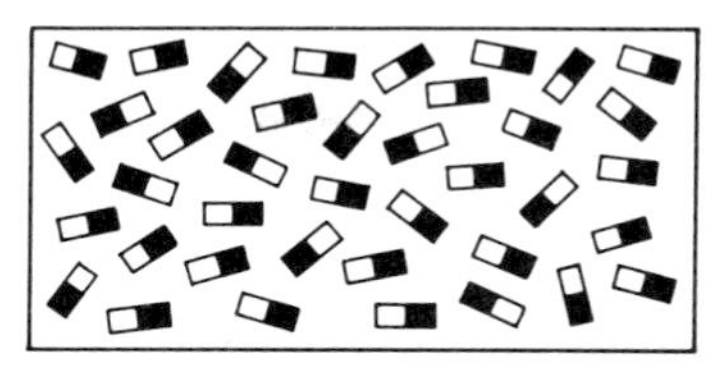

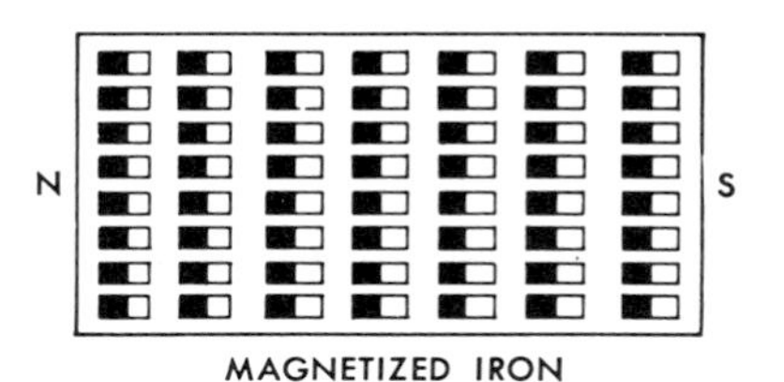

Fig. 16—First Theory of Magnetism—Particles Are Aligned

Theory No. 1 states that a magnet is made up of a very large number of small magnetized *particles.* When a bar of iron is not magnetized, the small magnetic particles are arranged in a random manner (Fig. 16). But when the bar of iron becomes a magnet, the magnetic particles are aligned so that their individual effects add together to form a strong magnet.

Theory No. 2 about magnetism concerns the *electron.* The electron has a circle of force around it, and when the electron orbits are aligned in a bar of iron so that the circles of force add together, the bar of iron is magnetized.

While iron is one of the better known magnetic materials, remember that some materials are non-magnetic since they never exhibit any of the properties of magnetism. Some of the non-magnetic materials are wood, paper, glass, copper, and zinc.

HOW MAGNETS ARE MADE

An ordinary iron bar can be converted into a magnet in a number of different ways.

One method is to stroke the iron with another piece of iron that has already been magnetized. The effect of inducing magnetism into the iron bar is called **magnetic induction.**

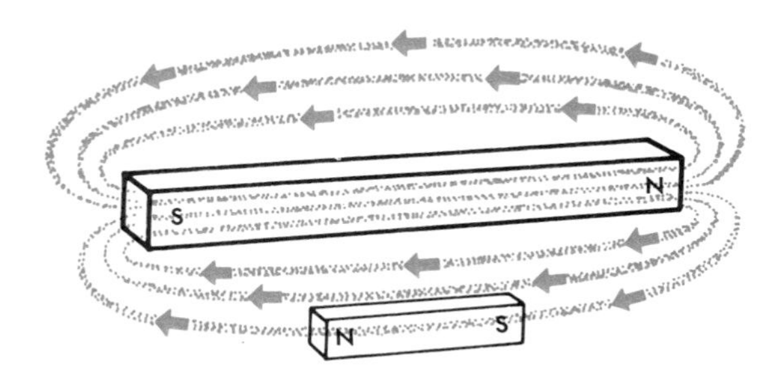

Fig. 17—Magnetic Induction of an Iron Bar

Another method of magnetic induction is simply to place an iron bar in a strong magnetic field (Fig. 17). The lines of force in the field passing through the iron bar will cause the bar to become a magnet as long as it is located in the field. If the bar is withdrawn from the field of force, and if its composition is such that it retains some of its induced magnetism, it is then said to be permanently magnetized and is called a **permanent magnet.**

Most permanent magnets are made of hard metals composed of alloys since soft metals will not retain much of their magnetism. Some of the more common alloys are nickel-iron and aluminum-nickel-cobalt.

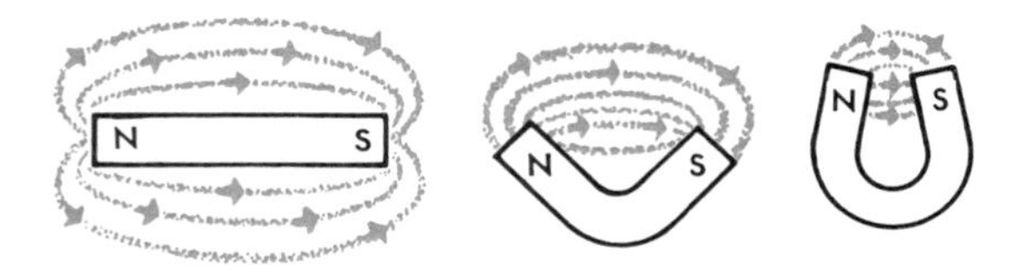

Fig. 18—Forming a Horseshoe Magnet

Permanent magnets are found in many shapes including the horseshoe magnet which concentrates the lines of force at the two poles in a small area (Fig. 18).

The most effective way of inducing a high level of magnetism in a material to form a permanent magnet is by the principles of electromagnetic induction. This principle is covered in a section which follows.

SUMMARY: MAGNETISM

In summary:

- **Every magnet has an N and S pole, and a field of force surrounding it.**
- **Magnetic materials are acted upon when located in a field of force.**
- **Unlike poles attract and like poles repel.**
- **An unmagnetized piece of iron can become a magnet through induction.**

ELECTROMAGNETISM

It was not until the year 1820 that *the relation between electricity and magnetism* was discovered. Before this time it was generally believed that magnetism existed only in the lodestone or iron ore found in nature, and there was no relationship at all between electricity and magnetism.

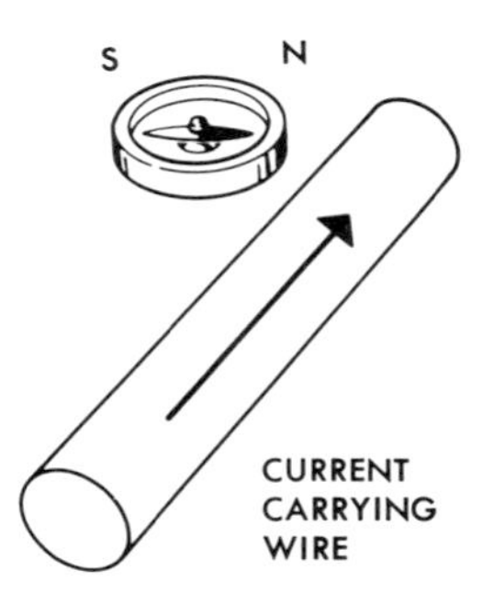

Fig. 19—Electric Current Creates Its Own Magnetic Field

An experiment with a compass and a wire carrying current revealed the connection between electricity and magnetism. When the compass was held over the wire, the needle turned so it was crosswise of the wire (Fig. 19). Since the only thing known that would attract a compass needle was magnetism, it was obvious that *the current in the wire created a magnetic field around the wire.*

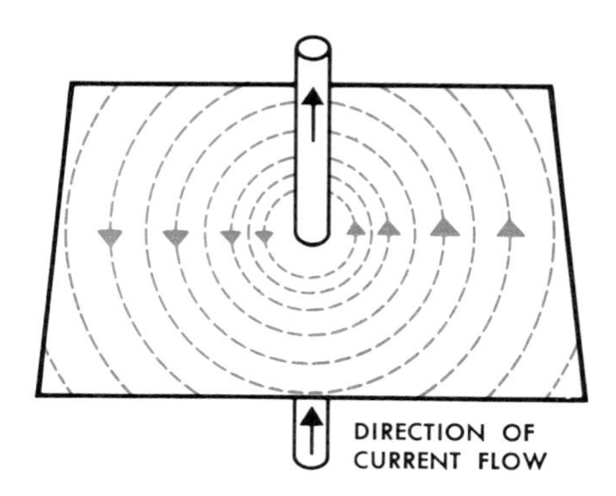

Fig. 20—Shape of Magnetic Field Around Wire Carrying a Current

The nature of the magnetic field around the wire is revealed when the current-carrying wire is run.

through a piece of cardboard, and iron filings are sprinkled on the cardboard. The iron filings align themselves to show a clear pattern of concentric circles around the wire (Fig. 20). The circles are more concentrated near the wire than farther away. Although the iron filings on the cardboard show only the pattern in one plane, remember that the concentric circles extend the entire length of the current-carrying wire.

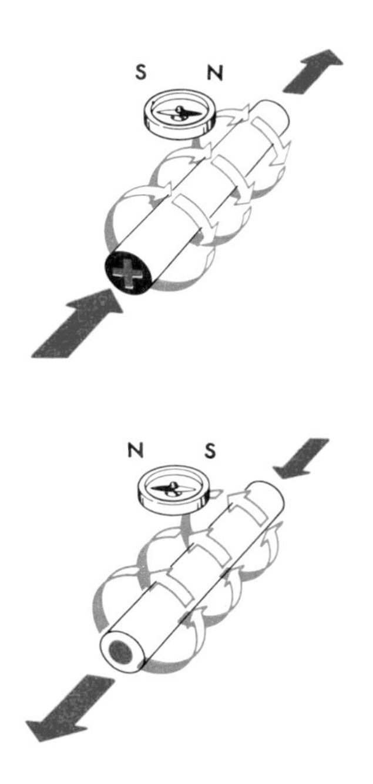

Fig. 21—Magnetic Lines Change Direction When Current Is Reversed

In Fig. 21, when current is flowing in a wire in the direction indicated by the cross, the N pole of a compass needle will always point in one direction. However, when current is flowing in the wire in the opposite direction as indicated by the dot, the north pole of the compass needle reverses and points in the opposite direction.

Since the needle always has a tendency to align itself so magnetic lines, or flux lines, enter its S pole and leave its N pole, we can conclude:

Magnetic lines have direction, and change direction when the current flow changes in the wire from one direction to another.

The Right Hand Rule for a Straight Conductor can be used to find the direction of the lines of force around the wire.

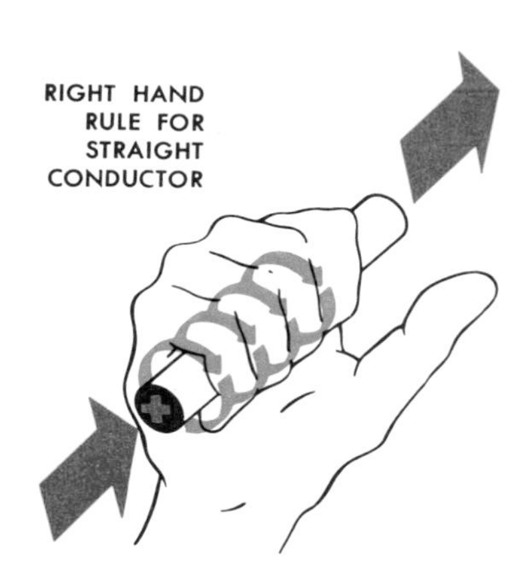

Fig. 22—Right Hand Rule Shows Direction of Lines of Force in a Straight Conductor

To apply the rule, grasp the wire with the thumb extended in the direction of conventional current flow (positive to negative); the fingers will then point in the direction in which the lines of force surround the conductor (Fig. 22). These lines of force are always at right angles to the conductor, and the compass needle confirms the direction as determined by the Right Hand Rule.

Unlike the flow of electrons in the conductor, which actually move, the magnetic lines of force do not move or flow around the wire; instead they merely have direction as indicated by their effect upon the compass needle.

The number of lines of force, or strength of the magnetism, increases as the current through the conductor is increased.

If a compass is moved farther away from the conductor, a point finally is reached where the compass is unaffected by the field (Fig. 23). If the current is then increased, the compass needle will be affected and will again indicate the direction of the magnetic field as shown.

The number of lines of force, and the area around the conductor which they occupy, increase as the current through the conductor increases.

In other words:

More current creates a stronger magnetic field.

 Litho in U.S.A.

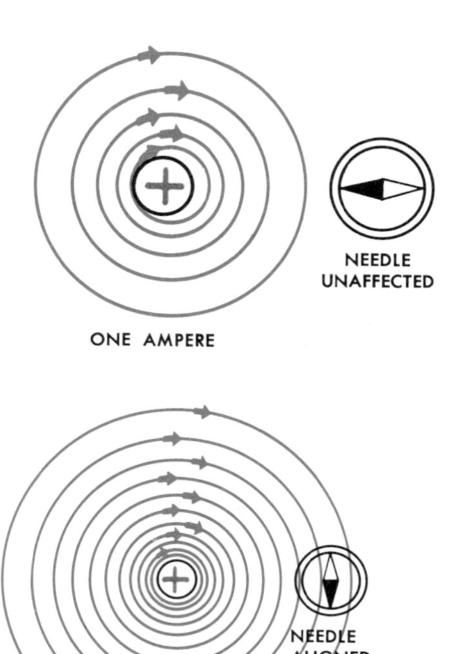

Fig. 23— More Current Creates a Stronger Magnetic Field

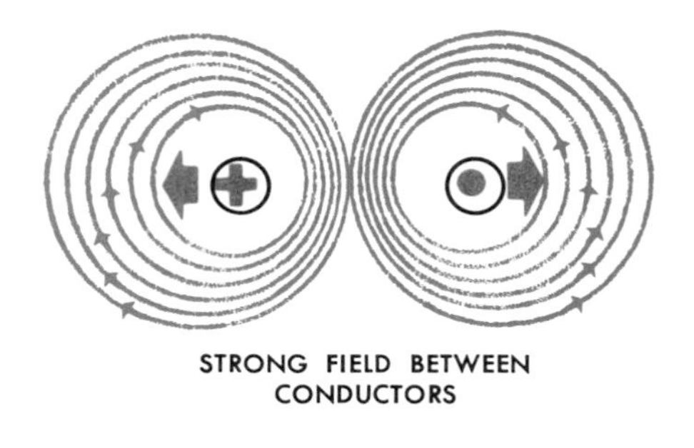

Fig. 24—How Conductors Are Affected By Strong Magnetic Fields

If two adjacent parallel conductors are carrying current in opposite directions, the direction of the field is clockwise around one conductor and counterclockwise around the other (Fig. 24). The lines of force are more concentrated between the conductors than on the outside of the conductors. The force lines between the two wires add together to form a strong magnetic field. Under this condition, the two wires willl tend to move apart, leading us to conclude:

A current-carrying conductor will tend to move out of a strong field and into a weak field.

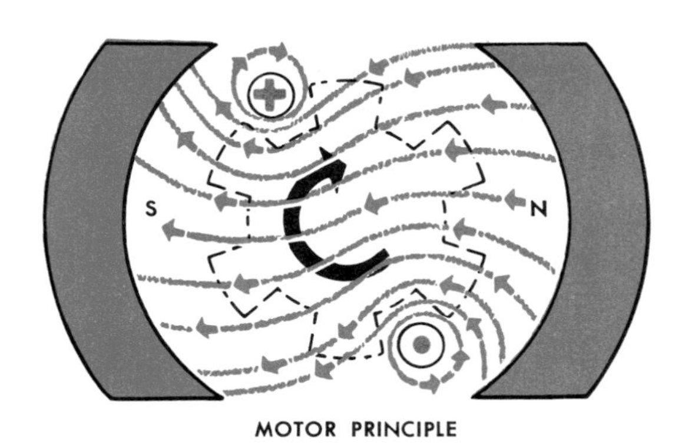

Fig. 25—Principle of the Starting Motor

In Fig. 25 two conductors are placed on an armature located between a strong N and S pole, and the conductors are made to carry current in opposite directions. The result is that a strong and a weak field are formed on opposite sides of each conductor as shown.

By the Right Hand Rule, current flowing into the top conductor will form magnetic lines on the underneath side of the conductor that add to the lines of the N and S poles. The conductor will then tend to move upward or clockwise into the weakened field.

Similarly, current flowing out of the lower conductor forms a strong field on top and a weak field underneath, causing the conductor to move downward or clockwise.

Thus, a rotation is caused by the current flowing in the conductors. This is the principle of the starting motor (Fig. 25). For more detail on starting motors, see Chapter 7, Starting Circuits.

 Litho in U.S.A.

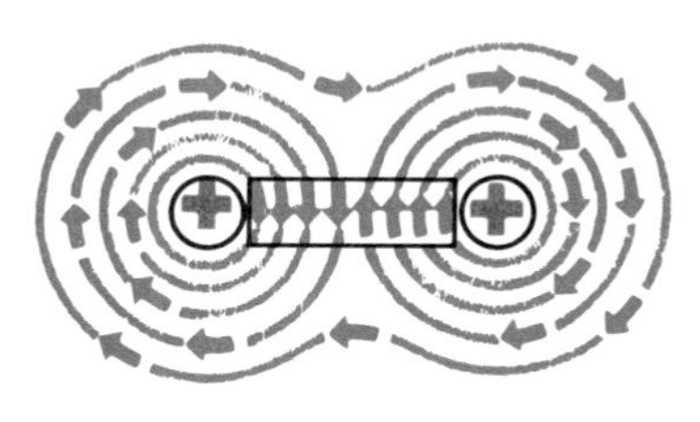

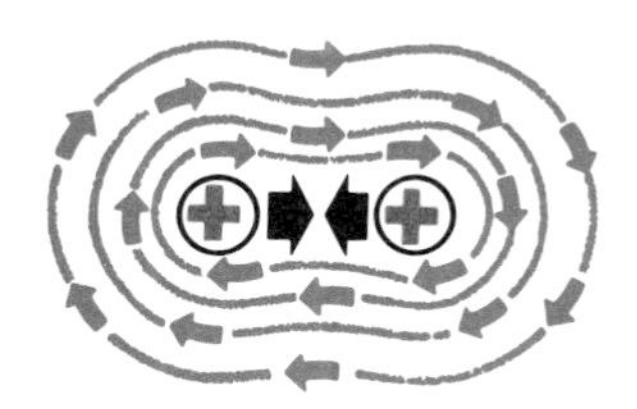

Fig. 26—How Conductors Are Affected By Weak Magnetic Fields

A different condition exists when two parallel conductors are carrying equal currents in the same direction (Fig. 26). A magnetic field, clockwise in direction, will be formed around each conductor, with the magnetic lines between the conductors opposing each other in direction. The magnetic field between the conductor is canceled out, leaving essentially no field in this area. The two conductors will then tend to move toward each other; that is, from a strong field into a weak field.

Two conductors lying alongside each other carrying equal currents in the same direction create a magnetic field equivalent to one conductor carrying twice the current (Fig. 27).

When several more conductors are placed side by side, the magnetic effect is increased as the lines from each conductor join and surround all the conductors.

Using the Right Hand Rule, we can see that all the lines of force enter the inside of the loop of wire on one side, and leave the other side as shown.

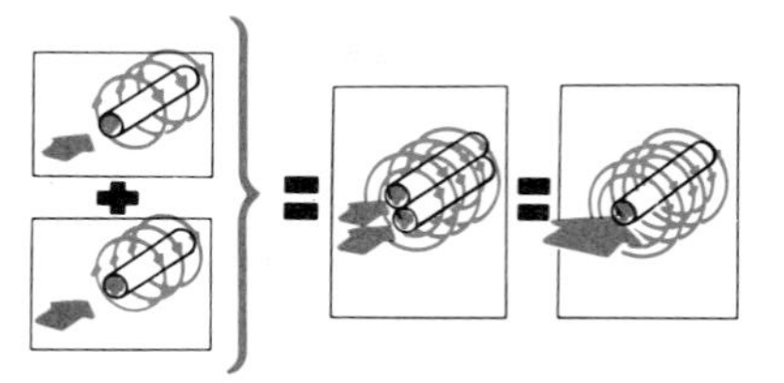

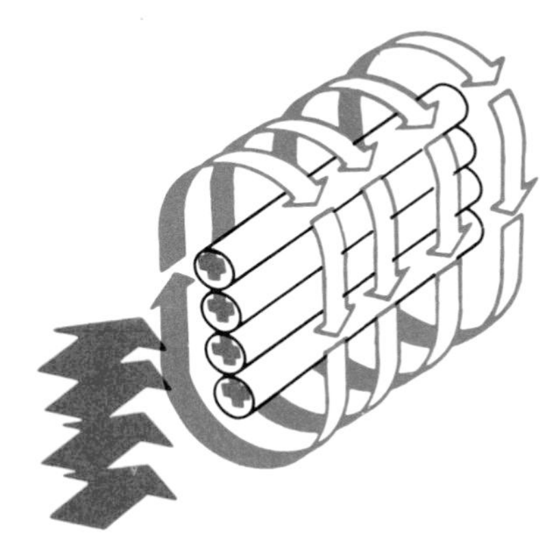

Fig. 27—How Two or More Adjacent Conductors Increase the Magnetic Field

The lines of force are concentrated inside the loop. A single loop of wire carrying current is called a **basic electromagnet.**

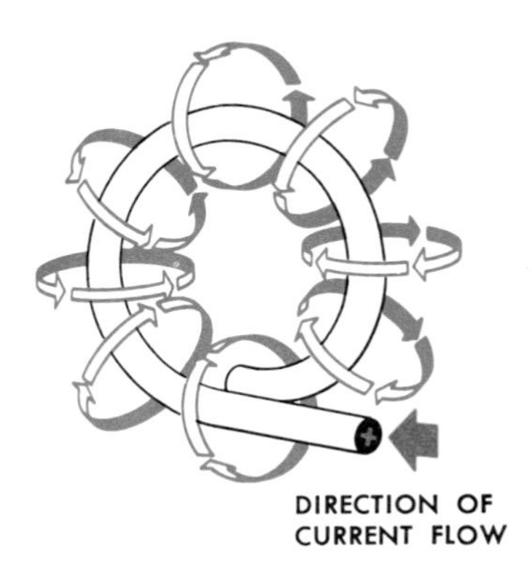

Fig. 28—Conductor in a Single Loop Has No Increase in Magnetic Field

A straight current-carrying wire when formed into a *single* loop has the same magnetic field surrounding it as when it was straight (Fig. 28).

HOW ELECTROMAGNETS WORK

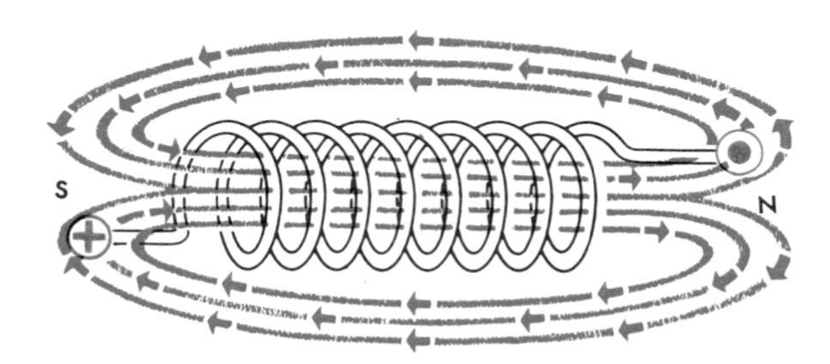

Fig. 29—Conductor in Several Loops Multiplies the Magnetic Field

But what happens when a current-carrying wire is wound into a *number* of loops to form a coil as shown in Fig. 29? Now the resulting magnetic field is the sum of all the single loop magnetic fields added together, since this is the same as several conductors lying side by side carrying current in the same direction.

With lines of force leaving the coil at one end and entering at the other end, a north and south pole are formed at the coil ends the same as in the bar magnet.

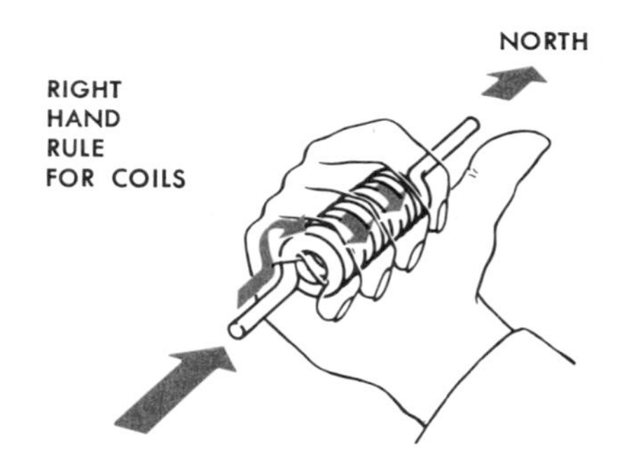

Fig. 30—Right Hand Rule for Coils

To find polarity of the coil ends, apply the Right Hand Rule for Coils by grasping the coil with the fingers pointed in the direction of current flow; the thumb will then point toward the N pole of the coil as shown in Fig. 30. If the current direction through the coil is reversed, the polarity of the coil ends will also reverse.

When a coil is wound over a core of magnetic material such as iron, the assembly becomes a usable **electromagnet** (Fig. 31).

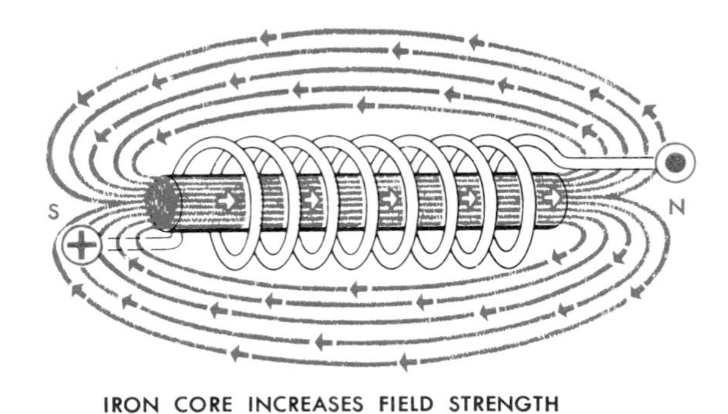

Fig. 31—Use of Iron Core to Increase Field Strength of Coil and form an Electromagnet

The strength of the magnetic field at the N and S poles is increased greatly by adding the iron core. The reason for this increase is that air is a very poor conductor of magnetic lines, while iron is a very good conductor. Relatively speaking, the use of iron in a magnetic path may increase the magnetic strength by 2500 times over that of air.

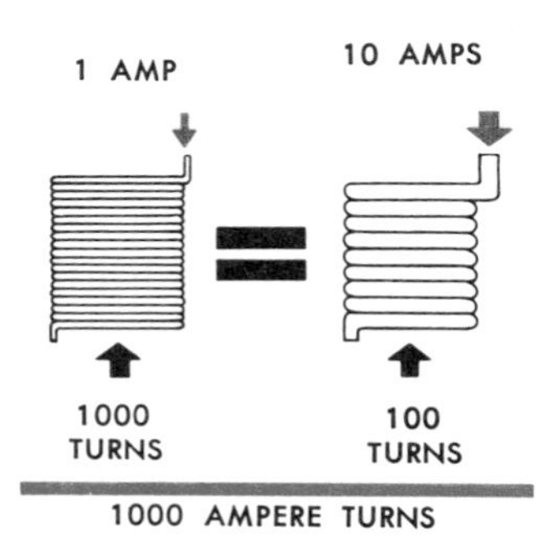

Fig. 32—Strength of Electromagnet Depends Upon Turns of Coil

The strength of the magnetic coils in an electromagnet is directly proportional to the number of turns of wire and the current in amperes flowing in the coil as shown in Fig. 32.

An electromagnet having one ampere flowing through 1000 turns and another electromagnet having 10 amperes flowing through 100 turns will each create 1000 ampere-turns, which is a measure of the magnetic field strength. The attraction on magnetic materials located in the magnetic field of each of these electromagnets will be the same.

 Litho in U.S.A.

Fig. 33—Electromagnet Picking Up Junk Metal

Just as electric current flows through a closed circuit, so do the lines of force created by a magnet occupy a closed magnetic circuit. Since the same number of lines that come out of the N pole must also enter the S pole, a complete circuit must be present for each magnetic field.

The resistance that a magnetic circuit offers to lines of force, or flux, is called **reluctance.** The reluctance is comparable to resistance in an electrical circuit.

There is an equation for an electromagnetic circuit that is similar to Ohm's Law for the electric circuit. This equation is:

Number of Magnetic Lines is Proportional to:

$$\frac{\textit{Ampere-Turns}}{\textit{Reluctance}}$$

Two facts related to this equation are important to us here:

1) The number of magnetic lines, or strength of the field, is directly proportional to the ampere-turns. In an electromagnet, more current through the coils means greater field strength.

2) The number of lines or field strength is inversely proportional to the reluctance; that is, if the reluctance increases the field strength decreases. Since most magnetic circuits consist of iron and short air gaps, the reluctance of such a series circuit is equal to the iron reluctance added to the air gap reluctance.

The effect of an air gap on the total reluctance of a circuit is very pronounced. This is true because air has a much higher reluctance than iron.

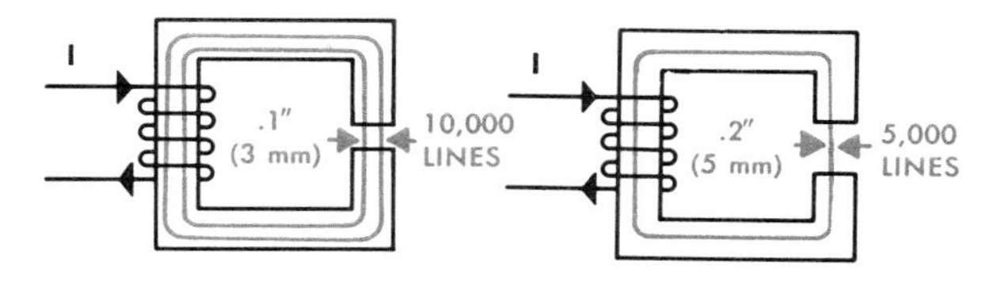

Fig. 34—Effect of Air Gap on Reluctance of a Circuit

To illustrate this fact, consider a magnetic circuit with a short air gap that has a field of strength of 10,000 lines of force (Fig. 34). If the length of the air gap is doubled, the reluctance will almost double, and the field strength will be reduced to approximately 5,000 lines of force. Although the air gap represents only a very short segment of the total magnetic path, increasing the air gap from, say, 0.1 inch (3 mm) to 0.2 inch (5 mm) may cut the field strength almost in half.

SUMMARY: ELECTROMAGNETISM

In summary:

- **Electricity and magnetism are related, because a magnetic field is established around a conductor that is carrying current.**
- **An electromagnet has an N pole at one end and an S pole at the other end of the iron core, much like a bar magnet.**
- **Every magnetic field has a complete circuit that is occupied by its lines of force.**
- **An electromagnetic field gets stronger as more current flows through its coils.**

ELECTROMAGNETIC INDUCTION

When a conductor is moved across a magnetic field, a voltage is induced in the conductor. This principle is called **electromagnetic induction,** and is defined as *the inducing of voltage in a conductor that moves across a magnetic field.*

HOW VOLTAGE IS INDUCED

To show this, move a straight wire conductor across the magnetic field of a horseshoe magnet (Fig. 35). Connect a sensitive voltmeter to the ends of the wire and the needle will register a small voltage as the wire is moved across the magnetic field.

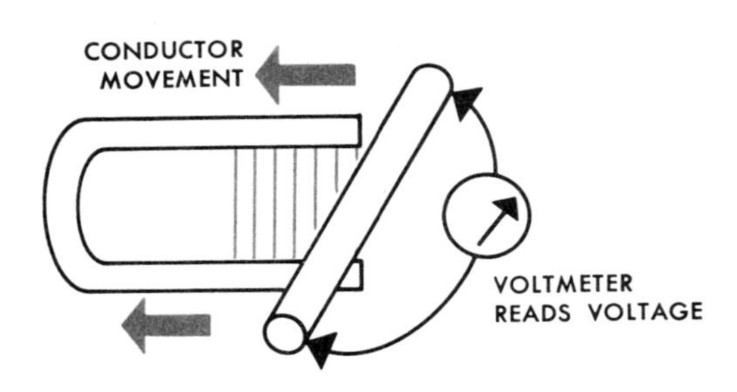

Fig. 35—Moving Conductor Across Magnetic Field—Voltage Is Induced

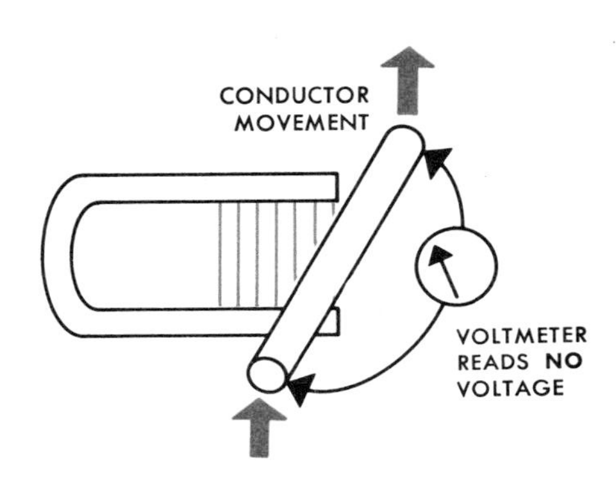

Fig. 36—Moving Conductor Parallel to Magnetic Field—No Voltage Is Induced

However, if the wire is moved *parallel* with the lines of force, no voltage will be induced (Fig. 36). *The conductor must cut across the lines of force in order to induce a voltage.*

We have observed that voltage has polarity; that is, positive and negative poles. We have also stated that current flows from the positive terminal of a voltage source through the external circuit and then back to the negative terminal of the source.

Now we can also see that a wire cutting across a magnetic field also becomes a source of electricity, and must have a positive and negative end, just like a battery.

However, unlike the battery, we will now see that the polarity at the ends of the wire can change. This polarity depends upon the relative direction of wire movement and the direction of the magnetic field.

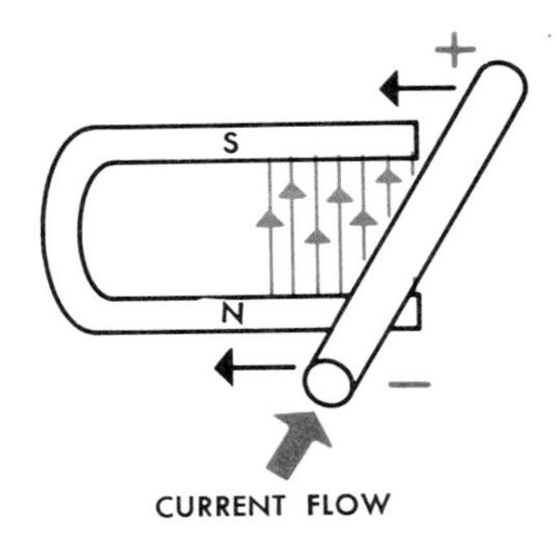

Fig. 37—Finding Polarity at Ends of a Conductor

To determine the polarity at the ends of a conductor and the consequent direction of current flow, consider a straight wire moving to the left across a magnetic field as shown in Fig. 37. With this direction of motion, the magnetic lines are striking the wire on the left side, and this side of the wire is called the leading side.

By applying the Right Hand Rule for an Induced Voltage, the voltage polarity and current flow direction can be determined as follows: Grasp the conductor with the fingers on the leading side of the wire, and pointed in the direction of the magnetic lines of force. The thumb will then point in the direction of current flow.

In Fig. 37, current is seen to flow into the page, or away from the reader, as indicated. This means that the polarities at the wire ends must be as shown in order to meet the condition that current flows from the positive side of a source through the external circuit and returns to the negative side of the source.

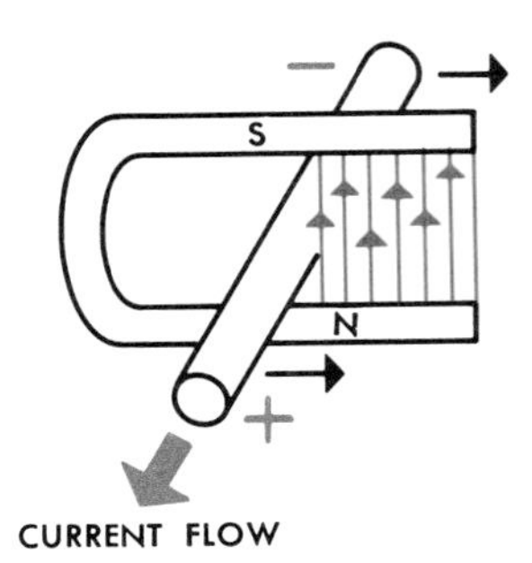

Fig. 38—Reversed Polarity in a Conductor

When the direction of motion of the conductor is changed to move toward the right, the right side of the conductor becomes the leading side (Fig. 38). By applying the Right Hand Rule, the current is seen to reverse its direction from Fig. 37, and to flow out of the page or toward the reader. This means that the voltage polarities at the wire ends have reversed.

In the previous examples, if, instead of moving the wire to the left, we move the magnetic field to the right across a stationary conductor, the same voltage and current flow will be induced in the wire. The same holds true for moving the field to the left across the conductor, because in each case the leading side of the conductor and the magnetic field direction are unchanged. Therefore, we can conclude: *A voltage will be induced in a conductor cutting across a magnetic field when there is relative motion between the two. Either the conductor can move, or the magnetic field can move.*

MAGNITUDES OF INDUCED VOLTAGE

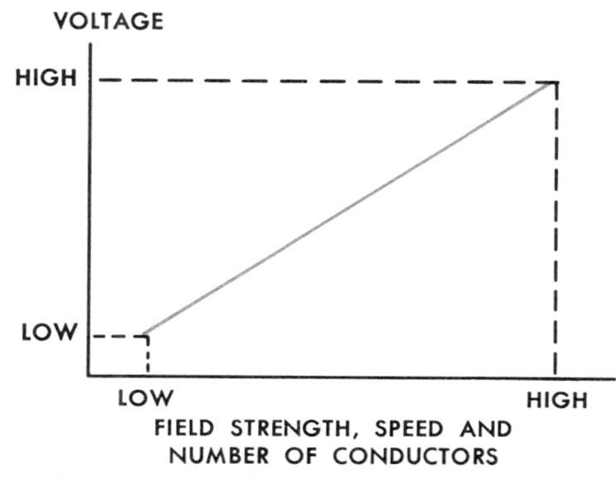

Fig. 39—Factors Which Determine the Magnitude of Induced Voltage

Now that we have observed the factors that determine the polarity of the induced voltage and the direction of current flow, let's consider the factors that determine the magnitude of the induced voltage (Fig. 39). These factors are:

1. *The strength of the magnetic field.*

2. *The speed at which lines of force are cutting across the conductor.*

3. *The number of conductors that are cutting across the lines of force.*

If the magnetic field is made stronger, such as by using a larger horseshoe magnet, more lines of force will be cut by the conductor in any given interval of time and the induced voltage will be higher.

If the *relative motion* between the conductor and magnetic field is increased, more lines of force will be cut in any given interval of time and so the voltage will be higher.

If the straight wire conductor is wound into a coil which is then moved across the field, all the loops of wire are in series and the voltage induced in all the loops will add together to give a higher voltage.

To summarize:

- **Stronger magnetic field = more induced voltage**
- **Faster relative motion = more voltage**
- **More conductors in motion = more voltage**

METHODS OF INDUCING VOLTAGE

There are three ways in which a voltage can be induced by electromagnetic induction:

- **Generated Voltage**
- **Self-Induction**
- **Mutual Induction**

Let's discuss each form of induction.

GENERATED VOLTAGE

A direct-current generator operates by moving conductors across a stationary magnetic field to produce voltage and current.

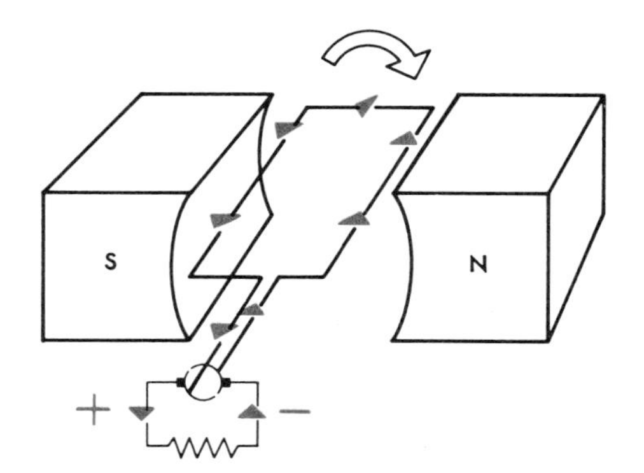

Fig. 40—Basic DC Generator

To show this, take the most basic type of DC generator where a single loop of wire is rotating between the N and S poles of a magnetic field (Fig. 40).

By applying the Right Hand Rule for Induced Voltage to both sides of the wire loop, current is seen to flow in the direction indicated, and the voltages induced in the wire loop give a coil voltage that appears at the two commutator segments attached to the wire ends. The current then flows through brushes riding on the commutator to the external circuit. The voltage polarities are as shown.

Another application of the principle of generated voltage is the alternating-current generator, or alternator, where the magnetic field is made to cut across stationary conductors in order to produce voltage and current.

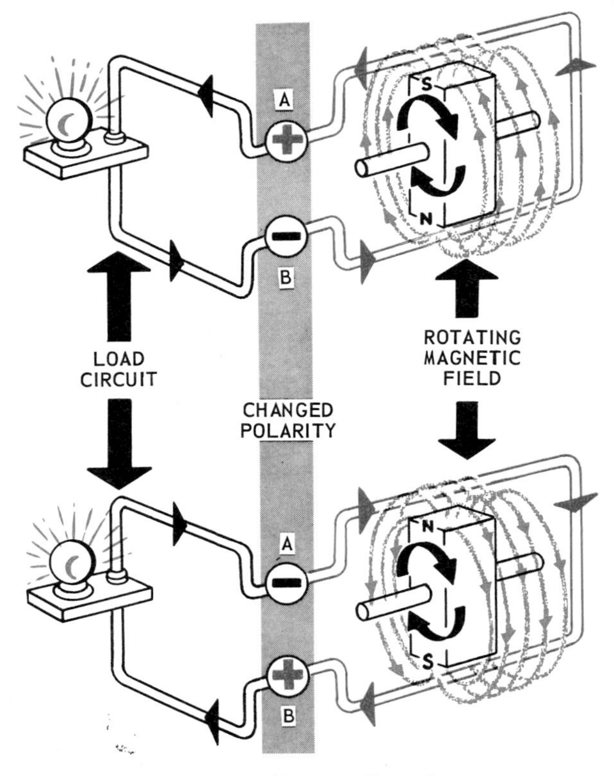

Fig. 41—Basic Alternator Operation

Fig. 41 shows the most basic type of alternating current generator, with a rotating magnetic field cutting across stationary conductors that are mounted on the generator frame.

By applying the Right Hand Rule, with the rotating magnetic field as shown, current flow through the conductors will alternate, thus causing an *Alternating Current* output.

The voltage induced in a conductor by physically moving the conductor or the field is referred to as *Generated Voltage.* This principle is used in DC generators and alternators, both of which are covered in detail in Chapter 6.

SELF-INDUCTION

Self-induction is the induction of a voltage in a current-carrying wire when the current in the wire itself is changing.

Earlier in this chapter we used a separate magnetic field provided by a horseshoe magnet to generate voltage in a conductor. In self-induction no separate field is used; instead the magnetic field created by a changing current through the wire itself is seen to induce a voltage in the wire. Hence, the voltage is **self-induced.**

The reason that a voltage is induced in a wire carrying a changing current is this: Since the current creates a magnetic field in the form of concentric circles around the wire which expand and contract as the current increases and decreases, these magnetic circles cut across the conductor and thereby induce a voltage in the conductor. Since there is relative motion between the field and conductor, the condition necessary for inducing a voltage has been met.

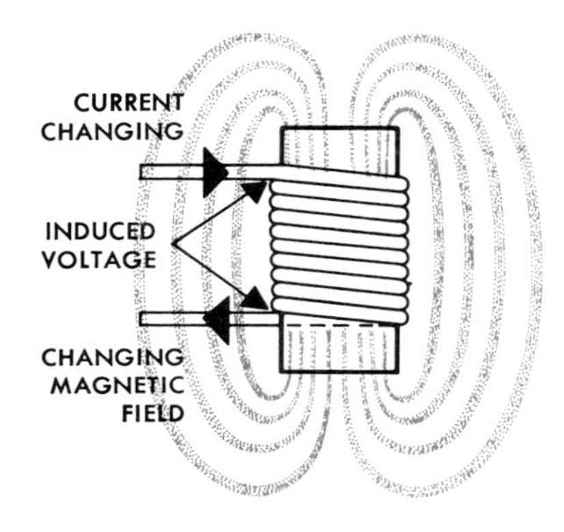

Fig. 42—Self-Induction in a Coil

Let's take a coil of wire with the turns wound tightly together over an iron core (Fig. 42). When the current increases in one loop the expanding magnetic field will cut across some or all of the neighboring loops of wire; thus inducing a voltage in these loops. The coil of wire wound over an iron core is often called an inductor, and possesses the property of inductance which causes a voltage to be induced in the coil when the current is changing.

Polarity of Induced Voltage in a Coil

Now, let's make a statement that determines the voltage polarity of the self-induced voltage in a conductor or coil of wire, and then explain this statement more fully.

The polarity of an induced voltage will oppose a change in the current that produced it.

"Change in current" refers to current which is either increasing or decreasing in value.

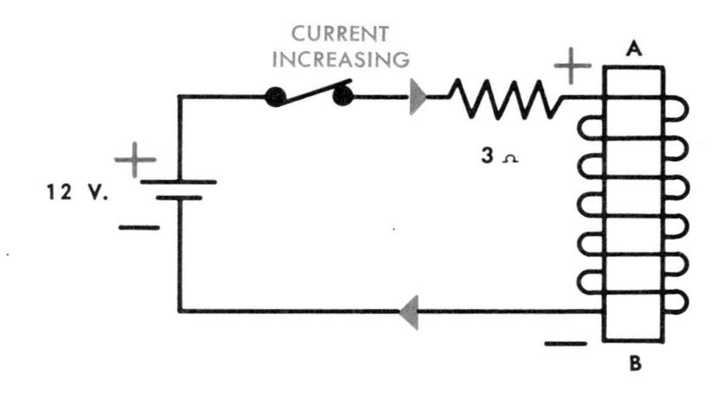

Fig. 43—Self-Induction in a Circuit When Current Increases

Fig. 43 shows a circuit containing a coil of wire (inductor).

After the switch is closed, the current increases from zero to its maximum value of, say, four amperes. During this time a voltage will be induced in the inductor in a direction opposing the increasing current; *the inductor itself becomes a source of voltage that attempts to prevent the current from increasing in the circuit.*

To oppose the increasing current, the inductor will have to generate a voltage in a direction opposite to the battery current, hence the polarity at A is positive (+) and at B is negative (−). The induced voltage opposes the change in current; that is, the induced voltage tries to maintain the "status quo" and keep the battery current at zero when the switch is closed.

The induced voltage polarities at the coil are therefore as shown.

However, the battery current in time overcomes the inductive effect of the coil, and reaches its final steady value of four amperes.

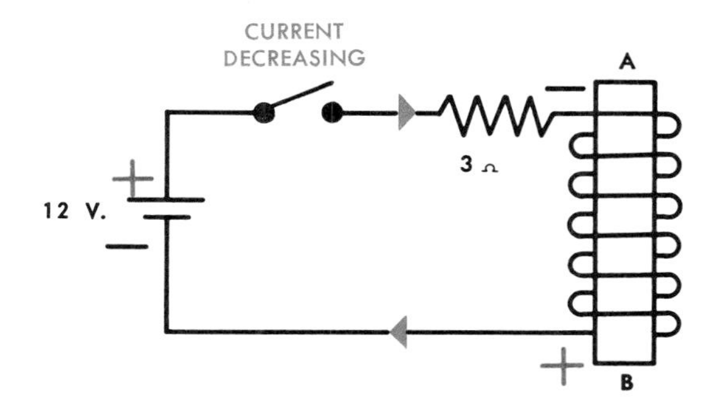

Fig. 44—Self-Induction in a Circuit When Current Decreases

When the switch is opened (Fig. 44), the current decreases from four amperes to zero. This changing current induces a voltage in the coil that again tries to maintain the "status quo" or to keep the current flow at four amperes. The polarity of the induced coil voltage, therefore, must be as shown, because the coil attempts to supply current in the same direction as originally supplied by the battery. It attempts to keep the current flow at the four-ampere value, and this may cause the switch to arc when it is opened.

Note that the induced voltage polarity for any direction of current flow is determined by whether the current is increasing or decreasing. For example:

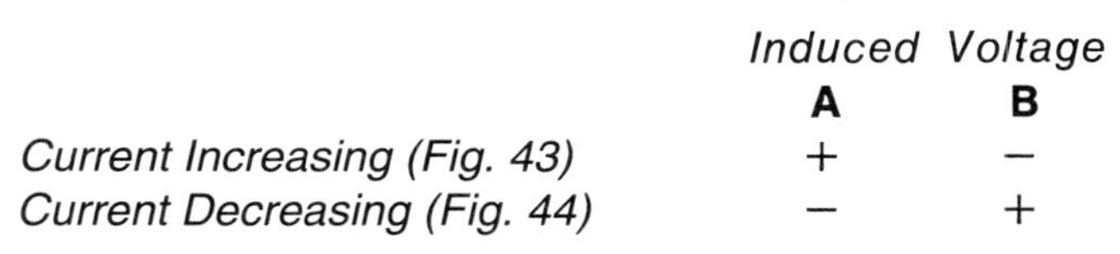

	Induced Voltage	
	A	**B**
Current Increasing (Fig. 43)	+	−
Current Decreasing (Fig. 44)	−	+

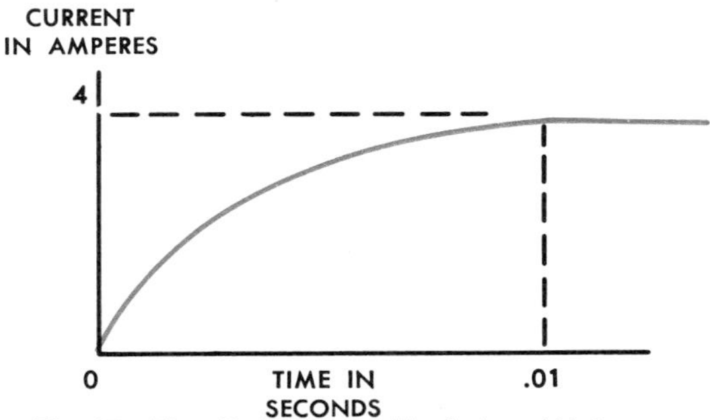

Fig. 45—Time Delay Caused by Induced Voltage

Although the inductive voltage tries to prevent any change in current value, the effects of the battery voltage and the closed or open switch in time cause the current to reach a constant value. The induced voltage, however, does cause a time delay while the current reaches its final value after the switch is closed or opened (Fig. 45).

Consider first the case when the switch is closed. Due to the inductive effect of the coil the current slowly rises to its maximum value of four amperes. When the final current of four amperes is reached, there is no changing magnetic field, no induced voltage, and the resistor alone acts to establish the final current value.

There is a certain amount of energy stored in an inductive coil when current is flowing through it. This energy is directly related to the amount of current (I) and the inductance of the coil, whose symbol is (L). The inductance of any coil is determined primarily by the number of turns of wire, their spacing, and the type of material used in the core of the coil. The amount of energy stored in a coil is given by the following equation:

$$\text{Coil's energy} = \frac{\text{Inductance} \times \text{current} \times \text{current}}{2} = \frac{L \times I \times I}{2}$$

This equation shows that the higher the inductance and the higher the current, the greater will be the energy stored in the coil.

Use of Self-Induction in Ignition Circuit Coils

A standard ignition circuit operates on the principle of energy stored in the primary winding of an ignition coil. When the distributor contacts open, the current suddenly drops to zero, and from the energy equation the energy in the coil suddenly drops to zero. Some of this energy is transferred by mutual induction (see the next section) to the secondary winding of the ignition coil, and the energy is dissipated in the form of an arc across the spark plug. In an ignition circuit, the time delay in build-up of primary winding current when the distributor contacts close is very important.

If the contacts open before the final maximum value of current is reached, the energy stored in the coil (see the energy equation) is reduced, making less energy available to fire the plug.

Although the inductance of the ignition coil may cause a time delay of only a fraction of a second, this interval of time must be closely correlated with the time the distributor contacts are closed. (See Chapter 8 for more details on ignition circuits.)

MUTUAL INDUCTION

If a changing magnetic flux created by current flow in one coil cuts across the windings of a second coil, a voltage will be induced in the second coil.

This induction of voltage in one coil because of a changing current in another coil is called **mutual induction.**

Mutual Induction in Coils

Fig. 46 illustrates the principle of mutual induction in a circuit where the blue winding (the secondary) is wound over an iron core, while the red winding (the primary) is wound over the blue winding.

When the switch is closed, current will increase in the primary, and the expanding lines of force will cut across the secondary, causing a voltage to be induced in the secondary.

Mutual Induction in Coils

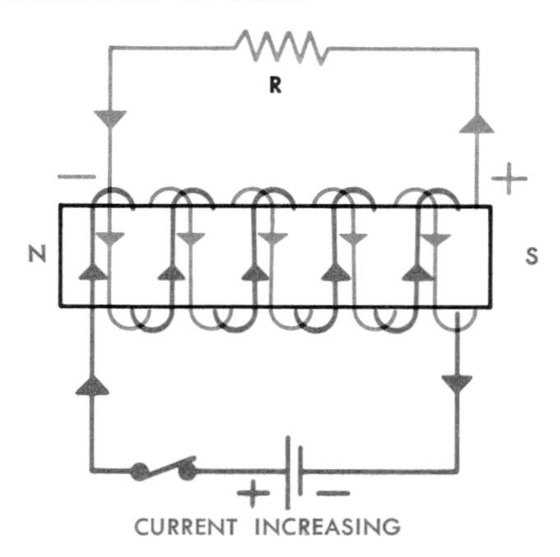

Fig. 46—Mutual Induction in Primary and Secondary Coils

Similarly, when the switch is opened, the sudden decrease in current in the primary will induce a voltage in the secondary. The blue secondary winding then becomes a source of voltage, and will supply current to resistor R.

Finding Polarity of Induced Voltage in Secondary Coil

The polarity of the induced voltage in the secondary can be determined in a number of different ways.

One of the simplest methods is to observe the direction of current in the primary, and note that the current direction in the secondary must oppose any change in the primary current. Thus when the primary current is increasing, the secondary current must flow in the opposite direction around the core in order to oppose the increase, and the secondary voltage polarity is established as shown in Fig. 46.

However, if the primary current is decreasing, the secondary current must flow in the same direction around the core in order to oppose the change; that is, to attempt to keep the flux in the core from changing. The secondary polarity is then as shown in Fig. 47.

An alternate method of finding the secondary induced voltage polarity is to use the Right Hand Rule for an Induced Voltage. Taking a lengthwise cross-sectional view of the assembly, when current increases in the primary, the circular lines of force expand and strike the secondary on the top side.

By using the Right Hand Rule for an Induced Voltage, the current flow direction is determined as shown in Figs. 48 and 49.

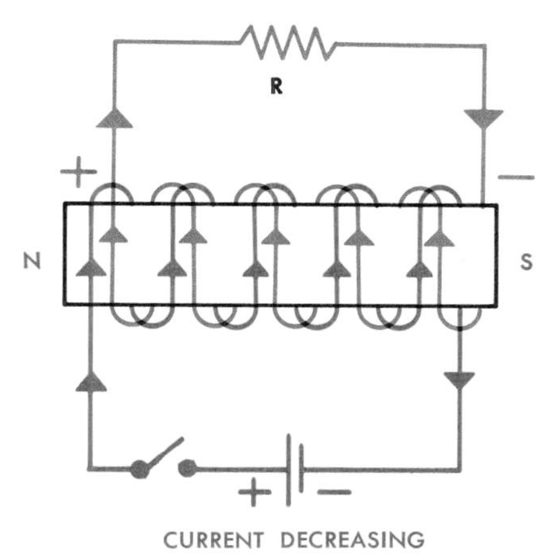

Fig. 47—Polarity in Coils with Primary Current Decreasing

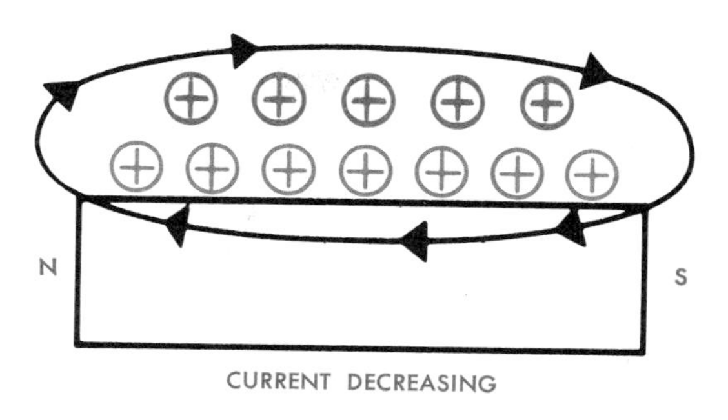

Fig. 49—Finding Polarity of Secondary Voltage When Current Is Decreasing

As stated, the two coils carry current in opposite directions around the core when the primary current is increasing (Fig. 48).

When the primary current decreases (Fig. 49), the circular lines of force strike the secondary windings on the underneath side, and the current flows in both coils in the same direction around the core. The voltage polarity is determined accordingly, with current coming out of the secondary positive terminal and returning to the negative terminal.

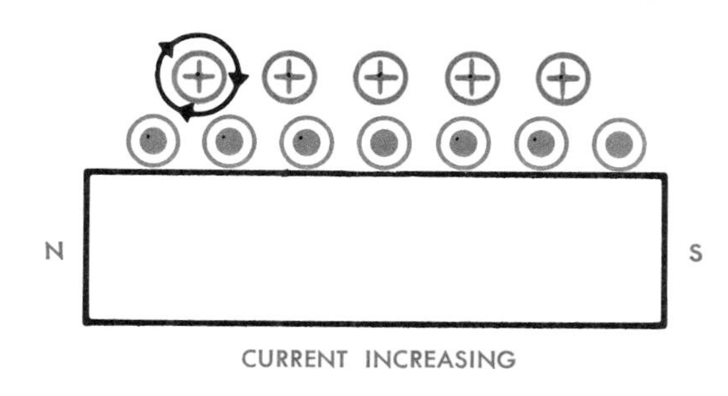

Fig. 48—Finding Polarity of Secondary Voltage When Current Is Increasing

The magnitude of the voltage induced in the secondary winding is determined primarily by the number of turns in the primary and in the secondary.

The ignition coil uses the principles of mutual induction in its primary and secondary windings.

SUMMARY: ELECTROMAGNETIC INDUCTION

In summary:

- **Electromagnetic induction is inducing voltage in a conductor that moves across a magnetic field.**
- **Conductor must cut *across* the field, not move parallel to it.**
- **Conductor and field must be moving in relation to each other.**
- **Faster relative motion = more voltage induced.**
- **More conductors in motion = more voltage.**
- **Stronger magnetic field = more voltage.**
- **Three ways of inducing voltage are generated voltage, self-induction, and mutual induction.**
- **Generated voltage by relative motion is used in generators and alternators.**
- **Self-induction creates its own voltage by a change of current in the conductor (as in the primary of ignition coils).**
- **Mutual induction occurs when changing current in one coil induces voltage in a second coil (as in the two windings of ignition coils).**

 Litho in U.S.A.

TEST YOURSELF

QUESTIONS

1. Electricity is the flow of ________________ from atom to atom in a ____________ .

2. Elements whose atoms have less than four electrons in their outer rings are generally good ______________ .

3. Like charges ____________ , while unlike charges ______________ .

4. The flow of electrons through a conductor is called a current. The *rate* of flow of the current is measured in ______________. The *force* of the current is measured in _________________. The *resistance* of the conductor to this current is measured in __________.

5. Electronics is the control of ______________ and the study of their ___________ and ___________ .

6. Resistance is the ___________ of electron flow in a conductor.

7. Every magnet has a ___________ of __________ surrounding it..

8. True or False? When electrons flow through a conductor a magnetic field is developed around that conductor.

9. An electromagnetic field gets __________ as more ____________ flows through its coils.

10. What are the two parts of a basic electromagnet?

11. Match the method of inducing voltage below with the component which uses this method:

a. Generated voltage	1. Primary winding of ignition coil
b. Self-induction	2. Secondary winding of ignition coil
c. Mutual induction	3. Alternator

(Answers on page 18 at the end of this book.)

 Litho in U.S.A.

MEASUREMENT OF ELECTRONS / CHAPTER 2

INTRODUCTION

Electronics would be nothing without measurement, numbers, and electronic circuits. In this chapter we will cover the following:

- **Ohm's Law**
- **Basic Electronic Circuits**
- **Electronic Circuits — Three Types**
- **Power (Watts)**
- **Pulses, Waves, Frequency, and Signals**
- **Multimeters**
- **Basic Testing of Circuits**
- **Other Types of Electronic Test Equipment**
- **Wire Diagrams and Schematics**

OHM'S LAW

In 1827 a mathematic reasoning to electronics was established by a German, Georg Simon Ohms. At that time, electromotive force (Volts) and electron flow (Amperes) through a conductor had been established, but nothing had been established for the resistance within a conductor. Ohm assumed that if the electromotive force and amperes were established from a certain length and material of a conductor, doubling the length of the same conductor would allow a passage of electrons at half the rate of the shorter conductor. This is known as Ohm's Law, whereby, the ratio of the electromotive force (E) and amperes (I) can be taken as a measure of resistance (R) within a conductor. Thus, R=E/ I, where E is the electromotive force (volts), I is the electron flow or current in amperes, and R is the resistance in ohms.

The formula for Ohm's Law can be expressed in three different ways (Fig. 1). When any two quantities are known, the third can be calculated as shown.

VOLTS (E) = AMPERES (I) X OHMS (R)

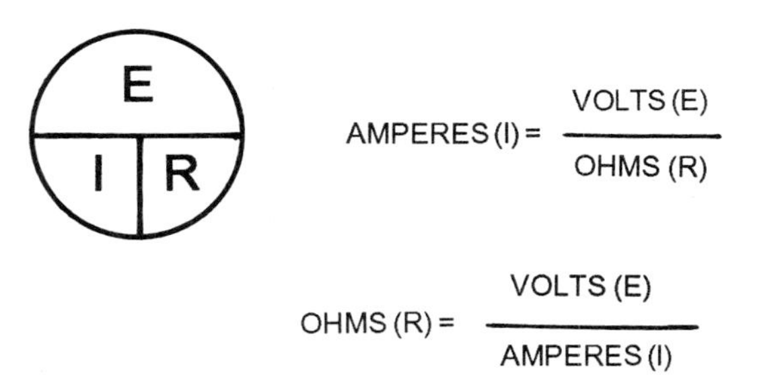

Fig. 1 — Ohm's Law—The Three Formulas

BASIC ELECTRONIC CIRCUITS

An electronic circuit is a schematic of electronic components that permits the flow of electrons. It is also the same as and can be called an electrical circuit. This manual will refer to the circuits as *electronic circuits.*

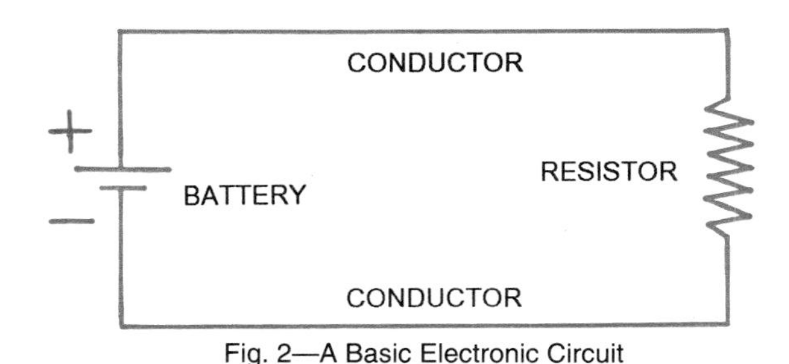

Fig. 2—A Basic Electronic Circuit

A basic electronic circuit consists of three parts:

- **A Voltage Source such as a battery**
- **A Resistor such as a light bulb.**
- **Conductors such as copper wires to connect the circuit together.**

 Litho in U.S.A.

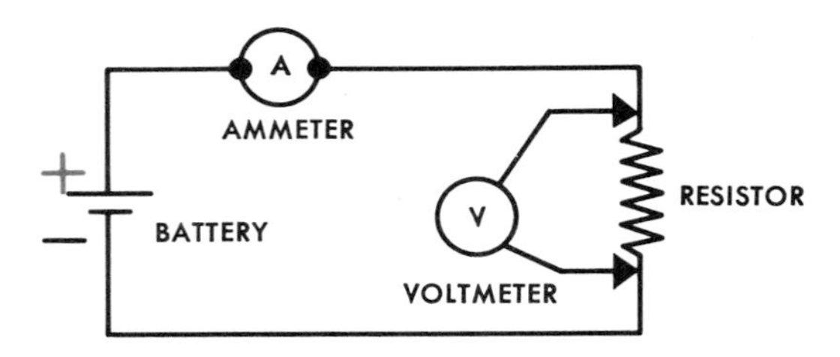

Fig. 3—Basic Gauges for Testing a Circuit

The basic gauges for testing an electronic circuit are: An *Ammeter* to measure current (flow), and a *Voltmeter* to measure voltage (pressure) between any two points in a circuit. (These will be covered in more detail later in this chapter.)

ELECTRONIC CIRCUITS—THREE TYPES

The three types of electronic circuits are:

- **Series Circuits**
- **Parallel Circuits**
- **Series—Parallel Circuits**

These circuits are compared in Fig. 4.

Series Circuits have several resistors connected so that current can flow along only one path.

Parallel Circuits have more than one path for current to flow. The resistors are side-by-side and provide separate routes for current.

Series-Parallel Circuits have some resistors connected in series and some in parallel.

SERIES CIRCUITS

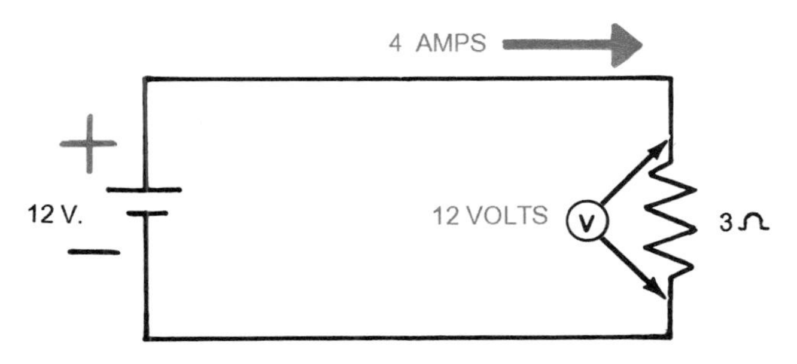

Fig. 5—A Basic Series Circuit

A **basic series** circuit may have a three-ohm (3Ω) resistor connected to a 12-volt battery. See Fig. 5.

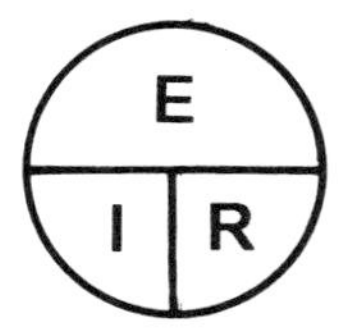

Fig. 6—Ohm's Law Formula Circle

To find the current, use Ohm's Law Formula Circle (Fig. 6), where I=E/R or 12/3=4 Amperes or 4 Amps. The voltage across the three ohm resistor would be E=I x R=4 x 3=12 volts.

Another series circuit is shown in Fig. 7. This circuit has a two-ohm resistor and a four-ohm resistor connected to a 12-volt battery.

In a series circuit, the total circuit resistance is equal to the sum of all the resistors. In this circuit, the total circuit resistance is 4+2=6 ohms. The current from Ohm's Law is: $I=\frac{E}{R}=\frac{12}{6}=2$ amperes.

SERIES

X.1347

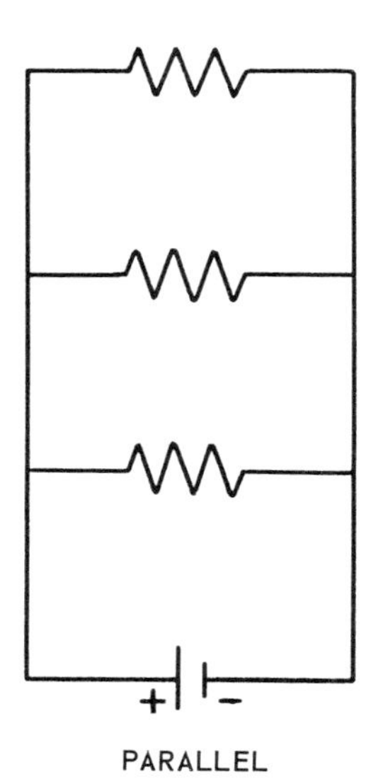

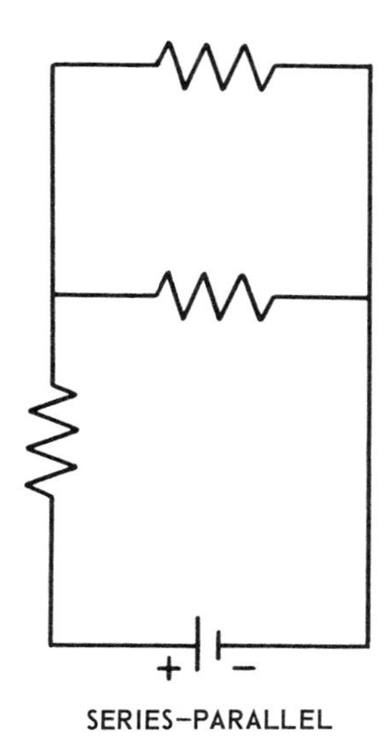

Fig. 4—Three Types of Electronic Circuits

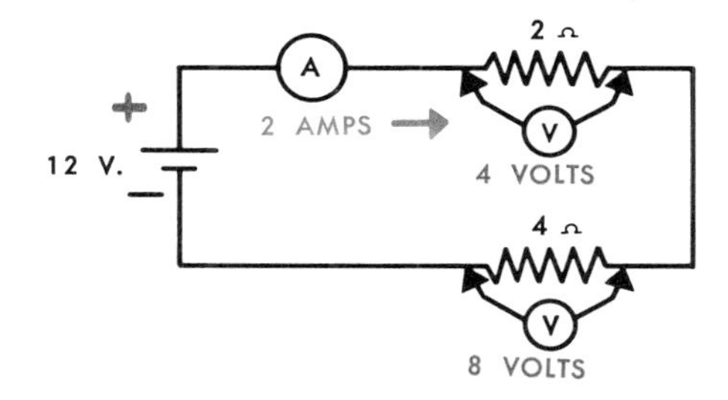

Fig. 7—Series Circuit with Two Resistors

Voltage across the two-ohm resistor can be figured using Ohm's Law: thus E = IR = 2 x 2 = 4 volts. For the four-ohm resistor, E = 2 x 4 = 8 volts. These values are called the **voltage drops,** and the sum of all voltage drops in the circuit must equal the source voltage, or 4 + 8 = 12 volts.

You will note that in order to calculate the voltage drops across the resistors you had to first calculate the current (amperes) of the circuit. Voltage drops are also known as *Voltage Division.*

The **voltage division** rule allows us to calculate the voltage across one or a combination of series resistors without first having to solve for the current. The basic formula for the voltage division rule is:

$$\frac{R_1 \times E}{R_1 + R_2} = V_1$$

Where R_1 is the resistor where the voltage division is to be found, E is the battery source. R_2 is the other resistor in the series circuit. If there are more resistors in the series they would have to be added into the formula too. V_1 would be the voltage division result. Using this rule let's find the voltage division of the four-ohm resistor in Fig. 7.

$$\frac{4 \times 12}{4 + 2} = \frac{48}{6} = 8 \text{ Volts (divided)}$$

Voltage Division is used extensively in the field of electronics to supply lower voltages to other circuits. Let's calculate the divided voltage outputs of each of the connections in a series circuit with three resistors (Fig. 8).

Using the **voltage division rule,** the first voltage output between the two-ohm and three-ohm resistors would be:

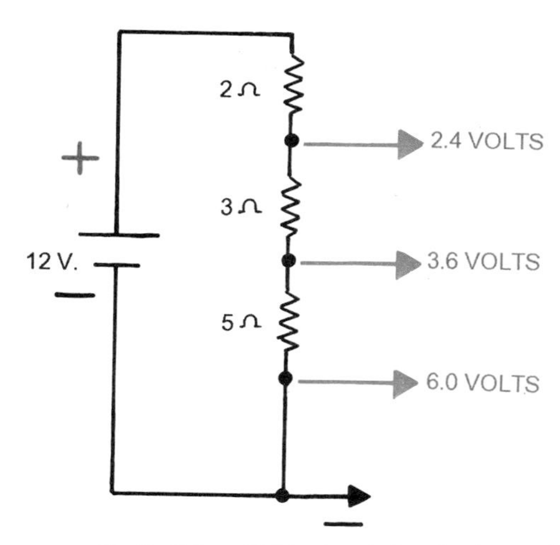

Fig. 8—Voltage Division in a Series Circuit

$$\frac{2 \times 12}{2 + 3 + 5} = \frac{24}{10} = 2.4 \text{ Volts}$$

The second voltage output would be:

$$\frac{3 \times 12}{2 + 3 + 5} = \frac{36}{10} = 3.6 \text{ Volts}$$

The third voltage division output would be:

$$\frac{5 \times 12}{2 + 3 + 5} = \frac{60}{10} = 6.0 \text{ Volts}$$

What would be the current (amperes) for each of the voltage outputs?

In a *series circuit,* the total resistance is equal to the sum of the resistors. Thus, using Ohm's Law:

$$I = \frac{E}{R} = \frac{12}{2 + 3 + 5} = \frac{12}{10} = 1.2 \text{ Amperes}$$

for each of the divided voltage outputs.

 Litho in U.S.A.

In summary, **series circuits** have the following features:

1. The current through each resistor is the same.

2. The voltage drops across each resistor will be different if the resistances are different.

3. The sum of the voltage drops equals the source voltage.

PARALLEL CIRCUITS

In a **parallel circuit,** the voltage drop across each resistor is equal to the potential of the current source since there is a separate path for current to flow through each resistor. This means:

1. The voltage across each resistor is the same.

2. The current through each resistor will be different if the resistance values are different.

3. The sum of the separate currents equals the total current in the circuit.

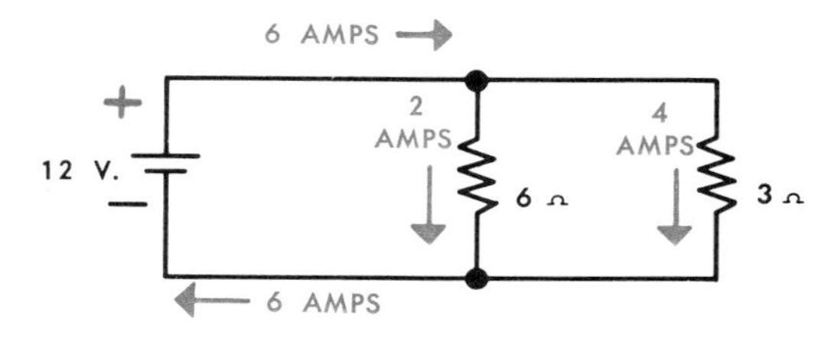

Fig. 9—Parallel Circuit with Two Resistors

The parallel circuit in Fig. 9 has a six-ohm and a three-ohm resistor connected to a 12-volt battery. The resistors are in parallel with each other, since the battery voltage (12 volts) flows across each resistor.

The current through each resistor or branch of the circuit can be figured using Ohm's Law. For the six-ohm resistor, $I = \frac{E}{R} = \frac{12}{6} = 2$ amps. For the three-ohm resistor, $I = \frac{12}{3} = 4$ amps. The total current supplied by the battery is 2 + 4 = 6 amps. The equivalent resistance of the entire circuit has to be two ohms, since $R = \frac{E}{R} = \frac{12}{6} = 2$ ohms. You will note that before the equivalent resistance of the entire circuit could be found the amperage had to be calculated across each resistor. The Current-Divider Rule for parallel circuits allows us to calculate the equivalent resistance of the circuit without having first to solve for the amperes. The formula for **two** resistors is:

$$R = \frac{R_1 \times R_2}{R_1 + R_2}$$

Where R_1 and R_2 are the resistors in the circuit and R is the equivalent resistance of the circuit. Let's calculate the equivalent resistance of the parallel circuit in Fig. 10 using the current-divider rule.

$$R = \frac{6 \times 3}{6+3} = \frac{18}{9} = 2 \text{ ohms}$$

But the current-divider rule formula changes when there are **three** or **more** resistors within a parallel circuit. The formula changes to:

$$R = \frac{1}{1/R_1 + 1/R_2 + 1/R_3 \,.\,.\,.\,.\,.}$$

The complication of calculating this formula makes the calculating of the amperes for each resistor easier.

SERIES-PARALLEL CIRCUITS

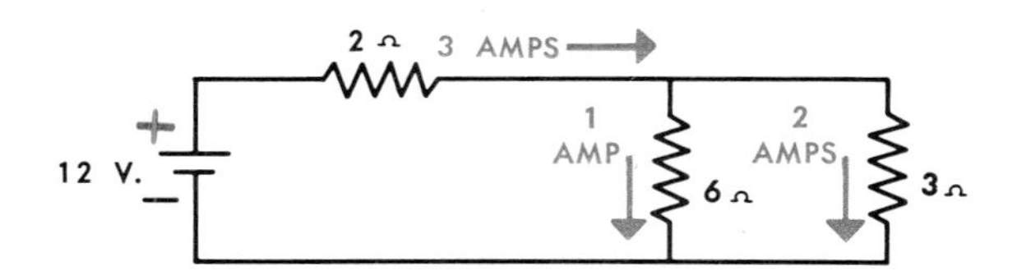

Fig. 10—Series-Parallel Circuit

A *series-parallel circuit* is shown in Fig. 10. Note that the 2Ω resistor is in series with a parallel combination (the three- and six-ohm ones).

Since there are two resistors in parallel, we can use the current-divider rule to calculate the resistance of the two resistors.

 Litho in U.S.A.

$$R = \frac{6 \times 3}{6+3} = \frac{18}{9} = 2 \text{ ohms}$$

Add this two ohms to the other two-ohm resistor because it is in series, for a total circuit resistance of four ohms. The total current of the circuit, using Ohm's Law is:

$$R = \frac{E}{I} = \frac{12}{4} = 3 \text{ amps}$$

With three amps flowing through the two-ohm resistor nearest the battery, the voltage drop across this resistor is E = IR = 3 x 2 = 6 volts, leaving six volts across the parallel six- and three-ohm resistors.

The current through the six-ohm resistor is $I = \frac{E}{R} = \frac{6}{6} =$ 1 amp, and through the three-ohm resistor is $I = \frac{6}{3} = 2$ amps.

Total current is the sum of these two current values or 1 + 2 = 3 amps.

CURRENT FLOW IN SERIES AND PARALLEL CIRCUITS

We have learned from Ohm's Law that there is a definite relationship between current, voltage, and resistance in an electrical circuit.

Now let's see how this applies to series and parallel circuits with a given resistance.

Each type of circuit shown in Fig. 11 has three four-ohm resistors. What is the total current flow in each case?

In the *series* circuit there is only one path for the current to flow, so the total resistance is 12 ohms. Therefore, current $= \frac{\text{volts}}{\text{ohms}} = \frac{12}{12} = 1$ amp for the circuit.

In the *parallel* circuit there are three different paths for current, each with 12 volts of force. Since current for each resistor $= \frac{\text{volts}}{\text{ohms}} = \frac{12}{4} = 3$ amps, the total current is 3 (amps) × 3 (resistors) = 9 amps for the circuit.

In the *series-parallel* circuit the total resistance is 4 ohms plus the resistance of the parallel resistors $\frac{(4 \times 4 = 16)}{(4 + 4 = 8)} = 2\,\Omega$ or a total of 4 + 2 = 6 ohms.

Therefore, current $= \frac{\text{volts}}{\text{ohms}} = \frac{12}{6} = 2$ amps for the circuit.

What does this mean for the design of electrical circuits?

- **Series circuit = high resistance**
- **Parallel circuit = low resistance**
- **Series-parallel circuit = medium resistance**

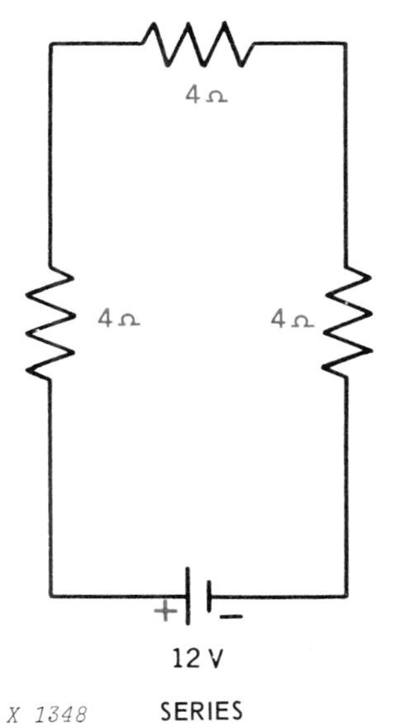

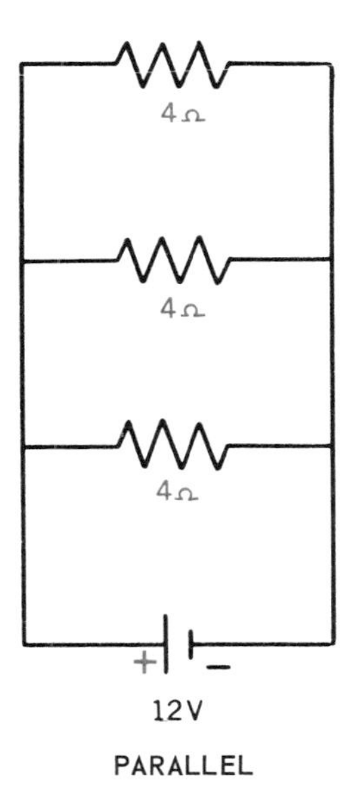

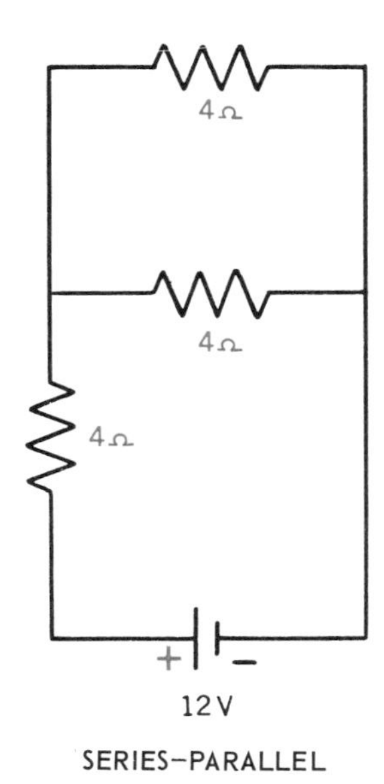

Fig. 11—Current Flow in Series and Parallel Circuits

 Litho in U.S.A.

POWER (WATTS)

In any system, *power* is a measure of the rate of energy conversion of that system. In an engine, the output horsepower rating is a measure of its ability to do mechanical work. In electronics, power is the measure of the rate at which electrical energy is converted into heat by the resistive elements within a conductor. This power is represented by the letter P and its measurement is in *Watts.* The formula for **power (watts)** is P = I times E, where I is amperes and E is the electromotive force (volts). To incorporate the resistive elements (ohms) into the formula, we can substitute the factor of I from Ohm's Law. The power formula can now be expressed as P = E/R times E or $P = E^2/R$. Many electronic components are labeled with watts or their capacity to withstand heat, which we will talk about in the next chapter.

PULSES, WAVES, FREQUENCY AND SIGNALS

PULSES

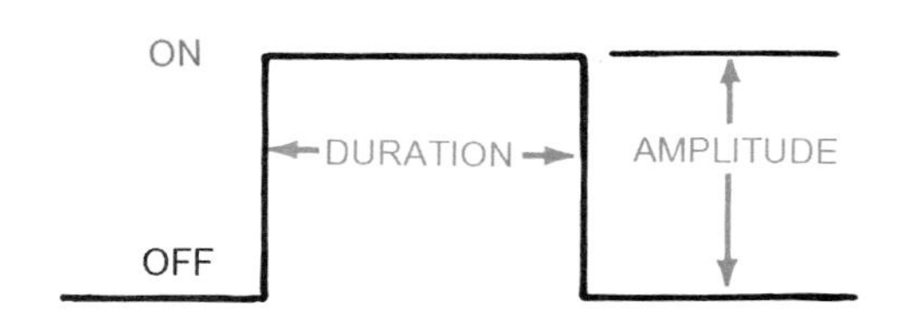

Fig. 12—A Pulse

A *pulse* is a sudden **On** and **Off** of direct current flow within a circuit (Fig. 12). In its basic form, an on-off switch will cause a pulse of electron flow within a circuit when the switch is turned on and off.

The width of a pulse is the length of time the switch was on. The height of the pulse is known as the amplitude and is determined by the amount of the voltage. An early electronic device that generated pulses was the telegraph key. Voltage (amplitude) generated a sound by manually turning a contact switch on and off (duration) to create a code of dots and dashes that were used to convey information.

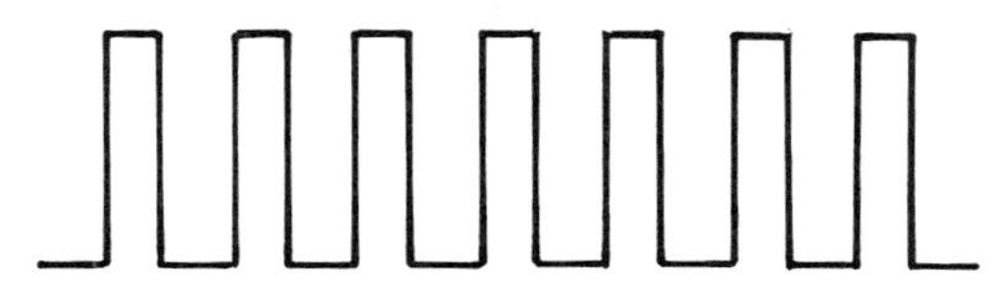

Fig. 13—A Typical Controlled Pulse Output

Today pulses (Fig. 13) are processed, generated and controlled by *Logic* or *Digital* electronic circuit devices, discussed in the next chapter.

WAVES

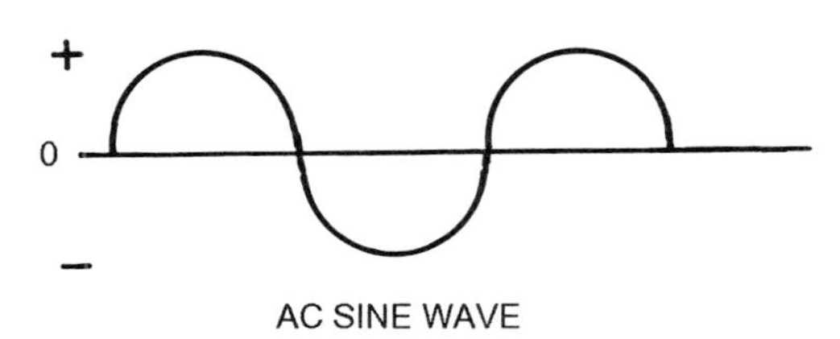

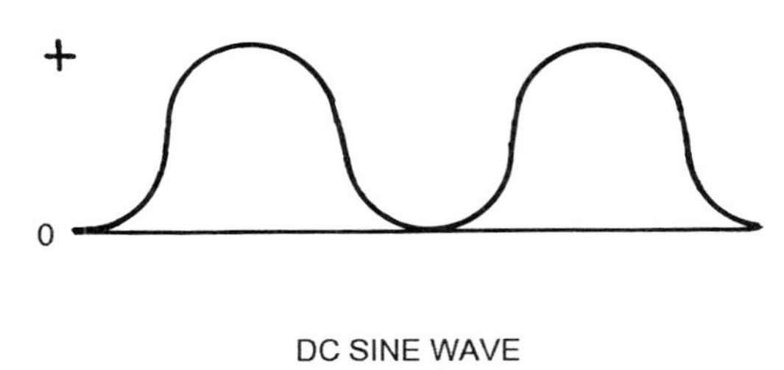

Fig. 14—Some Typical Waveforms

Unlike pulses created by the on-off flow of electrons, waves are created by varying a continuous flow of electrons within a circuit. This variation of current is accomplished by different types of devices within a circuit. In its basic form a manual variable resistor, known as a potentiometer (See Chapter 3), can vary the current flow and create waves within a circuit. Such a circuit is called an *Analog* or *Linear* circuit and the process that generates the waveform is known as **Modulation.** Obviously, manually varying the electron flow to control a constant waveform would be too slow to be of any use, so electronic devices are used to control and vary the electron flow. Waveforms can be created by either direct or alternating current to form signals (Fig. 14).

FREQUENCY

Controlled pulses and waves require a certain amount of time for one cycle to be completed. *Frequency* is the number of cycles occurring in one unit of time, generally one second. These cycles per second (cps) are measured in Hertz (Hz). The alternating current supplied to our homes is 60 Hz—a frequency of 60 cycles per second.

SIGNALS

The periodic waveforms (frequencies) caused by controlled variations of current can convey information in *Signals.* These signal waveforms can be modulated from devices designed to monitor various functions of a modern engine such as voltage, temperature, oil pressure, etc. These signal waveforms "trigger" digital circuits that display visual information to the operator. The information contained in the signal waveforms also can be stored in a computer chip and recalled later for diagnostic purposes when an engine system malfunctions. The use of signals is covered in more detail in the next chapter.

MULTIMETER

Just as circuit components in the field of electronics have become miniaturized, so too current flows have become smaller, so that the volts, ohms, and amperes within a circuit often total less than one. When this is the case, their names and values carry a "milli" prefix: millivolts, milliohms, and milliamps. To test for these small amounts in an electronic system, the very least you will need is a *Multimeter.* There are two types of multimeters that can be used to test electronic circuits.

ANALOG MULTIMETER

The Analog Multimeter (Fig. 15) gives less-precise value readings. It's also very hard to accurately gauge "milli" readings on the dial. It is best used for observing the trend of a slowly changing voltage, current, or resistance.

DIGITAL MULTIMETER

The Digital Multimeter (Fig. 16) is commonly referred to as **DMM**. It is highly accurate and easier to read than analog-type multimeters for certain signals. It is used to find the precise value of any type of voltage, current, or resistance.

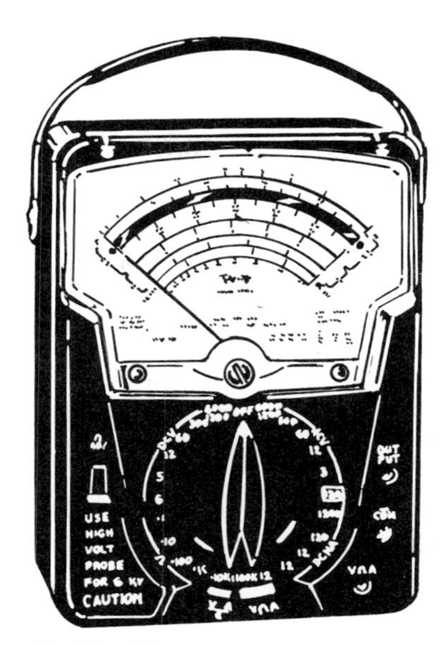

Fig. 15—A Typical Analog Multimeter

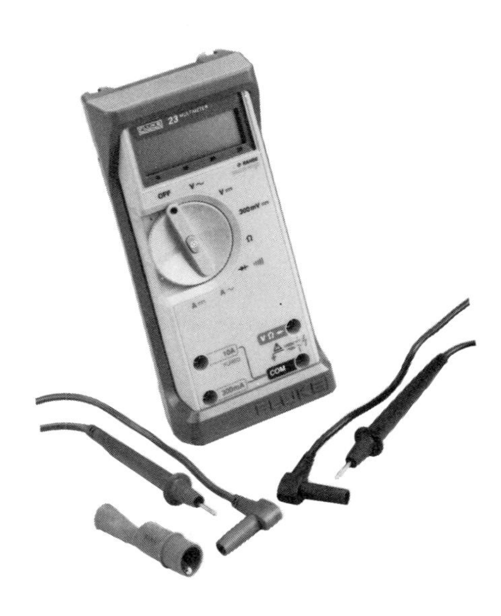

Fig. 16—A Digital Multimeter

Test Light vs. Multimeters

Both test lights and multimeters may be used to check the voltage in a circuit. Both operate by drawing current from the circuit that is being tested. A typical multimeter (digital or analog) draws 0.05 milliamps to operate and a test light draws 250 to 300 milliamps. This makes the current draw of the test light 5,000 times higher than the multimeter. The test light can be used for measuring battery voltage within a circuit, but for circuits with voltage lower than battery voltage a multimeter should be used for testing.

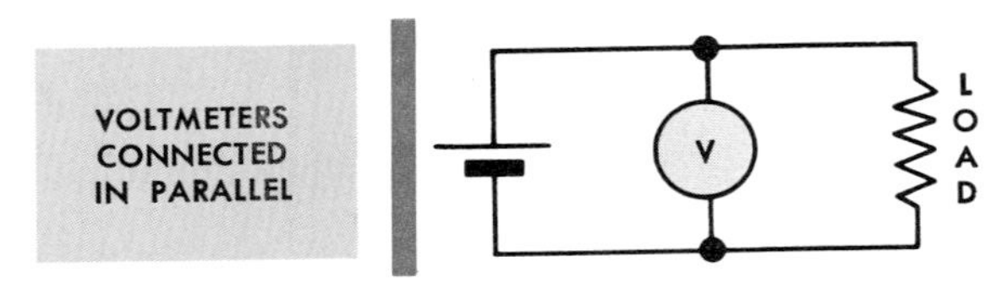

Fig. 17—Connect Voltmeter in Parallel

Whenever using these DMMs or any other test instrument, always follow the applicable instruction manual. However, we will make two points here. When using the DMM to measure voltage, connect the DMM in parallel (Fig. 17) by connecting the red lead to the high side of the circuit and the black lead to the grounded side. When using the DMM to measure current flow (amperes), connect it in series with the circuit (Fig. 18).

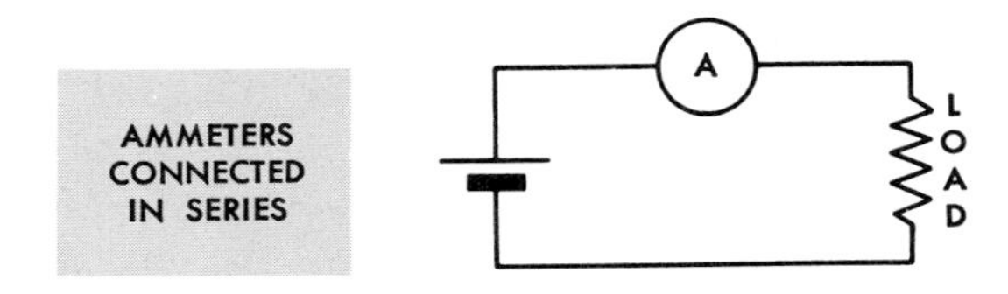

Fig. 18—Always Connect Ammeter in Series

SAFETY PRECAUTIONS

When using a DMM, double check the switch setting and lead connection before making measurements. Make sure you follow all the instructions in the manual. Disconnect the DMM or turn off the power before changing switch positions. Do not connect to circuits with voltage present when the switch or push buttons are in any ohms or current positions.

BASIC FAILURES OF CIRCUITS

Meters can be used in a circuit (Fig. 19) to locate four basic failures:

- **High Resistance**
- **Open**
- **Ground**
- **Short**

A HIGH RESISTANCE within a circuit can result in slow, dim, or complete failure of the component to operate. High resistance is caused by loose, corroded, dirty or oily terminals, or by broken strands within a circuit wire that reduce the capacity of the wire to carry current. Generally, circuits are protected from high resistance by circuit breakers and/or fuses, but high resistance may cause improper operation of a component and still not trip a circuit breaker or cause a fuse to fail.

An OPEN is a break in a circuit wire that results in a complete failure of the component to operate. An open in a circuit is caused by a failed protective device such as a fuse or circuit breaker, a disconnected terminal, a broken wire, or another failed component.

A GROUND in a circuit also results in component failure. A ground in a circuit is caused by non-insulated wire connections or frayed insulation on wire that accidentally or unintentionally comes in contact with the grounded frame of the unit.

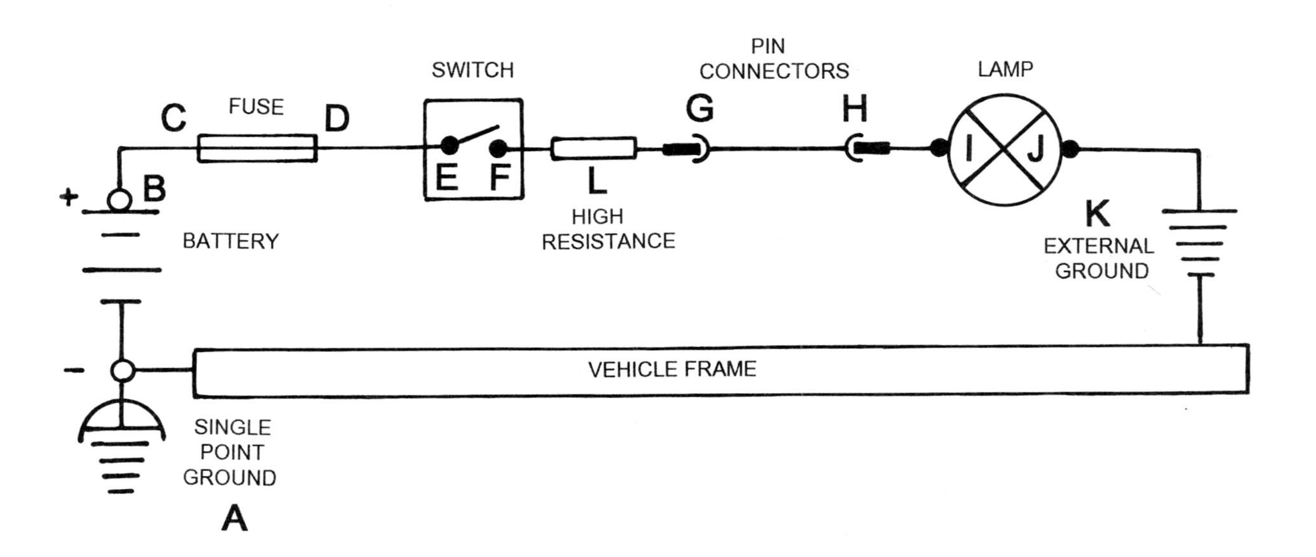

Fig. 19—A Typical Circuit for Testing

 Litho in U.S.A.

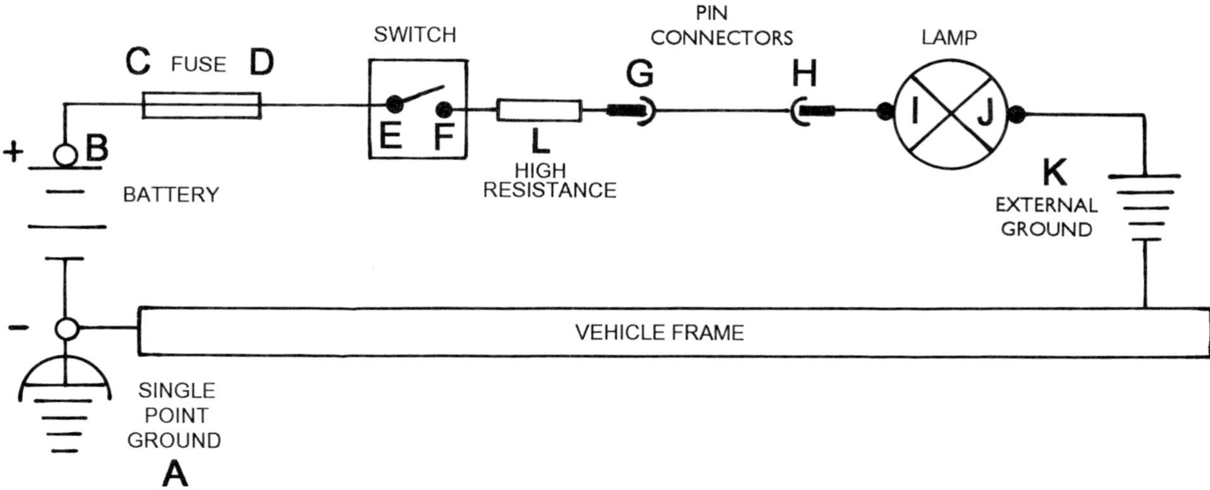

Fig. 20—A Typical Circuit with High Resistance

A SHORT usually results in two components operating when only one of two switches is turned on. A short happens when two wires from two different circuits inadvertently touch each other and make contact electrically.

BASIC TESTING OF CIRCUIT FAILURES

A High Resistance Circuit Failure

A high resistance failure in a circuit can result in an improper or component non-operation. In testing a circuit with high resistance (Fig. 20) we will do the following:

Remove the fuse from the circuit and visually inspect it for being warped or open. If the fuse is warped or cannot be visually inspected, check the fuse for continuity with a meter. Set the meter for the ohm or resistance position. Place each lead of the meter to each end of the fuse. The reading of the meter should read zero if the fuse is good. If any resistance is read on the meter the fuse should be replaced. If the meter reads infinite the fuse is open and should be replaced. If the fuse is found to be open, some high resistance within the circuit has caused the fuse to fail to protect the circuit. Replace the fuse if it is found to be open but do not operate the circuit until after the next step, otherwise you may have to replace the fuse again.

Visually inspect connections A, B, C, D, G, H, and K for being loose, corroded, or dirty. Clean and tighten terminals as necessary. If the components of the circuit are exposed to the environment, visually inspect connection E, F, I, and J for corrosion. Clean and/or replace components if necessary. Generally, in a circuit with high resistance, cleaning all connections within a circuit will solve the problem and complete the testing, but if the corrosion has worked its way under the wiring insulation, as in our example at point L, further testing must be done.

Disconnect connection I from the circuit. We start at point I because it is the beginning of the component that causes the load in the circuit. The connection is disconnected from the circuit to prevent the fuse from failing while testing for voltage. Set the meter to read DC voltage and connect the red lead to the wire at I and the black lead to the vehicle frame. Turn the controlling switch on and read the meter.

If voltage is less than battery voltage move toward the battery source within the circuit by disconnecting connections and checking for battery voltage, separately, with the meter at points H, G, F, E, D, and C. In our example, battery voltage would be found at point F, indicating the wire between points F and G should be replaced.

If the battery voltage was found at point I, you would have to reconnect the wire and move toward the ground source. Check voltages, separately, at the disconnected points of J and K until low voltage is found. In our example circuit, either the ground wire or lamp would have to be replaced.

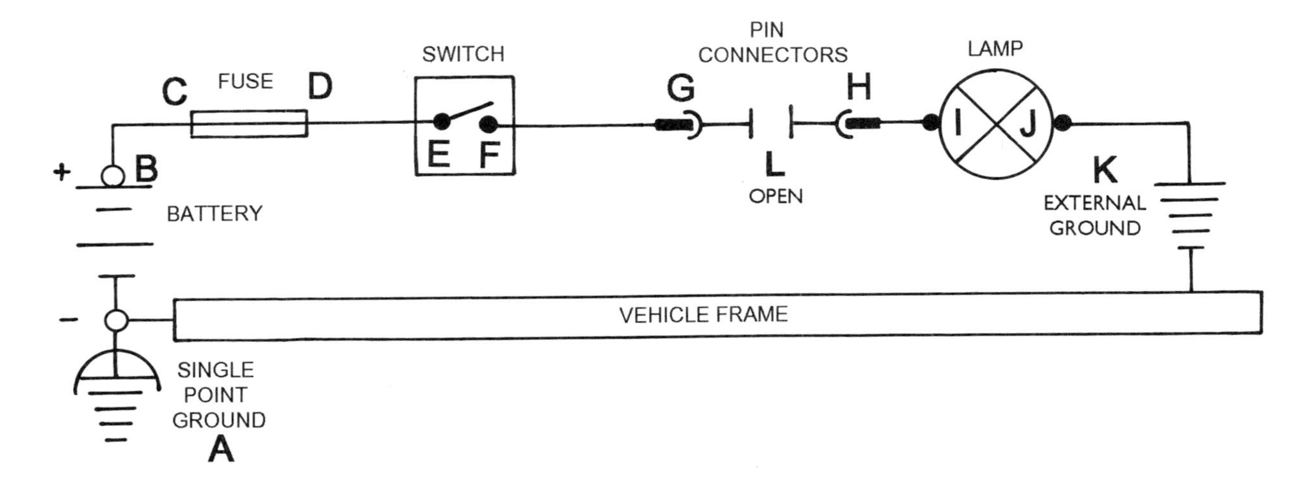

Fig. 21—A Typical Circuit With An Open

An Open Circuit Failure

An open circuit will result in component non-operation. In testing for an open in the circuit (Fig. 21) we will do the following:

Remove the fuse from the circuit and visually inspect it for being open. If the fuse cannot be visually inspected, check the fuse for continuity with a meter. Set the meter for the ohm or resistance position. Place each lead of the meter to each end of the fuse. The reading of the meter should read zero if the fuse is good. If the meter reads infinite the fuse is open and should be replaced. If the fuse is found to be open some high resistance within the circuit has caused the fuse to fail to protect the circuit. Replace the fuse if it is found to be open but do not operate the circuit until further testing.

Visually inspect connections A, B, C, D, G, H, and K for being loose, corroded, or dirty. Clean and tighten terminals as necessary.

Set the meter to read DC voltage and connect the red lead to any wire connection that is easily accessible between the points F and I and connect the meter's black lead to the vehicle frame. Turn the controlling switch on and read the meter.

In our circuit example, the meter will read zero voltage at points H and I and battery voltage at points F and G. This would indicate an open in the wire between points G and H and the wire should be replaced. When replacing a wire in a circuit make sure the proper wire gauge is used.

If battery voltage was read at all points between F and I then move to the next point toward ground until zero volts is found. Replace any component or wire prior to the point where the zero volts was found.

Fig. 22—A Typical Grounded Circuit

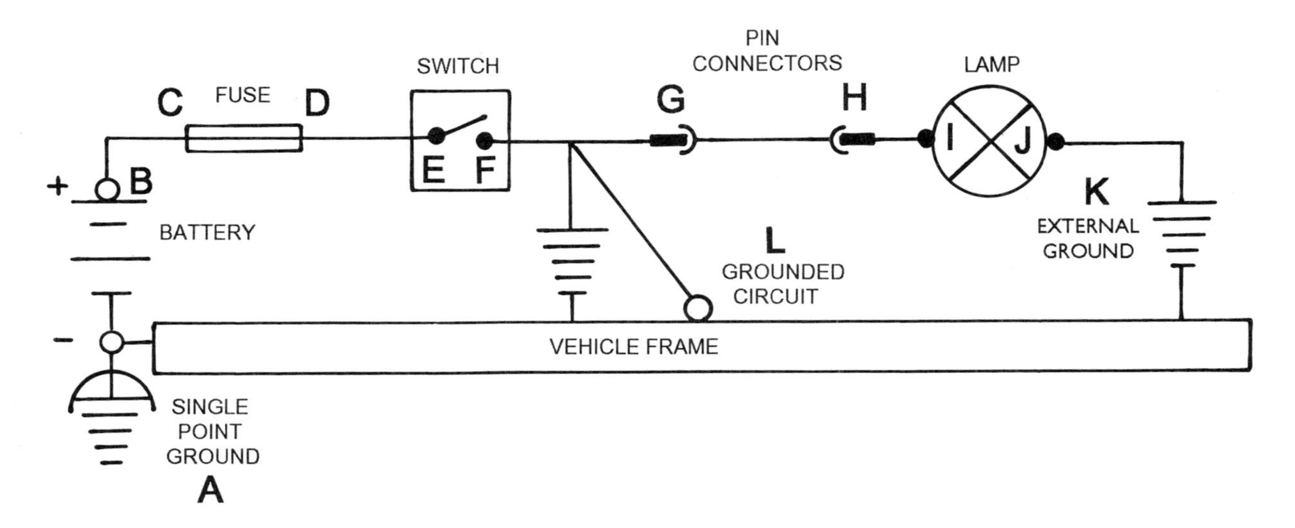

 Litho in U.S.A.

If zero voltage were read at all points between F and I then move to the next points toward the battery source until battery voltage is found. Replace any component or wire after the point where battery voltage was found.

A Grounded Circuit Failure

A grounded circuit results in component non-operation. If the battery or power wire is grounded, as it is in our example (Fig. 22), the fuse will be opened or failed.

Remove the fuse from the circuit and visually inspect it for being open. If the fuse cannot be visually inspected, check the fuse for continuity with a meter. Set the meter for the ohm or resistance position. Place each lead of the meter to each end of the fuse. The reading of the meter should read zero if the fuse is good. If the meter reads infinite the fuse is open and would indicate that the power wire is grounded. Do not install a replacement fuse until the grounded wire is found.

With the fuse removed, disconnect point K or external ground connection. Make sure point K connection does not come into contact with the frame during testing. Turn controlling switch of circuit on (closed). Set meter to read resistance or ohms. Check continuity of the circuit by touching the red lead to any point between D and I that is easily accessible. Touch the black lead to the frame of the unit or ground. In our example the reading on the meter will be zero, indicating the circuit is grounded. If the meter reading showed infinite the circuit would not be grounded.

Turn switch off or open and check continuity between point D or E and frame of unit. In our example the reading will be infinite, indicating circuit wire between D and E was not grounded. Disconnect connections at Point G. Check continuity of both connections of point G to frame or ground of unit. The blade or switch side of the wire will read zero, indicating a grounded wire between F and G. The reading at the other connection at G or to the lamp side will read infinite. The wire F-G will have to be replaced. If there were no connections in the circuit, the entire harness between points F and I would have to be replaced. After harness or wire is replaced, recheck circuit for continuity to ground.

A Shorted Circuit Failure

A shorted circuit will result in the operation of two or more circuits when either of the circuits are put into operation. To isolate the shorted circuit (Fig. 23) we will do the following:

Remove the fuses or circuit protection devices from both circuits. Depending upon the load of the circuits it may be possible that a fuse from either or both circuits may have failed during operation. Remove the

Fig. 23—Two Typical Circuits With A Short

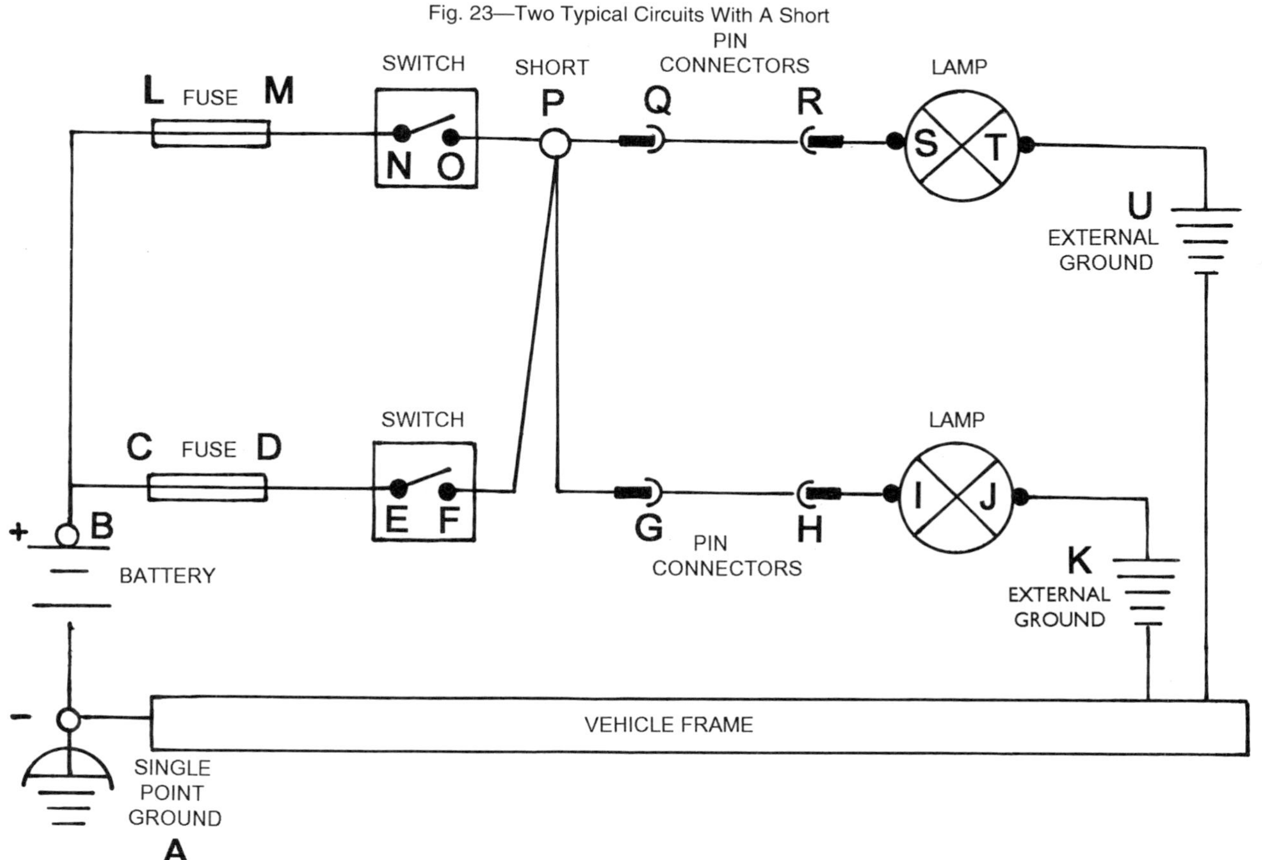

fuses from the circuits and visually inspect them for being open or warped. Replace any warped or failed fuse but do not install fuses until shorted circuit is found.

With fuses removed, disconnect point K and leave point U connected. Set meter to read resistance or ohms and close switch E-F. Touch the red lead of the meter to any points between D and I. Touch the meter's black lead to the frame or ground. In our example, the meter will read close to or at zero, indicating the circuit is shorted to the other circuit. Open switch E-F and check continuity at point D or E to frame or ground. The reading will be infinite indicating that wire (D-E) is not shorted to the other circuit. Disconnect any connections, separately, between points F and I and check continuity between each connection to ground. In our example, the switch side of connection G will read close to or at zero, indicating the wire F-G is shorted to the other circuit. The component side of connection G will read infinite on the meter.

Reconnect external ground K to frame and disconnect external ground U from the frame. Check continuity of points M through S to frame to locate shorted wire the same way we did in the lower circuit. In our example, the switch side of connection Q will have the zero or close to zero reading, indicating the shorted location. Be sure to use proper gauge of wire when replacing wires.

Note: It can be possible that both circuits can have the same external ground. If this exists, remove either points T or J and treat them as external grounds.

The service technician who has a good knowledge of electrical fundamentals and knows how to use test meters will soon find that this job becomes much easier.

OTHER TYPES OF ELECTRONIC TEST EQUIPMENT

Circuits in modern agriculture and industrial equipment have become so complex that using a multimeter to diagnose and test such circuits can be time-consuming. Because of this, test equipment has been designed that taps directly into the harness of the unit.

The test equipment shown here will give you a balanced group of aids for testing and troubleshooting complex electronic circuits. Shown are a few examples of the many models available.

ELECTRICAL DIAGNOSTIC RECEPTACLE KIT

Fig. 24—Electrical Diagnostic Receptacle Kit

The electrical diagnostic receptacle kit (Fig. 24) works with any digital, analog, or analog/digital meter. It is used for making measurements while a system is operating (as a "T" would be used in a hydraulic test kit).

SOLENOID TESTER

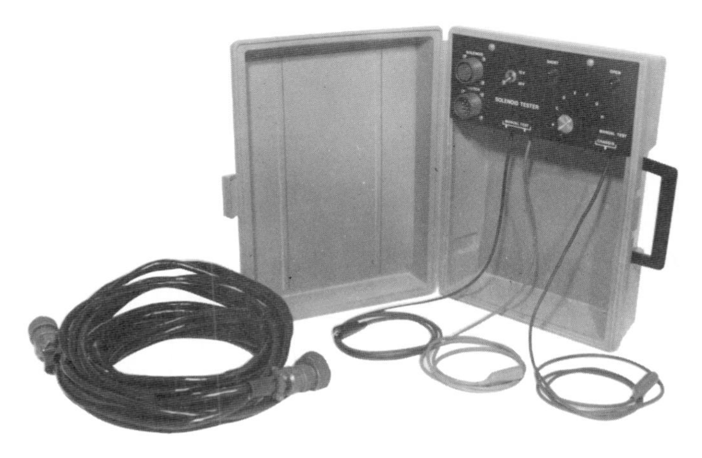

Fig. 25—Solenoid Tester

A solenoid tester (Fig. 25) tests solenoid-actuated poppet-type control valves. It determines whether a service problem is electrical or hydraulic.

COMBINE LEVELING CONTROL TESTER

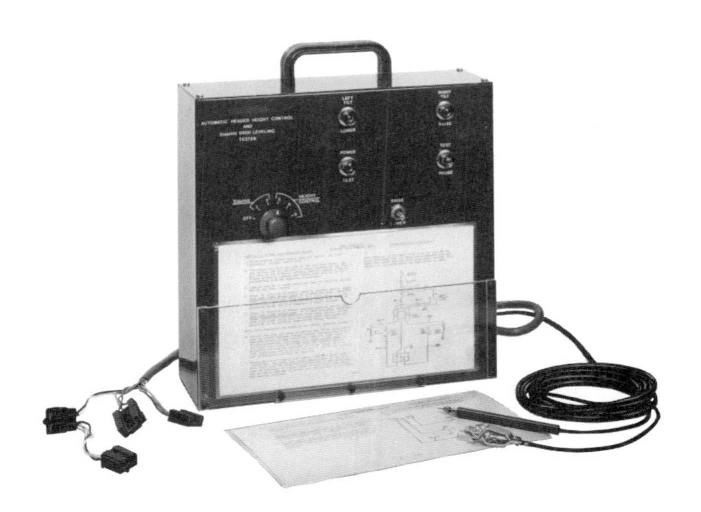

Fig. 26—Leveling Control Tester

The leveling control tester (Fig. 26) is used to test and diagnose automatic header height control and leveling control systems.

MOTOR GRADER HYDRAULIC FRONT-WHEEL DRIVE TESTER

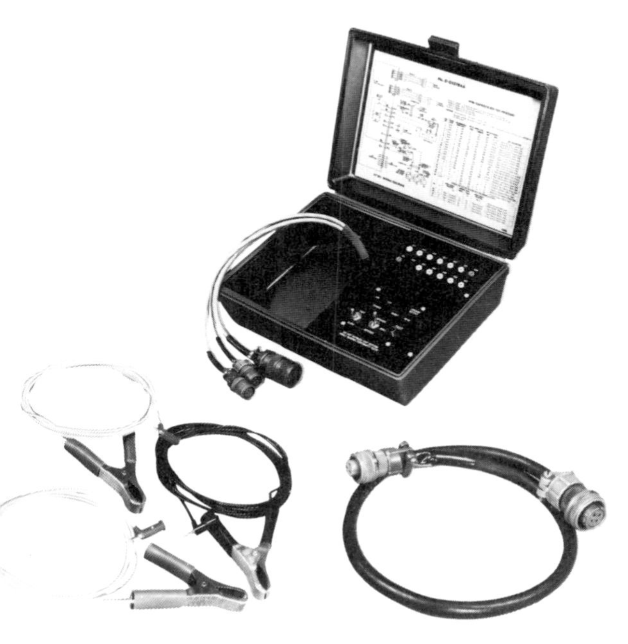

Fig. 27—Hydraulic Front-Wheel Drive Tester

The hydraulic front-wheel drive tester (Fig. 27) diagnoses problems in the electrical control system of front-wheel drive units. It checks for shorts and open circuits, and produces test signals to check the basic operation in both forward and reverse.

COMBINE MONITOR SYSTEM TESTER

The monitor system tester (Fig. 28) diagnoses performance monitoring systems both on and off the machine. It will isolate and diagnose individual system components.

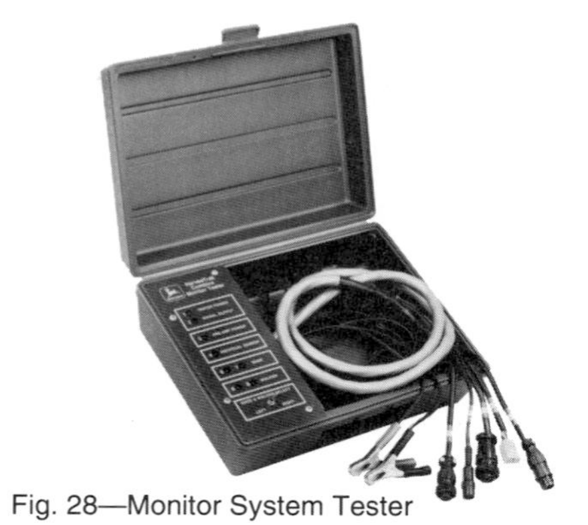

Fig. 28—Monitor System Tester

AUTOMATIC BLADE CONTROL SYSTEM TESTER

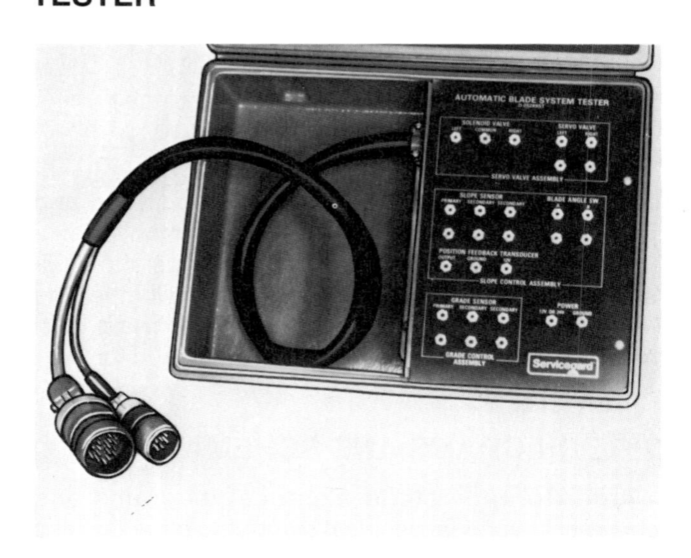

Fig. 29—Automatic Blade Control System Tester

The automatic blade control system tester (Fig. 29) works in conjunction with a multimeter. It checks servovalves, grade control sensor, sloper sensor, position feedback transducer, and blade angle switch for proper operation.

 Litho in U.S.A.

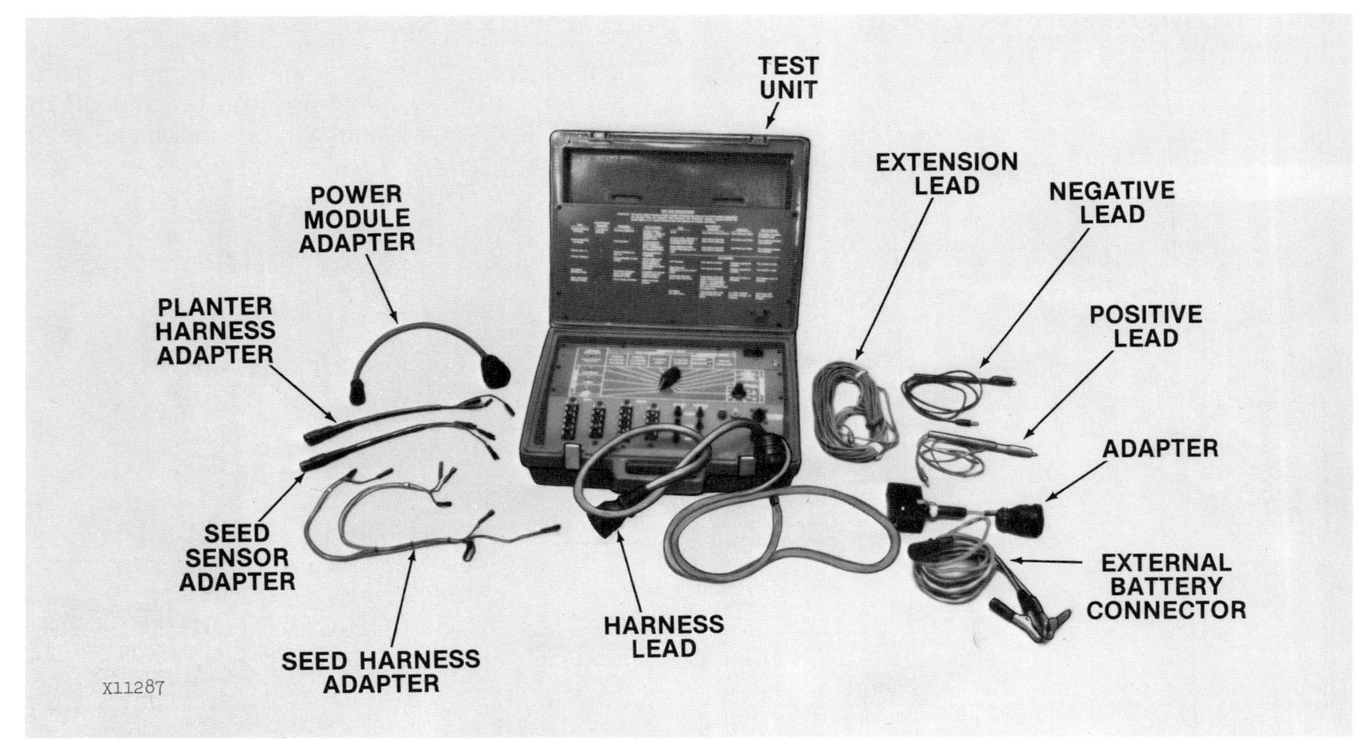

Fig. 30—Planter Monitor System Tester Kit

PLANTER MONITOR SYSTEM TESTER

Planter monitors use highly sophisticated seed and distance detection systems. The planter monitor system tester (Fig. 30) simulates seed and distance signals to allow the technician to determine whether the electronic components of the main system are performing properly.

WIRE DIAGRAMS AND SCHEMATICS

In order to understand electrical diagrams and schematics, various electrical standards organizations have tried to make diagrams and schematics more uniform. These groups include The International Electrotechnical Commission (IEC), International Standards Organization (ISO), Institute of Electrical and Electronic Engineers (IEEE), and American National Standards Institute (ANSI).

There are five electrical schematics and diagrams available to provide information to service and understand the operations of the electrical system of a machine. They are:

- **System Functional Schematic**
- **Subsystem Functional Schematic**
- **System Wiring Diagram**
- **Component Location Drawing**
- **Subsystem Diagnostic Schematic**

SYSTEM FUNCTIONAL SCHEMATIC

The System Functional Schematic is an electrical diagram of the complete machine and is made up of several foldouts of circuits divided into subsections. Each subsection is an electrical subsystem that contains one or more electrical circuits and is indicated by a letter/number and circuit description. When these subsections are laid out side by side they show a logical sequence of the relationship between all the various electrical devices and shows how they are connected to one another. The purpose of the System Functional Schematic is to show the operation, function, and interaction of each electrical subsystem of a machine. Each wire is identified by a number and/or color and all electrical devices are identified by a letter/number designation and description, and are represented by an international schematic symbol. When applicable, a device will also be represented by an SAE (Society of Automotive Engineers) pictorial symbol. The System Functional Schematic contains no harness or connector information.

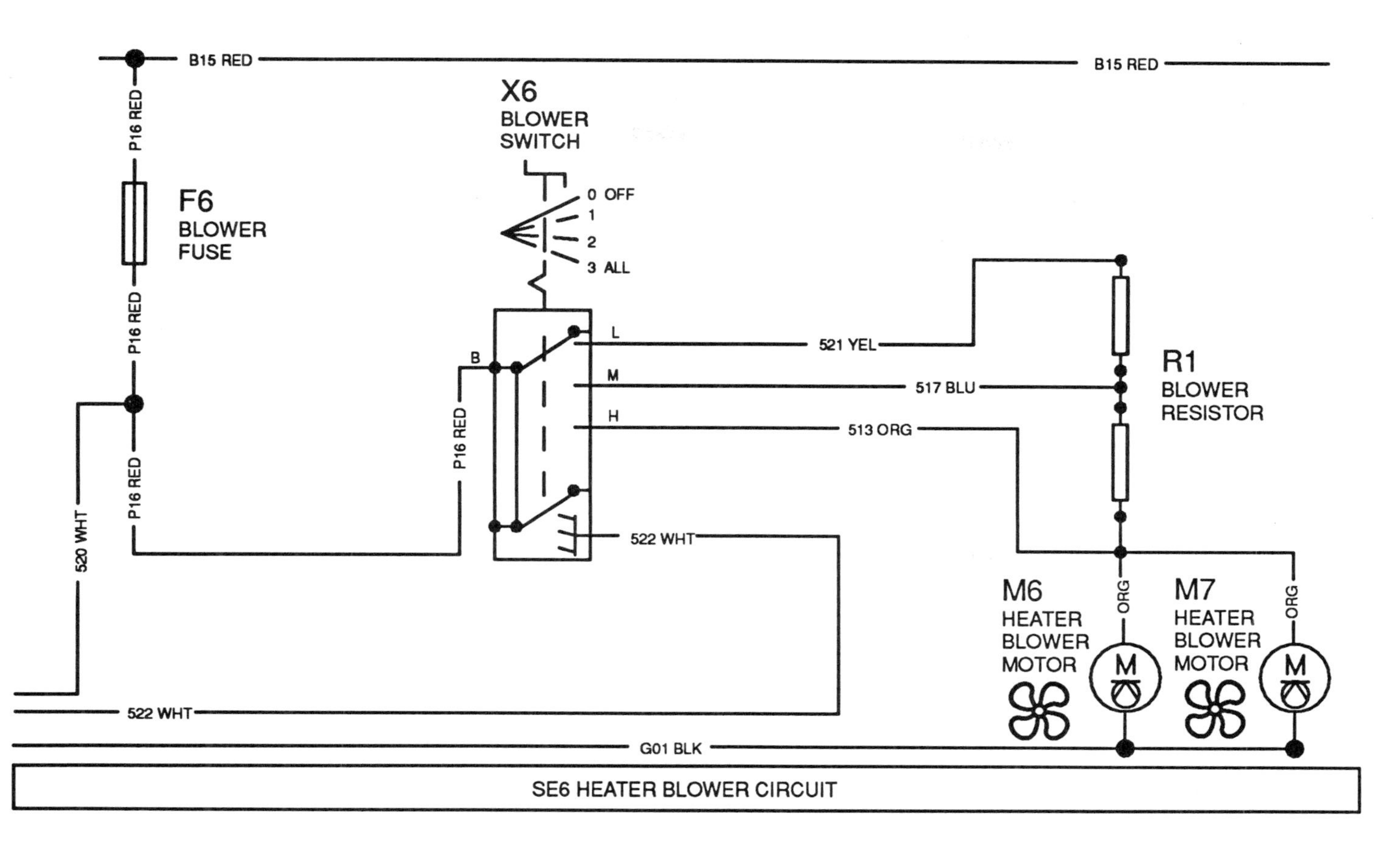

Fig. 31—A Subsystem Functional Schematic

SUBSYSTEM FUNCTIONAL SCHEMATIC

The Subsystem Functional Schematic (Fig. 31) is a sectional division of the System Functional Schematic and shows the same letter/number designations of wires, as well as component symbols. The section division circuit is identified in a rectangle at the bottom of the schematic. In our example, a heater blower circuit is shown and it is the sixth section (SE6) of the System Functional Schematic. All power supply wires are shown across the top of the drawing; the ground wires are shown across the bottom, with the components shown in between. The pictorials of the fans in our example are SAE symbols that indicate the function (blower fans) of the components (electrical motors).

SYSTEM WIRING DIAGRAM

A System Wiring Diagram (Fig. 32) is an illustration of the actual wire harness flattened out (dark lines) with extension of wires and connections to components. The wiring harness is laid out to allow identification of each connector, pin (male or female), wire numbers and/or color, and terminal location. Final terminal locations show the component symbol, letter/number identifier, and name of the component. A wire size given in mm^2 if it deviates from 1 mm^2 or 16-gauge wire. Specific wiring harnesses are identified in bold lettering (W15) at main connectors to allow one to follow a circuit through a diagram to additional diagrams. In our example (Fig. 32), note that the actual harness has wires and connections to the air conditioning pressure switch which were not shown in the Subsystem Functional schematic. The SAE symbol for air conditioning is a snowflake.

COMPONENT LOCATION DRAWING

A Component Location Drawing (Fig. 33) is a pictorial view of a harness showing location of all electrical components, connectors, harness main ground locations, and harness band and clamp locations. Each component is identified by the same identification letter/number and description used in the Subsystem Functional Schematic. When applicable, components also are shown with an SAE symbol.

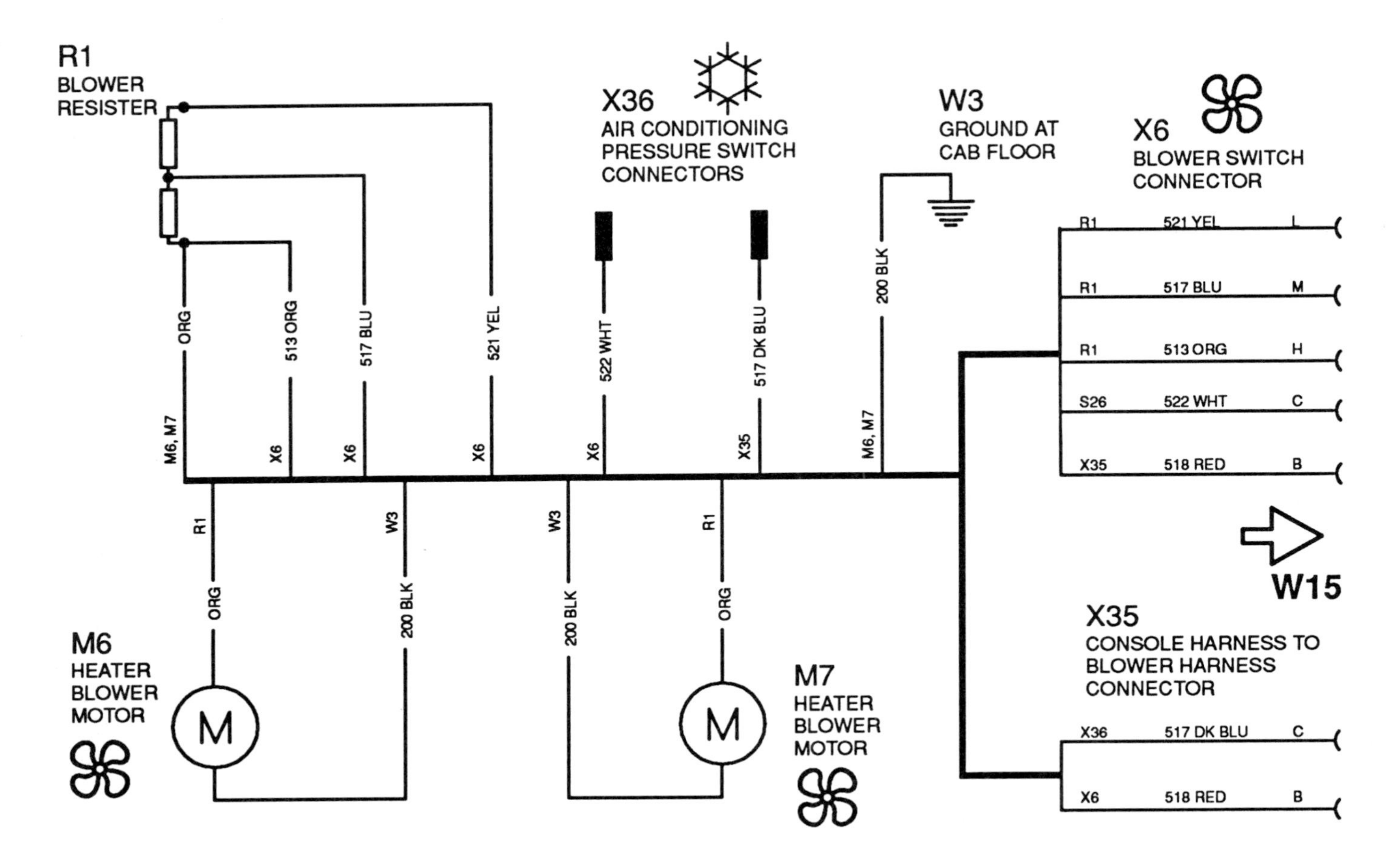

Fig. 32—A System Wiring Diagram

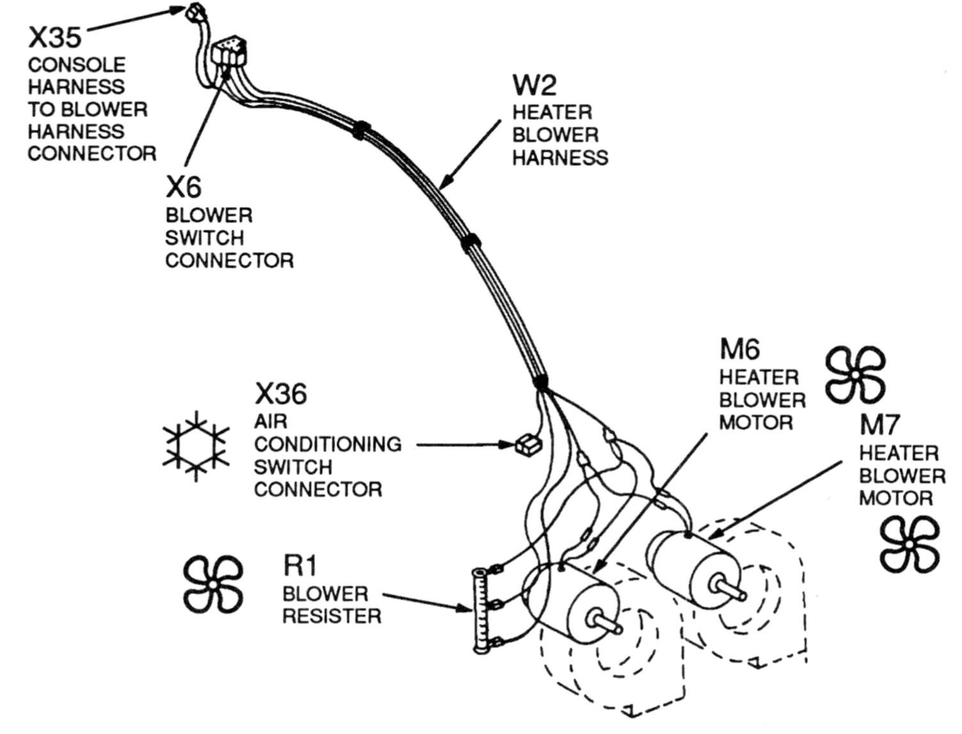

Fig. 33—A Component Location Drawing

Litho in U.S.A.

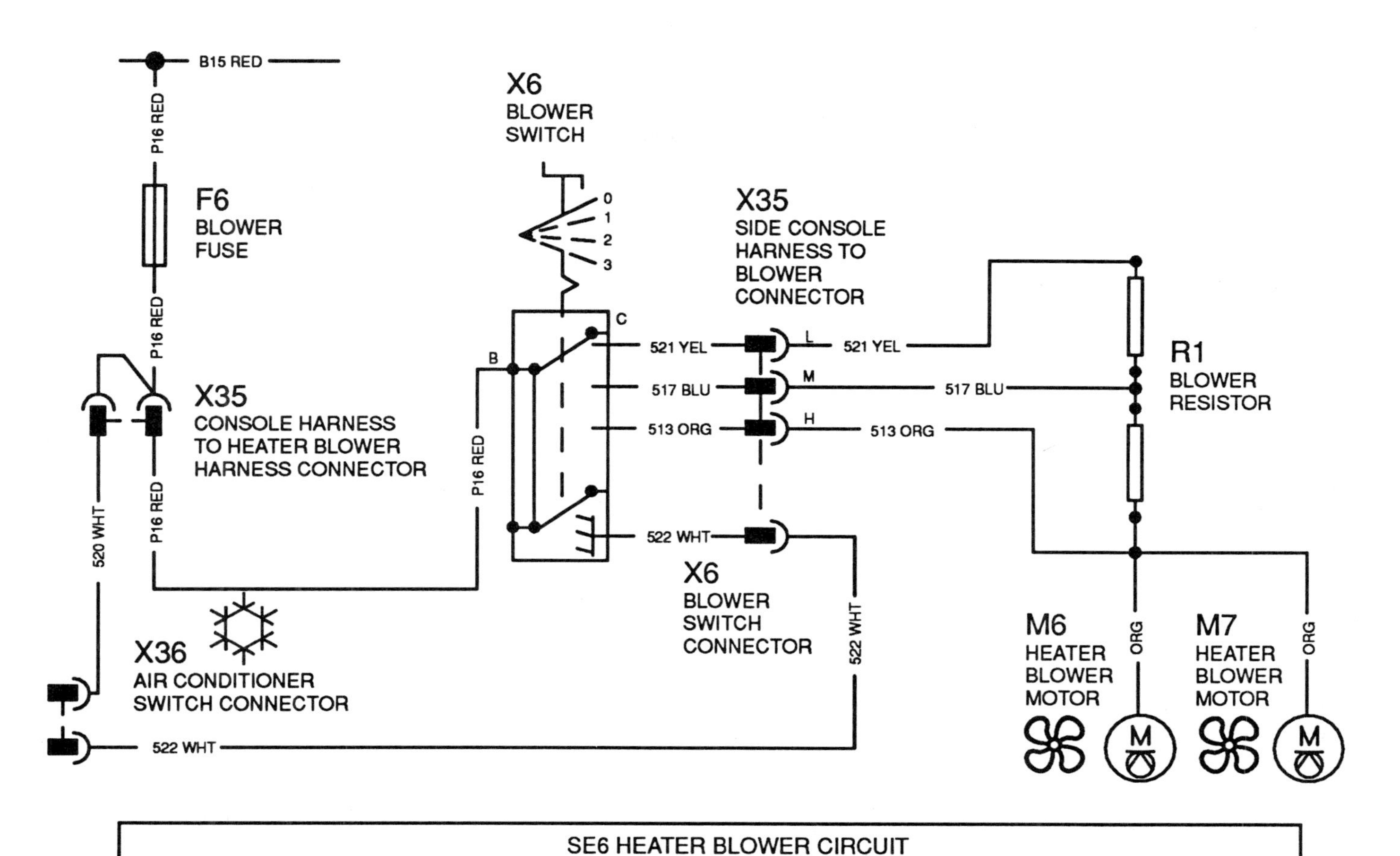

Fig. 34—A Subsystem Diagnostic Schematic

SUBSYSTEM DIAGNOSTIC SCHEMATIC

The Subsystem Diagnostic Schematic (Fig. 34) is a diagram that combines the Subsystem Functional Schematic with all harness connectors and pin locations to aid in diagnosing subsystems.

COMPONENT IDENTIFICATION LETTERS

Component identification letters have been developed by standard organizations to comply to all electrical components. Each electrical component and main harness connections will have an identification letter assigned to it. A number is added to the letter to separate and indicate the total components and main connections within that letter group. The letters I, O, and Q are not used. The following is a list of identifying letters and some examples of what they represent.

Identification letter		Examples
A	—	ABS control units, radios, and control units.
B	—	All types of sensors, horns, and microphones
C	—	Condensers and capacitors
D	—	Digital devices, pulse counters, and integrated circuits
E	—	Heaters, air conditioning, lights, distributors, and spark plugs
F	—	All protection devices such as fuses and circuit breakers
G	—	All power supply such as batteries, alternators, and generators
H	—	Any signal device such as alarms, buzzers, or signal lights

Identification letter	Examples
L —	Any inductor device such as coil windings
M —	Any electrical motor
N —	Regulators
P —	Any measuring instruments such as ammeters and tachometers
R —	Resistors
S —	Any switch
T —	Ignition coil or any transformer
U —	Converters and modulators
V —	Any semiconductors such as diodes
W —	All conductors of an electrical path
X —	All electrical connection devices
Y —	Any electrically-actuated mechanical device
Z —	Any electrical filter or suppressor device

An alphabetical listing of devices and their identifying letters are listed on page 22 in the back of this book.

WIRE NUMBER AND COLOR CODES

There is no set standard for wire number codes. Most major manufacturers of machines create their own number codes and list the information and show how to use these codes in their technical manuals, but most all manufacturers use standard color abbreviations to identify wires in a circuit. Red-colored wires are generally used for power source wires that can be traced back to the battery. Black-colored wires are generally used for grounds. Listed here are some of the abbreviations for colors.

Color	Abbreviation
Black	BLK or BK
Brown	BRN or BN
Red	RED or R
Orange	ORG or O
Yellow	YEL or Y
Green	GRN or G
Dark Green	DK. GRN or DG
Light Green	LT. GRN or LG
Blue	BLU or B
Dark Blue	DK. BLU or DB
Light Blue	LT. BLU or LB
Purple	PUR or P
Gray	GRY or GY
White	WHT or W

If the wire is bi-colored, both abbreviations are used, separated with a slash. BLK/WHT means a black wire with a white tracer.

DIAGRAM AND SCHEMATIC SYMBOLS

Standards organizations such as the American National Standards have developed international symbols for components that are used on electrical diagrams and schematics worldwide.

The Society of Automotive Engineers (SAE) also has established symbols to identify the operation and function of components on diagrams.

International Symbols

International electrical symbols (Fig. 35) have been established by standards organizations for components used for diagrams and schematics intended for worldwide distribution.

Fig. 35—International Electrical Symbols

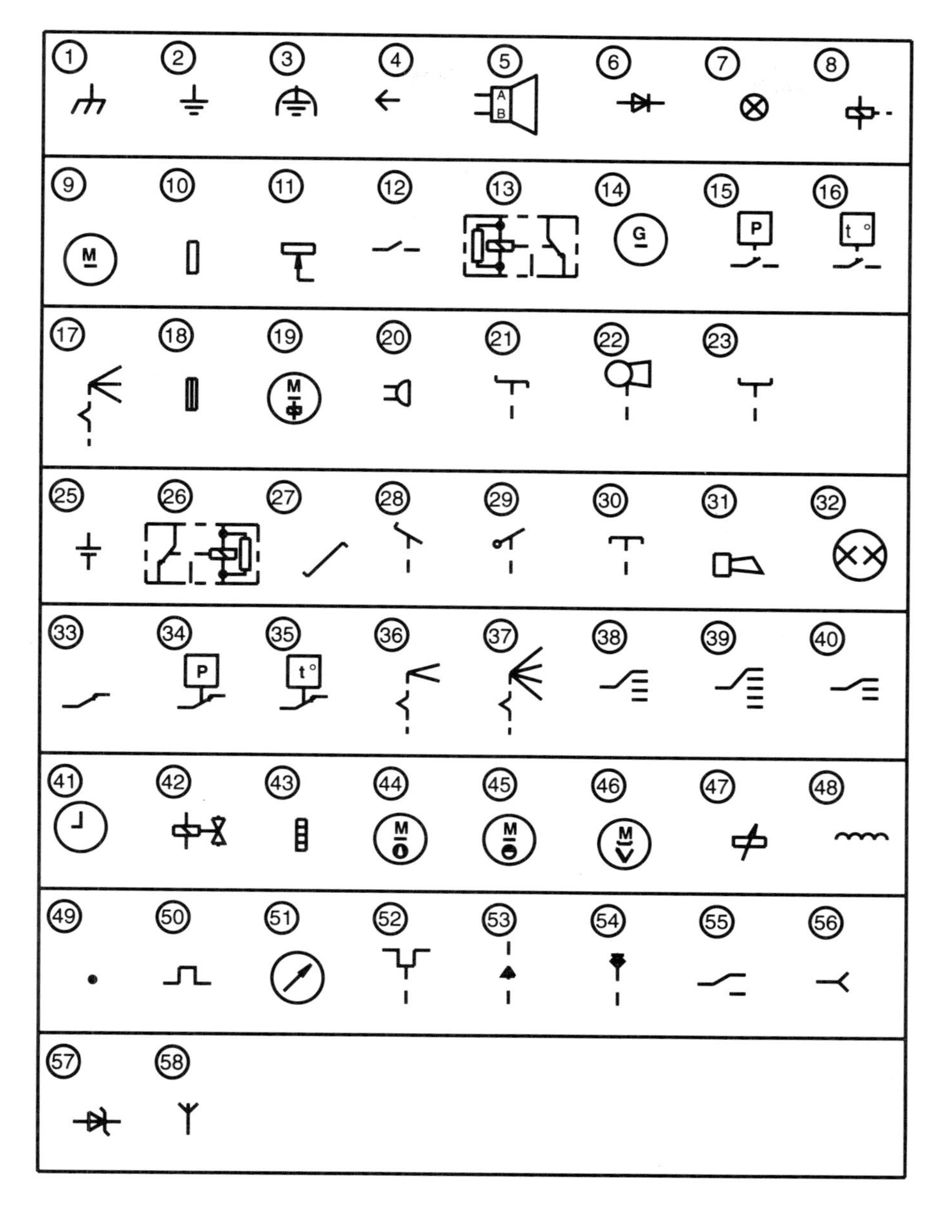

1. Frame ground
2. Ground
3. Shielded ground, (Single point ground)
4. Pin, male
5. Loudspeaker
6. Rectifying junction, (Semiconductor diode)
7. Lamp with single element
8. Actuator with one winding, (Solenoid)
9. Direct Current Motor
10. Fuse
11. Variable resistor
12. Make contact, normally open
13. Relay, normally closed
14. Direct current generator
15. Pressure activated switch, normally open
16. Temperature activated switch, normally open
17. Three position detent
18. Fuse
19. Direct current motor with solenoid
20. Buzzer
21. Operated by turning
22. Operated by key
23. Operated by pulling
25. Battery
26. Relay, normally closed
27. Twisted pair of wires
28. Operated by pedal
29. Operated by lever
30. Operated by pushing
31. Horn
32. Lamp with two filaments
33. Break contact, normally closed
34. Pressure activated switch, normally closed
35. Temperature activated switch, normally closed
36. Two position detent
37. Four position detent
38. Four position contact
39. Five position contact
40. Three position contact
41. Clock
42. Solenoid valve
43. Heating element
44. Direct current motor with fan
45. Direct current motor with pump
46. Windshield wiper motor
47. Variable resistor
48. Windings, (Coil of wire)
49. Connection point
50. Thermal effect
51. Analog meter
52. Thermal activated
53. Automatic reset
54. Operated by touching
55. One position contact
56. Socket, female
57. Zener diode
58. Antenna

Litho in U.S.A.

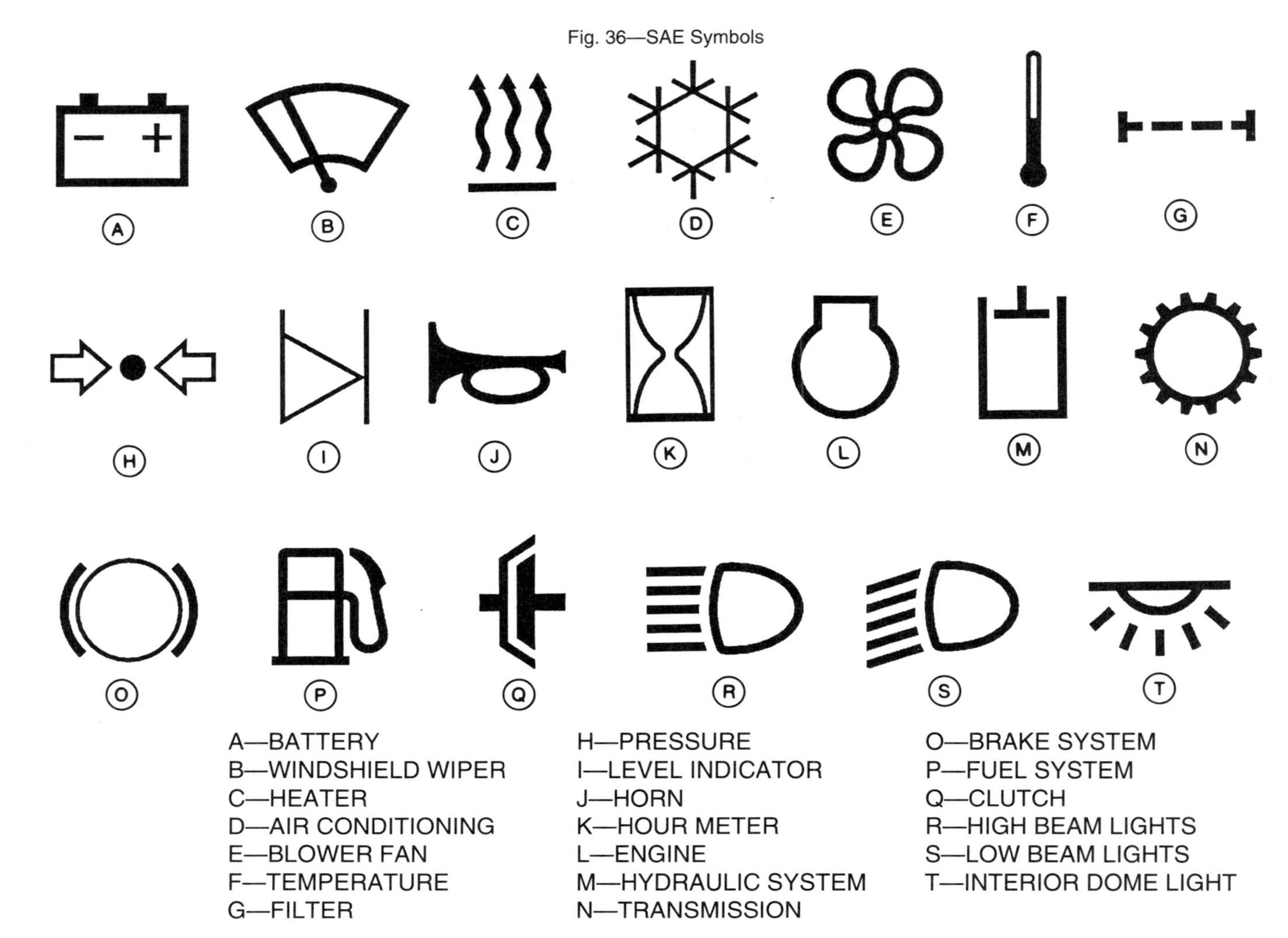

Fig. 36—SAE Symbols

SAE Symbols

SAE symbols are bold-faced symbols designed by the Society of Automotive Engineers. These symbols can easily be identified when reading an electrical schematic. Some of these symbols that may appear on electrical diagrams and schematics are shown in Fig. 36.

SUMMARY: MEASUREMENT OF ELECTRONS

In summary:

- **The measurement of volts, amperes, or resistance is calculated using Ohm's Law.**
- **The three types of electronic circuits are Series, Parallel, and Series-Parallel Circuits.**
- **Power (watts) is a measure of the rate at which electrical energy is converted into heat.**
- **Pulses are processed, generated, and controlled by Logic or Digital Circuits**
- **Waves are processed by varying a continuous flow of electrons within a circuit.**
- **Frequency is the measurement of time for one pulse or wave cycle and is measured in Hertz.**
- **The two types of Multimeters are Analog and Digital Multimeters.**
- **The basic failures of circuits are High Resistance, an Open, a Ground, or a Short.**
- **The five types of diagrams and schematics are System Functional Schematic, Subsystem Functional Schematic, System Wiring Diagram, Component Location Drawing, and Subsystem Diagnostic Schematic.**

TEST YOURSELF

QUESTIONS

1. A basic electronic circuit consists of a ___________ source, a ___________ such as a light bulb, and a ___________ such as copper wire.

2. What are the three basic types of electronic circuits?

3. Which of the basic circuits is used if a high resistance is required? Which is used if a low resistance is required?

4. In a parallel circuit each resistor receives the same ________________.

5. A pulse is a sudden ___________and_________ of direct ___________ flow within a circuit.

6. The variation of current within a circuit generates ________________.

7. Frequency is the number of ___________occurring in one second and is measured in ____________.

8. The two types of multimeters are ___________ and ________________.

9. Voltmeters are connected into a circuit in ________ while ammeters are connected in ___________.

10. Match the three basic failures of circuits with the correct descriptions.

a. Open or break	1. A wire touching metal
b. Ground	2. Two bare wires touching
c. Short	3. A loose wire

11. What should you make sure of before checking resistances within a circuit?

(Answers on page 18 at the end of this book.)

 Litho in U.S.A.

 Litho in U.S.A.

ELECTRONIC COMPONENTS / CHAPTER 3

INTRODUCTION

There are many different types of electronic components used to control the flow of electrons. This chapter will cover how they are made, how they work, and what they do within a circuit.

CONDUCTORS

Most metals are good conductors. The most widely used metal for a conductor is copper because of its cost and availability. Copper is one of the top six metals that have the least resistance compared to other metals (Fig. 1).

SILVER	0.936
COPPER	**1.000**
GOLD	1.403
CHROMIUM	1.530
ALUMINUM	1.549
TUNGSTEN	3.203

Fig. 1—Relative Resistance of Copper Compared to Other Metals

As you can see in Fig. 1, the relative resistance of copper is between silver and gold, two precious metals, and so is the best conductor available in terms of cost and availability.

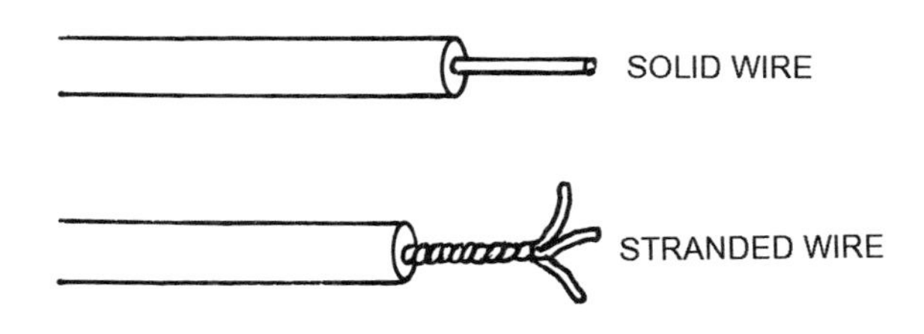

Fig. 2—Types of Wiring

Conductor wiring is solid or is stranded (Fig. 2). Solid wire is made with a single conductor and is generally called heavy-duty. Household wiring is a solid type of wire conductor. Stranded wire is two or more twisted or braided conductors. All conductor wiring is insulated with a covering of plastic or rubber.

There are many sizes of wires. The larger the number of the wire, the smaller it is. Therefore, No. 12 wire is smaller than No. 10 wire. This numbering system is known as the American Wire Gage (AWG) or Brown and Sharpe Wire Gage (B&S). The unit of measure for wire is the circular mil (0.001 in. dia.). No. 12 gage wire is 101.9 mil (0.10 in. dia.) and No. 10 gage wire is 80.8 mil (0.08 in. dia.).

All conductors offer some resistance to the flow of electrons. Wire must be selected which has an acceptable amount of resistance for the needs of a circuit. Let's show the effect of wiring resistance on the operation of a circuit.

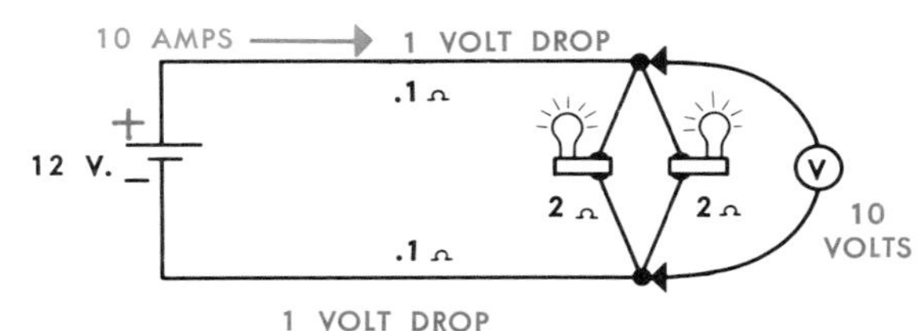

Fig. 3—Resistance in a Conductor (Copper Wire)

Fig. 3 shows a circuit which has two headlamps connected to a 12-volt battery with two copper wires each having a resistance of 0.1 ohms.

Each headlamp has a resistance of two ohms and are in parallel with the circuit. Using the current-divider rule for parallel circuits, the total resistance of the two headlamps would be:

$$\frac{R_1 \times E}{R_1 + R_2} = \frac{2 \times 2}{2 + 2} = \frac{4}{4} = 1 \text{ ohm}$$

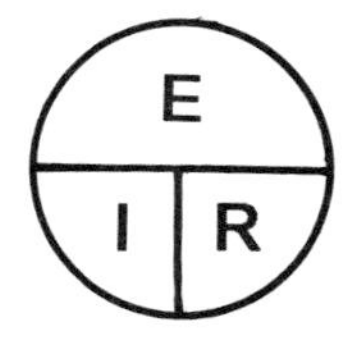

Fig. 4—Ohm's Law Formula Circle

The total resistance of the wires and headlamps in the circuit would be 0.1 + 1 + 0.1 or 1.2 ohms. To find the current of the circuit, use Ohm's Law Formula Circle (Fig. 4), where I = E/R or 12/1.2 = 10 amperes.

The voltage drop in each wire is E = IR = 10 x 0.1 = 1 volt, or two volts total for both wires.

The voltage left to operate the headlamps, or load, is 10 volts, since the sum of the voltage drops must equal the source voltage, or 1 + 10 + 1 = 12 volts.

The copper wire "robs" the headlamps of two volts, leaving 10 volts across the headlamps to provide illumination.

Litho in U.S.A.

The wiring used in any circuit must allow sufficient voltage across the load for proper operation. The smallest wire that will not create too much voltage drop is normally used.

Resistance in a wire depends upon:

1) the length of the wire.

2) the cross-sectional area of the wire.

3) the temperature of the wire.

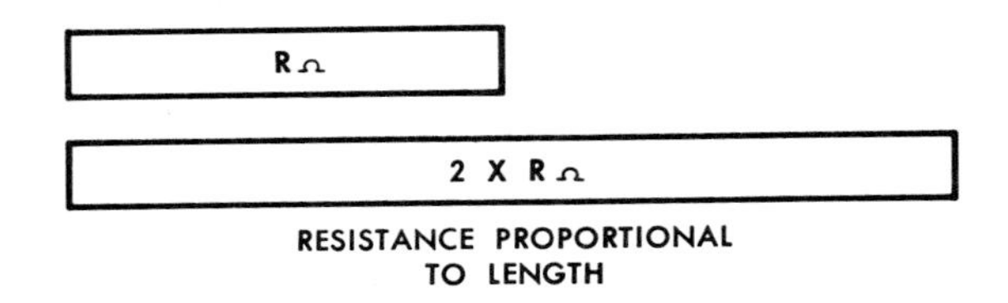

Fig. 5—A Longer Wire Contains More Resistance

If the length of a wire is doubled, the resistance between the wire ends is doubled (Fig. 5). In other words, the longer the wire, the greater the resistance between the wire ends.

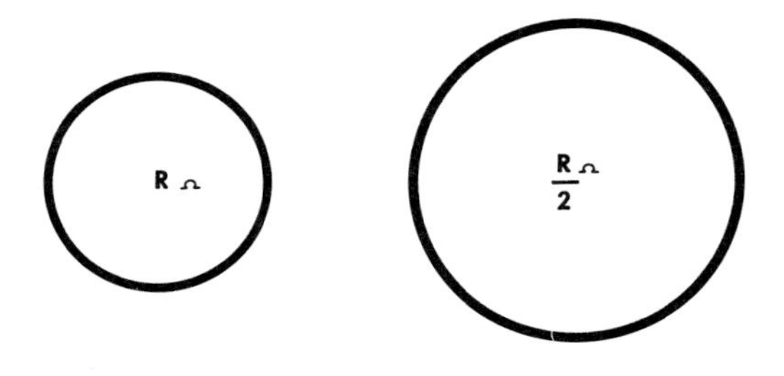

Fig. 6—A Smaller Wire Contains More Resistance

If the cross-sectional area of a wire is reduced by half, the resistance for any given length is doubled (see Fig. 6). That is, the smaller the wire, the more will be the resistance, and the larger the wire, the less will be the resistance.

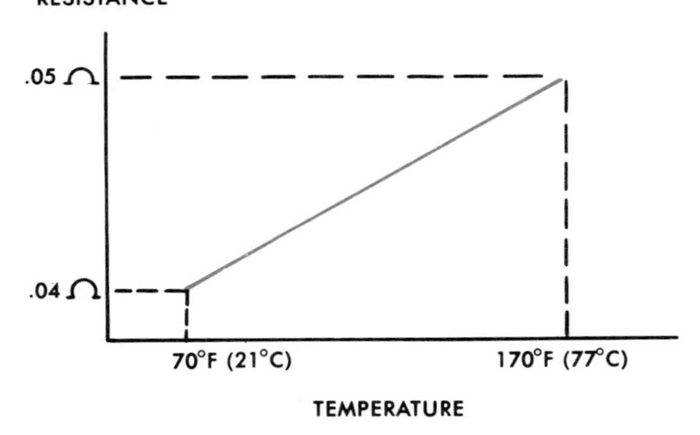

Fig. 7—A Hotter Wire Increases Resistance

As the temperature rises in a wire, the resistance increases.

Fig. 7 shows an example of a 10-ft. (3-meter) length of wire having a resistance of 0.04 ohm at 70° F (21° C).

At 170° F (77° C), the resistance is 0.05 ohm, or an increase of 25 percent.

Excessive resistance in the wiring from normal heat can hinder the performance of electrical equipment. This is why the selection of the proper wiring is so important.

Heat is developed in any wire carrying current because of the normal resistance in the wire. If the wire gets too hot, the insulation will be damaged.

In summary:

- **The longer the wire the greater the resistance.**
- **The smaller the wire the greater the resistance.**
- **The hotter the wire the greater the resistance.**

SWITCHES

A **switch** is an electronic device which opens and closes the flow of electrons or current. When a switch is turned off it causes an *open* within a circuit, stopping the flow of current. When a switch is turned on or *closed,* current is then directed to various other components within a circuit.

Fig. 8—Symbol for a Single-Pole Single-Throw Switch

The simplest switch is the single-pole single-throw (SPST) switch (Fig. 8).

Other types of switches (Fig. 9) are single-pole double-throw (SPDT), double-pole single-throw (DPST), and double-pole double-throw (DPDT).

There are many different ways of actuating switches:

1. A manually-actuated switch is controlled by an operator. It can be of a simple single-pole, single-throw (SPST) type or use multiple poles and positions.

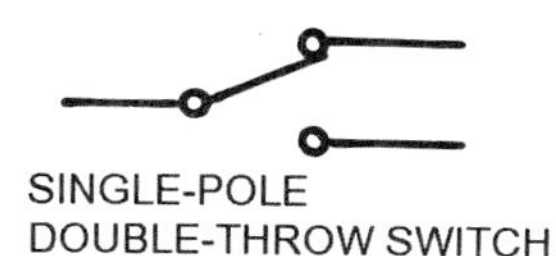

SINGLE-POLE
DOUBLE-THROW SWITCH

DOUBLE-POLE
SINGLE-THROW SWITCH

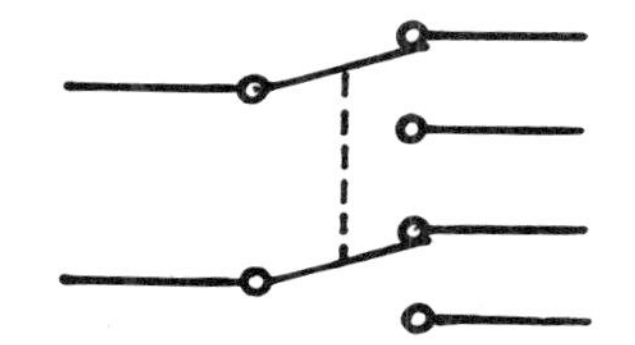

DOUBLE-POLE DOUBLE-THROW SWITCH

Fig. 9—Symbols for Other Types of Switches

Fig. 10—A Neutral Start Switch

2. A mechanically-operated switch is actuated by a mechanical device. These switches also are called "motion complete" or "limit" switches. Such a switch is a neutral start switch (Fig. 10). The neutral start switch (A), generally located on the transmission, is between the key switch and starter circuit. It is mechanically operated from the transmission shifter. The switch is closed only when the transmission is in the neutral position, completing the circuit to the starter. In any other position the switch remains open.

Fig. 11—Air Compressor Cycle Switch

3. Pressure switches use a pressure change to open or close switch contacts. An example is an air compressor cycle switch (Fig. 11). This switch is pressure-activated and is generally open until low pressure closes the switch, which then completes the current flow to the pump clutch, causing pressure to increase.

Fig. 12—A Magnetic Switch

4. A magnetic switch (Fig. 12) is activated by a magnetic field that opens or closes the switch.

Switches are used for:

1. Controlling circuits—Opening and closing current pathways to turn a circuit on or off.
2. Selection—Multi-position switches can provide a variety of possible paths and combination of paths for current to flow.
3. Sensing—These switches send a signal to some other type of device to relay information.

Some of the other major types of switches (Fig. 13) used are:

- **Ignition Key**
- **Toggle**
- **Push-Pull**
- **Cutout**
- **Pressure**
- **Limit-Position Switch**
- **Multiple Contact**
- **Push Button**
- **Twist**
- **Rotary**
- **Reed**
- **Mercury**

 Litho in U.S.A.

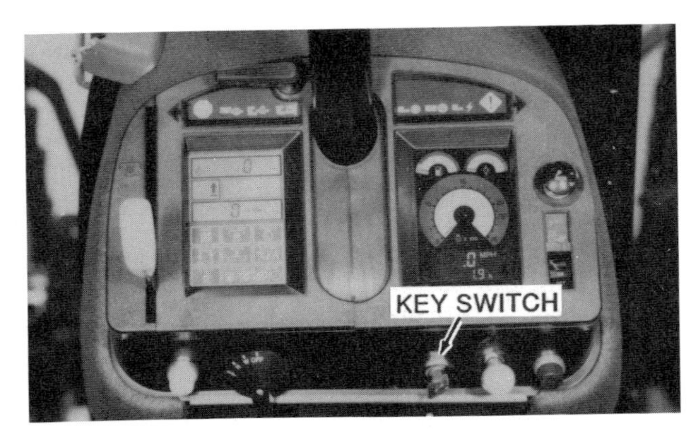

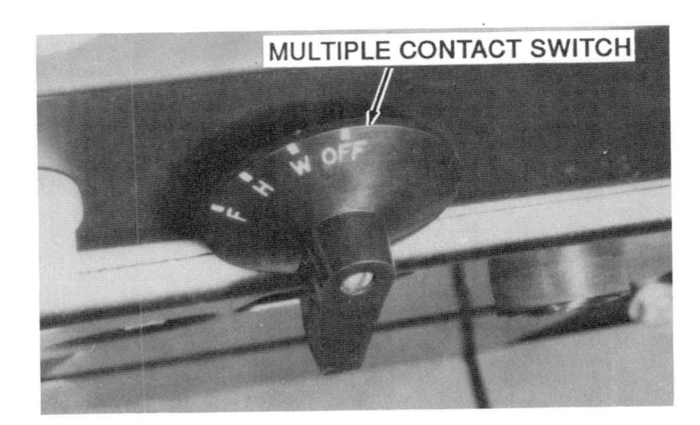

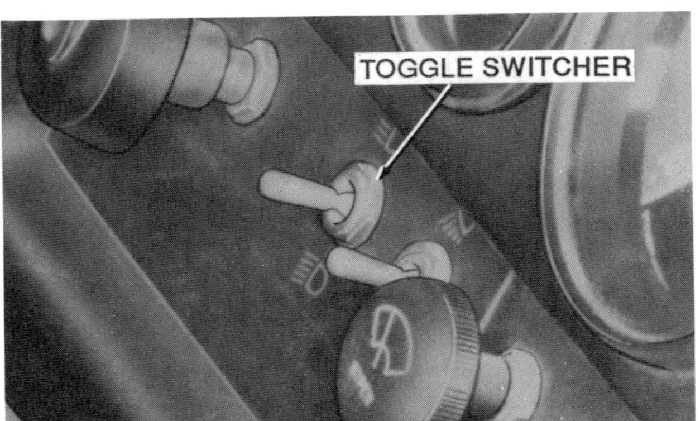

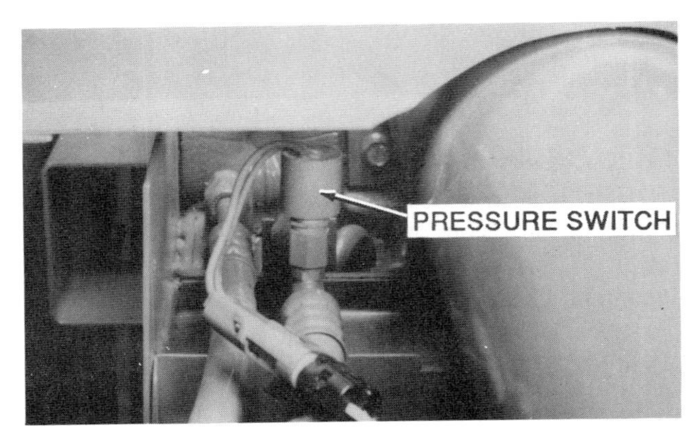

Fig. 13—Other Types of Switches

IGNITION KEY SWITCHES must have a key inserted before they can be operated. Used for starting switches because of safety and to avoid theft of the machine.

TOGGLE SWITCHES are simple on-off flip switches. Used to control small auxiliary circuits.

PUSH-PULL SWITCHES are usually on-off switches. Used for two-way control as of a simple lighting circuit.

CUTOUT SWITCHES are used to break an electrical circuit as during emergencies. They may be actuated manually or automatically.

MULTIPLE CONTACT SWITCHES may use a knob or a key which is turned to various positions to make or to break contact with different circuits. Used for complex circuits such as combination lighting or for variable speed control of a fan or heater.

PUSH BUTTON SWITCHES are momentary switches which are moved in only one direction to open or close a circuit. Used for simple jobs such as sounding a horn or buzzer.

TWIST SWITCHES give the same on-off control as push-pull types, except that they give a rotary action. Used for two-way or two-speed control of simple circuits such as windshield wipers.

 Litho in U.S.A.

PRESSURE SWITCHES are operated by an outside force from oil, water, air, or gas. Usually they are spring-type units which open or close a circuit automatically in response to pressure. Often these switches are used as sending units for oil pressure lamps, etc.

ROCKER SWITCHES (Fig. 14) are usually two-way control switches. They may or may not have LED lighting.

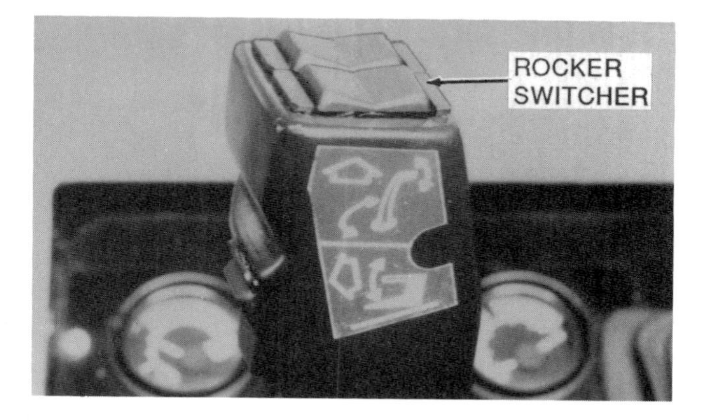

Fig. 14—A Rocker Switch Used to Lower a Harvesting Unit

A variation on the twist switch is the ROTARY SWITCH. This switch is often used with a fan to complete a circuit when the fan reaches a certain speed. It also shuts off when the fan drops below this speed.

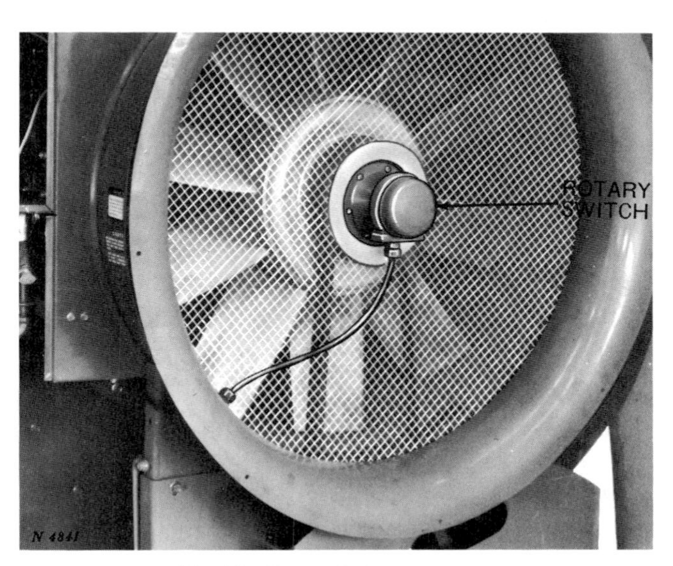

Fig. 15—Rotary Switch on a Fan

A good example of a rotary switch is one that is used with the fan on a crop dryer (Fig. 15). When the fan reaches a predetermined speed, the rotary switch will activate, completing an electrical circuit to the gas burner. This assures that the fan is always running before the burner starts. When the fan slows down, it will again activate the switch, shutting off the burner. Thus the switch is a protective device against possible failure of the fan and consequent heat build-up.

Fig. 16—Thermostatic Temperature Control Switch

A rotary switch is also used as a thermostatic temperature control switch in air conditioning systems (Fig. 16). This is a rotary-type switch with a gas filled temperature sensing tube inserted in the evaporator core. The switch end of the sending tube uses a diaphragm to control two external contacts wired to the compressor clutch. When the cab air needs to be cooled (to a preselected temperature setting inside the cab), the gas in the sensing tube expands the diaphragm completing the circuit in the switch and engaging the compressor clutch. The compressor continues to operate until the preselected cab temperature is reached.

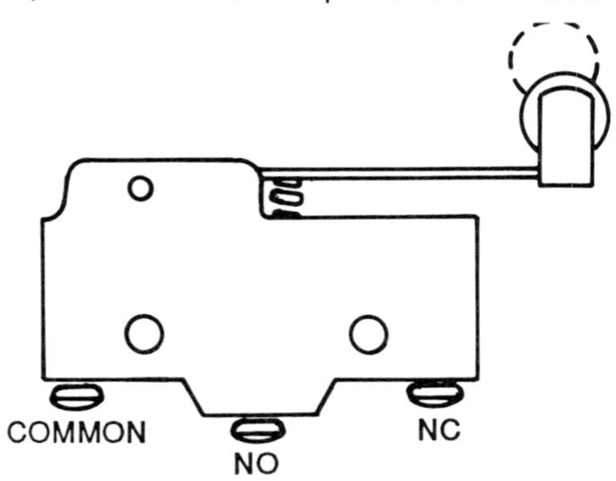

Fig. 17—Bale Size Limit-Position Switch

Another special type of switch is the LIMIT-POSITION SWITCH (Fig. 17), sometimes referred to as a microswitch. It is a miniature unit activated by a light pressure with a small amount of travel. It usually has three terminals: a "C" common terminal to which one wire is connected; a "NO" or normally open terminal, to which the other wire is connected if the circuit is open except when the bale is depressed; and an "NC" or normally closed terminal for another wire if the circuit is closed except when the bale is depressed.

 Litho in U.S.A.

Another switch that is smaller yet is the REED SWITCH. In this switch, the contacts are tiny reeds of metal that are actuated by a magnetic field to close or open the circuit. The reed switch may be a sealed unit and the magnetic windings may also be enclosed in the switch, making it a tiny relay about one-eighth inch in diameter and three-eighth inch long.

MERCURY SWITCHES are glass tubes which contain hermetically-sealed stationary electrodes and mercury. When the tube is tilted, the mercury pool moves which makes or breaks contact between the electrodes. These switches are practically silent in operation and have long lives even when used with high current loads.

SERVICING OF SWITCHES

If a switch is defective, always replace it with a new switch. **Do not void the safety of a defective switch by disassembling and trying to repair it.** It is usually cheaper and safer to replace it.

Important: Always be sure to replace a defective switch with one with the same electrical rating. A switch with too low a rating can be burnt out.

RELAYS

Relays are really electrically controlled switches that allow a small current to control a larger current.

Fig. 18—A Relay with Cover Removed

A relay (Fig. 18) is a device that has a coil winding around an iron core, a movable armature with a contact that is mounted over the core and winding, and another stationary contact that is aligned with the contact on the armature. When the windings are energized, the core becomes magnetized, causing the armature and contact to move to or away from the stationary contact. Thus, a small current to the windings and core can control (open or close) a much larger current to some other circuit through the contact points.

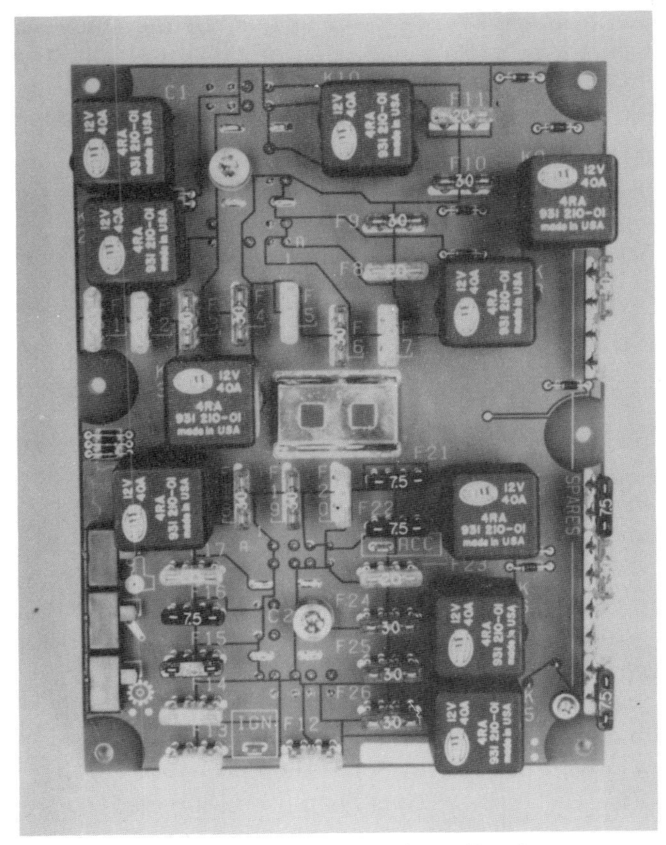

Fig. 19—Relays on a Load Center Board

Relays are operated by switch devices and are generally mounted to a load center board (FIg. 19). Some of the components that require a higher current and are controlled by relays are lights, blower fan motors, fuel shutoffs, wiper motors, etc. Note that the relays in Fig. 19 are labeled as 12-volt and 40-amp relays.

Fig. 20 shows the terminals at the bottom of a relay and a relay diagram. Most relays will have their terminals numbered. In our example, the diagram of the relay shows that when a small current is directed to terminal 85 the coil of the relay is energized, magnetizing the core and moving the armature and contact connected to terminal 30 to the stationary contact at terminal 87. Terminal 30 is generally connected directly to the battery (+) and terminal 87 is connected to the component.

To check a load center board relay, remove the relay from the load center board and connect battery voltage to terminals 85 and 86. Take care to read a wiring diagram so you know which terminal is connected to ground. In our example (Fig. 20), terminal 86 is the ground terminal, so the battery ground (—) wire should be connected to this terminal and the battery (+) wire should be connected to terminal 85. If the

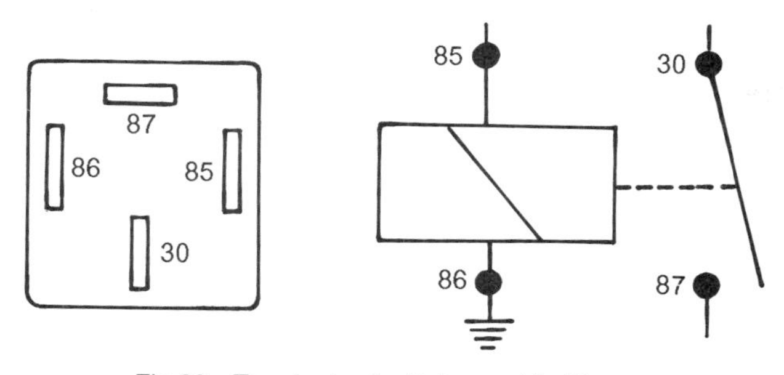

Fig 20—Terminals of a Relay and its Diagram

polarity of the battery were reversed, the magnetism of the core would be on the opposite end of the core and the armature will not move to a closed position. Once the battery is correctly connected to terminals 85 and 86, check the resistance across terminals 30 and 87 with a multimeter set to read resistance. The resistance should read zero. If the reading were infinite the relay would have to be replaced.

So far, we have only talked about one kind of relay that controls one circuit of one component. Some relays are designed with multi-high current contacts that allow one relay to control several devices simultaneously.

There are many different types of relays. Some of these types will be covered here:

- **Starter Relays**
- **Cut-out Relays**
- **Horn Relays**
- **Tell-tale Relays**

Let's discuss each relay in detail.

STARTER RELAYS

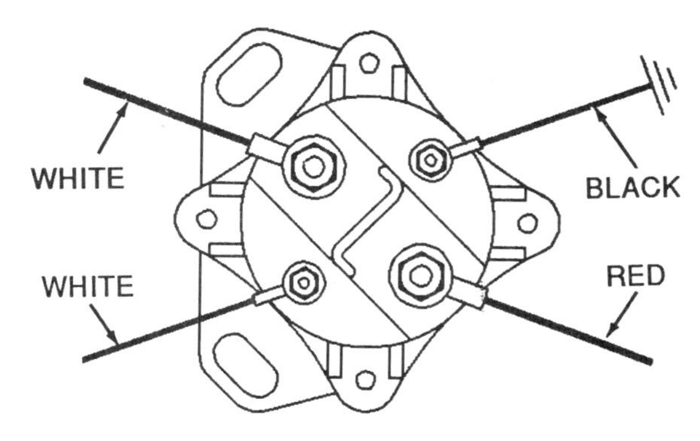

Fig. 21—A Starter Relay

A starter relay (Fig. 21) provides high current to the starter for starting an engine only if the transmission is in neutral.

Operation of Starter Relays

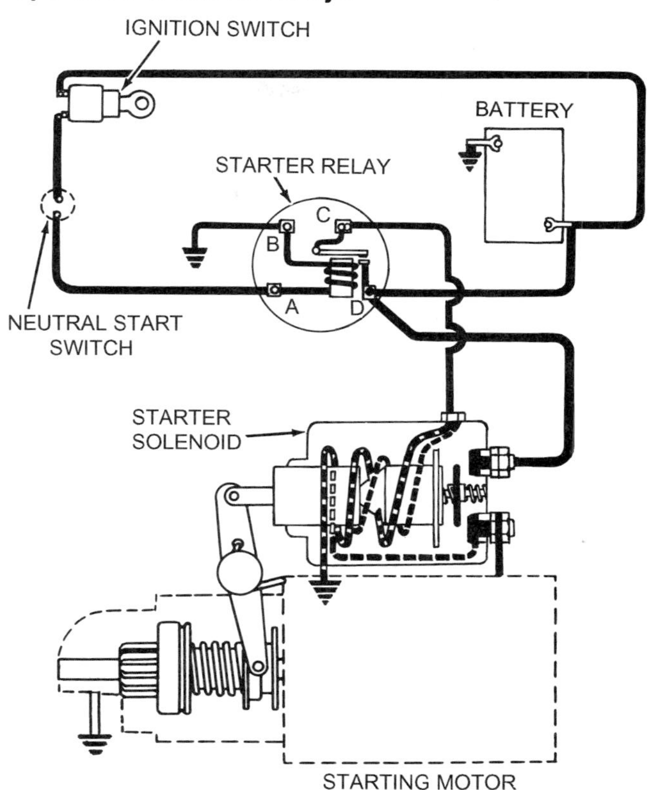

Fig. 22—Starter Relay Operation in a Typical Starting Circuit

The coil windings of the starter relay (terminal A) cannot be activated until the transmission is in neutral and the ignition switch is in the start position (Fig. 22). With the transmission in neutral, the neutral start switch is closed, closing the circuit from the ignition switch to terminal A of the starter relay. When the ignition switch is turned to the start position, the coil windings of the starter relay is energized, causing the relay contacts to close. With the contacts closed, battery voltage is delivered to the starter motor solenoid windings, causing the solenoid plunger to move and close the main battery connection of the starter solenoid to the starter motor. This completes the circuit and causes the starting motor to operate. Once the engine is started, the ignition switch is released from the start position to the run position, disabling the starter relay and starter motor.

Checking the Starter Relay

1. Remove the wire connection from terminal C (Fig. 23) of the starter relay to prevent accidental engagement of starter motor during checkout.

2. Using a multimeter set to read DC voltage, check voltage at terminal D by placing the red lead of the meter on terminal D and the black lead of the meter to

 Litho in U.S.A.

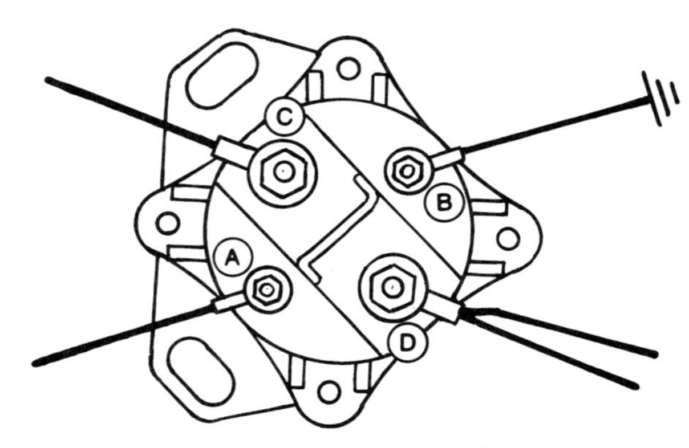

Fig. 23—Checking a Starter Relay

ground or frame of the unit. Reading should be battery voltage. If no voltage is noted, there is an open or high resistance in the wire to the battery. Clean connections at the battery and starter relay or replace wire.

3. If voltage is OK in step 2, place a jumper wire to terminals A and D. Check voltage at terminal C to ground with a multimeter. Reading should be battery voltage, indicating the contacts of the starter relay have closed properly. If no voltage is noted, the starter relay is defective and should be replaced.

Operation of Starter Relays (Early)

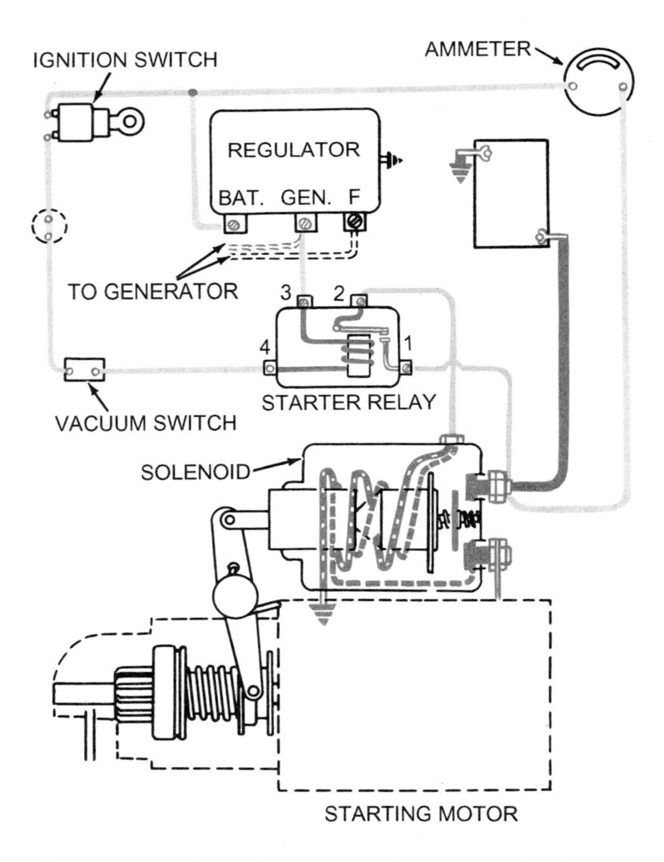

Fig. 24—Starter Relay Operation in a Typical Early Starting Circuit

On some early machines, starter relays were connected to the alternator regulator circuit to prevent the starter motor from operating after the engine was started.

When the ignition switch is turned to the start position, current flows from the battery through the ammeter, ignition switch, neutral start switch, vacuum switch, starter relay winding, and then to the regulator of the generator or alternator and to ground (Fig. 24). The neutral start and vacuum switches are added protection and may not be used in some applications.

As current flows through the windings of the starter relay, the relay contacts close, directing current to the windings of the starter solenoid. The starter solenoid engages the battery terminal of the starter solenoid to the starter motor, completing the circuit and causing the starting motor to operate.

When the engine starts, the buildup of alternator voltage through the regulator opposes the battery voltage so that the current flow through the starter relay windings is reversed, causing its contacts to open. This cuts the current flow to the starter solenoid windings and the starting motor stops operating. This action occurs even though the ignition switch is in the start position, and so prevents the starting motor from operating while the engine is running.

Checking the Starter Relay (Early)

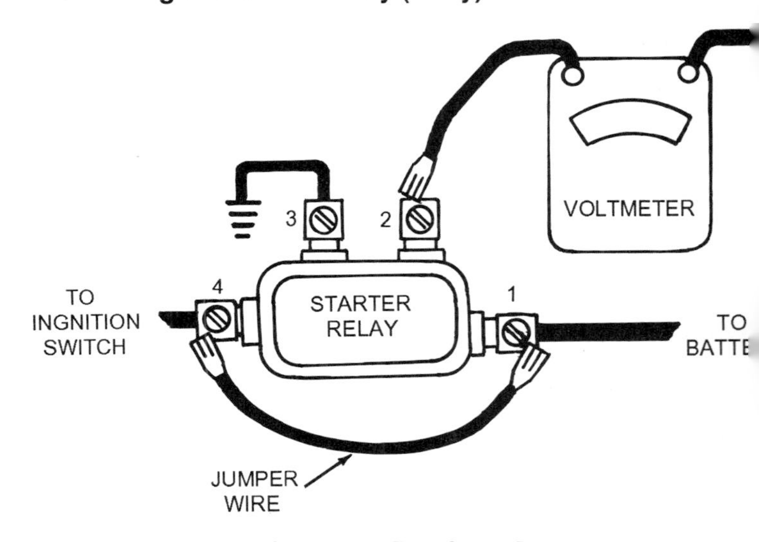

Fig. 25—Checking an Early Starter Relay

Refer to figures 24 and 25 for terminal numbers of the starter relay.

1. Disconnect wire from terminal 2 of the starter relay to prevent accidental engagement of starter motor during check.

2. Disconnect wire to terminal 3 of the starter relay from the regulator and connect it to ground or frame of the unit.

3. Place a jumper wire between terminals 1 and 4 of the starter relay.

4. Using a multimeter set to read DC voltage, check voltage at terminal 2 of the starter relay by placing the red lead of the meter to terminal 2 and the black lead of the meter to frame of unit or ground. Reading should be battery voltage. If no voltage is noted, the starter relay is defective and should be replaced.

Cutout Relays

The cutout relay is used to open the circuit between the generator and battery when the generator slows (idle) during operation. The cutout relay closes when engine speed increases and the battery is being charged. Since the cutout relay is a part of the charging circuit, it is covered in detail in Chapter 6 under "Generator Regulators."

HORN RELAY

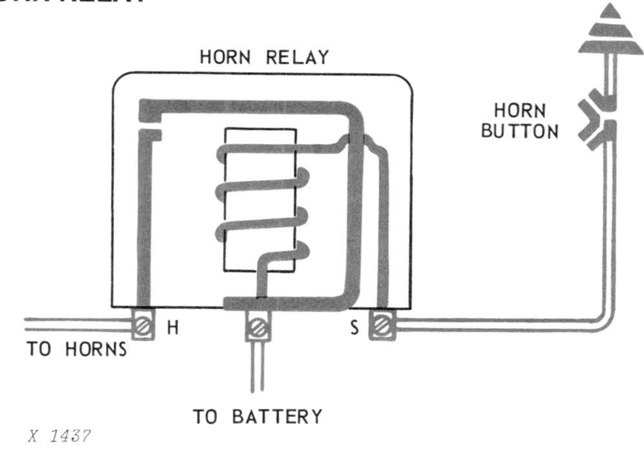

Fig. 26—Horn Relay Circuit

The function of the horn relay is to close the circuit between the horn and the battery when the horn button is pressed. A typical horn relay circuit is shown in Fig. 26.

When the horn button is pressed, the circuit from the battery is completed through the horn relay winding to ground and then back to the battery. The magnetism created in the winding by the current pulls the armature toward the core, closing the contacts. This completes the circuit between the horns and battery and causes the horn to operate. When the horn button is released, the flat spring on the back of the relay armature pulls the contacts apart.

NOTE: A light-duty horn relay may also be used as an indicator lamp relay. A normally-open sensor takes the place of the horn button and an indicator lamp is connected to terminal H of the relay. When the sensor closes (failure), the indicator lamp lights, telling the operator a failure has occurred and indicating what has failed.

Checking the Horn Relay

To check the horn relay (See Fig. 26), connect a voltmeter from the relay horn terminal H to ground, and use a jumper wire to terminal S and to ground.

The voltmeter will read battery voltage if the points are closed. If the voltmeter reads zero volts, the contacts are not closed, and the relay is defective and must be replaced.

TELL-TALE RELAY

Tell-tale relays are relays that operate with an electrical load of a circuit such as tail lights and/or turn signals. They are generally used on trucks and buses where indicator lamps are lighted on an instrument panel for normal operation and then go out when a failure occurs.

Tell-Tale Relay Operation

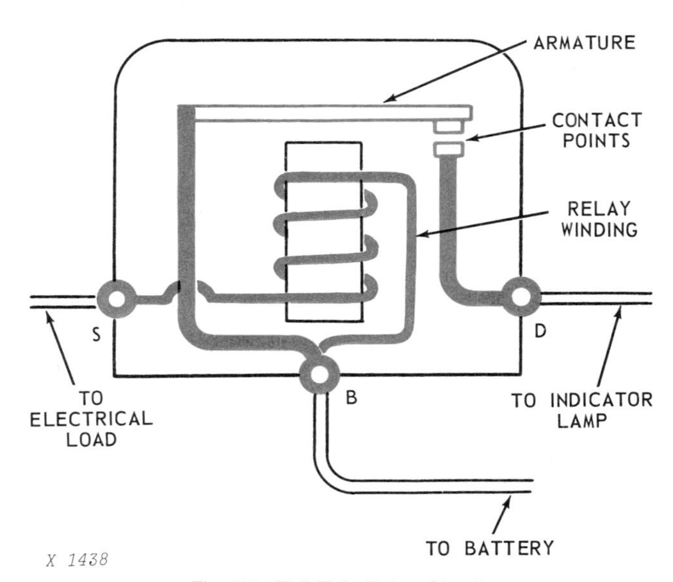

Fig. 27—Tell-Tale Relay Circuit

Fig. 27 shows a typical wiring circuit for a tell-tale relay.

When an electrical load of predetermined value is connected through the tell-tale relay winding to the battery, current flows through the relay winding. This creates a magnetic field which pulls the armature toward the core so that the relay contact points close. This connects the indicator lamp to the battery and so the lamp lights up.

If any part of the load circuit is not functioning properly such as a burnt out tail light, a reduced current flow through the relay winding allows the relay contact points to open. As a result the indicator lamp will go out, telling the operator that something is wrong in the circuit.

NOTE: Some tell-tale relays are designed so that the indicator lamp lights up when the electrical units are not working rather than when they are.

Some tell-tale relays require the following adjustments: air gap, point opening, and operating amperage of the contact points. The air gap and point opening are checked and adjusted the same as in the cutout relay (Chapter 6).

Testing the Amperage of Tell-Tale Relay

To measure the amount of current required to operate the tell-tale relay, connect an ammeter into the circuit at the terminal of the relay which is connected to the unit being operated. Also connect a variable resistance in series at the terminal of the relay which is connected to the battery. Close the signal lamp switch and slowly reduce the resistance until the relay operates, noting the ammeter reading.

If necessary, adjust the operating amperage by bending the armature spring post. Bend it up to increase the spring tension and the operating amperage and bend it down to decrease the amperage.

NOTE: Each model of tell-tale relay is designed to use with a specific number and size of bulb (or bulbs). Use of higher candle-power bulbs or more bulbs than specified will cause excessive contact point burning in the relay. The use of smaller candle-power bulbs or fewer bulbs than specified may result in failure of the relay points to close when these bulbs are turned on so that the indicator lamps will not burn. Find out the correct size and number of bulbs to be used with the relay for good operation.

SUMMARY: SWITCHES AND RELAYS

In summary:

- **Switches control circuits by opening and closing current pathways to turn circuits on and off.**
- **Most switches are manually and mechanically operated, but some switches are activated by pressure, temperature, a magnetic field, or a rotary motion.**
- **Relays are like switches, but they are activated automatically and can control several devices simultaneously.**
- **Relays allow a small current to control a much larger current.**

RESISTORS

Resistors reduce voltage output by resisting electron flow or current. Resistors create heat when current begins to flow through a circuit. They then transfer this heat to the surrounding area. Therefore, ambient temperature and humidity must be considered before putting a resistor in a circuit. Ambient temperature is that atmosphere immediately surrounding a given component. There are two types of resistors: fixed and variable.

FIXED RESISTORS

Fixed resistors can be made of wirewound resistance wire, metal film, carbon film, or molded carbon powder. Carbon resistors are labeled with a wattage value which indicates how much power they can dissipate.

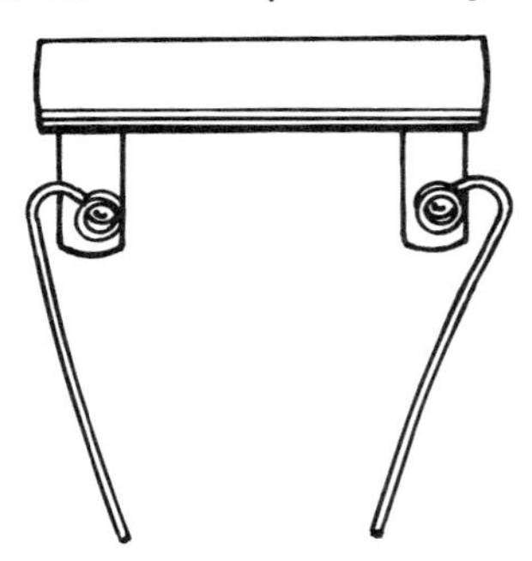

Fig. 28—A Wirewound Resistor

Wirewound resistors (Fig. 28) consist of a tubular form wrapped with coils of resistance wire. They can withstand lots of heat.

Metal film resistors use a thin film of metal or metal particle mixture to achieve various resistances.

Carbon film resistors use carbon film on a small ceramic cylinder. A spiral groove cut into the film controls the length of the carbon and thus the resistance.

Molded powder carbon resistors are known as carbon composition resistors. Such a resistor consists of carbon powder mixed with a glue-like binder and molded with a protective housing. The ratio of carbon powder and binder determines its resistance.

Resistors are marked with their value in ohms. For example: 5 ohms, 25K ohms (2500 ohms), or 3 meg ohms (3 million ohms). Some resistors are so small that they can't be marked with any numerical value. These small resistors are color coded with bands for their value.

RESISTOR COLOR CODES

Small carbon composition resistors are marked with color bands that indicate their resistance value (Fig. 29).

These bands are located on one end of the resistor. Always read the resistor values with the bands placed to your left. A resistor can have up to five color bands around it.

The first band shows the first digit of the resistor's value from zero to nine. The second band indicates the second digit of the resistor's value from zero to nine.

Fig. 29—Resistor Color Codes

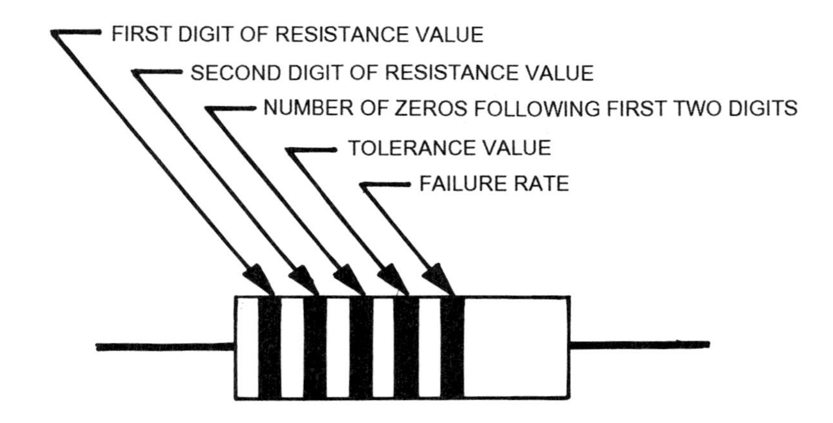

	BANDS				
COLOR	FIRST (DIGIT)	SECOND (DIGIT)	THIRD (MULTIPLIER)	FOURTH (TOLERANCE)	FIFTH (FAILURE RATE)
BLACK	0	0	1		
BROWN	1	1	10		1.0%
RED	2	2	100		0.1%
ORANGE	3	3	1000		0.01%
YELLOW	4	4	10,000		0.001%
GREEN	5	5	100,000		
BLUE	6	6	1,000,000		
VIOLET	7	7	10,000,000		
GREY	8	8	100,000,000		
WHITE	9	9	NONE		
GOLD				+ /-5%	
SILVER				+ /-10%	
NONE				+ /-20%	

The third band is a multiplier value that is multiplied by the first and second band digits together. The first three bands will give the resistors value in *Ohms.*

For example, what would be the resistance value of a resistor with bands of red, yellow, and red around it?

Looking at the chart (Fig. 29), the resistance digit for the color red is 2 and the resistance value for yellow is 4. The multiplier for red is 100. Therefore, the resistance is 24 x 100 or 2400 ohms resistance.

The fourth band, which is not always included on the resistor, indicates tolerance or how close the resistor actually comes to its rated value. This band will be either gold or silver in color. The gold band indicates a plus or minus 5 percent value of its resistance.The silver band indicates a plus or minus 10 percent value of its resistance. If there is no fourth color band on the resistor it indicates a plus or minus 20 percent value of its resistance.

In our previous example, there was no fourth band, so the resistance tolerance value would be 2400 plus or minus 20 percent, or 2160 to 2640 ohms.

The fifth band, if present, indicates the tested percent of failure per one thousand hours of use. A brown band would be 1.0%, a red band would be 0.1%, an orange band would be .01%, and a yellow band would be .001%.

The symbol for fixed resistors is shown in Fig. 30.

Fig. 30—Symbol for a Fixed Resistor

VARIABLE RESISTORS

Variable resistors (Fig. 31) are devices that control voltage output by changing their resistances. They are constructed with a resistance material, an input and output, and a means of adjustment.

There are many kinds of variable resistors with different names, including potentiometers, rheostats, trimmer resistors, thermistors, and photoresistors.

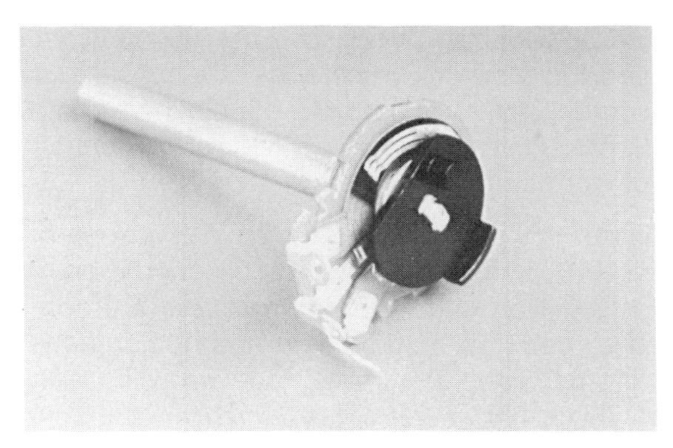

Fig. 31—A Variable Resistor

Potentiometers allow two current paths and act as sensing and voltage-sending devices. They are controlled manually (Fig. 32) or mechanically (Fig. 33).

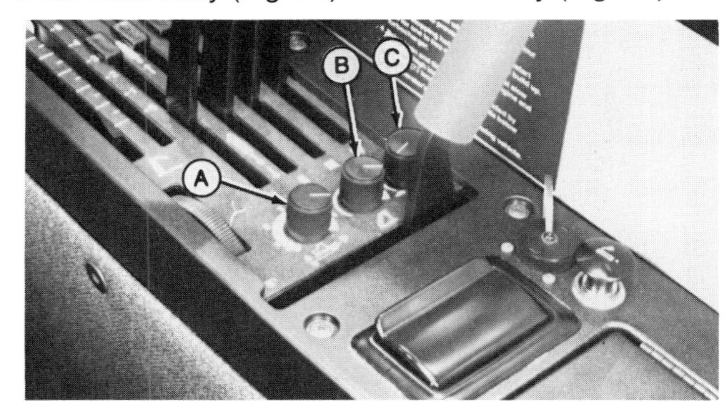

Fig. 32—Manually-Operated Potentiometers for Hitch Control

Fig. 33—Mechanically-Operated Potentiometer for Clutch Position

Rheostats allow one current path and are used to control varying current to a component such as a switch. It is also used to control the brightness of instrument lights.

Trimmer resistors are small variable resistors that are equipped with a thumbwheel or a slot for a small screwdriver. They are designed for occasional adjustment only and are found on circuit board assemblies.

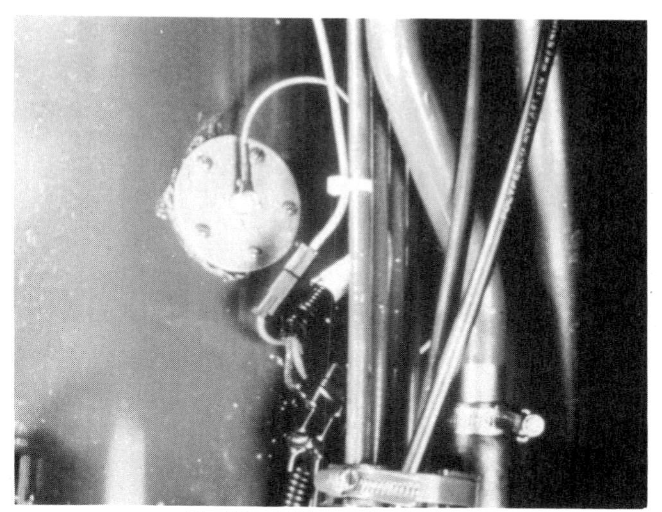

Fig. 34—A Thermistor Resistor

Thermistors (Fig. 34) are temperature-sensitive resistors. They are used mainly as sensors and are located at the temperature source that is to be monitored.

Photoresistors are light-sensitive resistors that change resistance when exposed to a light source. These devices are used as sensors. For example, a space heater uses a photoresistor as a safety device to shut down the unit when the light of a flame is not sensed.

The symbol for a variable resistor is shown in Fig. 35.

Fig. 35—Symbol for a Variable Resistor

CAPACITORS

Capacitors come in many sizes and shapes (Fig. 36), but they all do the same thing: they store electrons. How this is done will be explained later on in this section.

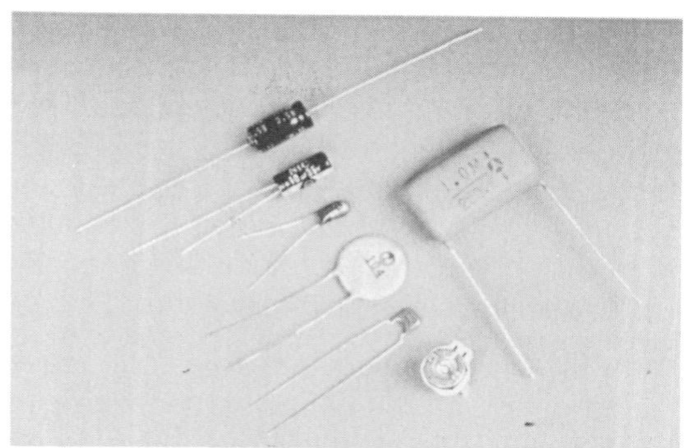

Fig. 36—Capacitors

HOW CAPACITORS ARE MADE

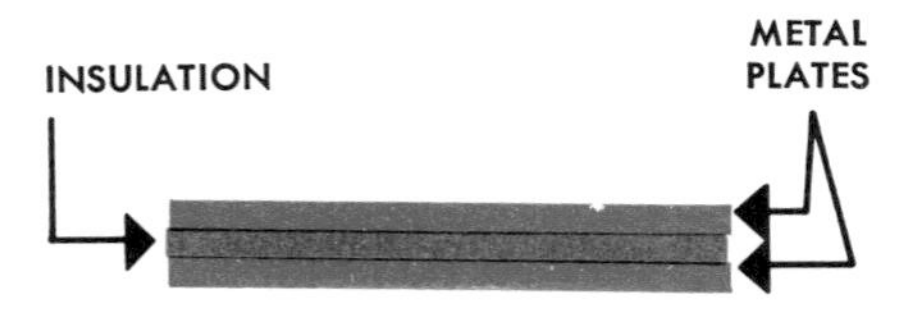

Fig. 37—Capacitor Construction

A capacitor is made by separating two metal plates with a thin insulating material called a die-electric. The die-electric can be paper, plastic, mica, or air. The metal plates can be aluminum foil or a thin film of metal (Fig. 37). They are then rolled into a cylinder or left flat.

When two metal plates are connected to a source of voltage, the capacitor exhibits the property of *capacitance* (stores up energy). The symbol for capacitance is *c,* and the unit of measure is *farad* (1F), *micro-farad* (0.000001 F), or pico-farad (0.000000000001 F). The greater the area of the plates, and the shorter the distance between them, the greater will be the capacitance.

HOW A CAPACITOR WORKS

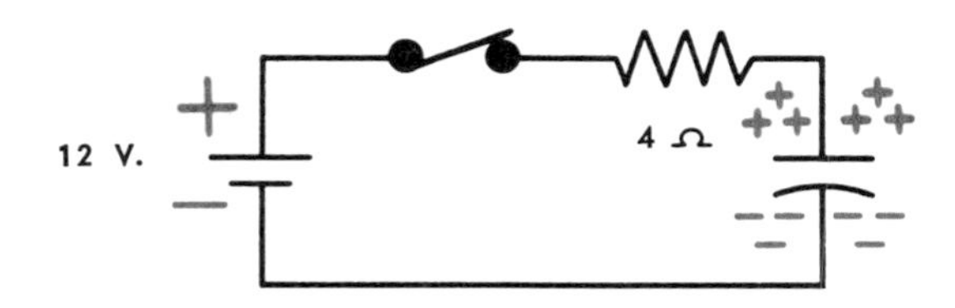

Fig. 38—Capacitor in a Circuit with Switch Closed

Fig. 38 explains the action of a capacitor in a circuit. Remember that the metal plates in all capacitors are insulated from each other. Since the plates are insulated, it would seem that no current will flow in this circuit.This is indeed the case, except at the instant of time when the switch is closed. When this is done, the voltage across the plates will suddenly change from zero to 12 volts. Let's see how this works.

When the switch is closed, electrons will leave the battery and gather on the capacitor negative plate as shown in Fig. 38. This plate will then have an excess of electrons. At the same time, electrons will leave the other plate and cause this plate to become positively charged. It is important to note that the electrons flow through the circuit but *not* through the insulating material separating the two plates.

As the two plates become negatively and positively charged, a voltage appears across the plates. Remember that a voltage between two points is the direct result of a difference in charge between the two points. As more and more electrons accumulate on the negative plate and leave the positive plate, the voltage across the plates approaches the battery voltage. When the capacitor voltage equals the battery voltage, the flow of electrons will stop. The equation giving the amount of charge on a capacitor is:

Charge = Capacitance x Voltage

Thus the higher the capacitance and applied voltage, the greater will be the charge on the capacitor.

The curves showing the flow of current into, and the voltage across, the capacitor plates are shown in Fig. 39.

At the very instant the switch is closed, the initial current is determined by the resistor R in the circuit, or $I = \frac{E}{R} = \frac{12}{4} = 3$ amperes. Initially the capacitor acts like a short circuit, since only the resistor R limits the initial current flow. As the charge on the plates increases, the capacitor voltage increases to oppose the battery voltage, and the current decreases with time. Finally the current flow stops completely when the capacitor voltage is equal and opposite to the battery voltage.

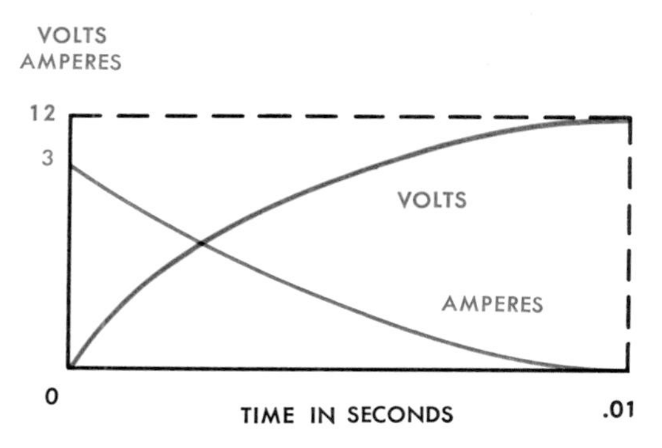

Fig. 39—Curves Showing Current and Voltage for Capacitor Plates

The time needed for the current to reach a zero value may be only a fraction of a second. Although this time interval may seem very short, it can be of major importance in electrical systems where voltages are changing at extremely rapid rates.

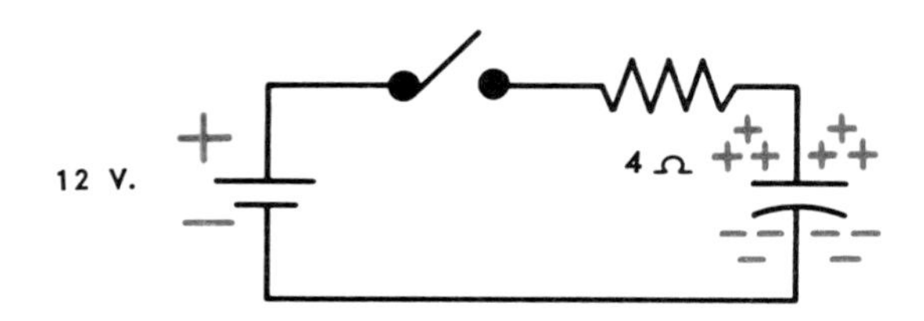

Fig. 40—Capacitor in Circuit with Switch Open

If the switch is suddenly opened, the charge of the capacitor will remain (Fig. 40). The accumulated charge represents stored energy, and a capacitor has the ability to store a charge. The amount of energy stored is represented by the formula:

$$\text{Energy} = \frac{\text{Capacitance x Voltage x Voltage}}{2}$$

Thus, the higher the capacitance and voltage the higher the energy stored in the capacitor. In time the charge will leak off the plates through the insulating material and surrounding air, reducing the charge to zero.

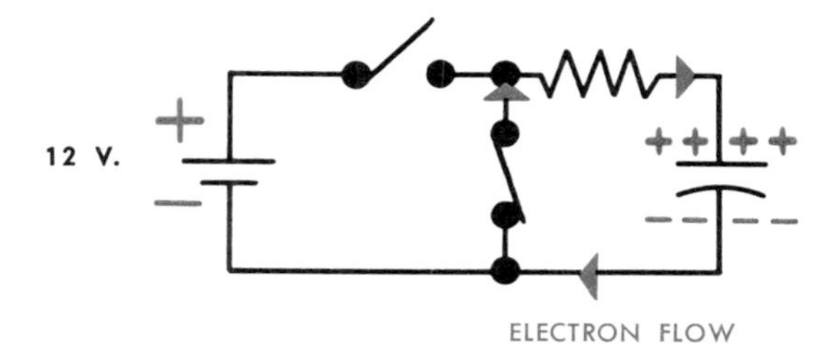

Fig. 41—Capacitor in Circuit with Two Switches

By adding another switch to the circuit, the energy stored in the capacitor may be used to send a current through the resistor (Fig. 41). When the switch is closed, a momentary surge of current will flow through the resistor until the charge on the two plates is equal. The initial current will be $I = \frac{E}{R} = \frac{12}{4} = 3$ amperes, since the capacitor voltage is 12 volts. As the electron flow continues, the voltage across the plates decreases, until finally the current value is zero. Again, the time interval may be only a fraction of a second.

The operation of capacitors can be summarized:

1. A capacitor as it builds up a charge develops a counter voltage across the plates.

2. The capacitor has the ability to store a charge, or to store energy.

3. Current flows only during the very short time when the capacitor plates are either charging or discharging.

USES OF CAPACITORS

Capacitors used in alternator electrical circuits usually consist of a roll of two layers of thin metal foil separated by a very thin sheet of insulating material. The assembly is then sealed in a metal can.

Capacitors of this type are used primarily to perform either one of two functions in an electrical circuit, as follows.

The characteristic of a capacitor to act initially like a short circuit when a sudden difference in voltage is applied across its plates makes it ideally suited to "trap" or temporarily store electrical energy that otherwise could damage electrical components.

To illustrate, consider an ignition circuit in which a capacitor is connected across the distributor contact points (Fig. 42).

When the contacts separate a high voltage is induced

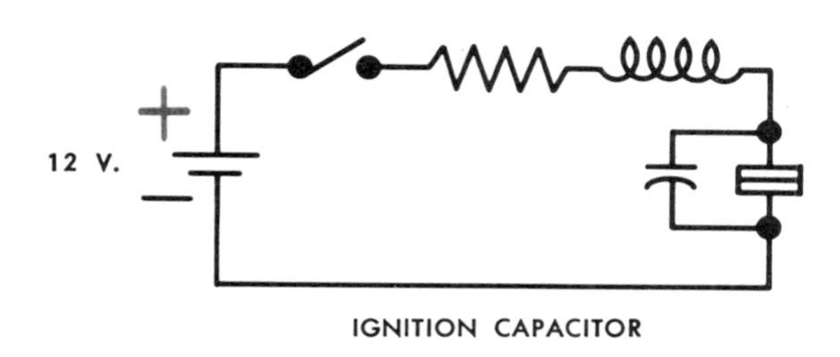

Fig. 42—Capacitor in Ignition Circuit

in the ignition coil primary winding because of self-induction. This high voltage causes the capacitor plates to charge when the contacts first separate; the capacitor acts initially like a short circuit and current flows into the capacitor to minimize arcing at the contacts.

High voltages originating when a circuit is opened may exist for a very short interval of time, such as 1/1,000,000 of a second, and are called **transient voltages.**

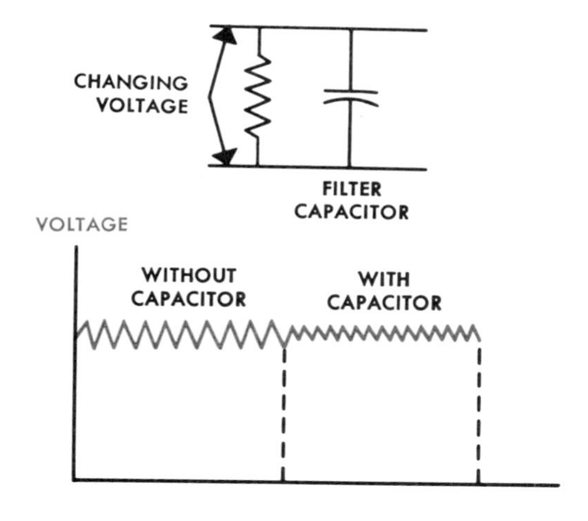

Fig. 43—Filter Capacitor

Another application where the capacitor is used to reduce the magnitude of changing voltages is in electrical circuits, where a constantly changing voltage needs to be "smoothed out" to a more constant voltage. A capacitor used in this way is connected across the changing voltage, and is called a **filter capacitor** (Fig. 43).

Besides reducing the magnitude of changing or transient voltages, capacitors are used in ignition amplifiers to store a charge of electricity until the charge can be transferred to another part of the circuit.

The uses of the capacitor in charging and ignition circuits are covered in more detail in Chapters 6 and 8 of this manual.

RESISTOR AND CAPACITOR CIRCUITS

There are two resistor and capacitor (R-C) circuits used to reshape incoming waves or pulses for signals. They are called the integrator circuit and the differentiator circuit. The product of the R-C in these circuits is called the RC time constant.The RC time constant is the time in seconds for a charging or discharging of a capacitor to go through 63.3 percent of the change in charge and is dependent upon the capacitor type.

Integrator Circuit

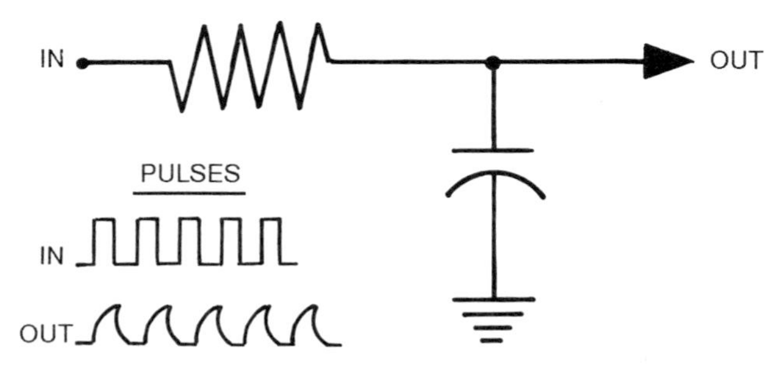

Fig. 44—The Integrator Circuit

In the integrator circuit (Fig. 44), if the input pulses are speeded up, the capacitor in the circuit charges and discharges and its RC time constant causes the output pulses to not reach their full height (amplitude). Therefore, the integrator circuit can function as a filter which can pass signals only below a certain frequency.

Differentiator Circuits

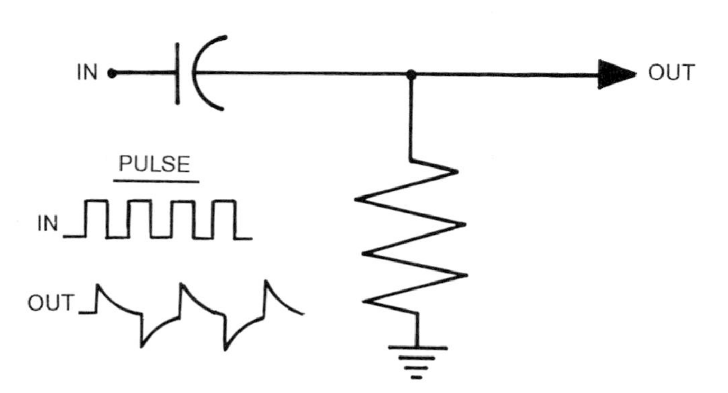

Fig. 45—The Differentiator Circuit

You will note that in the differentiator circuit (Fig. 45) the resistor and capacitor are positioned opposite in the circuit compared with an integrator circuit. The differentiator circuit has a DC input; operation of the capacitor produces filtered output waves with sharp positive and negative peaks (AC). It is used to make narrow pulse generators and to trigger digital logic circuits.

SUMMARY: RESISTORS AND CAPACITORS

In summary:

- **Resistors reduce voltage output by resisting current flow.**
- **There are two types of resistors: fixed and variable.**
- **A variable resistor controls voltage output by changing its resistance.**
- **A capacitor (or condensor) is made of two metal plates separated by thin insulation.**
- **A capacitor can store energy.**
- **A capacitor can build up a counter voltage across its two plates.**
- **Current flows between the plates only at the moment when the plates are either charging or discharging.**
- **At other times the plates are insulated.**
- **Capacitors can be used to store harmful energy for a moment (as to prevent arcing of distributor points).**
- **Another use is to "smooth out" changing voltages (as in regulators).**
- **A resistor and capacitor circuit that controls frequency is called an integrator circuit.**
- **A resistor and capacitor circuit that transforms DC input to AC output is called a differentiator circuit.**

SOLENOIDS

Solenoids are devices which use the strength of a magnetic field generated in a coil to move a metal core. The moving core along with linkage can transfer its motion to other mechanical devices.

Examples include a fuel shutoff solenoid (Fig. 46), fuel injection pump solenoid, and leveling valve solenoid as used on hillside combines (Fig. 47).

Solenoids of this type are basically electrical contacts and windings around a hollow cylinder containing a movable metal core or plunger.

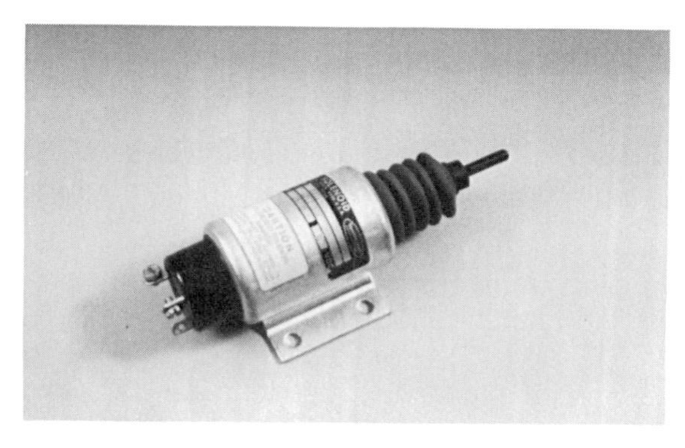

Fig. 46—A Fuel Shutoff Solenoid

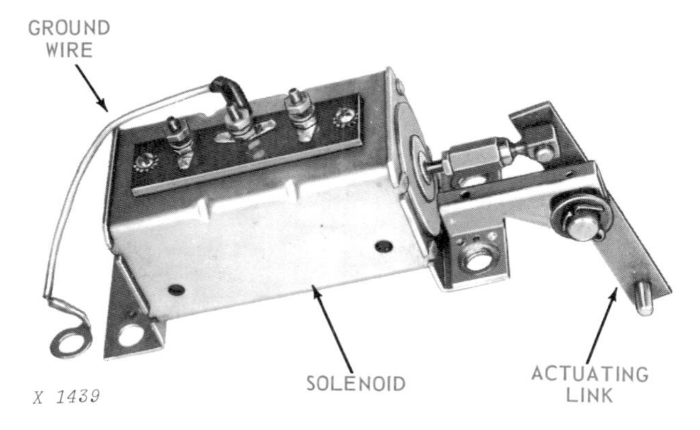

Fig. 47—Solenoid for Hillside Combine Leveling Valve

When the winding of the solenoid is energized by battery voltage through an external circuit, a magnetic field is created around the hollow cylinder, causing the plunger to be pulled or drawn inward. This movement is then transferred to switches and/or other mechanical devices. (See Chapter 7 on "Starting Circuits" for more detail on solenoid operation.)

Most solenoids and solenoid switches generally operate the plunger in only one direction. The plunger is returned by a spring when current is shut off from the winding. However, in the case of the solenoid on a hillside combine leveling valve (Fig. 47), the solenoid has two windings which operate the plunger in both directions.

Solenoids are also used to control the hydraulic spools on hydraulic valves to direct hydraulic flow to other hydraulic devices. These solenoid-controlled hydraulic valves are known as **Electro-Hydraulic Valves** (Fig. 48). (See Hydraulic Systems FOS Manual for operation details.)

Fig. 48—An Electro-Hydraulic Valve

When solenoids for electro-hydraulic valves are operated through a microprocessor controller and sensors instead of by direct battery voltage, hydraulic pressure can be controlled by varying the voltage to the solenoids. Such a system is used on modern tractors to control hitch operation. These valves are known as **Microprocessor-Based Electro-Hydraulic Valves.**

Let's look at the functions of a typical microprocessor-based electro-hydraulic valve for a hitch (Fig. 49).

Draft is sensed by strap (1), which uses a strain gauge sensor. The hitch position is sensed by potentiometer (3) attached to one of the rockshaft lift arms. The outputs of the sensors are read directly by hitch controller (2) and are summed by the hitch algorithm, which uses the setting of mix potentiometer (4) to determine the relative weight of each. The rockshaft control lever is attached to rotary potentiometer (5), which provides the lever command to the algorithm. Rate-of-drop potentiometer (7) and raise limit potentiometer (8) have also been added.

The lever command, the mix setting and the draft and position feedbacks are used to determine the command to send to the appropriate (pressure or return) valve solenoids (6). The electro-hydraulic hitch control algorithm is defined by the engineer in the design stage, using predetermined equations.

TESTING OF SOLENOIDS

Solenoids are tested by removing the solenoid from the circuit and testing the windings for resistance with a multimeter. The proper resistance readings of a solenoid will vary and will be listed in the technical manual of the machine.

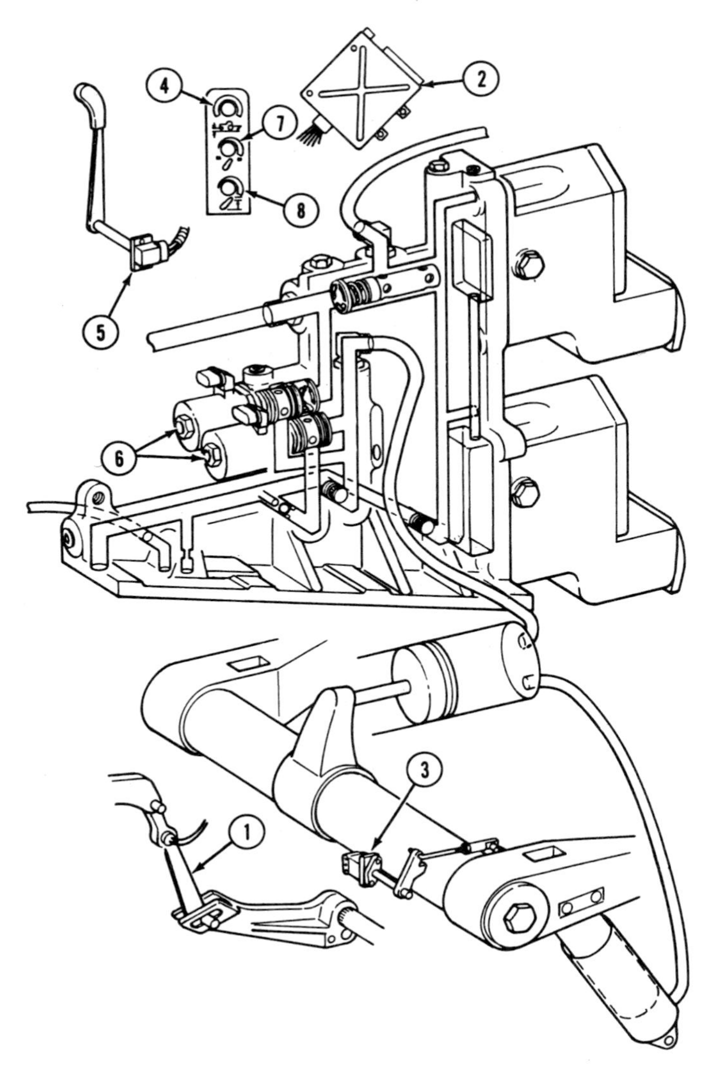

Fig. 49—Pictorial Diagram of Electro-hydraulic Hitch

SUMMARY: SOLENOIDS

In summary:

- **Solenoids are electromagnetic devices used to move a metal core and linkage.**
- **Some solenoids are used to control hydraulic valves. Such valves are called electro-hydraulic valves.**
- **When controlled by a microprocessor solenoid, voltage is varied to provide hydraulic control to a linkage.**

TRANSFORMERS

Transformers are devices with two or more windings around a metal core. They use induction to transform incoming voltage and current to a higher or lower voltage and current output.

In order to transform voltage and current from one winding to another, the current must be fluctuating or AC. A steady (DC) current will not transform voltage and current from one winding to another.

Transformers do not create an output power change from nothing, but are dependent upon the voltage and current input and the number of windings of each coil. The output of a transformer cannot exceed the power of its input.

If the input windings are less then the output windings or stepped-up, the output voltage is increased from the input voltage, but the output current is decreased from the input current.

If the input windings are more than the output windings or stepped-down, the output voltage is decreased and the output current is increased.

HOW A TRANSFORMER WORKS

A transformer steps-up or steps-down voltage and current by the amount of windings in each coil. The incoming voltage and current is connected to the winding known as the primary winding. The output winding is called the secondary winding.

Step-up Voltage Transformer

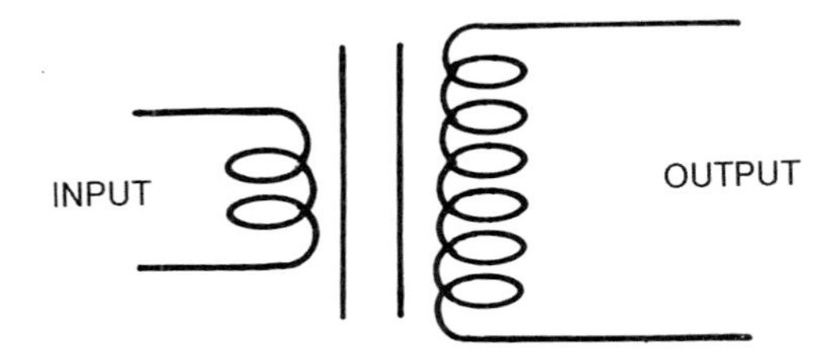

Fig. 50—Step-up Voltage Transformer Symbol

In a step-up voltage transformer (Fig. 50) the windings of the output or secondary windings are greater than the input or primary windings. Thus, the voltage output is increased by the turn ratio of the primary coil to the secondary coil and the current is decreased. For example, if the turn ratio of the coil is 1:5, then for every turn of the primary coil there are five turns at the secondary coil. If the incoming voltage is 12 volts, the output voltage would be 60 volts (12 x 5). In mobile machinery a good example of a step-up transformer would be an ignition coil (Fig. 51). (See Chapter 8 on "Ignition Circuits" for more information.)

 Litho in U.S.A.

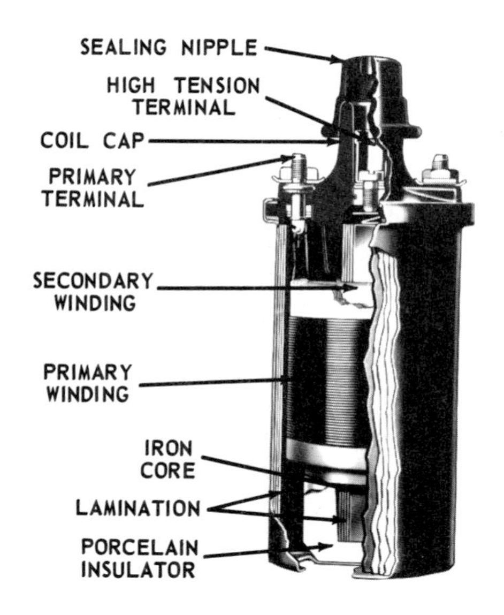

Fig. 51—Ignition Coil

Fig. 52—Transformer for Ignition on a Crop Dryer

Another type of step-up voltage transformer is used in crop dryers (Fig. 52), where it is used to increase the voltage to jump a spark across electrodes to ignite the dryer's gas. This type of transformer is also known as an ignition transformer.

Step-down Voltage Transformer

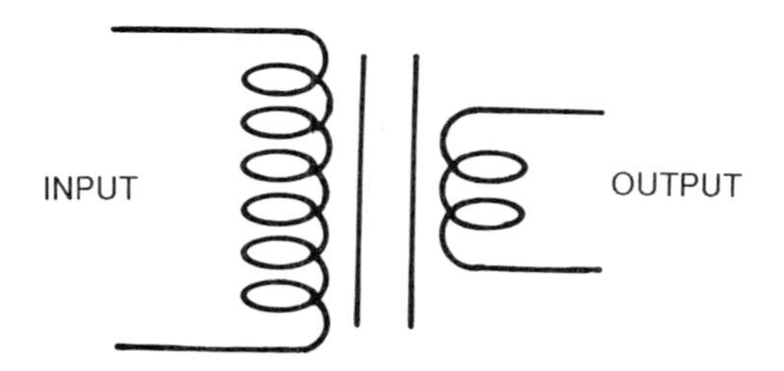

Fig. 53—Step-down Voltage Transformer Symbol

In a step-down voltage transformer (Fig. 53) the secondary windings (output) are less than the primary windings (input). Thus, in a 5:1 turn ratio the voltage will drop 60 volts at the primary windings to 12 volts at the secondary windings. Most step-down transformers are used by utility companies to reduce power line voltage and increase current to a usable level within our homes.

SUMMARY: TRANSFORMERS

In summary:

- **Transformers have the ability to transform incoming AC voltage and current to a higher or lower output AC voltage and current.**
- **Transformers will not work with DC voltage.**
- **The step-up voltage transformer has more secondary windings than primary windings.**
- **The step-down voltage transformer has more primary windings than secondary windings.**

SEMICONDUCTORS

In Chapter 1 we learned conductors have less than four electrons in the outer rings of their atoms, while insulators have more than four electrons in their outer rings.

Semiconductors are elements which have just four electrons in the outer rings of their atoms. They are neither good conductors nor good insulators. The most common semiconductor elements are silicon and germanium. Silicon is the most common element used for semiconductors. Since silicon makes up 27.7 percent of the earth's crust, availability is high and cost is low.

Semiconductors are used to make **Diodes** and **Transistors.**

Let's first see how the basic silicon element is made into semiconductors, then we'll talk about diodes and transistors.

HOW SEMICONDUCTORS ARE MADE

Silicon crystal for semiconductors is made by **covalent bonding** (Fig. 54). This means that the electrons in the outer ring of one silicon atom join the electrons of other silicon atoms so that the atoms share electrons in their outer rings. Now each atom really has eight electrons in its outer ring as shown. This creates a very good insulator since there are now more than four electrons in the outer ring.

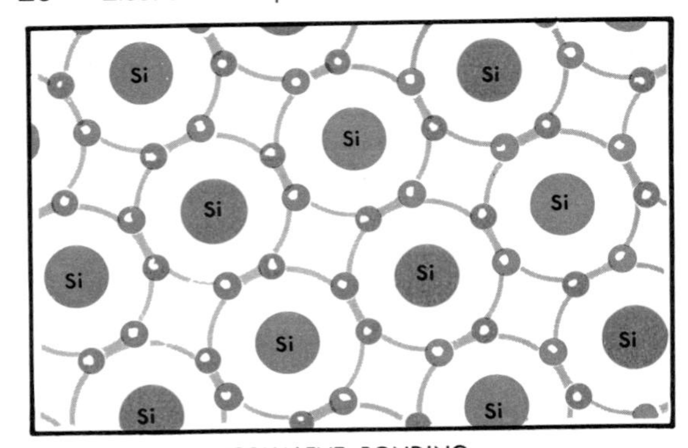

Fig. 54—Silicon Formed as Insulator by Covalent Bonding

The silicon crystal is then "doped" (selectively contaminated to control its conductivity) by adding other materials.

Two elements commonly used to dope the silicon are phosphorous and antimony. Both of these elements have five electrons in their outer ring. Covalent bonding occurs but there is one electron left over (see Fig. 55). This electron is called a "free" electron which can be made to move through the material very easily. Any material having an extra electron is called a negative or **"N" type material.**

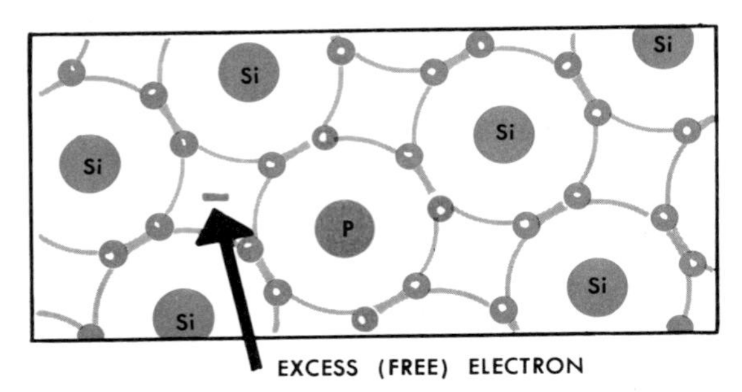

Fig. 55—Use of Phosphorous to "Dope" Silicon for "N" Type Material

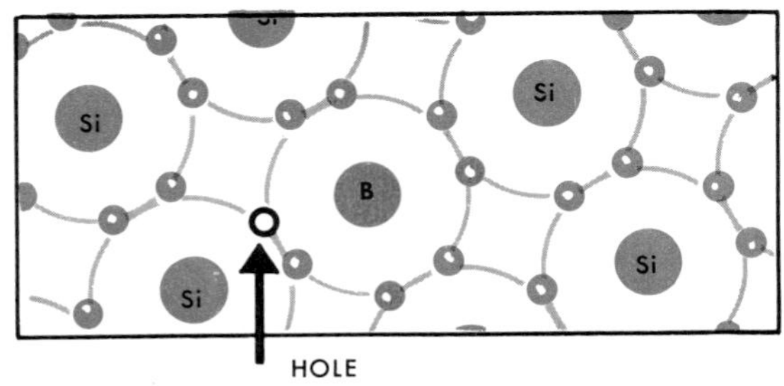

Fig. 56—Use of Boron to "Dope" Silicon for "P" Type Material

Two other elements commonly used to dope the silicon crystals are boron and indium. These elements have only three electrons in their outer ring. Covalent bonding occurs but there is a shortage of one electron for complete bonding. The resulting void is called a *hole* (Fig. 56). This hole can be considered as a positive charge of electricity. Materials lacking this electron and having this hole are called positive or **"P" type material.**

To understand semiconductors, thing of this hole as a positive (+) current carrier, just like the electron in a negative current carrier. The hole can move from atom to atom, just as an electron can move from atom to atom.

SUMMARY: SEMICONDUCTORS

In summary:

- **Semiconductors are made by covalent bonding.**
- **This is joining atoms which then share electrons.**
- **The result is a good insulator.**
- **"Doping" silicon crystals results in a free electron ("N" type material) or a voided electron, or hole ("P" type material).**
- **Either "N" or "P" material is a current carrier, moving electrons from atom to atom.**
- **Semiconductors have special uses in diodes and transistors.**

HOW SEMICONDUCTORS OPERATE

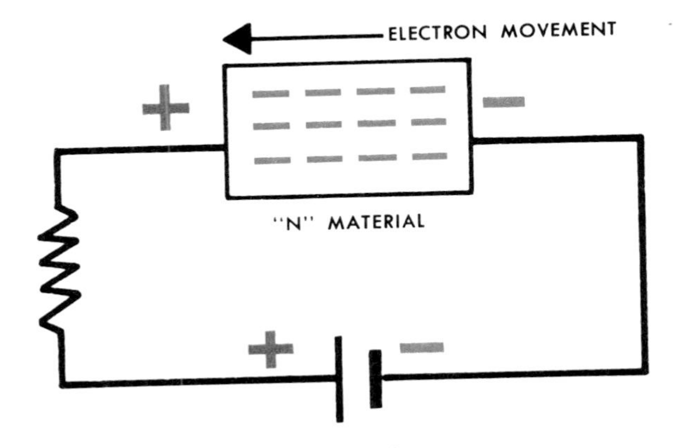

Fig. 57—Electron Movement in Circuit with "N" Type Material

The current flow in "N" type material is shown in Fig. 57. By connecting a voltage source such as a battery to the material, an electron current will flow through the circuit. This current is the movement of the excess of "free" electrons through the material and is very similar to what occurs in a copper wire.

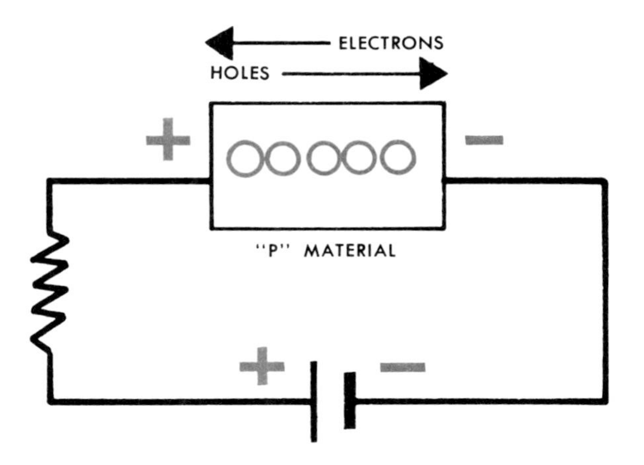

Fig. 58—Movement of Holes in Circuit with "P" Type Material

The current flow in "P" type material is shown in Fig. 58. But here the current is a movement of the positively charged holes. This hole movement works as shown in Fig. 59.

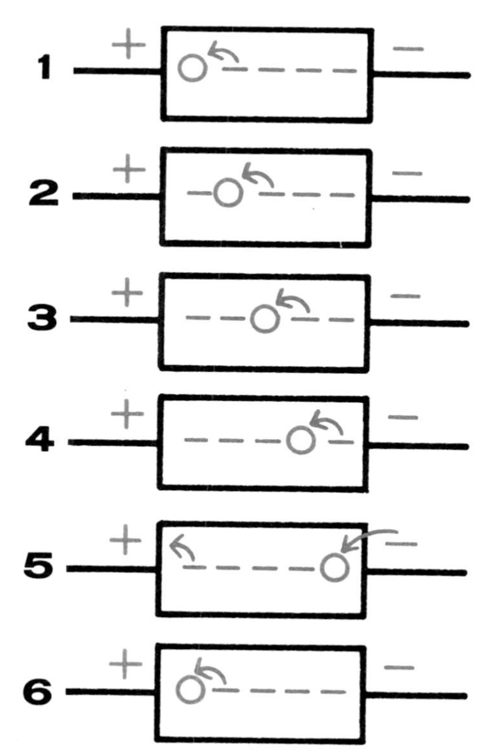

Fig. 59—Hole Movement in "P" Type Material

Notice how the (+) battery terminal in No. 1 attracts the (–) electrons in the material (unlike charges attracting). Similarly, the (–) battery terminal will repel the electrons. So an electron from one of the covalent bonds will move to the left toward the (+) terminal, and will fill one of the holes near the terminal. This movement of an electron leaves behind a hole. The positively charged hole, then, has moved to the right, toward the (–) battery terminal. This process continues and the hole keeps moving to the right until it nears the (–) connection at the semiconductor. At this time, the hole is filled by an electron which leaves the (–) wire connected to the semiconductor, and the (+) wire removes an electron from the semiconductor at the other end. (See No. 5 in Fig. 59.) The process is then ready to repeat itself.

The continuous movement of holes from the (+) terminal to the (—) terminal can be looked upon as current flow in "P" type material, and occurs when the battery voltage causes the electrons to shift around in the covalent bonds.

The hole movement occurs only *within* the semiconductor, while electrons flow through the entire circuit.

The hole movement theory will help us to understand how diodes and transistors operate, which follows.

In summary:

- **Outside voltage causes a current flow in the "N" or "P" material of semiconductors.**
- **In "N" material, current flow is the movement of "free" (—) electrons.**
- **In "P" material, current flow is the movement of (+) charged holes.**

Next we'll show how semiconductors are used in diodes and transistors.

DIODES

A diode is an electrical device that will allow current to pass through itself in *one direction only.*

How Diodes Are Made

A diode is formed when two semiconductor materials are joined, one of "N" type material, the other of "P" type. In diodes, the "N" material is usually phosphorus-doped silicon, while the "P" material is usually boron-doped silicon. (See Fig. 60.)

 Litho in U.S.A.

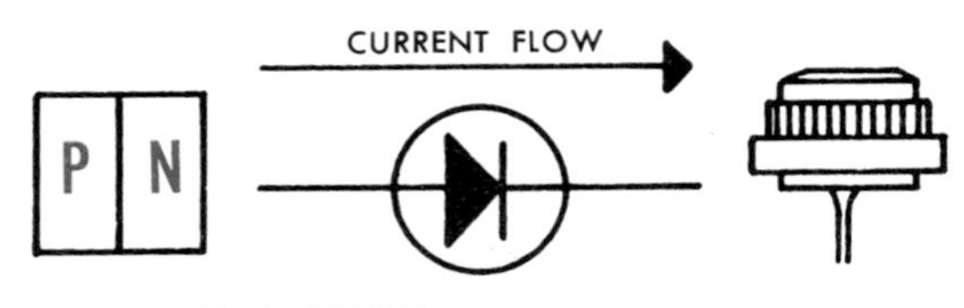

ELEMENT	ATOMIC NUMBER	NUMBER OF PROTONS	NUMBER OF ELECTRONS	VALENCE RING ELECTRONS
Boron (B)	5	5	5	3
Silicon (Si)	14	14	14	4
Phosphorus (P)	15	15	15	5

Fig. 60—Diode

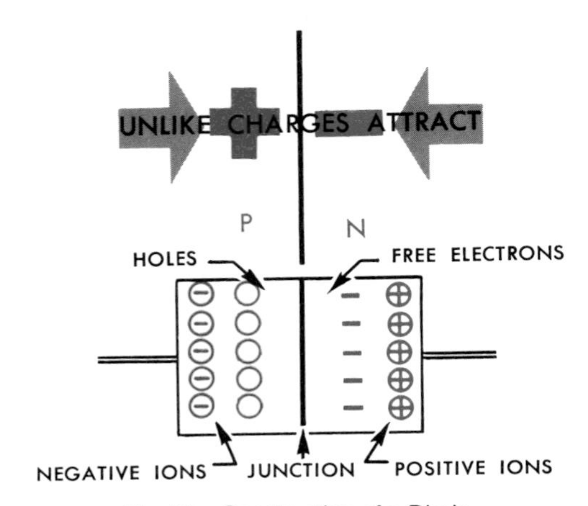

Fig. 61—Construction of a Diode

The basic construction of a diode is shown in Fig. 61.

"N" and "P" materials attract each other but are kept stabilized by positive and negative ions on each side. (An ion is an atom having a shortage or an excess of electrons.) The ions "pull back" on the free electrons and the holes to prevent them from crossing the junction.

The net result is a stabilized condition with a deficiency of electrons and holes at the junction area.

How Diodes Operate

Now let's activate the diode by connecting a battery to it (Fig. 62). The negative battery voltage will repel the electrons in the "N" material, while the positive battery voltage will repel the holes in the "P" material. With sufficient voltage, electrons will move from the negative terminal of the battery across the junction to the positive battery terminal and so create a flow of current. Also, the positive holes will move through the "P" material and through the junction as described above.

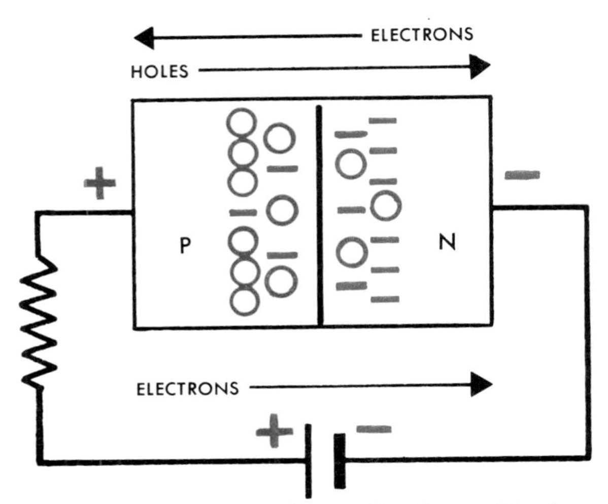

Fig 62—Diode Allowing Current Flow (Forward Bias)

The battery maintains the current flow, but for current to flow through the semiconductor, there must be holes present at the junction into which electrons can move.

Fig. 62 shows a forward bias connection of the semiconductor. This (–) to "N" and (+) to "P" connection creates the repelling action of the battery voltage which causes electrons and holes to congregate at the junction in large numbers—necessary for current flow through a diode.

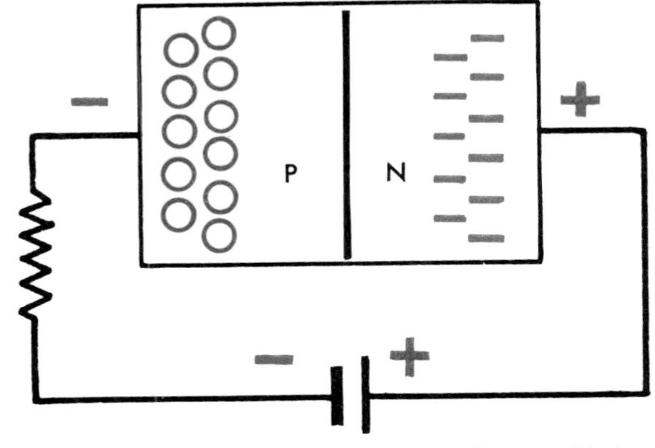

Fig. 63—Diode Blocking Current Flow (Reverse Bias)

Now let's see what happens when we reverse the battery connections (Fig. 63). As shown, the (+) of the battery attracts electrons away from the junction, while the (–) of the battery attracts the holes away from the junction. The result: no current flow.

This type of battery connection is called reverse bias, which causes the diode to block current flow.

In summary:

1. The diode will allow current to flow if the voltage across the diode causes electrons and holes to congregate at the junction area. (Forward Bias)

2. The diode will *__not__* *allow current to flow if the voltage across the diode causes the junction area to be void of electrons and holes. (Reverse Bias)*

Let's repeat a statement already made in this chapter. For electrons to move into the "P" material, there must be holes present in the "P" material near the junction *into which the electrons can move.* The reasons for the hole theory now becomes more apparent, as this theory provides a convenient means of explaining how a diode blocks or prevents current flow.

A complete description of diodes as used in alternators is given in Chapter 6.

Diode Leakage Current

When a reverse bias voltage is connected to a diode, it may be true that a small current will flow through the diode in the reverse direction, but the reverse current is very, very small.

If the voltage across the diode is increased, a value eventually will be reached called the *maximum reverse voltage* of the diode. At this voltage, the covalent bond structure will break down and a sharp rise in reverse current will occur. If the reverse current is sufficient in magnitude and duration, the diode will be damaged due to excessive heat.

Diodes are selected, of course, with an adequate maximum reverse voltage rating so that damaging reverse currents will not normally occur during operation.

Diode Types and Uses

Zener Diode

The zener diode (Fig. 64) is a specially designed type of diode that will conduct current in the reserve direction at a particular voltage. The primary feature of this type of diode is that it is very heavily doped during manufacture—the large number of extra current carriers (electrons and holes) allows the zener diode to conduct current in the reverse direction without damage if proper circuit design is used. The zener diode symbol is shown in Fig. 65.

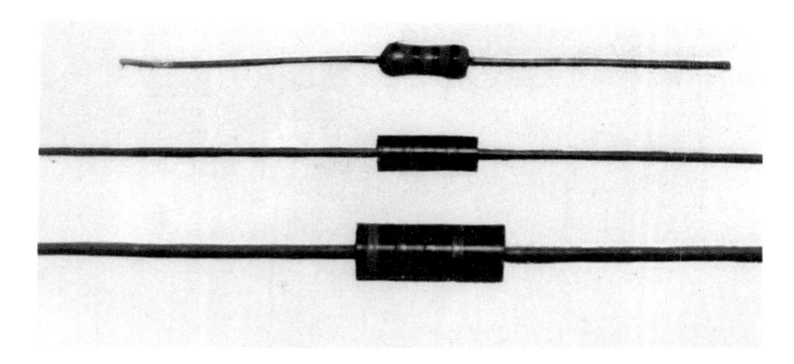

Fig. 64—Zener Diodes

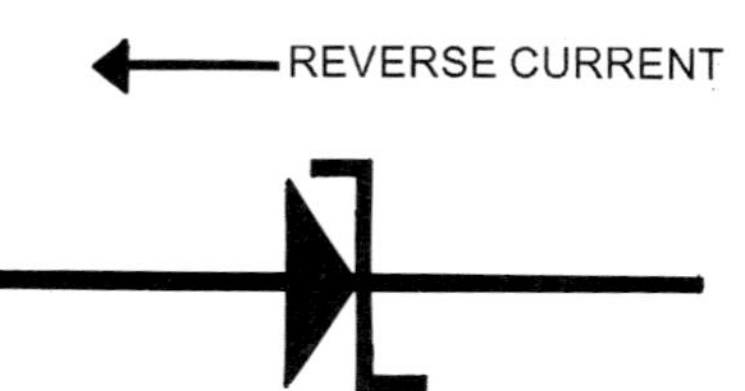

Fig. 65—Zener Diode Symbol

What makes the zener diode unique is that it will not conduct current in the reverse direction below a certain predetermined voltage (called reverse bias voltage). As an example, a certain zener diode may not conduct current if the reverse bias voltage is below six volts, but when the reverse bias voltage becomes six volts or more, the diode suddenly conducts reverse current. This type of diode is used in control and protection circuits.

POWER RECTIFIER DIODES

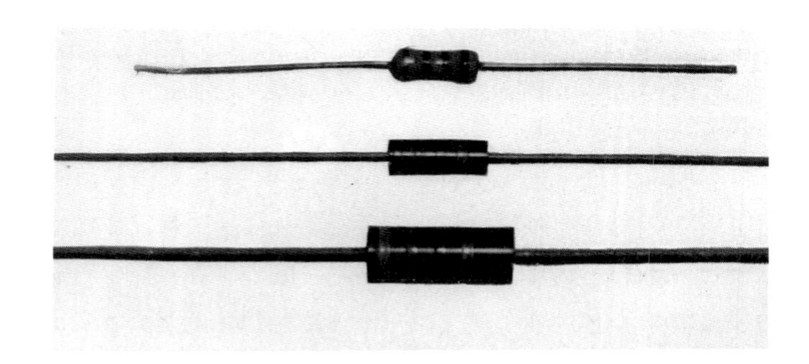

Fig. 66—Power Rectifier Diodes

Power rectifier diodes (Fig. 66) can handle high current. They are insulated in metal packages that act as heat sinks to dissipate excess heat. They are used mainly in power supplies such as alternators.

Litho in U.S.A.

SMALL SIGNAL DIODES

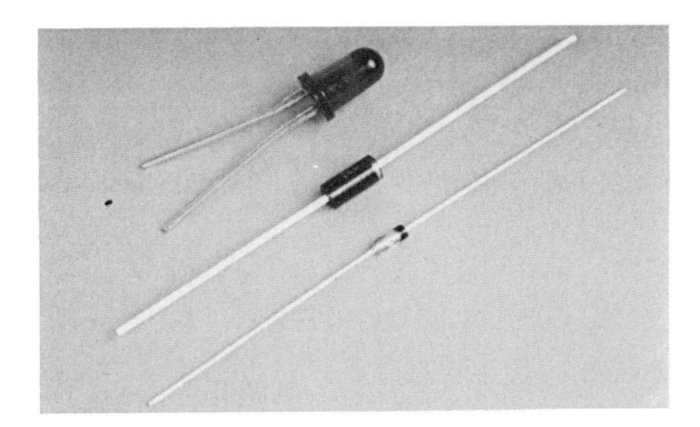

Fig. 67—Light-Emitting and Small Signal Diodes

Small signal diodes (bottom of Fig. 67) are used to transform low alternating current to direct current and absorb voltage spikes within a circuit. These diodes are generally part of a circuit board assembly and are used for control circuits.

Light-Emitting Diodes

All diodes emit some electromagnetic radiation when forward-biased. Diodes made from certain semiconductors like gallium arsenide phosphide emit considerably more radiation than silicon diodes.These diodes are called light-emitting or LEDs (see top of Fig. 67). They are used as visual signalling devices in instrument panels.

One use of diodes is for protection in electric clutch applications. When power is cut off to the clutch, a reverse voltage spike is created by the collapse of the coil magnetic field.

This reverse voltage spike travels back in the opposite direction, toward the system or switch. A diode is used to short this spike to ground, thus protecting the system and switch.

SUMMARY: DIODES

In summary:

- **Diodes control current by allowing it to pass through in one direction only.**
- **They are made by joining "N" and "P" semiconductor materials.**
- **Forward bias = allowing current to flow.**
- **Reverse bias = stopping flow of current.**
- **Zener diodes are special devices that will conduct current in reverse—above a certain voltage.**

TRANSISTORS

A *transistor* is a solid-state electronic device that is used in circuits to control the flow of current. It acts like a relay where a small current controls a larger current load and operates by either allowing current to flow or not allowing it to flow.

HOW TRANSISTORS ARE MADE

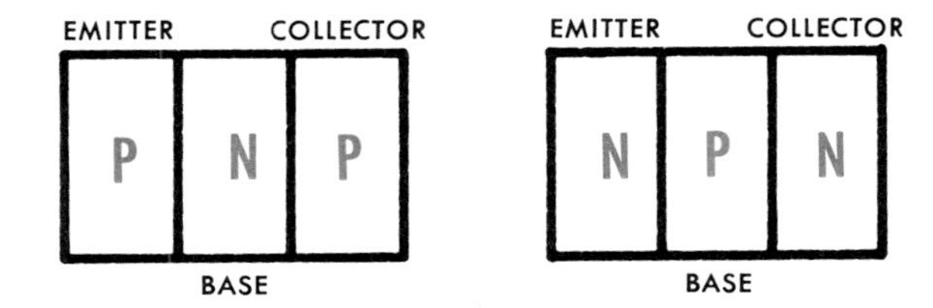

Fig. 68—Basic Parts of Transistors

A transistor is usually formed by adding a second section of "P" type material to the PN junction used for diodes. This results in the PNP transistor (Fig. 68). This type of transistor is known as bipolar.

The "P" material on the left is called the emitter, the "N" material in the center is the base, and the "P" material on the right is called the collector.

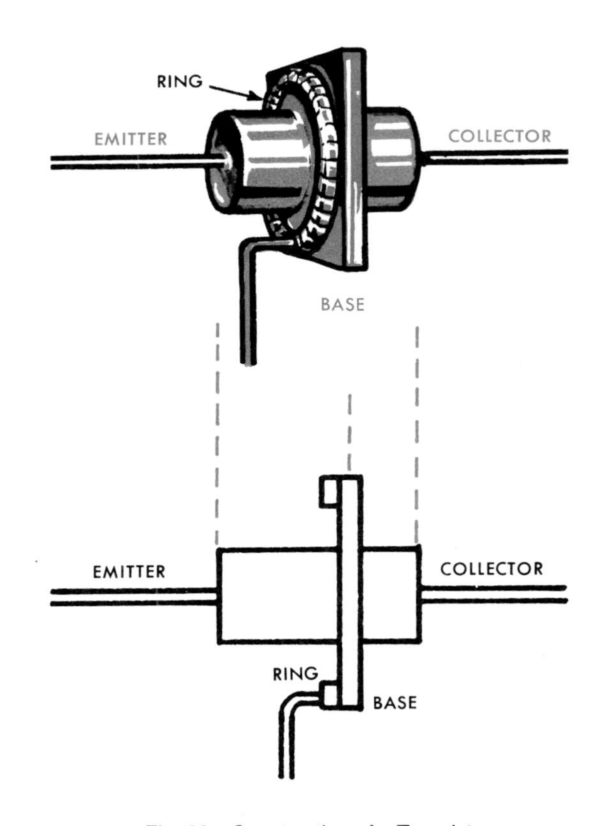

Fig. 69—Construction of a Transistor

The base of a transistor is very thin (Fig. 69). A metallic ring is attached around the base and this is connected to the circuit. By this design, the distance between the emitter and the collector is shorter than the distance between the emitter and the base ring. This gives the transistor its unusual operation as we'll see now.

HOW A TRANSISTOR OPERATES

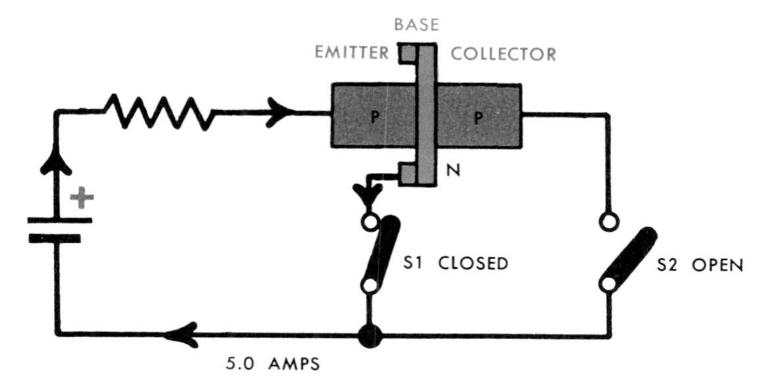

Fig. 70—Transistor in Circuit with Switch S2 Open

In Fig. 70, we have connected a battery to a PNP transistor.With switches S1 closed and S2 open, current will flow through the emitter-base of the transistor. With switch S2 open, the collector will not operate, and the circuit is a simple PN junction diode (through emitter-base) connected to the battery in the forward bias direction (as in Fig. 62).

At this point, we should recall a definition of current flow—the movement of electrons from atom to atom in a conductor. Under this definition, the electrons in this circuit flow from the negative battery terminal through the base-emitter back to the positive side of the battery. However, to better understand the operation of a PNP transistor, we must accept hole movement as a theory of current flow, and apply this theory to the PNP transistor.

Using the hole movement theory, the current flow in the transistor is a movement of holes through the "P" material to the "N" material. This movement of holes can be looked upon as current flow, and this theory simplifies the explanation of how a transistor works.

Let's assume that the emitter-base current is five amperes.

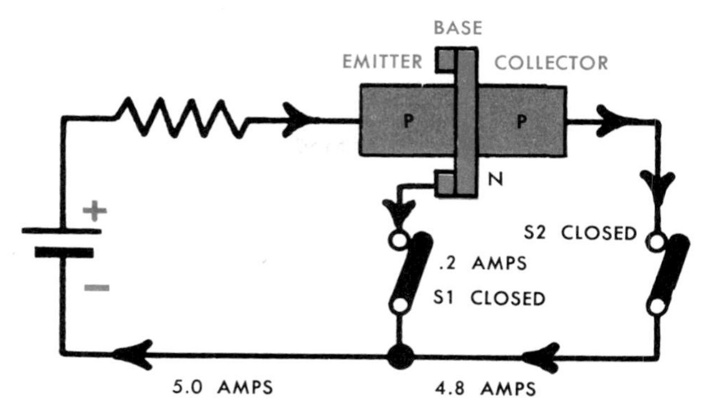

Fig. 71—Transistor in Circuit with Switches Closed

When switch S2 is closed, a rather startling thing happens (Fig. 71). The total current remains at five amperes, but now most of the current leaves the transistor through the collector circuit. The current through the collector is 4.8 amperes, and the base current has been reduced to 0.2 ampere. The reasons for this are as follows:

Because the transistor is arranged so that the emitter-collector are closer together than the emitter-base ring, most of the holes that are injected into the base by the emitter travel on into the collector due to their velocity. Also, the negative potential at the collector attracts the positive holes from the base into the collector.

In the example shown, the collector current is 24 times the base current. This factor is called the current gain.

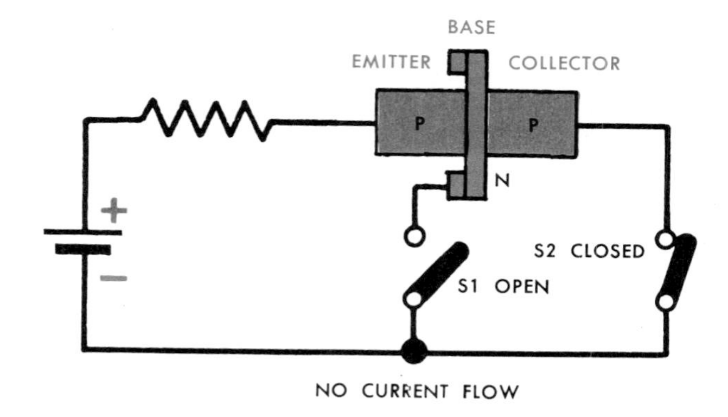

Fig. 72—Transistor in Circuit with Switch S1 Open

An important observation is that with switch S2 closed and switch S1 open, no appreciable current will flow (Fig. 72). The reason for this is that with the base circuit open, there are no holes being injected into the base from the emitter, and so there are no holes in the base which can be attracted by the negative battery potential into the collector. Furthermore, the negative battery potential at the collector attracts the holes in the collector away from the base-collector junction area and the resistance across the base-collector junction becomes very high. Although the emitter and collector are joined, opening switch S1 effectively "shuts off" the transistor so that no appreciable current flows.

An NPN transistor operates in the same way as a PNP transistor, with current flow consisting of a movement of electrons (instead of holes) from the emitter to the base and collector.

The significant thing about a transistor is that by controlling a small base current, a much larger collector current can also be controlled.

BIPOLAR TRANSISTORS

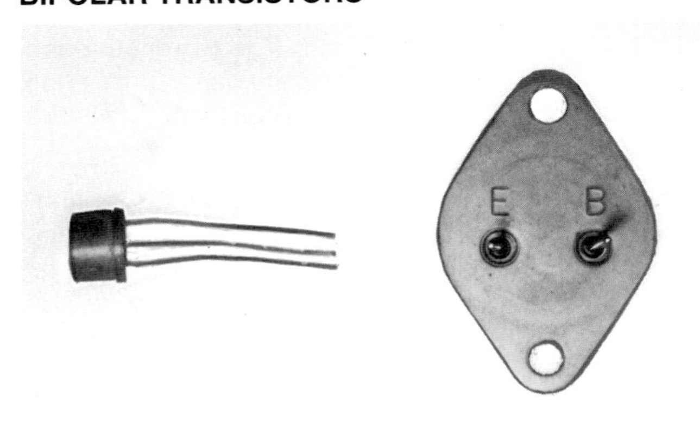

Fig. 73—Signal and Power Bipolar Transistors

There are many types of bipolar transistors (Fig. 73). Small signal transistors are used to amplify low-level signals or one can be made to act as an on-off switch. Some small transistors can be made to both amplify and switch on and off within the same unit.

The bipolar power transistor (Fig. 73) is used in high-power applications. Its large size and exposed metal casing act as a heat sink to keep it cool. In machine operations power transistors are used in transistorized regulators. (For more details see Chapter 6 on "Charging Circuits.")

Symbols for Bipolar Transistors

The symbols for bipolar resistors are shown in Fig. 74. The line with the arrow is the emitter, the heavy line is the base, and the line without an arrow is the collector. Note that the arrow points in the direction of conventional current flow; that is, from positive to negative in the external circuit.

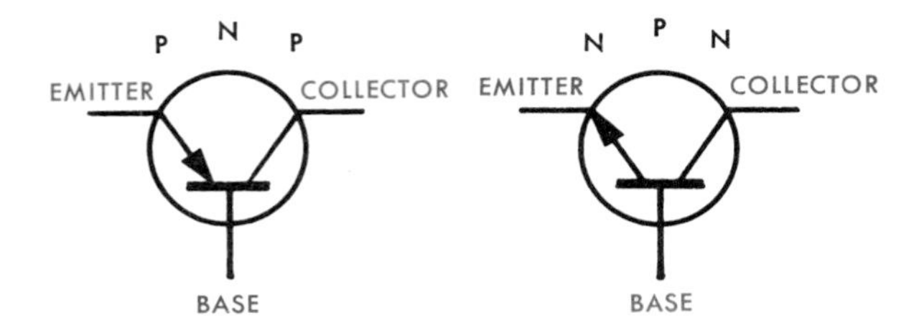

Fig. 74—Symbols for Bipolar Transistors

As we said, it is convenient to look upon current flow in the PNP transistor as a movement of holes, and in the NPN transistor as a movement of electrons. Although the electrons move against the arrow in the NPN transistor, this is not contradictory as it is easier to visualize the current carriers (electrons) as being emitted by the emitter into the base and collector.

FIELD EFFECT TRANSISTOR

Field effect transistors (FETs) use a small amount of voltage to vary a high flow of current. Connections are the *source* (input), *drain* (output), and *gate* (control). There are two major types of FETs in use today; the Junction and Metal Oxide Semiconductor.

Junction Field-Effect Transistor

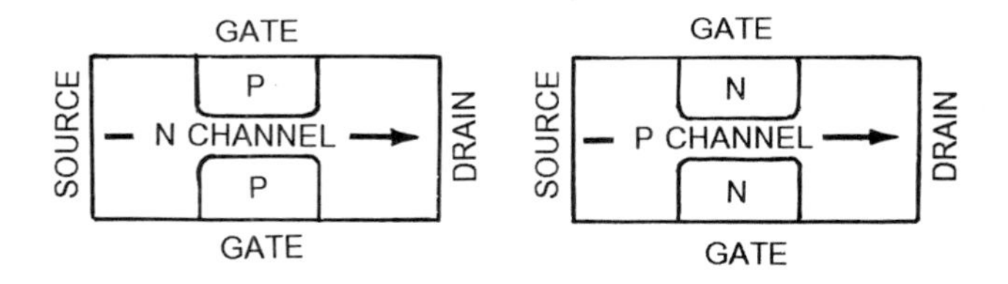

Fig. 75—N-Channel and P-Channel Type Junction FETs

There are two types of junction FETs. They are N-channel and P-channel (Fig. 75). The channel acts like a silicon resistor that conducts current, when in operation, from the source to the drain. At zero volts to the gate, maximum current is obtained from the source to the drain. As voltage is applied to the gate two high-resistance regions or *fields* are created around the channel, slowing the current flow. As more voltage is applied to the gate the fields will completely block the current flow. This gate-channel resistance is very high because in operation the gate-channel junction is a reverse-biased diode. Voltage to the gate generally vary from zero to one volt for an operational range.

Junction FETs are used at the input stage of amplifiers to provide a high-resistance input. They can produce high-frequency signals and can also be used as switches. The symbols for junction FETs are shown in Fig. 76.

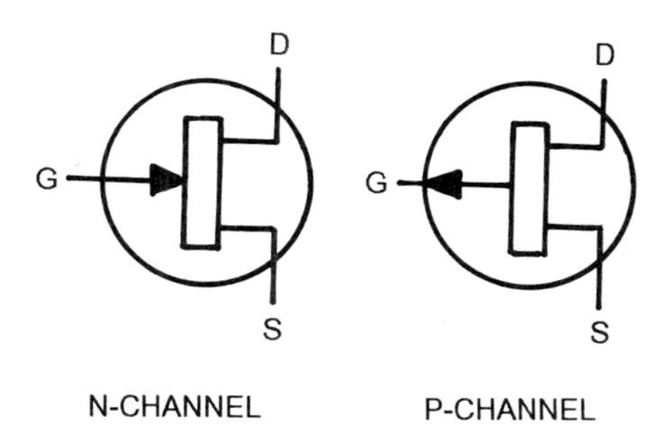

Fig. 76—Junction FET Symbols

Metal Oxide Semiconductor Field-Effect Transistors

Metal oxide semiconductor field-effect transistors (MOSFETs) have become an important transistor in use today. Most microcomputers and memory-integrated circuit devices are made up of thousands of MOSFETs on a very small piece of silicon. This is because their design is easily produced in integrated form. Integrated circuits are discussed later in this chapter.

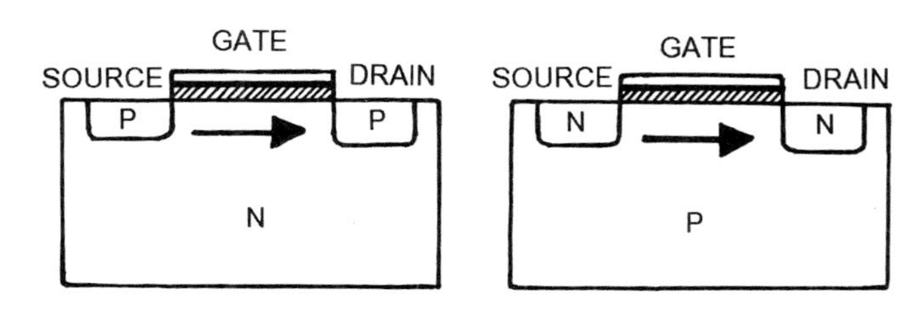

Fig. 77—N- and P-Type MOSFETs

All MOSFETs are N-type or P-type FETs (Fig. 77). Unlike the junction FET, the gate of the MOSFET does not have any electrical contact with the source and drain. An insulating silicon oxide glass-like layer separates the metal contact of the gate from the channel. Applying voltage to the gate causes electrons to be attracted to the region below the gate, creating a thin N- or P-type channel in between the source and drain, allowing current to flow through the channel. The amount of gate voltage determines the resistance of the channel.

MOSFETs have almost infinite gate-channel resistances. They use virtually no current and can switch at very high speeds. They are, however, very sensitive to static electricity. High voltage can easily pierce the insulation layer. MOSFETs are sometimes referred to as Insulated Gate Field-Effect Transistors or IGFETs. The symbols for MOSFETs are shown in Fig. 78.

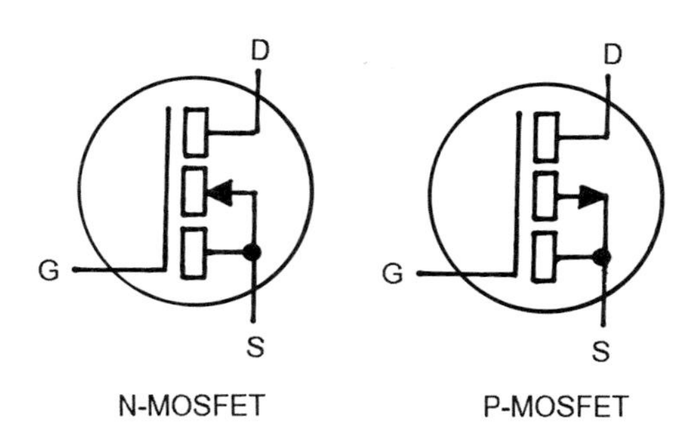

Fig. 78—MOSFET Symbols

SUMMARY: TRANSISTORS

In summary:

- **Transistors control the current in a circuit.**
- **A transistor uses a small current (base) to control a larger current (emitter).**
- **Transistors either allow current to flow or stop it.**
- **Transistors can be destroyed when the Emitter-base current flow is connected in forward bias. (See Chapter 6 on "Transistorized Regulators" as an example.)**
- **Field-Effect Transistors are used in microcomputers and memory-integrated circuits.**

INTEGRATED CIRCUITS

An integrated circuit (IC) (Fig. 79) is a device that contains circuits composed of resistors, diodes, transistors, and capacitors or any other electronic component. Sometimes referred to as "chips," they are unique because of their small size and the amount of work performed. They can contain a few components to form a simple circuit or can be made into a complex circuit with up to hundreds of thousands of components. It is not uncommon for an IC to have over 250,000 transistors on a silicon chip that is only 1/4-inch square! ICs have made possible video games, digital watches, affordable personal computers, and most importantly, microcomputers for use on agriculture and industrial machines.

 Litho in U.S.A.

Fig. 79—An Integrated Circuit Assembly

There are two major types of integrated circuits, *analog* or *linear* and *digital* or *logic.*

ANALOG IC

Analog ICs are circuits composed to produce, amplify, or respond to *variable voltages.* They include many kinds of amplifier circuits that involve analog-to-digital conversion and vice versa, timers, oscillators, and voltage regulators. Analog ICs are known as OPERATIONAL AMPLIFIER CIRCUITS or Op-Amps.

DIGITAL IC

Digital ICs are composed of circuits that produce voltage signals or pulses that have only two levels that are either ON or OFF. They include microprocessors, memories, microcomputers, and many kinds of simpler chips.

Some ICs are manufactured to combine both analog and digital functions on a single chip. We will discuss how each type of integrated circuit works, but first let's see how integrated circuits are made.

HOW INTEGRATED CIRCUITS ARE MADE

All IC components are assembled on pure crystal silicon of N or P type known as wafers (Fig. 80). Grooves are made in the wafers and filled with "doped" silicon (silicon with impurities). Pure crystal silicon is an extremely high-resistive material. Therefore, to allow electron flow, differing concentrations of impurities and silicon (P or N type) are added to the grooves to form components.

Previously in this chapter we became familiar with how these conventional electronic components were made. Let's look at how these components are made on silicon wafers to make integrated circuits.

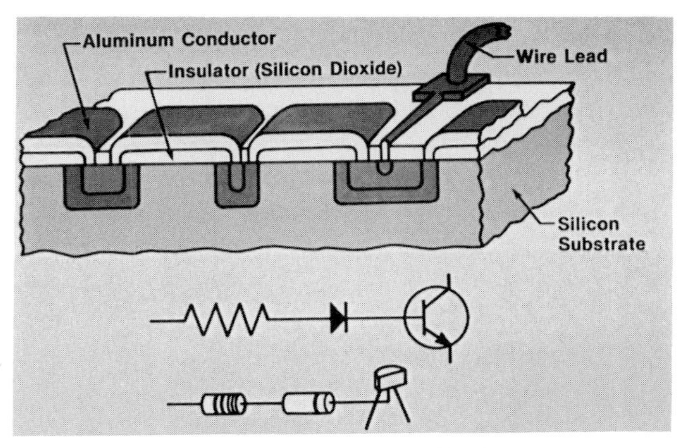

Fig. 80—An Integrated Circuit

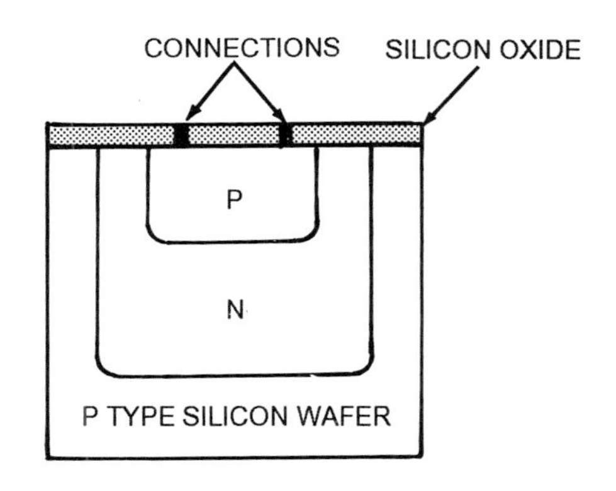

Fig. 81—An IC Resistor

IC RESISTOR

An IC resistor (Fig. 81) is made by adding a strip of P-type "doped" silicon that is arranged between two metal connections over a silicon wafer (P type). The P-type silicon wafer and N-type silicon sections seen in Fig. 81 serve primarily for physical reason only because, as we will soon see, other components are made of the same kind of wafer. The top of the resistor is covered with an insulator such as silicon oxide.

IC DIODE

An IC diode (Fig. 82) is made by layering two strips of "doped" silicon, one P type and the other N type, over a silicon wafer. Each section will have its own connection and the diode is covered with an insulator. You will note in comparing the IC diode to the IC resistor, the only thing different are the connection locations.

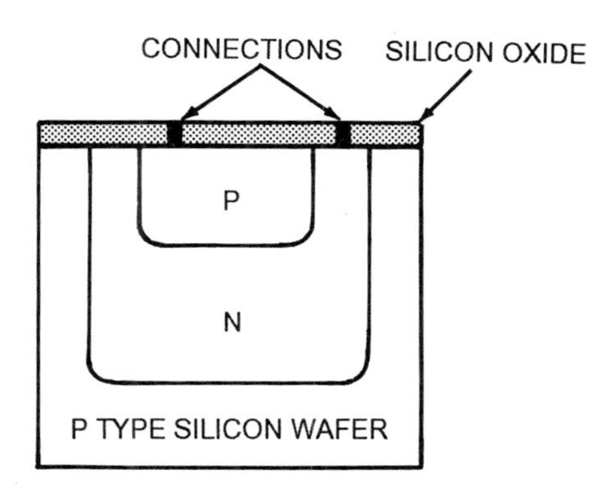

Fig. 82—An IC Diode

IC TRANSISTOR

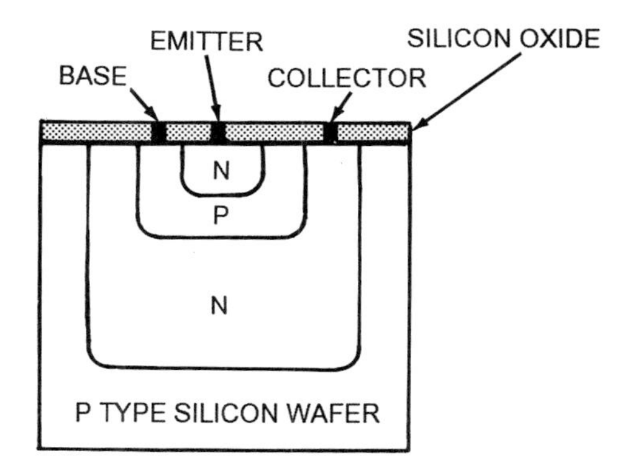

Fig. 83—An IC Transistor

An IC transistor (Fig. 83) can be made by adding another strip of N-type silicon over the P-type strip that could have been used for the resistor or diode. Rearranging the connections to each strip makes an NPN transistor. Note that the emitter, base, and collector connections are at the top of the wafer.

IC CAPACITOR

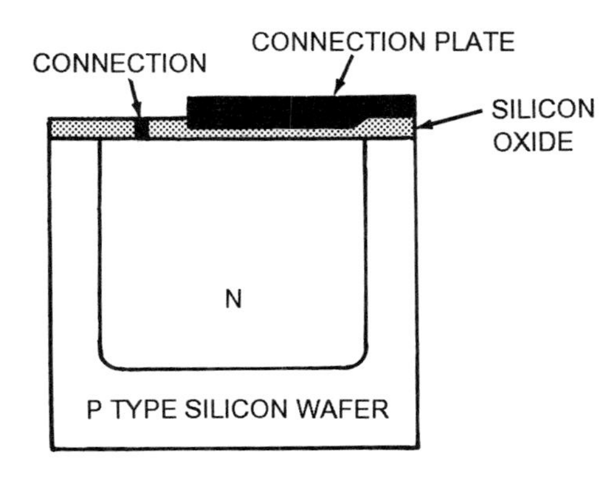

Fig. 84—An IC Capacitor

An IC capacitor (Fig. 84) is made when an N-type silicon is "doped" to act as a plate. The silicon oxide covering acts as the insulator. The other plate of the capacitor is a larger metal connection mounted above the insulator. Needless to say, a capacitor of this type can only be constructed with values above a few hundred pico-farads.

Such IC components are then connected together, usually with aluminum conductors, to make up circuits. However, within a P-type silicon wafer, the process could have been repeated with different types of silicon to form circuits of various diodes, resistors, transistors, and capacitors into a single unit.

HOW DIGITAL ICs WORK

The bases of memory or logic in a digital IC is like a group of on/off switches or latching relays connected together. In other words, the only thing each component of a digital IC "knows" is that it is ON or OFF. In order to process data electronically a binary number system is used in digital ICs. This binary number system is sometimes known as "machine language."

The decimal number system we learned in school is a base ten system. That is, there are ten different digits (0 through 9) we can use. To represent decimal numbers with the use of electricity would require a different voltage level for each of the ten different digits. Because of the high cost of switching circuits, which could handle the ten different voltages, all digital ICs use binary or base two numbers.

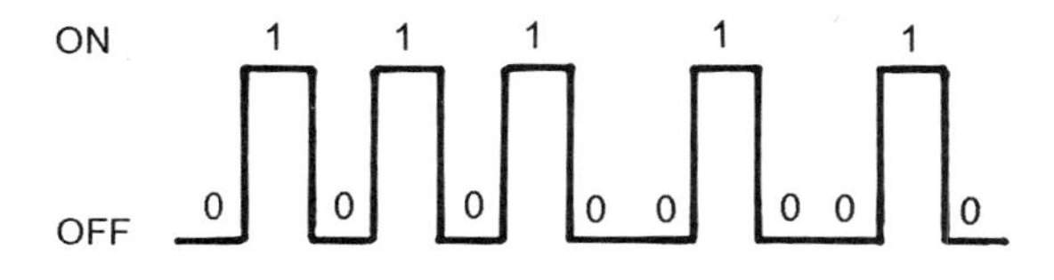

Fig. 85—Pulses and Binary Numbers

In Chapter 2, we discussed that a pulse was created by a sudden on/off of direct current. These pulses are used to process data electrically in a digital IC. The simple ON and OFF voltages (pulses) within a digital IC can be represented by two digits (binary) of 1 and 0, respectively (Fig. 85). With this binary system all numbers can be represented as a decimal system, but a lot more digits are required to do so.

 Litho in U.S.A.

DECIMAL VS BINARY NUMBER SYSTEMS

In the decimal system, the first (right-hand) digit represents ones, the next digit represents 10s, the next 100s, then 1000s, etc. Each digit is ten times as large as the next digit to the right. In the binary system, the first digit represents ones, the next digit twos, the next digit fours, the next digit eights, then 16s, 32s and 64s, etc. Each digit is two times as large as the next digit to the right.

DECIMAL	BINARY
1	8 4 2 1
0	0
1	1
2	10
3	11
4	100
5	101
6	110
7	111
8	1000
9	1001

The number 125 in the decimal system represents one 100, plus two 10s, plus five ones.

In the binary system, the number 125 is written as 1111101, which represents one 64, plus one 32, plus one 16, plus one 8, plus one 4, plus one one.

DECIMAL	BINARY
100 10 1	64 32 16 8 4 2 1
1 2 5	1 1 1 1 1 0 1

Each binary digit is called a "bit," and by grouping the digits into eight bit "bytes," handled in timed sequence, all characters can be represented and processed. Obviously, the binary system would be very cumbersome to use because of the great amount of time required to handle the number of digits involved, but given the speed of an electronic circuit, this is not a problem.

Adding binary numbers is just like adding decimal numbers, except in decimal numbers we carry a digit to the next column if the sum of the numbers added exceeds nine, whereas with binary numbers we carry a digit to the next column if the sum of the numbers exceeds one.

To electronically add numbers and process data requires logic circuits. The building blocks of logic circuits are called **gates.** A gate is a device which makes a YES or NO (one or zero) decision (output) based on two or more inputs.

GATES

There are three basic types of gates used in digital ICs to perform logic from input pulses. They are the AND gate, OR gate, and EXCLUSIVE OR gate.

AND GATE

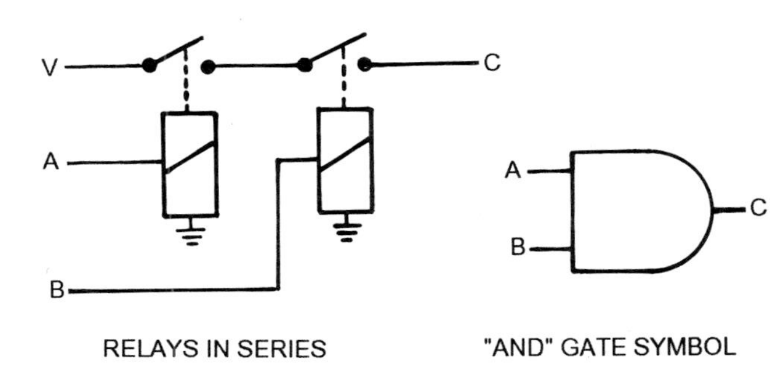

Fig. 86—Relay AND Gate

The "AND" gate functions like two normally-open relays connected in series with inputs at A and B (Fig. 86). Incoming voltages or pulses will only arrive at C if both A "and" B are on, or "1."

The logic on/off or 0/1 sequence to A and B with result C is known as a *truth table.* The symbol for an AND gate and its truth table are shown in Fig. 87.

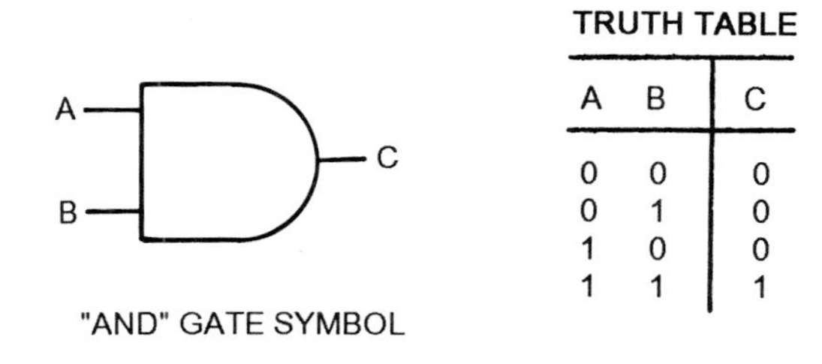

TRUTH TABLE

A	B	C
0	0	0
0	1	0
1	0	0
1	1	1

Fig. 87—AND Gate Symbol and Truth Table

OR GATE

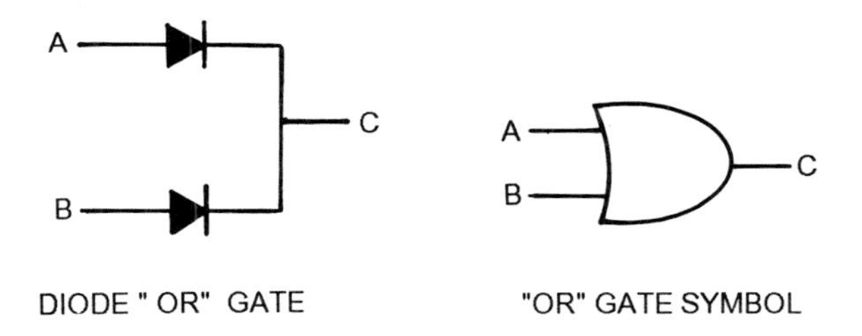

Fig. 88—Diode OR Gate

The "OR" gate functions like two diodes connected in parallel to one outlet (Fig. 88). This gate will give a 1

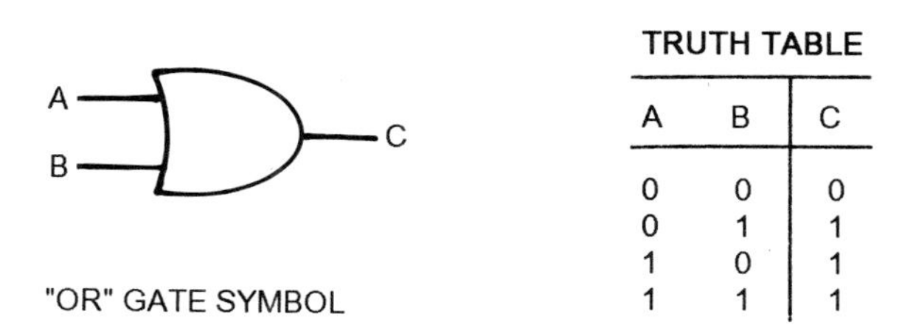

TRUTH TABLE

A	B	C
0	0	0
0	1	1
1	0	1
1	1	1

Fig. 89—OR Gate Symbol and Truth Table

or ON at C if either A "or" B is 1 or ON. The symbol for an OR gate and its truth table is shown in Fig. 89.

EXCLUSIVE OR GATE

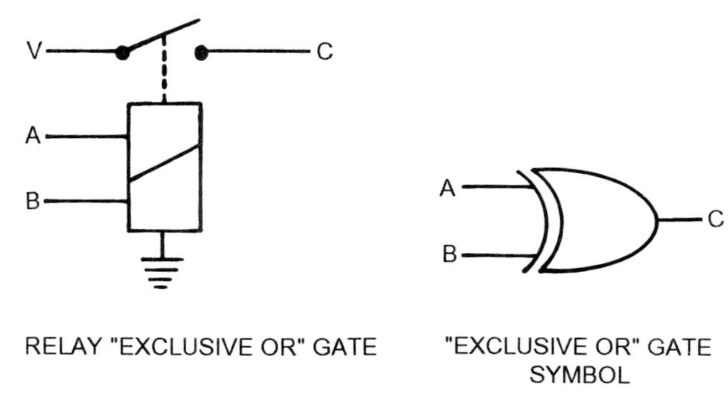

Fig. 90—Relay EXCLUSIVE OR Gate

The last basic type of gate is the EXCLUSIVE OR gate. It operates like a normally-open relay with inputs at A and B attached to the relay coil (Fig. 90). When both A or B are 1 (ON) the relay is "exclusively" Off, or 0. The symbol for an EXCLUSIVE OR gate and its truth table are shown in Fig. 91.

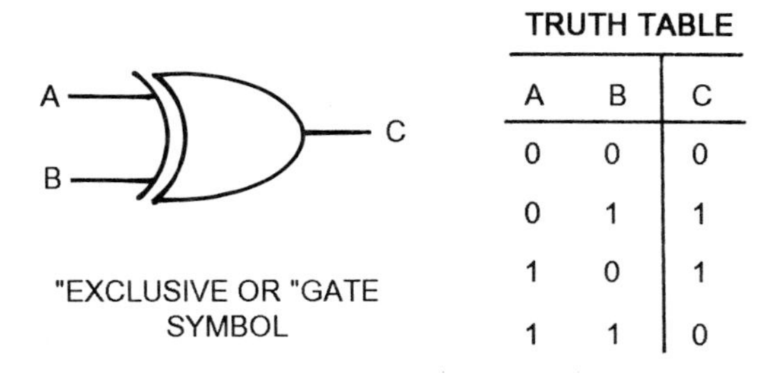

TRUTH TABLE

A	B	C
0	0	0
0	1	1
1	0	1
1	1	0

Fig. 91—EXCLUSIVE OR Gate Symbol and Truth Table

INVERTER

An inverter is a device with only one input A and one output B. An inverter functions like a normally-closed relay (Fig. 92) which reverses any input at A. The symbol and truth table for a inverter is shown in Fig. 93.

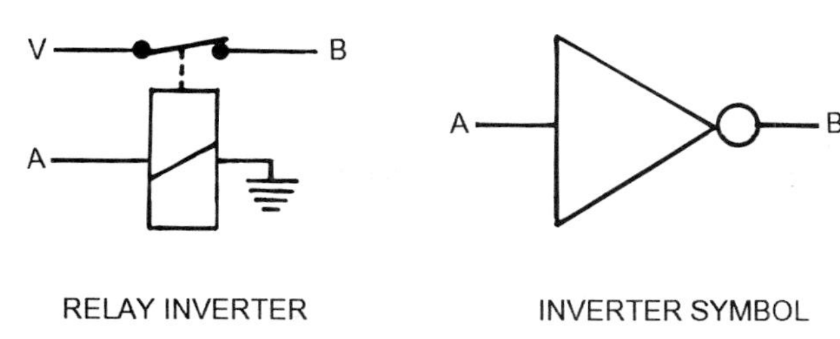

Fig. 92—Relay Inverter

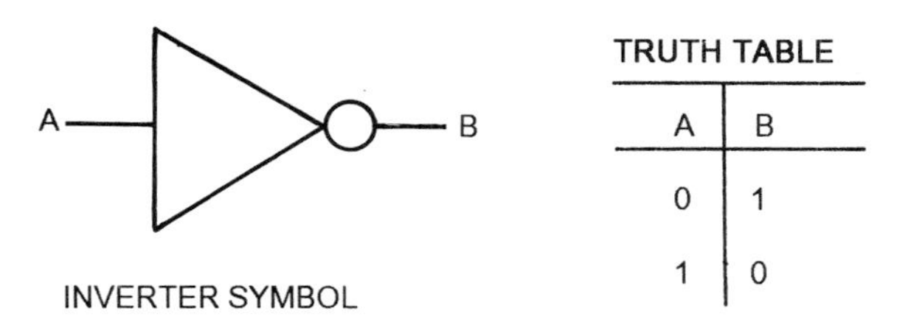

TRUTH TABLE

A	B
0	1
1	0

Fig. 93—Inverter Symbol and Truth Table

NAND GATE

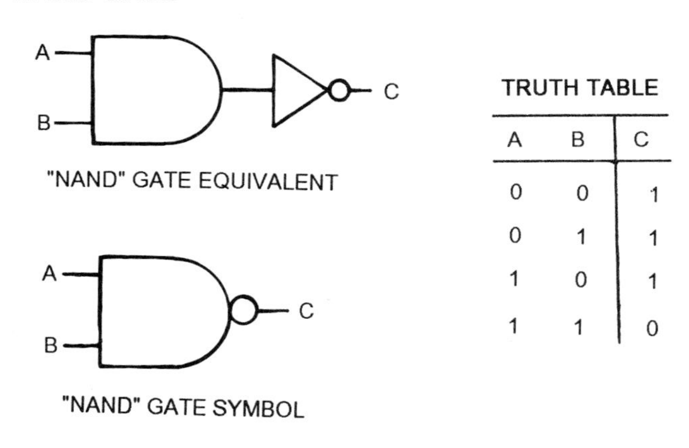

TRUTH TABLE

A	B	C
0	0	1
0	1	1
1	0	1
1	1	0

Fig. 94—NAND Gate Equivalent, Symbol, and Truth Table

The inverter can be used with any gate device to provide the required logic combinations. By combining an AND gate with an inverter (Fig. 94) the device is known as a NAND gate.

Looking at the truth table for a NAND gate in Fig. 94 you will note that it is the reverse of or the inverse of the truth table for an AND gate (Fig. 87).

NOR and EXCLUSIVE NOR GATES

NOR and EXCLUSIVE NOR gates are OR and EXCLUSIVE OR gates combined with an inverter. As you would expect, the outputs of their truth tables (Fig. 95) are just the opposite of the OR and EXCLUSIVE OR gates.

The NAND and NOR gates are important in making logic circuits, as we will now see.

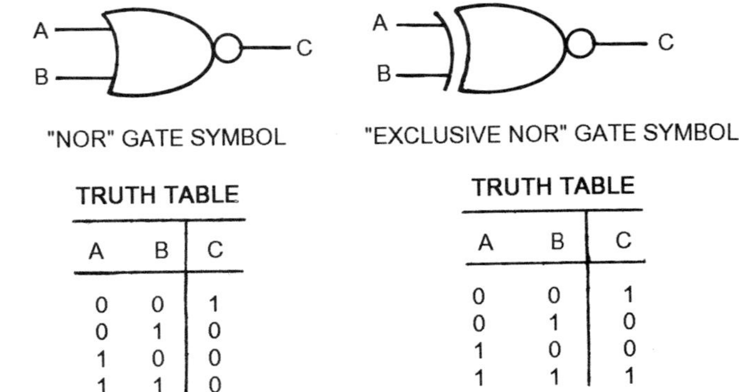

TRUTH TABLE

A	B	C
0	0	1
0	1	0
1	0	0
1	1	0

TRUTH TABLE

A	B	C
0	0	1
0	1	0
1	0	0
1	1	1

Fig. 95—NOR and EXCLUSIVE NOR Gates and Truth Table

HOW GATES ARE USED

Gates are used individually or connected together to form networks of gates to create logic circuits. Most logic circuits are one of two types, *Combinational* or *Sequential.*

COMBINATIONAL LOGIC CIRCUITS

Combinational gate circuits respond to incoming pulses immediately without regard to earlier events. They can be very simple or immensely complicated to form desired logic circuits. The NAND and NOR gates are important because in combination they can make virtually any type of gate.

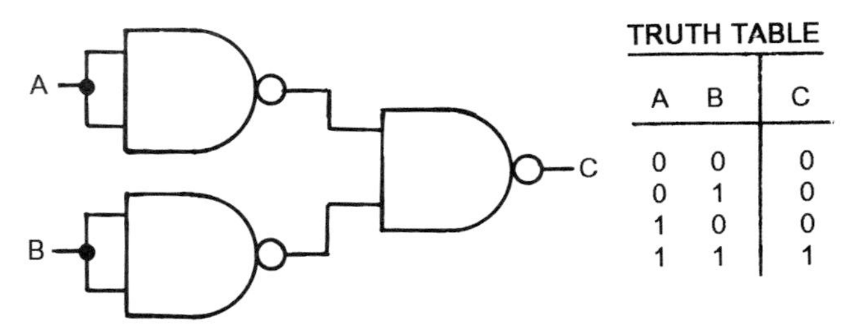

TRUTH TABLE

A	B	C
0	0	0
0	1	0
1	0	0
1	1	1

Fig. 96—Circuit of NAND Gates to make an OR Gate

For example, when three NAND gates are combined together, as in Fig. 96, they make an OR gate, as the truth table verifies.

Different types of gates can be combined together. In Fig. 97, two different types of gates form a logic circuit that converts a two-bit binary number (A and B) to its decimal equivalent (0, 1, 2, and 3). This is known as a decoder circuit.

As you can see from Fig. 97, this simple combination circuit becomes quite complicated when you follow the input pulses, operation of the gates, and obtaining the results. Because of this, a complicated circuit of gates uses a box for a symbol and is named to show what it does. Fig. 98 shows the box symbol for Fig. 97.

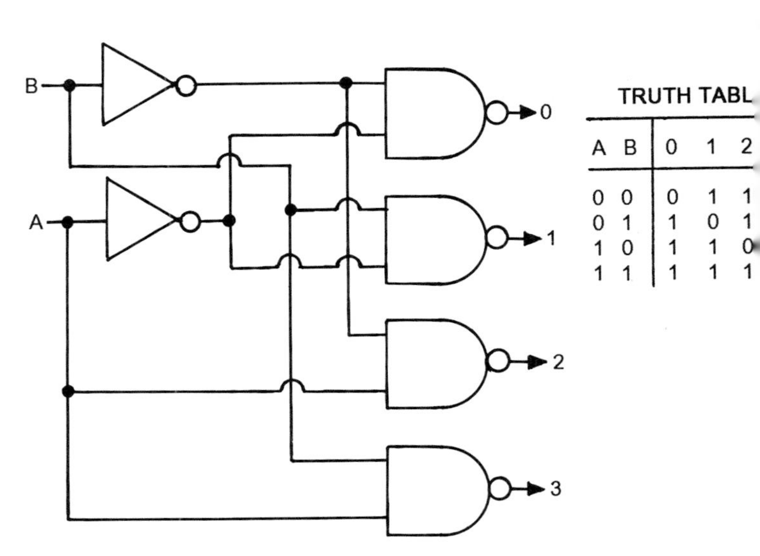

TRUTH TABL

A	B	0	1	2
0	0	0	1	1
0	1	1	0	1
1	0	1	1	0
1	1	1	1	1

Fig. 97—Decoder Circuit of NAND and Inverter Gates

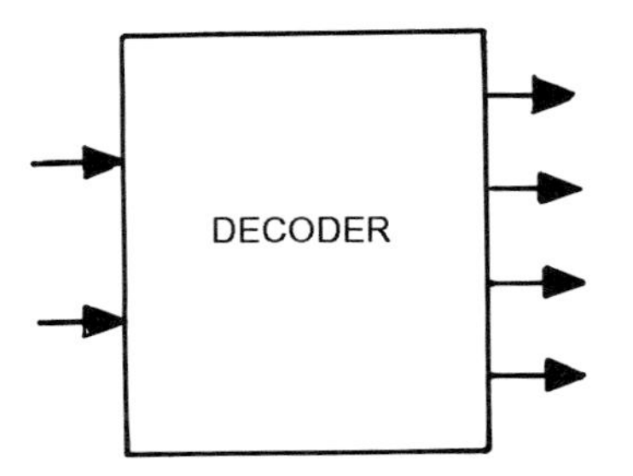

Fig. 98—Symbol for a Decoder Circuit

SEQUENTIAL LOGIC CIRCUITS

The output of a sequential logic circuit is determined by the previous state of the input (pulses). "Bits" of data move through sequential circuits step by step. Data (memory) advances a step each time a steady stream of pulses occur at the input. This "steady stream" of pulses is known as a *Clock.* The building block (gates) of a sequential logic circuit is called a *Flip-Flop Circuit.*

RESET-SET FLIP-FLOP CIRCUIT

The basic circuit of all sequential circuits is the RS (Reset-Set) flip-flop circuit (Fig. 99). In this circuit the two outputs always are opposite of each other and two inputs of 0 (off) is not allowed. If there are two 1 (on) inputs the outputs do not change. This circuit is sometimes called a *Latch.*

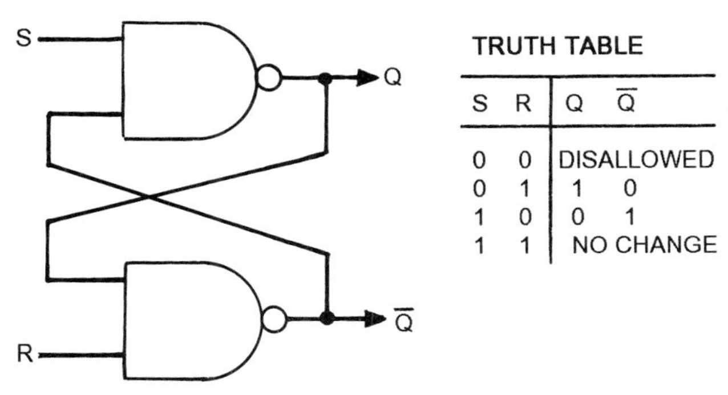

S	R	Q	Q̅
0	0	DISALLOWED	
0	1	1	0
1	0	0	1
1	1	NO CHANGE	

Fig. 99—Basic Reset-Set Flip-Flop Circuit

CLOCK LATCH CIRCUIT

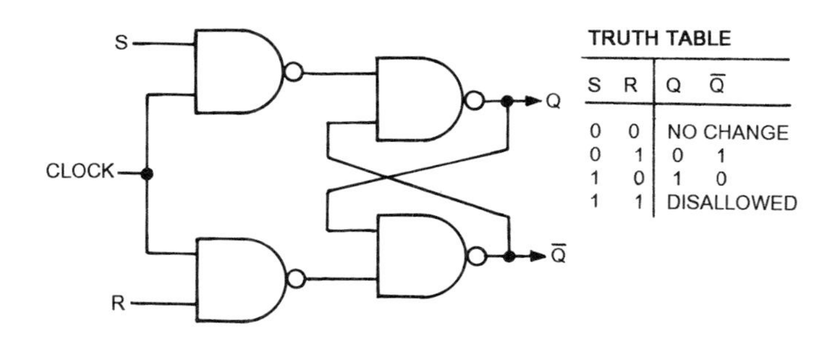

S	R	Q	Q̅
0	0	NO CHANGE	
0	1	0	1
1	0	1	0
1	1	DISALLOWED	

Fig. 100—A Clocked Latch Circuit

In a clocked latch (RS flip-flop) circuit (Fig. 100), the latch ignores any data input at S and R until a stream of pulses (clock) "triggers" the operation of the circuit. This circuit will not allow S and R to be both on (1). This circuit is the basis for *Random Access Memory* (RAM), discussed later in this chapter.

DATA LATCH CIRCUIT

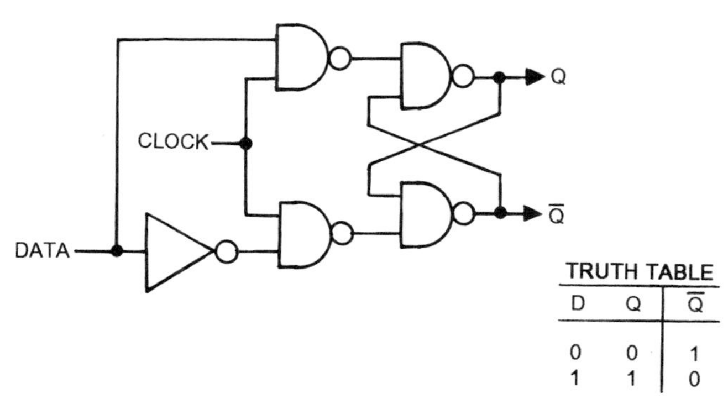

D	Q	Q̅
0	0	1
1	1	0

Fig. 101—A Data Latch Circuit

In a data latch circuit (Fig. 101) the latch "stores" the present outputs between the clock pulses. The data input is either on (1) or off (0). This circuit is the basis for a self-diagnosis system used on machinery.

JUMP-KEEP LATCH CIRCUIT

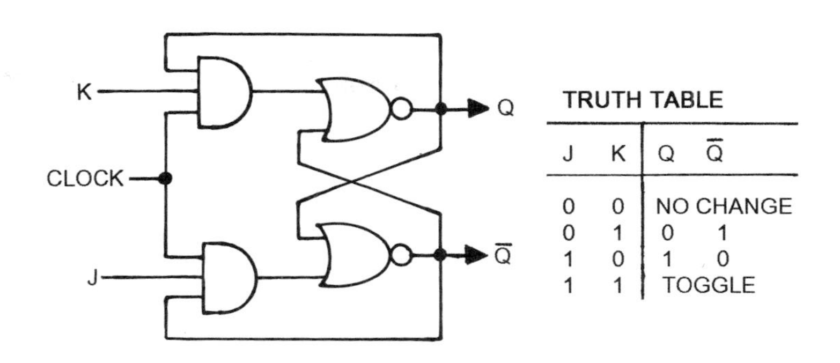

J	K	Q	Q̅
0	0	NO CHANGE	
0	1	0	1
1	0	1	0
1	1	TOGGLE	

Fig. 102—A Jump-Keep Latch Circuit

In a jump-keep (JK) circuit (Fig. 102) the outputs "keep" the same on/off as the inputs and the latch ignores the clock impulses. When both inputs (J and K) are on (1), then the outputs of the latch change state or "jump" with each clock pulse. This "jump" is also known as a "toggle." This circuit acts like a switch to control clock pulses and is used mostly as a basis for a counter circuit. Note that this circuit is also made up of two NOR gates and two AND gates.

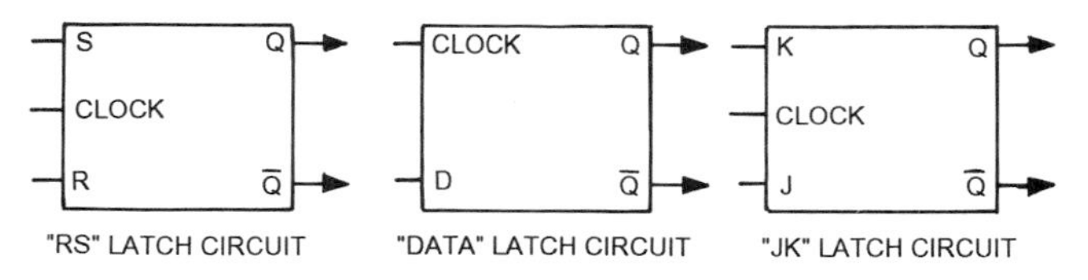

Fig. 103—Latch Circuit Symbols

As with the combination circuits, when the sequential circuits become complicated, a named box is used as a symbol. Fig. 103 shows the box symbols for the RS latch, the D latch, and JK latch circuits.

TOGGLE LATCH CIRCUIT

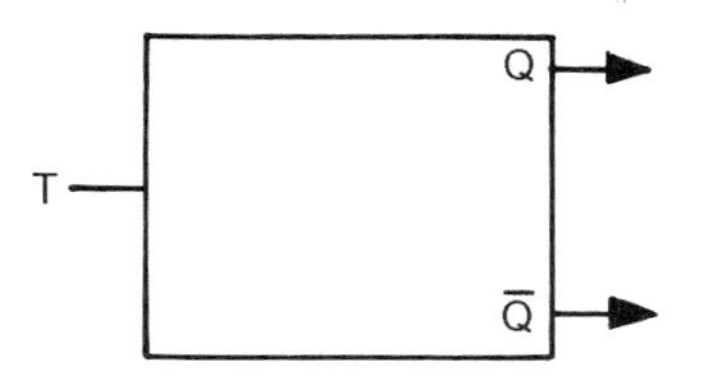

Fig. 104—Toggle Latch Circuit Symbol

A toggle (T) latch circuit (Fig. 104) has its outputs change state with every input pulse and has no restrictions. Toggle circuits can be made from any of the three latch circuits previously shown by changing the connections (Fig. 105).

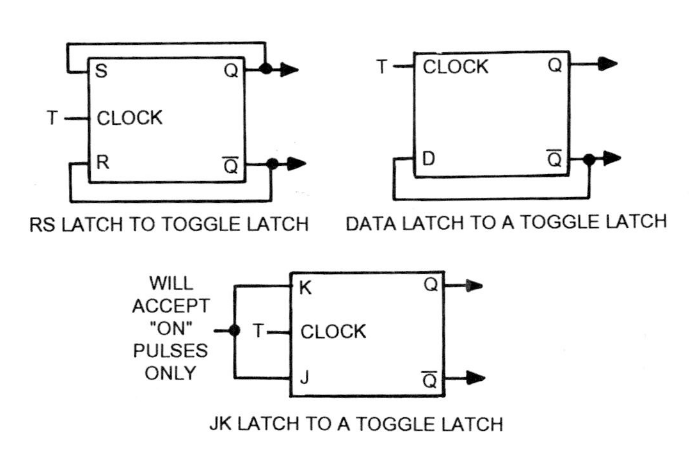

Fig. 105—Toggle Latch Circuits

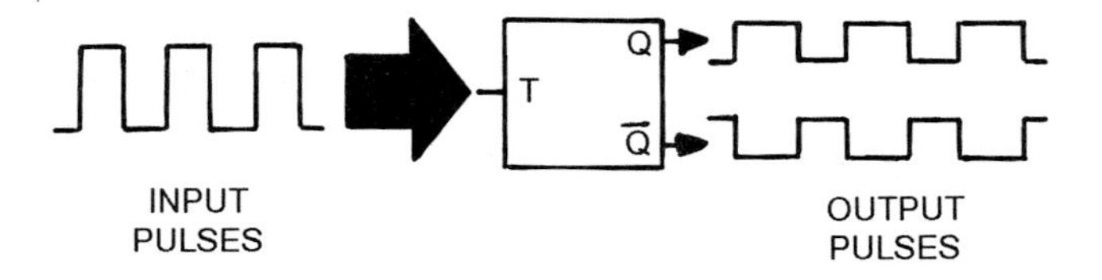

Fig. 106—Toggle Latch Input and Output Pulses

For every other input on/off pulse on a toggle latch circuit, the outputs change their state. In other words, the output of a toggle latch is one half of its input pulse (Fig. 106). Therefore, the input pulses are divided by two.

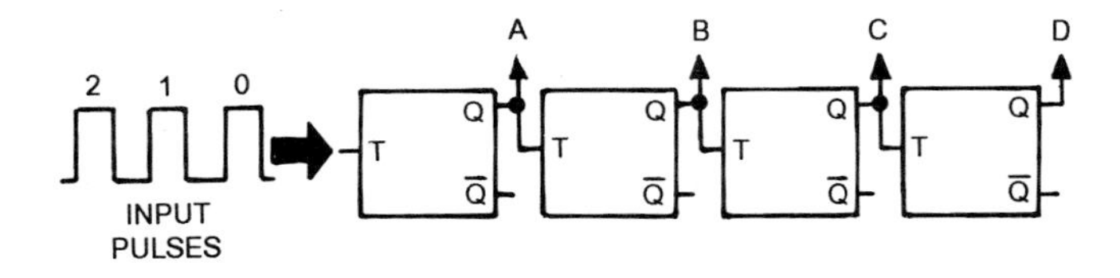

Fig 107—A Four-Bit Binary Counter

Toggle latch circuits can be combined to create a binary counter. Let's look at how four Toggle or "T" latch circuits can be formed to create a four-bit binary counter (Fig. 107). Each "T" latch circuit divides the incoming pulses by two.

The results, looking at the truth table (Fig. 108), is a four-bit binary count of 0000 to 1111. The count will recycle after the sixteenth incoming pulse and restart. This type of circuit can count pulses up to 20,000,000 times per second! There are many types of IC counters, most of which have special features that count up, count down, add, subtract, divide, rest, etc.

	INPUT PULSE COUNT															
OUTPUTS	0	1	2	3	4	5	6	7	8	9	10	11	12	13	14	15
D	0	0	0	0	0	0	0	0	1	1	1	1	1	1	1	1
C	0	0	0	0	1	1	1	1	0	0	0	0	1	1	1	1
B	0	0	1	1	0	0	1	1	0	0	1	1	0	0	1	1
A	0	1	0	1	0	1	0	1	0	1	0	1	0	1	0	1

Fig. 108—Truth Table for a Four-Bit Binary Counter

OPERATIONAL AMPLIFIER CIRCUITS

Analog ICs are circuits composed to produce, amplify, or respond to variable voltages and are called OPERATIONAL AMPLIFIER CIRCUITS.

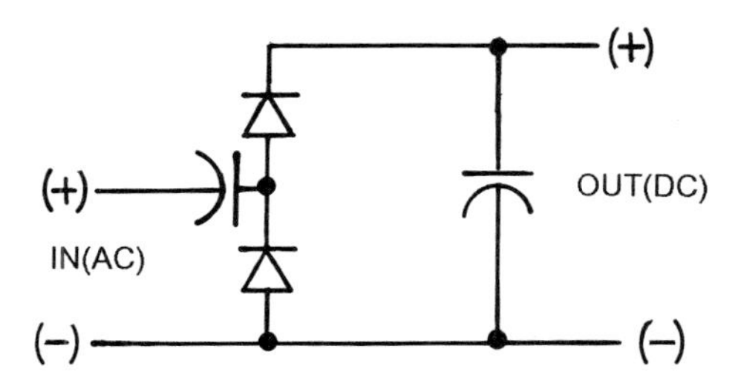

Fig. 109—An Operational Conversion/Amplifier Circuit

Fig. 109 shows a basic conventional circuit of diodes and capacitors that converts an incoming alternating current (AC) to direct current (DC) at its output, but in the process the outgoing DC is doubled in voltage from the incoming AC voltage. Such a circuit is called an Operational Amplifier (Op-Amp) Circuit.

Integrated Op-Amps (analog ICs) are made up of IC diodes, IC capacitors, IC transistors, and IC resistors. The amount of amplifier circuits in an analog IC is virtually unlimited. The symbol for analog IC Op-Amps is shown in Fig. 110.

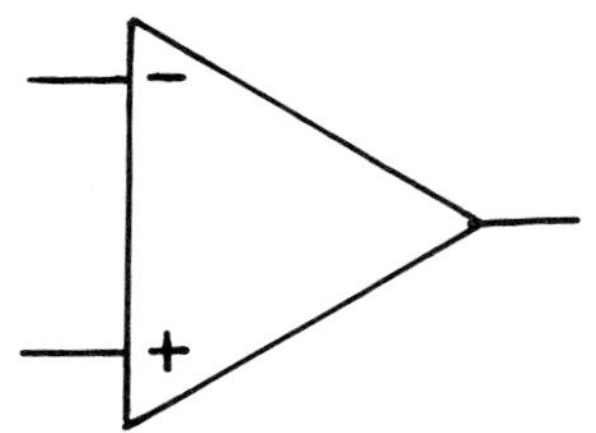

Fig. 110—Symbol for an Op-Amp

HOW OP-AMPS WORK

Op-Amps amplify the difference between voltages or signals applied to their two inputs. The input signals can be either AC or DC. The voltage output will be amplified if the voltage is applied to one of the input connections and the other input connection is grounded or maintained at some voltage level.

Depending upon the polarity connection of the Op-Amp to the incoming voltage, the Op-Amp has an *Inverting* or *Non-Inverting* operation.

INVERTING OPERATION

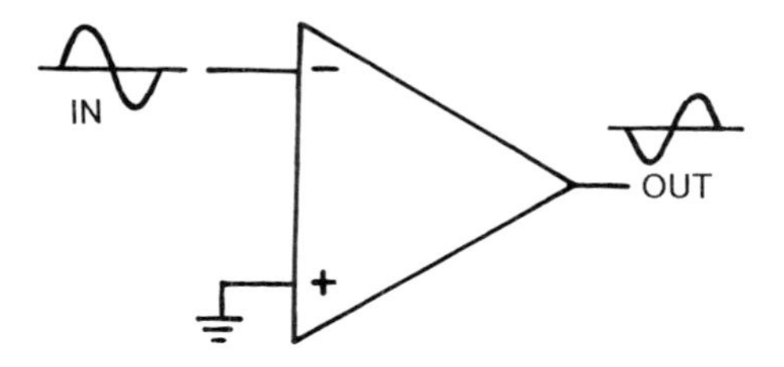

Fig. 111—An Inverting Op-Amp

If the incoming voltage (AC) is connected to the negative connection of an Op-Amp and the positive connection is grounded, the output voltage will be amplified and the signal will be inverted (Fig. 111).

NON-INVERTING OPERATION

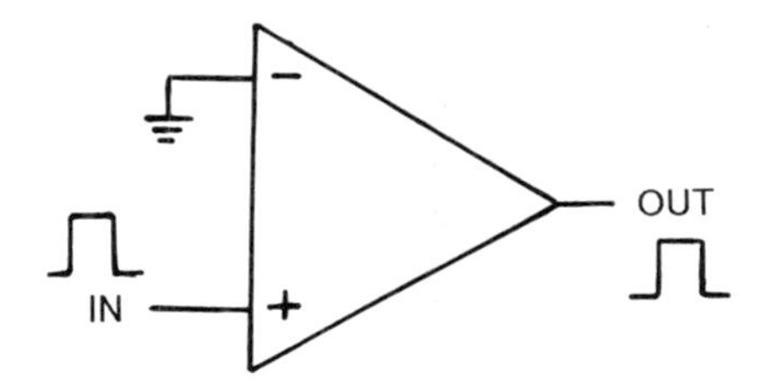

Fig. 112—A Non-Inverting Op-Amp

If the incoming voltage (DC) is connected to the positive connection of an Op-Amp and the negative connection is grounded, the output voltage will be amplified and the signal will not change (Fig. 112).

In both inverting and non-inverting operations the maximum amplification level or *gain* is obtained. The output voltage will swing from full ON to full OFF or vice versa depending upon the input. This on/off operation of an OP-Amp is called an *Inverter Buffer.*

OP-AMP OUTPUT CONTROL

With the operation of both inverting and non-inverting Op-Amps, the gain or voltage output, in most cases, is controlled by resistors.

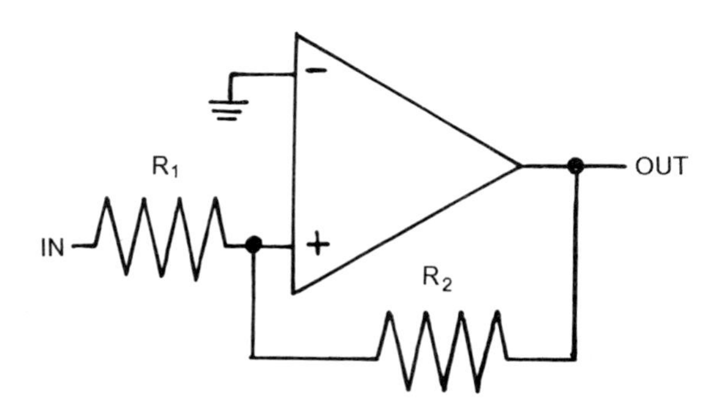

Fig. 113—A Non-Inverting Op-Amp with Resistors

Fig. 113 shows a non-inverting Op-Amp with resistors R_1 and R_2. The R_2 resistor feeds some of the output back to the input, reducing the gain. The formulation for this gain control is:

GAIN = R_2 / R_1 VOLTS OUT = VOLTS IN x (R_2 / R_1)

The output voltage control can be controlled manually by replacing resistor R_2 with a potentiometer.

MEMORY

On machinery with on board computers, memory is used as a self-diagnosis. There are two basic types of memory. They are Random Access Memory (RAM) and Read only Memory (ROM). ROM is permanently stored or programmed within integrated circuits and cannot be changed or effected by normal operational inputs. RAM is memory which is available for the operation of the unit and collects input from the operational components of the unit.

When operation of the unit is not normal, and the RAM inputs match the stored ROM, the ROM "tells" the microprocessor to "turn on" a warning indicator light for the operator or if the operation is critical, to "shut down" the unit to prevent damage. RAM can be recalled, after a problem occurs, in the form of codes that tell what component is at fault.

Other types of memory are Programmable Read Only Memory (PROM), Electrical Programmable Read Only Memory (EPROM), and Non Volitive Random Access Memory (NVRAM).

Most failures in integrated circuits occur during initial usage when they are new. For this reason, manufacturers do an operational "burn in" on new ICs to isolate "bad" components before they are sold. After this initial "burn in," very few IC failures occur until environmental factors start to take effect after a number of years.

PRINTED CIRCUIT BOARDS

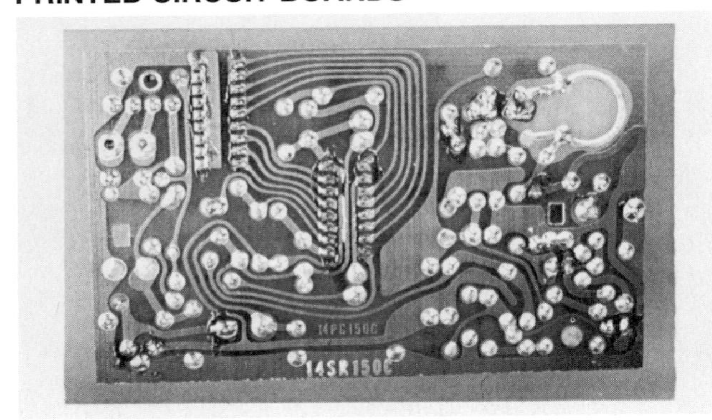

Fig. 114—Lines of a Printed Circuit Board

Printed circuit boards (PCs) are used to hold components in place and to provide current paths from component to component. The lines on a printed circuit board never touch each other (Fig. 114). If they did they would cause a short in the circuit.

Printed circuit boards are manufactured by a photographic process combined with acid etching of copper pathways.

After the boards are manufactured, components are added and soldered into place.

Printed circuit boards are relatively inexpensive for large numbers of devices and eliminate mistakes which could occur with hand wiring.

Problems include difficulty in repairing damaged pathways and great expense to modify circuit design once in production.

SUMMARY: INTEGRATED CIRCUITS

In Summary:

- **Integrated circuits are small electronic components that are assembled on pure silicon wafers.**
- **There are two types of integrated circuits.**
- **Integrated circuits use the binary number system to produce logic.**
- **The building blocks of logic circuits are called gates.**
- **There are two types of logic circuits composed of gates: Combinational and Sequential.**
- **Analog integrated circuits or Operational Amplifier (Op-Amp) circuits amplify the difference between voltages or signals applied to their inputs.**
- **Two types of memory are Random Access Memory (RAM) and Read Only Memory (ROM). There are many others such as Programmable Read Only Memory (PROM), Electrical Programmable Read Only Memory, and Non Volitive Random Access Memory (NVRAM).**
- **Integrated circuits are assembled on printed circuit boards.**

DISPLAY DEVICES

ANALOG GAUGE

Fig. 115—An Analog Fuel Gauge

The most familiar type of display device is the **analog gauge** (Fig. 115). A varying signal causes a mechanical change in the position of a needle.

LIGHT-EMITTING DIODE

A **light-emitting diode (LED)** is a solid-state display device. LEDs are self-illuminating, but have relatively high energy consumption, making them undesirable for some applications.

LIQUID CRYSTAL DISPLAY

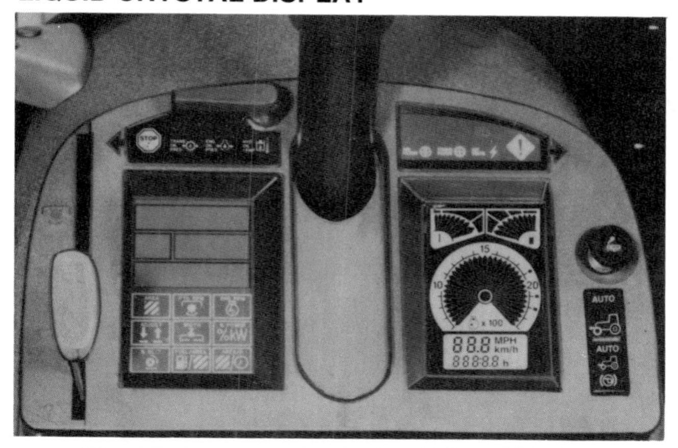

Fig. 116—Instrument Panel of Liquid Crystal Displays

Liquid crystal displays (LCDs) are used on modern machinery instrument panels to show data output from integrated circuits (Fig. 116).

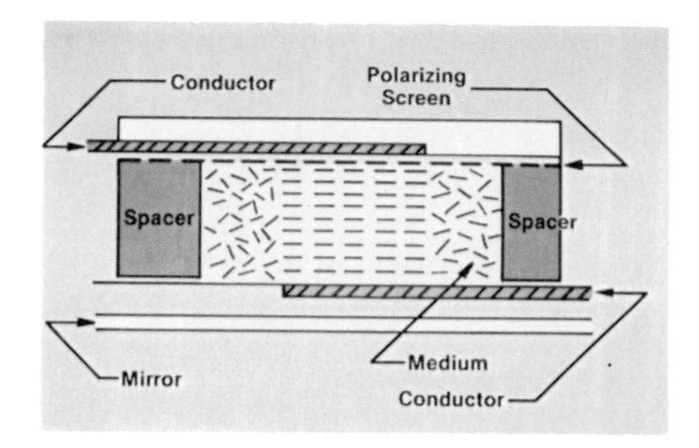

Fig. 117—Side View of a Liquid Crystal Display

LCDs use a special fluid medium to allow segmented displays (Fig. 117). When no current is applied to the conductive plates below and above the crystal fluid medium, the crystal elements float in random orientation.

Since LCDs do not provide their own light, only the relatively low power required to orient the crystals is consumed by their operation. For night operation, LCDs require some kind of illumination from behind.

There is a polarizing screen above the crystal layer. Light can travel freely through the screen and the crystal, and be reflected by the mirror at the bottom of the display back to the viewer.

When a signal turns on the plates for a segment, the crystals snap into alignment, the polarized light is polarized again at a 90° angle, and light no longer reaches the mirror. A dark segment forms.

Since LCDs do not provide their own light, only the relatively low power required to orient the crystals is consumed by their operation.

VACUUM FLORESCENT DISPLAYS

Fig. 118—A Panel Displaying a Vacuum Florescent Display

Vacuum florescent displays (VFDs) are instrument displays that give off their own light (Fig. 118). They work in the manner of neon lights, where output data from integrated circuits directs streams of electrons to strike phosphorescent segments (Fig. 119). Displayed information is controlled by turning on and off anodes below the phosphorescent layer.

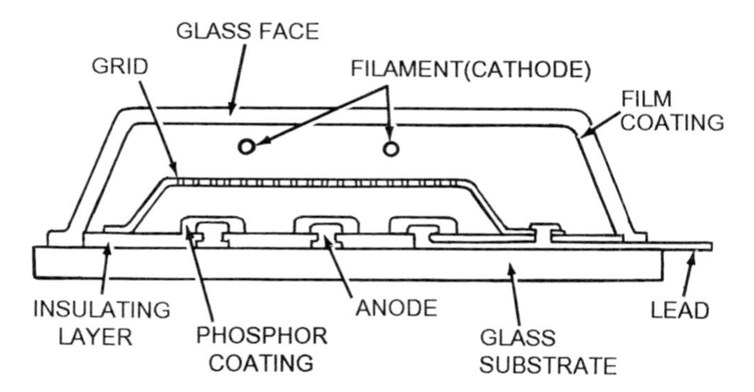

Fig. 119—Side View of a Vacuum Florescent Display

IN SUMMARY: DISPLAY DEVICES

In Summary:

- **An analog gauge operates by varying signals.**
- **Light-emitting diodes are self-illuminating, but have high energy consumption.**
- **Liquid crystal displays do not provide their own light and use low power to operate.**
- **Vacuum florescent displays give off their own light and operate in the manner of neon lights.**

MAGNETIC PICKUPS

Magnetic Pickups use the motion of a magnetic field past a coil or the motion of metal past a magnetic field to generate a signal.

MOTORS

Motors use the action of induced magnetic fields against opposing fields to cause a rotary motion.

GENERATORS AND ALTERNATORS

Generators and **alternators** use rotary mechanical motion to move magnetic fields and coils relative to each other. Current flow is induced within the coils. Alternators and generators are detailed in Chapter 6.

FUSES AND CIRCUIT BREAKERS

Both **fuses** and **circuit breakers** are protection devices.

A fuse is a conductor of electricity. However, when current passing through the fuse is too high, a thin element made of aluminum and other metals will overheat and melt. This results in a "blown" fuse and opens the circuit, protecting wiring and other components downstream of it. Fuses using blades to plug into or connect to a circuit are used in automotive electrical systems. When a fuse is blown, it must be replaced.

A circuit breaker performs the same function as a fuse. However, high current will cause a circuit breaker to "trip." Then, after the cause of the overcurrent has been eliminated, it can be reset manually like a light switch, or automatically by removing the excess load. Circuit breakers can be reset many times.

TEST YOURSELF

QUESTIONS

1. What metal is less resistive than copper?

2. Name two electronic components which use semiconductors.

3. Diodes allow current to pass in ________________ ________________ only.

4. Transistors control current in a circuit by allowing current to ______________or by ______________it.

5. Switches are used for __________ __________and ________________ .

6. Relays allow a ________________ current to control a ______________ current.

7. Resistors reduce voltage by resisting __________.

8. The bands on some resistors indicate their _______ value as measured in ____________.

9. Capacitors are electronic components that _______ electrons.

10. A resistor and capacitor circuit that controls frequency is called an ____________ circuit.

11. Electromagnetic devices which use the strength of the magnetic field generated in a coil to move a metal core are called ______________.

12. Integrated circuits use the _____________number system to produce ______________.

13. What are the building blocks of logic circuits called?

(Answers on page 19 at the end of this book.)

 Litho in U.S.A.

ELECTRICAL SAFETY / CHAPTER 4

SAFETY

Safety is too expensive to learn by accident. Hospital bills, doctor bills, medical supplies, and rehabilitation costs can be a big financial burden on both individuals and companies.

Because of recent dramatic increases in these costs, both company-provided and private insurance plans have begun to shift more of this burden onto the insured parties.

Accidents also result in lost time from work and more importantly, may cause a permanent handicap or a loss of health that affects the injured party's family and earning ability. Unsafe practices also result in property damage.

Accidents are reduced by incorporating safe work practices into shop management programs. Safety must be thought of as a normal part of the management process just as supplies, personnel and overhead costs are.

Electricity has brought advanced technology into our homes and workplaces. From light bulbs to microcomputers, it plays a significant role in our everyday lives.

It is easy to take this unseen and somewhat mystical convenience for granted and to forget that, as wonderful as its gifts, it is a powerful and dangerous force capable of causing property damage, serious bodily injury and even death.

Fig. 1—Safety Precautions Prevent Battery Explosions

Electricity and batteries can cause fires and explosions (Fig. 1). And, of course, the hazard of electrocution is always present.

People become especially complacent when working around low voltages.

However, keep in mind that a current of only one milliampere (one thousandth of an ampere) can be felt, a current of 25 milliamperes can kill and a current of 100 milliamperes probably will kill.

If certain conditions exist, a current of only 0.006 ampere can electrocute a healthy person in less than a second.

A typical battery may have two to five amperes of current flowing across its terminals (Fig. 2). Remember that this amount of current is enough to kill. Therefore, safety must be your first and most important consideration when working around and with electrical equipment.

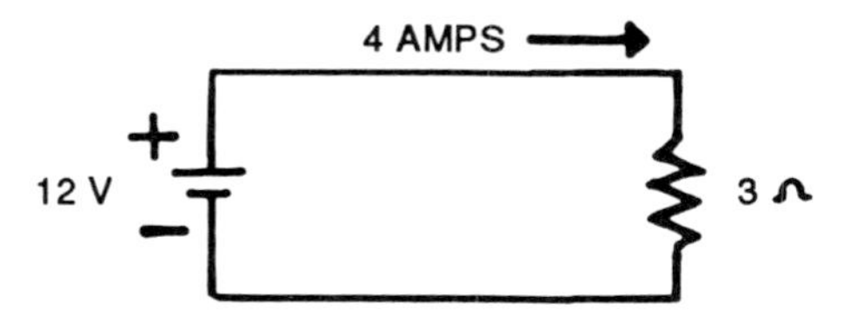

Fig. 2—A Basic Series Circuit of a Battery

Body resistance varies from person to person. It ranges from approximately 1000 to 500,000 ohms. The reasons for such variance depend on many factors: weight, height, body chemistry, etc.

An individual's own resistance may be lowered by perspiration, weather, wet ground conditions, and other variables.

When you learn to measure resistance, you will be able to measure your own body resistance by holding the end of a probe in each hand and taking an ohmmeter reading.

 Litho in U.S.A.

Electrical current follows the path of least resistance. Accidents happen when the human body becomes such a path.

ELECTRICAL SYSTEM

Avoid these hazards when servicing an electrical system:

- **Fires**
- **Short-Circuit Start**
- **Bypass Start Hazard**
- **Battery Explosions**
- **Acid Burns**
- **Electric Shock**

Fires

Electrical systems can cause fires if not properly maintained. One of the purposes of the energy stored in the battery is to start the engine. But if a bare wire of the start system touches a metal part of the machine it will become extremely hot or may even spark and could cause a fire in dust, chaff, leaves, or oil-covered wires. Most machinery fires do not result in personal injury, but every fire is a potential source of injury. Inspect electrical systems regularly. Make sure wires are properly insulated and clean dust, chaff, leaves, and oil off wires.

Every self-propelled machine should have an all-purpose ABC dry chemical fire extinguisher on board to cover all type of fires. Everyone involved with the machine should know how to use it and its charge should be checked annually.

Short-Circuit Start

If insulation on electrical wires is cracked or worn, a short circuit can occur. Electricity could flow to the cranking motor and start the engine when no one is around.

If the positive and negative terminals of a cranking motor are accidentally contacted by another metal object, the current will flow between the two terminals, and accidentally start the engine.

Bypass Start Hazard

Bypass starting of tractors and other farm equipment is a very serious safety concern.

Never short across the starter terminals with a screwdriver or other devices to start a tractor. You bypass the neutral start switch by doing so and if the tractor is in gear when the engine starts, it could suddenly lurch forward and crush you (Fig. 3). Many people have died doing it. DON'T TRY IT!

Never bypass start any tractor or other self-propelled machines and never start it while standing on the ground. Start tractors and self-propelled machines only from the operator's station and with the transmission in neutral or park.

Fig. 3—Avoid the Short-circuit Starting Hazard

Neutral-start switches keep the engine from starting when the transmission is engaged or when the clutch is engaged. Check them periodically to make sure they are working properly. Neutral-start switches can be located so that starting is only possible when:

A. The clutch or inching pedal is depressed.

B. The shift lever is in neutral or park position.

C. Any combination of the above.

These switches should prevent engine cranking when the rear wheels are engaged with the engine. If they don't, they must be adjusted or replaced.

Battery Explosions

Batteries contain sulfuric acid and explosive mixtures of hydrogen and oxygen gases.

When charging and discharging, a lead-acid storage battery generates hydrogen and oxygen gas. Hydrogen will burn, and is very explosive in the presence of oxygen.

A spark or flame near the battery could ignite these gases, rupturing the battery case and splattering acid on property, clothing, skin and eyes.

Special care must always be used around batteries:

Wear eye protection. Also wear rubber gloves and a rubber apron when working around electrolyte.

Keep sparks and flames away from the battery.

Never smoke around a battery.

Always work on batteries in places which are well-ventilated.

If the battery has been on charge or is being charged during a test procedure, blow away gases before continuing testing.

Do not break electrical circuits near the battery top as a spark could start an explosion.

To prevent battery explosions:

1. Maintain the electrolyte at the recommended level(Fig. 4). Check this level frequently.

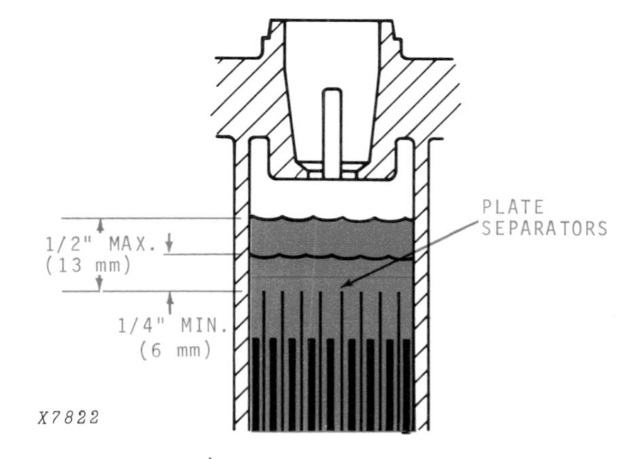

Fig. 4—Keep Electrolyte at the Proper Level to Prevent Explosions

When the level is properly maintained, less space will be available in the battery for gases to accumulate.

2. Put only distilled water in the battery.

3. Use a flashlight to check the electrolyte level. Never use a match or lighter. These could set off an explosion.

4. Do not short across the battery terminals by placing a metal object between them.

5. Do not charge a frozen battery. Warm the battery to 16 °C (60 °F).

6. Remove and replace battery clamps in the right order. This is very important.

   If your wrench touches the ungrounded (usually positive) battery post and the machine chassis at the same time, the heavy flow of current will arc across the terminals producing a dangerous spark.

   To prevent this from happening, follow these rules:

   a) When removing the battery, disconnect the grounded battery clamp first (Fig. 5).

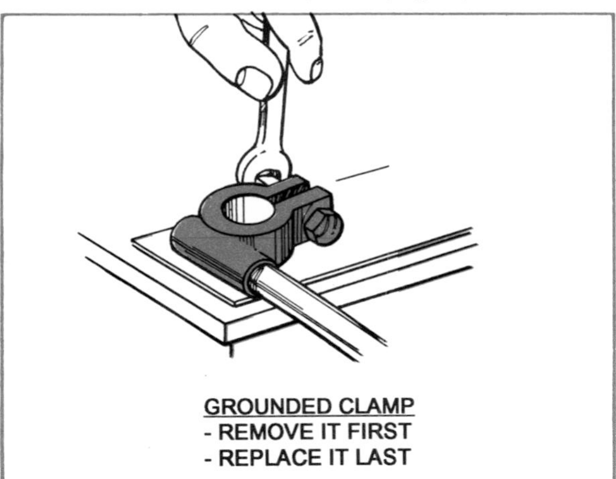

Fig. 5—Connect and Disconnect Battery Cables in the Proper Sequence

   The ground post lead will be connected to the engine block, frame or other metallic surface. The positive post lead will be connected to the starter relay.

   Some systems may have a positive ground. Although this is not common, always make sure you know which post is grounded.

   b) When installing the battery, connect the grounded battery clamp last.

7. Prevent sparks from battery charger leads.

   Turn the battery charge off or pull the power cord before connecting or disconnecting charger leads to battery posts (Fig. 6).

If you don't, the current flowing in the leads will spark at the battery posts. These sparks could ignite the explosive hydrogen gas which is always present when a battery is being charged.

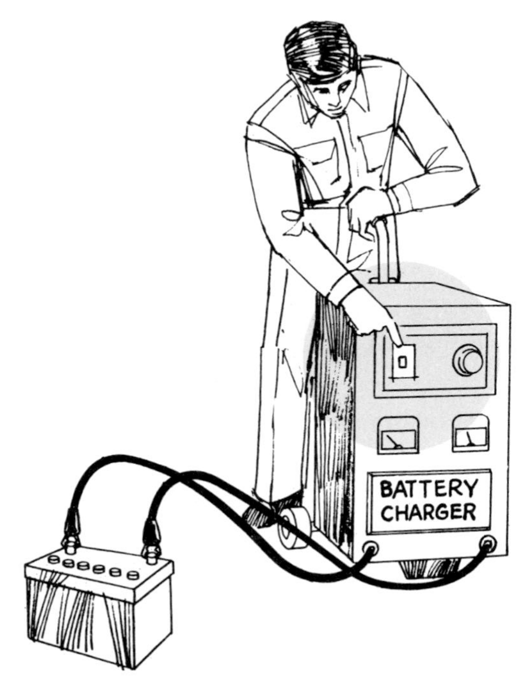

Fig. 6—Turn Off Charger Before Connecting or Disconnecting the Charger Leads

Connecting a Booster Battery

Improper connecting of a booster battery from one machine to the dead battery of another machine can be dangerous and can cause a battery explosion. Follow these procedures when connecting a booster battery from one machine to the dead battery of another machine.

1. Remove all cell caps of the dead battery (if so equipped).

2. Check to make sure the dead battery is not frozen. Never attempt to boost a battery with ice in its cells.

3. Be sure that booster battery and dead battery are of the same voltage.

4. Turn off all accessories and ignition of both machines.

5. Place gearshift of both vehicles in neutral or park and set the parking brake. Make sure vehicles do not touch each other.

6. Check the electrolyte level of the dead battery cells. Add distilled water to cells if low. Cover the vent holes with a damp cloth, or if caps are safety vent type, replace the caps before attaching jumper cables to the batteries.

7. Attach one end of one jumper cable to the booster battery positive terminal. Attach other end of the same cable to the positive terminal of the dead battery. Make sure of good, metal-to-metal contact between cable ends and terminals.

8. Attach one end of the other cable to the booster battery negative terminal. Make sure of good, metal-to-metal contact between the cable end and the battery terminal.

Caution: To prevent sparks and possible battery explosion, never allow ends of the two cables to touch while attached to the booster battery.

9. Connect other end of second cable to engine block or frame of the disabled vehicle as far away from the dead battery as possible. This is to insure that if a spark should occur at this connection, it would not ignite hydrogen gas that may be present above the dead battery.

10. Try to start the disabled vehicle. Do not engage the starter for more than 30 seconds or starter may overheat and booster battery will be drained of power. If the disabled vehicle will not start, start the vehicle with the booster battery and let it run for a few minutes with the cables attached. Try to start the disabled vehicle again.

11. Remove cables in exactly the reverse order from installation. Remove damp cloth and replace vent caps.

Acid Burns

Battery electrolyte is approximately 36 percent full-strength sulfuric acid and 64 percent water.

Even though it is diluted, it is strong enough to burn skin, eat holes in clothing and cause blindness if splashed into eyes.

Fill new batteries with electrolyte in a well-ventilated area, wear eye protection and rubber gloves, and avoid breathing any fumes from the battery when the electrolyte is added.

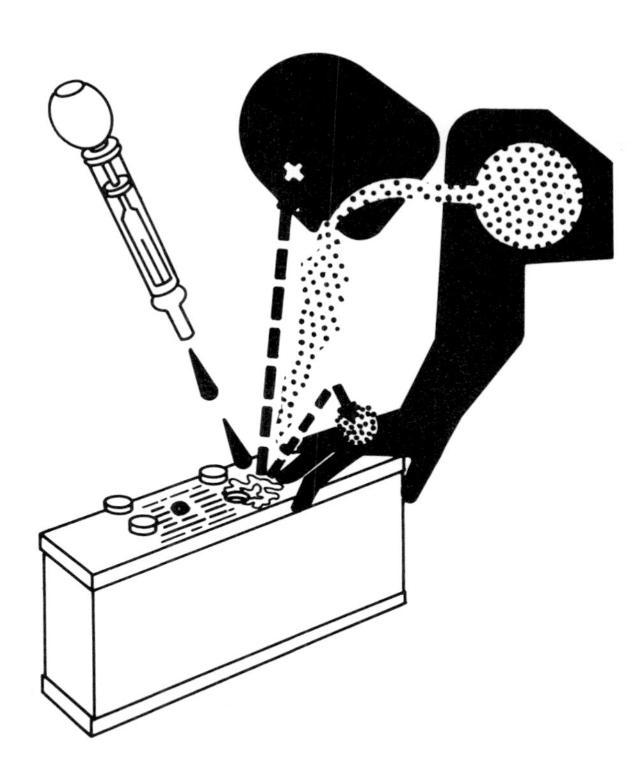

Fig. 7—Avoid Electrolyte Hazards When Using a Hydrometer

Avoid spilling or dripping electrolyte when using a hydrometer to check specific gravity readings (Fig. 7).

If you spill acid on yourself, flush your skin immediately with water for several minutes.

Apply baking soda or lime to help neutralize the acid.

If acid gets into your eyes, force the lids open and flood the eyes with running water for 15 to 30 minutes. Get medical attention immediately.

If the acid is swallowed, drink large amounts of water or milk, but do not exceed 2 quarts (2L). Get medical attention immediately.

Electric Shock

Injury from electric shock depends on the number of vital organs through which current passes.

Electricity travels at the speed of light which is 186,000 miles per second, leaving no reaction time.

Further, the hand muscles contract causing a firmer grip on a current-carrying wire or component.

To prevent a current path from flowing through your heart, always keep one hand away from the voltage source when working on a circuit. *Don't become part of the current path.*

The voltage in the secondary circuit of an ignition system may exceed 25,000 volts. For this reason, don't touch spark plug terminals, spark plug cables or the coil-to-distributor high-tension cable when the ignition switch is turned on or the engine is running (Fig. 8). The cable insulation should protect you, but it could be defective.

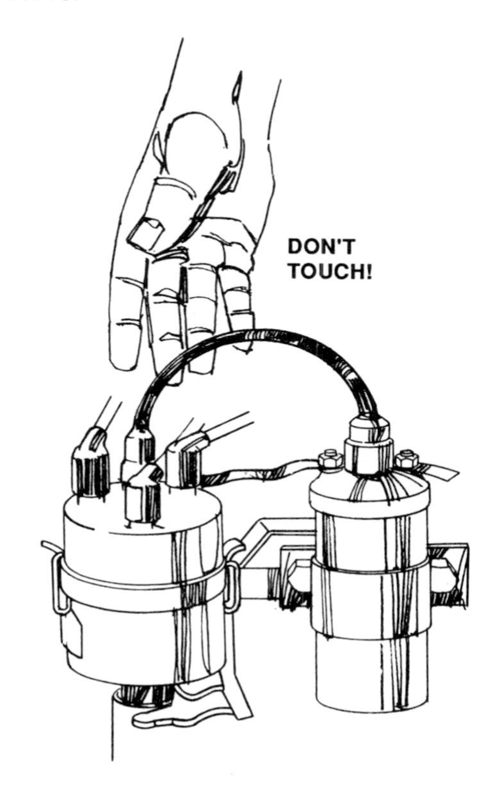

Fig. 8—Turn Off Ignition Switch Before Performing Repairs

Never run an engine when the wire connected to the output terminal of an alternator or generator is disconnected (Fig. 9).

If you do, and if you touch the terminal, you could receive a severe shock. When the battery wire is disconnected, the voltage can go dangerously high, and it may also damage the generator, alternator, regulator or wiring harness. Don't short across the battery terminals by placing a metal object between them.

Electric shock may cause unconsiousness and burns on the skin at the area of contact.

Use a dry rope or stick to move the victim to safety.

 Litho in U.S.A.

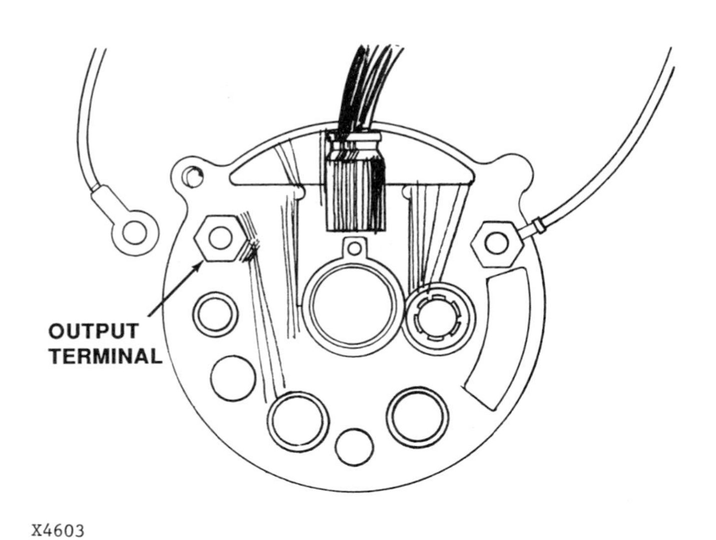

Fig. 9—Connect the Alternator or Generator Output Terminal Lead Before Running an Engine

De-energize the power source: open the switch or cut the cable or wire using an ax with a wooden handle.

Keep the victim lying down and still with clothing loosened around the neck, chest and abdomen.

If the victim is not breathing, apply cardiopulmonary resuscitation (CPR). Seek medical attention immediately.

SHOP PRACTICES AND WORK HABITS

The best way to minimize hazards associated with electricity and electrical equipment is to follow proven shop practices and employ good work habits:

- Wear eye protection. Plastic goggles protect eyes from impact from the front and sides. Unvented or chemical splash goggles also offer protection against chemical vapors and liquids (Fig. 10).
- Rest regularly to avoid the effects of fatigue. Never use drugs, alcohol or tobacco when working on equipment.
  ing on equipment.
- Avoid horseplay.
- Never work on dangerous equipment when you are ill, angry or anxious.

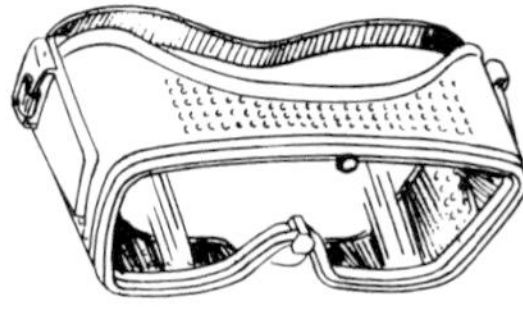

Fig. 10—Wear Unvented Safety Goggles Around Acids

- Be alert.
- Work only in adequately ventilated areas.
- Make sure the work area has sufficient light.
- Be aware of common machine hazards. Don't take shortcuts. Shortcuts shorten lives, and because of the problems that they cause, cost more in time and money than they save.
- Lock out or disconnect the electrical power source from the electrical system before you begin your service procedure.
- Know your limitations: age, weight and height all have a bearing on the jobs you are capable of doing.

Fig. 11—Keep a First Aid Kit Readily Available

- Learn the basic rules of first aid and keep a first aid kit readily available (Fig. 11).
- Apply immediate first aid to all injuries. If an injury appears severe, don't move the victim. Call a doctor and follow his instructions.

- Wear rubber gloves and a rubber apron when working around electrolyte.
- Know who to call for help. Keep emergency numbers for doctors, ambulance services, hospitals, and fire departments near the telephone.
- Never use water on electrical fires.

Keep a dry chemical fire extinguisher suitable for Class C fires close to but not in the fire hazard area.

Both Class B (burning liquids) and Class C (electrical equipment) fires require a pressurized dry chemical fire extinguisher of 20-pound capacity.

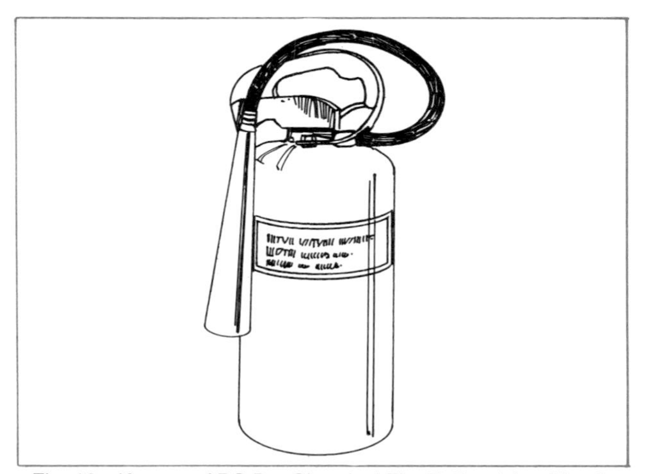

Fig. 12—Keep an ABC Dry Chemical Fire Extinguisher Handy

Shops should be equipped with at least one all-purpose ABC dry chemical fire extinguisher (Fig. 12). This extinguisher will put out Class A (combustibles like paper and wood), Class B, and Class C fires.

- Make sure that wires are properly insulated and clean and that electrical components are free of dust, chaff, leaves and oil.
- Always read, understand and follow the instructions and manuals provided with equipment (Fig. 13). Don't guess. Keep manuals in a clean, dry, readily available place.
- Read and understand labeling on the equipment. Replace labels that are damaged or worn. Heed the safety-alert symbols on the labels and in other instructional materials (Fig. 14).
- Pay attention to signal words:

  Signal words like *Danger, Warning* and *Caution* draw attention to potentially unsafe areas. Learn these signal words and let them become your "think trigger."

Fig. 13—Read and Follow Manufacturer's Instructions

Fig. 14—This Safety-alert Symbol Could Save Your Life

DANGER means that one of the most serious potential hazards is present. Exposure to these hazards would result in a high probability of death or a severe injury if proper precautions are not taken.

WARNING means the hazard presents a lesser degree of risk of injury or death than that associated with Danger.

CAUTION is used to remind of safety instructions that must be followed and to identify property damage hazards and hazards involving minor injuries.

Safety messages utilize colors as an aid to communication. Red and white are the colors used with the word Danger. Black and yellow are found on signs carrying the words Caution or Warning.

Fig. 15—Pictorial = Danger of Hazrdous Fumes or Dust

Pay attention to pictorial representations of safety hazards. A good pictorial should identify the hazard and portray the potential consequences of failure to follow instructions (Fig. 15).

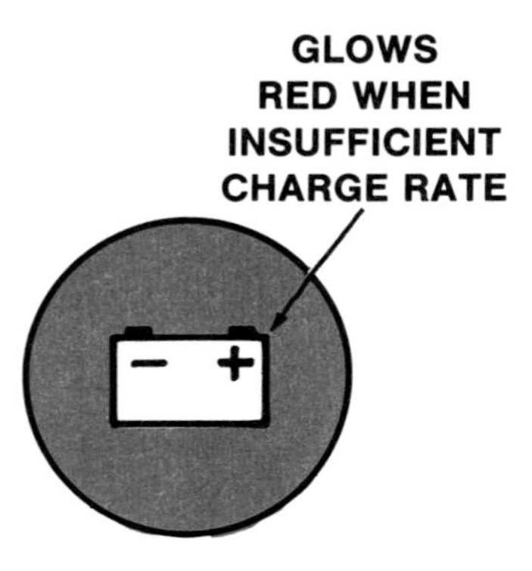

Fig. 16—Use of Color with Universal Ammeter or Generator Light

- Be able to identify universal symbols used to help identify controls. Color is often used with these symbols to indicate operating conditions (Fig. 16).
- Never work alone. Make sure someone knows you are working in the shop and will check on you and render aid if you are injured.
- Do not wear jewelry or other metallic objects when working on equipment. Keep metal parts of clothing, such as zippers and buttons, covered.
- Keep clothing, hands, feet and flooring dry. Make sure floor surfaces are clean.
- Use insulated tools whenever possible.

Fig. 17—Proper Use of Tools Prevents Injury and Property Damage and Extends Tool Life

Always select the right tool for the job and use it in the right way (Fig. 17).

Keep tools in good condition and store them safely when not in use.

Guard against eye injuries when cutting with pliers or cutters. Short and long ends of wire often fly or whip through the air when cut. Wear eye protection when cutting wire. Select a cutter big enough for the job. Keep the blades at right angles to the stock and don't rock the cutter to get a faster cut. Adjust the cutters to maintain a small clearance between the blades to prevent them from stiking each other when the handles are closed.

- Use non-metallic receptacles and funnels when working with electrolyte. Do not store electrolyte in a warm or sunny location.

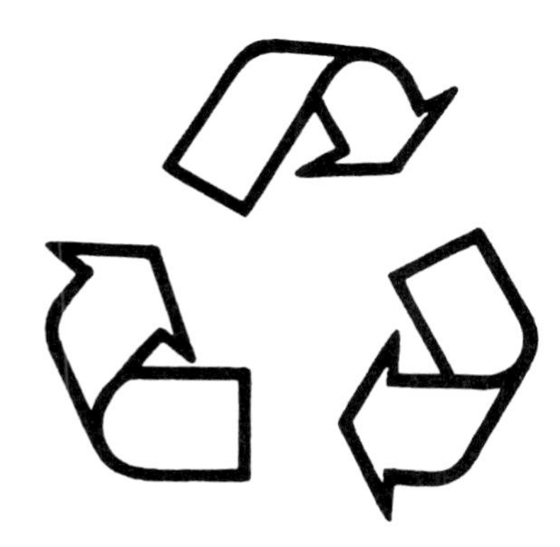

Fig. 18—Dispose of Fluids Properly

- Properly store and dispose of hazardous materials such as battery acid (Fig. 18). Do not pour battery acid down a drain or into a stream, pond or lake.

Improper disposal of fluids can harm the environment and ecology. Check with state environmental agencies for information concerning the proper disposal of battery acid.

Used batteries must be recycled. Battery retailers are required to take one old battery for each one you buy. Manufacturers recycle batteries to produce new ones.

- Remember that you and your coworkers contribute to each other's safety. Would you want to work around someone with unsafe working habits? Would you take your machinery to a repair shop known for unsafe working conditions and practices?

Fig. 19—Safety: Your Life Depends On It

When it comes to safety, be a leader, not a follower. Safety is everybody's business (Fig. 19).

SUMMARY: ELECTRICAL SAFETY

In summary:

- **Follow safe shop practices and work habits.**
- **Pay attention to signal words, like DANGER, WARNING, and CAUTION.**
- **Be aware of hazards that could occur while working on an electrical system.**
- **Even though battery acid is diluted, it can still cause severe burns. Always wear gloves and eye protection when servicing batteries.**
- **Always keep one hand away from the voltage source when working on a circuit.**

TEST YOURSELF

QUESTIONS

1. Shorting across starter terminals with a screwdriver is known as the __________ __________ hazard.

2. Batteries contain __________________ which make them very explosive.

3. Always wear ________________ when working on a battery.

4. After jump starting a dead battery, disconnect the ________________ jumper cable from the engine block first.

5. "Think" safety when you see signal words on machinery like ______________ , ____________ and ________________ .

6. When working on an electrical circuit, don't become part of the _____________ ______________.

7. If certain conditions exist, a current of only _______ __________ __________________ can electrocute a healthy person.

(Answers on page 19 at the end of this book.)

 Litho in U.S.A.

STORAGE BATTERIES / CHAPTER 5

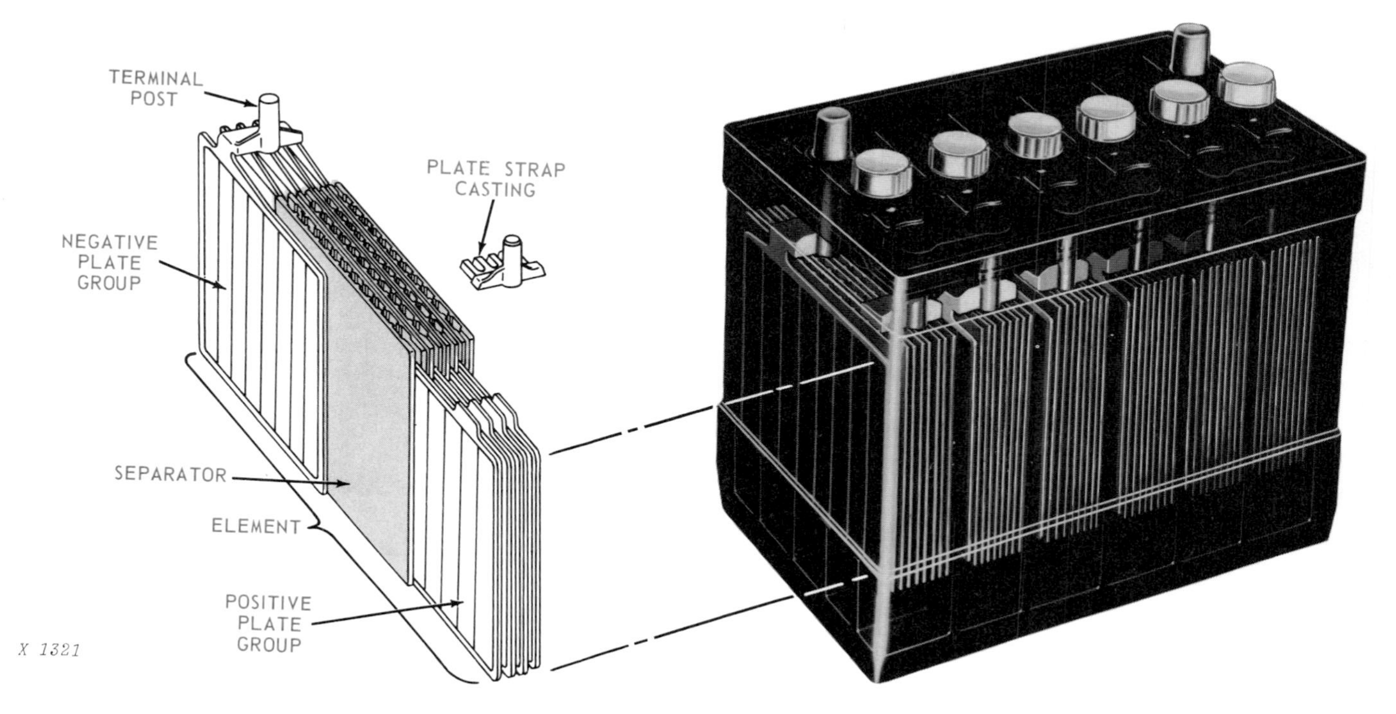

Fig. 1—Construction of a Storage Battery

WHAT A BATTERY DOES

The battery stores energy for the complete electrical system.

On demand, the battery produces a flow of direct current for the devices connected to its terminals.

Battery current is produced by a chemical reaction between the active materials of the *plates* and the sulfuric acid in the battery fluid or *electrolyte.*

After a period of use, the battery becomes discharged and will no longer produce a flow of current. However, it can be recharged by making an outside direct current flow through it in the opposite way from that which current flows out.

In normal operation, the battery is kept charged by current input from the generator or alternator.

For good operation, the battery must do three jobs:

- **Supply current for starting the engine.**
- **Supply current when the demand exceeds the output of the charging system.**
- **Stabilize the voltage in the system during operation.**

HOW A BATTERY IS CONSTRUCTED

The battery is made up of a number of individual **cells** in a hard rubber case (Fig. 1). The basic units of each cell are the **positive** and **negative plates.**

These plates hold the active materials in flat grids. Charged negative plates contain spongy lead (Pb) which is grey in color. Charged positive plates contain lead peroxide (Pb O_2) which has a chocolate brown color.

A **plate group** is made by welding a number of similar plates to a plate strap. (See Fig. 1.)

Plate groups of opposite polarity are interlaced so the negative and positive plates alternate. Negative plate groups normally have one more plate than the positive groups. This keeps negative plates exposed on both sides of the interlaced group.

Each plate in the interlaced plate group is kept apart from its neighbor by porous separators as shown in Fig. 1. The separators allow a free flow of electrolyte around the active plates. The resulting assembly is called an **element.**

After the element is assembled, it is placed in a cell compartment of the battery case.

On a soft-top battery, cell covers are installed next. Then the cell connectors are welded between the intermediate terminal posts of adjoining cells. In this way the cells are connected in **series.** Finally the top of the battery case is sealed.

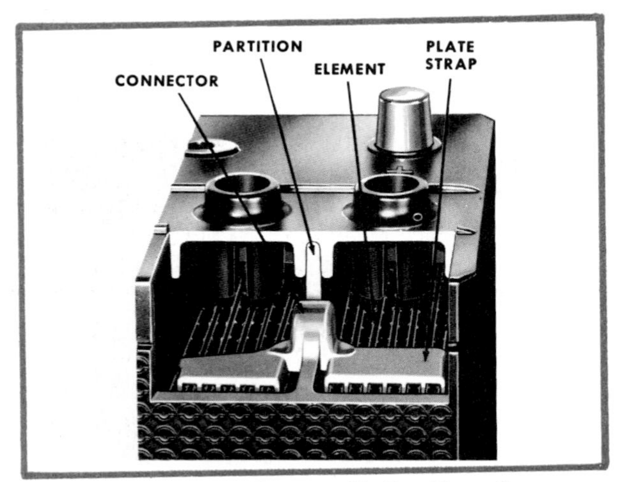

Fig. 2—Hard-Top Battery with One-Piece Cover

Hard-top batteries have one-piece cell covers which reduce the formation of corrosion on top of the case (Fig. 2). These batteries have cell connectors which pass through the partitions between cells. The connectors and partitions are sealed so that electrolyte will not transfer between cells. This improves battery performance, since the cell connections are shorter and the cover is more acid-tight.

The main battery terminals are the **positive** and **negative posts.** The positive terminal is larger to help prevent the danger of connecting the battery in reverse polarity.

Reversing the polaritiy may damage some components and wiring in the system.

There is usually a red cable connected to the battery positive post and a black cable connected to the battery negative post. The negative post cable will be connected to the engine block or other metal surface. The positive post cable will be connected to the starter.

Always disconnect the negative post cable first and connect it last. Otherwise a dangerous spark could occur. Never disconnect a battery with the key switch on or the engine running.

Do not lay metal tools or other objects across the battery as this may create a short circuit.

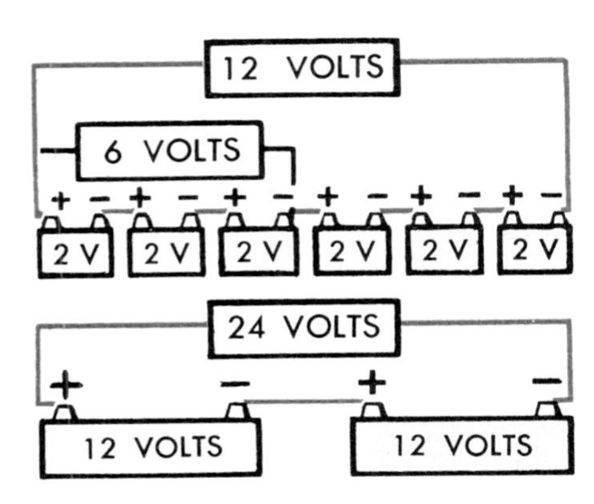

Fig. 3—Battery Cells and Batteries Connected in Series

Vent caps are located in each cell cover. The caps have two purposes: 1) They close the openings in the cell cover through which the electrolyte level is checked and water is added; 2) They provide a vent for the escape of gases formed when the battery is charging.

Electyrolyte can cause acid burns and the gases formed in batteries are very explosive. Review the battery safety information in Chapter 4 and be sure to follow proper precautions when working around or near batteries.

Each cell in a storage battery has a potential of about two volts. Six-volt batteries contain three cells connected in series, while 12-volt batteries have six cells in series (Fig. 3, top diagram).

For higher voltages, combinations of batteries are used. An example is where two 12-volt batteries are connected in series to serve a 24-volt system (Fig. 3, bottom diagram).

In summary:

- **The battery is made up of cells.**
- **Each cell has positive and negative plates.**
- **Similar plates are welded into plate groups.**
- **The plate groups are interlaced but separated.**
- **This allows a free flow of electrolyte around the active plates.**
- **The resulting assembly is an element—one for each battery cell.**
- **Each cell is connected in series.**
- **Main terminals — (+) and (−) — connect all cells.**
- **Six-volt batteries have three cells in series.**
- **Twelve-volt batteries have six cells in series.**
- **For higher voltages, two or more batteries are connected.**

HOW A BATTERY WORKS

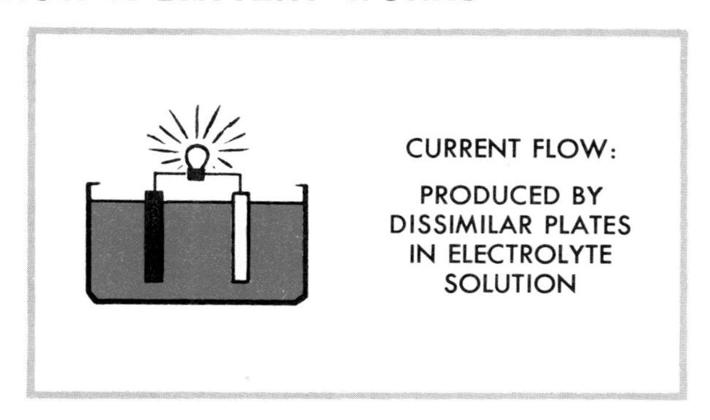

Fig. 4—How a Battery Produces Current Flow

The battery produces current by a chemical reaction between the active materials of the unlike plates and the sulfuric acid of the electrolyte (Fig. 4).

While this chemical reaction is taking place, the battery is **discharging.** After most all active materials have reacted, the battery is **discharged.** It must then be recharged before use.

We'll discuss these two cycles, but first let's see what the electrolyte is made of.

ELECTROLYTE SOLUTION

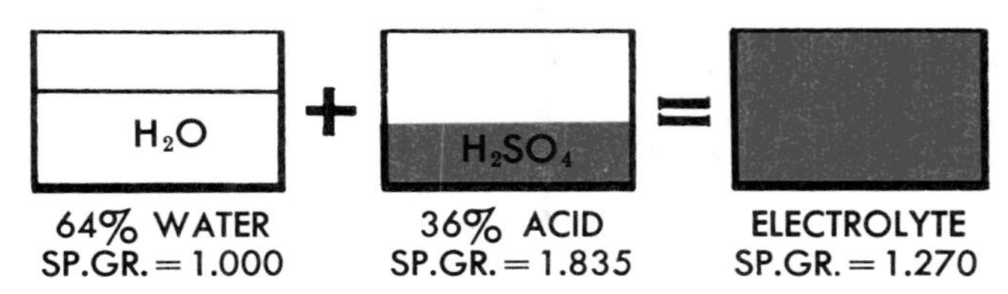

Fig. 5—Battery Electrolyte

The **electrolyte** in a fully charged battery is a solution of concentrated sulfuric acid in water (Fig. 5). Carefully follow the safety warnings in Chapter 4. It has a specific gravity of about 1.270 at 80° F (27° C) – which means it weighs 1.270 times more than water. The solution is about 36% sulfuric acid (H_2SO_4) and 64% water (H_2O) as shown.

The voltage of a battery cell depends upon the chemical difference between the active materials and also upon the concentration of the electrolyte.

DISCHARGE CYCLE OF BATTERY

When the battery is connected to a complete circuit, current begins to flow from the battery. The **discharge** cycle begins.

This current is produced by a chemical action as follows:

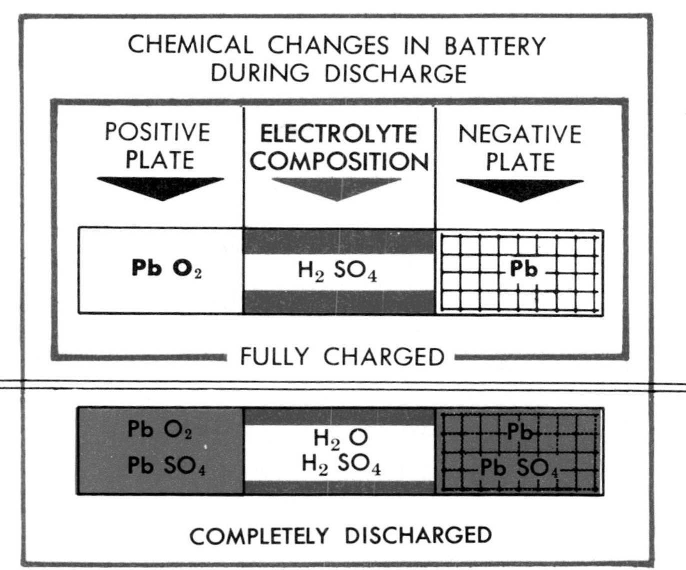

Fig. 6—Chemical Action of the Battery

The lead peroxide (PbO_2) in the positive plate is a compound of lead (Pb) and oxygen (O_2). Sulfuric acid is a compound of hydrogen (H_2) and the sulfate radical (SO_4) which in turn, is a compound of sulfur (S) and oxygen. Oxygen in the positive active material combines with hydrogen from the sulfuric acid to form water (H_2O). At the same time, lead in the positive active material combines with the sulfate radical, forming lead sulfate ($PbSO_4$). See Fig. 6.

A similar reaction takes place at the negative plate where lead (Pb) of the negative active material combines with the sulfate radical to form lead sulfate ($PbSO_4$). Thus, lead sulfate is formed at both plates as the battery is discharged, while the sulfuric acid in the electrolyte is replaced by water.

Note that the material in the positive plates and negative plates becomes chemically similar during discharge, as the lead sulfate accumulates. This condition accounts for the loss of cell voltage, since *voltage depends upon the difference between the two materials.*

As the discharge continues, dilution of the electrolyte and the accumulation of lead sulfate in the plates eventually brings the reactions to a stop. For this reason the active materials never are completely exhausted during discharge. At low rates of discharge the reactions are more complete than at high rates since more time is available for the materials to come in contact. When the battery can no longer produce the desired voltage, it is said to be discharged. It must be recharged by a suitable flow of direct current from some external source before it can be put back in service.

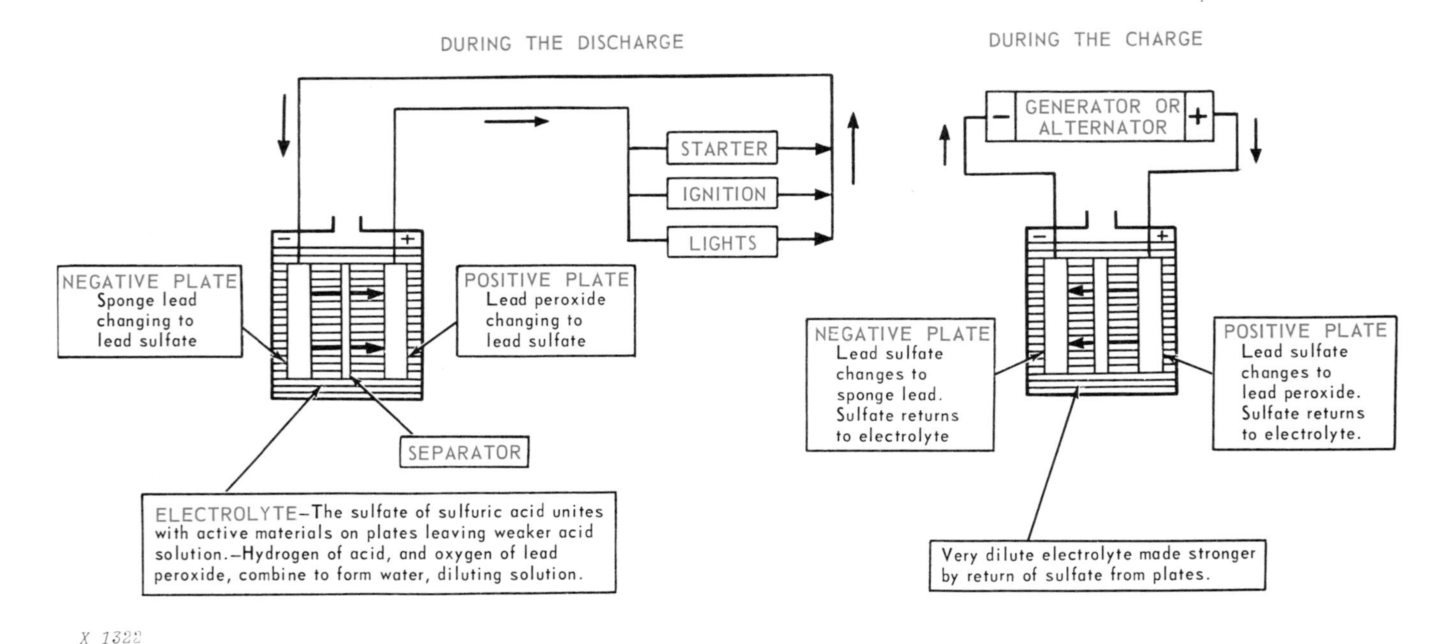

Fig. 7—Chemical Action in Battery During Discharge and Charge Cycles

In summary, refer to Fig. 7 for a diagram of the complete discharge cycle of the battery.

CHARGING CYCLE OF BATTERY

The chemical reactions which go on in the battery cell during charge are essentially the reverse of those which occur during discharge. See Figs. 7 and 8.

The lead sulfate on both plates is split up into Pb and SO_4, while water (H_2O) is split up to get hydrogen to form H_2SO_4 or sulfuric acid. At the same time the oxygen enters into chemical combination with the lead at the positive plate to form PbO_2 or lead peroxide.

These reactions demonstrate the important fact that water actually takes part in the chemistry of a lead-acid storage battery.

It is interesting to note that the specific gravity of the electrolyte decreases during discharge for two reasons—sulfuric acid (which is "heavier" than water) is used up and water is formed. Conversely, when the battery is charged, the specific gravity of the electrolyte increases—sulfuric acid is formed and water is used up.

When checking the specific gravity of the electrolyte after the battery has been fast charged, the reading may continue to rise for some time because the newly-formed acid requires time to diffuse from the plates into the electrolyte. Therefore,

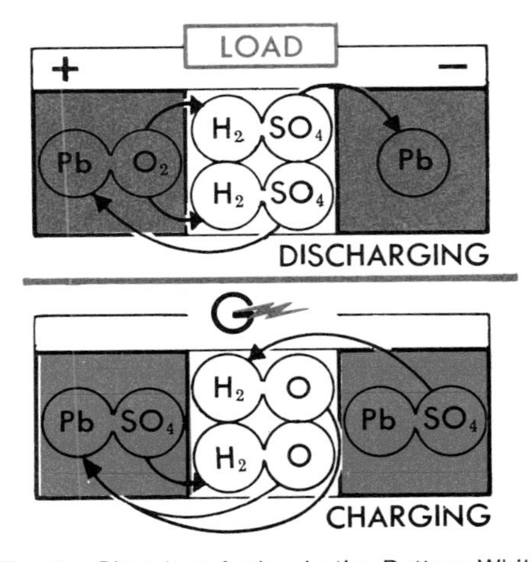

Fig. 8—Chemical Action in the Battery While Discharging and Charging

a specific gravity reading taken while the electrolyte is full of gas will be erroneously low. Conversely, higher specific gravity readings will be obtained as the gas is dissipated from the electrolyte. This situation often leads to the mistaken impression that specific gravity is still rising after charging has been discontinued, when this is not always true.

We have noted that water plays an important part in the chemical action of a storage battery. The purity of water for battery use has always been a controversial subject, but always resolves to the fact that distilled water is the best. Water with impurities hurts the life and performance of a battery.

 Litho in U.S.A.

Summary: How A Battery Works

In Summary

- **The battery produces current by chemical action.**
- **This chemical action discharges the battery.**
- **As chemical action fails, the battery discharges.**
- **Charging the battery reverses the chemical action.**
- **Chemical action is between lead plates and electrolyte fluid.**
- **Electrolyte is sulfuric acid in water. Be careful.**
- **More acid in electrolyte = higher charge.**
- **Higher charge = heavier electrolyte.**
- **"Specific gravity" measures weight of electrolyte.**

THE BATTERY AND THE CHARGING CIRCUIT

The battery is the heart of the electrical system. It plays its role in the operation of the starting, charging, ignition, and accessory circuits.

However, the battery is really part of the charging circuit.

In operation, the battery works in cycle with the generator or alternator (Fig. 9). This happens as follows:

1) The battery supplies current to the system, and becomes discharged.

2) The generator sends reverse current to the battery, recharging it.

3) The voltage regulator limits the voltage from the generator to a safe value which does not overcharge the battery at high speeds.

The charging cycle is different at various engine speeds.

When the engine is shut off, the battery alone supplies current for the accessory circuits.

At low speeds, both the battery and generator may supply current.

At higher speeds, the generator may take over and supply enough current to operate the accessories and also recharge the battery.

This charging cycle is more fully explained in Chapter 6, "Charging Circuits."

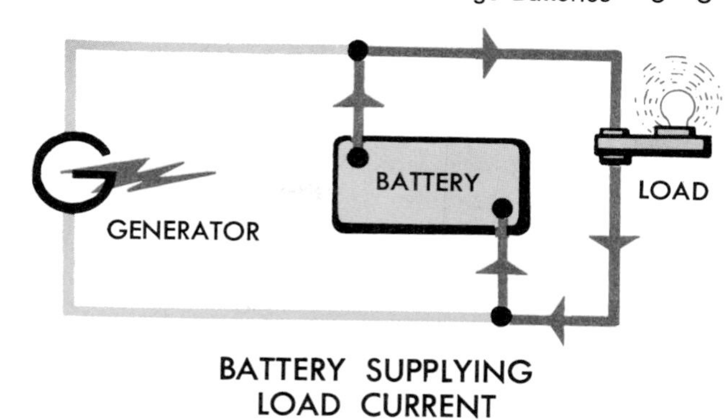

BATTERY SUPPLYING LOAD CURRENT

GENERATOR
BATTERY
LOAD

GENERATOR AND BATTERY SUPPLYING LOAD CURRENT

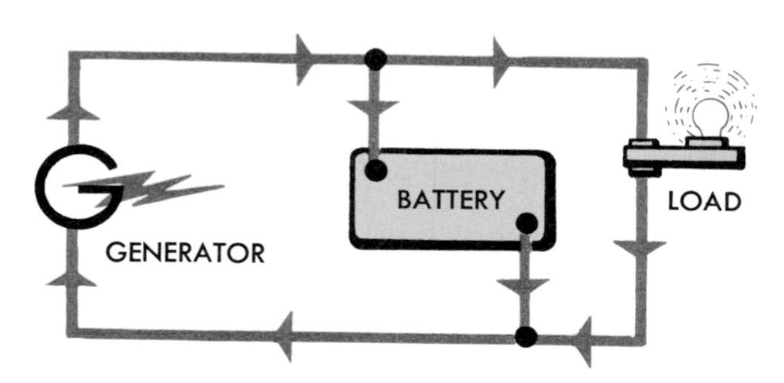

GENERATOR SUPPLYING LOAD CURRENT AND CHARGING BATTERY

Fig. 9—The Battery and the Charging Circuit

TYPES OF BATTERIES

There are two types of batteries:

- **Dry-charged**
- **Wet-charged**

The difference between the two depends upon the way they are sent out from the factory.

DRY-CHARGED BATTERIES

A dry-charged battery contains fully-charged elements. But it contains no electrolyte until it is activated for service in the field. Therefore, it leaves the factory in a dry state. Once activated in the field, it is essentially the same as a wet-charged battery.

At the factory, the battery elements are specially charged as follows: A direct current is passed through the plates while immersed in an electrolyte of dilute sulfuric acid. The fully-charged plates are then removed from the electrolyte, washed in water, and completely dried. The battery is then assembled.

A dry-charged battery retains its state of full charge as long as moisture is not allowed to enter the cells. If stored in a cool, dry place, this type of battery will stay factory-fresh and will not become "shelf worn" prior to use.

Activating Dry-Charged Batteries

The activation of a dry-charged battery is done in the field either by the warehouse or dealer.

To make sure the proper electrolyte is used and the battery is properly activated, many manufacturers furnish a packaged electrolyte for their dry-charged batteries along with instructions for placing the battery into service. These instructions must be carefully followed.

Under normal conditions, activate dry-charged batteries as follows:

1. Wear safety goggles and rubber gloves. Keep sparks and flame away from the battery. Make sure the work area is well-ventilated.

2. Remove the seals from the battery cell openings and remove the vent caps.

3. Carefully fill each cell with the approved electrolyte to the proper level (usually at the bottom of the filler neck). Examine the vent caps to be sure they are open and install the caps.

4. Check the specific gravity with a hydrometer and record the corrected reading. (See "Battery Testing.")

5. Using a date code ring, gently stamp the date code on the battery. See Fig. 22.

6. Allow the battery to stand for a few minutes, then recheck the level of electrolyte in each cell. If necessary, add electrolyte—not water.

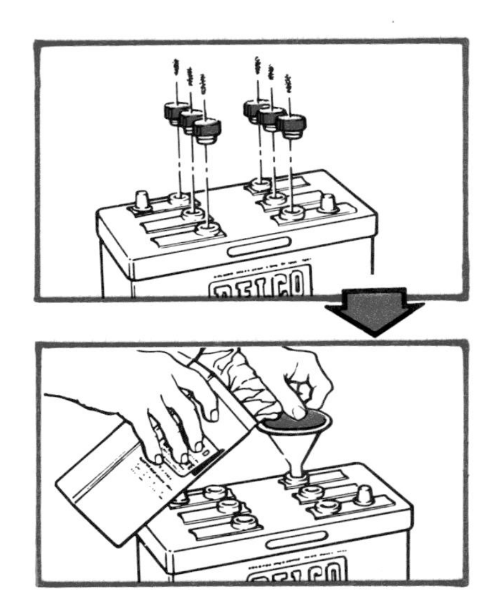

Fig. 10—Activating Dry-Charged Batteries

7. As a precaution, check the open-circuit voltage of the battery. (See "Battery Testing.") As a general rule, place a 12-volt battery in service if it tests 12 volts or more, charge it first if it tests 10-12 volts, and consider it defective if it tests less than 10 volts.

8. As a final test, check the specific gravity of the electrolyte again. If the reading shows more than a 30-point drop (0.030) from the previous reading, charge the battery.

9. Slow charge a freshly-activated battery to assure that the user receives a fully-charged battery. Always charge a newly-activated battery if the machine will not be run for at least one hour.

10. After the battery has been in service, add only approved water. **Do not add acid.**

11. Once in use, keep the battery serviced and charged just like a conventional wet-charged battery.

Dry-charged batteries should be stored in a cool, dry place with low humidity. Also be sure the temperature is between 60° and 90°F (16 and 32°C). Under these conditions, a battery can be stored for several years and keep a good charge. Under bad conditions, the battery may lose its charge in several weeks.

What is the advantage of dry-charged batteries? The prime advantage is that they do not sulfate and corrode during long storage like wet-charged batteries.

One disadvantage is that acids must be handled and it is difficult to give the battery a good quality check before shipment.

WET-CHARGED BATTERIES

Wet-charged batteries contain fully-charged elements and are filled with electrolyte at the factory. A wet-charged battery will not maintain its charged condition during storage, and must be recharged periodically.

During storage, even though the battery is not in use, slow reaction takes place between the chemicals inside the battery which causes it to lose charge. This reaction is called **self-discharge.**

The rate at which self-discharge occurs varies directly with the temperature of the electrolyte. A fully charged battery stored at room temperature of 100°F (38°C) will be almost completely discharged after a storage period of 90 days. The same battery, stored at 60°F (16°C) will be only slightly discharged after 90 days.

Wet-charged batteries, therefore, should be stored in as cool a place as available, as long as the electrolyte does not freeze.

A wet battery which is kept fully charged will not freeze, while a discharged battery can freeze. For example, a discharged battery with a specific gravity of 1.100 will freeze at 18°F (−8°C) whereas a fully-charged battery with a specific gravity of 1.260 is never in danger of freezing, unless it is −75°F (−59°C). See chart below.

CHART SHOWING WHEN ELECTROLYTE FREEZES AT VARIOUS SPECIFIC GRAVITIES

Condition of Battery	*Specific Gravity of Electrolyte*	*When Electrolyte Freezes (Temp.) (°F)*	*(°C)*
Discharged	1.100	+18°	− 8°
	1.140	+ 8°	−13°
	1.180	− 6°	−21°
	1.220	−31°	−35°
Fully-charged	1.260	−75°	−59°

Sulfated Batteries

Wet-charged batteries which are stored for long periods of time without recharging may be permanently damaged by the oxidation of the positive plate grid wires and the formation of lead sulfate crystals in the plates which become dense and hard.

If the sulfate crystals are not too dense and hard, the battery may be restored to normal service by applying a slow charge rate for a longer than normal period. However, if the sulfate crystals are excessively hard and dense, the battery can never be restored to a normal operating condition, regardless of the rate or time of the charge.

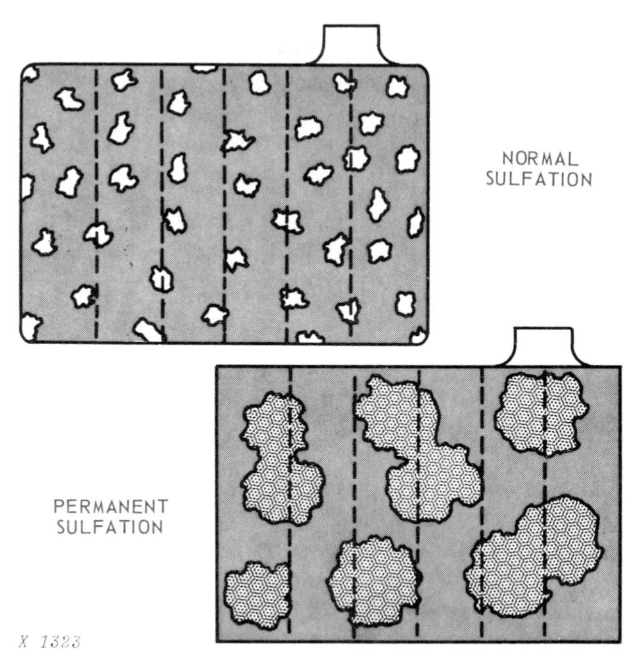

Fig. 11—Sulfation of Battery Plates

Sulfation is caused by the chemical reaction in the battery discharge and was discussed earlier in this chapter. Both plates become lead sulfate ($PbSO_4$) and the acid is converted to H_2O; no further reactions can take place and a discharged battery results.

If this condition is allowed to exist for a long enough period of time (30-90 days), the lead sulfate will harden and cause a sulfated battery (Fig. 11).

The hard sulfate deposit on the plates is extremely difficult to break down. Charging at high rates will only cause extreme heating, because the plates will tend to reject much of the current. Even the normal rate of charge is high for sulfated batteries. *For sulfated batteries, therefore, recharge at ½ the normal rate, or approximately ½ ampere per positive plate per cell.*

For example, a typical 6-volt battery has 12 negative and 11 positive plates per cell. If the battery is sulfated, recharge at a slow rate.

How long will it take to recharge a sulfated battery? This depends on the degree of sulfation, the state of battery charge, and the battery age. Charge at a rate which will not allow the electrolyte temperature to exceed 120° F (49° C), and charge until the specific gravity of all cells indicates a fully-charged battery.

This may take 60 to 100 hours when sulfation is extremely bad. If the specific gravity has not reached the normal full charge in this time, replace the battery.

While in storage, wet-charged batteries should be brought to full charge every 30 days, to guard against sulfation. Batteries removed from equipment during the winter should receive the same care or be stored under cool conditions.

Trickle-type charges have been developed to produce a charging rate measured in milli-amperes which is just sufficient to offset the losses due to self-discharge.

NOTE: Do not use trickle chargers for more than sixty days. Very low charging rates for long periods can cause permanent damage to the positive plate grids.

A word about "warm" storage of batteries: Remember, you don't do a wet battery a "favor" by keeping it warm during storage . . . in fact, the best bet is to store the battery in as cold a place as possible, as long as the electrolyte doesn't freeze. In general, *store batteries in a cool, dry place.*

CHECKING ELECTROLYTE LEVEL IN BATTERY CELLS

Carefully follow the safety warnings in Chapter 4.

Periodically check the level of the electrolyte in the battery cells. This should be done at least every week during steady operation of the system. Proper level is ¼ inch (6 mm) to ½ inch (13 mm) above plate separators (so that the tops of the battery plates are covered). (See Fig. 12.) *Do not overfill or acid may spew out of caps.*

Use only distilled water in the battery. If not available, be sure to use clean, soft water. *Avoid hard water.*

NEVER ADD ACID TO THE BATTERY unless electrolyte is lost by spilling.

Always wait until after checking specific gravity before you add water to the battery. This will assure a true reading. If level is too low to check specific gravity, add water, operate in circuit for a few minutes to mix water and electrolyte, then check.

In freezing weather, never add water to the battery unless it will be operated immediately to allow proper mixing of water with electrolyte.

MAINTENANCE-FREE BATTERIES

Maintenance-free batteries (Fig. 13) operate similarly to conventional style batteries. The use of lead-calcium plates instead of lead-antimony in their construction increases the ability of the battery to accept an overcharge, thus greatly reducing bubbling and gassing of the electrolyte. Less fluid is lost, eliminating the need to add water. Venting of gases from the cells is usually through a vent (Fig. 14) as most maintenance-free batteries do not have vent caps.

Fig. 12—Proper Level of Electrolyte in Battery Cells

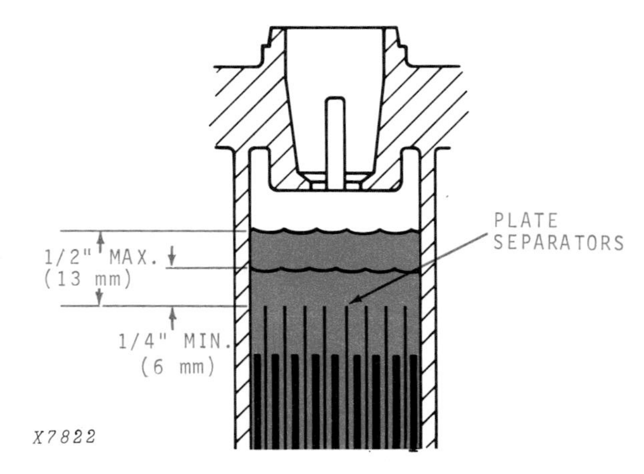

Fig. 13—Maintenance-Free Battery

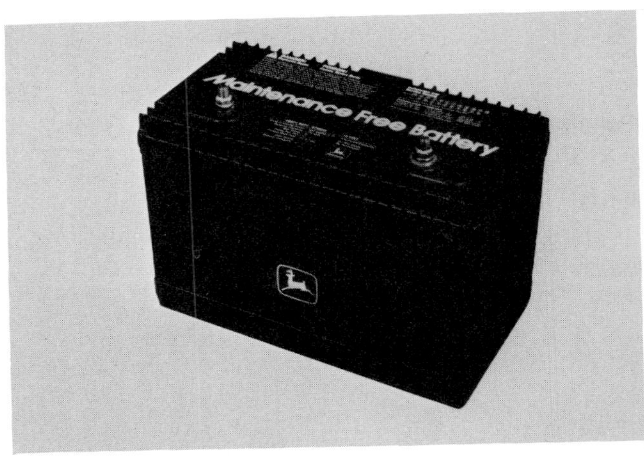

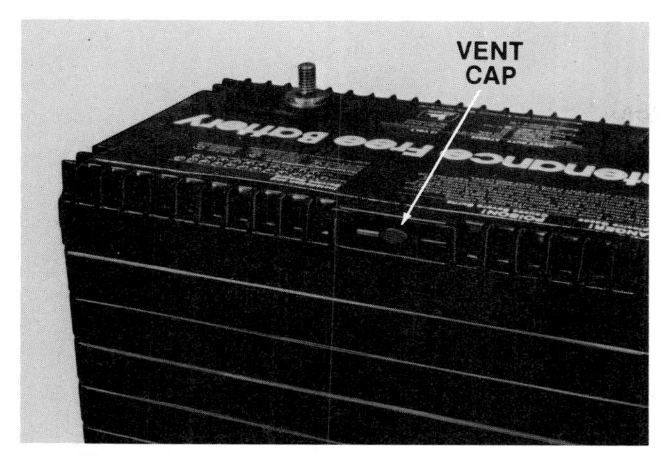

Fig. 14—Location of Vent on Maintenance-Free Battery

Most maintenance-free batteries are ready for service when they leave the factory. They have a very low rate of discharge and thus have a longer shelf life than conventional batteries—sometimes twelve months or longer depending on storage temperature.

VISUAL INSPECTION

A simple visual inspection of the battery will often prove valuable in locating a possible cause of failure.

For example, many conditions are easily noted, such as damage due to excessive tightness or looseness of the hold-down clamps, distorted, cracked, or leaky case or cell vent caps, evidence of overflowing electrolyte, or heavy deposits of dust on the top of the battery. Any of these can contribute to battery failure.

If necessary, check out your visual diagnosis with further battery tests.

SPECIFIC GRAVITY TEST

The state of battery charge is indicated by the specific gravity or weight of the battery electrolyte.

The strength of the electrolyte varies directly with the state of charge of each cell.

To find out how much energy is available from the battery, you need only find out what percentage of sulfuric acid remains in the electrolyte. One of the simplest and most reliable ways to do this is to measure the specific gravity or weight of the solution.

Specific gravity can be measured very quickly by means of a battery hydrometer with a thermometer for temperature correction.

Hydrometers are calibrated to measure specific gravity correctly at an electrolyte temperature of 80°F (27°C).

To determine a corrected specific gravity reading when the temperature of the electrolyte is other than 80°F (27°C): **Add** to the hydrometer reading four gravity points (0.004) for each 10°F (5.5°C) **above** 80°F (27°C). **Subtract** four gravity points (0.004) for each 10°F (5.5°C) **below** 80°F (27°C).

This compensates for expansion and contraction of the electrolyte at temperatures above or below the standard.

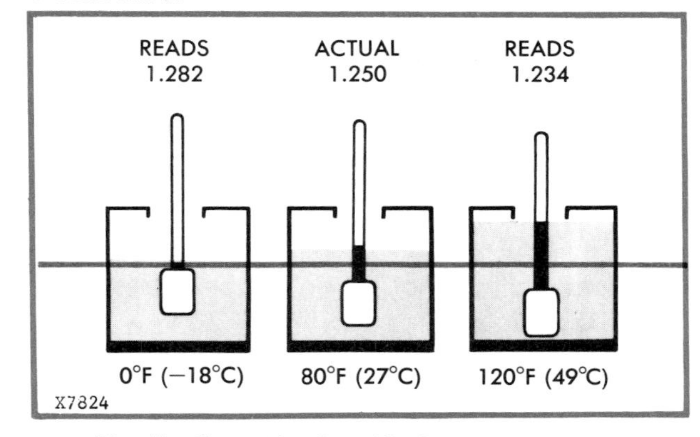

Fig. 15—Correcting Specific Gravity Readings to Allow for Temperatures

For example, a specific gravity reading of 1.234 is obtained at 120°F (49°C). Since this reading was taken with the electrolyte temperature 40°F (4°C) above the standard, a total of 16 (4×4) gravity points (0.016) is added, giving a corrected reading of 1.250. See Fig. 15.

As a further example, suppose a reading of 1.282 is obtained at 0°F (18°C). This reading was taken with the electrolyte temperature 80°F (27°C) below the standard. Therefore, a total of 32 (8×4) gravity points (0.032) is subtracted, giving a corrected reading of 1.250.

Batteries used in tropical or arctic regions are special cases. In tropical areas, use 1.225 as a full charge adjusted reading. In arctic regions, use 1.280 as a full charge reading. This allows for the special operating conditions and assures maximum performance under stress of heat or cold.

Using the hydrometer, test the specific gravity of each cell (Fig. 16). Be sure the hydrometer float is suspended freely in the liquid, not touching the walls, top, or bottom of the tube. Also be sure that your eye is at the level of the liquid when the reading is taken. Readings taken at a sharp angle are generally inaccurate.

If the liquid level is too low to check, add distilled or filtered water to the cells and charge the battery long enough to assure complete mixing of the water and electrolyte, then check with the hydrometer.

When checking specific gravity of a battery which has been gassing freely, allow sediment to settle or gas to escape from the sample before taking the reading.

- **Specific gravity should read from 1.225 to 1.280 (corrected for 80 °F (27 °C) electrolyte temperature).**
- **The variation in readings between cells should be not more than 0.050.**

If the readings are not within the specified range, do the following:

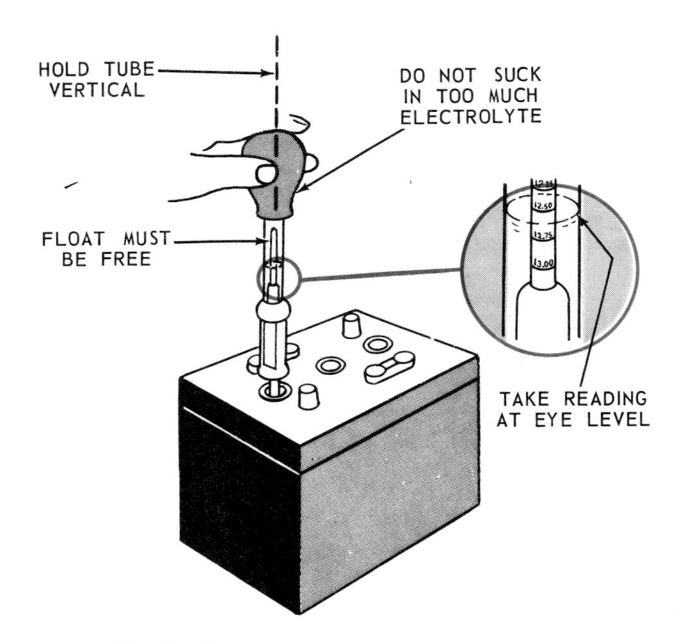

Fig. 16—Checking Specific Gravity of Battery with Hydrometer

If Specific Gravity Is Less Than 1.225

When the specific gravity reading is less than 1.225 (after correction for temperature), the battery may be in satisfactory condition although its state of charge is low. Charge the battery before making further tests. (See later in this chapter for details.)

If Specific Gravity Is Above 1.280

When the specific gravity reading is above 1.280 (after correction for temperature), the battery may be in satisfactory condition although it is above full charge. In use, its specific gravity should return quickly to the normal 1.225-1.280 range. Make further tests to be more certain of the battery condition.

If More Than 0.050 Variation Between Cells

A difference of more than 50 specific gravity points (0.050) between cells indicates an unsatisfactory battery condition. This may be due to unequal consumption of electrolyte in the cells caused by an internal defect, short circuit, improper activation, or deterioration from extended use. The battery should normally be replaced.

NOTE: Specific gravity readings do not always give a true indication of the state of charge of a battery. If water has been added recently or acid has been lost through accident or leakage, the reading will indicate a lower state of charge than is actually the case. Therefore, never use specific gravity readings alone to decide the condition of a battery.

Specific Gravity Reading (Adjusted)	State of Charge
1.260 Sp. Gr.	*100% Charged*
1.230 Sp. Gr.	*75% Charged*
1.200 Sp. Gr.	*50% Charged*
1.170 Sp. Gr.	*25% Charged*
1.140 Sp. Gr.	*Very Little Useful Capacity*
1.110 Sp. Gr.	*Discharged*

The table above shows the state of charge of a typical battery at various specific gravity readings.

PREDICTING BATTERY LIFE

When the battery is new, all cells are in good condition and the voltage difference between cells is about zero or is negligible.

But as the battery gets months or years of service, the voltage difference between cells is greater. This is normal, for all batteries are perishable.

However, when the maximum difference in cell voltages reaches 0.05 volts, the battery is worn out and should be replaced.

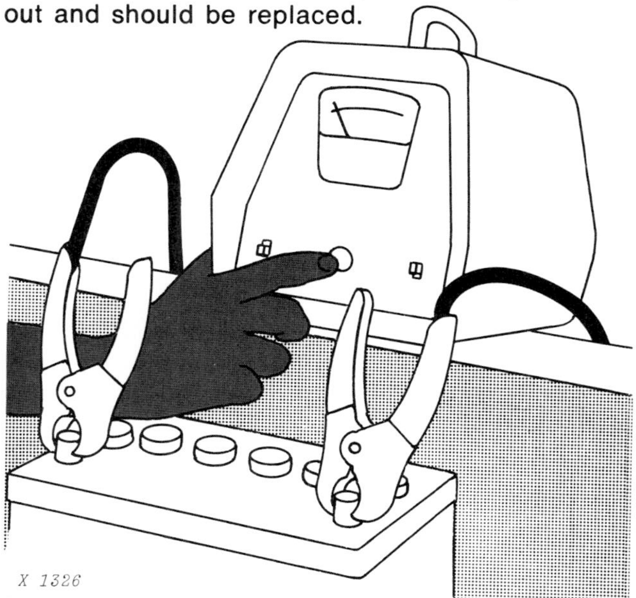

Fig. 17—High-Rate Discharge Battery Tester Connected to Battery

HIGH-RATE DISCHARGE TEST

When a battery is known to be in good condition, its discharge performance can be measured by the specific gravity test. However, this check gives no hint of other factors that may cause a battery to perform badly.

To be sure of a battery's ability to deliver current under load, give it a high-rate discharge test. This test shows the internal conditions that might not otherwise be detected.

The test is made by using a high-rate discharge battery tester (Fig. 17).

This instrument is primarily a high-capacity fixed or variable resistance through which the battery may be discharged at a known rate, voltmeter readings taken, and comparisons made. Terminal voltage under load is used as the standard of performance.

The following conditions must exist before this test is made:

(1) Battery specific gravity must not be less than 1.225 at 80 °F (27 °C). Otherwise, erratic or unreliable readings will result.

(2) Battery temperature must be between 70° and 90°F (21 and 32°C). Terminal voltage does not remain constant but actually decreases as the temperature of the battery drops below range. For example, the terminal voltage of a 12-volt battery at 80°F (27°C) under load might be as high as 10.8 volts, while the terminal voltage of the same battery under the same load at 0°F (−18°C) would be only 8.4 volts.

Note the ampere-hour capacity of the battery, which is normally printed or stamped on the case. For a 6-volt battery, apply a fixed load in the test of twice this capacity. For a 12-volt battery, apply a fixed load of three times this capacity.

Connect the high-rate discharge tester to the battery as shown in Fig. 17. Be sure to follow the manufacturer's instructions.

Discharge the battery under the fixed load for approximately 20 seconds, then read the terminal voltage.

If the battery is in satisfactory condition, the terminal voltage reading should remain above *4.8 volts* for a 6-volt battery or above *9.6 volts* for a 12-volt battery.

If the terminal voltage falls below this value, the battery is defective or it is not as fully charged as the specific gravity reading indicated in the specific gravity test.

To be sure of the battery condition, carefully charge it and repeat the test.

Replace the battery if it is defective.

The chart in Fig. 18 shows the basic test results from the specific gravity and high-rate discharge tests.

CHARGING THE BATTERY

The amount of electrical current a battery can produce is limited by the amount of chemical reaction which can take place within it.

When chemical reaction in a battery has ended through defect or long use, the battery is discharged and can no longer produce a flow of electrical current.

BATTERY TESTING CHART

TEST RESULTS	CONDITION	CORRECTIVE PROCEDURE
GRAVITY TEST GRAVITY BETWEEN 1.225 - 1.280	CHARGED	MAKE HIGH RATE DISCHARGE TEST
GRAVITY BELOW 1.225	DISCHARGED	RECHARGE; MAKE HIGH RATE DISCHARGE TEST
MORE THAN 50 GRAVITY POINTS (0.050) VARIATION BETWEEN CELLS	(a) SHORTED CELL (b) ACID LOST (c) OLD BATTERY	REPLACE
HIGH RATE DISCHARGE TEST* MINIMUM TERMINAL VOLTAGE: 4.8 VOLTS FOR 6–VOLT BATTERY 9.6 VOLTS FOR 12–VOLT BATTERY	(a) DISCHARGED (b) OLD BATTERY	RECHARGE REPLACE

*Ampere Load Should Equal 2 x Amp-Hr Rating for 6-Volt Batteries and 3 x Amp-Hr Rating for 12-Volt Batteries.

Fig. 18—Battery Testing Chart

The battery can be recharged, however, by causing direct current from an outside source to flow through it in a direction opposite to that in which it flowed out of the battery.

During discharge, current flows from the positive (+) terminal of the battery, through the circuit, and back into the battery at the negative (−) terminal (Fig. 19). Within the battery, current flow is from the negative to the positive terminal.

To recharge the battery, this flow of current is reversed as shown, restoring the chemicals in the battery to their active state.

The battery then becomes charged and ready to produce electrical current again. The chemical action which takes place within a battery during discharging and charging is explained earlier in this chapter.

Fig. 19—Current Flow During Discharge and Charge of Battery

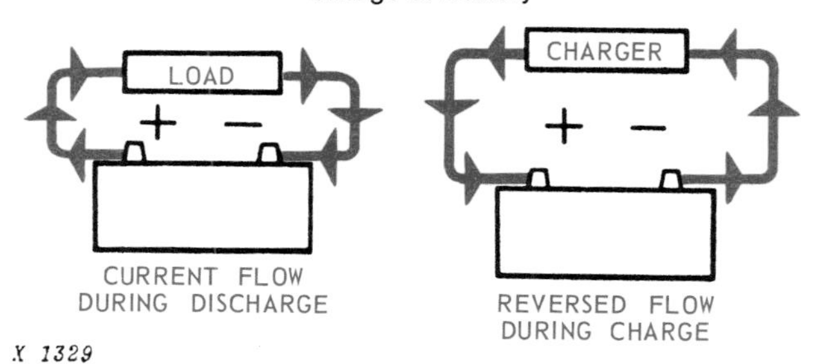

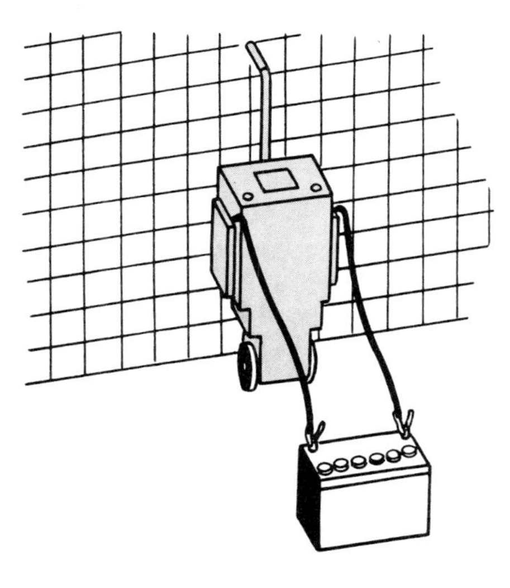

SLOW CHARGER

Fig. 20—Battery Charging

Batteries can be recharged in two ways:

- **Fast Charging**
- **Slow Charging**

A battery that is in satisfactory condition but requires recharging will accept a large amount of charging current without undesirable effect. This type of battery may be charged quickly at a high rate with a battery *fast charger*.

A battery that becomes sulfated, however, will not accept a high rate of charging current without possible damage. Its sulfated condition provides increased resistance to current flow within the battery. Flow of a high rate of charging current through this resistance creates heat, which can result in warping of the plates, boiling of the electrolyte, and eventual damage to the separators. Cell caps and covers and the battery case may be damaged or distorted.

A battery in this condition must be charged over a long period at a low rate. In this manner, sulfate formation on the plates will be gradually broken down and the battery returned to its normal charged state.

The reaction of the battery itself to fast charging will indicate the amount of charging current it can accept without damage. Never allow the battery electrolyte to heat up above 120°F (49°C).

If the battery is sulfated, use a fast charger and reduce the charging rate or *slow charge* it. For details, see "Sulfated Batteries" earlier in this chapter.

Wear safety goggles and rubber gloves when charging a battery. Keep sparks and flames away from the battery. Make sure the work area is well-ventilated. When charging and discharging, a lead-acid storage battery generates harmful fumes and gas. This gas is very explosive.

Batteries connected in series can be charged together if their voltages are within 0.1 volt for 6-volt batteries or 0.2 volt for 12-volt batteries, of if their specific gravity is within 20 points. Otherwise, batteries must be charged individually.

When charging several batteries in series, charge at the rate of the lowest capacity battery in the line.

Batteries connected in parallel must be disconnected and the surface charge removed before checking the voltage of each battery.

To remove the surface charge, ground each battery negative terminal one at a time and turn a machine light switch on for one minute. Then turn the light switch off and wait one minute and check battery for proper voltage using a voltmeter.

Batteries connected in parallel can be charged together without disconnecting them when the battery voltage is 6 volts or above for 6-volt batteries and 12 volts or above for 12-volt batteries.

If battery voltage is below these specifications, the batteries may be sulfated. Charging each battery individually will break down the oxide and revive the batteries quicker than charging all batteries together in parallel connection.

FAST CHARGING

Fast charging gives the battery a high charging rate for a short period of time. Never use a fast charger as a booster to start an engine.

Disconnect the battery ground cable. Then disconnect the battery positive cable. Remove the battery from the machine and, if necessary, fill the cells with distilled water to the level recommended by the load tester manufacturer.

Connect a charger or load tester of a 30 to 300 ampere rating to the battery. Be sure to follow the manufacturer's instructions for using the charger. A portable-type charger is shown in Fig. 20.

Set the charging rate at 30-60 amperes for a 6-volt battery or at 15-30 amperes for a 12-volt battery.

Start the charger at a slow or low charging rate. Increase the charging rate one selection at a time. Observe the ammeter after one minute at each selection for a 10-amp charging rate. If necessary, select boost.

After the charger has operated for at least three minutes, note the electrolyte. If it gasses excessively, the battery is sulfated and charging rate must be reduced to prevent possible damage. Reduce the charging rate until the electrolyte produces comparatively few bubbles but gassing has not stopped entirely. Replace the cell caps.

The maximum charging time at the boost selection is 10 minutes for one conventional battery and 20 minutes for a maintenance-free battery. Allow an additional five minutes charging time for each - 12°C (10°F). If the battery is not accepting the required 10-ampere charging rate by the specified time, replace the battery.

If the battery is accepting the charge, check specific gravity after 30 minutes for a conventional battery or 60 minutes for a maintenance-free battery.

Charge battery to a specific gravity reading of 1.230 to 1.265 points.

The charging rate for conventional batteries may require 2 to 4 hours. The charging rate for maintenance-free batteries may require 4 to 8 hours.

After replacing a battery or after cleaning battery terminals, use an electrical sealant around the base of the terminals.

SLOW CHARGING

A battery that is badly sulfated will not accept fast charging without possible damage. So these batteries must be charged at a slow rate.

Be sure to follow the manufacturer's instructions for using the charger.

Charge the battery at a low rate (7% of the battery ampere-hour rating or less) for an extended period of time until fully charged.

A battery is considered fully charged when three consecutive hydrometer readings, taken at hourly intervals, show no rise in specific gravity.

The normal slow-charging period is from 12 to 24 hours.

If a battery's specific gravity has not reached the normal full-charge range (1.225 to 1.280) within 48 hours of slow charging, replace the battery.

Badly sulfated batteries, however, may take from 60 to 100 hours to recharge completely.

SUMMARY: CHARGING OF BATTERIES

Remember these key facts when charging batteries:

1. Batteries are charged by reversing their flow of current using an outside power source.

2. There are two ways to recharge batteries: fast charging and slow charging.

3. If a battery is sulfated, use a fast charger but reduce the charging rate.

4. If the battery is badly sulfated and will not accept a fast charge, use the slow charging method.

5. If fast charging is used, never exceed 30 amps for a 12-volt battery or 60 amps for a 6-volt battery. Avoid heating up the battery electrolyte above 120°F (49°C).

6. To assure a fully charged battery, follow the fast charge with a slow charge.

7. After charging, always make sure the electrolyte is at the proper level in each cell.

BALANCING BATTERIES (24-Volt Split-Phase Systems)

Four 6-volt batteries or two 12-volt batteries are sometimes connected in series to furnish 24 volts to the starting system. The lights and other electrical equipment then operate on two 12-volt circuits. Each circuit uses half of the batteries. These circuits are connected so the electrical load is nearly balanced between the batteries.

In normal service, the two sets of batteries should stay equally charged. However, if one side uses more water than the other, plug a small lamp bulb into an outlet socket in the circuit of the batteries using more water. This should put enough load on this side to remedy the problem.

Another unbalanced condition may occur when more than a normal load is imposed on one half of the batteries through an outlet socket.

The method used to connect the batteries places them "in balance"; that is, the rate of discharge from all batteries is approximately the same provided that accessories used with the outlet socket, such as auxiliary lights, do not draw more than, say, 3 amperes. Thus the life of all batteries will be approximately the same. Auxiliary lights or other electrical equipment which draw up to 7 or 8 amperes can be used for short periods of time without materially affecting the batteries.

However, if auxiliary electrical equipment which draws more amperes is used for long periods, the batteries will be thrown out of balance and the life of those on one side will be affected. This is because the current drawn from this side is greater than that from the other, which results in failure to maintain the batteries at proper charge.

One method of correcting this condition is to install an extra outlet socket to the other side.

The outside electrical load should then be alternated between the two outlet sockets as the battery charges demand. Or, when the first pair of batteries become low in charge, disconnect the outlet socket wire from one pair and connect it to the other pair.

Under these unusual conditions, pay strict attention to the charge of the batteries at all times so the load can be switched when one pair of batteries becomes low in charge.

CAPACITY RATINGS OF BATTERIES

New capacity ratings for batteries were adopted in 1971 by the Society of Automotive Engineers (SAE) and Battery Council International (formerly known as AABM).

These new simplified ratings are:

- **Cold Power Rating**
- **Reserve Capacity**

Let's look at each one.

COLD CRANKING AMPERE (CCA) RATING

This rating tells the power for starting on cold days when the going gets tough. It gives the *number of amperes the battery at 0°F (−18°C) can deliver over 30 seconds and not fall below a voltage of 1.2 volts per cell* . . . the minimum voltage required for dependable starting.

This is the most important rating because it tells how much power the battery can deliver for its No. 1 job . . . *starting.* The tougher the engine is to start, the more amperes it takes to get it started. Many low-priced batteries can deliver only 200 amps. The more powerful batteries will deliver 600 amps under the same condtions. *A battery capable of delivering 200 amperes isn't likely to start an engine that needs 400.*

RESERVE CAPACITY

The second new rating is called **reserve capacity.** It tells . . . *"the number of minutes a new fully charged battery at 80°F (27°C) will deliver 25 amperes while maintaining a voltage of 1.75 volts per cell."*

Fig. 21—How Cold Weather Affects the Battery and the Engine When Starting

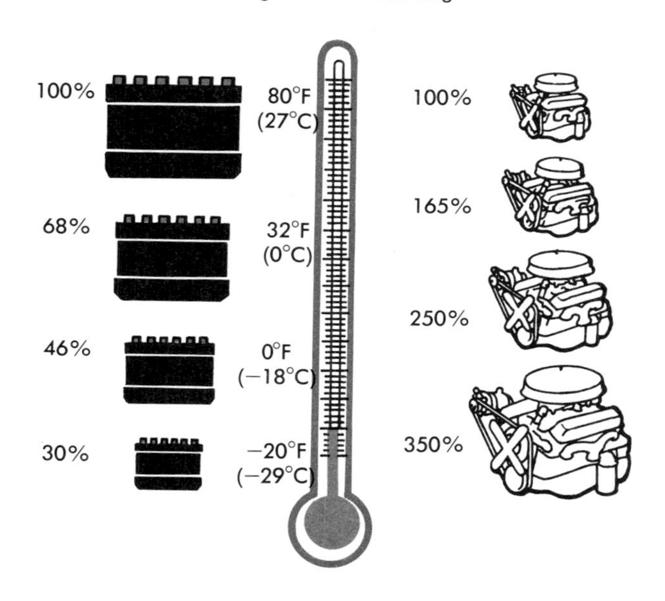

A simple translation is . . . if the charging system of your machine failed suddenly, how much time would you have to find help? The load of 25 amperes is equivalent to the needs of ignition, lights, and normal accessories.

In other words, 25 amperes is the power drain required to keep your machine operating. Reserve Capacity is always expressed in minutes . . . the time available to seek help. The greater number of minutes, the greater the margin of safety.

COLD WEATHER AND BATTERY SIZE

For *cold weather starting,* the battery must be big enough for the cranking job.

Fig. 21 shows how the load on the battery gets bigger when it is colder. At 20 below zero, the battery has only 30% of its full cranking capacity. At the same time, a greater load is put on the battery by the colder engine when starting.

In effect, *at cold temperatures the battery is "smaller" while the engine is "larger."*

If the replacement battery is smaller than the original, the engine will surely be harder to start in cold weather.

HOT WEATHER AND BATTERY SIZE

In southern areas, hot starts often require as much or more battery power than cold starts. They may occur hard during hot weather . . . after the machine has been worked and the engine is hot. Hot starts are more common with the big high-compression engines and are aggravated when the machine is equipped with air conditioning.

There are many times in the heat of summer when engines are just as hard to start as in the winter. Therefore, use the same system of matching battery power to the engine in the south as you do in the north.

OTHER FACTORS IN BATTERY SELECTION

When replacing batteries, be sure to replace the battery with one *at least equal* in size to the original.

A *larger* battery than the original may be needed if added accessories such as air conditioning put a larger load on the battery.

An extra-output generator may be the answer in cases where electrical loads are excessive or when operating mostly at idle speeds. This will help keep the battery charged and increase its service life.

The cheapest battery is not always the best buy for replacement. For example, three batteries in the same group size may vary in price, but they also vary in cold power rating, in construction, and in warranty period. Divide the price by the months of warranty, and you may find the most expensive batteries are really the cheapest, per month of expected service.

A final word on replacing batteries: One out of every four batteries returned for warranty have nothing wrong with them except that they are discharged. *Be sure to trouble shoot the battery before you replace it.*

BATTERY REMOVAL AND INSTALLATION

Remove the battery as follows:

1) Note carefully the location of the positive (+) terminal so that the battery is installed in the same way.

2) Disconnect the ground terminal first. Use only a box end wrench to loosen clamps on terminals. Remove clamps using a screw-type puller. Do not hammer on the battery posts.

3) Remove the battery and inspect the battery tray and hold-downs for dirt or corrosion. Clean any corroded part with ammonia solution or baking soda (¼ pound added to a quart of water).

4) Check cables for worn or frayed insulation. Replace cable or bolts if corroded.

Install the battery as follows:

1) Be sure the battery is fully charged.

2) On new batteries with a warranty tag (Fig. 22), punch the tag to show date of purchase and also when guarantee expires. For example, if battery is purchased in November 1990 and has a prorata warranty of 24 months, punch out dates as shown in Fig. 22.

3) On new batteries without a warranty tag, gently stamp the date on the top of the battery using a date coding ring (Fig. 22, top). As shown, the 12 letters in the outer part of the ring are for the months, while the ten figures in the inner part of the ring identify the year. Example: L0 = November 1990.

4) Set the battery in place, using a lifting strap if necessary. Make sure the battery is resting level in its tray.

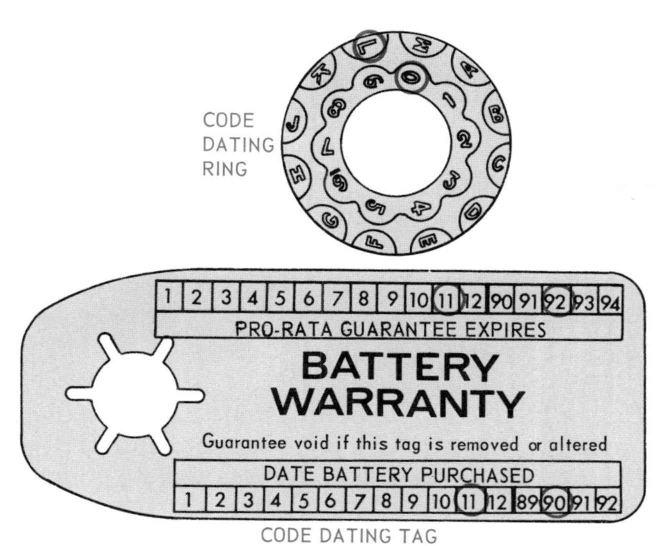

Fig. 22—Date Coding for new Batteries

5) Tighten the hold-down nuts **evenly** until the battery is secure. Do not overtighten as this will distort or crack the battery case.

6) Clean the battery terminals and cable clamps with a wire brush before attaching the clamps. This will assure a good contact. Coat the terminals with petroleum jelly to prevent corrosion. Never paint the terminal posts.

7) Check for the correct polarity of the battery. The ground may be to the positive or negative pole, depending upon the system. **Reversed polarity can damage the system.** Note that the positive battery terminal post is larger than the negative post—match it with the larger cable clamp. Wait until last to connect the grounded cable or strap to avoid short circuits. On alternator-equipped machines, before connecting the last cable or ground strap, momentarily touch it against the battery post. With all switches and accessories off, no spark should occur. If it does occur, check for reversed battery polarity, improper alternator connections, or defective electrical equipment.

Fig. 23—Installation of Typical Batteries

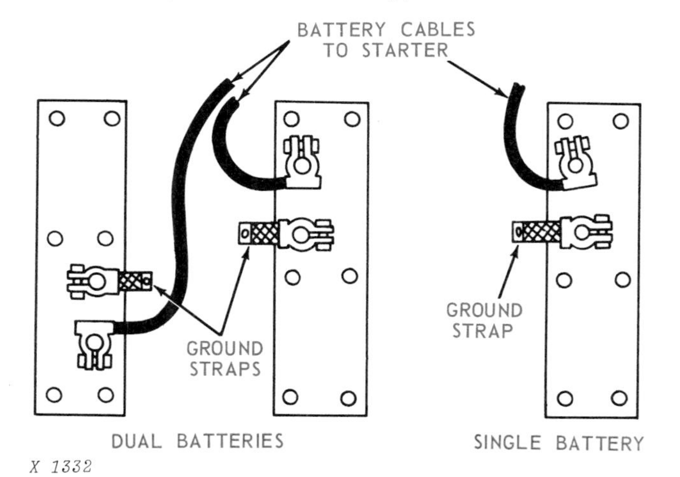

8) Tighten the clamps on the battery terminals. Use a box end wrench carefully to avoid twisting the battery terminal posts.

9) On generator-equipped machines, AFTER the battery is connected and BEFORE starting the engine, POLARIZE THE DC GENERATOR. (See Chapter 6.) NEVER ATTEMPT TO POLARIZE ALTERNATOR-equipped machines.

WHAT AFFECTS BATTERY LIFE

Periodic service has the best effect on battery life. In contrast, neglect and abuse will shorten the life of the battery.

Besides periodic cleaning of the battery top, posts, and cable clamps, the four key factors in battery life are:

- **Electrolyte Level**
- **Overcharging**
- **Undercharging**
- **Cycling**

Let's discuss each of these factors briefly.

ELECTROLYTE LEVEL

Keeping up the level of electrolyte in each battery cell is the most basic step in long battery life.

UNDERFILLING causes the electrolyte to become too concentrated, making the plates deteriorate more rapidly. The low level also exposes the tops of the plates, which harden and become chemically inactive.

OVERFILLING causes electrolyte to spill out, corroding the battery posts and cell covers.
Normally water is the only part lost from the battery electrolyte. This loss of water is due to evaporation, especially in hot weather and while charging the battery. Be sure to use distilled water if available. Otherwise, use only clean, soft water.

The correct level of electrolyte is always above the tops of the cell plates, which is usually to the bottom of the cell filler neck.

OVERCHARGING

Overcharging causes a loss of water in the cells by separating the electrolyte into hydrogen and oxygen gases. The gas bubbles wash active materials from the plates, and reduce the battery capacity.

If the battery uses too much water, check it for overcharging.

Overcharging also causes the battery to heat up inside and oxidates the positive plate grids, resulting in a loss of cell capacity and early failure.

UNDERCHARGING

A battery which stays undercharged will become sulfated. The sulfate normally formed in the plates will become dense, hard, and chemically irreversible if allowed to remain in the plates for long periods. The lowered gravity levels then make the battery more likely to freeze.

In cold weather, undercharged batteries often fail to crank the engine because of their lack of reserve power.

CYCLING

A cycle consists simply of a discharge and recharge. If operating conditions subject the battery to heavy and repeated cycling, its life will be shortened as cycling causes the positive plate's active material to shed and fall to the sediment tray in the bottom of the battery.

While the battery is perishable and will eventually wear out, be sure to remember that its life can be prolonged with a reasonable amount of care.

OTHER FACTORS IN BATTERY LIFE

If the battery hold-downs are loose, the battery will bounce around in its tray. This may cause the case to crack or cause internal damage to the elements.

Fig. 24—Things Which Affect the Life of the Battery

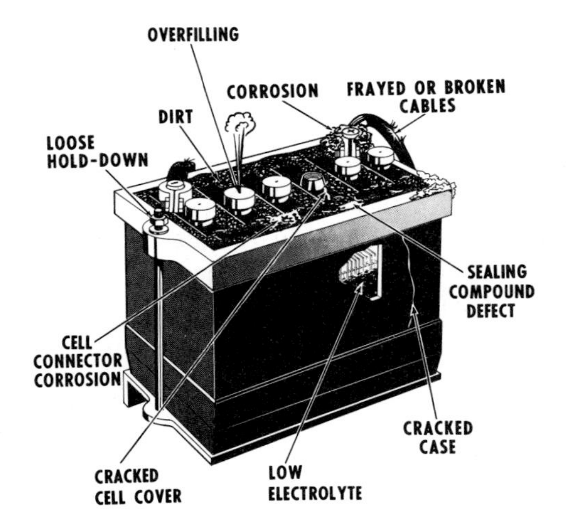

If the hold-downs are too tight, this can also damage the battery by warping the case.

Dirt and corrosion will damage the battery case and may cause a dangerous short circuit.

Frayed or broken battery cables may also create a short circuit.

BATTERY LOAD TESTING

Before load testing a battery, follow proper safety precautions and read the instructions provided by the load tester manufacturer. Wear safety goggles and rubber gloves. Keep sparks and flame away from the battery.

Always visually inspect the battery prior to load testing.

Batteries can be load tested in two ways:

1. Fixed load test.
2. Adjustable load test.

If the battery has been on charge, blow away gases before continuing testing.

Load test batteries one at a time.

After replacing a battery or after cleaning battery terminals, use an electrical sealant around the base of the terminals.

FIXED LOAD TEST

1. Take a hydrometer reading of each cell. If there are more than 50 points variation between cells, replace the battery. If less than 50 points variation, proceed to Step 2.
2. Test the specific gravity of the battery. It must be at least 1.225 @ 80° F (27° C). If specific gravity is lower than this, charge the battery. If the specific gravity does not come up to this level, replace the battery.

   If the battery tests above this level, test it for internal shorts. Do this by connecting a 300 ampere load for 15 seconds across the battery terminals. If the battery starts to smoke or gas, replace it.

   If the battery tests at 1.225, proceed to Step 3.
3. Make sure all electrical switches and accessories are off to reduce the possibility of arcing when the ground cable is disconnected. Disconnect the battery ground cable. Then disconnect the battery positive cable.

4. Clean corrosion from the battery and battery terminals.
5. Measure the temperature in the center cell of the battery.
6. Connect the voltmeter and load tests leads to the proper battery terminals.
7. Apply a test load equal to 50% of the cold cranking ampere (CCA) rating of the battery.

   Another way to determine the correct load to apply to the battery is to note the ampere-hour capacity of the battery which is normally printed or stamped on the case. For a 6-volt battery, apply a fixed load of twice this capacity. For a 12-volt battery, apply a fixed load of three times this capacity.

   Apply the load for 15 seconds. Read the battery voltage, then immediately remove the load from the battery.
8. If the test voltage is below the minimum required (4.8 volts or above for a 6-volt battery; 9.6 volts or above for a 12-volt battery), replace the battery. If test voltage is at or over the minimum required, return the battery to service.

ADJUSTABLE LOAD TEST

1. Perform Steps 1-6 of the adjustable load test.
2. Set the battery tester selector to the correct battery or engine size. The battery size selector must be set to a range which will include 50% of the cold cranking amp (CCA) rating of the battery. This range is three times the 20 ampere-hour capacity of the battery.
3. Connect the voltmeter and load test leads to the proper battery terminals.
4. Apply the load for 15 seconds. Read the battery voltage on the meter. Immediately disconnect the load from the battery.
5. Compare the battery voltage and temperature to the proper chart following.

LOAD TEST TEMPERATURE CHART
(6-volt battery)

°F	°C	Volts
70 and above	21 and above	4.8
60	16	4.75
50	10	4.7
40	4	4.65
30	-1	4.55
20	-7	4.45
10	-12	4.35
0	-18	4.25

LOAD TEST TEMPERATURE CHART
(12-volt battery)

°F	°C	Volts
70 and above	21 and above	9.6
60	16	9.5
50	10	9.4
40	4	9.3
30	-1	9.1
20	-7	8.9
10	-12	8.7
0	-18	8.5

6. If voltage is less than that shown in the chart, replace the battery. If reading is equal to or greater than that on the chart, the battery is good and can be returned to service.

CONNECTING A CHARGING BATTERY

Improper jump starting of a dead battery can be dangerous. Follow these procedures when jump starting a battery from a charging battery.

1. Check for a frozen battery. Never attempt to jump start a battery with ice in the cells.
2. Make sure the charging battery and dead battery are of the same voltage.
3. Turn off accessories and both machine ignitions.

4. Place the gearshift of each machine in neutral or park and set parking brake. Make sure the machines do not touch each other.

5. Check electrolyte level in both batteries. Add, if low. Make sure the vent caps are secure and level. Cover the vent caps with a damp cloth.

6. Attach one end of one jumper cable to the charging battery positive terminal. Attach the other end of the same cable to the positive terminal of the dead battery (Figs. 25 and 26).

   Make sure there is good, metal-to-metal contact between the cable ends and the terminals.

7. Attach one end of the other cable to the charging battery negative terminal.

   Make sure there is good, metal-to-metal contact between the cable end and the battery terminals.

   Never allow the ends of the two cables to touch while attached to the batteries.

8. Connect the other end of the second cable to the engine block or machine metal frame *below* the dead battery and as far away from it as possible.

   That way, if a spark should occur at this connection, it would not ignite hydrogen gas that may be present above the dead battery.

9. Try to start the machine with the dead battery.

   Do not engage the starter for more than 30 seconds or starter may overheat and the charging battery will be drained of power.

   If the machine with the dead battery will not start, start the other machine and let it run for a few minutes with the cables attached. Try to start the second vehicle again.

10. Remove cables in exactly the reverse order from installation, disconnecting the grounded cable from the engine block or other metallic ground first, then the other grounded cable from the charging battery.

    Disconnect cables at positive terminals last.

11. Remove the wet cloth from the vent caps.

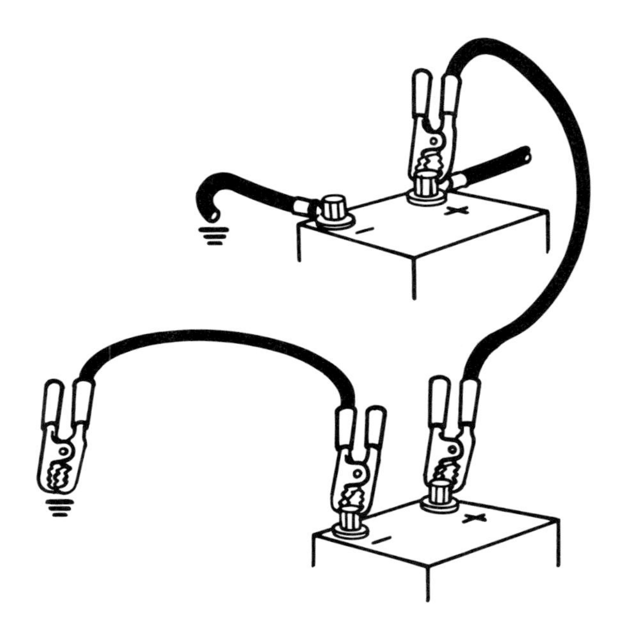

Fig. 25—Jump Starting 12-volt Batteries in a Single-Battery Application (Charging Battery is at Bottom)

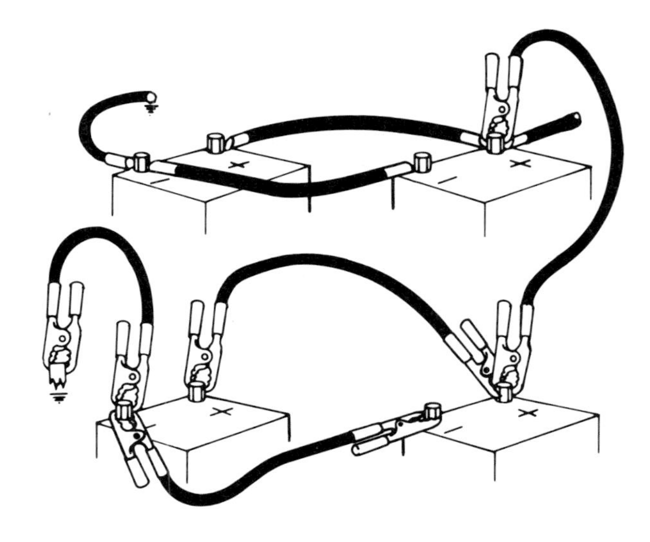

Fig. 26—Jump Starting 12-volt Batteries in a Two-Battery Application (Charging Battery is at Bottom)

Litho in U.S.A.

BATTERY AND BATTERY ACID DISPOSAL

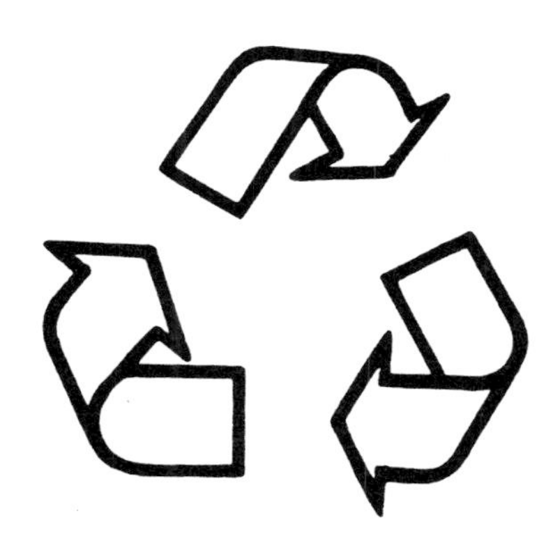

Fig. 27—Dispose of Fluids Properly

Improper disposal of batteries and battery acid can harm the environment and ecology. Check with environmental agencies for information concerning the proper disposal of these items. Do not pour battery acid down a drain or into a stream, pond or lake.

BATTERY TEST EQUIPMENT AND TOOLS

The testing equipment shown here will give you a balanced group of aids in testing and troubleshooting. But only one example from the many models available is shown. Also remember that this is not a complete listing; many other tools are useful for auxiliary tests.

HYDROMETER

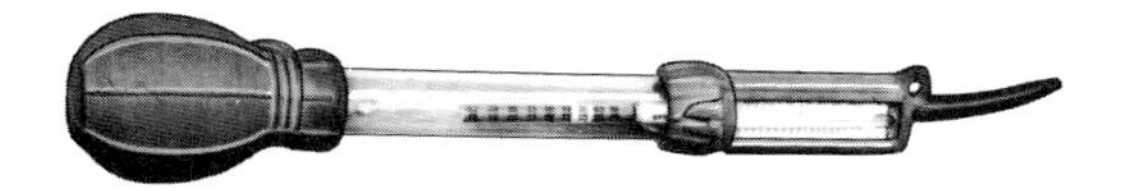

Fig. 28—Hydrometer

The hydrometer (Fig. 28) checks the specific gravity of the battery electrolyte. All the good hydrometers have a built-in thermometer.

BATTERY CHARGER

The battery charger in the Fig. 29 provides both fast and slow charges and is polarity protected to prevent damage if connected wrong. It can also be used as an engine starter for short periods to aid in cold weather starting.

Fig. 29—Battery Charger

BATTERY SERVICE KIT

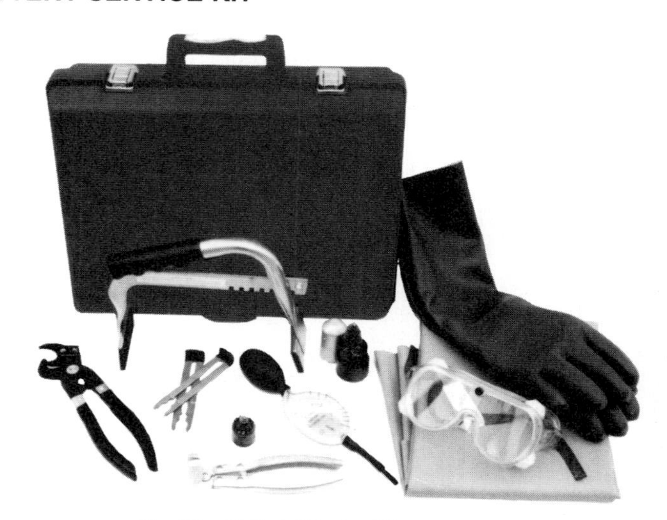

Fig. 30—Battery Service Kit

The battery service kit shown in Fig. 30 is equipped with safety goggles, rubber gloves, battery carrier, post/clamp cleaner, clamp remover, clamp spreader/cleaner, side terminal adaptor, hydrometer, and an apron. It services both side terminal and top post batteries. It is used to remove battery clamps, clean terminals and clamps, and test electrolyte.

BATTERY TESTER

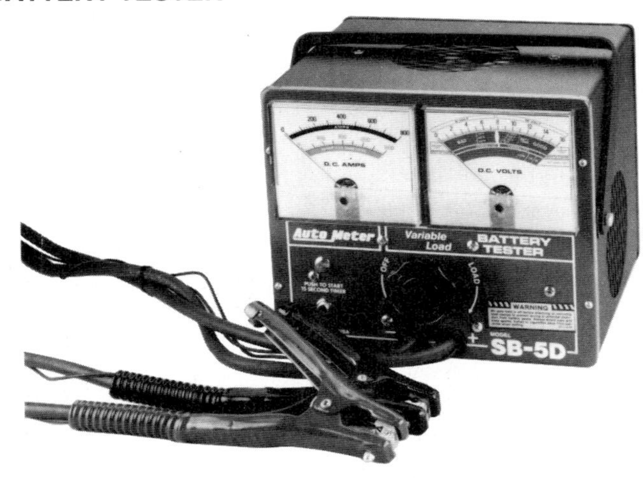

Fig. 31—Battery Load Tester

The solid state battery load tester shown in Fig. 31 electronically checks condition, power, and state of charge of both 6-and 12-volt batteries. It can also be used to compare batteries in multiple hookups and to check alternator regulated voltage.

CURRENT GUN

The current gun shown in Fig. 32 has an LCD readout which indicates both DC and AC currents. It measures current flow of each component from one location. It also measures alternator output, starting current, and battery charging levels.

Fig.32—Current Gun

TEST YOURSELF

QUESTIONS

1. (True or false?) "The specific gravity of a fully charged battery is the same regardless of the temperature of the electrolyte."

2. Battery electrolyte is made up of __________ and ______________________.

3. (True or false?) "Dry-charged batteries are activated at the factory."

4. If you test a battery and find that the specific gravity is 1.200 in all cells, what should you do?

5. If you test a 6-volt battery for light load voltage and get cell readings of 1.96, 1.93, and 1.94, what should you do?

6. The two methods of charging batteries are __________ charging and __________ charging.

7. Which method of charging should be used on badly sulfated batteries?

8. Batteries connected in ______________ can be charged together if their specific gravity is within _________ points.

9. When performing a fixed load test on a battery, apply a test load equal to ____________ of the ______________________ rating of the battery.

(Answers on page 19 at the end of this book.)

CHARGING CIRCUITS / CHAPTER 6

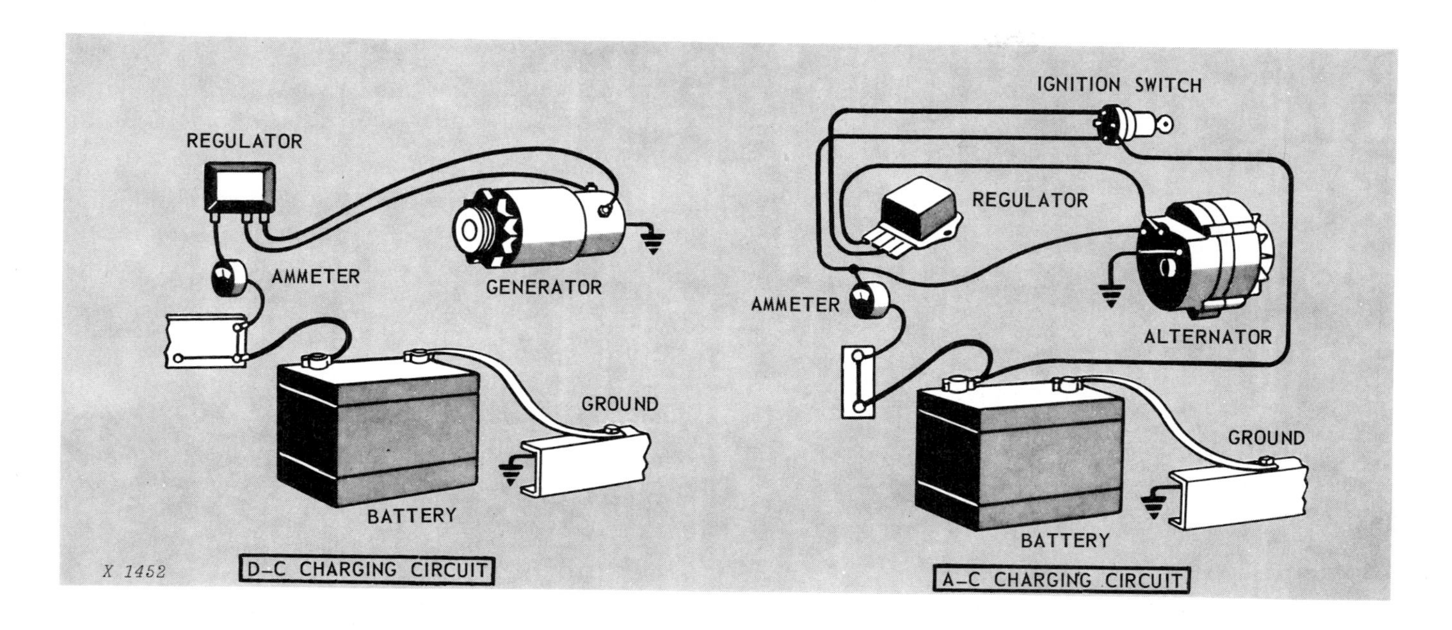

Fig. 1—Charging Circuits—Two Types

INTRODUCTION

The charging circuit does two jobs:

- **Recharges the battery**
- **Generates current during operation**

There are two kinds of charging circuits:

- **DC Charging Circuits (Use Generators)**
- **AC Charging Circuits (Use Alternators)**

Both circuits generate an alternating current (AC). The difference is in the way they rectify the AC current to the direct current (DC).

DC CHARGING CIRCUITS have a *generator* and a *regulator* (Fig. 1).

The *generator* supplies the electrical power and rectifies its current mechanically by using commutators and brushes.

The *regulator* has three jobs: 1) opens and closes the charging circuit; 2) prevents overcharging of the battery; 3) limits the generator's output to safe rates.

AC CHARGING CIRCUITS have an *alternator* and a *regulator* (Fig. 1).

The *alternator* is really an AC generator. Like the generator, it produces AC current but rectifies it electronically using diodes. Alternators are generally more compact than generators of equal output, and supply a higher current output at low engine speeds.

The *regulator* in AC charging circuits limits the alternator voltage to a safe, preset value. Transistorized models are used in many of the modern charging circuits.

OPERATION OF CHARGING CIRCUIT

All charging circuits operate in three stages:

- **During starting—battery supplies all load current**
- **During peak operation—battery helps generator supply current**
- **During normal operation — generator supplies all current and recharges battery**

In both charging circuits, the *battery* starts the circuit when it supplies the spark to start the engine. The engine then drives the generator (or alternator) which produces current to take over the operation of the ignition, lights and accessory loads in the whole system.

The battery also helps out during peak operation when the electrical loads are too much for the generator (or alternator).

But once the engine is started, the generator (or alternator) is the "work horse" which gives current to the ignition and accessory circuits.

 Litho in U.S.A.

The generator supplies this current as long as the engine is speeded up and running. When the engine slows down or stops, the battery takes over part or all of the load.

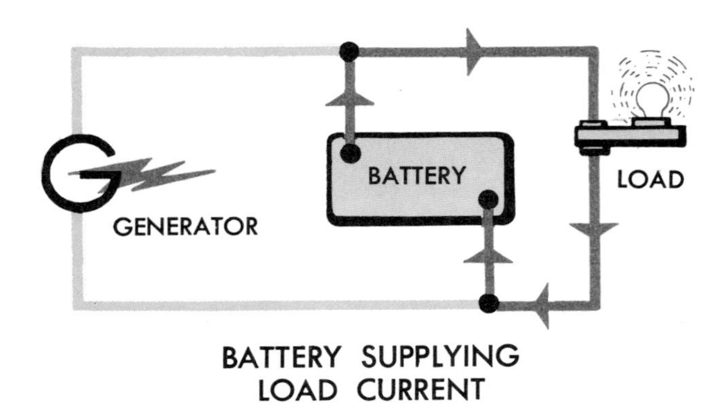

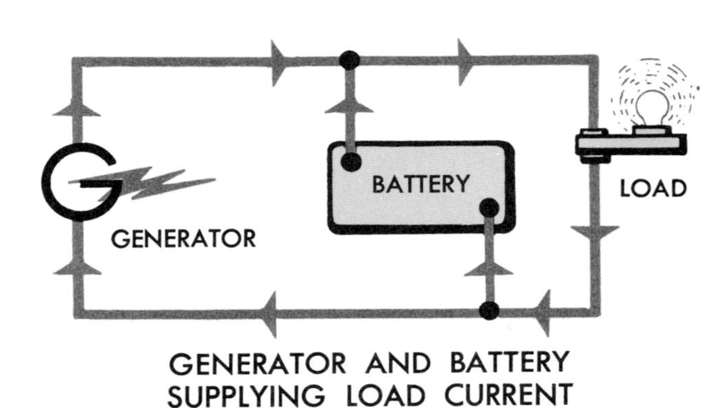

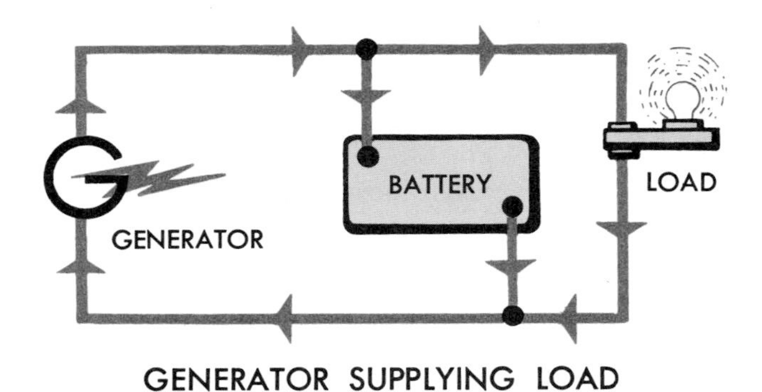

Fig. 2—Charging Circuit in Operation—Three Stages

Operation of the charging circuits during the three stages is illustrated in Fig. 2.

The top illustration shows operation while starting the engine.

The middle diagram shows what happens during peak electrical loads when the battery helps the generator.

The lower diagram shows the system during normal operation when the generator supplies all power for loads and also recharges the battery.

In the rest of this chapter, let's look at each charging circuit by itself, first the parts of **DC circuits.**

GENERATORS

The generator is the heart of the DC charging circuit.

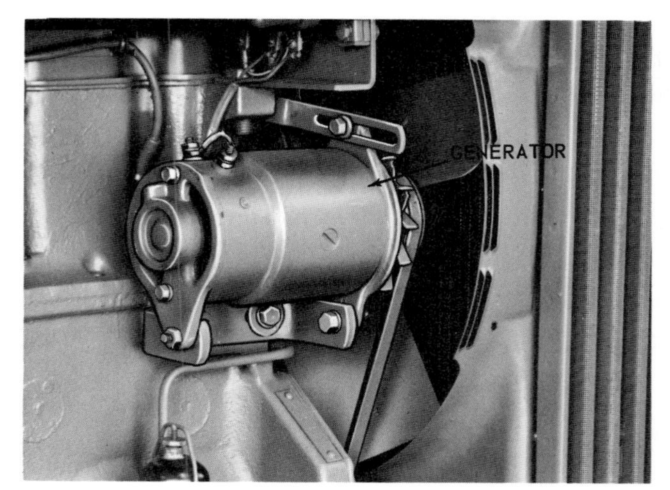

Fig. 3—Generator is the Heart of the DC Charging Circuit

BASIC OPERATION OF GENERATOR

The generator produces electrical power by means of *electromagnetic induction* (Chapter 1).

This is moving a conductor through a stationary magnetic field.

Let's build up a simple generator and explain each function.

The basic generator (Fig. 4) has two parts:

- **Armature—rotating wire loop (the conductor)**
- **Magnetic poles—stationary magnetic field**

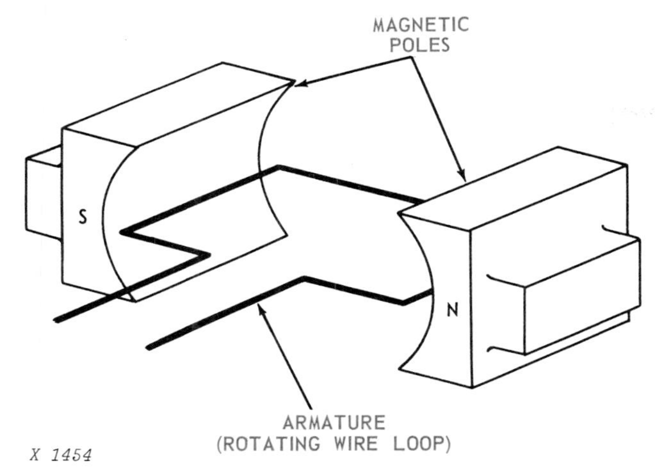

Fig. 4—The Basic Parts of a Generator

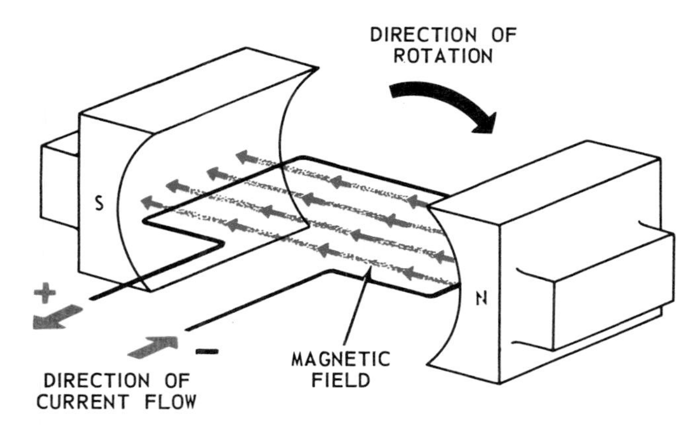

Fig. 5—Basic Generated Voltage

Now let's set the basic generator in motion (Fig. 5).

As the armature rotates through the magnetic field of the poles, voltage is generated.

Using the Right Hand Rule, we can see that the voltage comes toward us on the left side in Fig. 5, and flows away from us on the right side.

By the Conventional Theory, this means that the left end of the armature loop is positive (+) while the right end is negative (−).

For current to flow, we must add three more parts (Fig. 6).

Let's connect the ends of the armature loop to a split ring called a *commutator.*

Next we need some *brushes* to contact the commutator and some *wires* to connect the brushes to a load.

Now we have completed the circuit and current will flow.

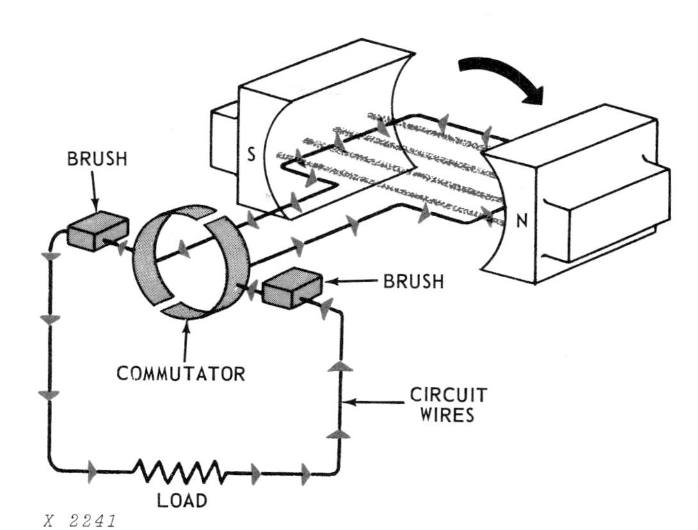

Fig. 6—Basic Current Flow in Generator

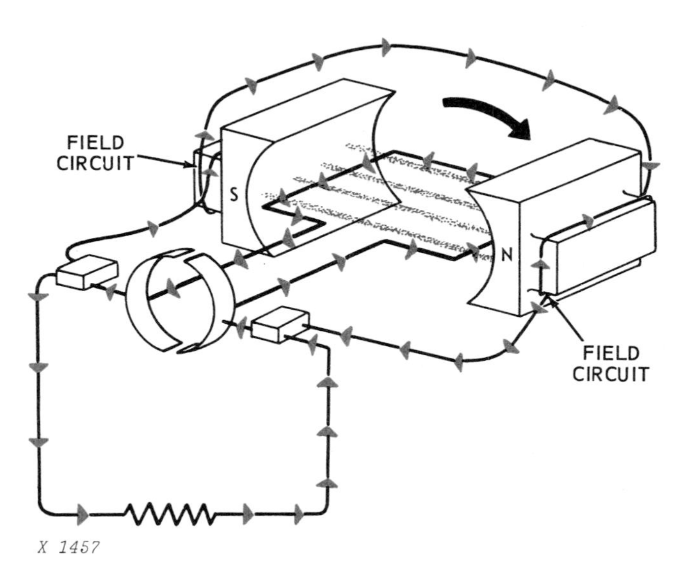

Fig. 7—Complete Parts of Basic Generators

To insure a strong current and proper flow, we must add one more feature (Fig. 7).

The magnets by themselves are weak and create a weak field. The result is that the voltage induced is low.

To remedy this, let's wind the wire conductors around the magnets as shown in Fig. 7.

Now by attaching the wires to the brushes, the current is used to strengthen the magnetic field between the poles. This wiring is called the **field circuit** of the generator.

This completes our basic generator.

HOW THE GENERATOR CONVERTS AC TO DC CURRENT

So far our basic generator has produced an alternating current.

This is because the armature **reverses** the polarity of the current and so changes the direction of current flow on each side of the loop as it rotates (Fig. 8).

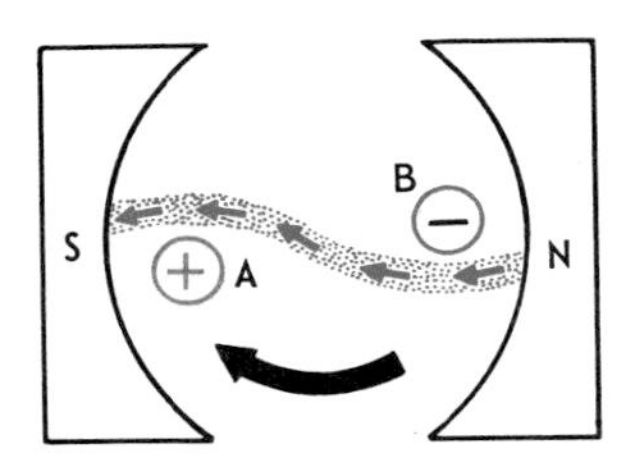

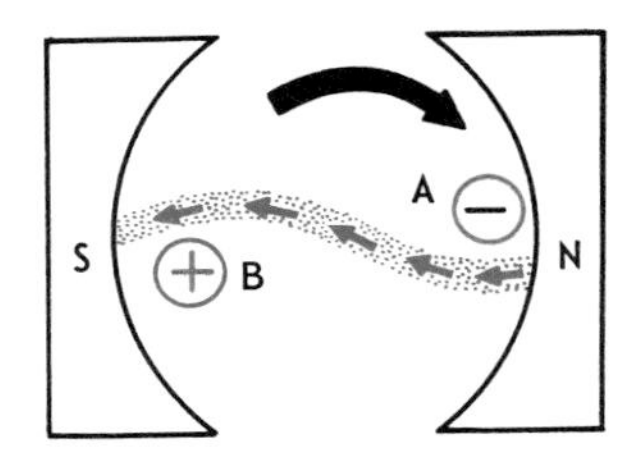

X 1458

Fig. 8—How the Polarity of the Armature Changes during Each Revolution

During the first half of its revolution in Fig. 8, the top of armature side A cuts through the magnetic field first, while the bottom of side B is first to cut the field. Using the Right Hand Rule, we find that current flows "toward" side A and "away from" side B. The Conventional Theory (+ to −) then gives us the polarities shown: (+) for A, and (−) for B.

During the second half of the revolution, the top of side B is the leading edge, while the bottom of side A is leading. Again using the rules, we find that B is now (+), while A is (−).

So we see that the *armature loop ends reverse polarity during each revolution.*

The result: **alternating current.**

How do we convert this AC current to DC?

Getting this AC current to flow to the load in the same direction (DC) is the job of the commutator and brushes (Fig. 9).

Twice during each rotation, the armature is vertical to the magnetic field as shown. Here the armature loop is not passing *through* the field and so no voltage is generated at this instant. This is the static **neutral point.**

The commutator is split into two parts with the open areas matching the neutral point of the armature as shown. This means that there is a gap as the commutator passes the brushes. Past this point the other half of the commutator contacts the brushes. Since the coil is in the same relative position as during the preceding one-half revolution, current flow to the brush stays in the same direction.

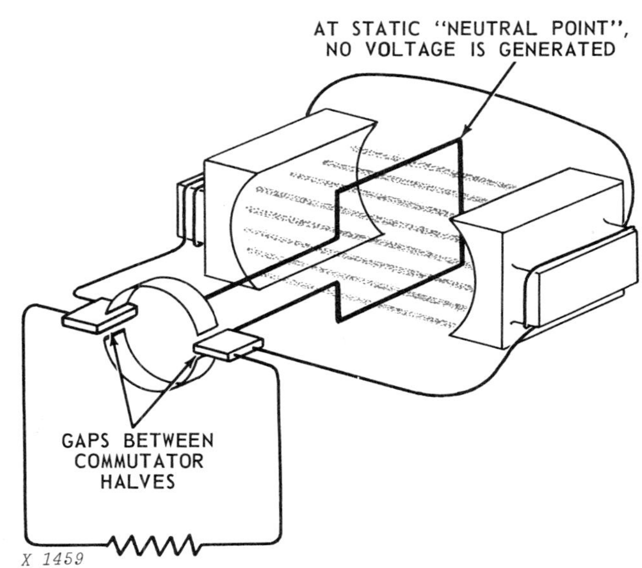

Fig. 9—How Generator Converts AC to DC Current

The result: **direct current (DC).**

This is how the generator converts AC current to DC current.

As we saw in Chapter 1, three factors decide how much voltage is generated:

1. *The strength of the magnetic field.*
2. *The number of wire conductors on the armature.*
3. *The speed of the armature.*

SUMMARY: HOW A GENERATOR WORKS

In summary:

- **Moving a conductor through a stationary field = basic generator.**
- **Basic generator = armature (rotating) + magnetic poles (fixed).**
- **Circuit is completed through commutator and brushes.**
- **Field circuit windings strengthen the magnetic poles.**
- **Commutator converts AC to DC current.**

Now let's look at the generator parts in more detail.

 Litho in U.S.A.

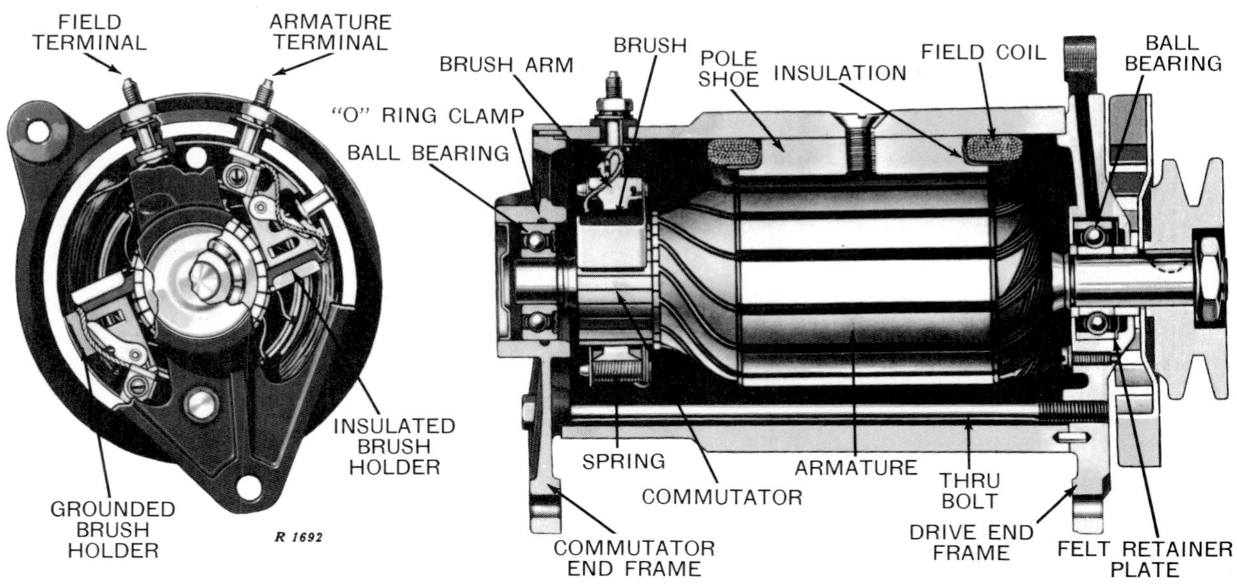

Fig.10—DC Generator in Cutaway View

ARMATURE

The armature is not just one wire loop or conductor, but many wire conductors. This means that a greater voltage is developed.

These conductors are wound around a core of soft laminated iron sections as shown in Fig. 10. The sections are then attached to the armature drive shaft.

The laminated iron sections are used in place of a solid core of iron to reduce heat. A solid iron core would generate unwanted voltage within itself. This would result in current flow called "eddy currents" which create excessive heat.

COMMUTATOR

The commutator must have a section to match each wire section of the armature.

So the commutator ring is composed of many sections or bars. (See Fig. 10.) Mounted on the end of the armature drive shaft, each section or bar is held together and separated from the adjacent bar by an insulating material. The ends of each wire conductor are connected to two adjacent commutator bars.

BRUSHES

The brushes are made of various materials, depending on the output needs of the generator. Fixtures in the generator hold the brushes so that their ends rub or ride on the commutator ring. Usually, spring pressure forces the brushes against the ring as the brushes wear.

POLE SHOES

The pole shoes are permanent magnets, fixed to the inside of the generator housing. Set directly across from each other, the two opposing poles set up a weak magnetic field.

FIELD CIRCUIT

The field circuit is composed of one wire conductor wound around both field poles many times. One end of the wire is attached to the brush; the other end to the field circuit terminal.

HOUSING

All generator components are enclosed in a metal housing. Most generator housings have openings at both ends. This allows air to pass through and cool the generator. On the pulley end of the drive shaft, there is usually a fan to force the air to circulate.

AUXILIARY UNITS

Three external units are an important part of generator operation.

We will briefly discuss these units now and cover them more extensively later in this chapter.

Cutout Relay

As you know, the generator recharges the battery as well as supplies current to the rest of the electrical system. To recharge the battery, a circuit must run from the generator to the battery. However, if this circuit was complete when the generator was *not* operating, the battery would discharge through the generator.

To prevent this, an automatic switch called a *cutout relay* is installed in the circuit. While the generator is operating, the switch is closed and the circuit is complete. When the generator is stopped, the switch opens the circuit.

Voltage Regulator

Voltage induced by the generator will go as high as necessary to overcome any resistance in the circuit. If resistance is high, voltage will be high, while low resistance means low voltage. But if voltage is too high, the field and load circuits may be damaged.

The generator cannot control the amount of voltage it produces. Therefore, an external unit called a *voltage regulator* is used in the field circuit. It has a shunt coil and contact points to control the strength of the magnetic field, thus limiting the voltage generated.

Current Regulator

Excessive current flow is caused by too little resistance and can also cause heat damage to the armature.

A *current regulator* is installed in the load circuit to control this current flow. This unit is very similar to the voltage regulator above.

Both the voltage regulator and current regulator are used, but while one is working the other is not. They never work at the same time.

All three units—cutout relay, voltage regulator, and current regulator—are usually housed together in one assembly.

ARMATURE REACTION

We have said that commutation takes place when the armature is in the "neutral" position. This is true of a one-loop armature (Fig 9).

However, an actual armature has more than one loop. In this case, "one loop at a time" is in the neutral position. Therefore, commutation occurs between the commutator sections connected to that loop. The other loops are still inducing voltage.

In describing the "neutral point," we said that it is perpendicular to the path of the magnetic field. It is obvious, then, that if the path of the field should change, the neutral position would also change. Therefore, the brush position would also have to be changed.

To see why, let's return to the basic operation for a moment.

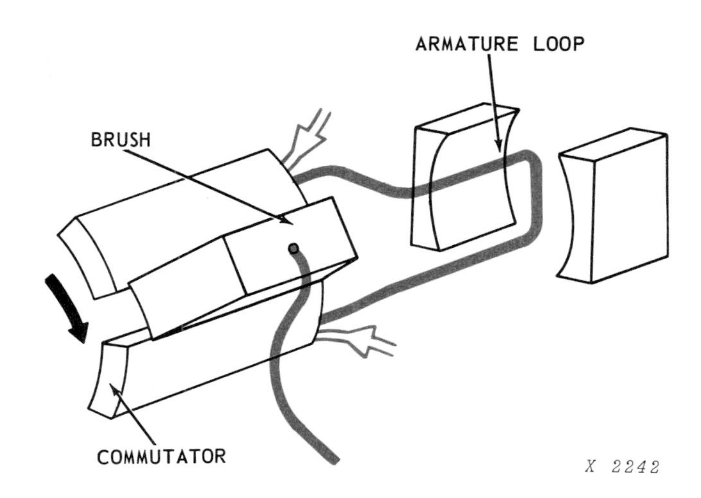

Fig. 11—Brush-Commutator Circuit (Normal Operation)

Remember that when the generator is producing voltage and current, the brush is in contact with the commutator section. This contact is like a completed circuit.

To get commutation, the commutator section beneath the brush is replaced by the section following it. In a sense, the circuit between the first section and the brush is broken—an open circuit.

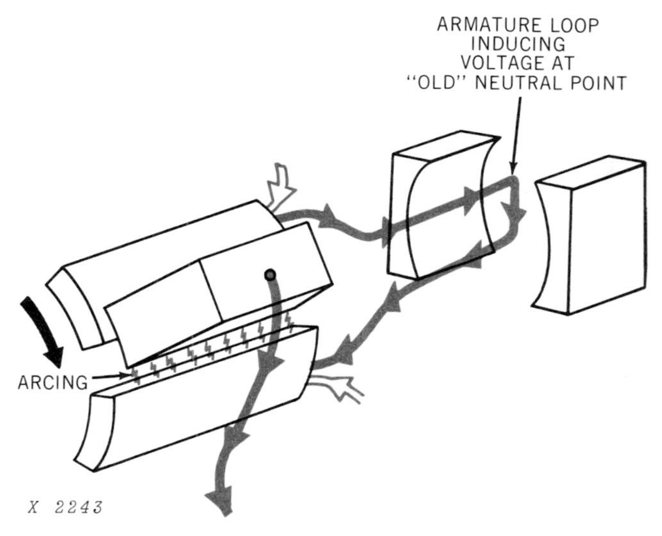

Fig. 12—Arcing Occurs When Neutral Point Is Changed

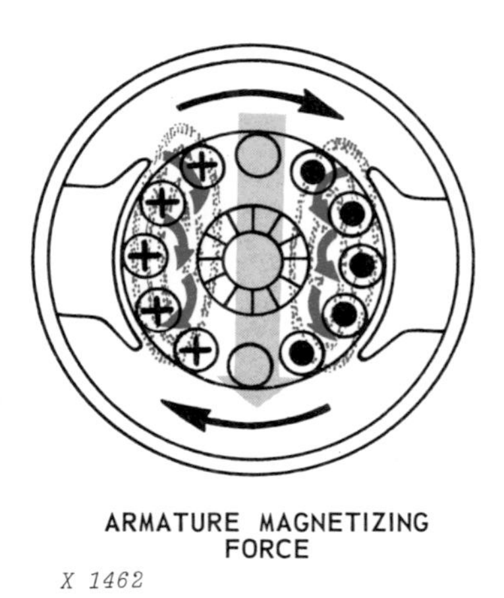

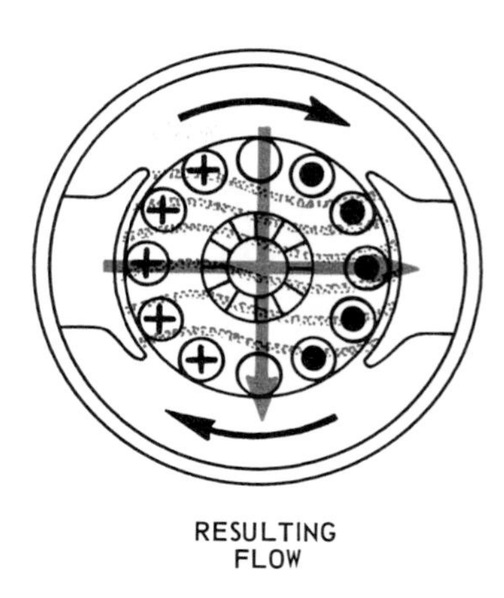

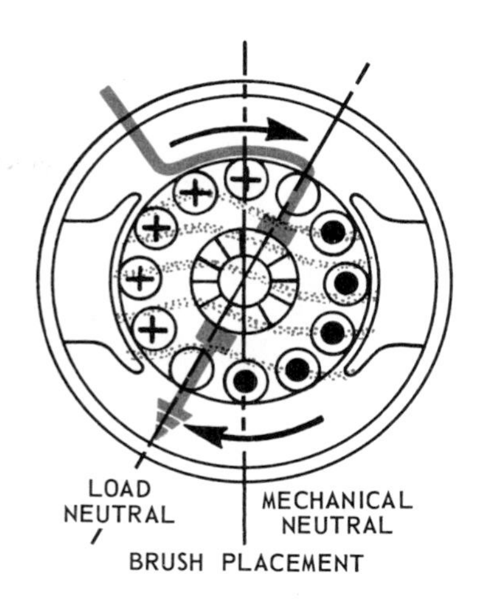

Fig. 13—Armature Reaction

During normal operation (Fig. 11), no voltage is induced in the loop attached to the commutator section at the time the connection is broken. However, if the voltage and current flow are still in that loop at this time, a spark or arc will appear between the brush and commutator section (Fig. 12).

This arcing is caused by the current which is attempting to cross between the section and brush. The duration of the arc is quite short because of the low voltage and rapidly expanding distance between brush and section. However, severe arcing will damage the brushes and commutator.

However, the brushes are set directly over the open area at the moment when the armature loop is in neutral position (Fig. 12).

But if the magnetic field path is distorted, the armature loop will conduct current during the commutation. The result: **arcing.**

Distortion of the magnetic field as shown really exists. This distortion is caused by the magnetic field set up around the armature conductors acting with the magnetic field of the poles. It is called **armature reaction.**

Fig. 13 shows the sequence of the armature reaction which distorts the field and how the brushes are placed to counteract it.

The illustration at left shows the magnetic field surrounding the coils of the armature that results when current flow is established in these coils. Remember that all the load current flows through the conductors of the armature, and the greater the current flow the greater will be the strength of the surrounding magnetic field.

The center diagram in Fig. 13 shows the magnetic field formed after combining the magnetic field of the armature with the field of the pole pieces. The change in the path of the magnetic field changes the neutral position to a new position under load.

To counteract this, the brushes must be located at the **load neutral** rather than at the **mechanical neutral.** This is shown at right in Fig. 13. The load neutral is located after the mechanical neutral during rotation as shown. At a constant speed and load, the new load neutral is the ideal commutating point. However, with varying speeds and loads, the load neutral point is constantly changing. For this reason a brush position is selected which will be the *best average location* that will create the least arcing at the brush under normal operating conditions.

TYPES OF GENERATORS

Basically, all generators are alike. They all use the basic parts we have described.

However, there are variations, each using the basic components in different ways or quantities.

The different types of generators are:

- **Shunt—standard for most uses**
- **Third Brush—needs no current regulator**
- **Interpole—provides better commutation**
- **Bucking Field—for changing loads and speeds**
- **Split Field—for low speeds but high loads**

Let's discuss each type of generator.

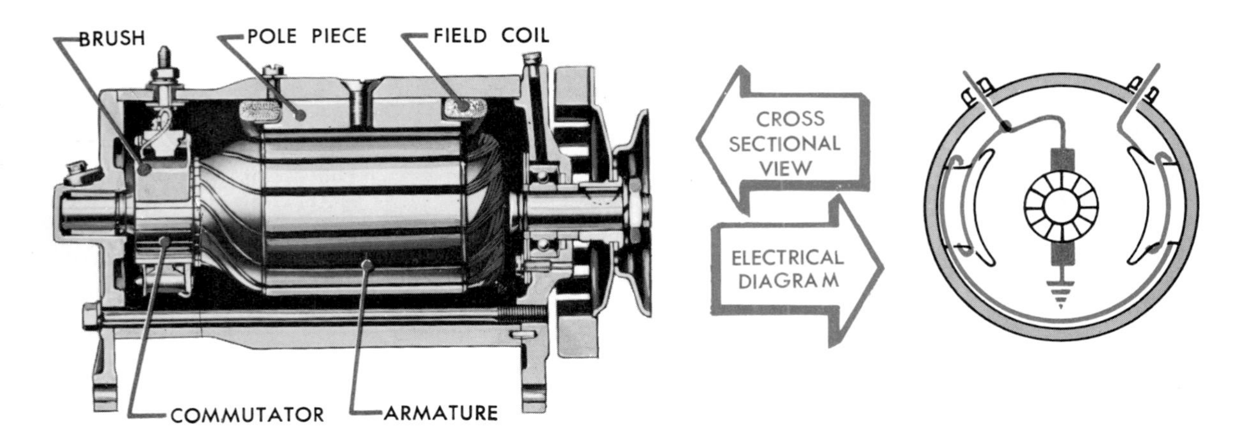

Fig. 14—Shunt Generator

SHUNT GENERATOR

The **shunt generator** is similar to the basic generator we described earlier. It is a two-pole, two-brush unit with an armature, a commutator, and field and load circuits (Fig. 14).

Operation is also the same: Voltage is induced by an armature rotating through a magnetic field, current is sent through the commutator into the field and load circuits, via the brushes.

A shunt generator uses an external voltage regulator to control the field circuit. The name "shunt" refers to the field coil which is in parallel with the armature.

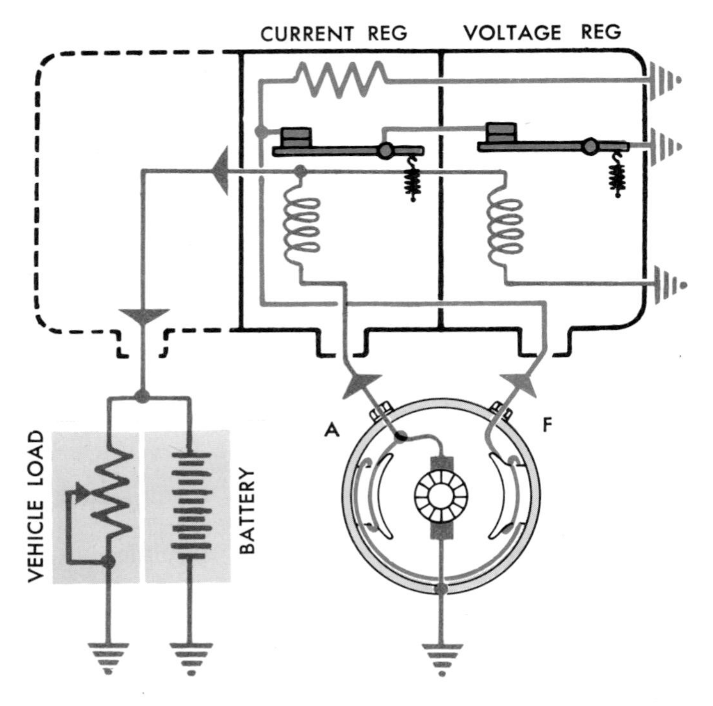

Fig. 15—Generator Field Circuit

The generator field circuit is shown in Fig. 15.

In this circuit the voltage regulator points are located *after* the field coils. The field circuit is grounded on one end at the generator regulator. The other end is attached to the insulated brush inside the generator. This circuit is called an "A" circuit.

NOTE: Some generators have field circuits in which the regulator points are located before the field coils and the field coils are grounded inside the generator. These models are called "B" circuit generators. They are not used on farm and industrial machines at the present time. In this chapter we will cover only the common "A" circuit generators.

THIRD BRUSH GENERATOR

The **third brush generator** has three brushes instead of two, and uses another means of controlling the flow of current.

The two main brushes, insulated and grounded, are positioned at the neutral point on the commutator to obtain the maximum voltage. The third brush is placed between the other two which means that it picks up less than maximum voltage.

As you can see in Fig. 16, the two main brushes (shown in red) are in the load circuit. The third brush (shown in blue) is connected to the field coil.

On most units the third brush position is adjustable as shown. By moving the brush toward the insulated brush, voltage is increased across the field circuit. Moving it away from the main brush decreases the voltage.

When a third brush is used, current is controlled in two ways:

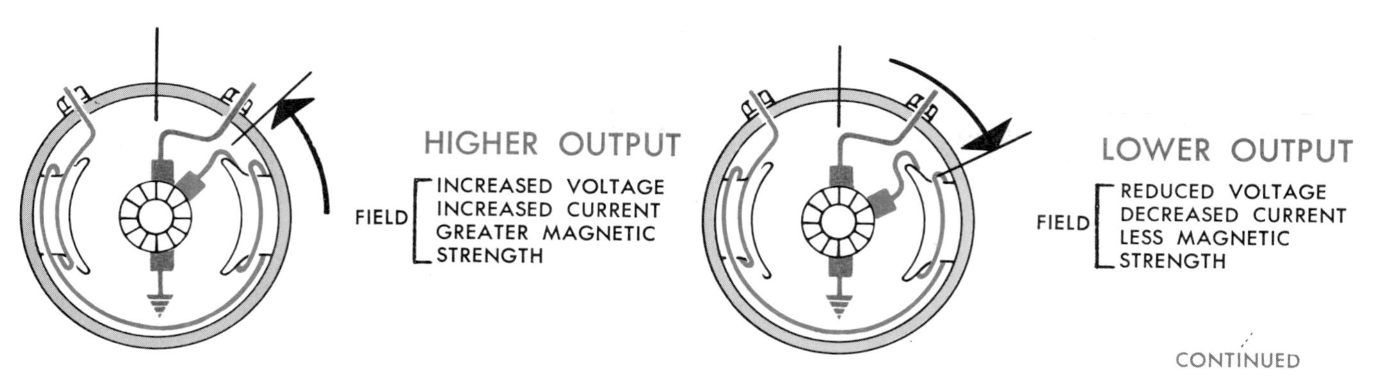

CONTINUED

Fig. 16—Third Brush Generator Output Adjustments

1) The third brush picks up less voltage; therefore, less current flows through the field circuit. This means a weaker magnetic field.

2) At greater loads and speeds, the magnetic field is more distorted so fewer lines of force are cut by the armature. This means that less voltage and current are developed in the field circuit and coils.

The interpole generator is quite similar to the shunt model, but has an extra magnetic pole shoe (Fig. 17). This changes the operation and solves a problem of locating the brushes at load neutral on the shunt generator.

At one time, the third brush generator was widely used in the automotive industry. It is relatively simple in design and by shifting the third brush, its output can be controlled. It can also reach its peak output at medium speed.

Today, however, the demands of electrical power have increased. Often the third brush generator cannot meet these demands. Though it reaches its peak at medium speed, power begins to decrease at higher speeds. For this reason it is now used primarily on slow speed equipment that has a low power requirement.

INTERPOLE GENERATOR

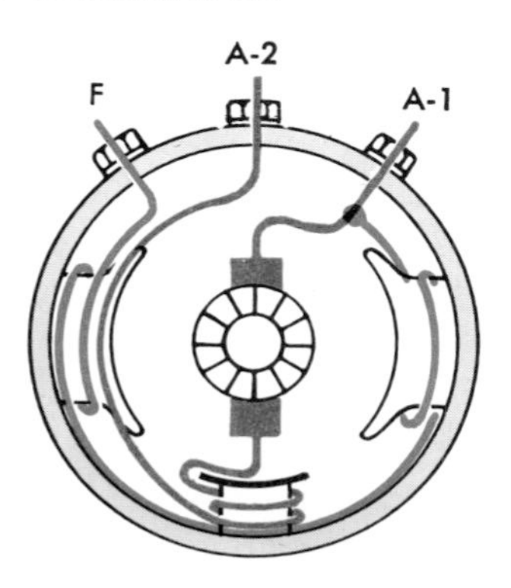

Fig. 17—Interpole Generator

The **interpole generator** is quite similar to the shunt model, but has an extra magnetic pole shoe (Fig. 17). This changes the operation and solves a problem of locating the brushes at load neutral on the shunt generator.

Earlier we explained that the best commutation took place when the armature was at the "neutral point." We also told how, because of magnetic field distortion, an ideal commutation point for all loads could not be reached.

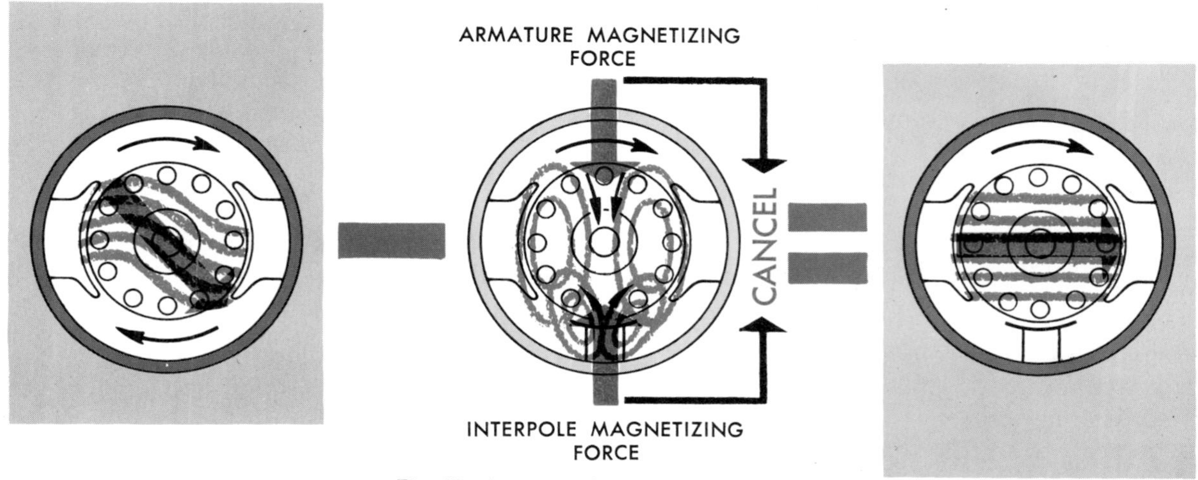

Fig. 18—Interpole Generator in Operation

This is true except when using the interpole generator. By adding another pole shoe, the distortion of the magnetic field is corrected.

By installing this extra magnet between the other two shoes, *the distorted magnetic field created by the current flow through the armature is neutralized.*

The interpole is wound with heavy copper wire since all armature current flow goes through this coil. The number of coil turns is calculated to produce enough ampere turns in the opposite direction to offset the magnetic field created by the armature.

Now the magnetic field lines are again in a straight line as shown in Fig. 18. The best commutation point is again at the "neutral" or "**mechanical** neutral point" and the brush position is matched to this.

Besides providing the ideal commutation point, the interpole generator can greatly increase the brush life over a non-interpole generator.

BUCKING FIELD GENERATORS

The **bucking field generator** is another variation on the standard shunt generator. But it can develop the voltage for high current at very low speeds.

Actually, getting a high current at low speeds is comparatively easy. (The problem is to control excessively high voltage at high speeds.)

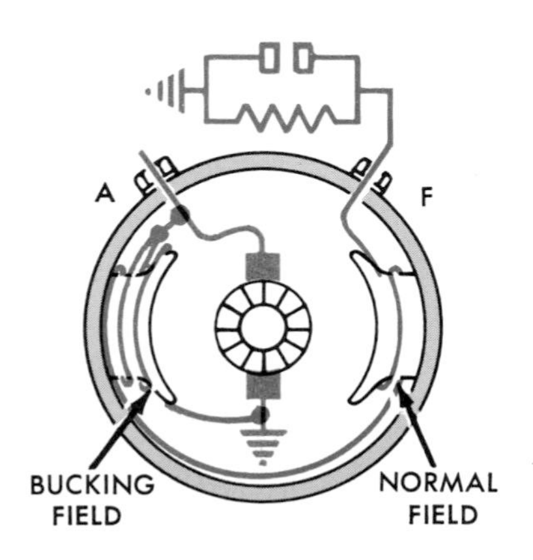

Fig. 19—Bucking Field Generator

To control the high voltage, a "bucking" field coil is used (Fig. 19). This coil is a high-resistance conductor wound around one pole shoe and connected across the brushes as shown. The coil is connected in reverse of the main field circuit. This produces a magnetic effect which *opposes* the normal magnetic field.

At low speeds, the bucking field has little or no effect on the magnetic field and the generator is able to produce high voltage and current.

As speed increases, the bucking field magnetism becomes greater and reduces the magnetic field lines of force. With a reduced field, the armature's voltage is lowered and the voltage output of the generator drops.

In effect, the bucking field circuit helps the voltage regulator to control voltage. It reduces the residual magnetism in the pole shoes while the regulator controls the magnetism and current flow in the field circuit.

Although the bucking field generator is able to provide adequate electrical power at low speeds, its primary use is in a system that has a wide variation in speed.

SPLIT FIELD GENERATOR

Some generator uses require a high power output at slow speeds for long periods of time.

These systems are usually on buses, farm machines, etc., that operate at low speeds or sit with the engine idling.

The **split field generator** works for these operations.

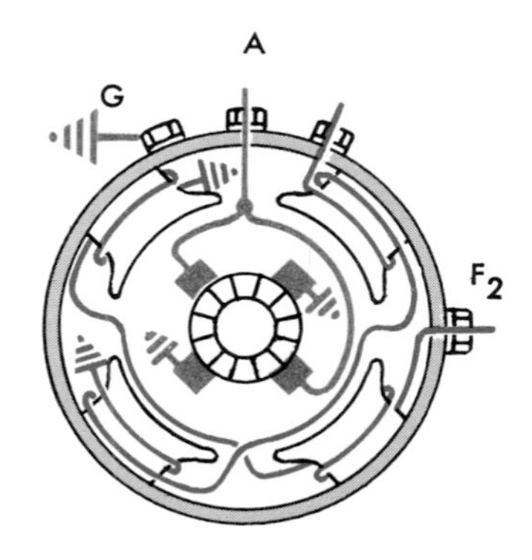

Fig. 20—Split Field Generator

The split field generator has an extra set of pole shoes and brushes (Fig. 20).

By doubling the shoes and brushes, a stronger magnetic field is created. This gives the extra voltage for low-speed charging and load current.

Generator output is doubled by separating the two fields and using a voltage regulator for each field.

Yet, each field circuit receives only the normal voltage and current so no damage is done to the regulators.

Summary: Types of Generators

Let's review the types of generators:

1. **Shunt**—used as a standard generator for most normal operations.

2. **Third Brush**—eliminates the use of a current regulator. Is relatively easy to change third brush position and control the output. Used in systems with low speed and low load requirements.

3. **Interpole**—provides a better commutation point and extends brush life.

4. **Bucking Field**—used where there is a wide variation of load and speed requirements.

5. **Split Field**—used in systems with low speed, but high load requirements.

USES OF GENERATORS

Why is one generator used for one application, but not in another?

We have already touched on the uses of generators when discussing the various types.

Now let's go a little deeper and see the *why* of generator applications.

Here are the key factors in selecting a generator:

- **Power requirements**
- **Operating conditions**
- **Service and maintenance needs**
- **Drive ratio—pulley to generator**

Below we'll discuss each of these factors.

Power Requirements

One of the most important factors to consider is the power requirements. The best rule is to select a generator that can provide 10 to 20 percent more than the total load requires. The extra output can be used to recharge the battery even when full load output is required. Also, the total load requirements would not take into consideration momentary load demands such as cranking motor, cigarette lighter, and other accessories.

Operating Conditions

Operating conditions are another prime consideration in generator selection.

Extremely dirty or damp conditions often require that the generator be fully enclosed, and the output of an enclosed generator is sharply reduced compared to a ventilated one of the same size.

High temperatures around the generator can add to the heat developed within the generator and reduce the output as well.

Service and Maintenance Needs

The type of service required of the generator and the type of maintenance on it are important factors, too.

The generator may be subject to long use at high, low, or variable speeds and loads.

The cost of repair parts, the length of service they give, and the ease in which they can be replaced are also important.

Drive Ratio—Pulley to Generator

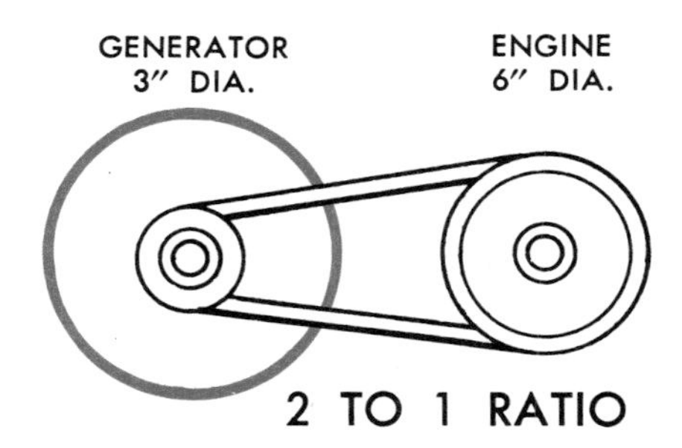

Fig. 21—Example of Drive Ratio—Engine to Generator

The drive ratio between the engine pulley and generator is also quite critical (Fig. 21). Armature speed is determined by this ratio.

A high armature speed may cause damage by developing more voltage than the regulator can handle. A low speed may not provide the necessary voltage output.

The drive ratio also affects the operation of the cutout relay switch. Due to varying idle speeds, the relay should be actuated at speeds of 100 rpm below or above idle speeds.

If the relay is set to operate at a *precise* idle speed, the relay points will constantly open and close due to this variation. The relay points would soon burn up in this kind of operation.

Summary: Uses of Generators

In summary:

- **Total load + 10-20 percent = size of generator required.**
- **Dirty or damp conditions = enclosed generator required.**
- **Heat conditions = ventilated generator required.**
- **Engine-to-generator pulleys = critical drive ratio.**
- **Drive ratio = speed of generator rotation.**
- **Too fast rotation = too much voltage for regulator.**
- **Too slow rotation = not enough voltage output.**

TESTING AND SERVICING OF GENERATORS

Like any other piece of equipment, generators are subject to failures and disorders.

To discover just what in particular ails them, there are several tests that can be made on the basic generator.

First, you should always test the complete circuit before removing the generator to the bench. This will insure that the generator is actually the "sick" member of the charging circuit team.

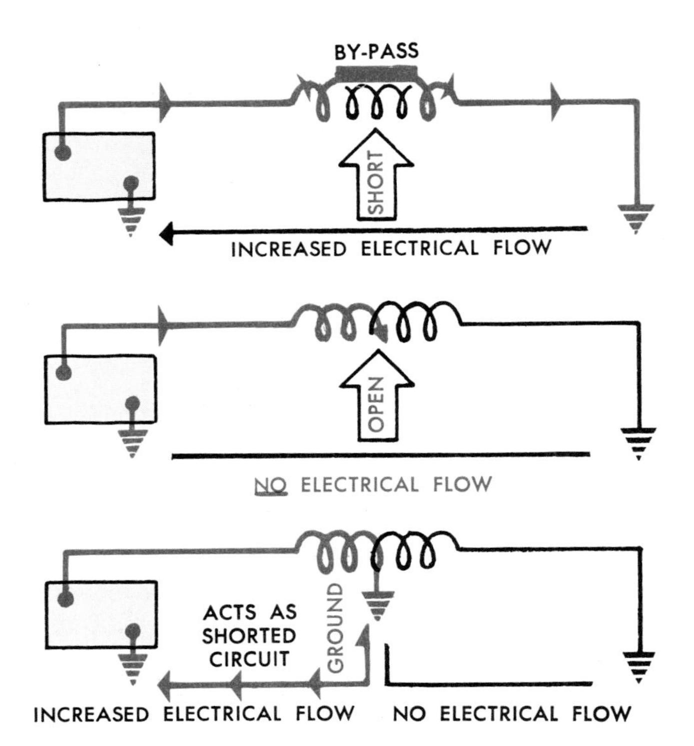

Fig. 22—Basic Failures of Electrical Circuits

Types of Failures

The four basic electrical failures are:

1. *Short Circuits*—These are unwanted connections, usually copper-to-copper, that allow current to bypass all or part of the circuit. (See Fig. 22, top.)

2. *Open Circuits*—These are breaks in the circuit which cause extremely high resistance. Usually no current will flow through an open circuit.

3. *Grounded Circuits*—These are unwanted connections that bypass all or part of the circuit from the insulated side to the grounded side of the circuit; usually a copper-to-iron connection.

4. *High-Resistance Circuits*—These are usually caused by poor or corroded connections, and frayed or damaged wires, all of which create greater resistance in the circuit. See Fig. 23.

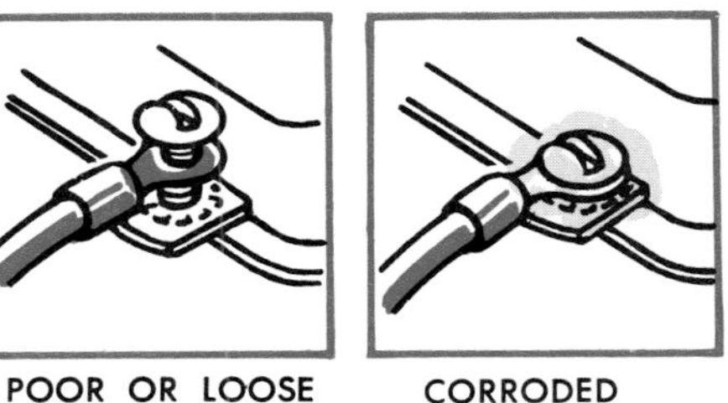

Fig. 23—Causes of High Resistance

After you have isolated the cause of the problem as a "sick" generator, test it thoroughly out of the circuit. The tests can tell you which component in the generator is failing.

To perform these tests, several types of testing units are necessary. Each test may require just one or a combination of these units.

However, one item is necessary for nearly all the tests—the manufacturer's Technical Manual for the machine. This manual is kept up-to-date and lists all the generator specifications plus servicing and repair procedures. If in doubt, always consult the Technical Manual. We will only give general specifications here.

Generator Output Test

A generator output test is an overall test which generally tells which internal component is malfunctioning.

To test the output, connect the units as shown in Fig. 24.

1. Connect an ammeter and switch in series with a battery to the generator output terminal.

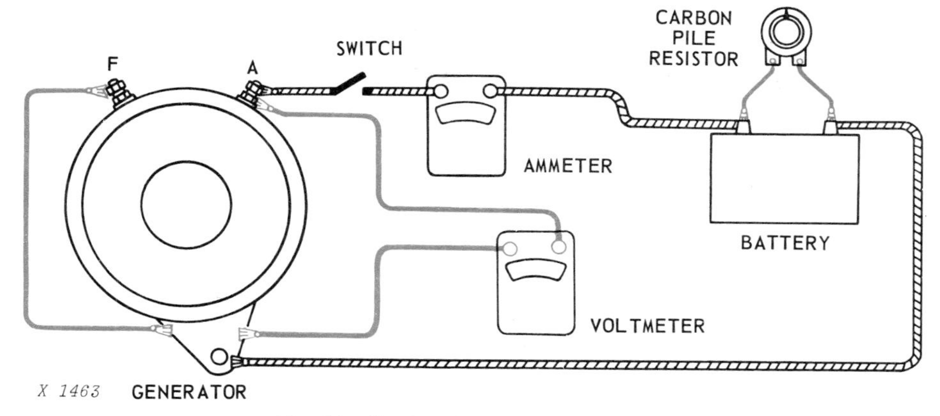

Fig. 24—Testing the Generator Output

2. Connect a voltmeter from the generator output terminal to ground as shown.

3. Connect a carbon pile across the battery.

4. Connect a jumper lead to the generator field terminal.

5. Operate the generator to obtain battery voltage and close the switch.

6. Speed up the generator to its rated value, adjust the carbon pile to obtain the specified voltage, and compare the current output with the generator specifications.

7. If the generator output is below par, disassemble it for further testing.

NOTE: The output test shown in Fig. 24 is for the common "A" circuit generator. For "B" circuit models, a different test hookup is used.

Generator Troubleshooting Chart

Below is a general list of generator failures and their possible causes.

NO OUTPUT

Sticking brushes

Dirty or corroded commutator

Loose connections

Grounded, shorted, or open armature

Grounded, shorted, or open field circuit

Grounded terminals

EXCESSIVE OUTPUT

Grounded or shorted field circuit

VARIABLE OR LOW OUTPUT

Loose or worn drive belts

Low brush spring tension

Dirty or burned commutator

Eccentric or worn commutator

Partial grounded, shorted, or open armature

Partial grounded, shorted, or open field circuit.

NOISY GENERATOR

Loose mounting

Loose pulley

Worn or dirty bearings

Improperly seated brushes

From this list, make the easy checks first. A good visual check of all the components may help you to catch the more obvious causes of failures such as:

Loose or broken lead wires—dirty or worn brushes — improperly seated brushes — broken brush springs — worn or glazed commutator — loose pulley — copper-to-copper contact of conductors in either the armature or field circuit (shorted) — broken conductors in the armature or field circuits (open) — copper-to-iron contact of conductors in either the armature or field circuit (grounded).

If these failures are not visible or cannot be corrected, then make detailed tests on each component.

These component tests are given on the following pages.

Armature Tests

Four possible failures of the armature are:

1. *Open armature circuits.*
2. *Shorted armature coils.*
3. *Grounded armature coils.*
4. *Dirty or worn commutator.*

Let's review each of these failures.

OPEN ARMATURE CIRCUITS

An open circuit in the armature coil will cause severe arcing at the brushes and commutator. Another indication is when it requires twice the rated speed to attain the specified voltage and current.

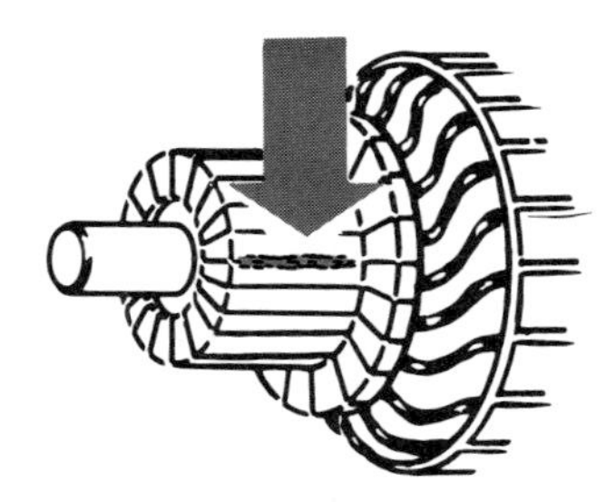

Fig. 25—Checking for Open Circuit in Armature

The broken winding can be found by inspecting the commutator (Fig. 25).

The trailing edge of the bar attached to the open loop will usually be badly burned as shown.

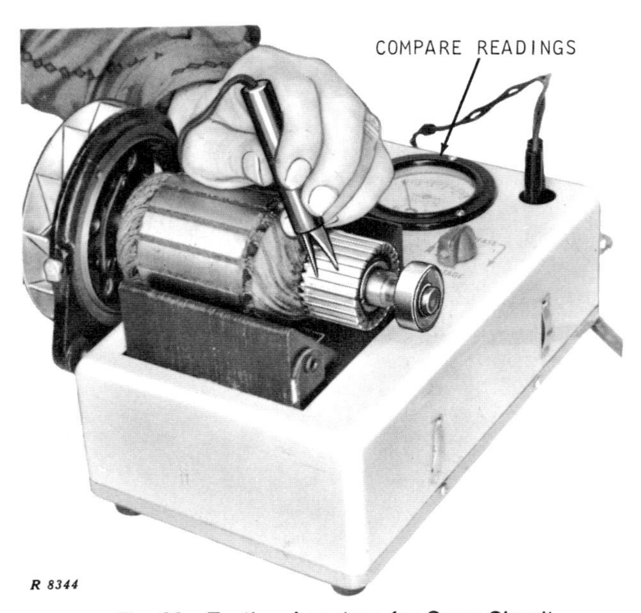

Fig. 26—Testing Armature for Open Circuit

Another method of finding the broken loop is with a combination test unit and growler (Fig. 26).

Place the test prods of the unit on two adjacent commutator bars. Rotate the armature to get the highest meter reading and record the reading.

Then turn the armature slightly and check two other bars. If one reading is lower than the others, that circuit is open.

SHORTED ARMATURE COIL

Another cause of low generator output is a short-circuited armature coil. If the armature loops are touching each other, they set up a closed circuit and very little current can get to the external load circuit. The more loops shorted, the higher the speed required to obtain voltage and current. Long use of a shorted armature can create damage from overheating.

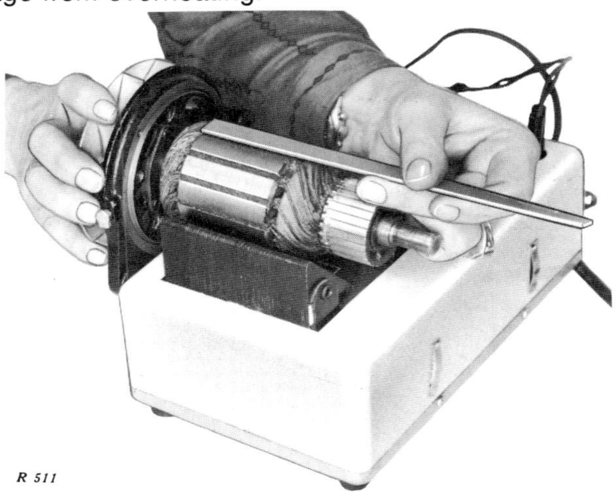

Fig. 27—Testing for Shorted Windings

Shorted windings can be found with the growler test unit (Fig. 27). The growler has an oscillating magnetic field which cuts across the armature loops, creating voltage and current flow through the shorted loops.

By holding a strip of metal such as a hacksaw blade over the armature and slowly rotating the armature in the growler, the magnetic field set up by the shorted windings will attract and release the metal strip. This will tell you that the armature coil is shorted.

GROUNDED ARMATURE

A grounded armature is another source of current loss. The armature is grounded when the winding is in contact with the ground. This bypasses the external load circuit and, since the grounded connection has low resistance, the remaining voltage is very low.

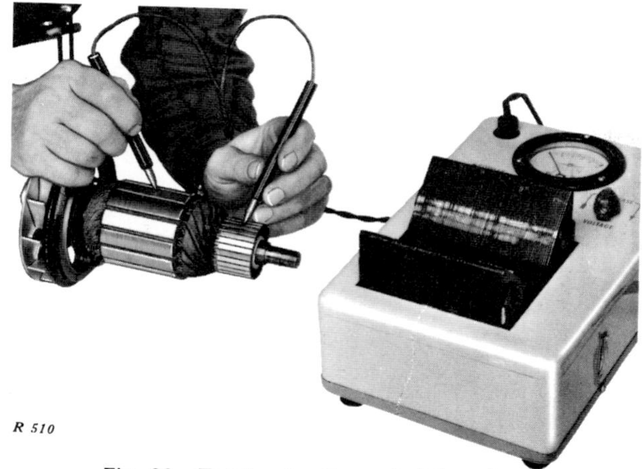

R 510

Fig. 28—Testing for Grounded Armature

To find a grounded armature, use the test lamp on the growler, if equipped (Fig. 28). Or use a separate test lamp.

Place one test prod on the metal of the armature and the other prod on the commutator. If the lamp lights, the winding is grounded.

DIRTY OR WORN COMMUTATOR CAUSING ARMATURE FAILURE

A dirty or oxidized commutator can cause symptoms that appear as if the armature were shorted or grounded.

Foreign material packed between the commutator bars can offer a path for current flow, thus shorting the armature loops.

A corroded commutator can prevent brush contact, thus prohibiting voltage development.

RECONDITIONING THE COMMUTATOR

To correct this, turn down the commutator with an armature turning tool (Fig. 29).

Also be sure to undercut the material between each bar as shown. This will prevent rapid brush wear due to bouncing or arcing brushes.

Use a strip of No. 00 sandpaper to polish the commutator. *Do not use emery cloth.*

Armature Repair

The commutator can be turned down with the proper tools. The armature, however, is seldom rewound in cases of shorts, opens, or ground.

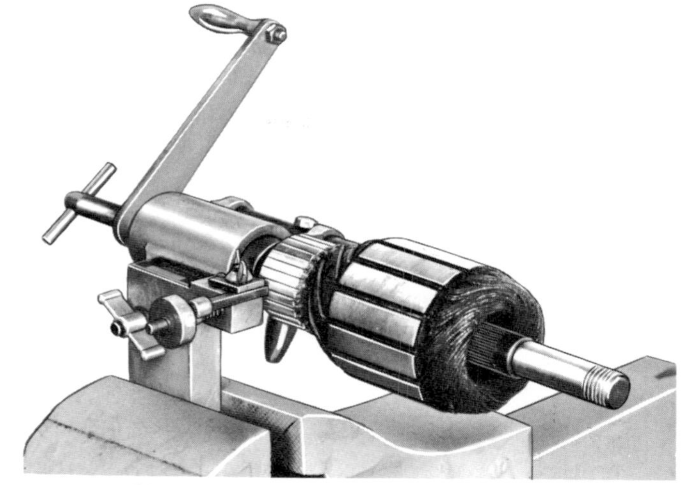

X 1464 TOOL FOR TURNING DOWN COMMUTATOR

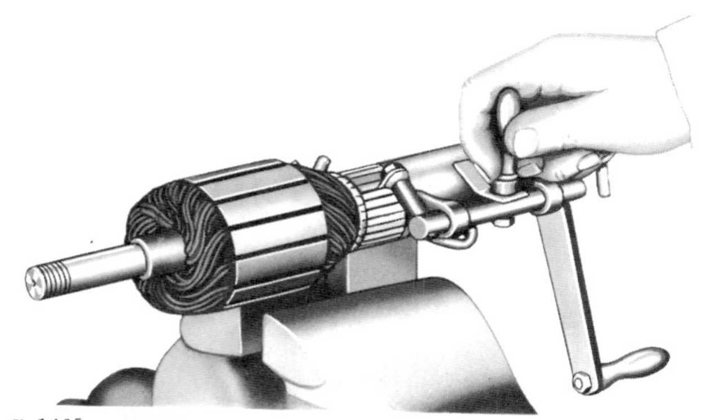

X 1465 UNDERCUTTING MICA ON COMMUTATOR

Fig. 29—Reconditioning the Commutator

Normally, it is replaced with another armature. Of course, if the proper tools and qualified personnel are available, then it can be repaired.

Field Circuit Tests

Like the armature, the field circuit can fail due to open, grounded, or shorted coils. It can also fail because of high resistance.

OPEN FIELD CIRCUITS

A broken coil or *open circuit* in the field circuit stops current flow.

This means the magnetic field cannot be strengthened and the amount of developed voltage is dependant on residual magnetism. This voltage is too low to supply the battery.

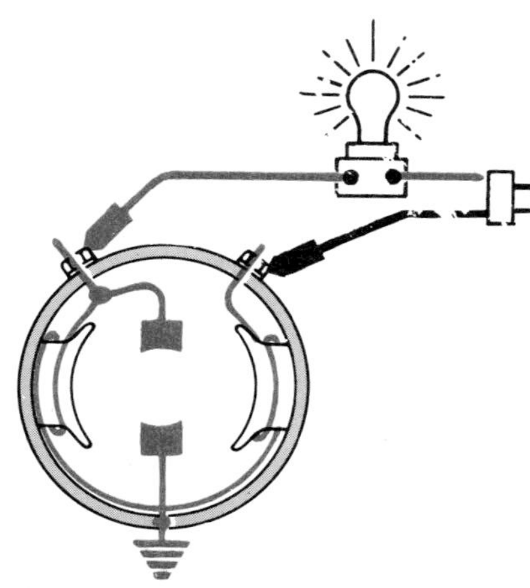

Fig. 30—Testing for an Open Field Circuit

Use a test lamp to find an open field circuit. Touch the prods of the lamp to the generator as shown in Fig. 30. If the lamp does not light, the field circuit is open. In this case, look for a broken wire in the coils or leads.

GROUNDED FIELD CIRCUIT

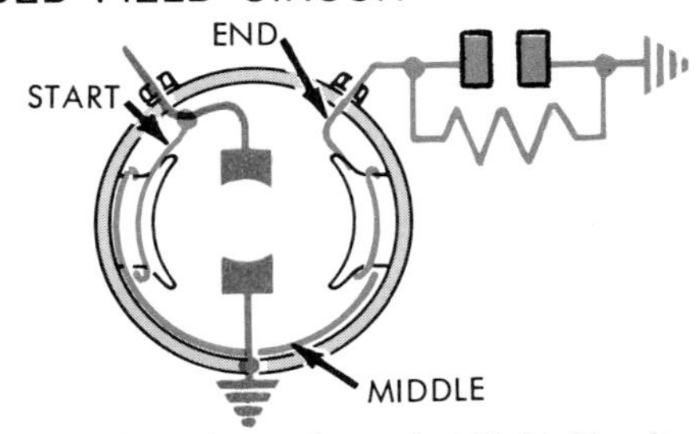

Fig. 31—Locating a Grounded Field Circuit

A *grounded field circuit* will have a different effect on generator output depending on where the ground is located (Fig. 31).

Let's assume that all circuits start at the insulated brush.

If the circuit is grounded before the coils, there will be no current flow through them. Therefore, the magnetic field will not be strengthened and no voltage will develop. In this case the armature will also be grounded since the field circuit is connected to the armature circuit.

If the field is grounded at the middle, the circuit will still build up voltage. However, with less resistance in the field circuit, more current will flow. The ground will bypass the regulating points and regulator resistance and there will be no control over voltage or current.

If the circuit is grounded after the coils, the circuit will react similar to the ground at the middle and there will be no control over voltage.

TESTING FOR A GROUNDED FIELD CIRCUIT

A test lamp is used to find out if the field circuit is grounded. All *intended* ground connections of the field circuit must be disconnected.

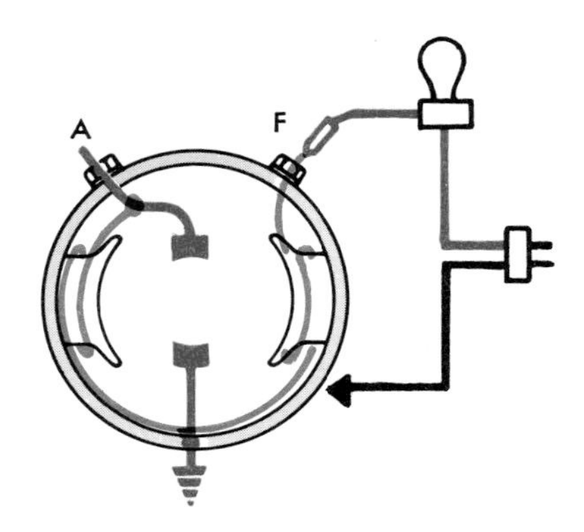

Fig. 32—Testing for Grounded Field Circuit

Place the prods of the test lamp as shown in Fig. 32.

If the lamp lights, the field circuit is grounded. Add this information to what is known about the effect a grounded circuit has on generator output, and you should be able to locate the grounded wire.

For a grounded circuit, either replace the field windings or repair them by reinsulating, revarnishing, and retaping.

SHORTED FIELD CIRCUIT

A *shorted field circuit* may cause excessive current flow. The number of shorted coils cuts resistance and so creates a higher current flow. This higher current can burn and oxidize the regulator points, creating an open circuit.

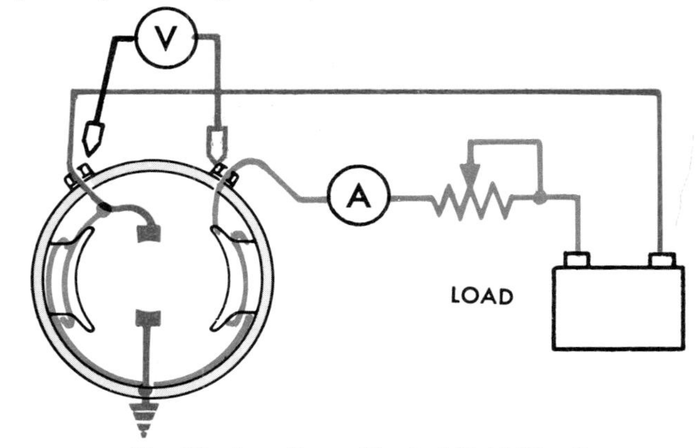

Fig. 33—Locating a Shorted Field Circuit

Use a volt-ampere tester with a 100-amp variable resistance unit for this test. Refer to the machine Technical Manual for the proper test readings.

1. Follow the test unit manual for the proper connections. (Fig. 33 shows a general hookup for the two types of field circuits.)

2. Adjust the variable resistance unit so that the voltmeter reads the specified voltage as given by the technical manual specifications.

3. Measure the current flow by the ammeter. It should be the same as the Technical Manual specifications. Any difference between the reading and specifications, either way, indicates a bad field circuit.

If the current flow is more than specified, the field circuit is shorted.

If the current flow is less than specified, there is high resistance (a damaged wire or loose connection) in the field circuit. This high resistance cuts the magnetic field strength and so reduces the generator output.

In either case, repair or replacement of the field circuit coils is required. Fig. 34 shows a typical set of generator field coils.

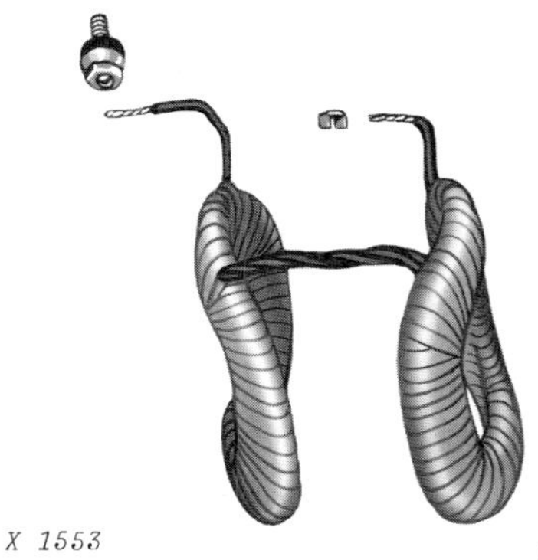

Fig. 34—Field Circuit Coils

Brush Testing

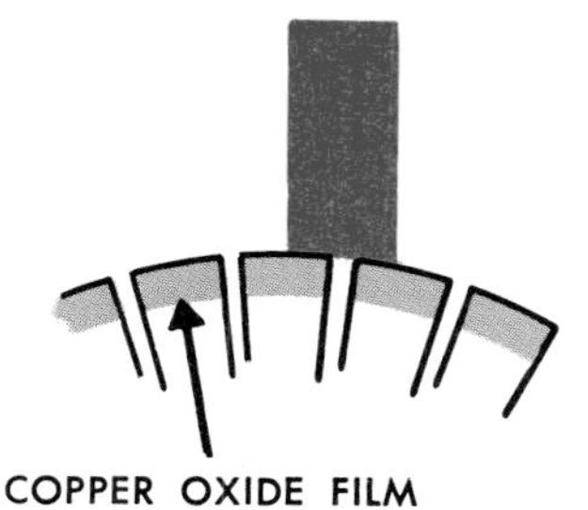

Fig. 35—Copper Oxide Film on Commutator

During normal operation a copper oxide film is formed on the commutator (Fig. 35). This film reduces the friction between the brushes and commutator. At low or no current loads, the film can be worn away by the brushes and the brushes will then chatter.

Continual brush chatter can loosen the brush lead, forming a high-resistance connection. Current will then follow an easier path through the brush surface and brush holders. This can eventually lead to burned brushes, brush holders, and brush arms.

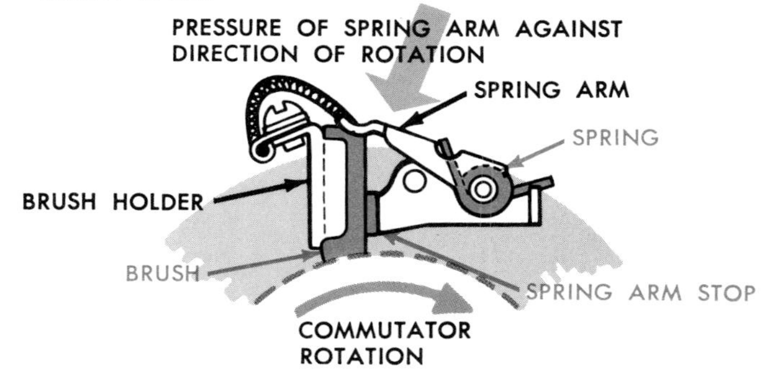

Fig. 36—Brush Arm Stop

A brush arm stop prevents the brush arm from pressing on the brush when the brush is worn away. However, the main purpose of the stop is to prevent the arm from scoring the commutator. But if the brushes are checked regularly for wear, this should never happen.

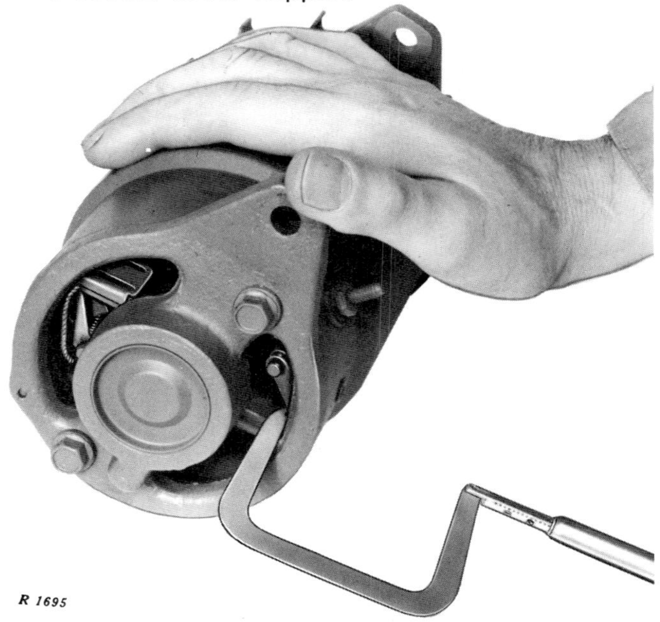

Fig. 37—Checking Brush Spring Tension

Brush arm spring tension should be checked during generator repair (Fig. 37). The Technical Manual can give you the proper tension specifications. Be sure to measure tension at a point as close to the middle of the brush as possible.

Too little tension will cause the brushes to bounce and arc at high speeds. Too strong a tension will cause excessive friction and brush wear.

 Litho in U.S.A.

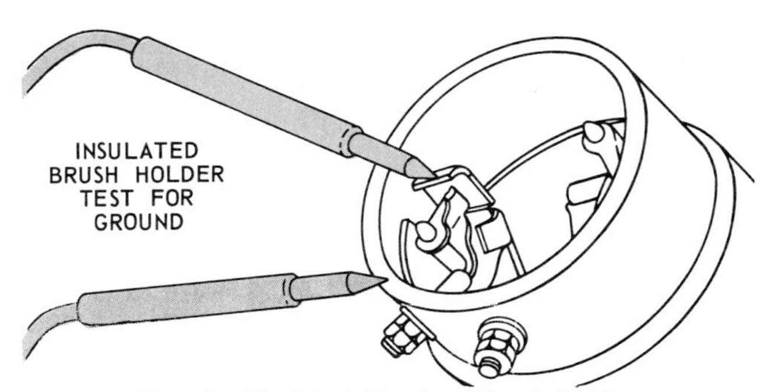

Fig. 38—Electrical Check of Brush Holders

To check for grounds, or opens on the brush holders, use a test lamp (Fig. 38).

Place the test prods across the *grounded* brush holder and the frame to check for opens. The lamp should light. If it does not, the brush holder is insulated from the frame and the circuit is open.

Place the lamp across the *insulated* brush holder and frame to check for grounds. The lamp should not light. If it does, a ground is indicated at the frame and the brush holder is grounded at this point.

Replacing Brushes

Replace the brushes when they have worn to one-half of their original length. When replacing brushes, be sure that a brush seat is obtained across the thickness of the brush. *A 25 percent contact area is satisfactory.*

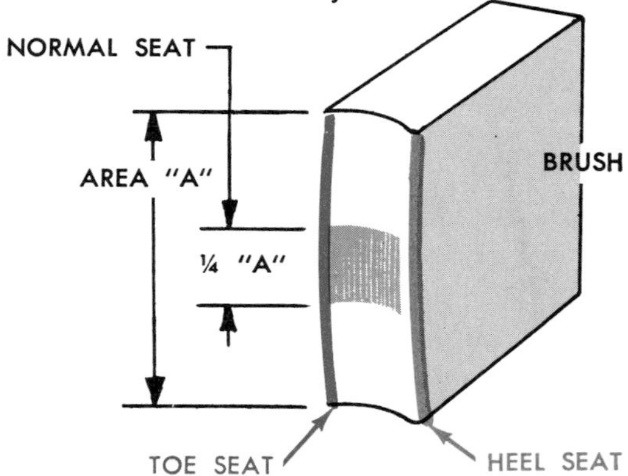

Fig. 39—Seating of Brush for Good Contact

If the brush seats only on the edges, this is not satisfactory regardless of whether it is a 25 percent contact.

Fig. 39 shows in color the bad heel and toe seats at the edge of the brush. The grey area at the center of the brush is a good seat. Seating on the edges will change the neutral position and severe arcing can occur.

Use seating stones or compounds to get a perfect fit between the brushes and the commutator.

POLARITY OF THE GENERATOR

Polarity is the direction of current flow through the generator. It is determined by the magnetic pole shoes. If the polarity of the pole shoes is changed, current flow will also change, regardless of whether other conditions change or not. Therefore, by changing pole polarity the generator can supply load current in either direction.

Pole shoe polarity is determined by the magnetism of the field coils the last time current passed through the coils. Anytime a current flows through the field coils, it sets the pole shoe polarity until the next time current flows.

This sets up a potential trouble spot. When working on or testing the generator you can accidentally change pole polarity. Even a slight current through the coils can do it.

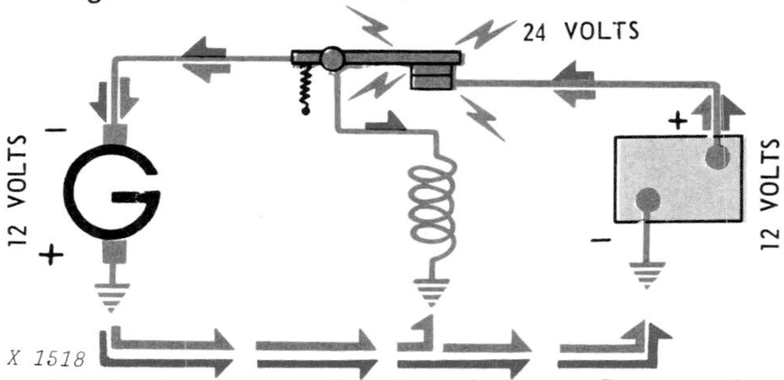

Fig. 40—How Reverse Polarity in Generator Damages the Charging Circuit

The damage does not occur until the generator, with a polarity opposite that of the battery, is put back into the charging circuit (Fig. 40).

When this is done, the generator builds up voltage and closes the cutout relay points. This, in effect, puts the battery in series with the generator and their voltages are added together. This high voltage across the points (about twice the system voltage) can create high current and enough heat to weld the points together as shown.

However, this damage does not happen immediately. The instant the points close, voltage is about the same on both sides of the relay coil. Since there is little difference in voltage, very little current will flow and spring tension opens the points. But generator voltage will again close the points and the cycle is repeated. Because this action repeats very rapidly, heat and arcing will finally cause the points to weld together.

When the points do weld together, the battery and the generator are connected at all times. The low resistance in the generator allows the battery to continue to discharge into the generator. This develops a high current which creates enough heat to burn the armature, ruining it.

In summary, **after any service, polarize the DC generator.**

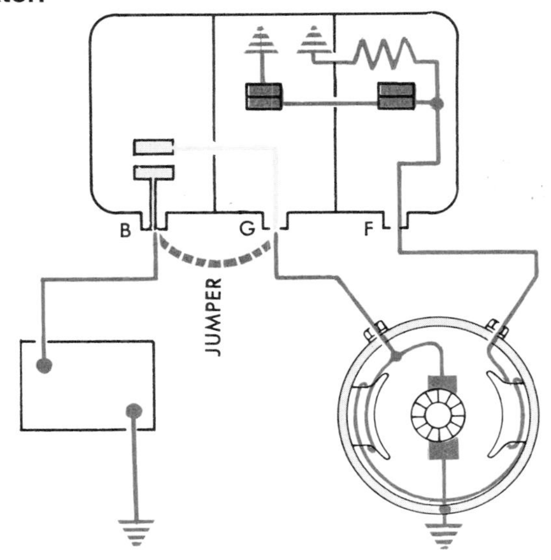

Fig. 41—Polarizing the Generator

Polarize the generator by connecting a jumper lead from the insulated side of the battery to the armature terminal as shown in Fig. 41. The battery, generator and regulator grounds must be connected. Just a touch of the generator lead between the "BAT" and "GEN" terminals of the regulator will cause current to flow in the proper direction in the field coils. A flash or arc will be noted when the lead is removed.

On high-voltage generators it is best to insulate the brushes. This will allow current to flow only through the field coils.

Summary: Polarity of Generators

The correct polarity of the DC generator is very important. A good rule to follow is to pass a current through the field coils in a direction that will have the grounded side of the coils connected to the grounded side of the battery.

GENERATOR REGULATORS

The **DC regulator** is the *control for the generator.* Otherwise, the generator would produce too much current and voltage at high speeds. Or the battery would discharge back through the generator at low speeds.

Fig. 42—Contact Point Generator Regulator on a Tractor

The "regulator" is really a combination of three functions:

- **Cutout Relay**
- **Voltage Regulator**
- **Current Regulator**

The CUTOUT RELAY (Fig. 43) *prevents battery discharge.* It closes the circuit between the generator and battery when the generator starts charging and breaks the circuit when the generator stops charging.

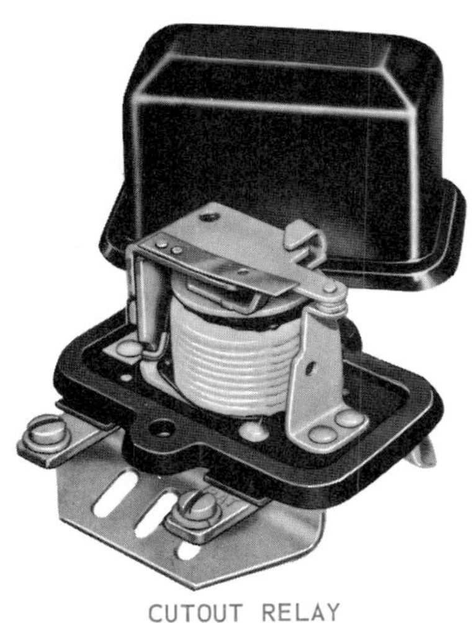

CUTOUT RELAY

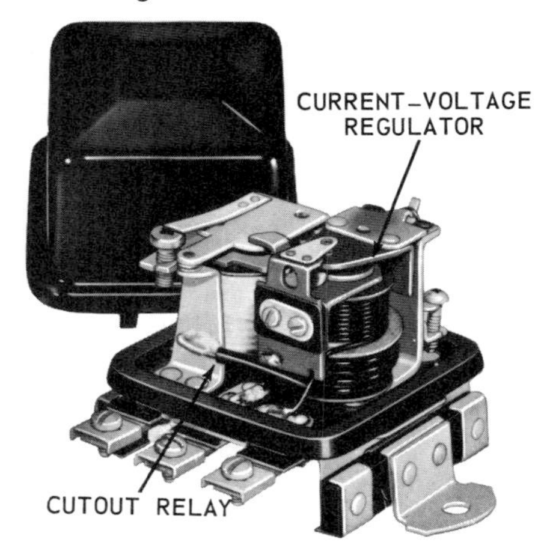

CUTOUT RELAY CURRENT-VOLTAGE REGULATOR

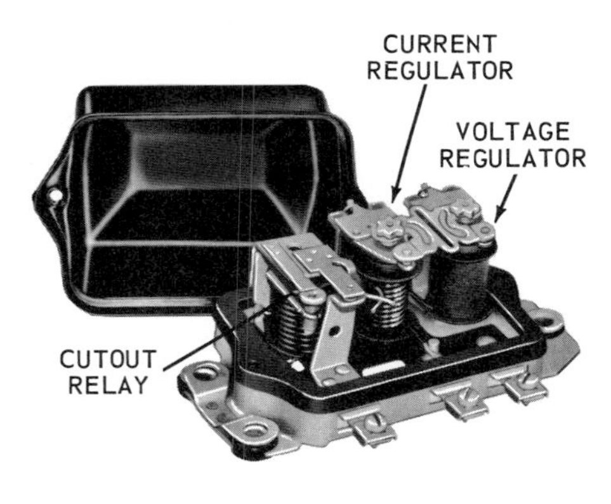

CUTOUT RELAY, CURRENT REGULATOR AND VOLTAGE REGULATOR

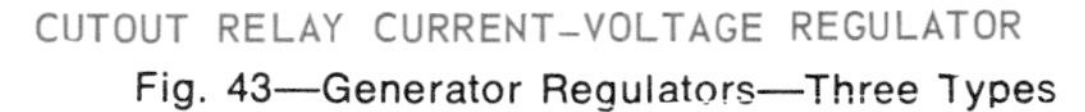

Fig. 43—Generator Regulators—Three Types

The last two regulators work at separate times as we'll see later. Both control the generator's output by varying the strength of its magnetic field.

Regulators may have one, two or three of these features combined in one unit (Fig. 43).

All three of these devices operate by electromagnetism and are very similar in operation and appearance.

Let's see how each specific function works in the charging circuit.

CUTOUT RELAY

The main purpose of the cutout relay is to open and close the circuit between the generator and the battery.

The cutout relay operates, as do all the regulator units by **electromagnetic induction.** What happens is that current flow through a coil of wire wound around an iron core magnetizes the core. The strength of this magnetism is determined by the strength of the current and the number of turns in the coil.

NOTE: The voltage and current regulators also operate on this same principle.

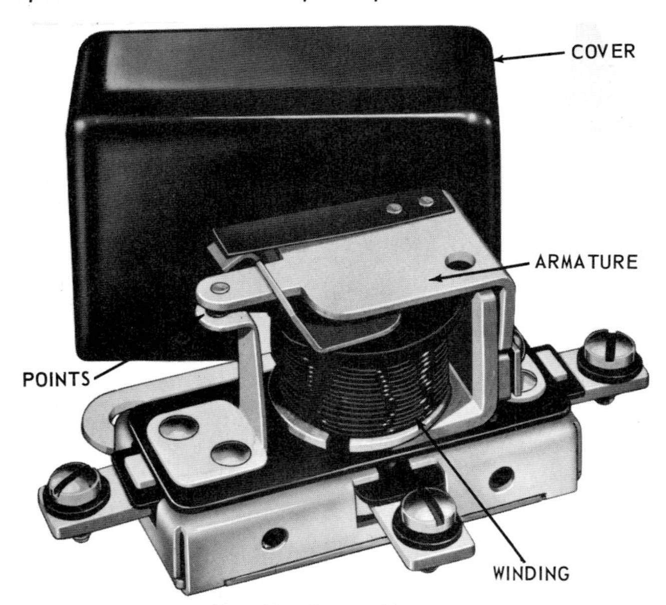

Fig. 44—Cutout Relay

Fig. 44 illustrates the physical features of the cutout relay. The relay consists of: two windings, one heavy and one fine, one iron core, a hinged flat metal piece called the armature and a stationary contact point.

The **heavy** winding, called the current or **series winding,** is wound around the core. One end is attached to the relay armature and the other end is connected to the charging circuit.

The **fine** winding is the voltage or **shunt winding.** One end is also attached to the relay armature. The other end is connected across the generator so that generator voltage is impressed on it at all times.

The relay armature is mounted above the core, but does not touch it. It is hinged at the back and is held away from the core by spring tension.

A contact point on the armature is aligned directly above the stationary contact point, which is connected to the battery by wire.

Operation of Cutout Relay

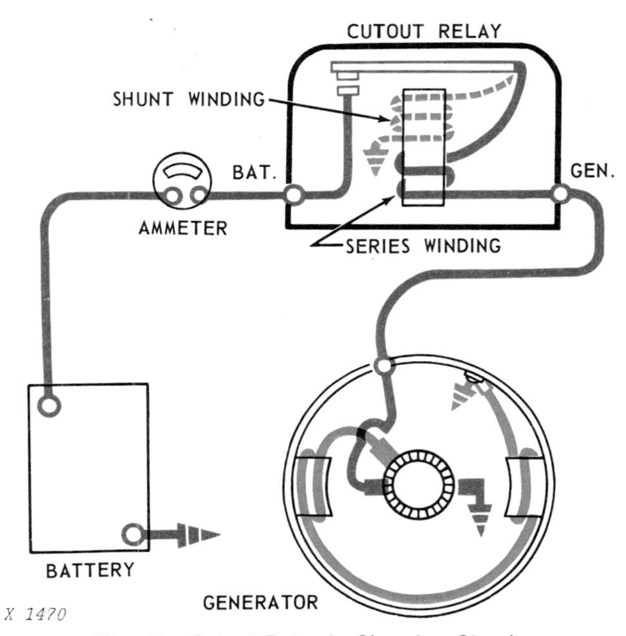

Fig. 45—Cutout Relay in Charging Circuit

As the generator builds voltage, current flows through the cutout windings, creating a magnetic field in the core. This magnetism pulls the armature toward the core and the two contact points close. The circuit between battery and generator is completed. Current then flows through the series winding to the battery (Fig. 45).

When the generator slows or stops, battery voltage becomes greater than generator voltage. Current then begins to flow from the battery to the generator.

With current flow reversed, polarity of the series winding is reversed. The shunt winding polarity remains the same, however.

The two opposing fields reduce the strength of the magnetism and the armature point is released

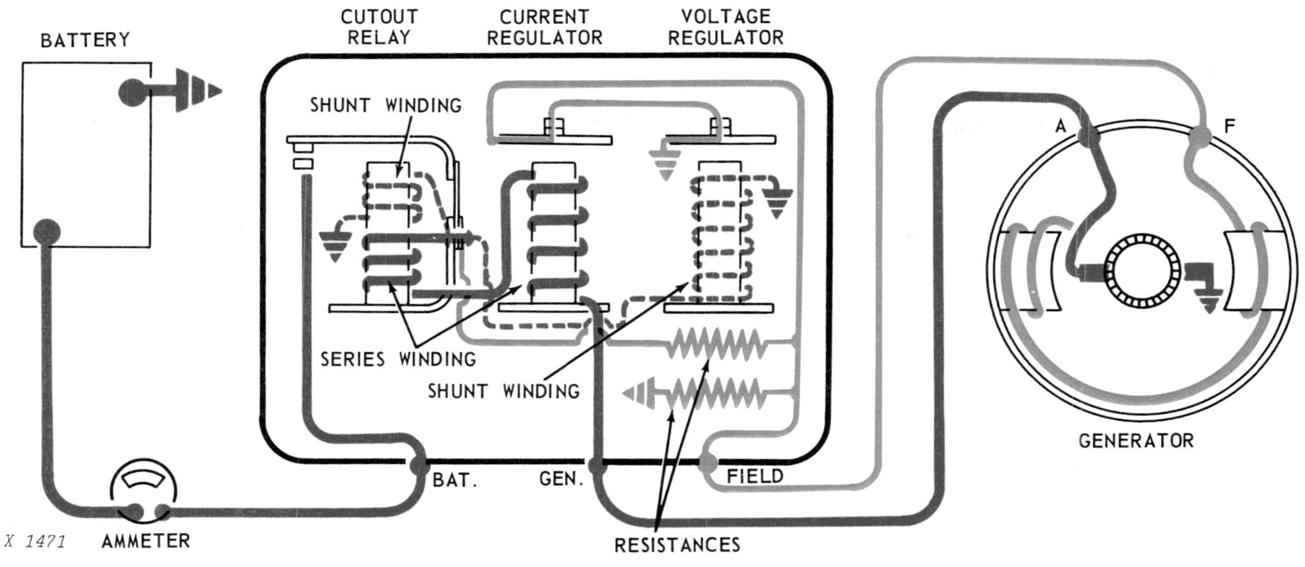

Fig. 46—Voltage Regulator with Single Winding

from the stationary contact point by spring action. This opens the circuit and stops current flow from the battery to the generator, and so prevents battery discharge.

VOLTAGE REGULATOR

As we said earlier in this chapter, the voltage regulator's main function is to regulate the generator magnetic field. In this way, the voltage build up in the generator is controlled.

The appearance of the voltage regulator is similar to the cutout relay. Like the relay, it has an iron core, flat hinged armature, and stationary contact point.

However, the position of some of these components is different, as are the type and number of windings on some models.

The armature is hinged on top, and the armature contact point is directly below the stationary contact point. Spring tension holds the points together when the regulator is not operating, and is provided by an adjustable spiral spring at the back of the regulator.

The hinge on the voltage regulator armature is made of two thermostic materials to allow for changing temperatures. This causes the regulator to regulate at a higher voltage when cold.

Types of Voltage Regulators

Two types of windings are used on standard voltage regulators.

- **Single Winding**
- **Accelerator Winding**

The SINGLE WINDING type has only a *shunt* winding of fine wire (Fig. 46.) This winding is connected across the generator. When the generator reaches a specified voltage as determined by the regulator, the magnetic field created by current flow through the winding draws the armature contact point away from the stationary contact. This opens the contact circuit and inserts resistance into the generator field circuit. The result is a reduction of voltage and current output in the generator.

Reducing the voltage and current reduces the magnetic attraction in the shunt winding. This allows the spring to pull the armature point back into contact with the stationary point. The generator field circuit is then grounded and voltage and current increases to the preset point.

The cycle then starts over again and is repeated many times a second. This is the method used in regulating generator voltage to a predetermined value.

The second type of regulator is the ACCELERATOR WINDING type (Fig. 47). It is equipped with the same components as the first type and has an added winding called an accelerator winding. This is a *series* winding of heavy wire.

Operation is basically the same as the first type except at a faster rate of speed.

In a sense, the series winding or coil, connected to the generator field circuit, helps the shunt winding create a stronger magnetic field faster. This draws the two points out of contact quicker.

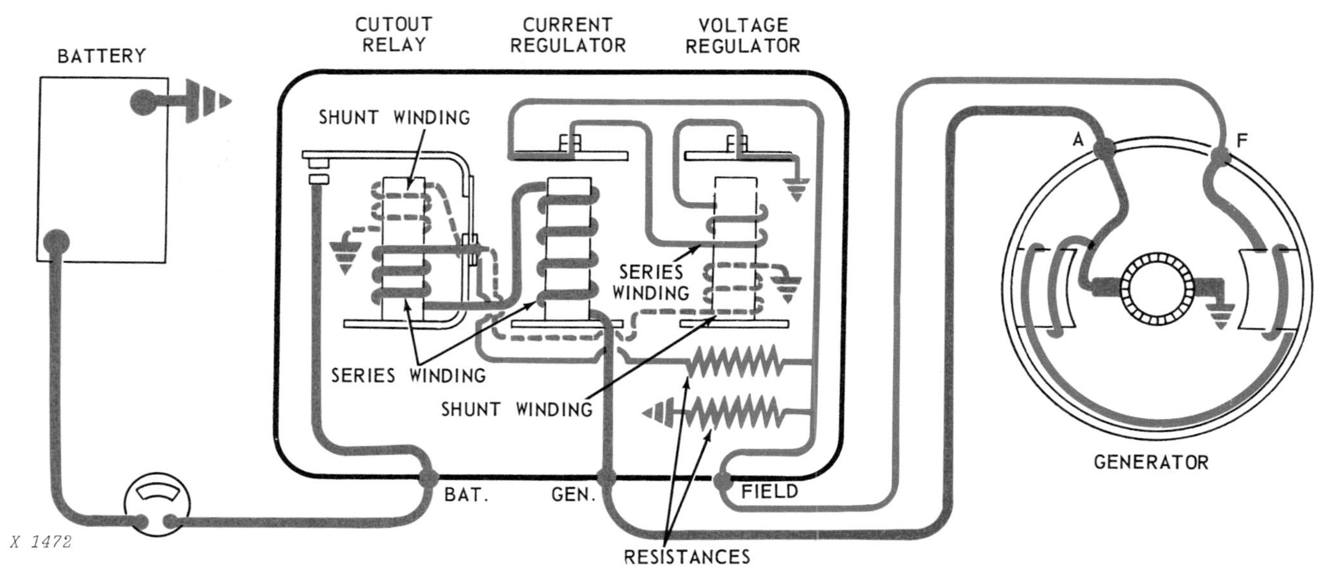

Fig. 47—Voltage Regulator with Accelerator Winding

The same speed causes the series coil magnetic field to collapse faster. When the armature point is suddenly released, it closes the circuit with the stationary point. The cycle is again repeated at a fast rate.

Voltage regulated in this manner allows the generator to supply varying amounts of current as needed by the battery and load circuits.

CURRENT REGULATOR

The **current regulator** controls the *current* output of the generator.

The current regulator looks the same as the single-winding voltage regulator. However, the current regulator winding is a *series* coil of heavy wire. The coil is connected into the generator armature or load circuit and carries the entire generator current output.

During a heavy load, such as a discharged battery and use of electrical accessories, voltage may not increase enough to actuate the voltage regulator. In these cases, the generator output will continue to increase to meet this demand. To limit this output to a safe point, the current regulator steps in.

As the output reaches a point as determined by the regulator, the magnetism of the regulator series coil draws the armature point toward the coil assembly. This breaks the contact between the two points. Resistance is then inserted into the generator field circuit and so the generator output is reduced.

This reduced output cuts the magnetism in the current regulator coil. Spring tension then draws the points back into contact and the generator field is directly connected to ground. The output again increases and the cycle is repeated as long as there is a heavy load.

As soon as the load is reduced (the battery is fully charged or the accessories are turned off) the voltage will increase. The *voltage* regulator then operates and gradually reduces the output. With the voltage regulator operating, there is no need for current regulation and the current regulator points remain in contact.

When one regulator is working, the other is not. Both will not operate at the same time.

HOW THE THREE REGULATORS WORK TOGETHER

Now that we have described the three generator regulators separately, let's take a look at the three operating as a unit.

Fig. 48 shows the three regulators in one unit operating in the charging circuit.

Shown in red are the series windings of the cutout relay and current regulator. In dashed red are the shunt windings in the cutout relay and voltage regulator. And in blue are shown the field circuit and the resistors.

When the generator begins to operate, it must recharge the battery. Two of the three units, the cutout relay and current regulator, perform in this operation.

The current flows from the generator armature into the series coil of the current regulator. From there it flows into the relay series coil, through the closed relay contact points and into the battery.

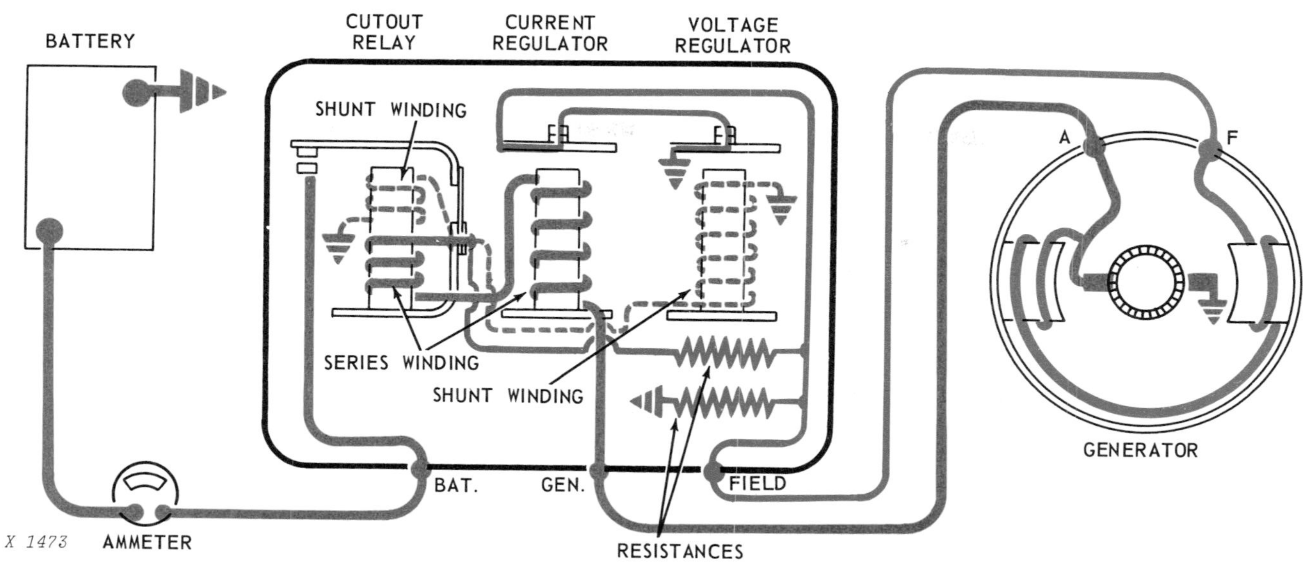

Fig. 48—Operation of All Three Generator Regulator Units

As the battery becomes charged, its resistance increases and the generator voltage rises. Then the voltage regulator goes into action, controlling generator field current and voltage.

When the engine stops, current from the battery flows back through the cutout series winding and reverses the polarity of its magnetic field. This causes the cutout points to open and disconnect the generator from the battery.

REGULATOR RESISTORS

The regulator also uses two common resistors. Both are fixed beneath the base of the regulator.

One resistor is inserted into the field circuit when either current or voltage regulator is used.

The second resistor is placed between the regulator field terminal and cutout relay frame.

The resistors are used to reduce a sudden voltage surge in the field coils caused by opening contacts. This prevents excessive arcing at the contact points.

VARIATIONS ON BASIC REGULATORS

As we said earlier, the three regulators are combined as units in some uses.

Let's discuss the major variations.

Current-Voltage Regulator

This unit combines the current and voltage regulators into one regulating unit (Fig. 49).

It is an economical means of controlling both generator voltage and output simultaneously.

However, the current-voltage regulator is limited to machines having low electrical load.

As you can see in Fig. 49, this regulator also has a cutout relay. Basically this is a standard type cutout relay we discussed earlier in this chapter.

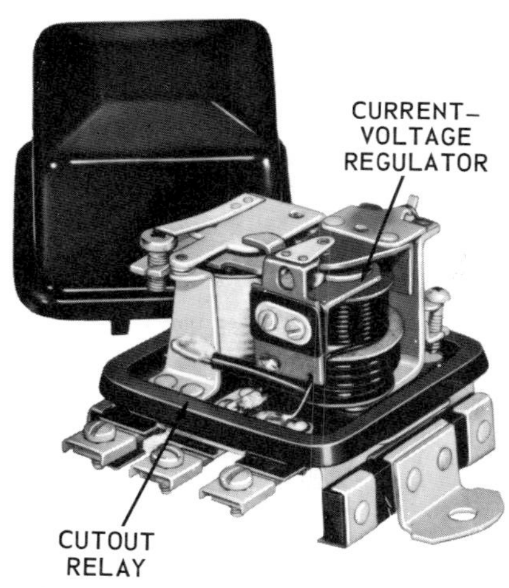

Fig. 49—Current-Voltage Regulator

In many respects, the current-voltage regulator unit of this regulator is similar to the voltage regulator in a standard 3-unit regulator. It is equipped with a hinged armature, iron core, stationary contact point, and adjustable spiral spring.

The major differences between this unit and other regulators are the number of windings and their operation.

The current-voltage regulator employs *three windings*–a shunt and *two* series (Fig. 50). The shunt winding is of fine wire and is connected across the generator to receive generator voltage at all times. One winding is of heavy wire and is con-

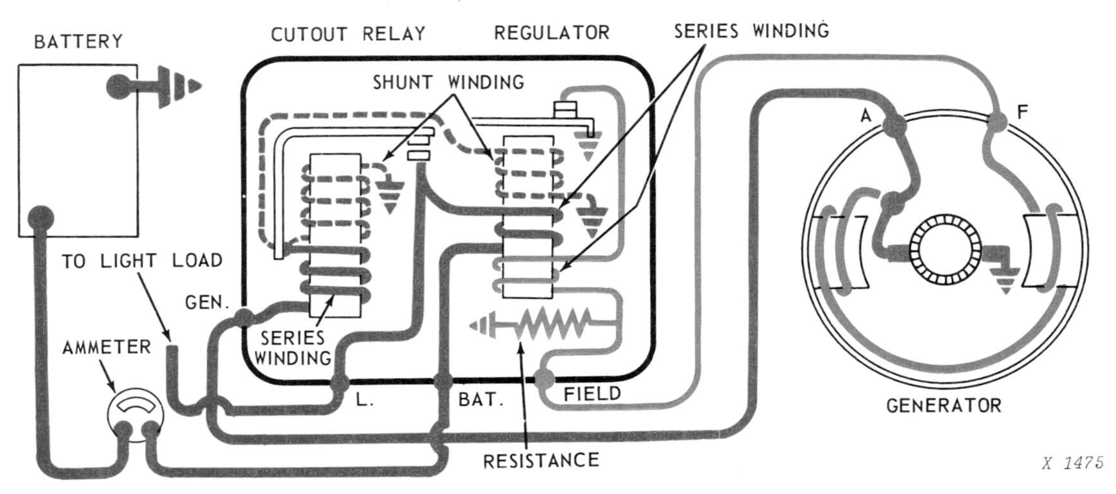

Fig. 50—Current-Voltage Regulator Operation

nected in series with the charging circuit. The other series winding is of a relatively heavy wire and is connected in series with the generator field circuit when the regulator contact points are closed.

In Fig. 50, the shunt winding in the cutout relay and the current-voltage regulator unit are shown in dashed red. The charging circuit series windings are shown in solid red. The field circuit series winding is shown in blue.

OPERATION OF CURRENT-VOLTAGE REGULATOR

As generator voltage builds, it causes current to flow through the cutout relay and into the regulator charging circuit winding and on into the battery. At the same time, voltage sensed in the regulator shunt winding causes some current to flow in it.

The two magnetic fields created by this current draw the regulator armature toward the iron core and opens the contact points. Field current is now diverted to ground through a resistor. This causes a drop in generator voltage and output.

A voltage drop, of course, reduces current flow through the windings and the magnetic field is weakened. Spring tension is then able to pull the armature point back into contact with the stationary point. This completes the field circuit and again the generator voltage and current output increase. This cycle is repeated many times a second as long as the generator and regulator are operating.

The field current series winding acts as an accelerator winding in that it helps to speed up the regulating cycle. When the regulator points are closed, field current flows through this winding, creating another magnetic field.

This added magnetic force helps to open the points faster. However, when the points open, field current flow stops and this field suddenly collapses. This, in turn, reduces the total magnetism enough to close the points quickly. The result is a faster regulator armature action rate.

In this way, the regulator controls *both* the generator voltage and current output.

One feature of current-voltage regulator operation is that as the charging rate increases, regulated voltage drops. This means that the higher the current flow to the battery, the lower the regulated voltage and vice-versa.

CURRENT-VOLTAGE REGULATOR TERMINALS

We have mentioned that this unit is limited to use in low-output, low load systems. An example is a diesel engine whose load requirements are only for battery recharging and occasional starting.

However, we don't want to rule out the ability of the current-voltage regulators to control more load than this. Because of another feature, it can handle some extra load such as lights or ignition.

The added feature is the extra terminal which this unit has (Fig. 51). It is a *load terminal,* mounted on the base of the regulator, where the battery terminal is on a 3-unit regulator. The "L" stamped on the terminal stands for "load."

The load terminal is connected to the stationary point of the cutout relay. It allows current from the generator to be diverted to a load, such as lights or ignition, without first passing through the current-voltage regulator. As a result, this current has no effect on the operating voltage and the regulator is only affected by current to and from the battery.

When the generator is not working and the lights are used, battery current will flow through the regulator winding and the "L" terminal to load. Any load that exceeds the generator output must be connected to the battery in the system.

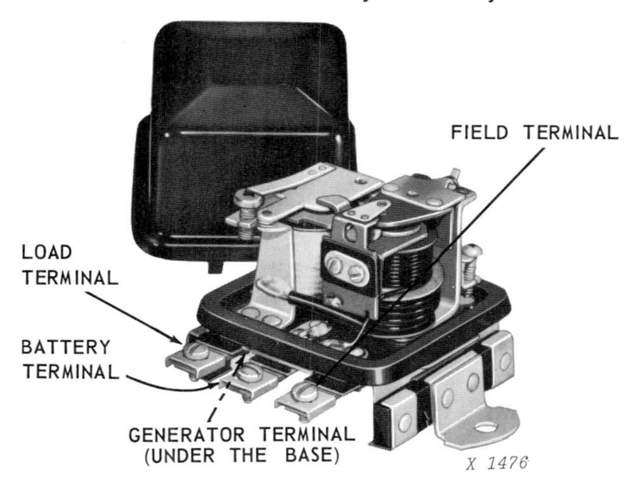

Fig. 51—Terminals on Current-Voltage Regulator

Heavy-Duty Regulators

The heavy-duty regulator is very similar to the 3-unit standard regulator. It is usually used with a high output or split field generator.

This type of installation is usually found on heavy equipment, such as trucks, where high loads in a wide range of speeds are required.

Heavy-duty regulators have special features such as double contact points, fiber insulating brackets on the stationary points, and added resistors for better control.

Double Contact Generator Regulator

The cutout relay and current regulator parts of this unit are the same as used on the standard 3-unit regulator.

However, the voltage regulator is different from the standard voltage regulator. It has an extra set of contact points. One of the points is on an extra armature which is over one of the stationary contact points. The other point is on the top of stationary contact point frame.

During normal operation the lower set of contact points operate like those on a standard voltage regulator.

But during high generator speeds and low loads, the voltage control shifts to the upper set of points. This connects the normally grounded end of the field circuit to the generator output circuit, shorting out the coils to control the generator voltage.

TESTING AND ADJUSTING THE DC REGULATOR

Before testing and adjusting the DC regulator, take these preliminary steps:

1) Check out the whole charging circuit to isolate the component which is failing.

2) Start with a **visual** *check of the circuit components, lead wires, and connectors.*

3) Then make an **electrical** *check, using a voltmeter, ammeter, tachometer, and resistance unit.*

NOTE: In many cases it is cheaper to replace a regulator rather than repair it. However, tests must be made to find out if the regulator is defective.

Preliminary Test Procedures

Once the problem has been traced to the regulator, take the following preliminary steps and precautions:

1. First make sure that the specified regulator is being used—one that is of the proper polarity. Most regulators are stamped on the base with a "P" for positive or "N" for negative.

2. Clean the contact points and make any minor adjustments outlined by the regulator or machine technical manual. Quite often this will correct the regulator problem.

3. Mechanical tests and adjustments should be made with the battery disconnected and the regulator out of the circuit.

CAUTION: Never close the cutout relay points by hand while the battery is connected. High current flow from the battery can damage the regulator.

4. Electrical checks and adjustments can be made with the regulator in or out of the circuit. However, the checks must be made with the regulator in operating position and the regulator cover in place.

5. Be sure to warm up the regulator to normal operating temperature before making the electrical tests. Do this by operating the regulator for 15 minutes with a ¼-ohm resistance in series with the battery. The regulator cover must also be in place. No electrical load other than ignition should be used during testing.

6. Speeds during the tests should be as specified by the Technical Manual.

7. The circuit must be polarized after each test or adjustment and before the engine is started. To do this, connect a jumper wire, momentarily, between the battery and generator regulator terminals.

8. Be sure to use the regulator service manual or the machine's Technical Manual. It will list all the necessary specifications and testing and adjustment procedures. A complete knowledge of the regulator components and operation is also necessary. Quite often an improper adjustment will do more damage than no adjustment at all.

9. Though the regulator can be electrically adjusted on the machine, it is better to remove it and its generator to the bench for servicing.

The first part of this chapter listed some general problems caused by a malfunctioning generator. Unfortunately, a faulty regulator can also create these same problems.

Let's assume that we have checked and found the generator and battery operating properly. Now we'll see how the *regulator* can cause these problems.

Regulator Trouble Shooting Chart

LOW CHARGING RATE—
FULLY CHARGED BATTERY

Regulator operation is normal.

LOW OR NO CHARGING RATE—LOW BATTERY

Low regulator setting. Burned or oxidized contact points, or open series circuit in the regulator.

HIGH CHARGING RATE—
FULLY CHARGED BATTERY

Improper voltage regulator setting. Defective voltage regulator unit, or grounded generator field circuit in the regulator.

HIGH CHARGING RATE—LOW BATTERY

Regulator operation is normal.

Sequence of Testing Regulator Units

The generator regulator should be tested and adjusted in the following sequence:

- *Check the voltage regulator first.*
- *Next check the cutout relay.*
- *Check the current regulator last of all.*

Always remember to bring the regulator units up to operating temperature. This gives the best testing results.

Voltage Regulator Tests and Adjustments

Two tests—air gap distance and voltage setting—are performed on the voltage regulator.

AIR GAP TEST

The *air gap* between the regulator armature and windings is an important mechanical adjustment. Remember that the air gap in a magnetic circuit acts as a *resistance.* Therefore, the greater the gap between armature and windings, the more magnetic force and so current needed to draw the armature down.

To measure the air gap, push down on the armature until the points are just touching. Then measure the gap between the armature and the core. The distance should be as specified by the Technical Manual.

STANDARD VOLTAGE REGULATOR

X 1477

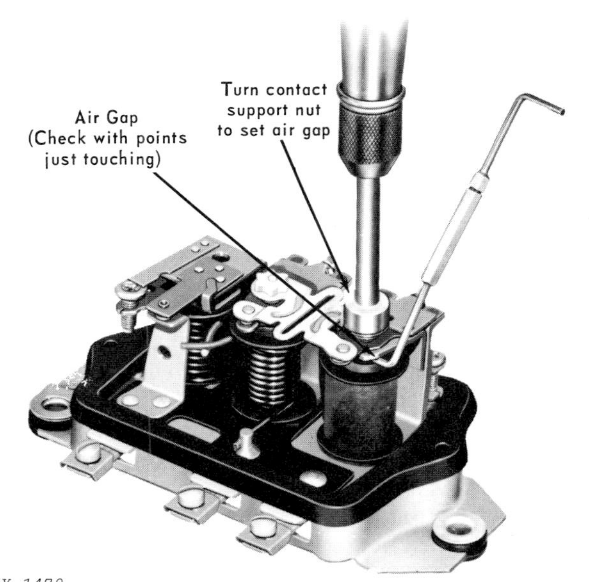

X 1478

POST-TYPE VOLTAGE REGULATOR

Fig. 52—Checking and Adjusting Air Gap in Voltage Regulators

Adjust the gap as shown in Fig. 52. On some regulators, a screw post at the top of the unit is adjusted to lengthen or shorten this distance.

VOLTAGE SETTING TEST

There are two methods for checking the voltage setting:

1) Fixed resistance method

2) Variable resistance method

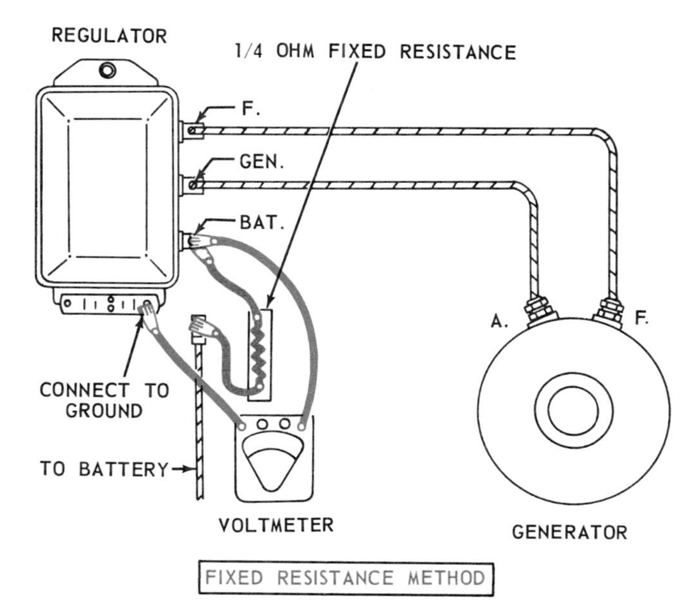

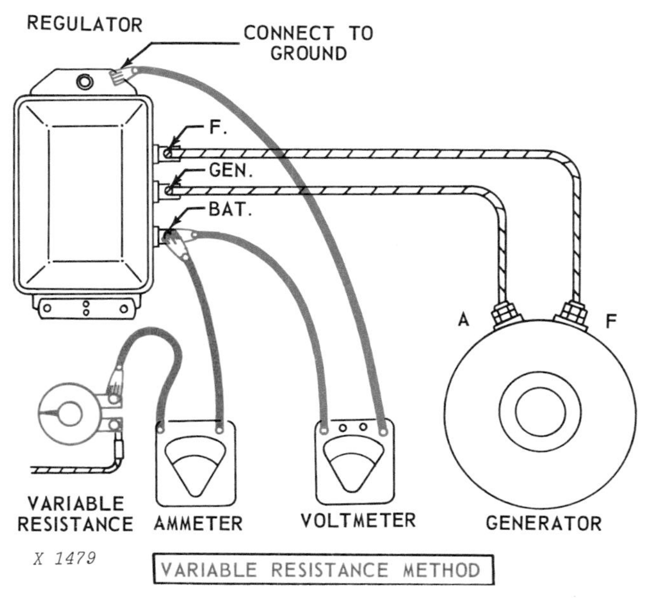

Fig. 53—Checking the Voltage Setting of Voltage Regulator

Fig. 53 shows how to make either check. Details are given below.

Fixed Resistance Method

1. Insert a ¼-ohm fixed resistor into the charging circuit at the battery terminal. (See Fig. 53, top.)

2. Connect a voltmeter from the battery terminal to ground.

3. Operate the circuit for 15 minutes at specified speed to warm it up. (See step 5 of "Preliminary Test Procedures.")

4. Cycle the generator by one of two methods: a) Slow it down until voltage drops to about ¼ of rated value. Then increase speed and note the voltage reading. b) Or cycle the generator by inserting a variable resistance into the field circuit. Slowly increase resistance until voltage drops to about ¼ of rated value. Decrease the resistance and note the voltage reading.

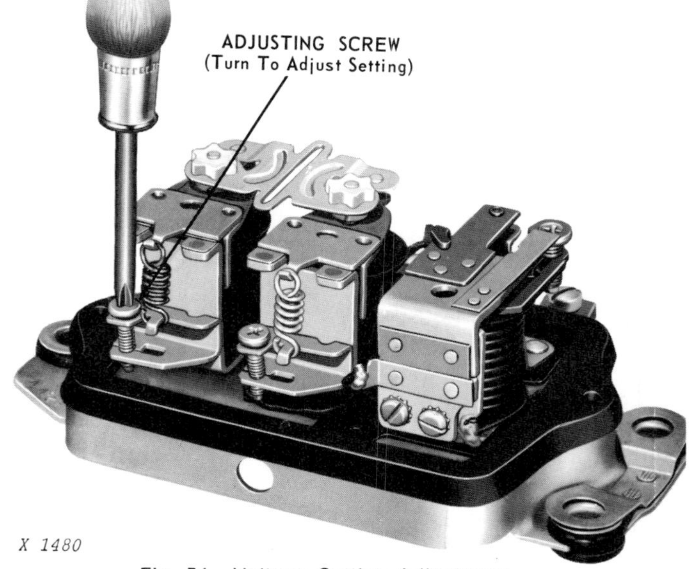

Fig. 54—Voltage Setting Adjustment

5. Adjust the voltage setting by turning the adjusting screw as shown in Fig. 54. If the adjusting screw is turned to its limits, it may be necessary to bend the spring support. However, do this very carefully.

Final adjustment should always be made by increasing the spring tension. If the setting is too high, adjust the unit below the specified value and then bring it back to this value by increasing the spring tension.

After each adjustment and before taking a reading, replace the cover and cycle the generator.

Variable Resistance Method

1. Connect a variable resistor and an ammeter into the charging circuit at the battery terminal. (See Fig. 53, bottom.)

2. Connect a voltmeter from the battery terminal to ground.

3. Start generator and adjust resistor to get a current flow of not more than 10 ampers. Operate the generator at specified speed to warm it up. (See step 5, "Preliminary Test Procedures.")

4. Cycle the generator as described in step 4 of fixed resistance method above.

5. Adjust the voltage setting as described in step 5 of the fixed resistance method.

ADJUSTING VOLTAGE FOR TEMPERATURE

The voltage reading should be corrected to allow for the ambient temperature around the regulator. See the machine Technical Manual.

A typical correction for a 12-volt regulator is 0.15 volts for every 10 degrees F (5.5°C) over or under 125°F (52°C). Add to the reading if over, or subtract if under 125°F (52°C).

For a 6-volt regulator, the correction factor may be 0.075 volts. On 24-volt regulators, 0.3 volts is a typical correction.

On voltage regulators with an accelerator winding or a series current windings, test at a specified amperage output because current flow in the windings affect the voltage of the test.

Cutout Relay Tests and Adjustments

There are three checks and adjustments on the cutout relay:

1) Air gap

2) Point opening

3) Closing voltage

The point opening and air gap tests must be made with the battery disconnected from the regulator.

AIR GAP CHECK

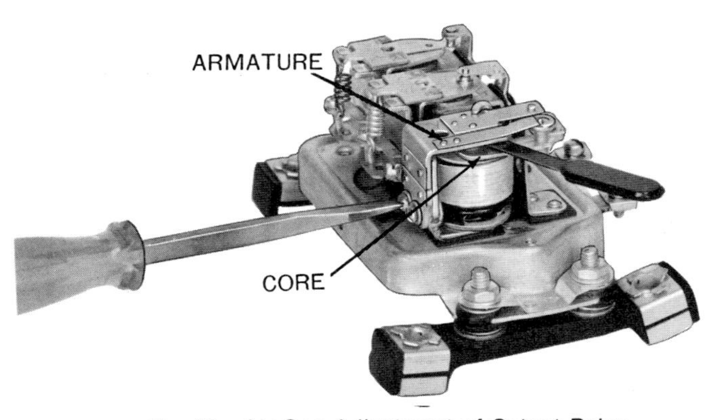

Fig. 55—Air Gap Adjustment of Cutout Relay

Push the cutout armature down until the points are just touching. Measure the air gap between the armature and the center of the core using a feeler gauge (Fig. 55). Adjust the air gap as shown. Raise or lower the armature as needed and make sure the points are aligned. Tighten the screws after adjustment.

POINT OPENING CHECK

Fig. 56—Point Opening Check

Check the point opening and adjust by bending the armature stop with a tool as shown in Fig. 56.

CLOSING VOLTAGE CHECK

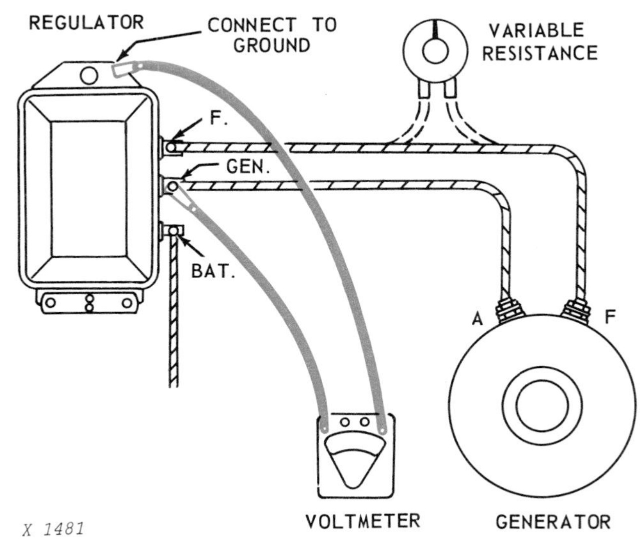

Fig. 57—Checking Closing Voltage of Cutout Relay

Connect a voltmeter between the generator terminal and ground (Fig. 57). Slowly increase the generator speed and note the relay closing voltage. Decrease the speed and make sure the points open before specified current flow is exceeded (with the battery connected).

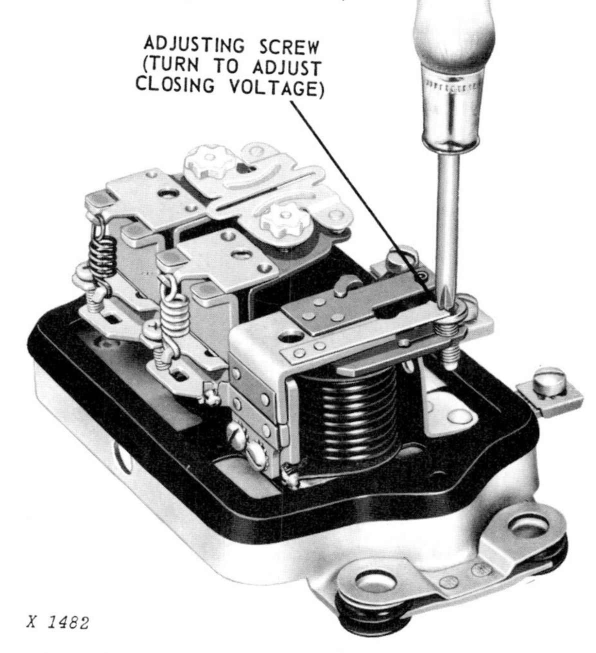

Fig. 58—Adjusting Closing Voltage of Cutout Relay

Adjust the closing voltage as shown in Fig. 58. Turn the screw clockwise to increase the setting.

Current Regulator Tests

Two checks are required on the current regulator:

1) Air gap

2) Current setting

AIR GAP CHECK

The air gap is tested and adjusted in the same manner as the voltage regulator air gap (see above).

CURRENT SETTING CHECK

Most current regulators have a temperature compensation. For these units, make the following test by the "load method."

1. Connect an ammeter into the charging circuit as shown in Fig. 59.

2. Turn on all accessories and connect an additional load across the battery (such as a bank of lights) to drop the system voltage about 1 volt below the voltage regulator setting.

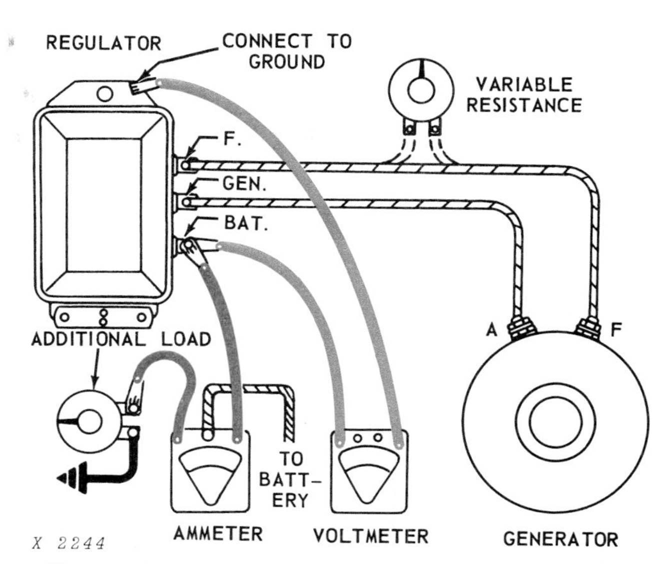

Fig. 59—Checking Current Setting of Current Regulator

3. Operate the generator at specified speed to warm it up. (See step 5 of "Preliminary Test Procedures.")

4. Cycle the generator and note the current setting.

5. Adjust the setting in the same way as for voltage setting. (See Fig. 54.)

Before slowing down the generator, be sure to remove the extra load. This will prevent overloading of the wiring.

Cleaning Contact Points

A great majority of regulator problems can be eliminated by a simple cleaning of the regulator points.

On positive-grounded regulators, the contact points that require the most attention are located on the current regulator armature and on the stationary contact point of the voltage regulator. *On negative-grounded regulators,* the other contact will require the most attention.

It is not necessary to have a flat surface on the points. However, it is necessary to remove all the oxidation.

Use a spoon or riffler file to remove the oxides until pure metal is exposed (Fig. 60).

Follow this with a thorough washing with trichlorethylene or a similar non-toxic solution. Wash by dampening a strip of lint-free cloth and drawing it between the points.

To clean the points it is necessary to remove the upper contact support as shown in Fig. 60.

Other smaller contact points of soft alloy do not oxidize, but they may be cleaned with a crocus cloth and then washed.

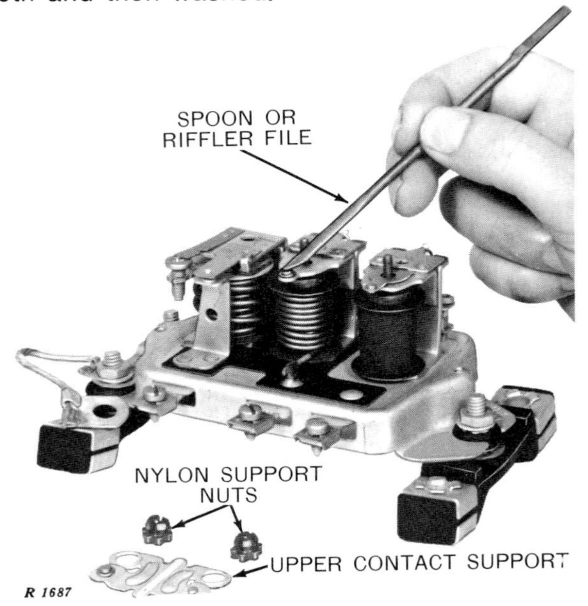

Fig. 60—Cleaning the Regulator Contact Points

Voltage Regulator Settings for Abnormal Operation

The perfect voltage regulator setting will keep the battery at or near full charge with a minimum use of water. The specified voltage value is usually satisfactory for average conditions. However, the operating conditions may be above or below the average. Therefore, it is sometimes necessary to adjust for abnormal conditions.

Two conditions indicate that special adjustment of the voltage regulator is necessary:

1. If the battery uses too much water at normal setting, reduce the voltage regulator setting 0.2 or 0.3 volt. Check to see if this is an improvement over a reasonable length of time. Repeat this adjustment until the battery remains charged with a minimum use of water. Remember that any reduction of the voltage requires a reduced cutout relay setting.

2. If the battery stays undercharged (¾ charge or less), increase the voltage regulator setting by 0.3 volt. Repeat until the battery is fully charged with a minimum use of water. When increasing the voltage setting, avoid a high setting that could damage lights or other electrical equipment during cold weather operation.

Always make sure that the battery is in top operating condition before making any of the above adjustments.

Now we must turn to the parts of the **AC charging circuit.**

ALTERNATORS

Fig. 61—Alternator (AC Generator) on a Modern Tractor

The **alternator** is the heart of the **AC** charging circuit.

Basically an alternator is like a generator. It converts mechanical energy into electrical energy.

We might say the alternator is an AC generator.

The difference is in the way the alternator rectifies its current to DC for the system. The alternator does this *electronically* using diodes.

Alternators are generally more compact than generators and can supply a higher current at low engine speeds.

In recent years, there has been more use of electrical accessories at low or idle engine speeds.

The alternator can best supply this output and for this reason AC charging circuits are used more today.

BASIC OPERATION OF ALTERNATOR

First let's compare the basic principles:

- **Generators—moving conductor through stationary field = induced voltage**
- **Alternators—moving field across stationary conductor = induced voltage**

Now let's see how the alternator works.

Fig. 62 shows a wire loop and a magnetic field, the same parts we saw earlier in the basic generator.

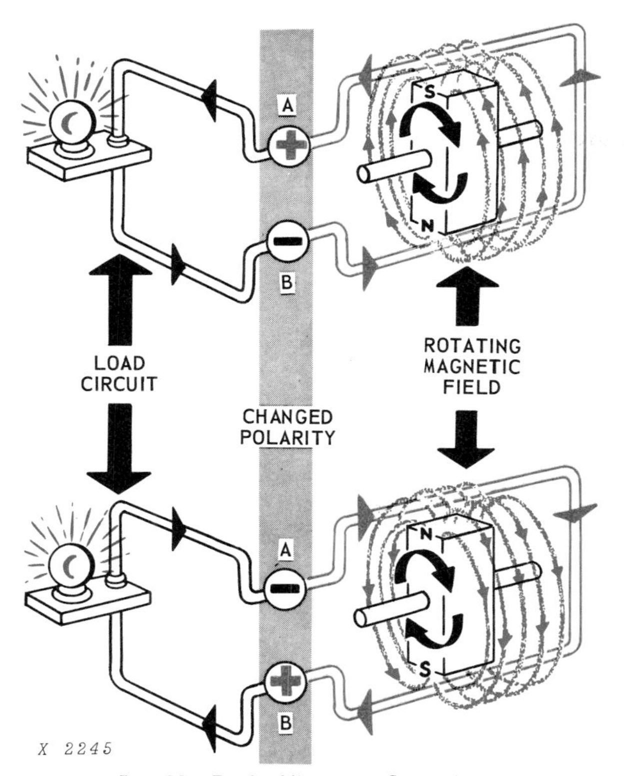

Fig. 62—Basic Alternator Operation

But now the wire loop is *stationary,* while the magnetic field is *rotating.*

The magnetic field (blue circles) is supplied by a bar magnet. (See N—S poles.)

The wire loop (in red) is connected to the charging circuit and its load (the light bulb.)

Let's start the operation. As the magnet rotates, its field cuts across the wire loop, inducing voltage. Since we have closed the loop circuit, current will flow. But in what direction?

This brings us back to another law of induction —magnetic lines of force leave the north pole of a magnet and enter its south pole.

In the upper diagram in Fig. 62, if the S pole is next to the upper part of the loop during that half-revolution, current will flow as shown by the red arrows. The same direction is induced by the lower N pole on that side of the loop.

Since current flow is from positive to negative, the end of the upper loop is (+) while the lower end is (−) as shown.

Now let's go to the second half of the revolution as shown in the lower diagram in Fig. 62.

The bar magnet has reversed poles and so the direction of current flow has reversed. This changes the polarity of the loop ends—the top one is now (−) while the bottom one is (+) as shown.

In summary: With each revolution, current flows from loop to load, first in one direction and then in the other. This is **alternating current** and the reason why the **alternator** is so named.

How Voltage Is Induced

However, an alternator made with a bar magnet rotating inside a single loop of wire is not practical, since very little voltage and current are produced.

The performance is improved when both the loop of wire and the magnet are placed inside an iron frame. The iron frame not only provides a place onto which the loop of wire can be assembled, but also acts as a conducting path for the magnetic lines of force.

Without the iron frame, magnetism leaving the N pole of the rotating bar magnet must travel through air to get to the S pole (Fig. 63).

Because air has a high reluctance to magnetism, only a few lines of force will come out of the N pole and enter the S pole.

Since iron conducts magnetism very easily, adding the iron frame greatly increases the number of lines of force between the N pole and the S pole. This means that more lines of force will be cutting across the conductor which lies between the bar magnet and the frame.

It is important to note that a very large number of magnetic lines of force are at the center of the tip of the magnet, whereas there are only a few lines of force at the leading and trailing edges of the tips. Thus, there is a strong magnetic field at the center and a weak magnetic field at the leading and trailing edges.

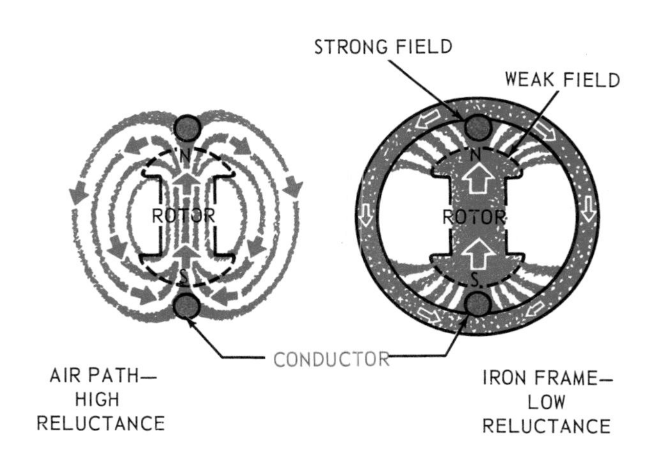

Fig. 63—Magnetic Lines of Force

 Litho in U.S.A.

This condition results when the distance, called the air gap, between the magnet and field frame is greater at the leading and trailing edges than at the center of the magnet.

The amount of the voltage induced in a conductor is proportional to the number of lines of force which cut across the conductor in a given length of time.

Therefore, if the number of lines of force is doubled, the induced voltage will be doubled (Fig. 64).

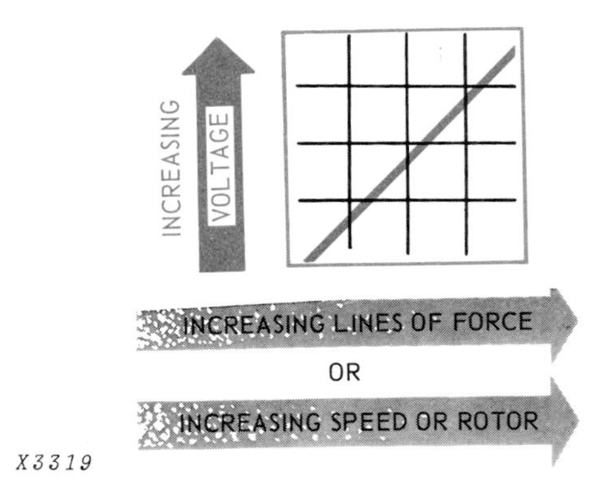

Fig. 64—Increasing Voltage

The voltage will also increase if the bar magnet is made to turn faster because the lines of force will be cutting across the wire in a shorter period of time.

Be sure to remember that either increasing the speed of rotation of the bar magnet, or increasing the number of lines of force cutting across the conductor, will result in increasing the voltage.

Similarly, decreasing the speed of rotation or decreasing the number of lines of force will cause the voltage to decrease.

As we saw earlier, the rotating magnet in an alternator is called the **rotor,** and the loop of wire and outside frame assembly is called the **stator.**

Fig. 65 shows the different positions of the rotor as it rotates at constant speed. At the top is a curve showing the magnitude of the voltage which is generated in the loop of wire as the rotor revolves.

The voltage curve shows the generated voltage or

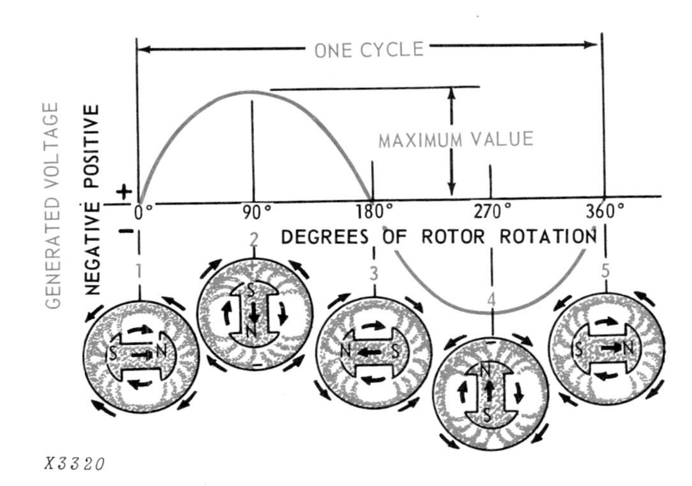

Fig. 65—Pattern of Generated Voltage During Each Revolution

electrical pressure which can be measured across the ends of the wire, just as voltage can be measured across the terminal posts of a battery.

With the rotor in the first position (1), there is no voltage being generated in the loop of wire because there are no magnetic lines of force cutting across the conductor.

As the rotor turns and approaches position (2), the rather weak magnetic field at the leading edge of the rotor starts to cut across the conductor, and the voltage increases.

When the rotor reaches position (2), the generated voltage has reached its maximum value, as shown above the horizontal line in the illustration.

The maximum voltage occurs when the rotor poles are directly under the conductor. It is in this position that the conductor is being cut by the heaviest concentration of magnetic lines of force.

Note that the magnitude of the voltage varies because the concentration of magnetic lines of force cutting across the loop of wire varies.

The voltage curve shown is not the result of a change in rotor speed, because in the illustration the rotor is considered to be turning at a constant speed.

By applying the Right Hand Rule to position (2), the direction of current in the loop of wire will be out of the top end of the conductor, and into the bot-

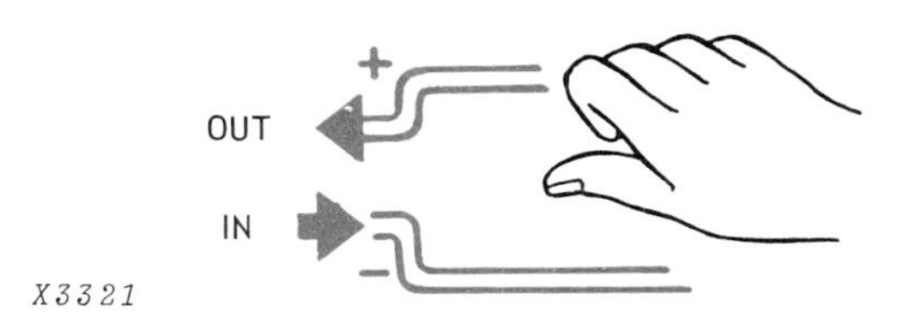

Fig. 66—Right-Hand Rule

tom end (Fig. 66). Thus, the top end of the conductor will be positive, and the bottom end negative.

The voltage curve which is shown above the horizontal line in Fig. 65 represents the positive voltage at the top end of the wire loop which is generated as the rotor turns from position (1) to position (3).

As the rotor turns from position (2) to position (3), the voltage decreases until at position (3) it again becomes zero.

As the rotor turns from position (3) to position (4), note that the N pole of the rotor is now passing under the top part of the wire loop, and the S pole under the bottom part. From the Right Hand Rule, the top end of the loop of wire is now negative, and the bottom end positive. The negative voltage at the top end of the loop is pictured in the illustration by the curve which is below the horizontal line.

The voltage again returns to zero when the rotor turns from position (4) to position (5).

The voltage curve in the illustration represents one complete turn or cycle of the rotor.

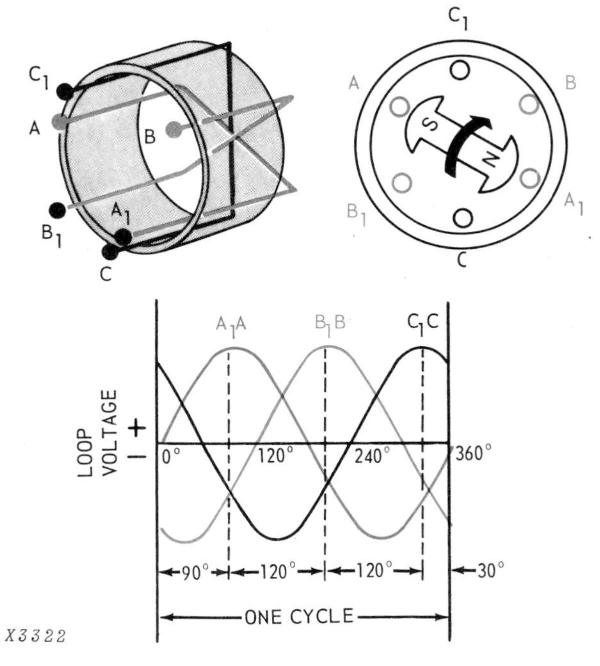

Fig. 67—Loop Voltage

With the rotor making 60 complete turns in one second, there will be 60 such curves, one coming right after the other, resulting in 60 cycles per second. The number of cycles per second is called the frequency. Since the generator speed often varies, the frequency also varies.

In Fig. 67, the single loop of wire acting as a stator winding, and the bar magnet acting as the rotor, show how an a.c. voltage is produced in a basic alternator. When two more separate loops of wire, spaced 120 degrees apart, are added to our basic alternator, two more separate voltages will be produced.

With the S pole of the rotor directly under the A conductor, the voltage at A will be maximum in magnitude and positive in polarity.

After the rotor has turned through 120 degrees, the S pole will be directly under the B conductor and the voltage at B will be maximum positive. Similarly, 120 degrees later, the voltage at C will be maximum positive.

This means that the peak positive voltages at A, B and C in each loop of wire occur 120 degrees apart. These loop voltage curves are shown in Fig. 67.

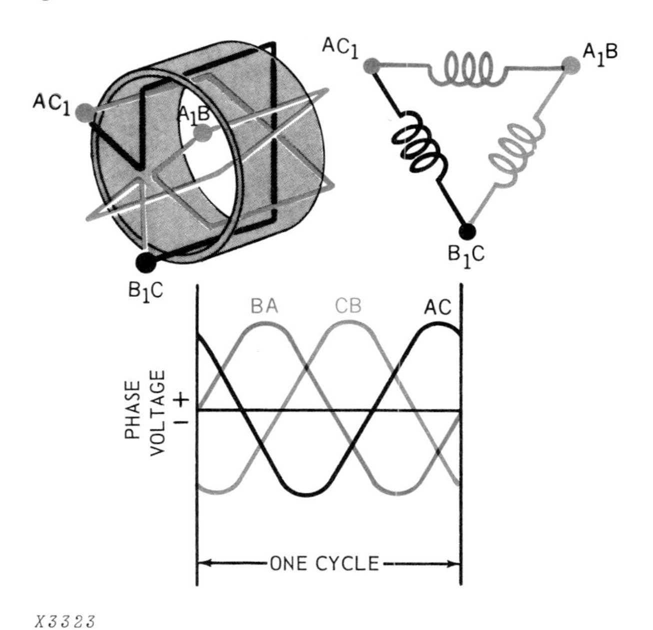

Fig. 68—Phase Voltage (Delta Stator)

When the ends of the loops of wire marked A_1, B_1 and C_1 (Fig.68) are connected to the ends marked B, C and A respectively, a basic three phase **"delta"-connected stator** is formed. The three AC voltages available from the delta-connected stator are identical to the three voltages previously discussed, and may now be denoted as the voltages from B to A, C to B, and A to C, or more simply BA, CB and AC. Inspect the illustration to see the logic of this. *Example:* the voltage formerly called A_1A may now be called BA.

When the ends of the loops of wire marked A_1, B_1 and C_1 are connected together, a basic three-phase **"Y"-connected stator** is formed (Fig. 69). The three voltages available from the "Y"-connected stator may be labeled BA, CB and AC.

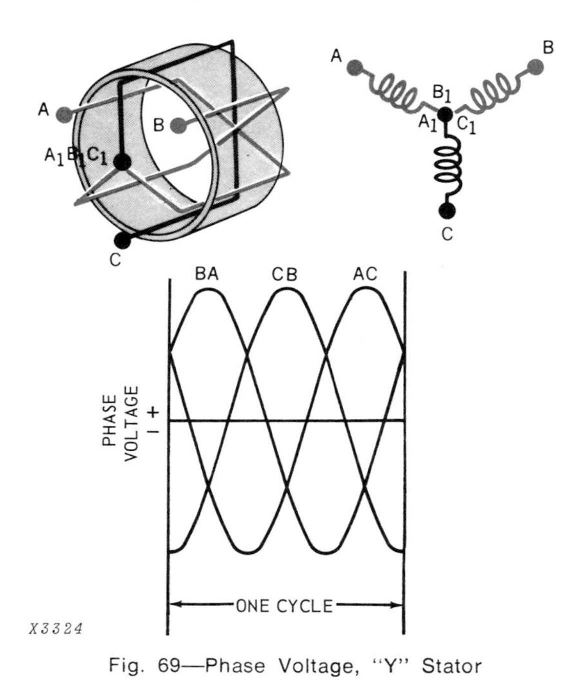

Fig. 69—Phase Voltage, "Y" Stator

From the illustration you can see that each of these voltages consists of the voltages in two loops of wire added together. For example, the voltage measured from B to A consists of the voltages in loops B_1B and A_1A added together. This addition yields a voltage curve BA similar in shape and form to the individual loop voltages, except that the voltage curve BA will be approximately 1.7 times as large in magnitude as an individual loop voltage.

Remember that three AC voltages spaced 120 degrees apart are available from the "Y"-connected stator, as illustrated. These voltage curves will be considered in more detail in the following sections.

We have now developed the two basic types of stator windings, and have shown how three separate complete cycles of AC voltage spaced 120 degrees apart are developed for each complete revolution of the rotor.

Now we will look at the *diode*, and will see how six diodes connected to the stator winding change the three AC voltages to a single DC voltage needed for the DC electrical system:

How Diodes Change AC to DC

The operating principles of diodes are covered in Chapter 1, "Electricity—How It Works."

Here we need know only that *a* **diode** *is an electrical device that will allow current to flow through itself in one direction only.* The diode is often pictured by the symbol in Fig. 70, and current can flow through the diode only in the direction indicated by the arrow.

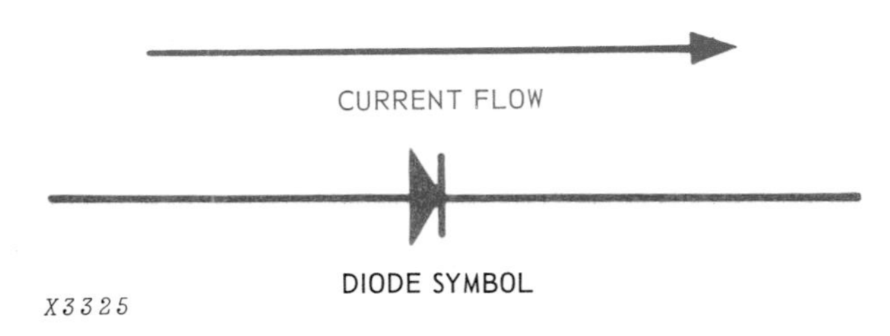

Fig. 70—Diode Symbol

When a diode is connected to an AC voltage source having ends marked A and B, current will flow through the diode when A is positive (+) and B is negative (–). The diode is said to be "forward-biased" (Fig. 71), and with the voltage polarity across the diode as shown, it will conduct current.

When the voltage at A is negative and at B is positive, the diode is said to be "reverse-biased" and it will not conduct current.

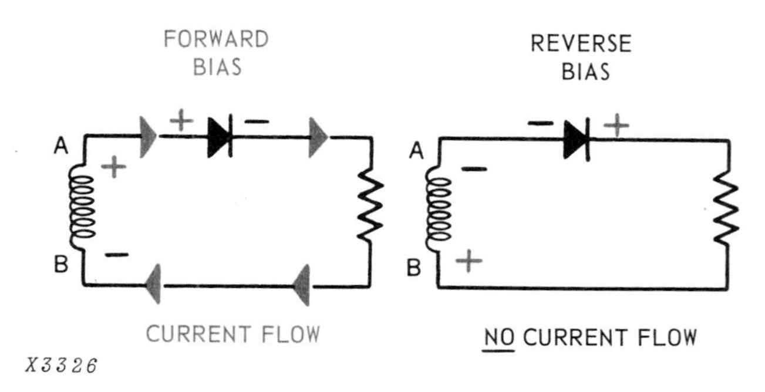

Fig. 71—Forward and Reverse Bias

The current flow that would be obtained from this arrangement is illustrated in Fig. 71.

Since the current flows only half the time, the diode provides what is called "half-wave rectification" (Fig. 72). A generator having only one diode would provide very limited output.

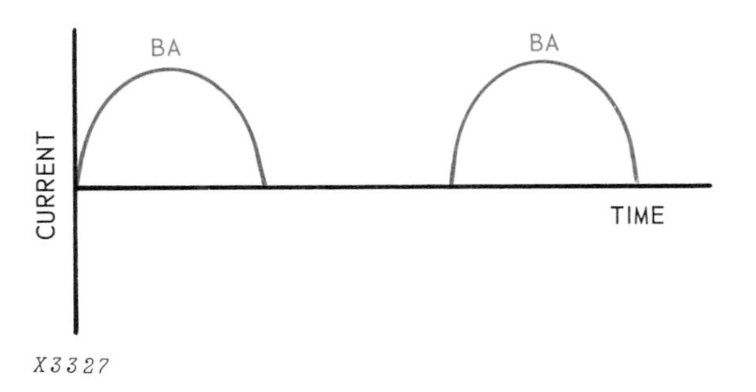

X3327

Fig. 72—Half-Wave Rectification

The output is increased when four diodes are used to provide "full-wave rectification" (Fig. 73). Note that the current is more continuous than with one diode, but that the current varies from a maximum value to a zero value.

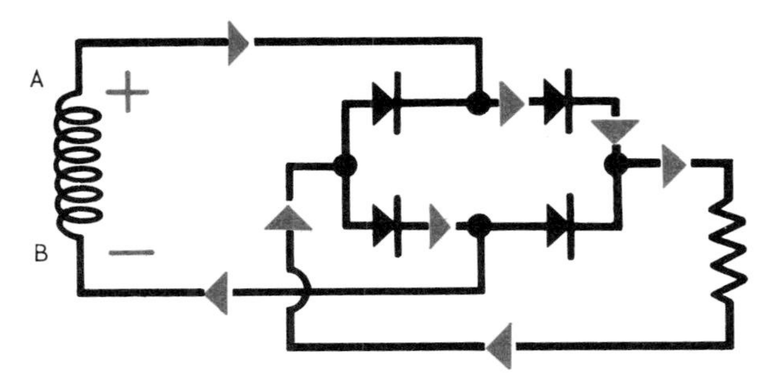

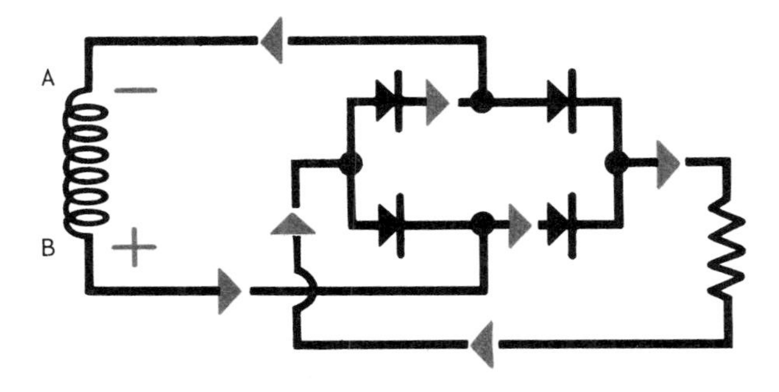

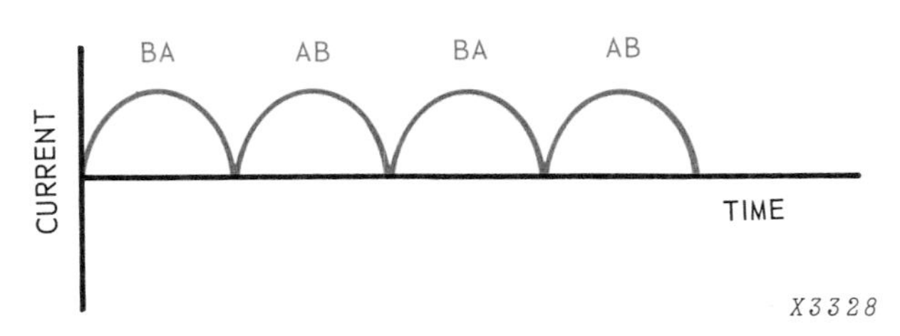

X3328

Fig. 73—Full-Wave Rectification

It is particularly important to observe that the current flow through the external load resistor is in one direction only. The AC voltage and current have therefore been rectified to a unidirectional or DC voltage and current.

This circuit arrangement could be used to charge a DC battery, but it does not produce the most output that can be obtained in a generator.

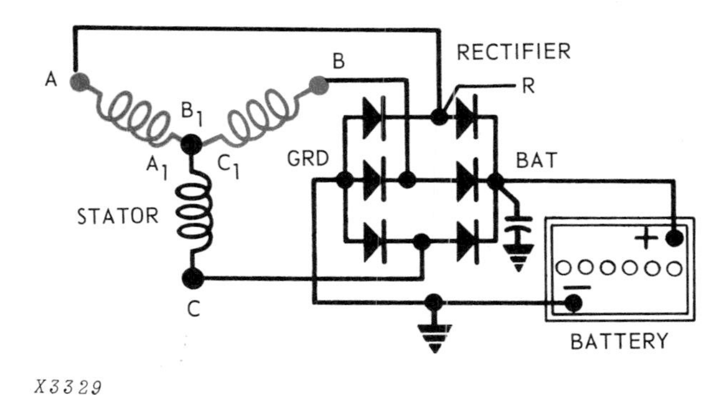

X3329

Fig. 74—Three-Phase, Full-Wave Rectification

In order to obtain a higher output and a smoother voltage and current, a three-phase stator is connected to six diodes which together form a "three-phase full-wave bridge rectifier" (Fig. 74).

The operation of the "Y"-connected stator will be illustrated first, then that of the delta-connected stator. A battery connected to the DC output terminal will have its energy restored as the alternator provides charging current. Note that the blocking action of the diodes prevents the battery from discharging directly through the rectifier.

To explain the direction of current flow in the stator-rectifier combination, we will review briefly our previous discussion concerning the three AC voltage curves produced in the "Y"-connected stator winding (Fig. 75). Our first reference was to the voltages developed in each loop. These loop voltage curves A_1A, B_1B and C_1C are reproduced here for reference. However, these individual loop voltages do not appear across the rectifier diodes, because the rectifier is connected only to the A, B and C terminals of the stator. Therefore, the voltages which appear across the rectifier diodes are the phase voltages BA, CB and AC.

 Litho in U.S.A.

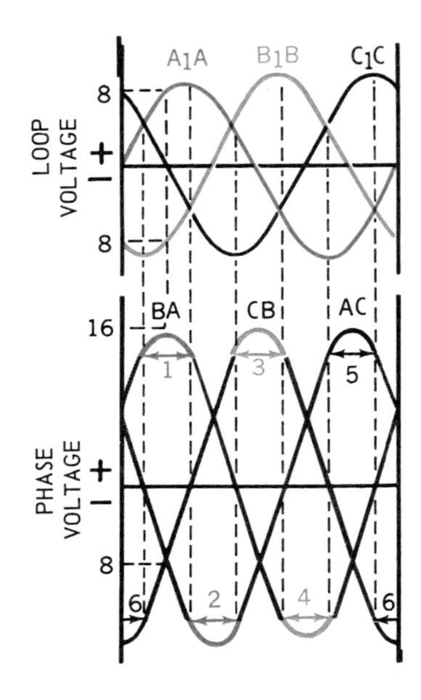

Fig. 75—Loop and Phase Voltage Curves

The phase voltage curves BA, CB and AC are also reproduced here, and are obtained as previously explained by adding together each pair of loop voltages (Fig. 75).

As an example, phase voltage BA is obtained by adding together the voltages in loops A_1A and B_1B. To obtain the phase curve BA, we add together the voltage from B to B_1, and the voltage from A_1 to A. Consider the instant when the voltage in curve BA is maximum in magnitude and positive in polarity. At this same instant the voltage B_1B is minus 8, or the voltage from B to B_1 is plus 8. This value added to the A_1A loop voltage of plus 8 volts yields a maximum positive voltage of 16 volts for curve BA.

By taking different instants of time, the entire curve BA and curves CB and AC can be obtained in this same manner.

Voltage Curves—"Y"-Connected Stator

For convenience, the three a.c. voltage curves provided by the *"Y"-connected stator* for each revolution of the rotor have been divided into six periods, 1 through 6. Each period represents one-sixth of a rotor revolution, or 60 degrees.

An inspection of the voltage curves during period 1 reveals that the maximum voltage being induced appears across stator terminals BA. This means that the current flow will be from B to A in the stator winding during this period, and through the diodes as illustrated.

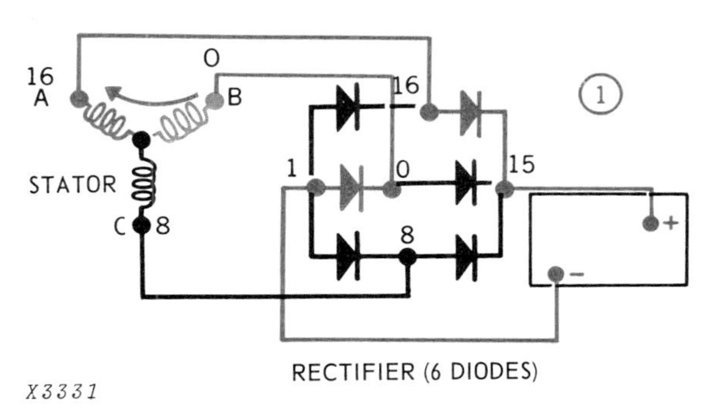

Fig. 76—"Y" Stator, Period 1

To see more clearly why the current flows during period 1, assume that the peak phase voltage developed from B to A is 16 volts (Fig. 76). This means that the potential at B is zero volts, and the potential at A is 16 volts.

Similarly, from the curves the phase voltage from C to B at this instant is minus 8 volts. This means that the potential at C is 8 volts, since C to B, or 8 to zero, represents a minus 8 volts.

At this same instant the phase voltage from A to C is also minus 8 volts. This checks, since A to C, or 16 to 8, represents minus 8 volts.

Neglecting voltage drops in the wiring, and assuming a one volt drop in the conducting diodes, the voltage potentials are noted on the rectifier.

Only two of the diodes will conduct current, since these diodes are the only ones in which current can flow in the forward direction. The other diodes will not conduct current **because they are reverse biased.**

For example, the lower right-hand diode is reverse biased by 7 volts ($15-8=7$), and the right-hand middle diode is reverse biased by 15 volts ($15-0=15$). **It is the biasing of the individual diodes, provided by the stator, that determines how current flows in the stator-rectifier combination.**

Throughout period 1 the current flows as indicated, because the bias direction across the diodes does not change from that shown. Although the voltage potentials across the diodes will vary numerically, this variation is not sufficient during period 1 to change a diode from reverse bias to forward bias and from forward bias to reverse bias.

Inspect the phase voltages curves in Fig. 75 and you will see that between periods 1 and 2 the maximum voltage being impressed across the diodes changes or switches from phase BA to phase

AC. This means that as the maximum voltage changes, the current flow will change from BA to CA.

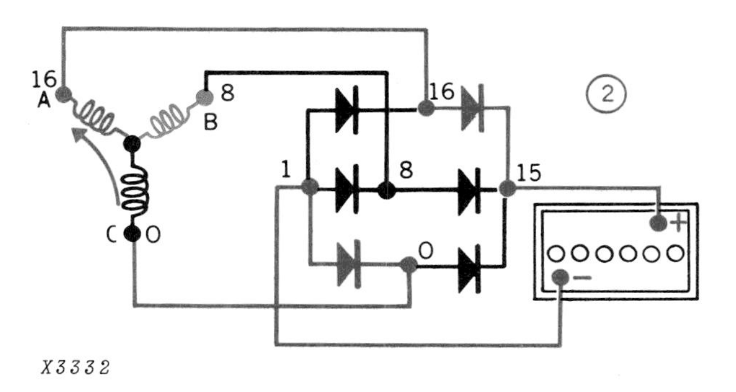

Fig. 77—"Y" Stator, Period 2

It is important to note in Fig. 77 that the maximum voltage being produced in the stator windings during period 2 appears across phase AC and that this voltage is negative from A to C.

Taking the instant of time at which this voltage is 16 volts, the potential at A is 16, and at C is zero (A to C, or 16 to 0, is a negative or minus 16).

Similarly, at this same instant, the voltage across phase BA is 8 volts, and across phase CB is 8 volts. This means that the potential at B is 8 volts, as shown. The direction of current flow during period 2 is illustrated.

Following the same procedure for periods 3-6, the current flows can be determined, and are shown in Figs. 78 and 79.

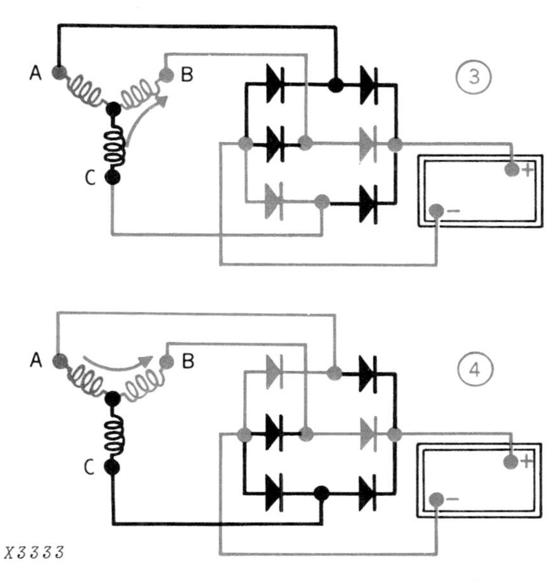

Fig. 78—"Y" Stator, Periods 3 and 4

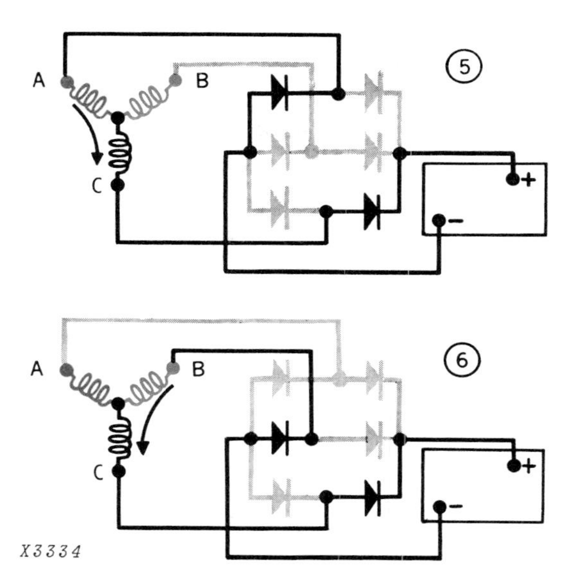

Fig. 79—"Y" Stator, Periods 5 and 6

These are the six major current flow conditions for a three-phase "Y"-connected stator and rectifier combination.

The voltage obtained from the stator-rectifier combination when connected to a battery is not perfectly "flat," but is so smooth (Fig. 80) that for all practical purposes the output may be considered to be a non-varying DC voltage. The voltage, of course, is obtained from the phase voltage curves, and can be pictured as shown.

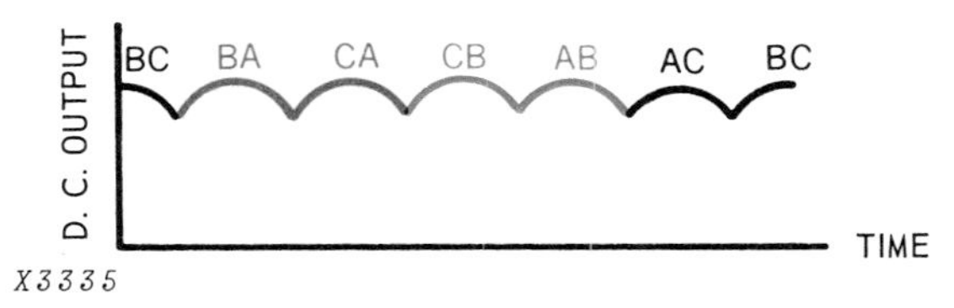

Fig. 80—DC Current Output

An alternate method of establishing the direction of current flow through the rectifier for a "Y"-connected stator is to refer to the illustration showing the loop voltage curves (Fig. 75).

During period 1 the two loop windings having the largest voltages are A_1A and B_1B, with the voltage in loop C_1C always being less than the voltages in the other two loops. Since the voltage in A_1A is positive, and in B_1B is negative (positive from B to B_1), the current will flow from B to A during period 1. The phase voltage curve BA, of course, is simply a picture of the actual voltage that the two loop voltages A_1A and B_1B added together impress across the rectifier diodes.

 Litho in U.S.A.

Referring again to the loop voltage curves, the two loop windings having the largest voltages during period 2 are A_1A and C_1C. Since the voltage in A_1A is positive, and C_1C is negative (positive from C to C_1), the current will flow from C to A during period 2.

In the same manner, the current flow directions can be determined for the remaining four periods.

Although this alternate method of using loop voltages can be used to determine the current flow directions, it cannot be used to explain why the current flows as it does through the stator-rectifier combination.

To explain why, it is necessary to determine the voltages that actually exist at the rectifier, because it is these voltages and the biasing of the diodes that determine the current flow directions. These voltages are represented by the phase voltage curves, which are the voltages that actually appear at the rectifier diodes.

Again, as we have already seen, the phase voltage curves are simply the loop voltage curves added together.

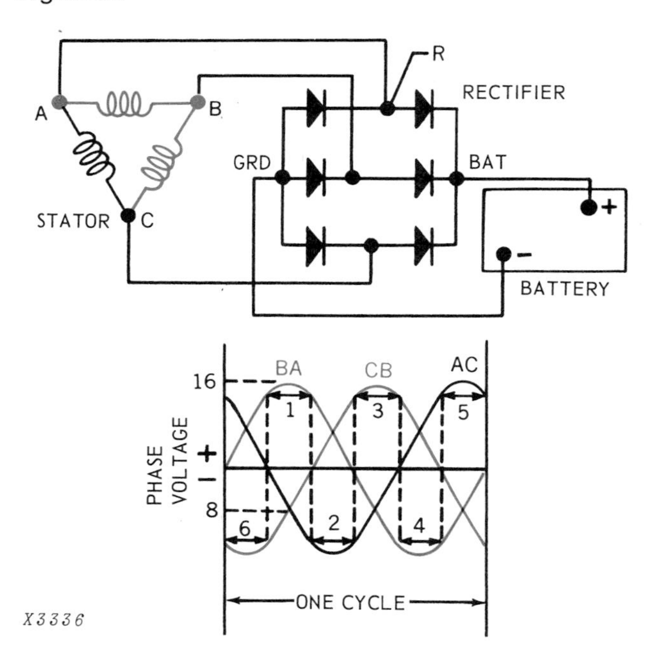

Fig. 81—"Delta"-Wound Stator

Voltage Curves—Delta-Wound Stator

A *delta-connected stator* wound to provide the same output as a "Y"-connected stator also will provide a smooth voltage and current output when connected to a six-diode rectifier (Fig. 81).

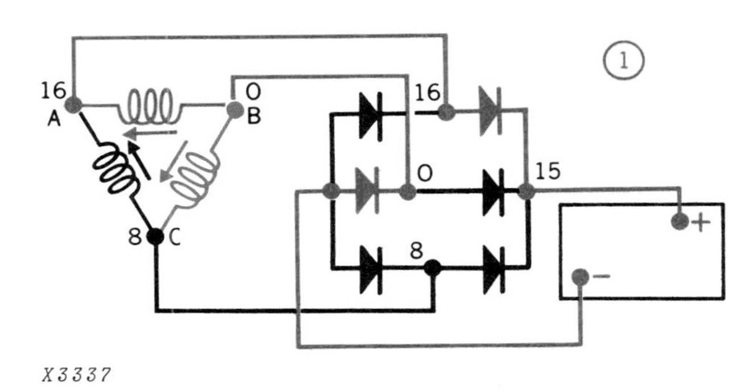

Fig. 82—"Delta" Stator, Period 1

For convenience, the three-phase AC voltage curves obtained from the basic delta connection for one rotor revolution are reproduced here and have been divided into six periods.

During period 1 (Fig. 82), the maximum voltage being developed in the stator is in phase BA.

To determine the direction of current flow, consider the instant at which the voltage during period 1 is a maximum, and assume this voltage to be 16 volts. The potential at B is zero, and at A is 16.

From the curve, it can be seen that the voltage of phase CB is a negative or minus 8 volts. Therefore, the potential at C is 8 (C to B or 8 to 0 is a minus 8 volts).

Similarly, the voltage of phase AC is minus 8 volts. This checks, since A to C, or 16 to 8, is a minus 8. These voltage potentials are shown in Fig. 82.

The current flow through the rectifier is exactly the same as for a "Y"-connected stator, since the voltage potentials on the diodes are identical.

An inspection of the delta stator, however, reveals a major difference from the "Y" stator. Whereas the "Y" stator conducts current through only two windings throughout period 1, the delta stator conducts current through all three. The reason for this is apparent, since phase BA is *in parallel* with phase BC plus CA. Note that since the voltage from B to A is 16, the voltage from B to C to A also must be 16. This is true since 8 volts is developed in each of these two phases.

During period 2 (Fig. 83), the maximum voltage developed is in phase AC, and the voltage potentials are shown on the illustration at the instant the voltage is maximum.

 Litho in U.S.A.

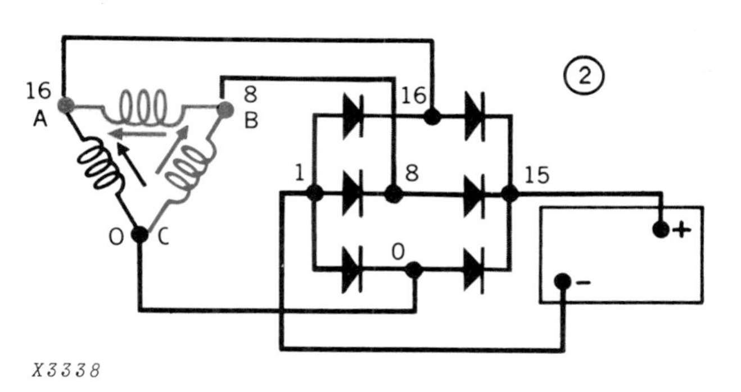

Fig. 83—"Delta" Stator, Period 2

Also shown are the other phase voltages, and again, the current flow through the rectifier is identical to that for a "Y" stator, since the voltages across the diodes are the same. However, as during period 1, all three delta phases conduct current as illustrated.

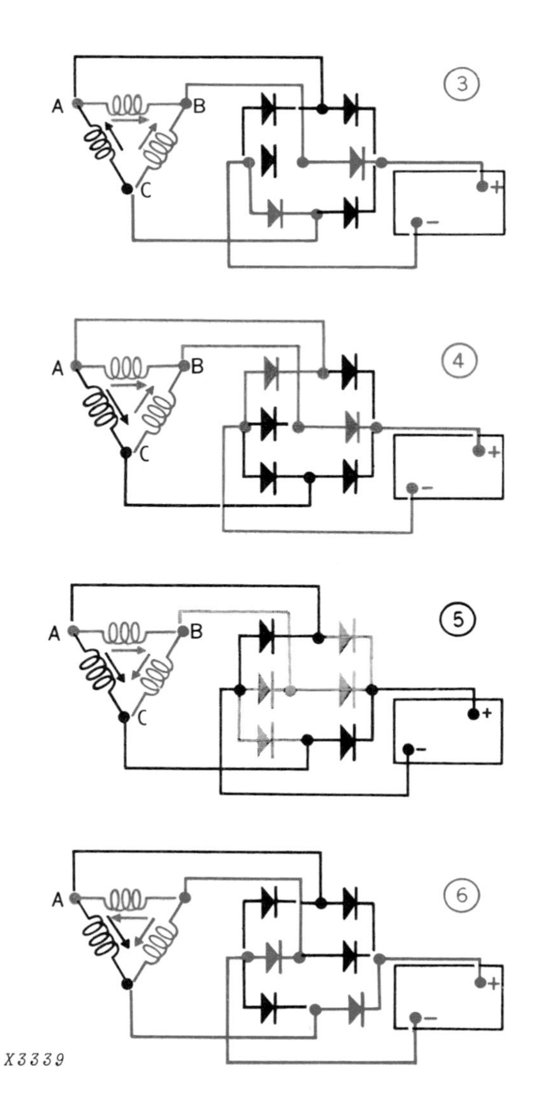

Fig. 84—"Delta" Stator—Periods 3, 4, 5, and 6

Following the same procedure for periods 3-6 the current flow directions are shown in Fig. 84.

We have seen the six major current flow conditions for a delta stator.

This concludes our study of the fundamental principles by which a basic alternator develops and rectifies three AC voltages into a single DC voltage and current flow.

CONSTRUCTION OF ALTERNATOR

Now let's look at the actual parts of a typical alternator.

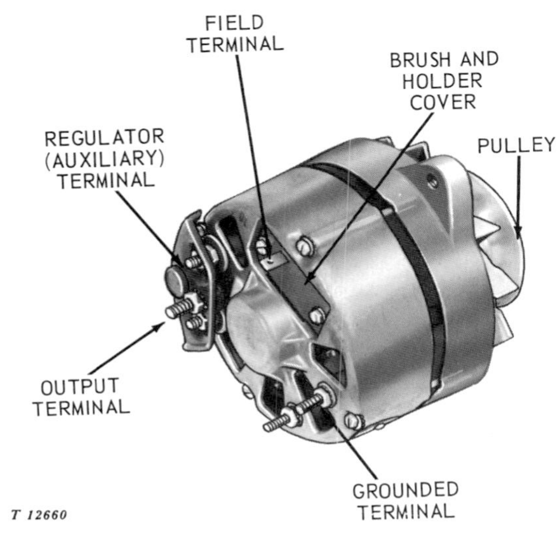

Fig. 85—Complete Alternator

Alternators have three main units:

- **Rotor Assembly—magnetic field which rotates.**
- **Stator Assembly—conductors which are stationary.**
- **Rectifier Assembly—diodes which change AC to DC current.**

Fig. 86 shows the parts of a typical alternator.

ROTOR ASSEMBLY

The **rotor assembly** does the same job as the field coil and pole shoe assembly in the DC generator. The difference is that the rotor assembly revolves while the field coil and pole shoe assembly is stationary.

The rotor assembly consists of a *wire coil* wrapped around an *iron core* and mounted on a *rotating shaft* (Fig. 86). The coil is enclosed between two interlocking soft iron sections. The ends of the coil are connected to two *slip rings* mounted on one end of the shaft.

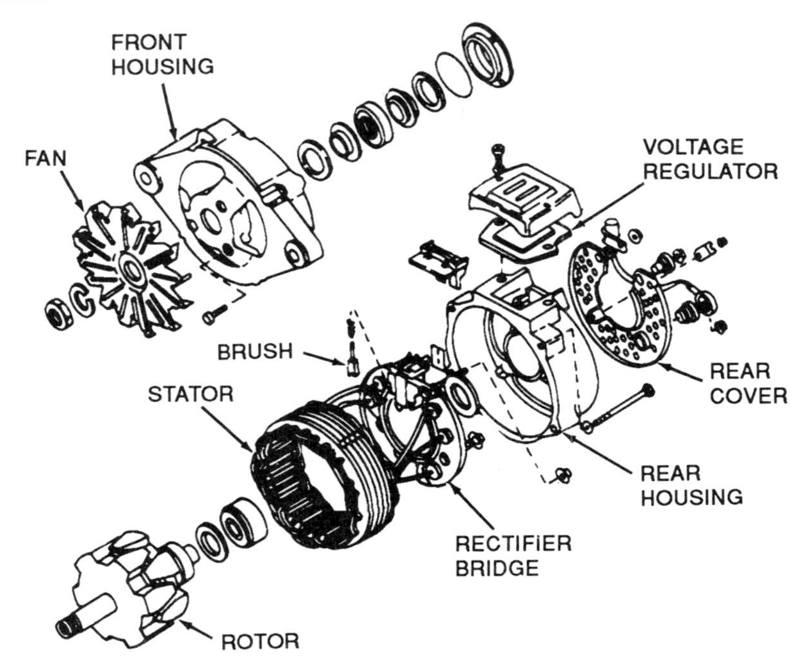

Fig. 86—Exploded View of a Typical Alternator

Small *brushes* ride on the slip rings. One of the brushes is connected to ground. The other is insulated and connected to the alternator field terminal. This terminal is connected through the regulator and the ignition switch to the battery.

When the ignition switch is turned on a small amount of current from the battery flows through the regulator to the insulated brush, through one of the slip rings, into the coil, out through the other slip ring and the other brush to ground.

The current through the coil creates a magnetic field which makes the rotor a rotating multi-pole magnetic field. Each finger on the iron section becomes a magnetic pole.

The rotor poles do not retain magnetism—unlike the generator's pole shoes. Direct current from the battery must flow through the rotor coil to magnetize the poles *before* the alternator will start to charge. The rotor or alternator field is, therefore, *externally* excited.

STATOR ASSEMBLY

The **stator assembly** does the same job as the armature in a DC generator. However, the stator is fixed while the armature rotates. An actual stator assembly is shown in Fig. 87.

The stator assembly is a laminated soft *iron ring* with three groups of *coils* or *windings* in the slots.

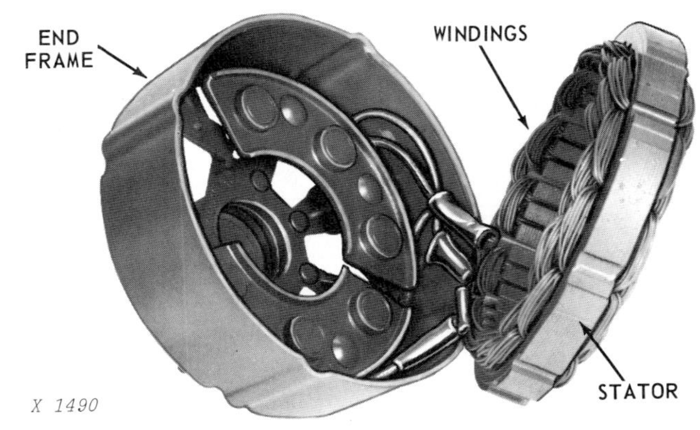

Fig. 87—Stator Assembly

Each group is made up of from 8 to 16 coils, depending on the design.

One end of each stator winding is connected to a positive and negative diode. (We will cover the diode installation later.)

As we saw earlier, the other ends of the stator windings can be connected by either one of two ways (Fig. 88):

- **"Y"-Connected Stator Windings**
- **Delta-Connected Stator Windings**

The delta-connected alternator may be used for heavy-duty operations where lower voltage, but higher amperage, is needed.

The "Y"-connected alternator usually provides a higher voltage and a moderate amperage.

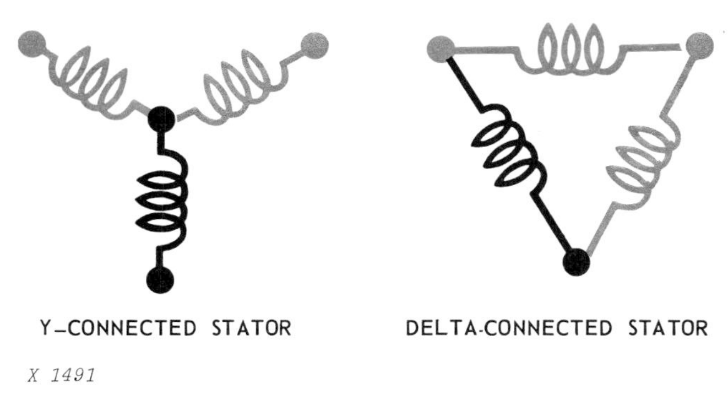

Fig. 88—Two Connections of 3-Phase Stator Windings

RECTIFIER ASSEMBLY (DIODES)

To convert the AC to DC current, the rectifiers or **diodes** are used.

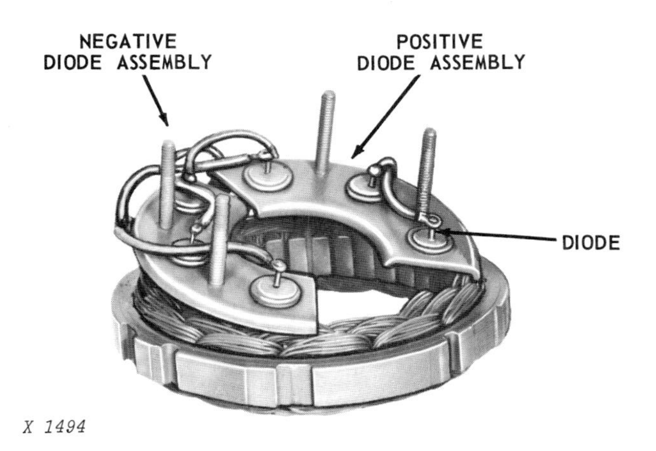

Fig. 89—Dioded Assemblies in Alternator

In a 3-phase alternator, six diodes are mounted at the slip ring end of the alternator housing (Fig. 89). Three negative diodes are mounted in the end frame or in a heat sink bolted to the end frame. Three positive diodes are mounted in the heat sink which is insulated from the end frame.

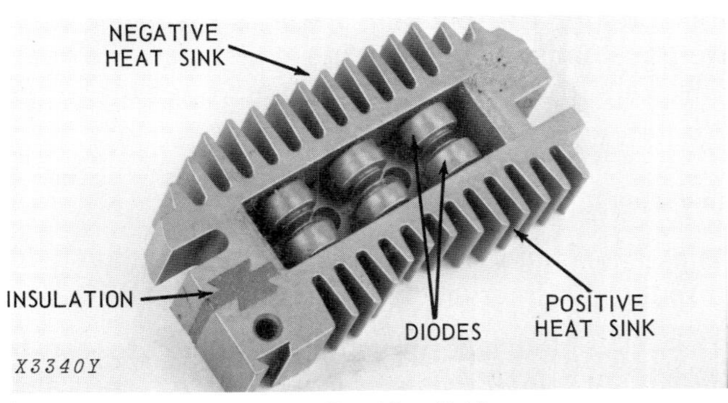

Fig. 90—Rectifier Bridge

Some alternators have all six diodes mounted in one assembly, called a *rectifier bridge* (Fig. 90).

NOTE: On negative-grounded alternators, the positive diodes are mounted in the insulated heat sink. On positive-grounded alternators, the negative diodes are mounted in the insulated heat sink.

Isolation Diode

Some alternators use an extra diode assembly in the circuit. This diode is called the *isolation diode.*

Its primary function is to act as an automatic switch between the battery and alternator. It will block any current flow from the battery back to the alternator and regulator when the alternator is not operating.

The isolation diode is mounted in a heat sink metal frame. (See Fig. 86.) (Two diodes in parallel are used on some high-amperage alternators.)

Because all output passes through this diode, high temperatures can be expected. For this reason, the diode and frame are mounted on the slip ring end frame to allow free circulation of air around it.

The charge indicator light is placed in parallel with the isolation diode and connected between the regulator terminal and the ignition switch.

The light supplies the initial current to the regulator to excite the rotor field. This is shown when it lights up while starting the engine.

Most alternators have a resistor in parallel with the light so that when the bulb burns out, the alternator will still be energized and charge the batteries.

The isolation diode and the charge indicator lamp operate as follows:

1. *With the ignition switch off,* there is no voltage at either end of the indicator lamp. The light is out and the isolation diode blocks current from the battery to the alternator.

2. *When the ignition switch is turned on,* voltage is applied to the one side of the indicator light. This will cause a small amount of current to flow through the regulator to the rotor field and this causes the indicator light to glow.

3. *Once the rotor field is excited and the engine starts rotating,* voltage is generated in the stator windings and the output through the rectifiers steadily increases.

4. *Soon the generated voltage exceeds the voltage potential of the battery.* At this time the indicator light goes out and the voltage regulator takes control of field voltage and current.

ALTERNATOR TYPES AND DESIGNS

The alternator has only one function in the electrical system—*to supply current to charge the battery and operate electrical accessories.*

Since each application makes its own special requirements on the alternator, there are many different types and designs.

Some of the factors which determine alternator design are: type of mounting, vibration, belt loading, rotor speeds, current output, service life required, and environmental factors such as dust and dirt.

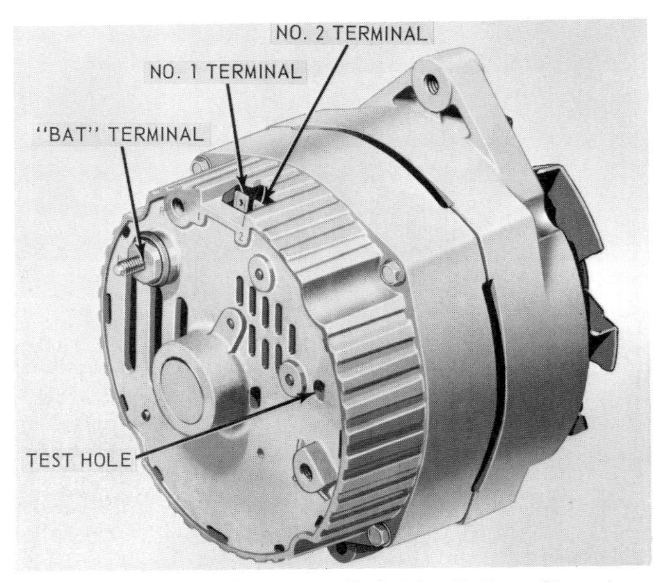

Fig. 91—Typical Alternator ("A" Circuit Type Shown)

Fig. 91 illustrates one type of alternator which is used in many applications in the off-the-road and automotive field. It is shown in cutaway in Fig. 92.

VARIATIONS ON ALTERNATOR DESIGN

While basically every alternator is the same, there are variations for different uses:

- **Housing—open or closed**
- **Field Circuit—A or B type**

Let's look at these designs briefly.

Open-Type Alternators

The open-type alternator has openings in its housing for better air circulation. A fan mounted on

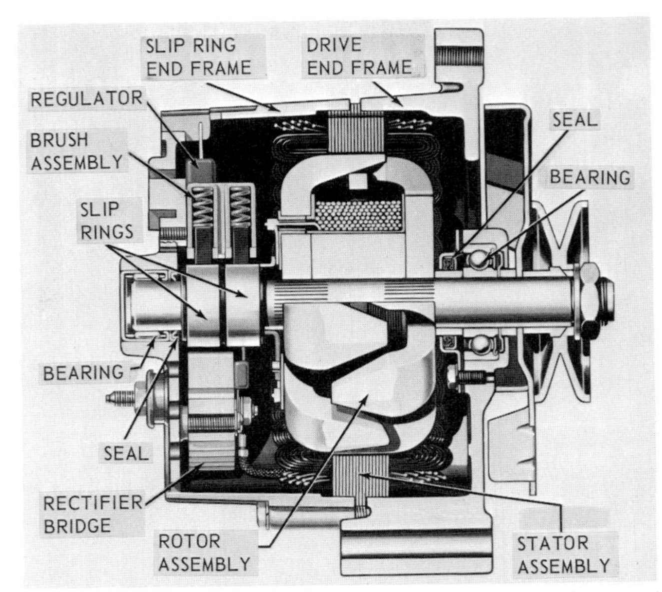

Fig. 92—Typical Alternator ("A" Circuit Type)

the pulley end of the rotor shaft helps circulate air to cool the inner parts of the alternator.

Closed-Type Alternators

The closed-type alternator is completely encased, except for the pulley and fan assembly and its terminals.

It is designed for use in extremely dusty conditions or in areas where a spark might ignite combustible material. However, it is still necessary to cool these units by either of two methods.

1) Air cooling

2) Oil cooling

AIR-COOLED ALTERNATORS

The first method of cooling is by circulating air. The fan is driven by the drive pulley and forces air around the sealed housing. Some alternators use two fans for cooling. The housing itself is made of lightweight aluminum which helps to dissipate the heat. The housing may also be ribbed or finned for better circulation of air.

OIL-COOLED ALTERNATORS

The other method of cooling is by circulating oil through the interior of the alternator.

The oil-cooled type is usually confined to use in stationary electrical systems. This design also eliminates the brush and slip ring assemblies.

Alternator Field Circuit Variations

There are two basic field circuit connections for an alternator—"A" circuit and "B" circuit. These circuits are similar to those described on page 4—8 for the generator.

"B" CIRCUIT FIELD

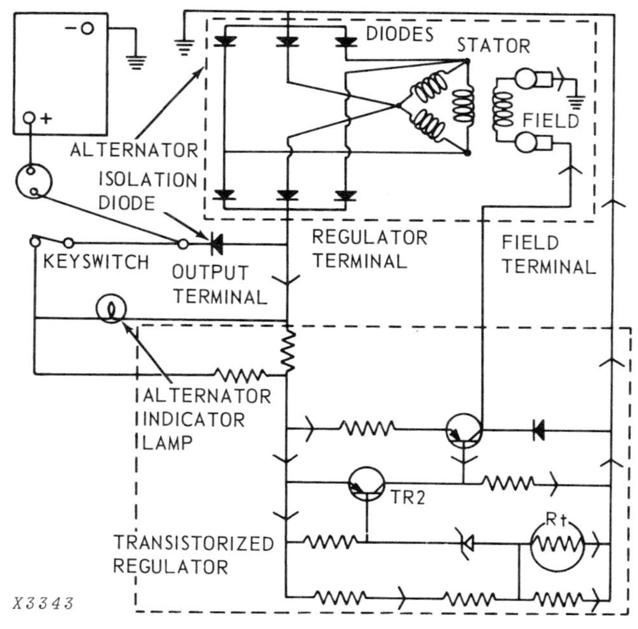

Fig. 93—"B" Circuit Field for Alternator

In a "B" circuit alternator (Fig. 93) the regulator is located before the field. The current flow is usually from the regulator terminal of the alternator (the output diode assembly, usually the positive diodes) to the regulator. After the regulator the current flows to the field coil in the rotor, then to ground, and finally to the negative or return diode assembly. Full alternator output is obtained by connecting the field terminal to the regulator terminal or output terminal.

"A" CIRCUIT FIELD

In an "A" circuit alternator (Fig. 94) the regulator is located after the field, between the field and the alternator ground or negative diodes. Full alternator output is obtained by grounding the field windings. One alternator has a tab in a test hole (see Fig. 91) so that the field is grounded by placing a screwdriver against the tab end and the alternator frame.

DIODE TRIO

To further isolate the field and regulator circuit, some alternators use a separate diode assembly to obtain field current from the stator (Fig. 94).

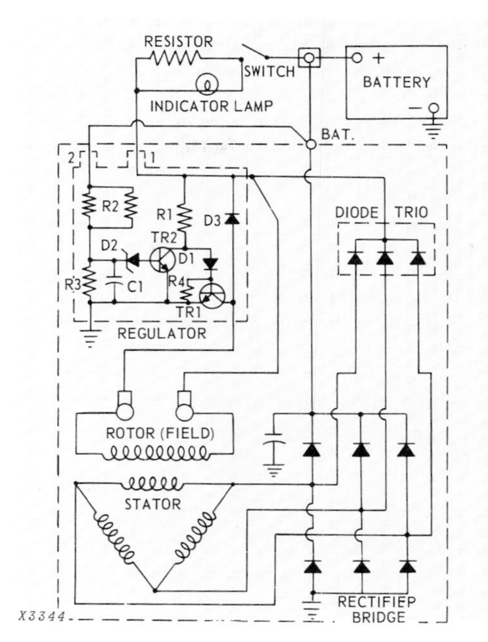

Fig. 94—"A" Circuit Field for Alternator

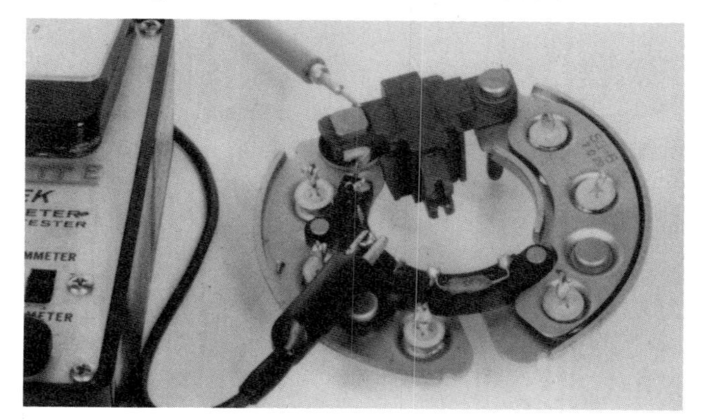

Fig. 95—Diode Trio

This diode assembly has three diodes and is called a diode trio (Fig. 95).

SUMMARY: OPERATION OF ALTERNATORS

In summary:

- **An alternator is an AC generator.**
- **It rectifies current to DC using diodes.**
- **Voltage is generated by a moving field in a stationary conductor.**
- **This is reverse of generator principle.**
- **Alternator = rotor (moving) + stator (fixed) + rectifier (diodes).**
- **Rotor-stator produces AC current.**
- **Rectifier diodes convert current to DC.**

TESTING OF ALTERNATORS

NOTE: The following discussion refers to the two particular types of alternators shown on pages 4—39 to 4—42. For information on other types, refer to the Technical Manual covering that model.

General Testing

Many of the general testing procedures and precautions that applied to the DC generator also apply to the alternator.

1. Make sure that all components, especially the alternator, are the exact models specified for the machine.

2. Make a complete *visual* check of all leads, wires, connections, etc., to find possible loose connections, broken wires, worn brushes and dirty slip rings.

3. Use a recommended testing procedure to make a complete *circuit check* of all the circuit components. This will isolate the problem.

4. In testing or repairing the alternator, follow the recommended testing and service procedures and specifications. These will be found in the alternator service manual or the machine technical manual.

5. Disassemble the alternator only as far as necessary to make adjustments or repairs.

6. The following list of tools and test equipment are generally used:

a. Machine or alternator Technical Manual for reference use.

b. Normal and special tools as noted by the Technical Manual.

c. Ammeter.

d. Voltmeter.

e. Ohmeter (1½ volt)

f. Variable carbon pile resistor.

g. Test lamp in series with 12-volt battery.

NOTE: A 110-volt test lamp can be used, but NOT to test diodes or any components while they are connected to the diodes. Use the lamp with extreme care.

Precautions for Alternator Testing

Below is a list of general precautions which must be taken prior to and during alternator tests:

1. Before disconnecting the battery or alternator from the circuit, satisfy yourself that the proper polarity is observed.

2. Before installing the battery or alternator, BE SURE THAT PROPER POLARITY IS OBSERVED. Turn off all switches and accessories. Then, before connecting the battery ground strap, momentarily touch it to the battery post. No sparks should occur.

3. Disconnect the battery ground strap when using or connecting a battery charger if you are not sure the charger is safe for an alternator charging circuit.

4. Never operate the alternator on open circuit. All three components of the circuit must be connected.

5. Never short or ground the alternator terminals unless instructed.

6. When making alternator tests in the circuit, be sure battery is in good condition and fully charged for control of alternator output with an external load.

7. IMPORTANT: NEVER ATTEMPT TO POLARIZE AN ALTERNATOR. To do so will damage the rectifying parts of the unit.

Failure to take the above precautions can result in damage to the regulator, the rectifier diodes, or other parts of the alternator.

General Diagnosis of Alternators

The following chart can help you to locate malfunctions within the alternator charging circuit.

At the end of this chapter we will relate these alternator problems to the whole charging circuit.

Below each problem is listed the most likely causes.

Alternator Troubleshooting Chart

ALTERNATOR FAILS TO CHARGE

1. Alternator belt loose.

2. Open or high resistance in charging circuit.

3. Worn or defective brushes.

4. Malfunctioning regulator.

5. Open isolation diode.

6. Open rotor field coil.

LOW OR UNSTEADY CHARGING RATE

1. Belt loose.

2. Open or high resistance in the charging circuit.

3. Open, worn, or defective brushes.

4. Faulty regulator.

5. Shorted or open rectifier diode.

6. Grounded or shorted rotor field coil.

7. Open, grounded, or shorted stator windings.

EXCESSIVE CHARGING RATE

1. Loose alternator or regulator connections.

2. Malfunctioning regulator.

NOISY ALTERNATOR

1. Defective or worn belt.

2. Misaligned belt or pulley.

3. Loose pulley.

4. Worn bearings.

5. Shorted rectifier diodes.

From this list, make the easy checks first. A good visual check of the charging components may help you to catch the more obvious causes of failures.

Alternator Bench Testing

As you can see, most of the problems listed in the chart can be caused by any one of the charging circuit components. This means that a *complete* circuit check must be made to isolate the faulty component.

Once you have located the problem in the alternator itself service the alternator on the bench.

Disassembly of Alternator

First disconnect the battery ground cable. Then disconnect the wires and drive belt from the alternator and remove the alternator from the machine. Select a clean, well-lighted work area for alternator service.

Prior to disassembly, take all the pre-test precautions. Also, remember that the alternator should be disassembled only as far as necessary to correct the problem.

For detailed procedures, refer to the alternator or machine technical manual. Here we will dismantle a standard alternator to give you only the *general* procedures.

REMOVING MAIN PARTS

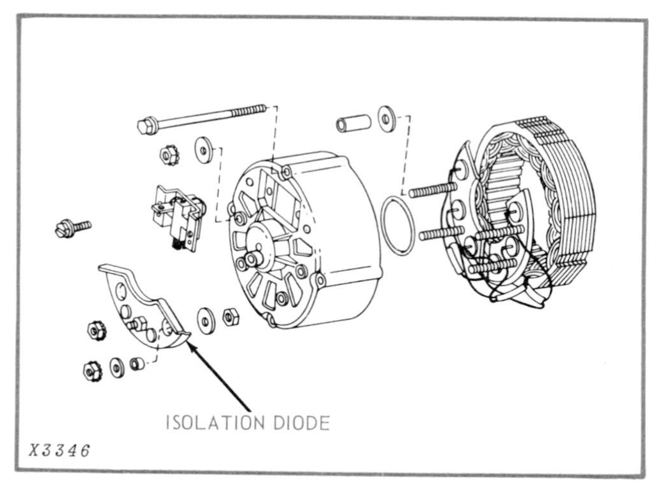

Fig. 96—Isolation Diode

1. On alternators equipped with an isolation diode, remove this assembly. (Fig. 96).

2. On some alternators, the regulator is mounted directly on the alternator housing. In this case, remove the mounting screws and wiring leads and lift off the regulator.

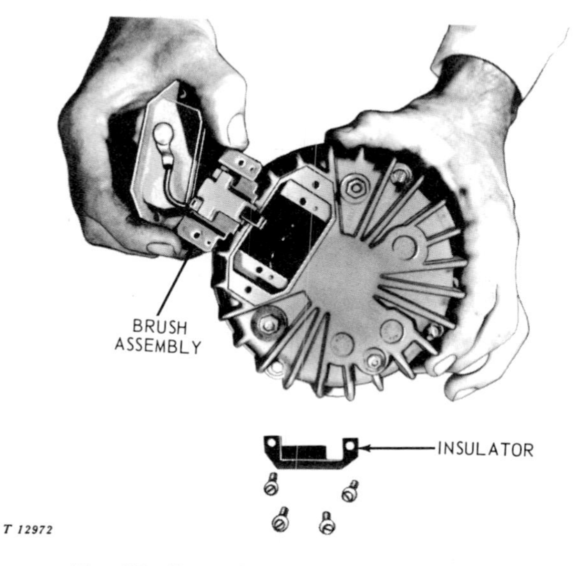

Fig. 97—Removing the Brush Assembly

3. The brush assembly may be either a separate unit or part of the slip ring end frame. If it is separate unit, remove the mounting screws and any wire leads. Then tilt it slightly at the top and slip it out of its cavity (Fig. 97).

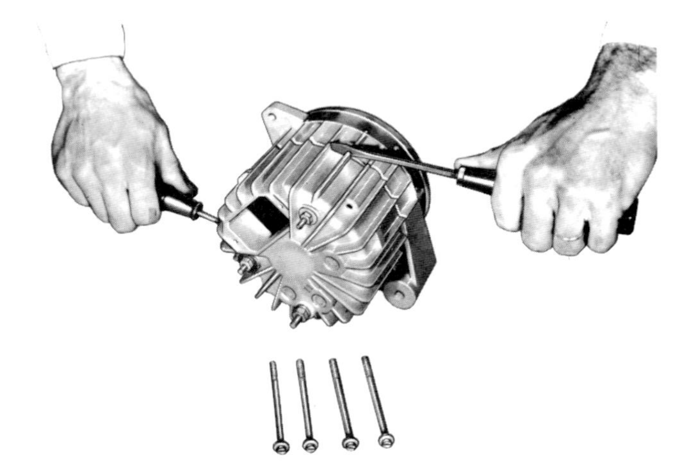

Fig. 98—Removing the Slip Ring End Frame

4. Separate the slip ring end frame, rectifier diode assembly, and stator from the drive end frame and rotor. Do this by removing the long through bolts. Slip screwdrivers in the slots between the drive end frame and stator (not more than 1/16-inch—any deeper may damage the stator windings). Then gently pry the stator loose (Fig. 98).

5. To remove the slip ring end frame from the diode and stator assemblies, remove the lock nuts, insulators, and insulating washers from the diode terminal posts.

IMPORTANT: Notice the position of the insulators before removing them. They must be replaced in the same order.

6. On closed-type alternators, separate the diode assembly from the stator by removing the stator leads from the screw terminals (Fig. 99).

On some open-type alternators, the stator leads must be unsoldered from the diodes to separate the two assemblies (Fig. 99).

SOLDERING AND UNSOLDERING DIODE LEADS (SOME OPEN-TYPE ALTERNATORS)

When soldering or unsoldering diodes, be sure to grip the diode lead with a pair of needle-nosed pliers between the connection and the diode. The pliers will act as a heat sink and prevent heat damage to the diode hermetic seal.

Also, avoid bending the diode lead if possible. If the lead must be bent, use two pliers. Place one between the diode and the bend and use the other pliers to bend the lead very carefully.

IMPORTANT: Diodes will be ruined if the hermetic seal at the diode lead is broken.

When soldering a diode lead, use a solder containing 60 percent tin and 40 percent lead.

IMPORTANT: Never use an acid-core solder. Use only a rosin-core solder.

REMOVING BEARINGS

The rear bearing on the slip ring of the rotor shaft need not be removed unless it must be replaced. The front bearing need not be pressed off the shaft unless the drive end frame is to be removed. Be sure to use the recommended pullers and presses for these jobs.

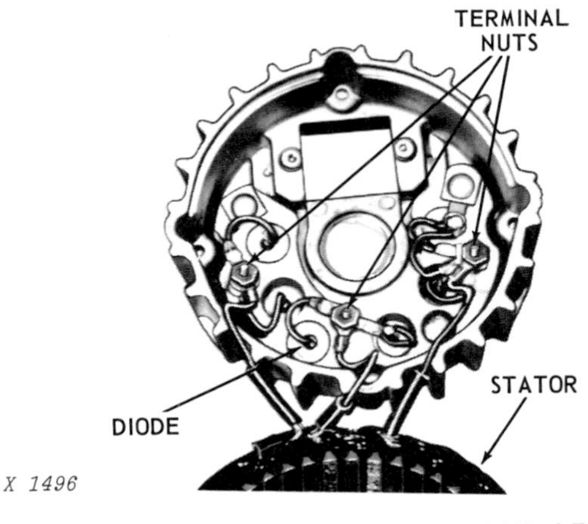

OPEN TYPE – UNSOLDER DIODE LEADS

CLOSED TYPE – REMOVE TERMINAL LEADS

Fig. 99—Removing Diode Assemblies

REPLACING DIODES

On some closed-type alternators, individual diodes may be replaced by pressing the faulty diode out of the plate and pressing a new one in (Fig. 100). On other alternators, replace the whole positive or negative diode plate assembly.

IMPORTANT: Be sure to replace a diode with one of the same type. Positive diodes are marked in red, while negative ones are marked in black.

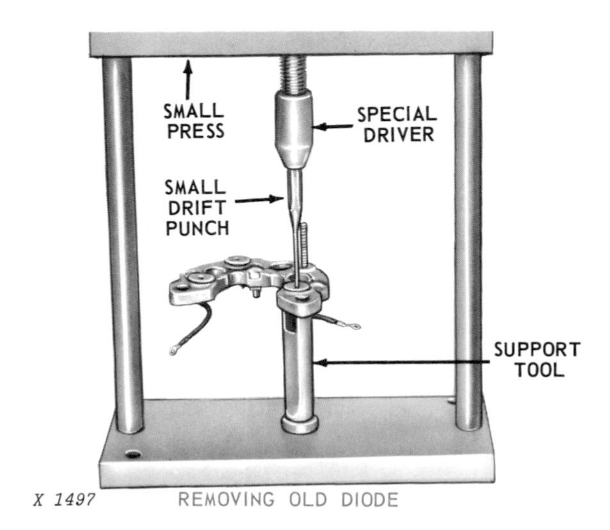

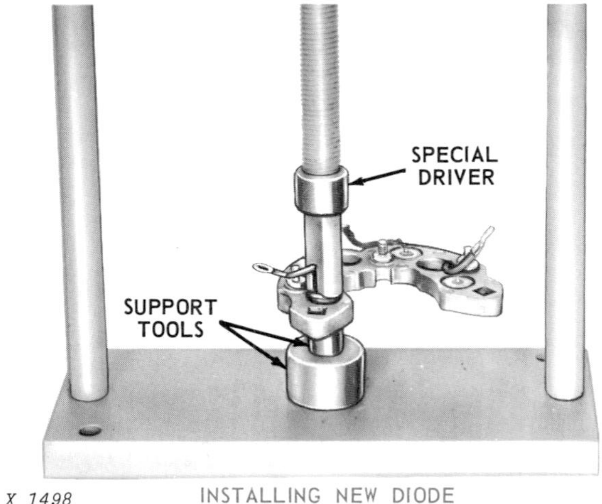

Fig. 100—Replacing Individual Diodes (Closed-Type Alternators Only)

Testing the Individual Components

The tests given here are only general outlines to give you an idea of what tests should be performed on actual electrical components. The alternator service manual or machine Technical Manual will give you the detailed tests and specifications.

Testing the Brushes

First make a visual check of the brushes and leads. If the brushes are worn down to an exposed length of ¼-inch (6 mm) or less or are otherwise defective, replace the entire brush assembly.

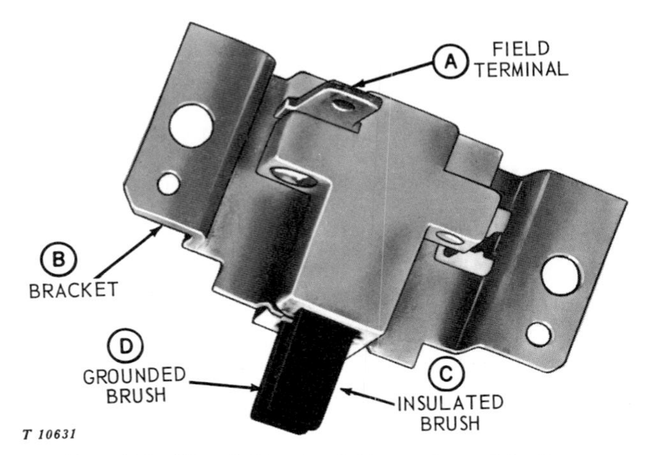

Fig. 101—Visual Check of Brush Assembly On 12-Volt Alternator

An *insulation test* can be made by connecting an ohmmeter or test lamp to the field terminal (A) and assembly bracket (B). (See Fig. 101.) Resistance should be infinite or the test lamp should not light. If not, brush assembly is shorted and must be replaced.

A *brush continuity test* is made by connecting an ohmmeter to the field terminal (A) and a brush (C). Also between bracket (B) and brush (D). The resistance should be zero in both cases.

NOTE: In some alternators, brush "D" is also insulated and connected to a separate terminal.

While making the tests, wiggle the brushes and leads to locate poor connections or intermittent grounds.

Testing the Diodes

Test the diodes in two steps:

1) Test the diodes before they are disconnected from the stator.

2) If a malfunction appears during this test, disconnect the suspected diode and retest the others.

The first indication of a faulty diode may be a humming or growling sound heard while the alternator is operating. At the same time, the alternator will usually have a low output.

Use a commercial diode tester, a 1½-volt ohmmeter or a test lamp connected in series to a

12-volt battery to test the diodes. *Never use a 110-volt test lamp to test the diodes.*

Some commercial diode testers will check the diodes in one step.

TESTING INDIVIDUAL DIODES WITH OHMMETER

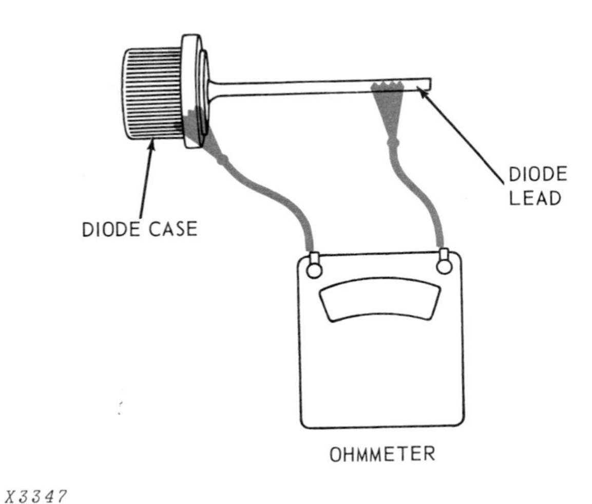

Fig. 102—Testing Diode with an Ohmmeter

Connect one ohmmeter lead to the diode case and the other lead to the diode lead (Fig. 102). Note the ohmmeter reading.

Now reverse the ohmmeter leads and note the reading. A good diode has a high and a low reading.

After testing each group of diodes, compare the ohmmeter readings. If the high and low readings on all the diodes are the same, the diodes are good. If even a slightly different reading is obtained for one diode, it is probably defective. Disconnect it and recheck the diodes.

TESTING DIODES IN THE RECTIFIER BRIDGE

Fig. 103—Testing Diodes in the Rectifier Bridge

Connect one ohmmeter probe to the metal base. With the other ohmmeter probe, touch each of the three diodes on one side of the rectifier bridge (Fig. 103). Note the ohmmeter readings.

Reverse the ohmmeter probes and make the same checks. Note the ohmmeter readings.

Repeat this same procedure on the diodes on the opposite side of the rectifier bridge.

Ohmmeter readings should indicate that each diode should have continuity in only one direction. A shorted diode would have continuity in both directions. An open diode would not have continuity in either direction.

TESTING DIODES WITH TEST LAMP

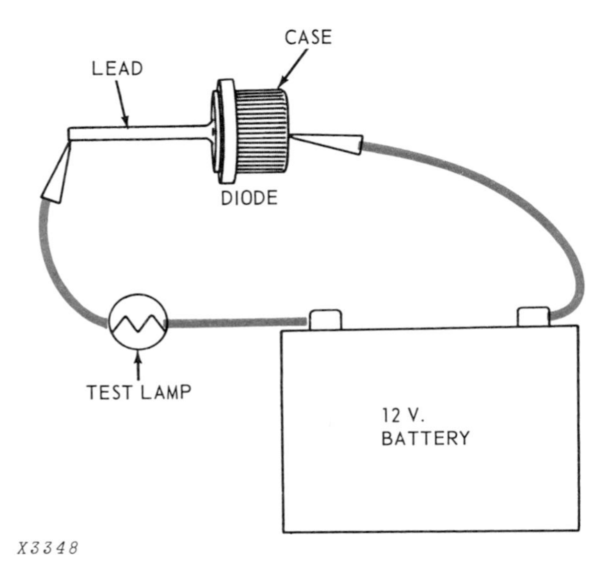

Fig. 104—Testing Diode with Test Lamp

If a test lamp in series with a 12-volt battery is used, place one test probe on the base of the diode and the other probe on the diode lead. Then reverse the probes (Fig. 104).

The lamp should light in one direction, but not the other. If the lamp lights in both directions, one or more of the diodes in the group is shorted. If the lamp does not light in either direction, all of the diodes are open.

Disconnect the diodes from the stator and test them one at a time to locate the shorted diode.

TESTING DIODES WITH ALTERNATOR DIODE TESTER

A commercial diode tester usually has its own directions. Follow that testing procedure when checking out diodes.

When replacing diodes, remember their polarity. **Positive diodes are marked in red, negative diodes in black.**

Testing the Stator Assembly

Again, a visual inspection may uncover some defects in the stator windings. For example, shorted windings are usually discolored.

Test stator as follows:

1. If a 110-volt test lamp or an armature tester is used, disconnect the diodes from the stator to prevent damage to the diodes.

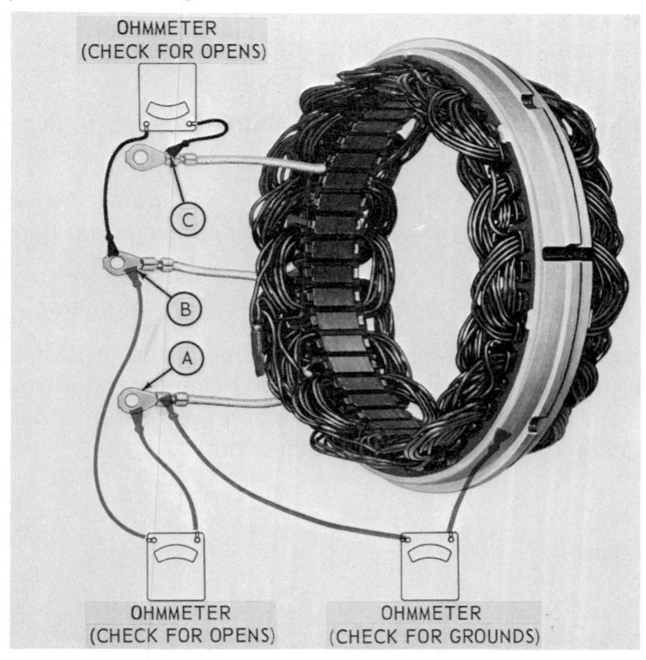

Fig. 105—Checking "Y"-Connected Stator with an Ohmmeter

2. Disconnect the stator leads from the diode assemblies.

3. *Check for a grounded winding* by connecting an ohmmeter to one stator lead and to the stator frame. The ohmmeter reading should be infinite.

4. Check a "Y"-connected stator for an *open circuit* by connecting an ohmmeter from point A to B and B to C (Fig. 105).

5. Some delta-connected stator leads must be welded to the terminals and cannot be checked for an open circuit except by a sensitive ohmmeter.

6. If a very sensitive ohmmeter is available, the following procedure may be used. However, it will not detect a short or open circuit that occurs only when the stator is hot.

7. To check for a short-circuited winding or, on a delta winding, and open circuit, carefully zero the ohmmeter and connect the leads to A and B (Fig. 105). A typical reading would be 0.1 ohm for a delta stator or 0.2 ohm for a Y stator. If on a delta connected stator, the reading was approximately 0.2 ohm and then a 0.1 ohm reading was obtained in step 3, the phase winding is probably open-circuited. A high reading indicates two delta phase windings are open-circuited. Now touch the A and B leads together several times. The meter pointer should deflect to zero. If there is no pointer movement, the windings are shorted.

8. Repeat the open circuit and short circuit tests between A and C and between B and C in Fig. 105.

NOTE: A single short-circuited winding of a delta-wound stator is very difficult to distinguish. Therefore, the accuracy of this test depends upon meter sensitivity to resistances of 0 to 1 ohm.

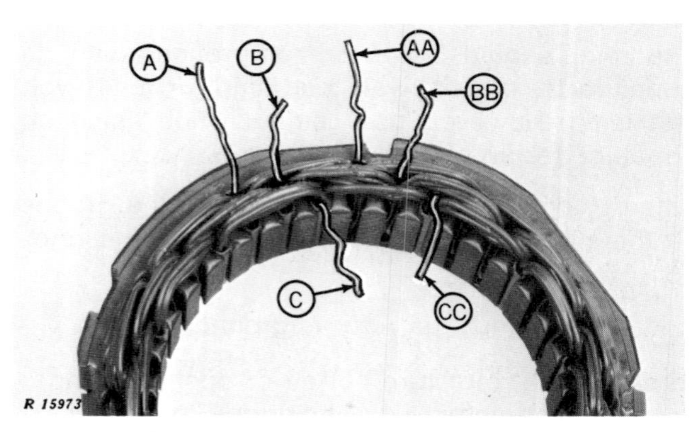

Fig. 106—Testing the Stator Assembly ("Delta" Connected Stator Shown)

9. If a sensitive ohmmeter is not available for the above, and the leads are not welded together, carefully disconnect the stator leads (Fig. 106). Stator leads may be brittle if they have been overheated or if they are old.

10. Each phase winding may now be checked for shorts to an adjacent phase winding (A to B, B to C and A to C). A delta-connected stator may be also checked for an open-circuited phase winding by connecting A to AA, B to BB, and C to CC.

11. After testing, connect stator leads again.

12. This completes the testing of the stator assembly.

Remember that a shorted stator is usually discolored and will smell. Replace the stator only after other components have proved to be satisfactory.

Testing the Rotor Assembly

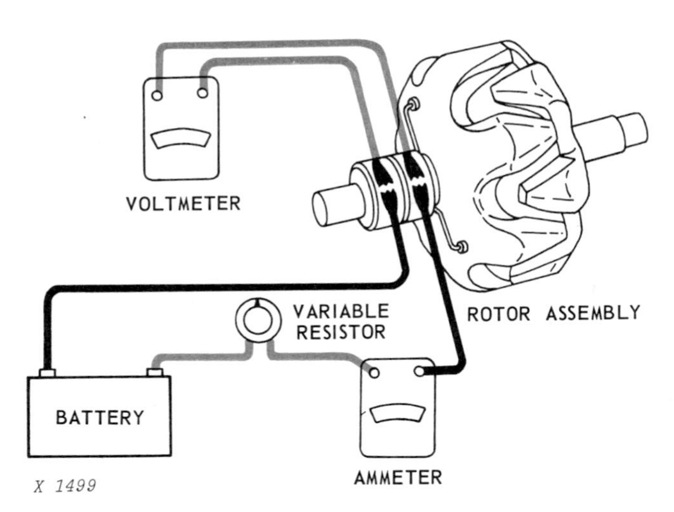

Fig. 107—Testing the Rotor Assembly

The rotor should be tested for grounds with an armature tester, a 12-volt test lamp, or a 110-volt test lamp. However, use caution when using an armature tester or a 110-volt lamp.

When using a test lamp, place one of the probes on the slip ring and the other probe on the rotor coil.

If the lamp lights, the rotor is grounded.

To check the rotor for shorted or open windings, connect a voltmeter to the slip rings (Fig. 107). Then connect an ammeter in series with a variable resistor to the slip ring and 12-volt battery as shown. Set the variable resistor to maximum resistance. Connect the other slip rings to the battery and adjust the resistor to obtain full battery voltage.

IMPORTANT: When removing the wire from the slip ring, it will cause an arc and may damage the slip ring surface, requiring clean-up.

Rotor field current draw should equal the amount specified for the applied voltage. For example, it should be between 2.0 to 2.5 amperes at 12.4 volts.

Shorted windings are indicated by excessive current draw, open windings by no current draw at all.

If the slip rings are scored, turn them in a lathe until they just clean up. If desired, polish them sparingly with No. 00 sandpaper or No. 400 grit silicon carbide paper.

Assembly of Alternator

Reassembly of the alternator is generally just the reverse of disassembly. However, observe the pretest precautions again and use the specific directions for assembly given in the machine Technical Manual. Always test the alternator after assembly to be sure it is operating properly.

ALTERNATOR REGULATORS

The AC **regulator** is the *control for the alternator.* Otherwise, the alternator would produce too much voltage.

The regulator does this by placing a resistance in the field circuit which reduces current flow to the alternator rotor. This is much the same principle as used by the regulator in DC charging circuits.

How is the AC regulator different from the DC regulator?

1. No current regulator is needed, since the alternator limits its output by setting its own opposing field during operation.

2. The AC regulator is often only a voltage regulator.

There are special cases where combination regulator or field relay units are used for AC circuits, but these are not basic to our present story. (See later under "Other Types of Alternator Regulators.")

TRANSISTORIZED VOLTAGE REGULATOR

The most popular AC voltage regular is **transistorized (solid state).** It is a fully electronic unit composed of resistors, diodes, Zener diode, transistors and thermistor. These components (Fig. 108) are usu-

Fig. 108—Transistorized Voltage Regulator
(For Circuit, see Fig. 110)

ally in a sealed case. This eliminates the need for adjustments or the chance of failure due to dirt or moisture.

For the basic operation of transistorized regulator components, see Chapter 1. Here we will only give a basic description:

1. **Resistors**—Devices made of wire or carbon which present a resistance to current flow.

2. **Zener Diode**—A diode which is connected in a reverse bias and below a certain voltage works like the usual diode. However, beyond a predetermined voltage, the Zener diode will conduct reverse current.

3. **Transistor**—Semi-conductors which control the flow of current in the circuit by either allowing it to flow or stopping it.

4. **Thermistor**—A temperature-compensated resistor. The degree of its resistance varies with the temperature. It controls the Zener diode so that a higher system voltage is produced in cold weather, when needed.

Operation of the Transistorized Regulator

Basically, the AC regulator has two jobs:

1. Allow battery current to excite the alternator field coils.

2. Control charging voltage at safe values during operation.

Figs. 109 and 110 show the circuits for two typical alternators.

Since all regulators are similar in operation, let's take the regulator in Fig. 110 and see how it operates.

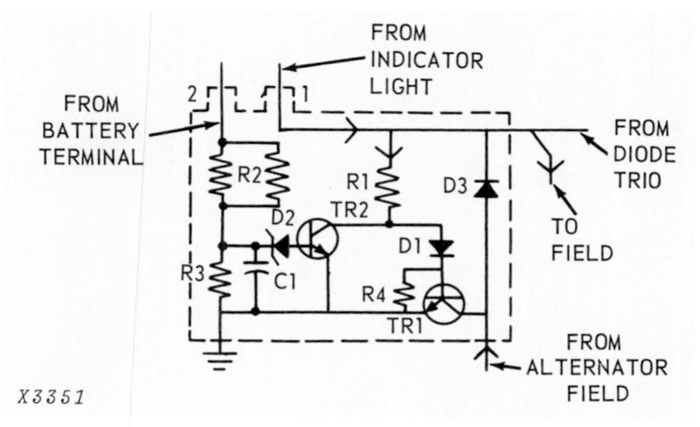

Fig. 109—Circuit of One Transistorized Regulator (With NPN Transistors)

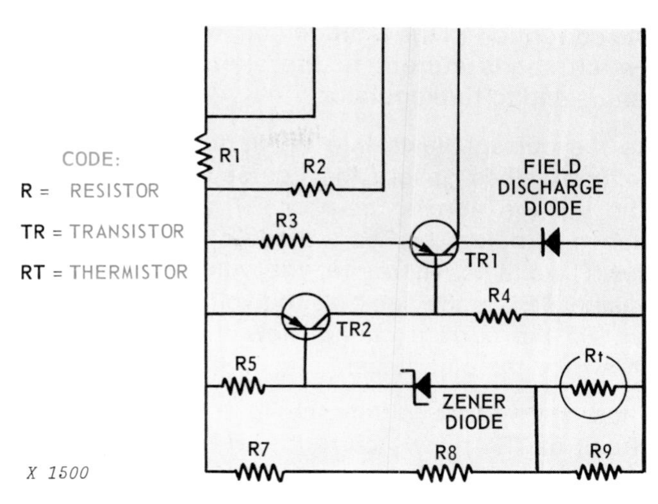

Fig. 110—Circuit of Another Transistorized Regulator

How The Regulator Works During Starting

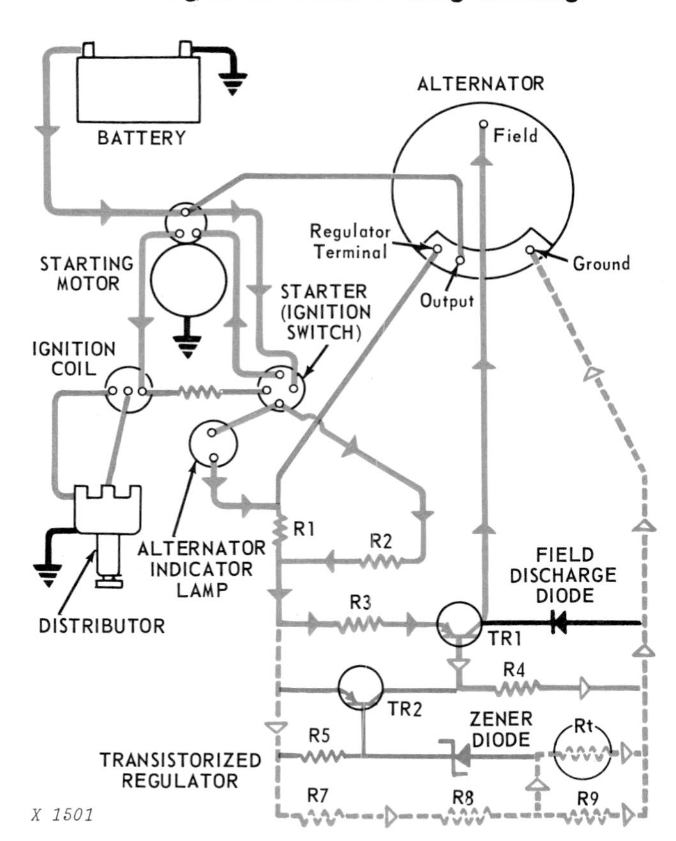

Fig. 111—Operation of Transistorized Regulator during Starting

When the starter switch is turned on, the circuit is completed (Fig. 111). Battery current flows to the starter solenoid and to the starter (ignition) switch as shown by the red lines. The starter solenoid sends the current on to the ignition coil, while the starter switch sends current to the alternator indicator lamps and to the regulator.

As the current flows into the regulator, different voltage values govern the course of the current. The voltage across resistors R7 and R8 for instance, is below the Zener diode critical or "break down" voltage. Therefore, the voltage felt at the base of TR2 is the same as the voltage at its emitter. So the current cannot flow through TR2 (as shown by the blue lines).

Thus the voltage difference in the emitter-base circuit of TR1 allows current flow from its emitter through its base and collector. The collector current then goes on to excite the alternator field (vertical red line). At the same time a slight amount of current flow travels to the alternator ground as shown by the dotted red line.

How The Regulator Works During Engine Operation

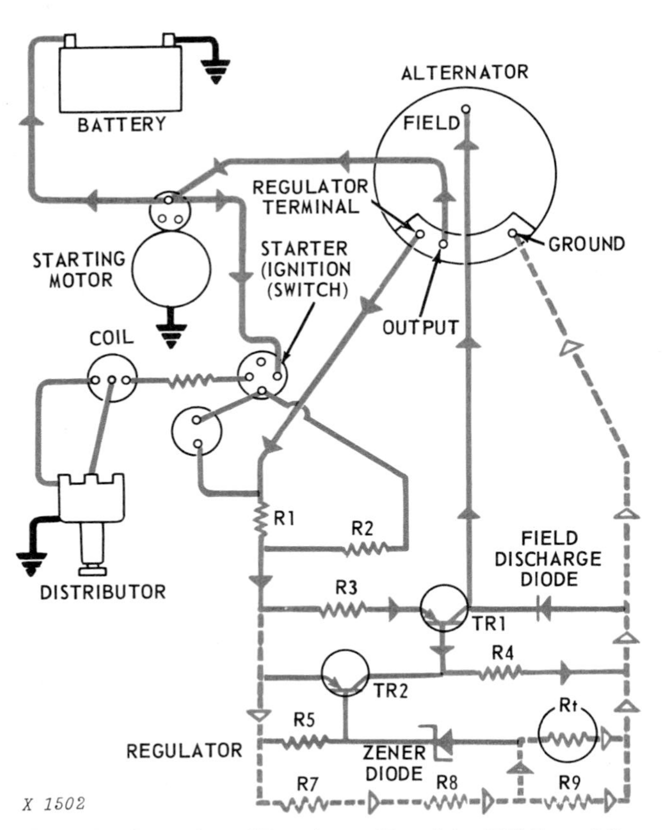

Fig. 112—Operation of Regulator—Transistor TR1 Turned On

The early part of engine operation (Fig. 112) is similar to the starting period above except that as the engine speeds up the alternator field around the rotor generates voltage which flows out to supply loads.

However, the voltage values are still the same and transistor TR1 still conducts the current to the alternator field as shown by the vertical red line.

As the engine operates and load requirements begin to decrease, the alternator voltage builds up (Fig. 113).

This causes the voltage across the resistors to also increase. Then the voltage across R7 and R8 becomes greater than the Zener diode critical voltage.

The Zener diode immediately "breaks down," allowing current to flow through in a reverse direction. This "turns on" transistor TR2 and so current is able to flow through TR2's emitter-base and collector.

When current flows through TR2, the voltage at the base of TR1 is equal to or greater than at its emitter. This prevents current from flowing through TR1 to the alternator field.

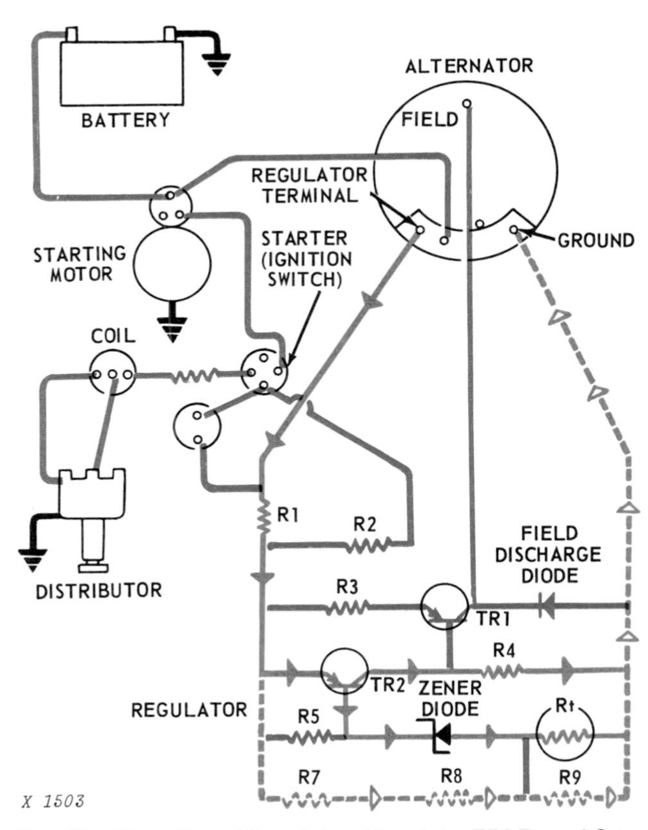

Fig. 113—Operation of Regulator—Transistor TR2 Turned On

This collapses the field and reduces the output of the alternator, protecting the circuit.

The system voltage then drops below the critical voltage of the Zener diode and it stops conducting. This turns off TR2 and turns on TR1 and current again flows to the alternator field. This operation is repeated many times a second.

In effect, the two transistors act as switches controlling the voltage and alternator output.

Note the *field discharge diode* in Fig. 113. This diode prevents damage to transistor TR1.

When TR1 turns off, the alternator field current cannot drop immediately to zero, because the rotor windings cause the current to continue to flow.

Before the current flow reaches zero, the system voltage and regulator start current flow again. However, the decreasing field current flow induces a high voltage and this can damage the transistor.

The purpose of the field discharge diode is to divert the high voltage away from the transistor.

Operation of Regulator With NPN Transistors

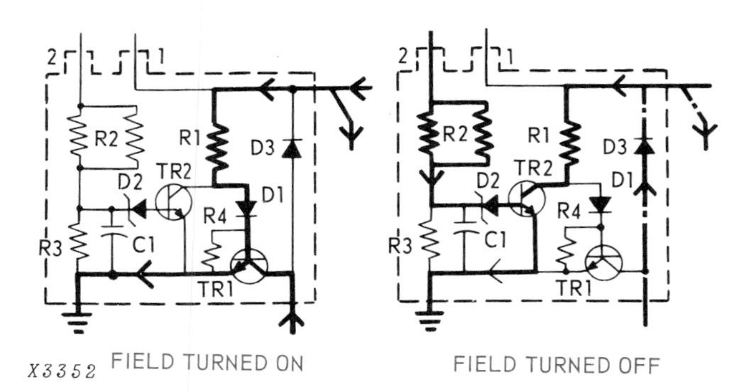

Fig. 114—Operation of Regulator with NPN Transistors

The second typical regulator we saw in Fig. 109 works as follows:

The alternator is generating current and supplying its own field current from the diode trio (Fig. 114). When voltage rises to the critical voltage of the zener diode (D2), the diode conducts and a *positive* voltage is applied at the base terminal of driver transistor "TR2" and turns it on. (The positive voltage at the base terminal turns an NPN transistor on.) This reduces the voltage at the base terminal of the power transistor (TR1), which turns the power transistor off.

Testing and Repairing the Transistorized Regulator

First make a complete circuit check to isolate the faulty component. Take all the normal precautions for safety.

For a general diagnosis, use the checks given earlier in the alternator testing section. Then make the more specific tests which follow.

For a particular test procedure, always follow the machine Technical Manual. The following test is an example of the type usually made on a transistorized regulator. Since the regulator is only used to control voltage, the voltage test is all that is required.

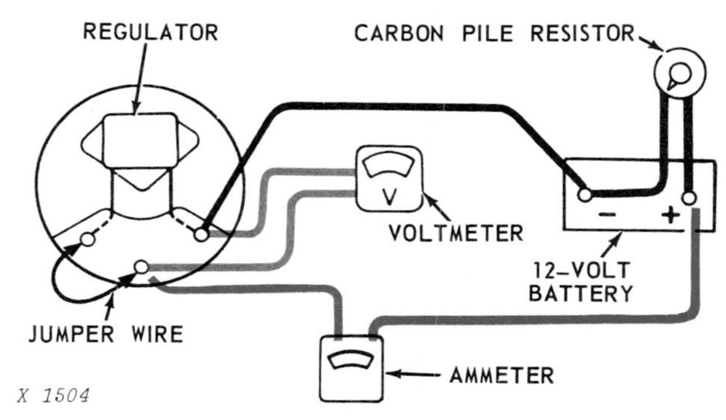

Fig. 115—Voltage Test for Transistorized Regulator

VOLTAGE TEST

1. This test can be performed either on or off the machine. Use an alternator that is known to be in good repair. Set up the test circuit shown in Fig. 115.

2. Connect a voltmeter to the alternator ground and output terminals as shown. Be sure to use a voltmeter with an accuracy within 0.1 volts.

3. Start the engine, momentarily connect jumper wire to excite the field, and apply a load of about 10 amperes (use lights, motors, carbon pile resistors, etc.).

4. Operate the circuit for about 15 minutes to stabilize the temperature of the regulators. Measure and record the temperature about one inch from the regulator case.

5. Compare the voltmeter reading with the voltage specifications listed in the machine Technical Manual. Do not forget to adjust the reading for the temperature recorded above.

ADJUSTMENT AND REPAIR

Some transistorized regulators have an adjusting screw for changing the operating voltage for different conditions.

Since most transistorized regulators are sealed units, they offer no means of repair. Therefore, they must be replaced if they are faulty. But at the same time, this type of regulator is usually more reliable than other kinds.

OTHER TYPES OF ALTERNATOR REGULATORS

As popular as the fully transistorized regulator is, it is not the only regulator used. Vibrating contact regulators, similar to the units used in DC charging circuits, are still in use today.

Transistor Vibrating Contacts Regulator

This combination of old and new, so to speak, has been effectively used in some AC charging circuits.

A typical model has a *circuit breaker* to protect the alternator in case of shorts, grounds or reverse polarity. A *voltage regulator* controls a *transistor* which in turn energizes the alternator field. *Diodes* are used as safety devices to protect the regulator, while *resistors* also aid in controlling the circuit.

DIAGNOSIS AND TESTING OF CHARGING CIRCUITS

In this chapter we have strongly suggested that you begin by testing the complete charging circuit. The checks given here should be done *before* any of the component tests given earlier in this chapter. A complete circuit check will isolate the failing component.

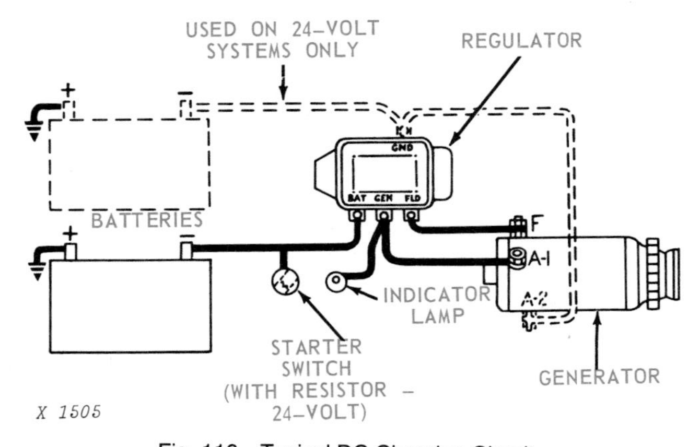

Fig. 116—Typical DC Charging Circuit

TROUBLE SHOOTING OF CHARGING CIRCUITS

When using this chart, assume that the battery is in good operating condition. (Battery problems are covered in Chapter 3.)

Problem	Possible Cause
1. Low Battery Output	a. Faulty Regulator—AC or DC
	b. Faulty Generating Unit—AC or DC
	c. Slipping Drive Belt—AC or DC
2. Low Generating Output	a. Slipping Drive Belt—AC or DC
	b. Poor Lead Connections—AC or DC
	c. High Circuit Resistance—AC or DC
	d. Cutout Relay Malfunction—DC
	e. Faulty Regulator Circuit Breaker—AC
	f. Regulator Diode Malfunction—AC
	g. Regulator Transistor Malfunction—AC
	h. Oxidized Regulator Points—AC or DC
	i. Open Isolation Diode—AC
	j. Shorted Field Circuit Windings—AC or DC
	k. Open Diodes—AC
3. Excessive Output	a. Faulty Regulator—AC or DC
	b. Grounded Field Terminal—DC
	c. High Voltage Regulator Setting—AC or DC
	d. High Temperature—AC or DC

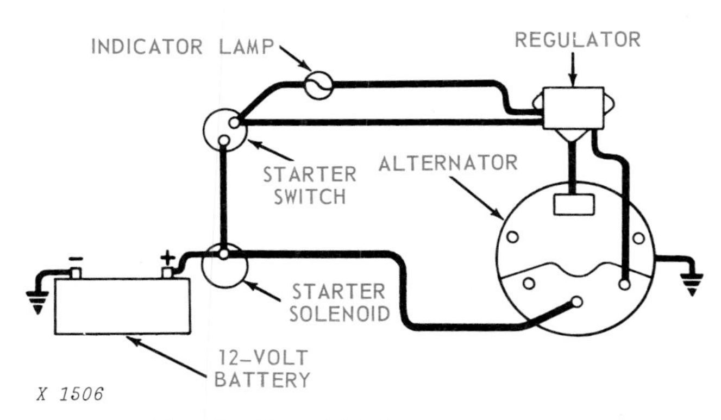

Fig. 117—Typical AC Charging Circuit

e. Shorted Regulator Transistor—AC

4. Noisy Generating Unit

a. Defective Bearings—AC or DC.

b. Loose Mounting—AC or DC

c. Shorted Diode—AC

d. Open Diode—AC

e. Brush Chatter—DC

5. Battery Uses Too Much Water

a. Regulator Setting Too High—AC or DC

b. Faulty Regulator—AC or DC

6. No Generating Output

a. Sticking Brushes—DC

b. Loose or Open Connections —AC or DC

c. Grounded, Shorted, or Open Armature—DC

d. Grounded, Shorted, or Open Field Circuit—DC

e. Grounded Terminals—DC

f. Dirty or Corroded Commutator—DC

g. Open Diodes—AC

h. Open Stator Windings—AC

i. Open Rotor Windings—AC

j. Faulty Regulator Circuit Breaker—AC

k. Faulty Cutout Relay—DC

l. Faulty Current Regulator—DC

m. Faulty Voltage Regulator—DC

n. Open Transistor—AC

o. Open Isolation Diode—AC

SAFETY RULES

Before we start the circuit tests, let's quickly review some of the precautions that have been mentioned in the previous sections of this chapter.

DC Circuits

1. Polarize the circuit after each test or adjustment.

2. Be sure the battery is in good operating condition before making any tests or adjustments. Review the battery safety messages in Chapters 4 and 5.

3. Disconnect the battery ground cable when removing the generator or battery.

4. Never immerse the circuit components in a cleaning solution.

AC Circuits

1. Never attempt to polarize the circuit.

2. Be sure the battery is in good operating condition before making any tests or adjustments.

3. Never operate the alternator in an open circuit, except when instructed in the Technical Manual.

4. Never short or ground the alternator terminals.

5. Do not disconnect the voltage regulator while the alternator is running.

6. Disconnect the negative battery cable first when removing the alternator or battery.

7. Do not use acid-core solder on the alternator terminals. Use only a rosin-core solder.

8. Never immerse the circuit components in cleaning solution.

TESTING THE DC CHARGING CIRCUIT

Remember that these are general tests performed on a circuit which has a standard shunt-type generator and a 3-unit regulator. The battery should be considered in good operating condition for these tests.

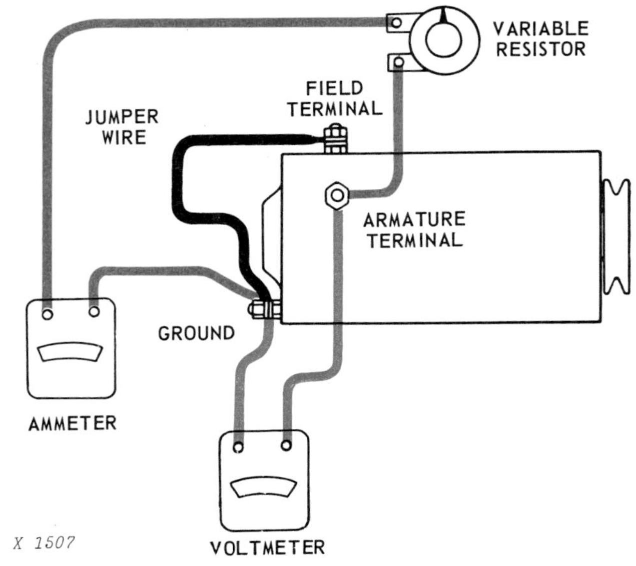

Fig. 118—Generator Output Test (12-Volt Circuit)

Generator Output Test

The test equipment you should have for this test are: *an ammeter, a voltmeter, a tachometer, a variable resistor and a jumper wire.*

You will notice that we have shown this equipment in its basic form. An electrical testing unit as shown in Chapter 2 may combine them all into one unit. If this is the case, follow the tester's instructions in making test connections.

TEST PROCEDURE

1. Make a visual check of all wire leads, connections, and brushes.

2. Disconnect the leads to the generator armature and field terminals.

3. Connect the ammeter and variable resistance in series with the generator armature terminal and ground (Fig. 118). Connect the voltmeter across the generator armature terminal and ground as shown.

4. Connect the jumper wire to the generator field terminal and to ground.

5. Adjust the variable resistor for the least resistance.

6. Operate the engine at the specified speed.

7. Adjust the variable resistor to obtain the specified voltage. (For example, 14-volts for a 12-volt circuit, or 28.5 volts in a 24-volt circuit.)

The generator output should be as specified. For example: 20 amps for a 12-volt circuit, or 10 amps for a 24-volt circuit.

If the results are not what is specified, the generator is faulty. Remove the generator from the circuit and make further tests.

If the test shows a good generator, reconnect the leads to the terminals and polarize the generator.

Resistance Tests

As is shown in Fig. 119, we are again performing this on a 12-volt circuit. Because these tests are given as examples of the type of tests used, you will need to consult the machine Technical Manual for particular testing connections, procedures, and specifications.

TESTING PROCEDURE

1. Connect the ammeter to the regulator battery terminal and to the wire disconnected from this terminal.

2. Connect a jumper wire to the generator field terminal and to ground.

3. With all accessories off, run the engine at a speed that will produce a 10-amp charging rate. *(DO NOT exceed 10 amps in a 24-volt circuit.)*

4. Measure the voltage at the following points:

a. From the generator armature terminal to a pin connector in the negative battery post.

b. From the generator frame to the regulator base.

c. From the generator frame to the grounded battery post.

5. Disconnect the jumper wire and turn on the lights. Continue the 10-amp charging rate. If you must lower the battery voltage to get the 10-amp charging rate, use a heavy-duty carbon pile resistor connected to the battery.

6. Check the voltage from the generator field to the regulator field terminal.

 Litho in U.S.A.

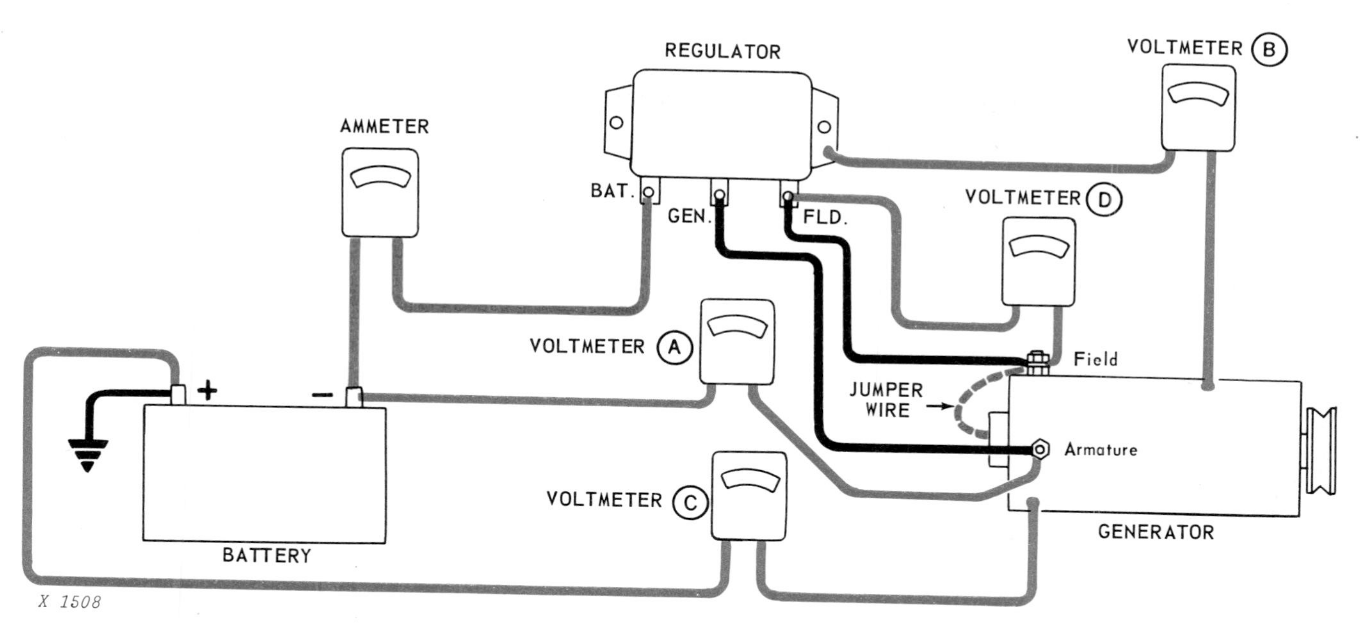

Fig. 119—Wiring Resistance Tests (12-Volt Circuit)

JUDGING THE TEST RESULTS

If the voltage readings are too high, check out each circuit component individually.

If the readings are normal, reconnect the regulator lead and polarize the generator.

Regulator Tests

OXIDIZED REGULATOR POINTS TEST

1. Connect the ammeter into the circuit as shown in Fig. 120.

2. Run the engine at a speed capable of producing 8 amps with all lights on.

3. Connect a jumper wire as shown in Fig. 120.

JUDGING THE TEST RESULTS

If the ammeter reading increases more than two amps, the regulator points are oxidized.

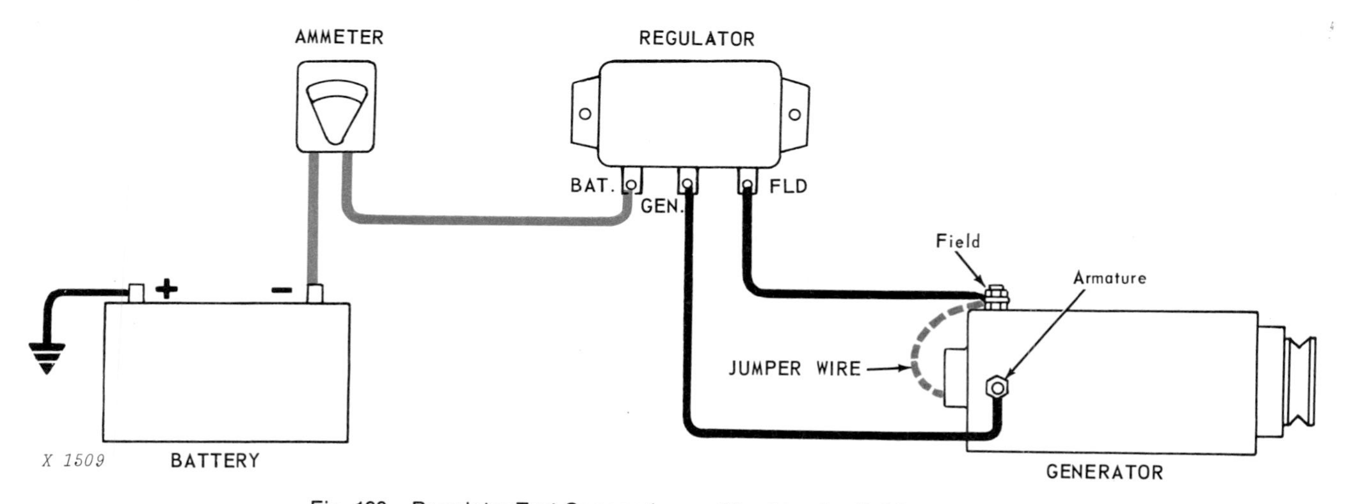

Fig. 120—Regulator Test Connections—Checking for Oxidized Points

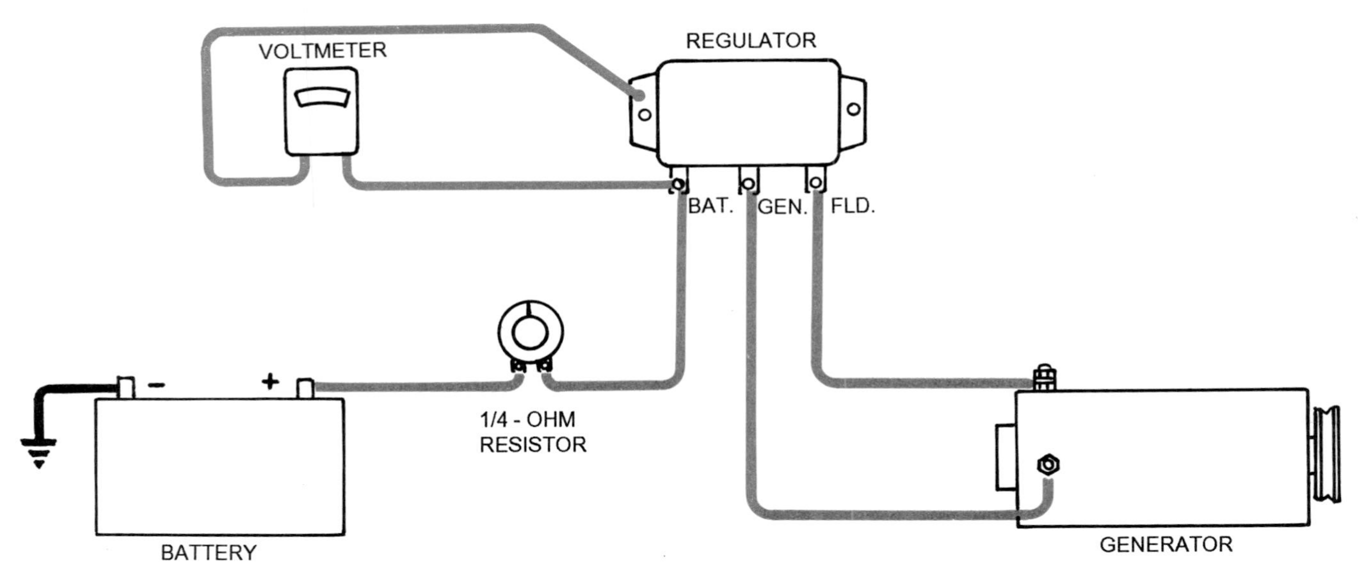

Fig. 121—Regulator Voltage Tests (12-Volt Circuit Connections)

REGULATOR VOLTAGE TESTS

1. Disconnect the battery wire from the regulator battery terminal.

2. Connect the voltmeter and a ¼-ohm resistor as shown in Fig. 121. (On regulators with accelerator windings, use the specified resistor only.)

3. Run the engine for 15 minutes to reach the regulator operating temperature.

4. Stop and restart engine to cycle the generator.

5. Run the engine at the specified speed (for example, 1900 rpm).

6. Check the voltmeter reading and the operating temperature of the regulator. Compare these readings with the specifications, such as the examples listed below:

Temperature	**Correct Voltage (12-Volt Circuit)**
85°F (29°C)	*14.2-15.0 volts*
105°F (41°C)	*14.0-14.7 volts*
125°F (52°C)	*13.8-14.5 volts*
145°F (63°C)	*13.5-14.1 volts*

JUDGING THE TEST RESULTS

If the specifications are not met, check the voltage regulator unit of the regulator more closely.

REGULATOR CURRENT TEST

Test instrument connections for this test are given in Fig. 121.

Test Procedure

1. Disconnect the battery wire from the regulator battery terminal.

2. Connect the voltmeter and ammeter as shown in Fig. 122.

3. Start the engine several times to lower the battery voltage.

4. Turn on all lights and run the engine at the specified speed.

5. The voltmeter reading should be 1 volt below the voltage regulator setting.

6. The ammeter reading should be as specified. For example, 18.5 to 21.5 amps on a 12-volt circuit.

JUDGING THE TEST RESULTS

Make further tests on the current regulator portion of the regulator if this test is unsatisfactory.

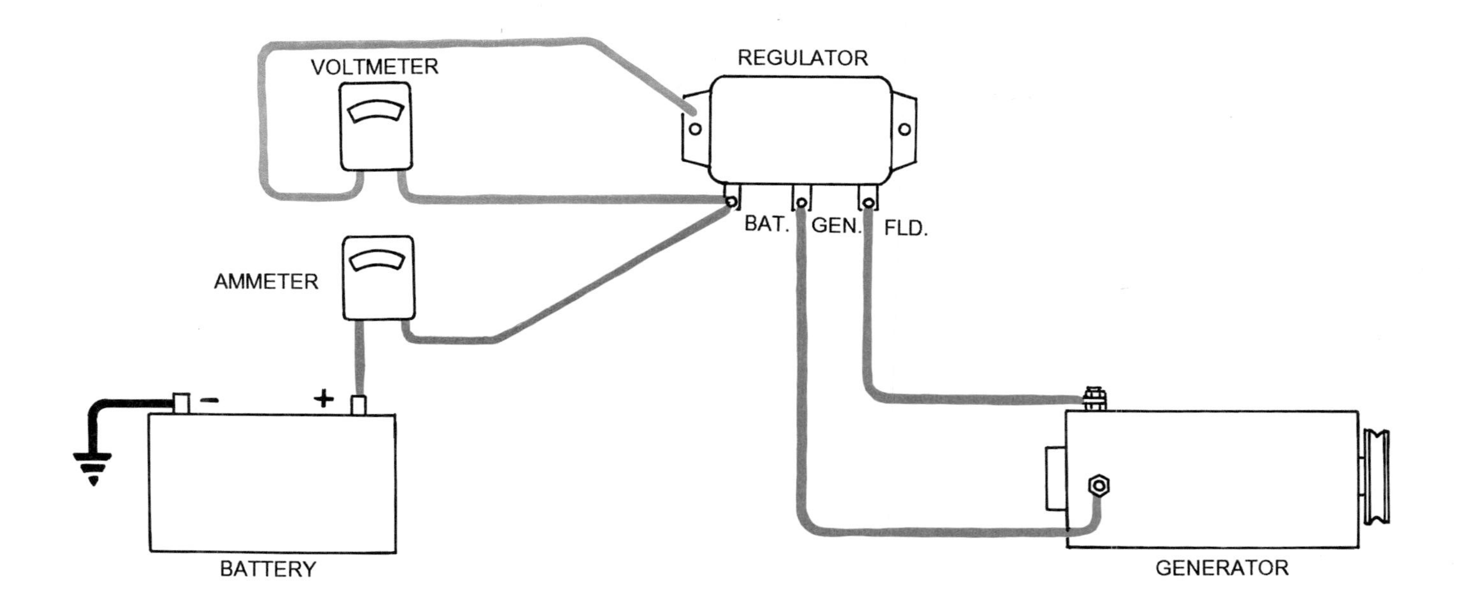

Fig. 122—Regulator Current Test (12 Volt Circuit Connections)

TESTING THE AC CHARGING CIRCUIT

In the following test examples, an open-type alternator and a transistorized regulator are used.

We remind you again to follow the precautions that we have given. To ignore them will only increase the chance of damaging the electrical components or injuring yourself.

The test equipment must be accurate and in good condition.

The instruments needed for these tests are:

1. *Voltmeter*

2. *Ammeter*

3. *Variable Resistor*

4. *Jumper Wire*

Before making any electrical tests, satisfy yourself that all leads are firmly connected and in good repair. Also check the alternator drive belt tension.

Test No. 1

1. Connect the voltmeter as shown in Fig. 123.

2. With the engine, ignition key switch, and accessories off, the voltmeter should read less than 0.1 volt.

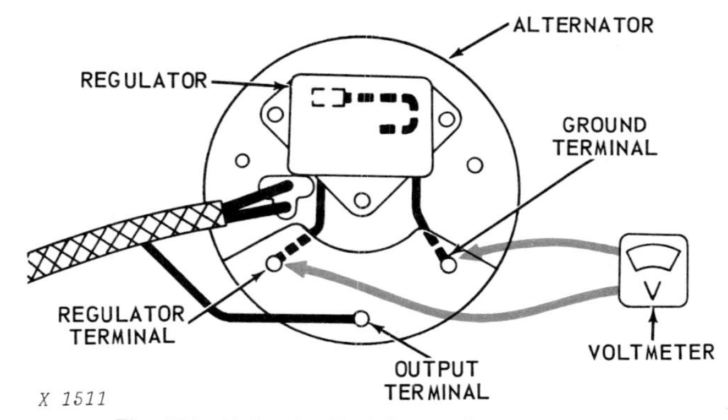

Fig. 123—Voltmeter Test Connections

JUDGING THE TEST RESULTS

A high reading indicates a shorted isolation diode or ignition key switch.

Test No. 2

1. Under the same conditions as Test No. 1, turn the ignition key switch on.

2. The voltmeter reading should be between 2 and 3 volts.

JUDGING THE TEST RESULTS

A high reading could be caused by a high resistance

in the alternator field, defective brushes, or a defective regulator.

A low reading might indicate a shorted alternator field, a defective regulator, or an open circuit.

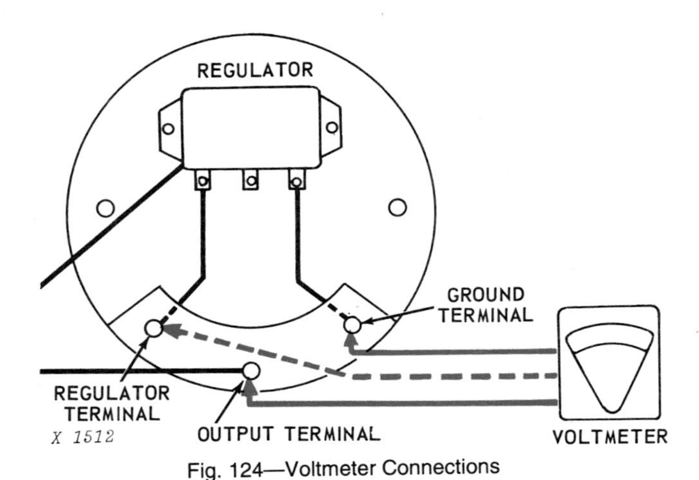

Fig. 124—Voltmeter Connections

Test No. 3

1. First, leave the voltmeter connected to the ground terminal and regulator terminal of the alternator (Fig. 124).

2. With the engine running and the key switch on but all accessories off, the voltmeter should read more than 15 volts.

3. Move the voltmeter lead from the regulator terminal to the output terminal. The voltmeter should read 1 volt less.

JUDGING THE TEST RESULTS

If the regulator terminal voltage is correct and the output terminal is the same as battery voltage, then the isolation diode is open.

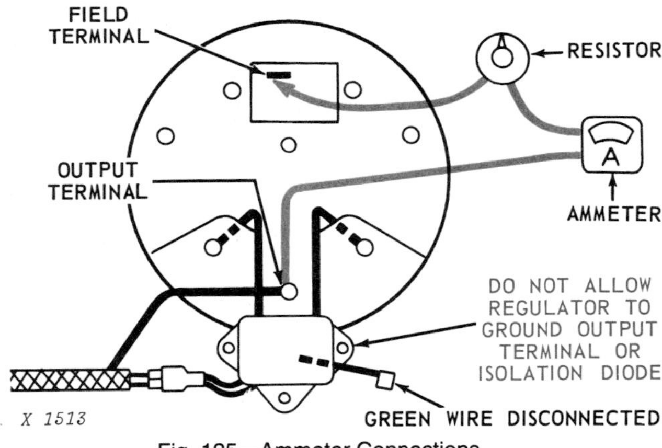

Fig. 125—Ammeter Connections

Test No. 4

This test is usually performed if Test No. 2 indicated a malfunction.

1. With the engine and switch off, disconnect the regulator-to-alternator field terminal wire (Fig. 125). Let the regulator hang on the wires connected to the regulator and ground alternator terminals. Be very careful in how the regulator hangs. Do not allow it to ground on the output terminal.

2. Connect the ammeter in series with a variable resistor to the field terminal and output terminal.

3. With all resistance eliminated, the ammeter should read 2.0 to 2.5 amps with the alternator cold.

JUDGING THE TEST RESULTS

A high reading indicates a shorted field winding or brushes.

A low reading means a high resistance in the brushes or slip ring, or an open circuit in the field windings.

Test No. 5

1. Connect the voltmeter and jumper wire as shown in Fig. 126.

2. Run the engine at a specified speed—for example, 800 rpm.

3. This should give a voltmeter reading of 15 volts. *Do not allow voltage to go above 16.5 volts.*

Overall Test Evaluation

If test No. 5 proved to be satisfactory, but test No. 3 voltage was below specifications, the regulator is probably at fault.

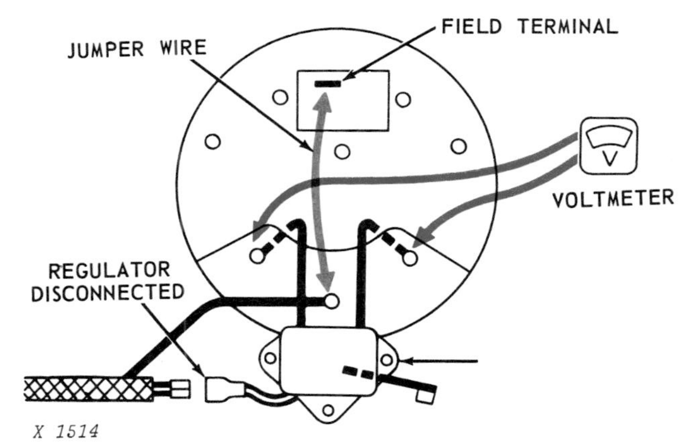

Fig. 126—Jumper Wire and Voltmeter Connections

If test No. 5 voltage was low, but tests No. 2 and 4 were satisfactory, the alternator is probably faulty.

Wiring Resistance Test

If the alternator and regulator operate properly, then check the wiring.

TEST PROCEDURE

1. Make a quick visual check of the lead connections and wires.

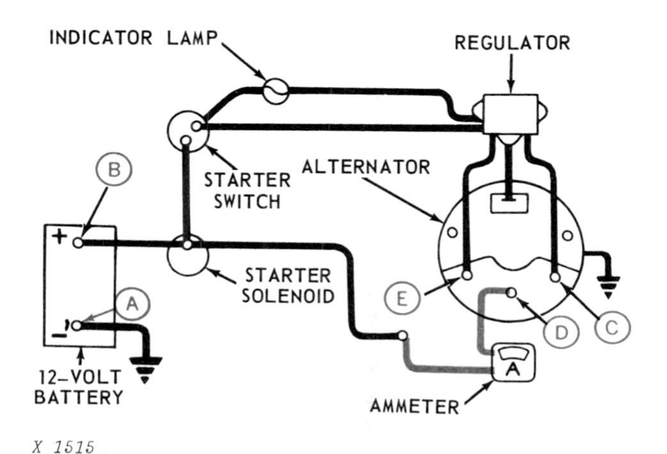

Fig. 127—Wiring Test Points

2. Disconnect battery ground cable. Then disconnect alternator output wire and connect ammeter as shown in Fig. 127.

3. Connect ground cable and run engine to obtain a 10-amp charging rate.

4. With a voltmeter, check the voltage at different points as illustrated in Fig. 127.

5. The voltage between these points should be as listed in the chart below.

Test Points	Max. Voltage
A-C	*0.3 volts*
B-D	*0.3 volts*
B-E	*1.3 volts*

6. Always disconnect battery ground cable to prevent accidental grounding while connecting the alternator output terminal wire. Then reconnect battery ground cable.

JUDGING THE TEST RESULTS

A high reading indicates a high resistance in the wiring or components.

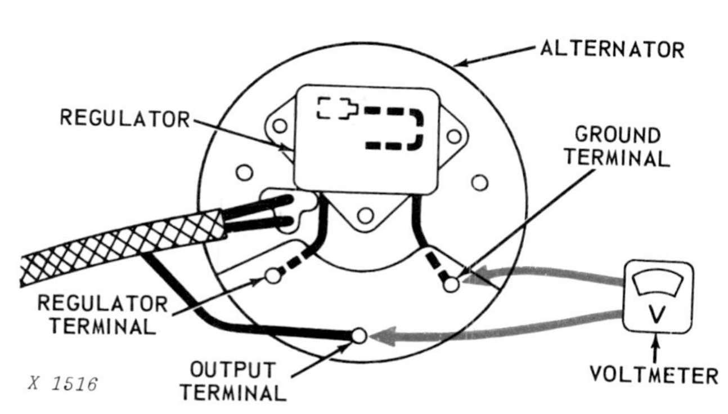

Fig. 128—Regulator Test Connections

Regulator Testing Procedure

1. Connect the voltmeter as shown in Fig. 128. Use an accurate voltmeter—one that will measure to within plus or minus 0.1 volts.

2. Run the engine to obtain a 10-amp charging rate for about 15 minutes to stabilize the regulator temperature.

3. Measure the regulator temperature about one inch from the regulator and check the voltmeter reading.

4. Compare the reading with the chart below.

Temperature	Correct Voltage (12-Volt Circuit)
40°F (4°C)	*14.4-14.9 volts*
60°F (16°C)	*14.3-14.7 volts*
80°F (27°C)	*14.2-14.6 volts*
100°F (38°C)	*14.0-14.4 volts*
120°F (49°C)	*13.8-14.3 volts*
140°F (60°C)	*13.6-14.1 volts*

JUDGING THE TEST RESULTS

If the voltage is not within limits, the regulator is faulty.

TEST YOURSELF

QUESTIONS

1. What are the two main jobs of all charging circuits?

2. True or false? "DC charging circuits generate a direct current while AC charging circuits generate an alternating current."

3. Match each item on the left below with the correct item on the right.

a. During starting	*1.* Generator supplies all current and recharges battery
b. During normal operation	*2.* Battery supplies all load current
c. During peak operation	*3.* Battery helps generator supply current

4. What are the two main parts of a basic generator?

5. How is the current generated by these two parts?

6. A moving field in a fixed conductor describes a(n) ____________________. A moving conductor in a fixed field describes a(n) ____________________.

7. Name the three main units of an alternator.

8. Which of these units contains diodes?

9. What is the purpose of the diodes?

10. True or false? "Never polarize an alternator."

11. What is the most popular type of alternator regulator?

12. In an "A" circuit alternator, the regulator is located ______________________________ the field.

13. To further isolate the field and regulator circuit, some alternators use a ____________________ to obtain field current from the stator.

(Answers on page 19 at the end of this book.)

 Litho in U.S.A.

STARTING CIRCUITS / CHAPTER 7

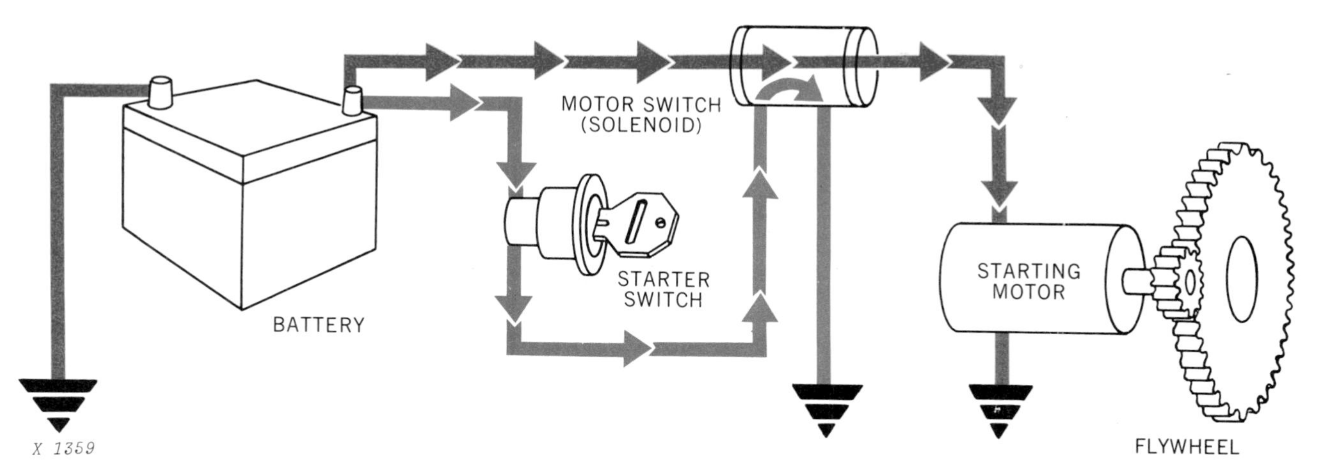

Fig. 1—Basic Starting Circuit

HOW THE STARTING CIRCUIT WORKS

The starting circuit converts electrical energy from the battery into mechanical energy at the starting motor to crank the engine.

A basic starting circuit has four parts:

1. The BATTERY supplies energy for the circuit.

2. The STARTER SWITCH activates the circuit.

3. The solenoid-operated MOTOR SWITCH engages the motor drive with the engine flywheel.

4. The STARTING MOTOR drives the flywheel to crank the engine.

How do these parts work together as a team?

The starting circuit is shown in operation in Figs. 2, 3, and 4.

When the starter switch is activated by the operator (Fig. 2), a small amount of electrical energy flows from the battery to the solenoid and back to the battery through the ground circuit.

As the solenoid gets this power from the battery, it moves the solenoid plunger and engages the pinion with the flywheel (Fig. 3). The plunger also closes the switch inside the solenoid between the battery and starting motor, completing the circuit and allowing a large amount of electrical energy to flow into the starting motor.

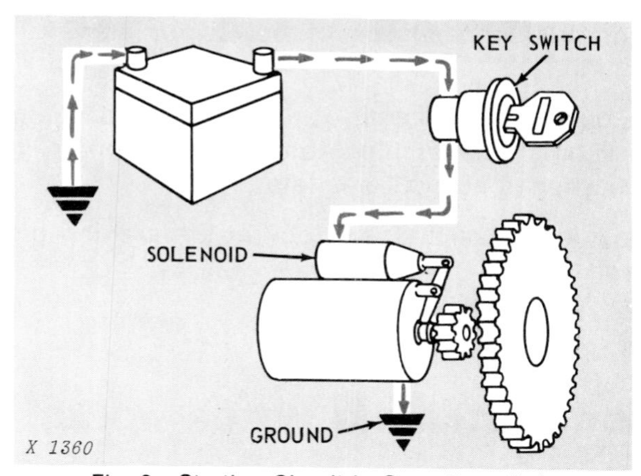

Fig. 2—Starting Circuit in Operation: 1) As the Starter Switch Is Activated

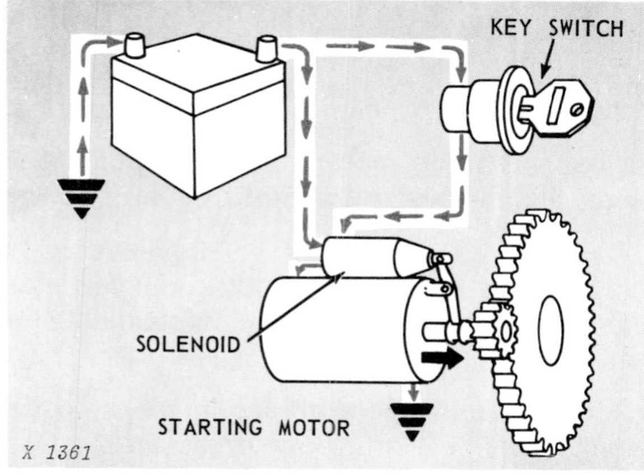

Fig. 3—Starting Circuit in Operation: 2) Starting Motor Engages Flywheel

Litho in U.S.A.

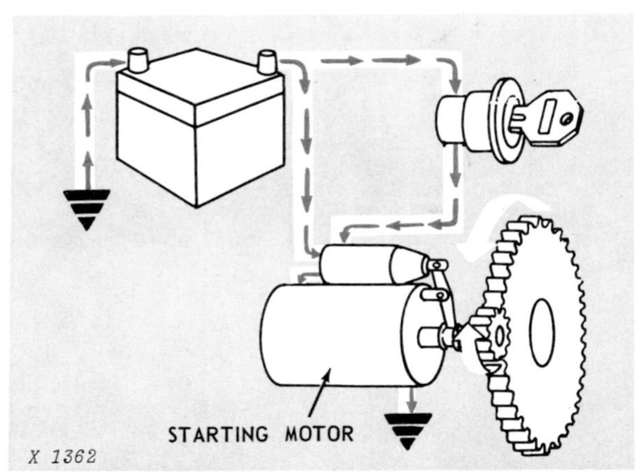

Fig. 4—Starting Circuit in Operation: 3) Starting Motor Cranks Engine

The starting motor takes the electrical energy from the battery and converts it into rotary mechanical energy to crank the engine (Fig. 4).

The battery is covered in Chapter 5 since it serves the whole electrical system. But remember that the battery is the source of power for starting the engine.

In our typical system, we have used a solenoid switch as an example. Motors can use other types of switches as we'll see later.

Now let's take a closer look at the starting motor itself.

HOW A STARTING MOTOR WORKS

The starting motor does the actual job of cranking the engine. It is a special type of electric motor:

1) It is designed to operate for short intervals under great overload.

2) It produces very high horsepower for its size.

The basic starting motor has a **solenoid,** a **field frame assembly,** an **armature,** and a **drive mechanism.**

Let's see how these parts work to convert electrical energy from the battery into mechanical energy to crank the engine.

First consider the **pole pieces** in the field frame assembly of the starting motor as the ends of a magnet (Fig. 5). The space between these poles is called the magnetic field.

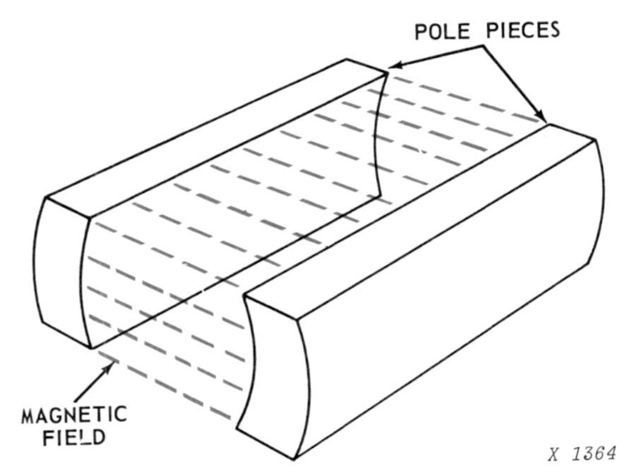

Fig. 5—Pole Pieces and Their Magnetic Field

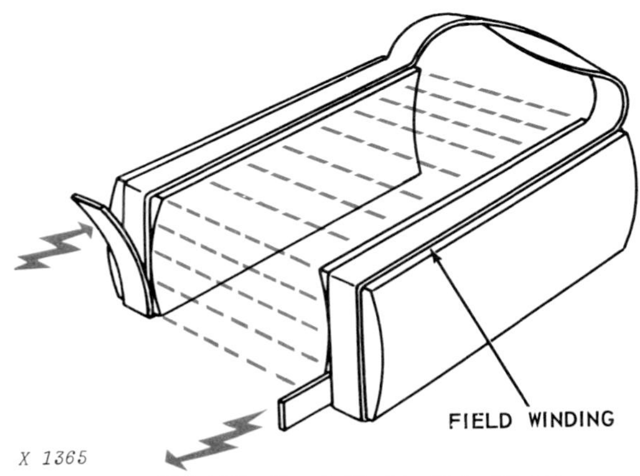

Fig. 6—Field Winding Added to Pole Pieces

If a wire, called a **field winding,** is wrapped around these pole pieces and current is passed through it, the strength of the magnetic field between the pole pieces is increased (Fig. 6).

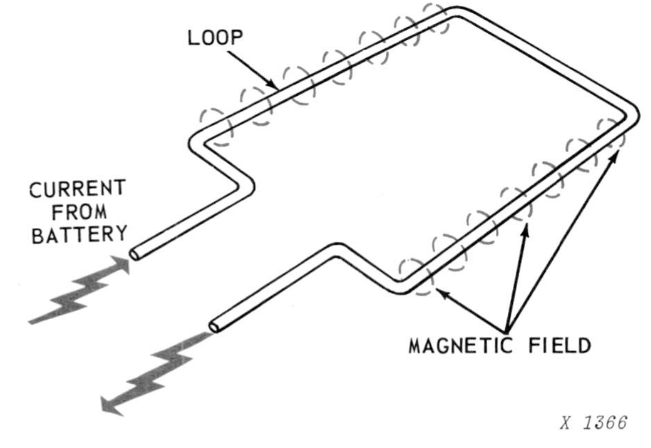

Fig. 7—Loop of "Live" Wire and Its Magnetic Field

Now let's consider a loop of wire (Fig. 7). When we feed electrical energy from the battery into this loop, a magnetic field is also formed around the wire.

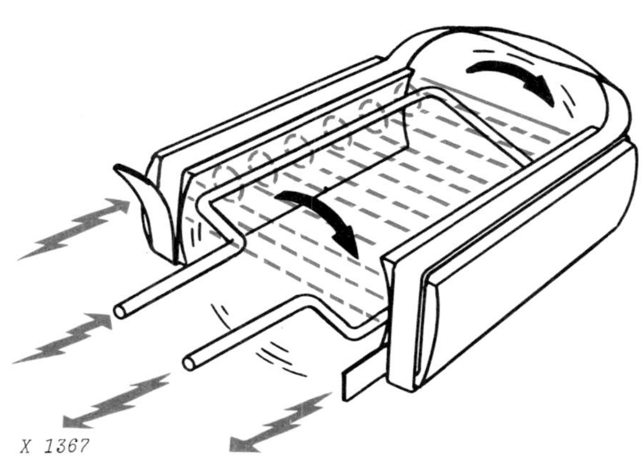

Fig. 8—Loop of Wire Placed in Field between Poles

If we place the loop of wire in the magnetic field between the pole pieces and pass current through the loop, we have the makings of a simple **armature** (Fig. 8). The magnetic field around the loop and the field between the pole pieces repel each other, causing the loop to turn.

In an actual armature we use more loops (Fig. 9). By attaching separate metal segments to the ends of each loop, we form a simple contact surface called a **commutator.** When we feed electrical energy to the commutator through sliding contacts called **brushes,** the repelling or kicking action causes a continuous rotation. All starting motors use this basic principle to develop useful mechanical energy.

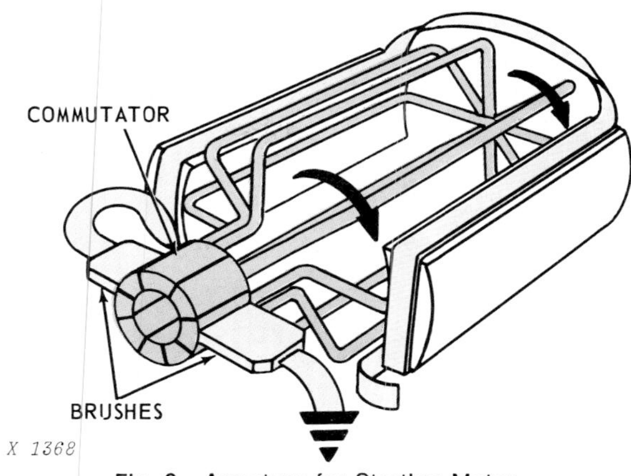

Fig. 9—Armature for Starting Motor

Every starting motor has brushes, an armature, field windings, pole pieces, and a drive mechanism (Figs. 10 and 11).

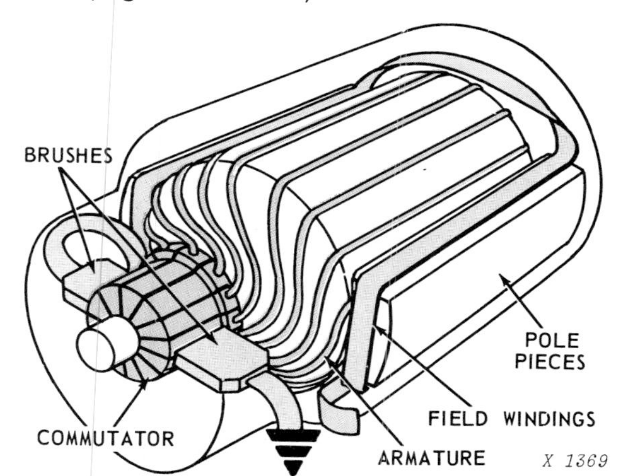

Fig. 10—Armature and Brushes in Starting Motor

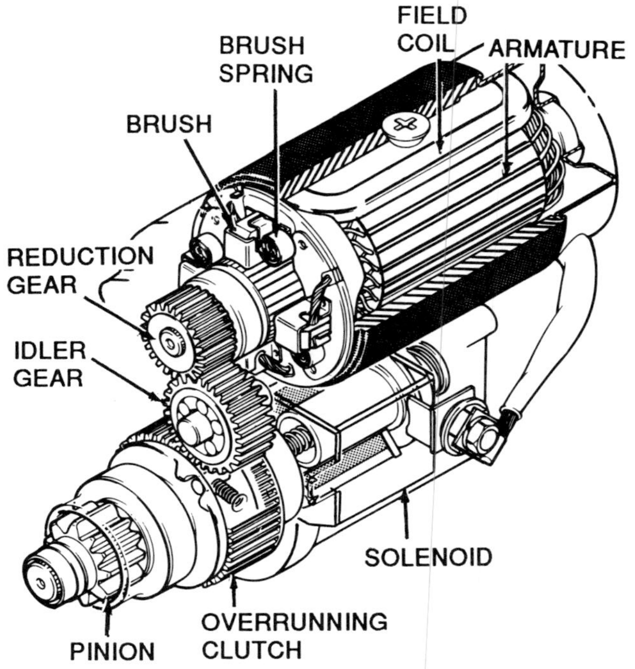

Fig. 11—Cutaway View of a Gear Reduction Starting Motor

 Litho in U.S.A.

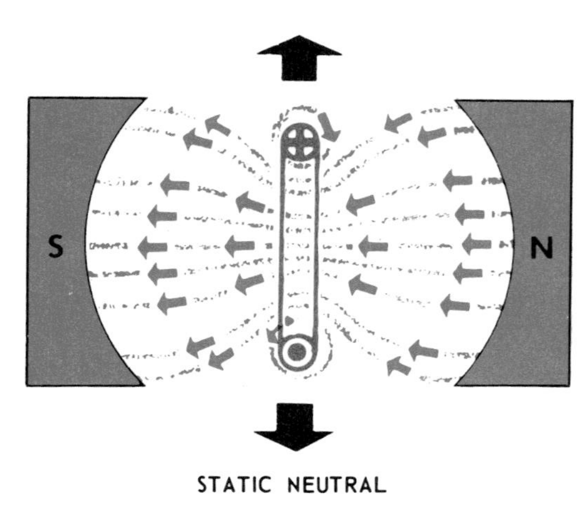

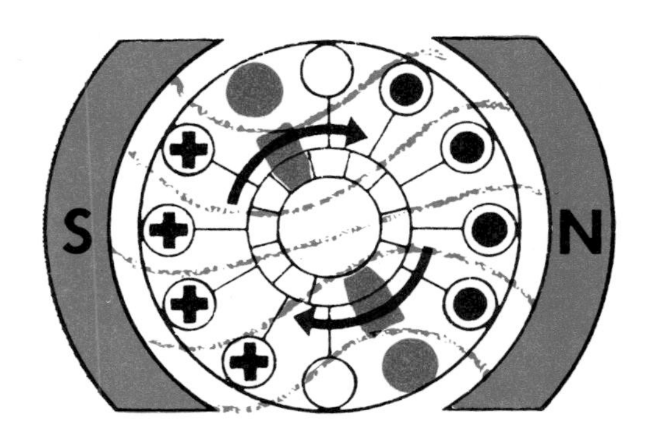

STATIC NEUTRAL

OPERATIONAL NEUTRAL

Fig. 12—Static Neutral Point of Rotating Armature

Let's build on the basics of Chapter 1 and explain more about the repelling action of the brushes and the armature which makes the starting motor operate.

In Chapter 1 we discussed the behavior of magnetic fields around a conductor when current is passing through. The principles we covered there are the basis for the operation of the starting motor.

As the armature rotates (Fig. 12) the sides of its loops reach a point shown when they are as far out of the magnetic field as possible. This is the "static neutral point" and is always halfway between the pole pieces of the motor. Current must be changed at this point to keep the same turning force.

The reversal of current is done every half-turn by the commutator. It works as follows: When the armature moves, so does the commutator. By the time the left-hand side of the armature has swung around to the north pole, the commutator segments will have reversed their connections with the brushes. Current will then flow in the *opposite* direction in the armature windings. This change of current flow would cause the armature to reverse, but since the windings have made a half-turn while the commutator changed connections, the force exerted on the armature will continue its rotation.

So, to keep rotating the motor, the current flow must be reversed every half-turn of the armature. This keeps the magnetic lines of force acting in the same direction.

The **static neutral point** is always *halfway* between the pole shoes and is the point where the direction of current must be changed to maintain a turning force in the same direction. This is true whether the motor has two, four or six poles. However, *when current flows through the armature windings* creating another magnetic field, *the normal field between the pole shoes is distorted.* Since lines of force may be assumed not to cross each other, the *neutral point is* therefore *shifted.* The motor brushes are shifted back from the static neutral point to an **operational** neutral (against the direction of rotation) to prevent excessive arcing and to obtain more efficient operation.

The main point is that the magnetic field of the armature DISTORTS the field of the pole shoes and SHIFTS THE NEUTRAL POINT TO A NEW POSITION. To match this, the motor brushes are mounted back at the new position.

SUMMARY: HOW A STARTING MOTOR WORKS

To summarize:

1) A current-carrying conductor, formed in a loop and mounted on a shaft, will cause the shaft to rotate when placed inside a magnetic field. The result: a basic starting motor.

2) If the direction of current flow in the loop is reversed as the loop passes the neutral position, the loop and shaft will keep on rotating. This is the way a starting motor is kept running.

3) It follows that to increase the power of the motor, more loops of conductors connected in series with an equal number of commutator segments are needed.

4) It also follows that increasing the strength of the magnetic field will affect the turning power of the motor, and will be directly related to the number of field poles and the number of ampere turns on each pole.

5) Basically the starting motor is a series-wound, direct-current electric motor designed to provide high power for a short time using current from a storage battery. Most starting motors have two, four, or six field poles with windings, a wound armature with a commutator, and two, four or six brushes.

TYPES OF STARTING MOTORS

Starting motors can be typed in several ways:

- **Type of Motor Circuit**
- **Type of Armature**
- **Type of Switch Control**
- **Type of Motor Drive**

Let's compare the types of starting motors as given above.

STARTING MOTOR CIRCUITS

All starting motors have a stationary member **(field)** and a rotating member **(armature).** The field windings and the armature are usually connected so that all current entering the motor passes through both the field and the armature. This is the motor circuit.

The brushes are a means of carrying the current from the external to the internal circuit—in this case from the field windings to the armature windings.

The brushes are carried in brush holders. Normally, half the brushes are grounded to the end frame while the other half are insulated and are connected to the field windings.

Starting motors on most off-the-road machines are considered to be series wound; that is to say, the field windings and the armature windings are *in series.*

However, starting motors have four main types of *field circuits:*

- **Series-wound**
- **Parallel-wound**
- **Series-parallel wound**
- **Compound-wound**

Let's look at the common series-wound circuit first. Then we will bring in the other three types.

SERIES-WOUND FIELD CIRCUITS

Four-Pole—Two-Coil Circuits

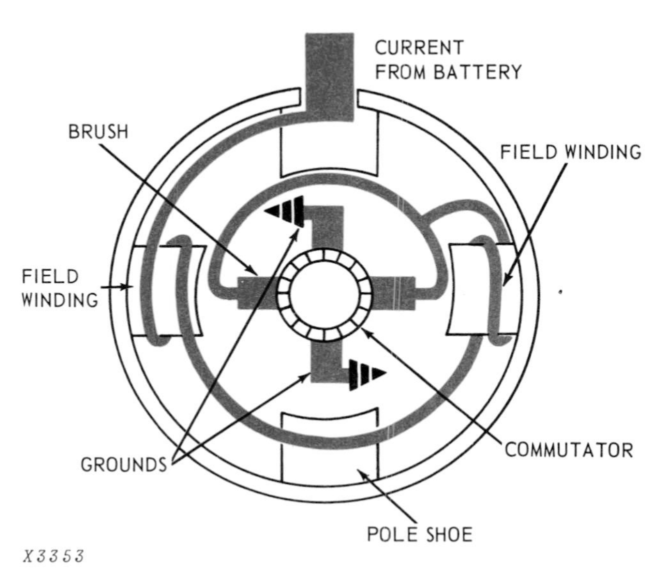

Fig. 13—Four-Pole—Two-Coil Series-Wound Motor

As shown in Fig. 13, some starting motors have four pole shoes and only two field windings.

Before looking at the circuit, we should understand the polarity of the pole shoes.

Using the right-hand rule for coils (see Chapter 1) we can see that when the starting motor is in operation, the face of the pole shoe with the field winding has a north polarity. The magnetic lines of force pass from this pole to the pole without windings, through the frame, and back to the original pole to complete the magnetic circuit.

In all starting motors the adjacent pole shoes are of opposite polarity, providing a north, south, north, south sequence around the frame. These magnetic lines of force passing from pole to pole cause four magnetic paths through the armature windings.

The field circuit in Fig. 13 is the series-wound type. This provides a four-pole action with only two field windings.

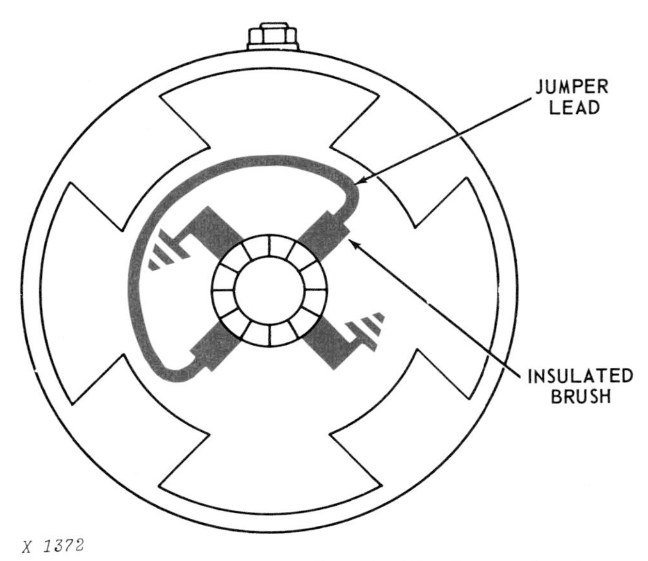

X 1372

Fig. 14—Motor Brushes Equalized by Jumper Leads

As a rule, all of the insulated brushes are connected together with jumper leads (Fig. 14) so that voltage in all the brushes is equalized. Without this, the result could be arcing and burning of commutator bars, insulating the brush contacts from the commutator surface. Once the brush contacts are insulated, the continuous current flow is interrupted and the motor stops.

Four-Pole—Four-Coil Circuits

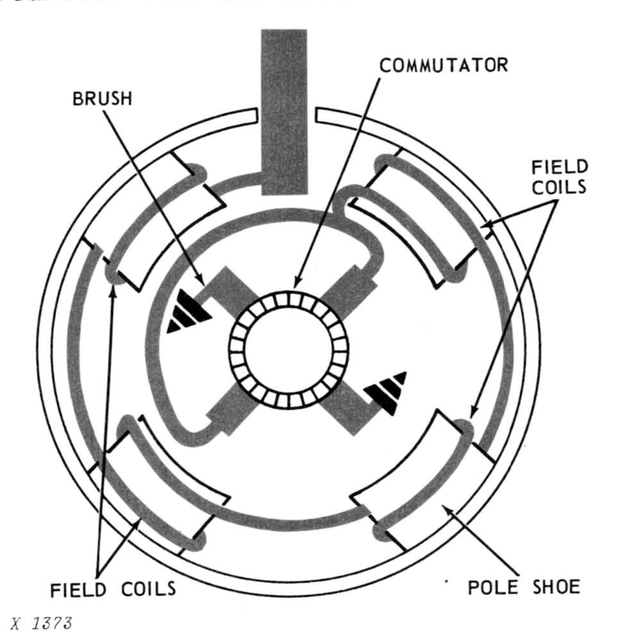

X 1373

Fig. 15—Four-Pole—Four-Coil Series-Wound Motor

Now that we have seen the basic series-wound starting motor, let's add field windings to the two bare poles. By doing so we create a four-pole, four-coil field circuit (Fig. 15).

By adding two field windings to the pole shoes we create more ampere turns of low resistance, resulting in a stronger magnetic field. Stronger fields produce greater torque, so we get more starting power from the four-coil motor than from the two-coil motor.

This field circuit has four field pole windings and four brushes. The field windings are in series so all current flow is through all of the windings before it flows through the two insulated brushes to the armature.

Six-Pole—Six-Coil Series Wound Circuits

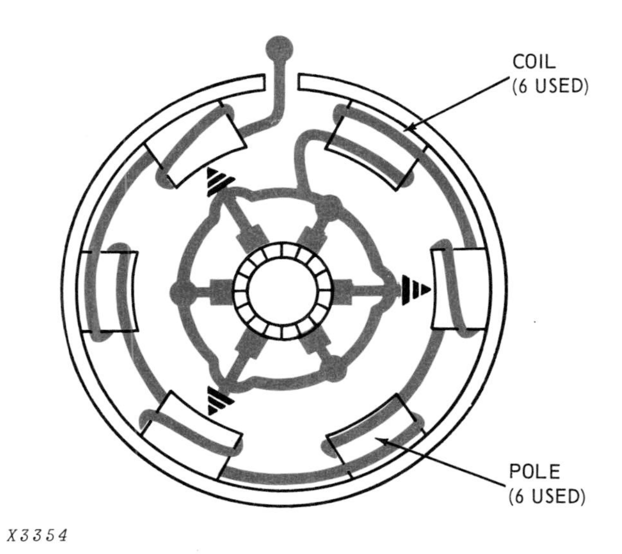

X3354

Fig. 16—Six-Pole—Six-Coil Series Wound Motor

Fig. 16 illustrates a six-pole—six-coil series-wound motor. As in the four-pole type the current flows through all the field coils before it flows to the armature. The addition of the two extra poles and windings adds to the torque of the motor.

PARALLEL-WOUND FIELD CIRCUITS

Most field windings on larger engines are PARALLEL-WOUND. That is, current flows through one field winding (Fig. 17) to the insulated brushes; also through the other field winding to the insulated brushes. Placing the field windings in parallel allows extra current flow to create greater torque or turning power.

Four-Pole—Four-Coil Field Circuit

The four-coil, four-pole field circuit (Fig. 18) is often used for smaller gasoline and diesel engines. The stronger four-coil fields produce the torque required to turn over these engines.

This field circuit has four pole windings and four brushes. The field windings are paired off so that

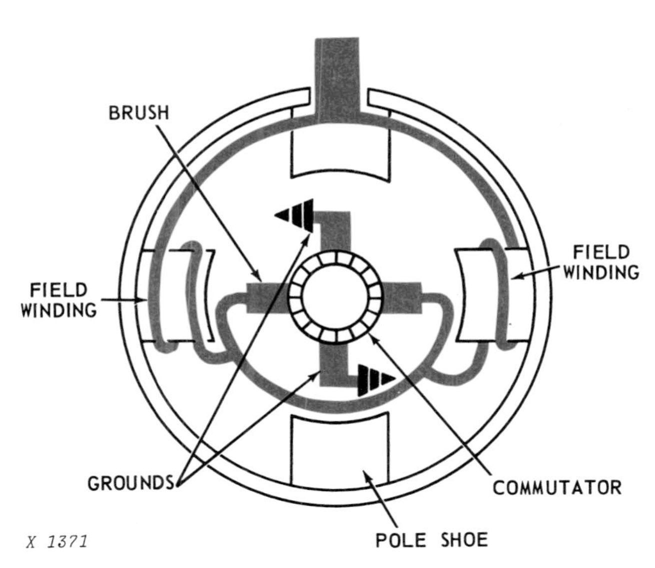

Fig. 17—Parallel Wound Field Circuit

half the current flows through one pair of field windings to one of the insulated brushes, and the current in the other pair of field windings flows to the other insulated brush.

The field circuit shown in Fig. 19 is a **four-pole**, four-coil circuit similar to those we have seen before. But this particular circuit has been developed recently for large diesel engines. In the past, large diesel engines required 24-volt starting motors for dependable starting. Because of 12-volt lighting and charging circuits, the 24-volt starting circuit required the installation of complicated wiring circuits to satisfy both needs. The new 12-volt

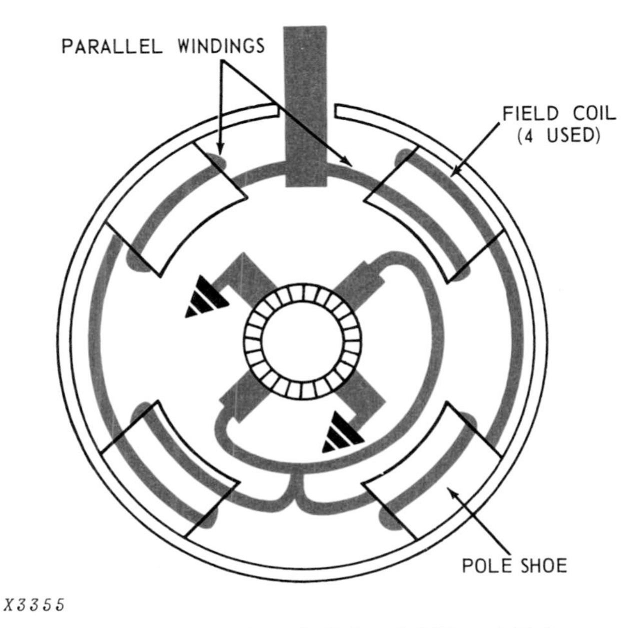

Fig. 18—Four-Pole—Four-Coil Parallel-Wound Motor

high output starting motor is the answer. It simplifies installation and maintenance since it permits operation of the whole starting circuit on the conventional 12-volt system.

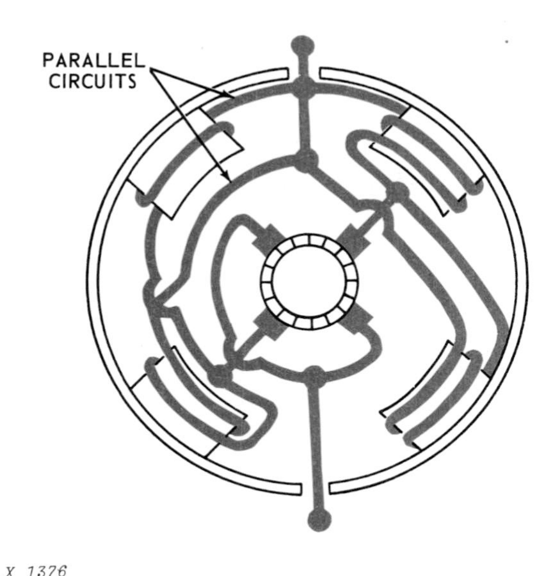

Fig. 19—12-Volt High-Output Starting Motor Circuit

However, the high-output starting motor circuit needs extra power from the batteries to produce the output required by large engines. Reducing the motor voltage from 24 to 12 volts means that the circuit must carry about twice the current to provide the same output. This is usually done by retaining the same number of batteries that were used in series on the 24-volt starting system, but connecting them in parallel. This gives the increased current at 12 volts.

How does this high-output starting motor work? Let's start with the fact that the power (watts) produced by a starting motor is equal to the product of voltage and current flow:

Watts = Voltage × Current

The high-output starting motor, which operates at twelve volts, must be able to carry the increased current to produce the same amount of power or watts as a 24-volt starting motor.

For example, if a 24-volt starting motor will carry 500 amps of current, the watts produced would be 12,000. If the high-output starting motor is to operate at a lower voltage (twelve volts), the current flow must be almost doubled to develop the same output as the 24-volt motor. For example, 12 volts and 1000 amps of current will produce 12,000 watts, the same power output as the 24-volt motor.

The increased current flow in the high output motor requires a different type of field winding circuit than the normal 12-volt motor. The four-coil, four-pole starting motor circuit in Fig. 18 is the high output circuit. It differs from the previous circuits shown in that *the field windings are in parallel instead of series-wound. This parallel circuit in the field windings permits the increased current flow needed at a reduced battery voltage.*

SERIES-PARALLEL FIELD CIRCUITS

This type of starting motor is designed for heavy-duty service.

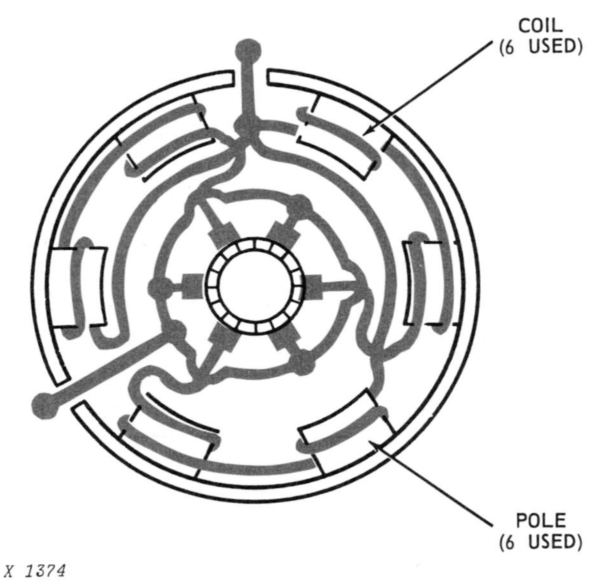

Fig. 20—Six-Pole—Six-Coil Series—Parallel-Wound Motor

A six-pole, six-coil field circuit is illustrated in Fig. 20. Here the current is split three ways instead of two, as in a four-pole circuit. The result is that one-third of the current flows through each pair of field windings to one of the three insulated brushes.

In this way we increase the number of circuits through the starting motor, keeping resistance low so that higher current can flow to develop more horsepower for starting the engine.

COMPOUND-WOUND FIELD CIRCUITS

In the previous field circuits, the starter field coils are connected in series with the armature. All of the current that flows through these coils also flows through the armature. These windings usually contain several turns of heavy copper ribbon. Free-running motor speed is quite high.

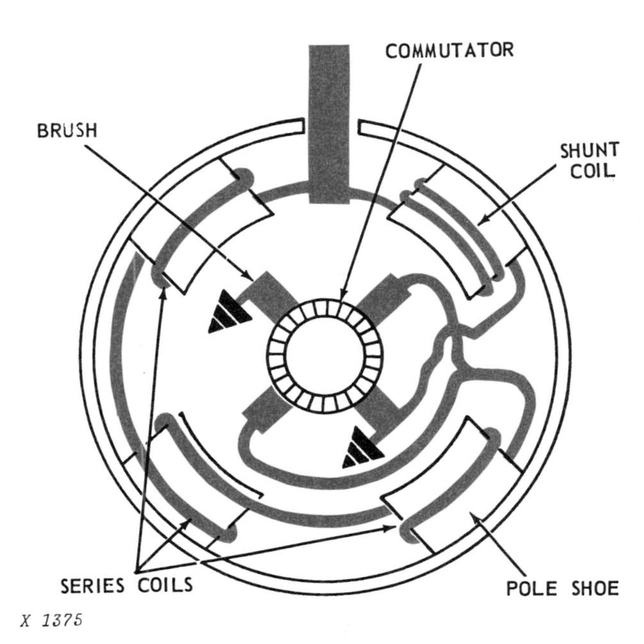

Fig. 21—Compound-Wound Motor

As the armature speed increases, a counter voltage (CEMF) in the armature windings also increases. The counter voltage is induced into the armature windings when they pass through the magnetic field of the field coils. We can now see that the faster the armature turns, the greater will be the counter voltage in the armature, until finally the armature counter voltage reduces current flow to a point where maximum free speed is reached.

In some cases excessively high free speed could result in the armature windings being thrown from their slots. However, to prevent excessively high free running speed, one or more of the poles is wound as a *shunt* winding. A shunt winding has many turns of small wire. The field circuit with a shunt winding as shown in Fig. 21 is known as a **compound-wound** field circuit.

The compound-wound starter shunt winding prevents overspeeding because the shunt winding is connected directly to ground. The current flow is at a high constant value of magnetic flux as determined by battery voltage and is not affected by the armature counter voltage. With a consistently high magnetic field from the shunt coil, the armature counter voltage is higher at lower armature speeds, thus reducing the series field winding and armature current flow as well as the armature free speed.

For an example of magnetic flux values, consider a typical compound-wound starter drawing 220 amps when cranking (200 amps for the series coils and 20 amps for the shunt coil) and 70 amps when at free running speed.

When cranking, the series coils drawing 200 amps and containing 11 turns of copper strap will have 2200 ampere-turns of magnetic flux. The shunt winding containing 110 turns and drawing 20 amps will also have 2200 ampere-turns of magnetic flux. Compare this to when turning at free running speed and the series coils will have about 550 ampere-turns each while the shunt will probably have more than 2200 ampere-turns (battery voltage rises).

ARMATURES FOR STARTING MOTORS

We have referred to the armature often in order to understand the field pole and field winding circuit.

The armature is the main drive of the motor and converts electrical energy into mechanical energy. The field windings and the armature windings produce mechanical *rotary* motion. In Fig. 22 we see the armature in the starting motor circuit.

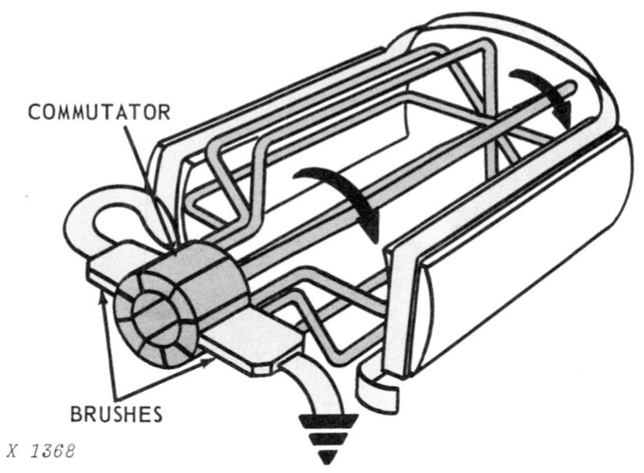

Fig. 22—The Armature in the Starting Motor Circuit

The armature must be wound to conform to the magnetic fields provided by the field poles. In a simple two-coil, two-pole unit, each armature winding is connected to one segment of the commutator. And all windings in the armature are connected in series.

Current flows into the typical armature as follows:

Two brushes are placed on opposite sides of the commutator as shown in Fig. 22. As current flows into the brushes, it flows through the armature windings except those shorted out by the two brushes. Since both ends of a winding contact a brush at the same time, the winding is shorted out in the commutating position. This provides two separate and parallel paths for the current to follow. Current flows in the same direction in all conductors passing before the pole pieces, but the direction of current flow changes at the neutral position as the armature turns up and out of the magnetic field between the pole pieces.

WAVE- AND LAP-WOUND ARMATURES

WAVE-WOUND armatures have only two paths for current flow.

LAP-WOUND armatures have as many paths as there are pole pieces.

The lap-wound armature is connected in series and divided into individual paths to permit large amounts of current to flow through it. This increased current capability is the reason the lap-wound armature is used in the high output starting motor we described earlier in this chapter.

SWITCHES FOR STARTING MOTORS

We learned earlier that starting motors convert electrical energy into mechanical energy. Now let's see how the **switch** performs a mechanical job electromagnetically, and completes the circuit from the battery to the starting motor.

Starting motors can use four types of switches:

- **Manual Switch**
- **Magnetic Switch**
- **Solenoid Switch**
- **Series-parallel Switch**

Let's talk about each one.

MANUAL SWITCHES

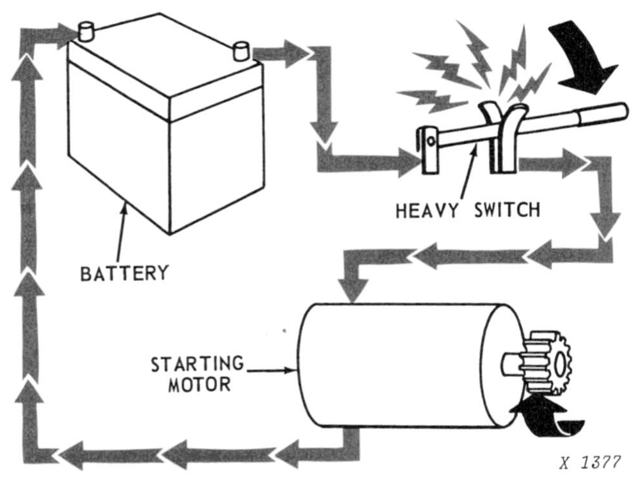

Fig. 23—A Manual Switch

A manual switch performs a mechanical operation by the simple closing of switch contacts. Operation of this type of switch closes a circuit or opens a circuit as in turning an electric light on and off. A simple manual switch is shown in Fig. 23.

The manually operated starting switch may be mounted where it is directly accessible to the operator, or it may be mounted on the starting motor and made accessible by various devices such as a hand lever.

MAGNETIC SWITCHES

To understand the magnetic switch and later on the solenoid switch, we must review the fundamentals of an electromagnet.

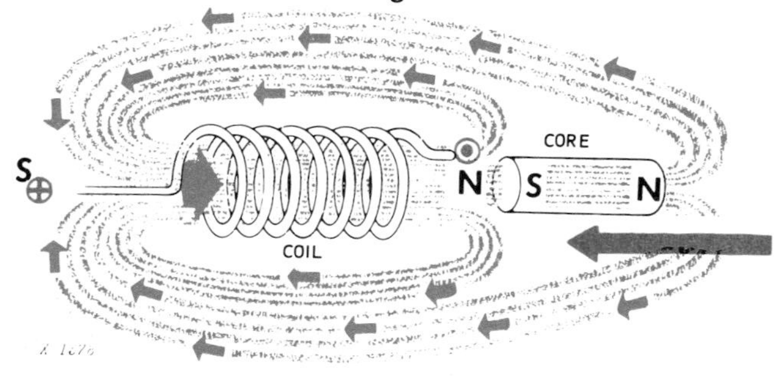

Fig. 24—Polarity of an Electromagnet

We learned in Chapter 1 that a magnetic field is made stronger by placing a soft iron core in a coil. We also learned that the polarity of the core is the same as for the coil. If the core has freedom to move, and is placed at one end of the coil in Fig. 24, it will also assume the polarity of the coil. Therefore, the adjacent poles are of opposite polarity, and the core is drawn into the center of the coil when current flows through the coil. As soon as current stops flowing, the field collapses and the core is free to move away from the coil. This is the principle of the magnetic and solenoid switches.

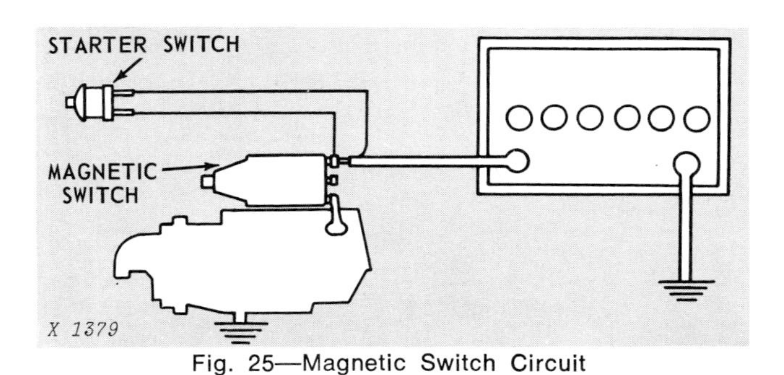

Fig. 25—Magnetic Switch Circuit

The *magnetic switch* is mounted on the starting motor frame like some manual switches. It is operated by a magnetic coil energized directly from the battery through a "start" control on the starter switch (Fig. 25).

The magnetic switch works as follows:

It has many turns of small wire wound around a hollow core. Floating in the core is a plunger with one end acting as a contact between the two main switch terminals. These terminals are connected in series with the starting motor. Normally, a small coil spring holds the plunger away from the main terminal contacts.

When the circuit to the coil is closed, a strong magnetic field is created in the core, causing the plunger to overcome the spring tension and complete the circuit between the terminal contacts. When the core contacts the terminals, the main circuit to the starting motor is completed and the engine is turned over.

When the control circuit is opened at the starter switch, the magnetic field collapses and the spring forces the plunger to its original position, opening the starting motor circuit. The motor then stops turning.

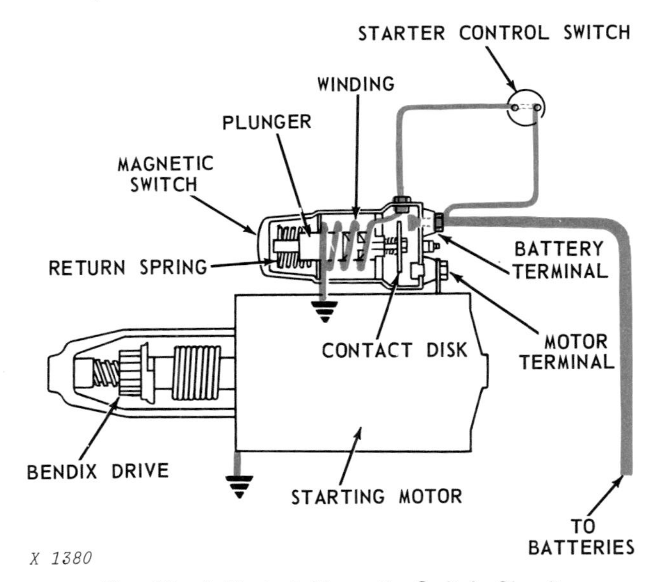

Fig. 26—A Typical Magnetic Switch Circuit

Magnetic switches are normally used with Bendix drives since there is no mechanical shifting action provided with this switch.

A Bendix drive with a magnetic switch is shown in Fig. 26.

SOLENOID SWITCHES

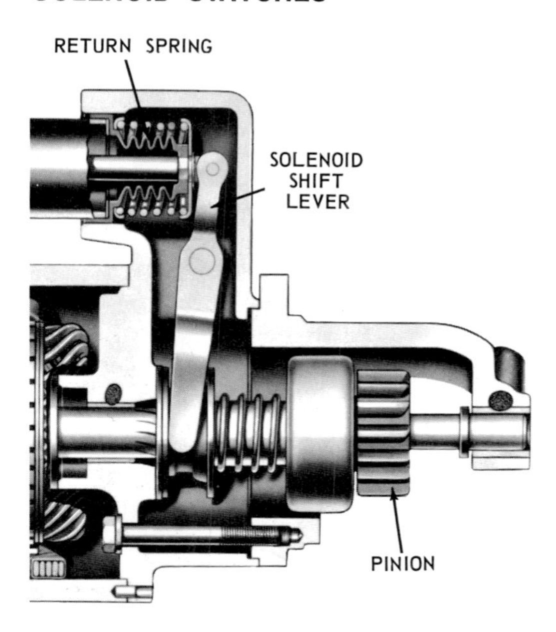

Fig. 27—Solenoid Shift Lever

The solenoid switch is very similar to a magnetic switch, but in addition to closing the circuit, the solenoid provides a mechanical means of shifting the starting motor pinion into mesh with the flywheel ring gear (Fig. 27).

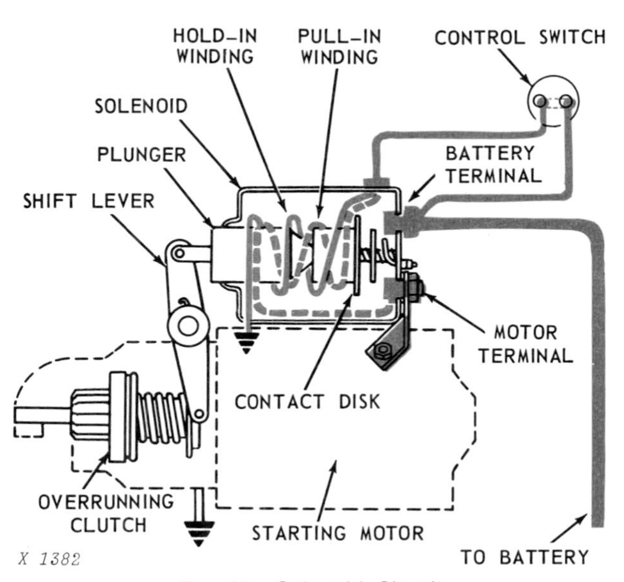

Fig. 28—Solenoid Circuit

The solenoid circuit is shown in Fig. 28.

The solenoid switch has two coils of wire wound in the same direction.

The **"pull-in" winding** is made up of *heavy* wire connected to the motor terminal of the solenoid and through the motor to ground.

The **"hold-in" winding** has an equal number of turns of *fine* wire with one end connected to ground.

These coils are energized directly from the battery through the "start" position on the starter switch.

In Figs. 29, 30, and 31 we will show how the solenoid switch operates from the moment the operator turns the starter switch.

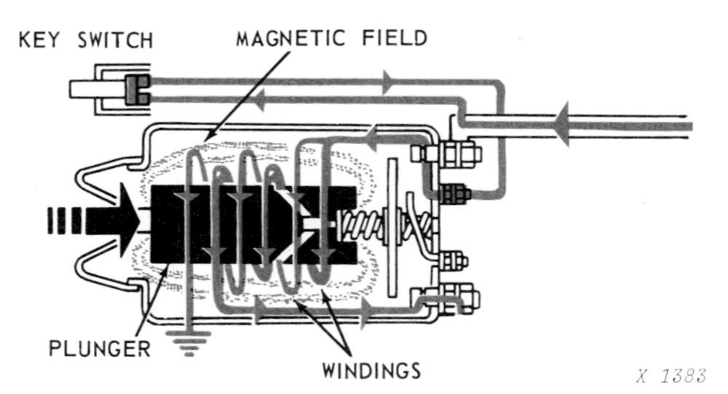

Fig. 29—Solenoid Plunger Starting to Move as Starter Switch Is Turned On

When the operator turns the starter switch, current flows to the starter solenoid (Fig. 29). Since the solenoid coils are wound in the same direction, current flows in the same direction, creating a strong magnetic field which pulls the plunger into the field as shown. The initial plunger movement engages the drive pinion with the flywheel ring gear.

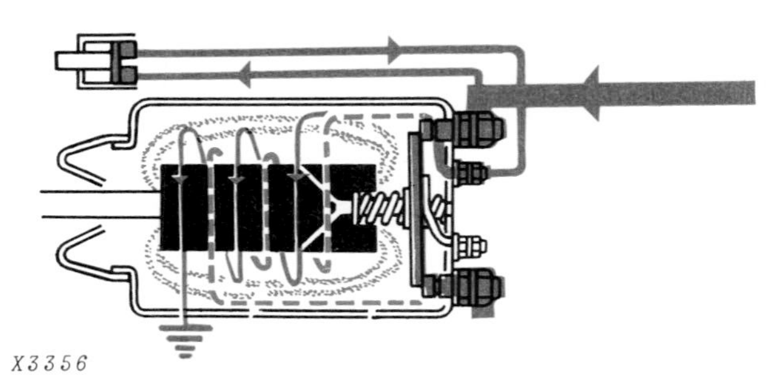

Fig. 30—Motor Engaged and Cranking the Engine

Further movement of the plunger closes the switch contact points within the solenoid (Fig. 30). This permits a heavy flow of current from the battery into the starting motor to crank the engine.

Inside the solenoid the closing of the points shorts out the heavy "pull-in" winding, leaving only the fine "hold-in" winding energized during the starting period. The initial flow of current through the "pull-in" winding is of very short duration. The flow of current through the "hold-in" winding continues as long as the control circuit is closed.

 Litho in U.S.A.

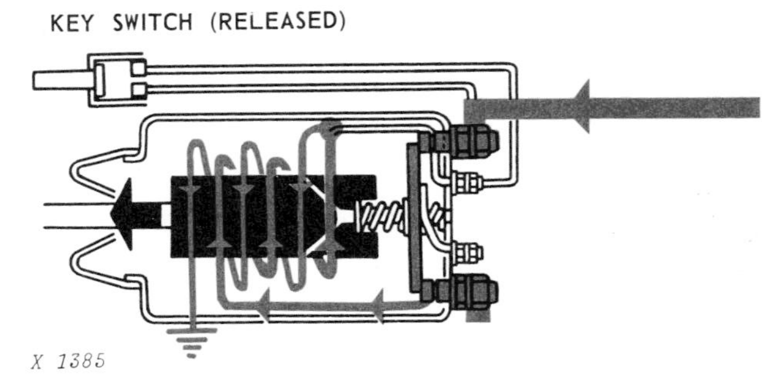

Fig. 31—Motor Disengages as Key Switch is Released

When the engine begins to run and the starter switch is released, several things happen quickly (Fig. 31).

First, current through the starter switch to the solenoid is cut off. Then a strong return spring pushes out the solenoid plunger, breaking the circuit from the battery to the starting motor and simultaneously pulling the pinion out of mesh.

Inside the solenoid, what has happened is this: When the starter switch opened the circuit, the two solenoid windings became connected in series and were energized from the motor terminal.

The current then flowed in a reverse direction in the "pull-in" winding, but continued in the same direction as before through the "hold-in" winding, since it is grounded. Since the number of turns in the two windings are equal, the ampere flow is equal, and the direction of flow is opposite, the magnetic field of one coil opposes that of the other, causing an immediate collapse of the magnetic field. Therefore, spring tension quickly moves the plunger back to its original position.

Solenoid Switch Mountings

Solenoid switches are mounted in two ways:

- **Coaxial Mounting—enclosed in the motor housing.**
- **External Mounting—"piggyback" on the motor housing.**

The solenoids do the same basic job with either mounting.

In the COAXIAL mounting, the solenoid switch is completely enclosed and the windings are around the armature shaft. (These solenoids have no "pull-in" windings.) Coaxial mountings are primarily used for adverse conditions or where space is not available for the bulkier "piggyback" arrangement. A coaxial starter is shown in Fig. 32. In the EXTERNAL mounting, an open or enclosed shift lever is used to actuate the pinion. An external solenoid model is shown in Fig. 33.

Fig. 32—Coaxial Starting Motor with Enclosed Solenoid

Fig. 33—Starting Motor with External Solenoid

SERIES-PARALLEL SWITCHES

A higher voltage in the starting circuit may be needed on some heavy-duty engines, especially diesel models.

Cold weather starting and other adverse conditions also draw more heavily on the starting circuit.

In some cases the high output starting motor can handle these conditions on a 12-volt circuit.

But in other cases, higher voltage starting motors must be used. By using a 24-volt battery supply, much higher starting speeds can be produced under heavy loads.

However, since lighting circuits have not yet been developed which meet highway regulations with 24-volt systems, special provisions for lighting must be made.

The **series-parallel switch** is one solution to this problem. This makes it possible to connect two 12-volt batteries in *parallel* for normal operation, but to connect the batteries in *series* through the switch during starting. This gives adequate power for starting the engine.

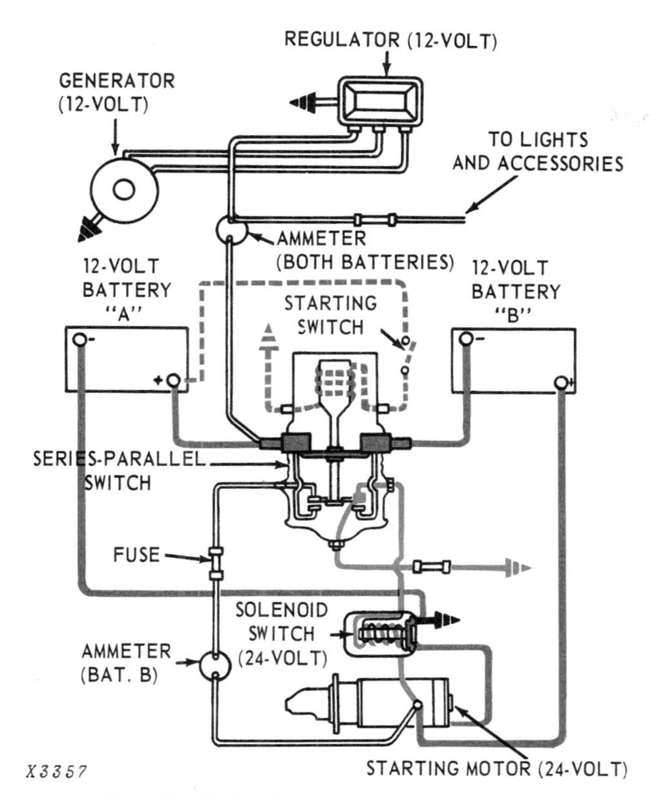

Fig. 34—Series-Parallel Switch for 24-Volt Starting Motor Operation

Fig. 34 shows a series-parallel switch for a 24-volt starting circuit. The series connection between the two batteries and the starting motor is shown in solid red. The starter solenoid circuit is shown in solid blue.

Operation during starting is as follows:

As the starting switch is closed, the solenoid coil within the series-parallel switch is energized (dashed red circuit) creating a magnetic force which attracts the series-parallel switch plunger. The plunger closes the two main switch terminals and connects the two batteries in series with the starting motor.

At the same time the solenoid circuit is completed by a set of points mechanically closed by the series-parallel switch plunger. This completes the battery-to-starting motor circuit and the starter turns over.

After the engine is started and the starting switch is released, the two batteries become connected in parallel when the series-parallel switch goes into a neutral position. (See Fig. 47.) This permits operation of the machine's electrical equipment at a normal system voltage of 12 volts.

STARTING MOTOR DRIVES

After electrical power is transmitted from the battery through a switch to the starting motor, some type of connection is needed to put this energy to work.

The last link in the starting circuit is the starting motor **drive.** The drive makes it possible to use the mechanical energy produced by the starting motor.

Let's learn more about these drive mechanisms and see what some of them look like.

INTRODUCTION

The starting motor armature revolves at a relatively high speed to produce turning power. Since the turning speed required to start an engine is comparatively slow, the starting motor is equipped with a small drive pinion which meshes with the teeth of the flywheel ring gear. The result is a gear reduction with the armature revolving as much as twenty times for every revolution of the flywheel. This permits the starting motor to develop high armature speeds and considerable power while turning the engine over at a lower speed.

But when the engine starts, it speeds up immediately and may soon reach as high as 2000 rpm. This is too fast for the starter and would result in damage to the armature.

To prevent this, starter drives are used. These are devices on the end of the armature shaft which mesh the drive pinion with the ring gear on the flywheel, and then prevent the starting motor from overspeeding after the engine has started.

TYPES OF DRIVES

There are two basic ways in which starter drives are engaged:

- **Inertia Drives**
- **Electromagnetic Drives**

On an INERTIA DRIVE, the pinion gear is weighted on one side to aid in its initial rotating motion. The Bendix drive is a type of inertia drive.

ELECTROMAGNETIC DRIVES are shifted in or out of mesh by the magnetic field of the switch. The Overrunning Clutch, Dyer Drive, and Sprag Clutch Drive are electromagnetic types.

 Litho in U.S.A.

BENDIX DRIVE

The Bendix drive depends upon inertia of the counterweight pinion and acceleration of the armature to move the pinion into mesh with the flywheel. This is shown in Fig. 35.

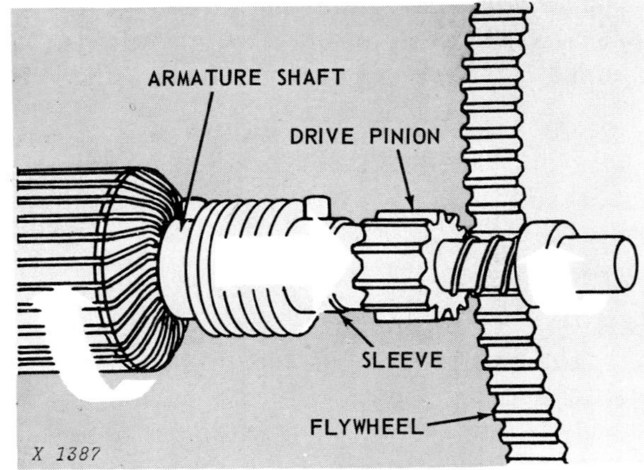

Fig. 35—Bendix Drive Engaging Flywheel as Engine Is Cranked

The Bendix drive is normally out of mesh and separated from the flywheel ring gear.

When the starting switch is closed and the battery voltage is fed to the motor, the armature shaft accelerates rapidly. The pinion gear, due to inertia created by the counter-weight, runs forward on the revolving screw sleeve until it meets or meshes with the flywheel gear (Fig. 35).

In other words, the Bendix drive uses inertia of the pinion and high starting speed of the armature shaft to move the pinion into mesh with the flywheel.

When the pinion becomes fully meshed, its forward motion stops, locking the pinion to the rotating armature shaft. The spring cushions the shock as the rotating armature starts to turn the flywheel.

This spring also acts as a cushion while cranking the engine against compression. In addition, it dampens the shock on the gear teeth during meshing or when there is a backfire of the engine.

All parts are now rotating as a unit, turning the engine over as shown in Fig. 36.

When the engine starts, the flywheel rotates faster than the armature shaft, causing the pinion to turn in the opposite direction on the screw and spin itself out of mesh (Fig. 37). This prevents the engine from driving the starting motor at an excessive speed.

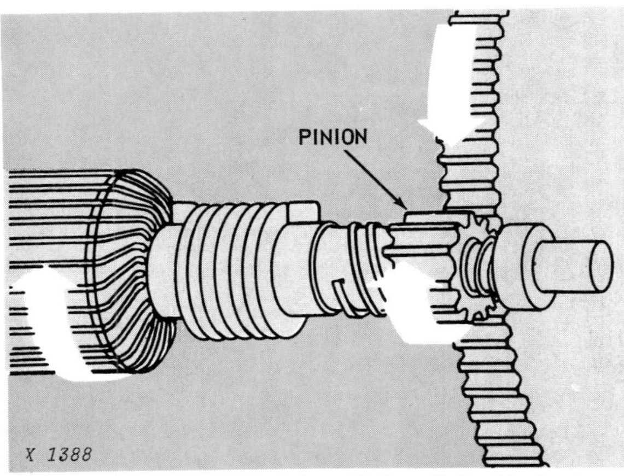

Fig. 36—Bendix Drive Pinion Fully Meshed

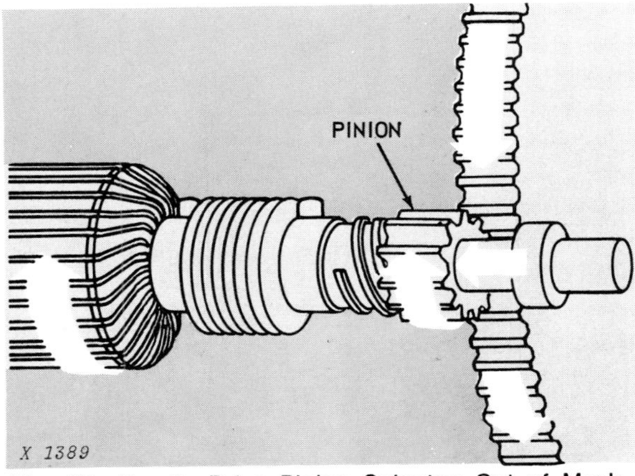

Fig. 37—Bendix Drive Pinion Spinning Out of Mesh After Engine Starts

The centrifugal effect of the weight on one side of the pinion, when spun from the flywheel, holds the pinion to the sleeve in an intermediate position until the starting switch is opened and the motor armature comes to rest.

For as long as the operator keeps the motor energized with the engine running, the motor will free speed. This is why the starter switch should be released immediately after the engine has started.

Certain precautions must be observed in operating a Bendix-type starting motor. If the engine backfires with the pinion in mesh with the engine flywheel, and the starting motor operating, a terrific stress is placed on the parts. This is because the motor armature attempts to spin the drive pinion in one direction while the engine, having backfired, turns the drive pinion in the opposite direction. This clash of opposing forces sometimes breaks or "wraps up" the Bendix spring.

Engine ignition timing should be checked and corrected to overcome backfiring.

Damage may also occur when the engine starts, throwing the Bendix drive pinion out of mesh with the engine flywheel teeth. When the engine is coming to rest it often rocks back, or rotates in reverse, for part of a revolution. If the operator attempts to re-engage the drive pinion at the instant the engine is rocking back, serious damage will result. As in a backfire, it may break the drive housing or "wrap up" the Bendix spring.

To prevent this, the operator should always wait at least five seconds between attempts to crank so that the engine stops turning.

Variations on Bendix Drives

Another Bendix drive known as the "folo-thru" has a detent pin which locks the drive in the cranking position to prevent disengagement on false starts. This pin is thrown out by centrifugal force when the engine runs and the pinion then disengages.

Some heavy-duty cranking motors use a *friction-clutch type* Bendix drive. This type of drive functions in much the same manner as other Bendix drives. However, it uses a series of spring-loaded clutch plates, instead of a drive spring, which slip momentarily during engagement to relieve shock.

OVERRUNNING CLUTCH DRIVE

Now let's take a look at one of the most widely used drive mechanisms, the *overrunning clutch* (Fig. 38). It allows positive meshing and demeshing of the drive pinion with the flywheel.

The overrunning clutch uses a shift lever to actuate the drive pinion. The pinion, together with the overrunning clutch mechanism, is moved endwise along the armature shaft and into, or out of, mesh with the flywheel. The shift lever may be either manual or operated by a *solenoid.* Operation is given below.

As we learned earlier, the drive pinion is normally out of mesh and separated from the flywheel ring gear. When the starting switch is closed, current flows to the solenoid, closing the switch circuit. As the solenoid switch closes, the shift lever moves the pinion into mesh and completes the circuit to the starting motor. If the pinion and the flywheel teeth meet, instead of meshing, the spring-loaded pinion rotates the width of one-half tooth and drops into mesh as the armature starts to rotate.

When the armature shaft rotates during cranking, small rollers become wedged against the clutch collar attached to the pinion. This wedging action locks the pinion gear to the armature shaft and causes the pinion to rotate with the shaft, as shown in Fig. 39.

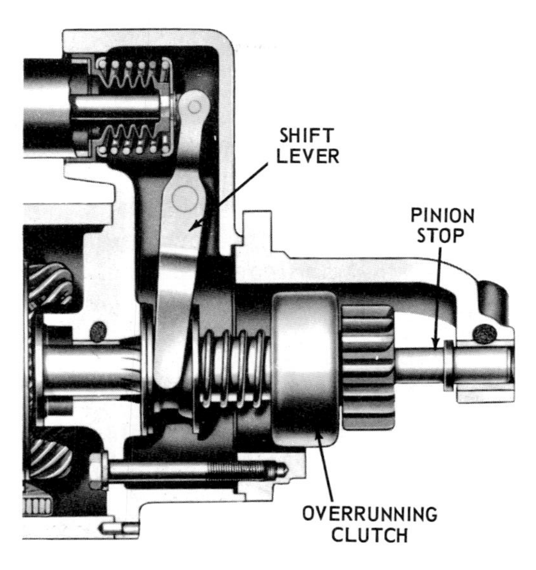

Fig. 38—Overrunning Clutch Drive

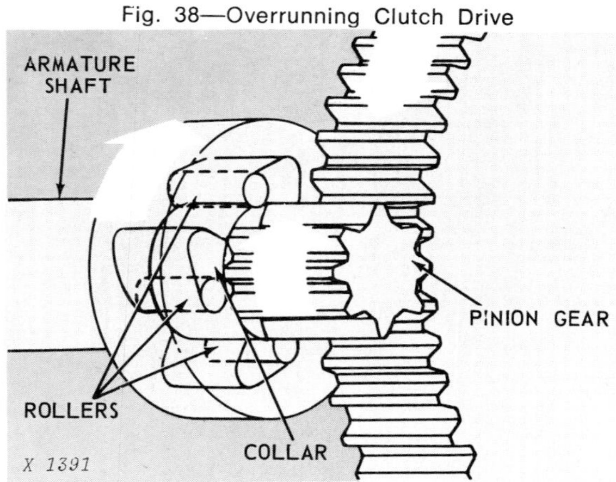

Fig. 39—Overrunning Clutch Drive Engaged

When the engine starts, the flywheel spins the pinion gear faster than the armature, releasing the rollers and unlocking the pinion from the armature shaft. The pinion, still meshed with the flywheel, overruns safely and freely until the switch is opened and the shift lever pulls the pinion out of mesh. This feature prevents the armature from being driven at excessive speed by the engine.

DYER DRIVE

The Dyer Drive is a special drive mechanism that provides positive meshing of the drive pinion with the flywheel, *before the cranking motor switch is closed and the armature begins to rotate.*

This action eliminates the clashing of pinion teeth with flywheel teeth, as well as the possibility of broken or burred teeth on either gear.

The Dyer drive is used on heavy-duty applications where it is important that the pinion be engaged before rotation begins. Engagement of the pinion while in motion is impossible because of the high horsepower developed and the acceleration of the armature when the starting circuit is completed.

The Dyer drive works as follows:

Movement of the motor shift lever forces the shift sleeve and pinion endways along the armature shaft so that the pinion engages the flywheel.

Further movement of the shift lever then closes the starting motor switch and the engine is turned over.

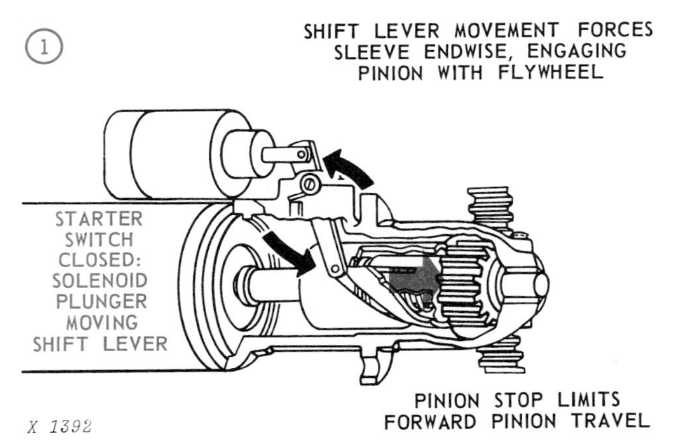

Fig. 40—Dyer Drive in Operation: 1) Solenoid Plunger Moving Shift Lever

The pinion, fitting loosely on the armature shaft splines, rotates freely while the pinion guide is forced to rotate forward on the spiral splines. The rotation of the pinion guide is transmitted to the pinion by two lugs on the pinion guide. The pinion rotates freely, without any forward movement, until alignment of the gear teeth takes place. Then it is thrust forward into mesh as shown in Fig. 40.

The pinion stop limits the forward movement of the pinion. As the shift lever completes its travel, it closes the starting motor switch, linked mechanically to the shift lever.

The motor armature then begins to rotate. This rotates the sleeve back, away from the pinion to its original position as shown in Fig. 41.

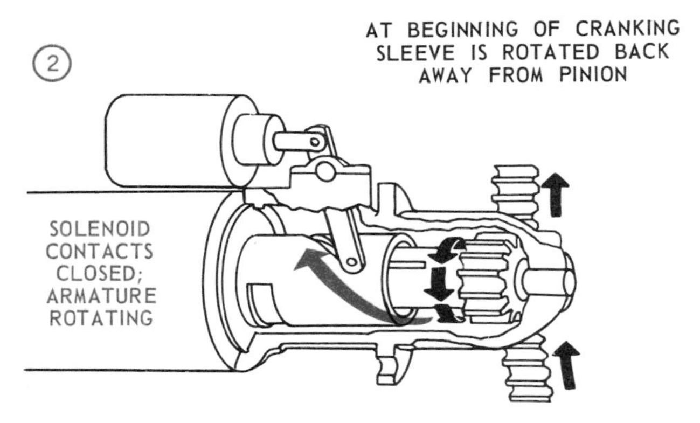

Fig. 41—Dyer Drive in Operation: 2) Armature and Pinion Rotating Flywheel

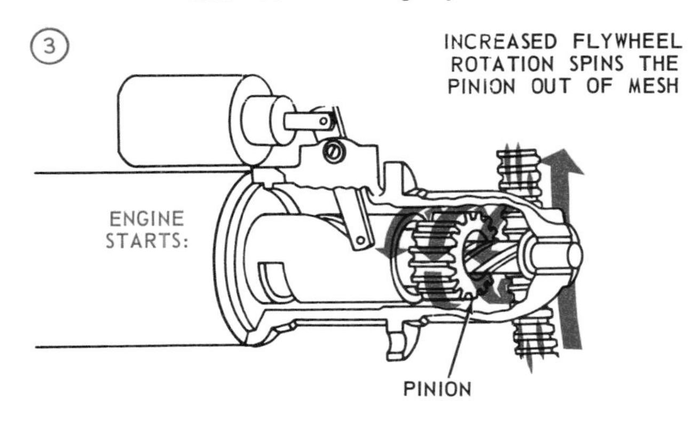

Fig. 42—Dyer Drive in Operation: 3) Flywheel Spinning Pinion out of Mesh

The instant that the engine starts, the flywheel attempts to drive the pinion faster than the armature is turning. Then both the pinion and the pinion guide are spun out of mesh with the ring gear (Fig. 42). The pinion guide drops into the milled section of the shaft splines, locking the pinion out of mesh.

It is impossible to start another cranking cycle without completely releasing the shift lever. The lever must drop all the way back to the "at rest" position.

On any automatic disengaging drive, it is always good policy to wait five seconds after a false start before attempting another start. In this period of time the engine will be at complete rest, and damage to the drive mechanism due to improper meshing is less likely to occur.

SPRAG CLUTCH DRIVE

The Sprag Clutch Drive is constructed and operates like the Overrunning Clutch Drive, except that a series of sprags, replace the rollers between the shell and sleeve. The sprags are held against the shell and sleeve surfaces by a spring. This assembly is then splined to the armature shaft with a stop collar on the end of the sleeve. See Fig. 43.

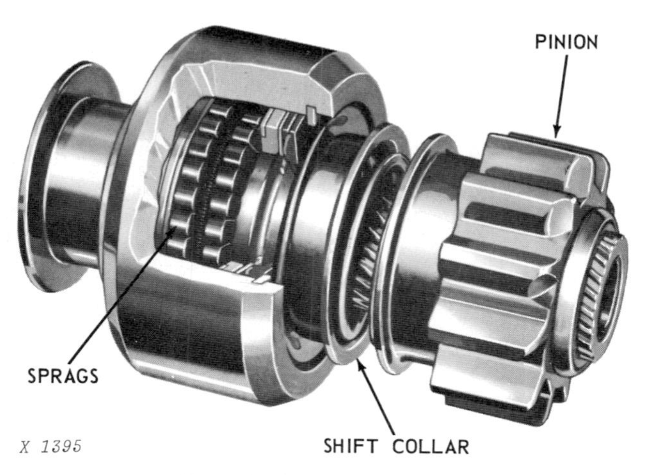

Fig. 43—Sprag Clutch Drive

Operation is as follows:

Movement of the shift lever against the collar causes the entire clutch assembly to move endways along the splined shaft, and the pinion teeth to engage the ring gear. If a clash of teeth should occur, continued movement of the shell and spiral splined sleeve causes the pinion to rotate and clear the teeth. The compressed meshing spring then forces the pinion into mesh with the ring gear. If the pinion does not clear before the two retainer cups meet, shift lever movement is stopped by the retainer cups and the operator must start the engagement cycle over again. This prevents closure of the switch contacts to the motor with the pinion not engaged, which would result in damage due to spinning meshes. On the second attempt, the pinion will engage in a normal manner.

With the pinion engaged and the switch closed, torque is transmitted to the pinion through the sprags. The sprags tilt slightly and are wedged between the shell and sleeve. When the engine starts, the ring gear drives the clutch faster than the armature, and the sprags tilt in the opposite direction to allow the pinion and sleeve to overrun the shell and armature. To avoid prolonged overrunning, the starting switch should be opened as soon as the engine starts.

The Sprag Clutch Drive is used primarily on larger starting motors to carry the high torque required to turn over high-compression engines.

STARTING CIRCUITS

A TYPICAL STARTING CIRCUIT

A typical starting circuit consists of a low resistance cable connection from the storage battery, through a control switch, to a direct-current starting motor.

The return circuit is to the battery, through the engine block and the frame of the vehicle.

Because of the large amount of current used to operate the starting motor, large cables are used. All connections must be clean and tight to prevent high resistance or voltage drop.

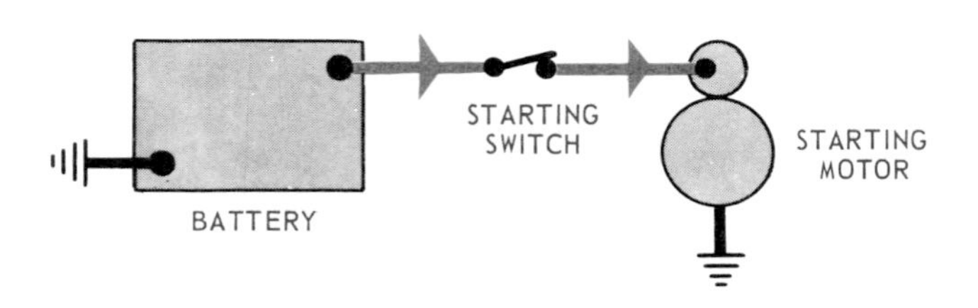

Fig. 44—A Typical Starting Circuit (12-Volt)

Closing the switch completes the starting circuit (Fig. 44).

Current will flow from the battery, through the cable, across the switch contacts, through the starting motor fields and armature, and back to the battery through ground.

SPLIT LOAD CIRCUIT

We learned earlier that diesel engines and similar heavy-duty engines require high-output starting motors. We also learned that regulation lighting systems have not been developed for 24-volt systems.

Let's look at two methods used to provide this extra starting power, as well as adequate lighting.

The first method is the 24-volt **split load** circuit illustrated in Fig. 45.

This circuit is called *split load* because the lights and accessories are split as equally as possible with each half connected to a 12-volt circuit.

The right-hand light circuit draws current from battery "A" and the left-hand circuit draws current from battery "B."

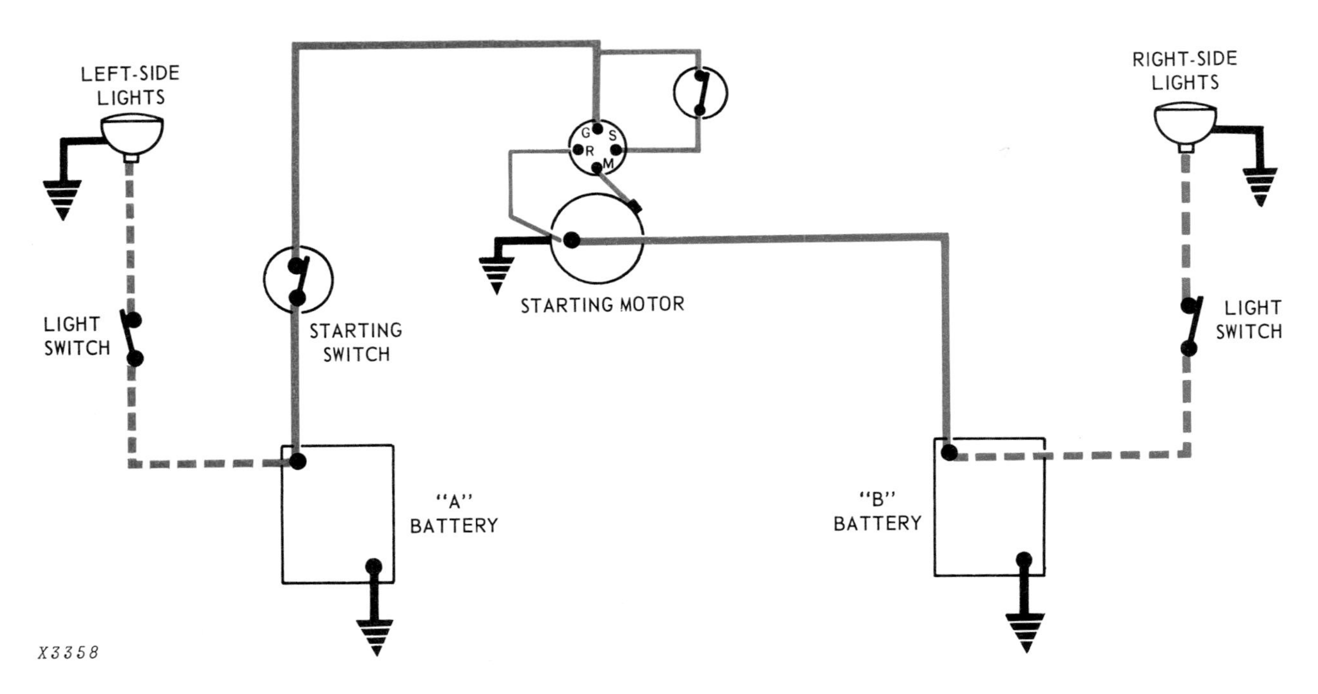

Fig. 45—Split Load Circuit (24-Volt)

The batteries and lights of each circuit are connected to a common ground, completing the circuit to the source. The vehicle frame and engine block are often used as the ground, and actually become part of the circuit.

When the switch in the starting circuit is closed, the vehicle frame becomes part of a series circuit between batteries "A" and "B" so that enough voltage and current is supplied to the 24-volt starting motor. The starting motor and the starting circuit are insulated so that none of the components in the circuit are attached to a common ground, allowing the circuit to be completed at the terminal of "B" battery.

Although the batteries are in series, the left- and right-hand lighting circuits remain split.

If the light switches are closed as shown, each light circuit will draw current from the battery to which it is connected. (The light switches are shown in the closed position for illustration purposes only. Of course, better starting performance is obtained by starting with the lights off.)

For details on the batteries used with split load circuits, see Chapter 3.

NOTE: Some split-load circuits have four six-volt batteries instead of two twelve-volt ones. In this case, two batteries operate each 12-volt circuit.

SERIES-PARALLEL CIRCUIT

Another method of using two 12-volt batteries to provide 24-volt starting is the **series-parallel** switch shown in Fig. 46.

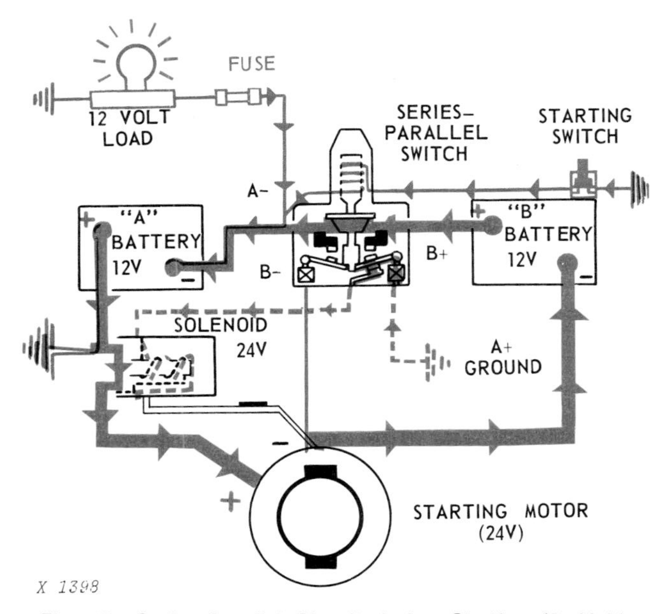

Fig. 46—Series-Parallel Circuit during Starting (24-Volt)

In Chapter 1 we learned about series and parallel circuits. We know that two 12-volt batteries connected in series will provide increased voltage while the current flow remains the same.

This applies directly to the series-parallel starting circuit. When the starting switch is closed, the series-parallel switch is actuated, connecting the terminals between "A" and "B" batteries as shown. We now have an insulated series circuit to the starting motor giving the 24 volts required.

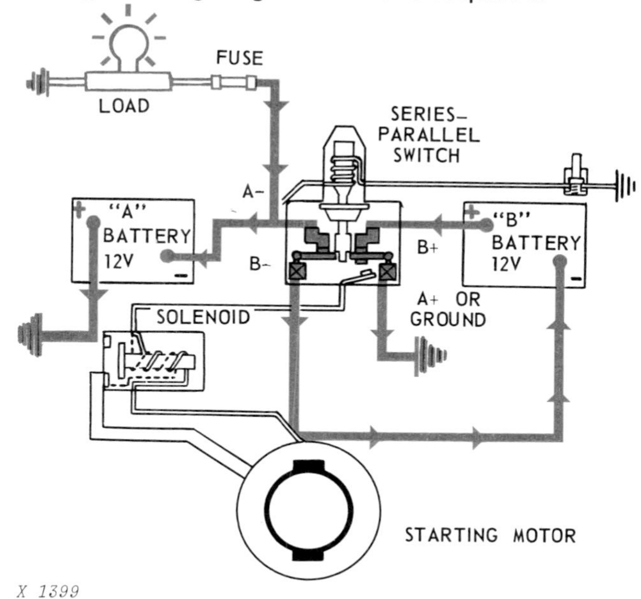

Fig. 47—Series-Parallel Becomes Parallel Circuit (12-Volts) After Starting

As soon as the engine starts and the starting switch is opened, the series-parallel switch also opens. This breaks the series connection between the batteries and provides a parallel circuit like the one shown in Fig. 47.

The starting switch operates as follows: The plunger moves up, opening the ground circuit for the 24-volt solenoid, which stops current flowing to the starting motor. As the plunger moves on up it opens the series connection.

Until recently, a 12-volt starting motor could not produce the torque required to start diesel and other high-performance engines. But the 12-volt *high-output* starting motor provides the features common to 24-volt starting motors, and may be used to replace the 24-volt systems we have covered. (The 12-volt high-output motor circuit is shown in Fig. 19.)

TESTING AND DIAGNOSING THE STARTING CIRCUIT

We have seen how each part of the starting circuit works and how the parts operate as a whole. Now we will look at the main troubles and how to test, diagnose, and remedy them. Let's start with the complete starting circuit on the machine. We should locate the trouble first, before we start removing components.

Several checks, both visual and electrical, can be made to isolate the trouble before removing any part of the circuit. Many times a component is removed from the machine only to find that it is not defective after making reliable tests.

Therefore, be sure to make the following tests before removing a component from a defective starting circuit.

TESTING THE STARTING CIRCUIT

Let's begin with the **battery.** The battery is the source of energy for all vehicle electrical systems. Therefore, it is also important to the starting circuit.

If the battery is less than three-quarters charged, accurate tests cannot be made in the starting circuit due to below par operation of the entire circuit. It is vital to the testing of starting circuit problems to have the battery fully charged and free of shorted or dead cells. Placing the test points of a voltmeter across the battery posts will indicate the charged condition of the battery while the engine is turning over. If the battery cannot provide the voltage required it must be recharged or replaced before the circuit can be checked. (For details, see Chapter 3. Also review the battery safety information provided in Chapter 1.)

CAUTION: When performing starter test, do not start the engine by shorting across starter terminals. The machine may start in gear if normal circuitry is bypassed.

If the battery is eliminated as a possible cause of starting circuit problems, inspect all clamps and **connections** in the circuit for corrosion and tighten them if necessary. Now we can continue our tests on the starting circuit.

When the starting switch in the circuit is closed, we can expect one of five things to occur if the starting circuit is defective:

1. Nothing happens—there is no "click" indicating that the solenoid contacts closed.

2. An audible "click" in the solenoid is heard, but the starting motor does not operate.

3. The starting motor is running but the engine does not turn over.

4. The starting motor turns over the engine slowly or erratically.

5. The engine starts but the starting motor drive does not disengage from the flywheel.

We can check out these five cases as follows:

Case No. 1: *We closed the starting switch and nothing happened.* This indicates that current is not reaching the solenoid and the switch contacts are not closing to complete the circuit. We can suspect then that the problem lies in that part of the circuit leading up to the solenoid, or in the solenoid.

To check this diagnosis, connect a jumper lead from the battery post to the switch terminal of the solenoid. (Be sure that the machine is not in gear.) If the engine starts the circuit is open somewhere between the battery and the solenoid. However, if the solenoid switch does not close, the solenoid is defective. Perform tests on the solenoid as covered later in this chapter.

Case No. 2: *The solenoid contacts "clicked" but the starting motor did not operate.* This indicates that the circuit problems lie within the starting motor. If the solenoid switch contacts close, and the switch begins to "chatter," there is low voltage at the starter because of low battery charge or high resistance in the circuit, or an open circuit exists in the hold-in winding of the solenoid. If low voltage at the starter is not the cause, the starting motor should be removed and tested as covered later in this chapter. (On 24-volt motors, a defective solenoid return wire could also cause the above symptom.)

Case No. 3: *The starting motor ran but did not turn over the engine.* We know that the starting motor is getting enough current to operate. The problem, then, is either in the shifting of the drive assembly into mesh, a broken armature shaft, or a dirty or faulty drive assembly. These causes require disassembly of the starting motor and proper service or repair.

Case No. 4: *The starting motor turned the engine over slowly or erratically.* The problem may be in the starting motor or in the drive assembly. Before removing the motor, a voltage drop test should be made. A voltage drop test will locate any high-resistance connections or shorted or grounded windings which would affect starting motor efficiency. This test is made with a voltmeter while the engine is turning over. With test points on the insulated BATTERY POST and the starting motor terminal, the voltage drop generally should not exceed 0.3 volts. With test points on the battery ground post and the starter frame, the voltage drop in the ground circuit usually should not exceed 0.1 volts. (On 24-volt motors, the return circuit is insulated and the voltage drop should be less than 0.3 volts.)

If no high-resistance connections are detected, the pinion drive and the flywheel ring gear can be inspected by removing the starting motor. If either gear is damaged, it must be replaced. If they are not damaged, the starting motor must be disassembled and tested.

Case No. 5: *The engine starts but the motor drive does not disengage from the flywheel.* This indicates a defect in the drive mechanism, the solenoid pull-in windings, solenoid contacts, or solenoid control circuit which will not allow the drive to disengage. The starting motor or the circuit should be serviced.

Summary: In each of these cases the trouble can be located without extensive circuit testing. Once the trouble has been found, follow the test procedures for the component as outlined later in this chapter.

Below is a brief list of general diagnosis tips for tests on the complete starting circuit.

HOW TO DIAGNOSE STARTING CIRCUIT FAILURES

Starting switch is on but no operation.

Look for an open circuit, defective starting switch, or starter safety switch, poor connections, open solenoid windings, stuck drive plunger or pinion.

Solenoid contacts close but starting motor does not operate.

Brushes are sticking, worn, or have weak spring tension. Commutator bars are dirty, burned, worn, pitted, or rough. Armature or field windings, armature bearings, or solenoid contacts are defective.

Starting motor operates but engine does not turn over.

Drive assembly not meshing. Dirty or faulty drive assembly. Drive pinion or flywheel ring gear damaged. Broken armature shaft.

Motor drive pinion does not move out of mesh.

Solenoid switch is defective. Drive assembly dirty or damaged. Armature shaft dirty or damaged.

SUMMARY: TESTING AND DIAGNOSING THE STARTING CIRCUIT

We have now covered testing and diagnosis of the complete starting circuit.

All of these checks should be made *before* removing any components for repair.

Next we will cover the testing and servicing of the various components of the starting circuit, once they are pinpointed as possibly defective.

Let's begin with the starting motor.

TESTING AND SERVICING STARTING MOTORS

Once you have tested the complete circuit and find that the starting motor may be defective, make the tests given here to confirm your diagnosis.

TESTING MOTOR BEFORE DISASSEMBLY

The tests below should be made before the starting motor is disassembled.

Preliminary Checks

Even before you remove the starting motor, inspect it.

Listen for the grinding of clashing teeth when the starting motor drive is engaged, and for a squealing or rattling noise when the motor drive is released from the flywheel. These are indications of a dry or worn drive mechanism.

Look at the starting motor. Are there any loose mounting bolts? Remove the commutator end frame and check for burned commutator bars, high mica, worn brushes, or an oily commutator and brushes. These can all cause inadequate starting and must be corrected before further testing of the motor.

Be sure to check the motor also for freedom of movement. Both the pinion gear and the armature must be free to move during the no-load test below.

Check the pinion for freedom of operation by turning the shaft. If the pinion turns freely, turn the pinion and check the armature operation. If the armature drags, look for tight, dirty, or worn bearings, a bent armature shaft, or a loose pole shoe screw. In this case, disassemble the starting motor immediately. The no-load test may damage the starting motor if the armature does not turn freely.

No-Load Test

The no-load test is the basic check of the starting motor's internal condition.

To perform a no-load test, connect the starting motor to a fully charged battery (Fig. 48).

Insert an ammeter capable of measuring several hundred amperes between the battery and the starting motor. (If the technical manual specification is for basic motor only, connect the jumper wire between the "S" terminal and the battery post so the solenoid current draw is not measured.) Connect the voltmeter to the starting motor terminal and frame. Connect a carbon pile resistor across the battery as shown. Place a jumper lead between the battery terminal and the "S" terminal on the solenoid to complete the test circuit. Place a tachometer on the end of the armature to measure armature speed.

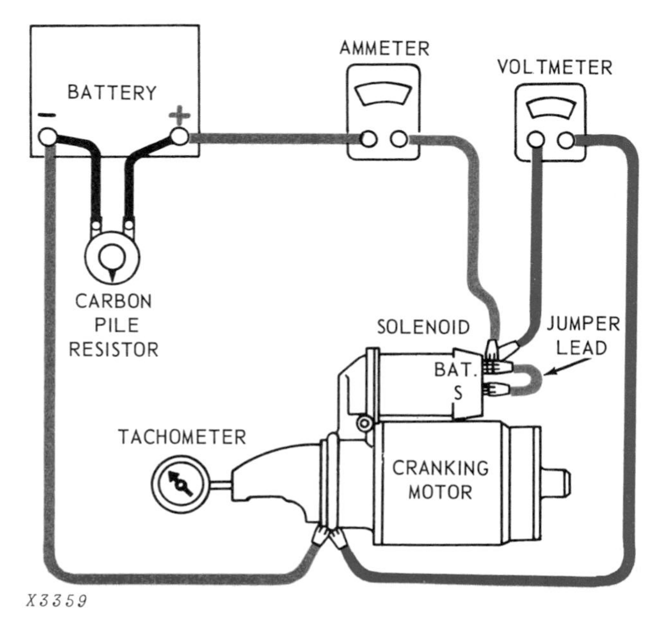

Fig. 48—No-Load Test Circuit

When the leads are connected to the battery terminals, current flows to the starting motor. The variable resistor is used to obtain the specified operating voltage of the starting motor. Do this by varying the resistor until the proper reading is on the voltmeter. When the specified voltage is attained, read the ammeter for the current drawn, and the tachometer for the armature speed. Compare these readings with the Technical Manual specifications for the starting motor being tested.

How to Interpret the No-Load Test Results

Take the no-load test readings and judge them by the possible results and causes given below.

NOTE: Open-circuited field coils will not be detected with this test on starters with series field coils which are connected in parallel with each other.

1. *Rated current draw and no-load speed* indicates a normal starting motor condition.

2. *Low free speed and high current draw indicates:*

a. Too much friction—Tight, dirty, or worn bearings, bent armature shaft or loose pole shoes allowing armature to drag.

b. Shorted armature—This can be checked on a growler after disassembly.

c. Grounded armature or fields — Check further after disassembly.

3. *Failure to operate with high current draw indicates:*

a. A direct ground in the terminal or fields.

b. "Frozen" bearings (this should have been determined when turning the armature by hand).

4. *Failure to operate with no current draw indicates:*

a. Open field circuit. This can be checked after disassembly by inspecting internal connections and tracing circuit with a test lamp.

b. Open armature coils. Inspect the commutator for badly burned bars after disassembly.

c. Broken brush springs, worn brushes, high insulation between the commutator bars or other causes which would prevent good contact between the brushes and commutator.

5. *Low speed and low current draw indicates:*

a. High internal resistance due to poor connections, defective leads, dirty commutator or an open field circuit.

6. *High free speed and high current draw* indicate shorted fields. If shorted fields are suspected, replace the field winding assembly and check for improved performance.

NOTE: The lock-torque test, requiring special equipment, can be used to diagnose motor malfunctions. However, a thorough no-load test will usually reveal these faults without extra testing.

DISASSEMBLY OF THE MOTOR

General

Now that we have completed the no-load test and interpreted the results, let's assume the starting motor does not perform per the specifications. This requires disassembly and further testing of the motor.

Normally the starting motor should be disassembled only as far as necessary to repair or replace the defective parts. But to show the procedures, we will completely disassemble and assemble *a typical starting motor with an externally-mounted solenoid switch.* As a precaution, wear safety glasses while doing this job.

As the starting motor is disassembled, clean each part and inspect it for excessive wear or damage. An overall inspection should be made during disassembly. Things which are easy to do during disassembly are: 1) checking bearings for proper clearance, roughness or galling, 2) removing oil and dirt from insulation, and 3) inspecting the condition of the insulation.

Disassembling the Motor

The first step in disassembling the starting motor is to remove the commutator end frame. Mark the position of the end frame and the main frame with chalk before removal to aid in aligning the parts when reassembled. While removing the end frame, note the location of the brush assembly.

Brushes may be located in the end frame—be careful during removal of these—or they may be located in the main frame. While removing the end frame, check the brushes to see if they slide freely in their holders and make full contact on the commutator. **Brushes worn to half their original length or less should be replaced.**

The second step in disassembly is to disconnect the solenoid (externally-mounted). After removing the fasteners to the main frame and the drive housing, the solenoid can be removed along with the main frame assembly.

Removal of the main frame will expose the armature and commutator. The commutator should be checked for evidence of excessive arcing, discoloration or excessive wear. If it is only slightly dirty, glazed or discolored, cleaning with No. 00 or 000 sandpaper will restore the commutator to a serviceable condition. NEVER USE EMERY CLOTH TO CLEAN A COMMUTATOR.

If the commutator is worn or rough, however, it should be turned on a lathe and the mica undercut if recommended. See Technical Manual. At this point the armature can also be inspected for rough bearing surfaces, and rough or damaged splines.

The solenoid shift lever must be disconnected from the drive housing to separate the armature and drive housing. In doing so, the complete armature shaft is exposed along with the drive mechanism. When the drive housing is removed, inspect it for cracks, and check the bearing for excessive wear. Remove any rust, paint or grease from the housing flange before reassembly.

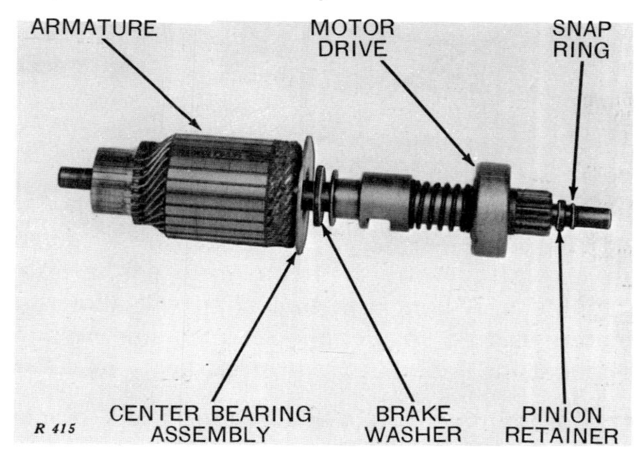

Fig. 49—Armature and Drive Assembly

With the overrunning clutch drive and armature shaft exposed as in Fig. 49, make a visual inspection for damage due to overheating.

Overheating occurs if the starting switch is not released as soon as the engine begins to operate, causing the drive pinion to remain in mesh and overrun the armature. Clutch mechanisms can withstand this for only a brief time before the lubricant gives out and they seize, causing excessive armature speeds. Look for galling of clutch bearings under the drive pinion, and bluing or deposits of bearing material on the armature shaft. If this is evident, replace the armature.

Checking the Overrunning Clutch Drive

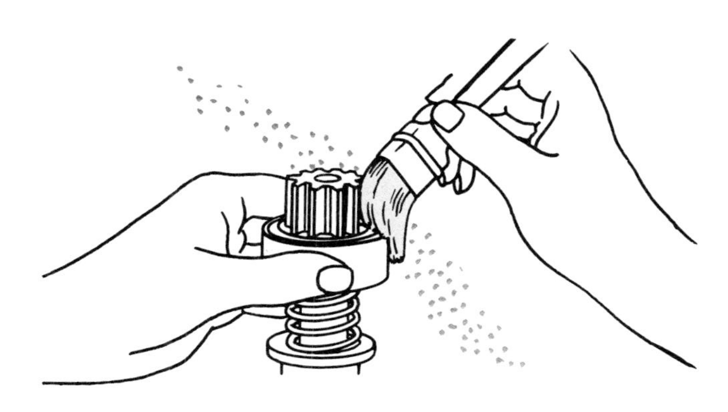

Fig. 50—Cleaning an Overrunning Clutch

After inspecting the motor, check the drive mechanism. Simply turn the clutch by hand to find out whether the drive slides freely on the splines. Replace the drive if it does not move freely. The pinion should turn smoothly in one direction and lock when turned *slowly* in the other direction. Replace the pinion if it is excessively worn or damaged.

Never clean the overrunning clutch by any high-temperature or grease-removing methods. This removes the lubricant originally packed in the clutch and causes rapid clutch failure. The overrunning clutch can be wiped or brushed clean with a solvent as shown in Fig. 50, but should never be submerged since it cannot be repacked with grease.

Checking the Bendix Drive

The spiral threads on the Bendix drive sleeve should be free of dirt and grease. If the pinion teeth are badly burred, chipped or otherwise damaged, replace the pinion and shaft assembly.

Distorted drive springs can cause breakage of the motor drive housing. When the drive spring is dis-

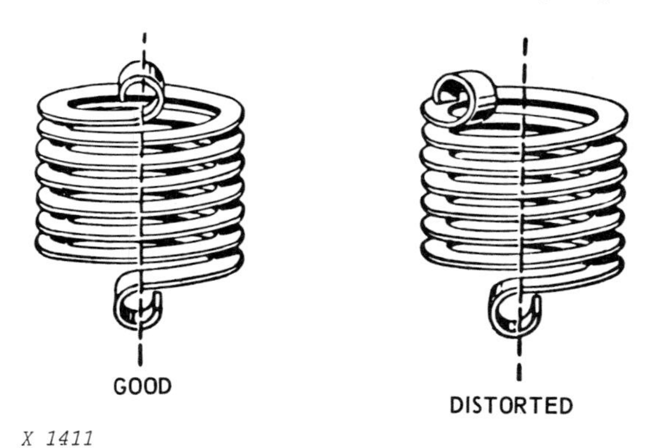

Fig. 51—Bendix Drive Springs

torted as shown in Fig. 51, replace it. When installing a new spring, always use new locks on the screws. Be sure to lock the screws securely and apply a thin coat of oil to the spiral threads.

Checking the Dyer and Sprag Drives

The Dyer and Sprag drives vary with the application. This means varied cleaning procedures, lubrication requirements, and adjustments. Consult the machine's Technical Manual before cleaning these drive mechanisms.

Removing the Drives

To separate the drive from the armature shaft, the pinion stop (Fig. 52) must be removed. When removing the drive assembly, use a pipe coupling or suitable metal cylinder to drive the pinion stop toward

Fig. 52—Removing Pinion Stop

the armature core and reveal the snap ring. After removing the snap ring, the drive assembly will slide off the armature shaft.

TESTING AND SERVICING THE MOTOR COMPONENTS

When the starting motor has been disassembled, inspect the motor parts in preparation for component testing.

General inspection helps to locate trouble spots which could cause later failures of the motor.

Mechanical failures, such as worn bushings, bent armature shafts, worn brushes or worn motor drive, are common troubles. Except for these, **starting motor failures are generally caused by one of the following:**

OPEN-CIRCUITED WINDING—This occurs due to a broken wire, poorly-soldered connection, or a disconnection at a terminal. The result is usually no operation or a decrease in starting ability.

GROUNDED WINDING—This occurs when insulation becomes defective and a bare winding wire contacts the metal of the armature, frame, or field poles. This usually results in failure of the motor to operate.

SHORT-CIRCUITED WINDING—This occurs when insulation becomes defective and the winding coils contact adjacent coils, resulting in decreased motor cranking power.

Testing and Reconditioning the Armature

When the no-load test shows a low armature speed and a high current draw, the armature should be tested for opens, grounds, and short circuits. If the armature is tested, perform all of the tests given below.

Fig. 53—Checking Armature Shaft for Straightness

Before testing, check the armature shaft for straightness. A bent armature shaft will allow the armature to drag, resulting in reduced starting performance and increased current draw.

To check, put the shaft in a lathe or rest in V-blocks and check the shaft with a dial indicator as shown in Fig. 53. Replace a bent or damaged shaft.

OPEN CIRCUIT TEST

If the armature shaft is straight, test it for open circuits in the commutator.

Open circuits are usually caused by too-long starting periods. The most likely place for an open circuit or loose connection to occur is at the commutator riser bars. Poor connections cause arcing and burning of the commutator bars as the starting motor is operated.

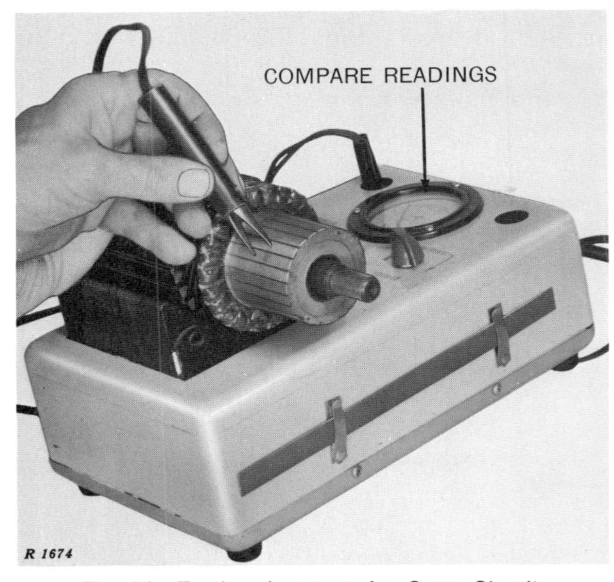

Fig. 54—Testing Armature for Open Circuits

If the commutator bars are not too badly burned, they can often be repaired by resoldering or welding the leads in the riser bars (using rosin flux) and turning down the commutator to remove the burned material. The insulation should usually be undercut. Refer to "Reconditioning Armature," later in this chapter.

A growler is shown in Fig. 54 with the armature mounted for testing open circuits. Use a two-pronged tester and span the adjacent commutator bars as illustrated. Slowly rotate the armature back and forth to get the maximum reading. If necessary, adjust the voltage control so that the meter hand rests about midway on the scale.

Voltage readings should read approximately the same as each commutator bar is tested with the adjacent bar. Wide deviations in readings will indicate an open circuit.

If an open circuit exists in the armature that cannot be located or repaired, the armature must be replaced.

GROUND TEST

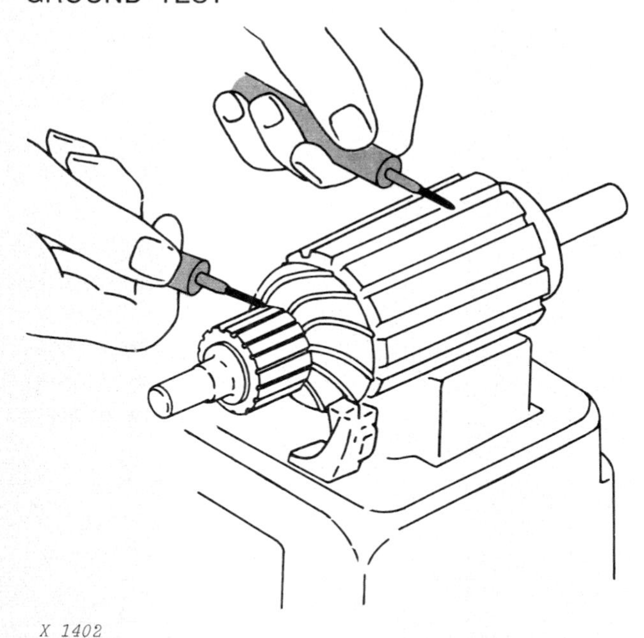

Fig. 55—Testing Armature for Grounds

When testing for grounds, place one test point on the iron core of the armature and the other test point on the copper commutator (Fig. 55). We are checking for connections between the copper and iron core which would cause a grounded armature. The test lamp should not light while making this test. If it does light, a winding is grounded and the armature must be rewound or replaced.

SHORT CIRCUIT TEST

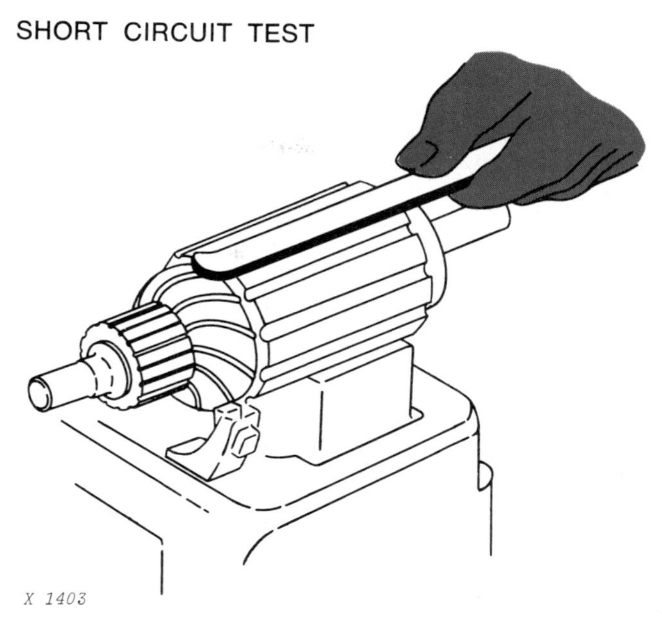

Fig. 56—Testing Armature for Short Circuits

An indication of a shorted armature is a burned commutator bar. To verify this, place the armature on a growler as shown in Fig. 56 and test it. With the growler turned on, place a thin strip of steel or a hacksaw blade on the armature as it is slowly rotated. If the metal strip vibrates over a winding, that winding is short circuited.

Short circuited windings are sometimes caused by metal in the commutator bridging the gap from one commutator bar to the next. By removing the bridged metal, this condition can be corrected. However, if this does not correct the short, replace the armature.

RECONDITIONING THE ARMATURE

If tests indicate that the armature is suitable for service, turn the commutator down and undercut it before the starting motor is assembled.

To turn down a commutator, use a suitable tool, such as the one illustrated in Fig. 57. If a tool specifically for turning commutators is not available,

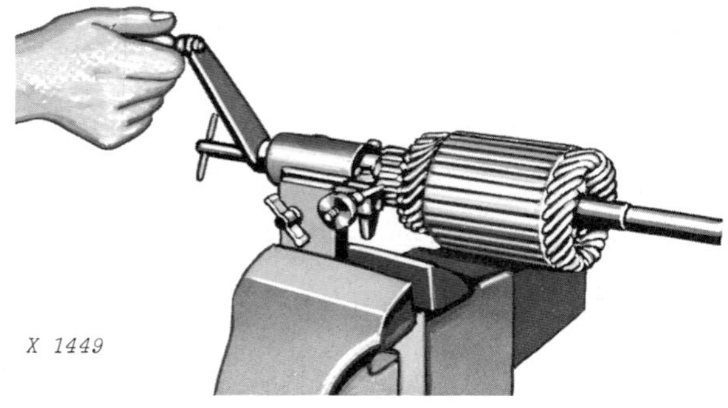

Fig. 57—Turning Down the Commutator

the commutator can be turned down on a lathe. In either case, remove only enough metal to "true up" the commutator.

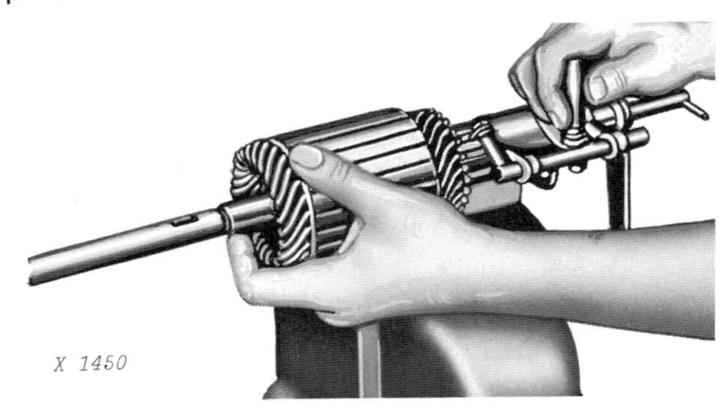

Fig. 58—Undercutting Mica on Commutator

On some starters, after "truing" the commutator, undercut the mica between the commutator bars. A tool used for this purpose is shown in Fig. 58. If a tool for undercutting is not available, use a hacksaw blade with the sides of its teeth ground to the same width as the distance between commutator bars. An undercutting tool is preferred, however, since it can provide more accurate and consistent reconditioning.

Always consult the starting motor specifications before reconditioning the armature. Most high-output starting motor armatures should *not* be undercut after the commutator is turned down.

Testing and Servicing Field Windings

Field windings should be checked for grounded wiring and open circuits while the armature is removed. Grounded wiring is caused by the copper wiring touching the steel case due to worn insulation or trapped metal between a winding and the case.

Before testing the field windings in a compound-wound starting motor, separate the shunt field winding lead from the series field winding lead. By doing this the shunt is eliminated from the circuit during the tests for grounds and open circuits.

GROUNDED CIRCUIT TEST

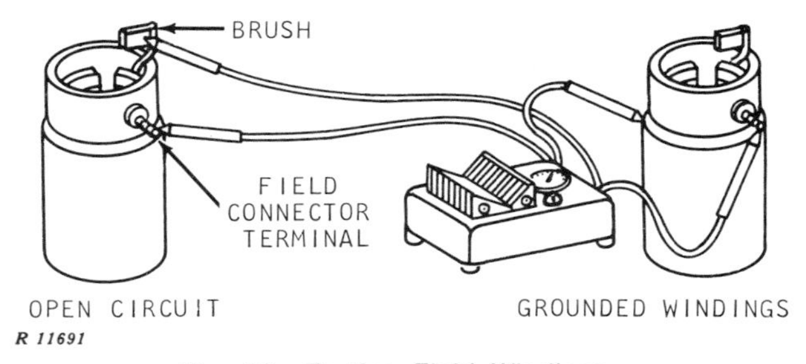

Fig. 59—Testing Field Windings

To test for a grounded circuit, first disconnect the field winding ground connections. Then connect one test point to the field frame and the other to the field connector (Fig. 59).

If the test lamp lights, the field windings are grounded. Field windings may be repaired or replaced if faulty.

OPEN CIRCUIT TEST

An open circuit test can also be made with a test lamp. If the test lamp does not light with the test points applied at the ends of the windings, the field winding circuit is open and current cannot flow through them.

When testing 24-volt starting motors, two sets of terminals are tested when checking for open circuits. Place one test point on the input terminal of the starting motor and the other test point on the winding as we did before. Test the second set of field windings by moving the test point from the winding to the insulated terminal.

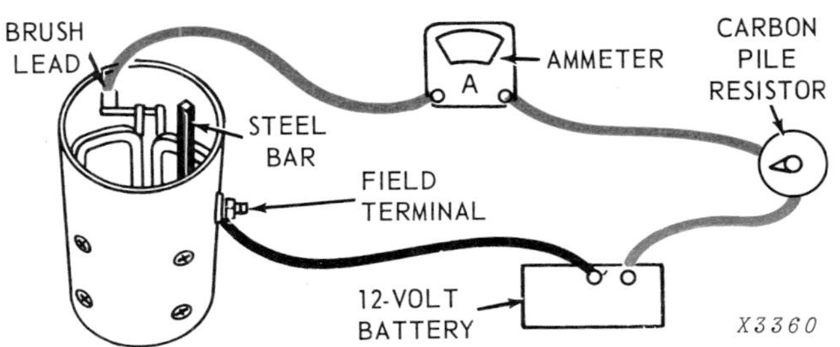

Fig. 60—Open Circuit Test in Parallel-Wound Field Winding

To check parallel-wound field windings for an open circuit, make connections shown in Fig. 60. Adjust carbon pile resistor to obtain 60 to 75 amperes. Place a steel bar against the pole shoe. No magnetism at pole shoe indicates an open-circuited field winding. Connect to other brush lead and check the other field windings.

SHORT CIRCUIT TEST

Because of the low resistance in the field windings there is no satisfactory test for short-circuited field windings. If the starting motor does not perform after all other tests have been made and no defects found, a short circuit can be suspected. It usually shows up in the no-load test as high ampere draw and high free speed.

REPLACING FIELD WINDINGS

If the field windings must be removed for repair or replacement, use a pole shoe spreader and pole shoe screwdriver. Be careful in replacing the field windings to prevent grounding or shorting them.

When the pole shoe has a long tip on one side, assemble it in the direction of armature rotation.

Testing and Servicing Brushes

Before assembling the starting motor, the brushes should be inspected and tested. To test the insulated brush holders, connect one test point to the brush holder and connect the other test point to the motor frame. The test lamp should not light. If all of the brush holders are insulated, the test lamp should light when connected to opposite brush holders but should not light when connected to adjacent brush holders or from any brush holder to the frame. Repair or replace the brush holder if tests indicate faulty brush holders. **Always use matched sets when replacing the brushes.**

On some motors, the brushes are soldered to the field coils instead of using mechanical fasteners. To replace these brushes, cut off the brush leads at the points where they are attached to the field coils. The ends of the coils must then be prepared for soldering on the new brush lead assemblies.

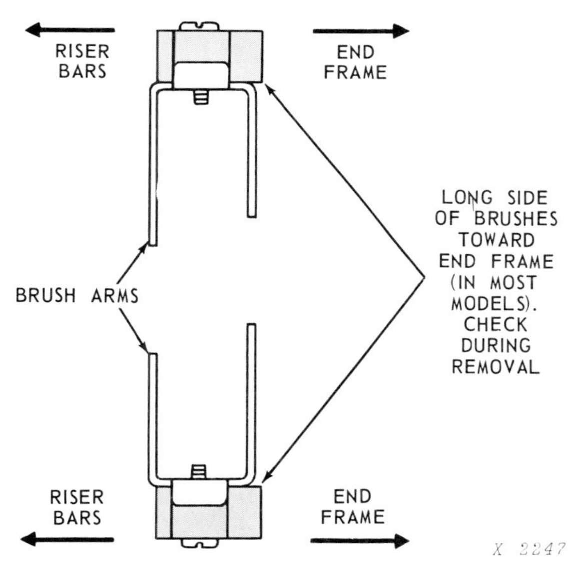

Fig. 61—Assembling Brushes to Brush Arm

Assemble all brushes to the brush arms in the proper direction. In many cases, the long side of the brush is toward the commutator end frame as shown in Fig. 61. Otherwise the brushes may come in contact with the riser bars. Always check the brush offset during removal.

When soldering leads, solder the **back** sides of the coils so that the connection will not rub the armature. Clean the ends of the coils thoroughly by filing or grinding off the old brush lead connections. Remove varnish only as far back as necessary to make the solder connections. Using rosin flux, the leads may then be soldered to the field coils, making sure that they are in the same position as the original brush holders. If the leads are overheated, solder will run on them and they will no longer be flexible.

To replace grounded brush assemblies, remove the old brush holders, the same as insulated brushes, and then attach the new assemblies to the frame. If the brush holder assemblies have retainers, peen the screws with a hammer so that the nuts cannot vibrate loose during engine operation.

When the field coil and brush assembly is reassembled in the frame, recheck the frame and field assembly with a test lamp to make sure that the soldered connection is not touching the frame and grounding the fields.

SPRING TENSION

The brushes should make good, clean contact with the commutator and should have the proper spring tension. They must also have freedom of movement so they can follow the commutator.

Never allow brush spring tension to fall below the specified limits. Starting motor brushes carry a high current, so good contact between the brush and the commutator will cut resistance. Most brush springs test above the specified tension but since the starting motor is used intermittently, this is not objectionable. Weak or distorted springs, or springs that show evidence of overheating, rust or excessive wear should be replaced. Check the spring tension of most brushes during assembly of the starting motor. (See later in this chapter.)

Servicing Bushings

COMMUTATOR END FRAME BUSHING

The bushing in the commutator end frame should be inspected for wear or damage. An excessively worn bushing will allow the armature to drag on the pole shoes.

The bronze bushings in the starting motor are oil-impregnated and should need no added lubrication. They should be lubricated when the motor is disassembled; at this time a few drops of light engine oil may be placed on each bushing before reassembly. *Be careful so that no oil reaches the commutator.*

In some cases, the bushing is lubricated by a felt wick. To replace a lubricated bushing, do the following.

REPLACING LUBRICATED BUSHINGS

Remove the oil plug so that the oil wick can be removed. Then drive the expansion plug out of the housing. Note the depth of the old bushing; drive it out, and press the new bushing in to the same depth as the original.

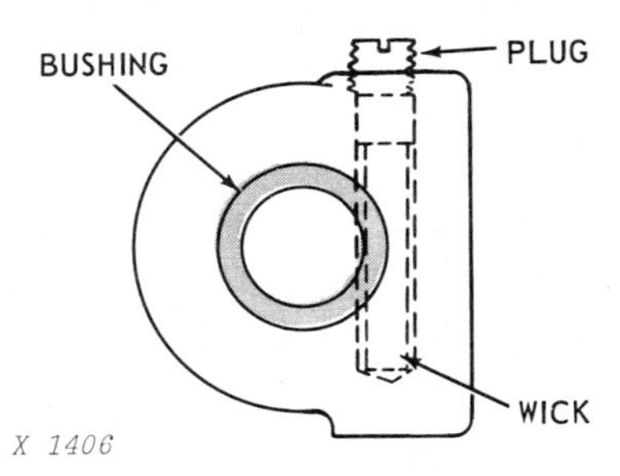

Fig. 62—Rebushing Diagram

After the new bushing is in place, use a drill the same size as the oil wick hole to drill the bushing (Fig. 62). Remove the burrs and clean out the oil wick hole before testing the bushing with the respective bearing surface of the armature. If the armature does not turn freely, ream the bushing.

The bushing can collapse when installing a prelubricated or absorbent bronze type. To prevent this, use the proper bushing arbor to obtain a correct bearing fit. The arbor is usually several ten-thousandths larger than the armature shaft. If the prelubricated bushing is undersized after installation, burnish it to size (polish by friction or rubbing). *Avoid reaming* (cutting excess metal away) if possible.

DRIVE HOUSING AND CENTER BEARING BUSHINGS

If the drive housing bushing or the center bearing bushing is worn, replace it. Loose-fitting center bearings are normal. The loose fit allows easier bushing alignment. When replacing the bushing, follow the same steps used to replace the commutator end frame bushing.

Before assembling the armature shaft with the drive mechanism, lubricate the armature shaft with a light coat of SAE 10 engine oil. Using a heavier oil may cause failure of the drive to mesh at low temperature.

When assembling motors having a center bearing, lubricate the center bearing in the same way and slide the bearing and washers onto the armature shaft ahead of the drive assembly. (Use a graphite-type starting motor lubricant as recommended.)

ASSEMBLY OF MOTOR

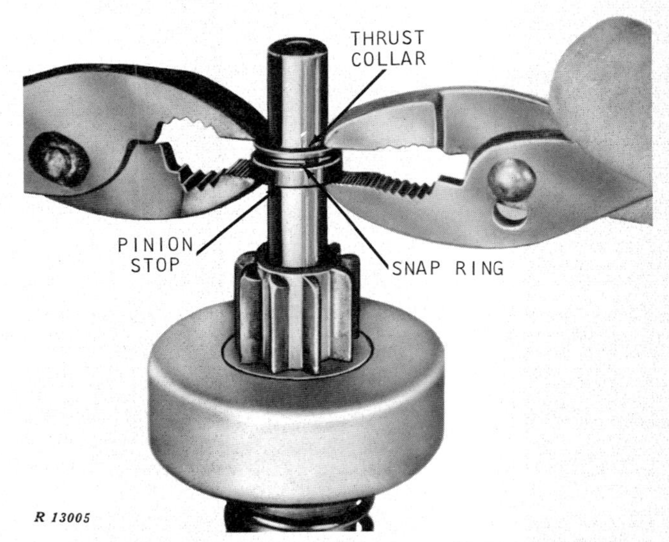

Fig. 63—Forcing Pinion Stop over Pinion Ring

Place the drive assembly on the armature shaft and slide the pinion stop onto the shaft with the cupped surface facing the open end of the shaft. Install the snap ring and squeeze it so it fits into the groove on the armature. Slide the thrust ring onto the shaft and position the pinion stop and washer next to the snap ring as shown in Fig. 63. By using two pairs of pliers, force the pinion stop over the snap ring.

With the thrust collar on the armature shaft, lubricate the bushing surface in the drive housing with the starting motor lubricant. Then guide the armature and solenoid shift lever into the drive housing. Secure the lever with the pivot screw and move the lever back and forth. The shift lever and drive assembly should move freely in both directions.

If the starting motor has a center bearing, secure it to the drive housing after checking the shift lever.

Refer to the machine Technical Manual for drive assembly tests on coaxial or heavy-duty starting motors.

Next, install the solenoid gasket, the return spring and the solenoid. Before attaching the solenoid to the drive housing, use a recommended sealing compound between the solenoid flange and the starting motor main frame.

Install the main frame and guide the brushes onto the commutator, being careful not to damage the brushes or the brush holders. After the main frame is installed, hold down the brush holders to align the brushes with the commutator and tighten the brushes.

Seating the Brushes

If new brushes were installed while servicing the starting motor, they may require seating before completing assembly.

To seat the brushes, wrap a piece of No. 00 or 000 sandpaper around the commutator so that the exposed end is trailing when the armature is turned in the running direction. The sandpaper should be one-eighth inch wider than the commutator and one inch longer than needed to reach around the commutator. Turn the armature clockwise as viewed from the drive end of the starting motor. Do not turn the armature for too long. A few turns will seat the brushes without too much brush wear. Be sure to remove all dust and abrasive particles.

Checking Brushes

We have learned that improper contact of the brushes with the commutator can cause faulty motor performance. Therefore, on most motors it is good practice to make one final check before completing assembly. That test is for spring tension on the brushes. (On some heavy-duty motors, this test is not recommended.)

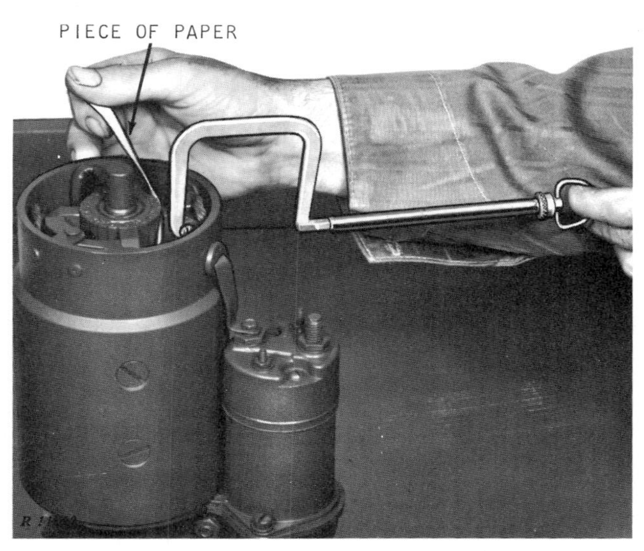

Fig. 64—Testing Brush Spring Tension

Measure brush spring tension using a spring scale (Fig. 64). By pulling the spring scale on a line parallel to the brush, we can obtain a reading at the point where the spring just leaves the brush holder. Consult the motor specifications in the technical manual, since spring tension ratings vary with different starting motors. To adjust the tension of the brushes, bend the springs or holders.

Completing Assembly of Motor

To complete assembly, lubricate the bushing in the commutator end frame with a starting motor lubricant and slide it onto the armature shaft. Arrange the brush leads so that the thru bolts will clear them and align the register marks on the main frame and the commutator end frame. Install the thru bolts and tighten the commutator end frame. Lubricate the oil reservoirs with a light SAE 10 engine oil.

Checking Pinion Clearance

The pinion clearance test is made to check for proper assembly of the starting motor and excessive wear in the shift linkage. Many starting motors have no provision for pinion adjustment. As a result, always refer to the machine technical manual for details on pinion adjustments. If the pinion is not adjustable, the drive shift assembly must be replaced when it becomes worn or faulty.

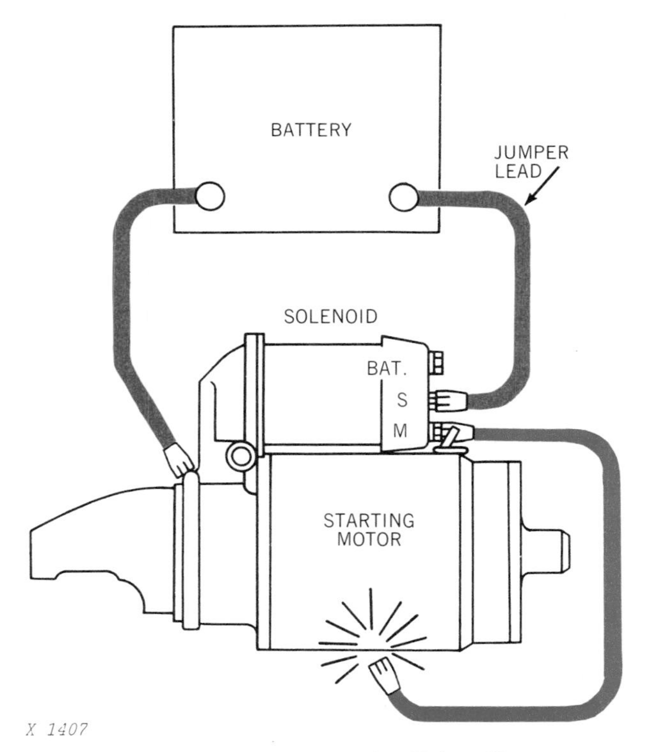

Fig. 65—Circuit for Testing the Pinion Clearance

To check pinion clearance, disconnect the field coil connector from the solenoid motor terminal (Fig. 65). Then connect a battery, of the same voltage as the solenoid, from the solenoid switch terminal to the solenoid frame or ground terminal. Momentarily connect a jumper lead from the solenoid motor terminal to the solenoid frame or ground terminal to shift the drive into starting position. It will remain there until the battery is disconnected. Now the pinion clearance can be checked as given below.

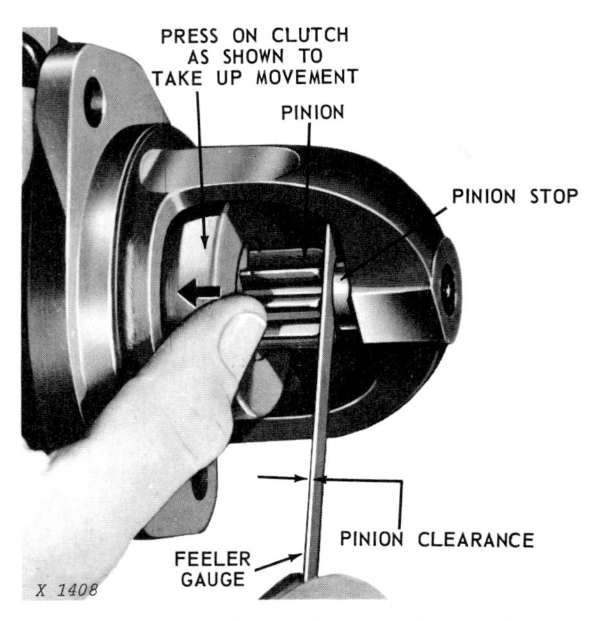

Fig. 66—Checking Pinion Clearance (Intermediate-Duty Clutch Motor Shown)

Check the pinion clearance by pressing the pinion or drive towards the commutator to take up slack movement and measuring the clearance between the pinion and pinion stop as shown in Fig. 66.

The correct pinion clearance will vary for different motors. Consult the machine Technical Manual for specific readings and procedures.

Final Assembly Check

It is good practice to give an overhauled starting motor a no-load test. If current draw is too high, two or three raps with a rawhide hammer will often help align the bearings and free up the armature.

PERIODIC MAINTENANCE OF STARTING MOTORS

Check the starting motor periodically during operation on the vehicle. This will help to eliminate failures due to neglect. Abnormal service which requires many daily starts, operation in dusty or very humid climates, or operation in arctic or tropical temperatures, all put an added strain on the starting circuit and tend to wear the parts more rapidly. Under these conditions, inspect the motor more frequently.

Keep the battery serviced to be sure of current for starting. Check the specific gravity of the battery electrolyte at regular intervals. Inspect the cables and connections and keep them clean and tight.

To prevent overheating, NEVER operate the starting motor for MORE THAN 30 SECONDS AT A TIME without pausing a few minutes to cool it off. (Always consult the machine technical manual for the exact safe period.)

Thrown solder is an indication that the motor has been overheated by too-long operation. This abuse may cause open circuits to develop at the commutator bars, resulting in burnt-out bars. Each time an open-circuited bar passes under a brush, severe arcing occurs and the bar soon becomes badly burned. If the bars are not severely burned when the starting motor is serviced, the commutator may be repaired by resoldering the leads and turning down the commutator.

Always keep the starting motor mounting tight and be sure the drive is in good condition. The condition of the drive can be checked by operating the starting motor two or three times while noting the action of the drive. The motor may have to be removed to actually examine the drive.

On clutch-type motors, operating the motor also serves as a check on the freedom of shift lever operation.

The high output starting circuit, especially, must be kept in good condition to get maximum performance from the high output motor. Refer to the machine Technical Manual for circuit maintenance and specifications when servicing these circuits.

As a final step in periodic maintenance, lubricate the starting motor by adding a few drops of light engine oil to the visible hinge cap oilers. The bearings in many starting motors are of the oilless type, but they require oiling when the motor is reassembled.

TESTING OF SWITCHES

No starting motor service is complete unless the switch is checked out.

The allowable resistance (or voltage drop) of the switch contacts will vary with the machine. See the Technical Manual.

TESTING MANUAL SWITCHES

A general inspection will indicate whether the manual switch is serviceable or faulty.

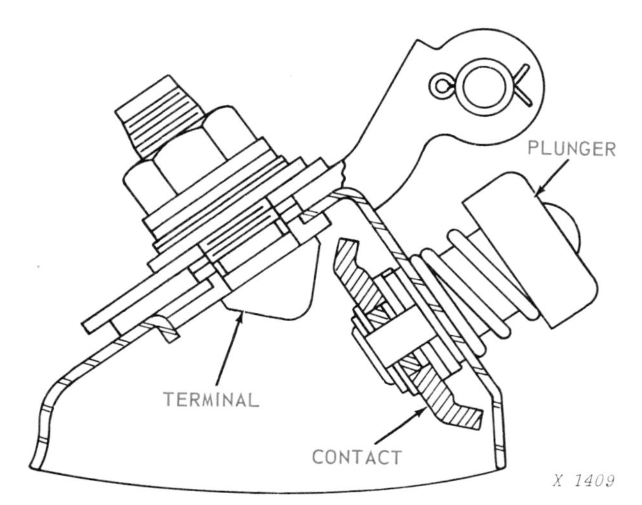

Fig. 67—Manual Switch

If the switch has a bypass ignition terminal, check it to be sure that contact is made as the switch button is pushed in. The ignition terminal should be in good condition, and if there is doubt it should be replaced. If the terminal stud is burned or bent, it should also be replaced; however, if it is not, remove any burrs with a file, and polish the contacts with No. 00 sandpaper. Install new insulators when the switch is replaced on the starting motor.

TESTING SOLENOID SWITCHES
(Starter Removed)

To prevent damage, the solenoid must be on the starter when testing the windings. Remove terminal cover and field coil connector. If equipped, disconnect shunt field winding lead from terminal on field frame.
Replace the solenoid if it fails any of the following tests.

Three tests are usually made on solenoid switches:

- **No load test**
- **Return test**
- **Pull-in winding test**

NOTE: Refer to your machine Technical Manual for the correct solenoid test procedure. Some solenoids are tested using the alternate procedure given on page 5—32 rather than the following typical procedures. If the machine manual recommends testing of both windings at once, use the procedures recommended in the machine manual. Note the ammeter reading at the end of its swing upward.

Testing Starter Solenoid (No Load Test)

Make connections as shown in Fig. 68 and use an ammeter capable of measuring several hundred amps.

Measure current draw and compare with the specifications for the starting motor being tested.

If speed and current draw are slightly low, connect a voltmeter between motor terminal and frame. Observe voltage during test. Voltage may be reduced because of high current draw on battery.

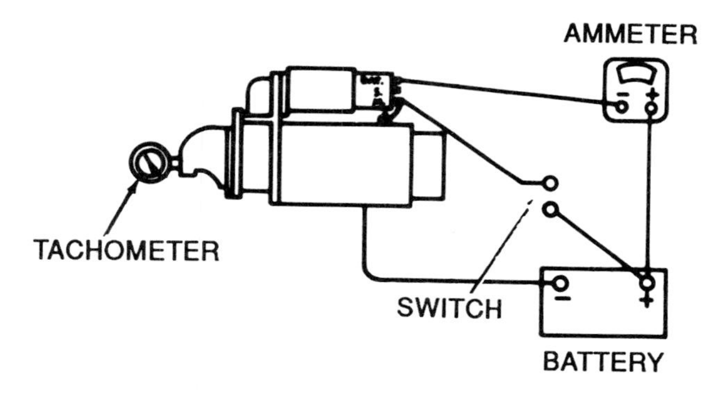

Fig. 68—Solenoid No Load Test

Evaluating No Load Test

Fails to Operate — Low Current Draw
Open series field circuit.
Open armature coils.
Defective brush contact with commutator.

Fails to Operate — High Current Draw
Grounded terminal or fields.
Seized bearings.

Low Speed — Low Current Draw
High internal resistance.
Defective brush contact with commutator.

Low Speed — High Current Draw
Excessive friction.
Shorted armature.
Grounded armature or fields.

High Speed — Low Current Draw
Open shunt field circuit.

High Speed — High Current Draw
Shorted series field coils.

Testing Magnetic Switches

A magnetic switch can be checked by connecting the test leads to the main switch terminals.

If the switch is in good condition, the plunger will close the main switch contacts.

If the action is slow or the plunger does not operate, take an ammeter reading of the current drawn by the coils.

1. Ammeter readings higher than specified mean a shorted or grounded wire.

2. A low reading means excessive resistance.

3. No reading means that there is an open circuit in the coils.

Most magnetic switches are sealed, so any of the above conditions mean that the switch must be replaced.

Testing Starter Solenoid (Alternate Method)

Some machine technical manuals use the following solenoid tests. Make these tests with the solenoid on the starter and the leads disconnected.

PULL-IN WINDING

Make connections as illustrated in Fig. 69.

Adjust the carbon pile to obtain 8 volts.

When the jumper wire is connected, the solenoid should move the pinion out to the stop.

When the jumper wire is disconnected, the pinion should remain out at the pinion stop.

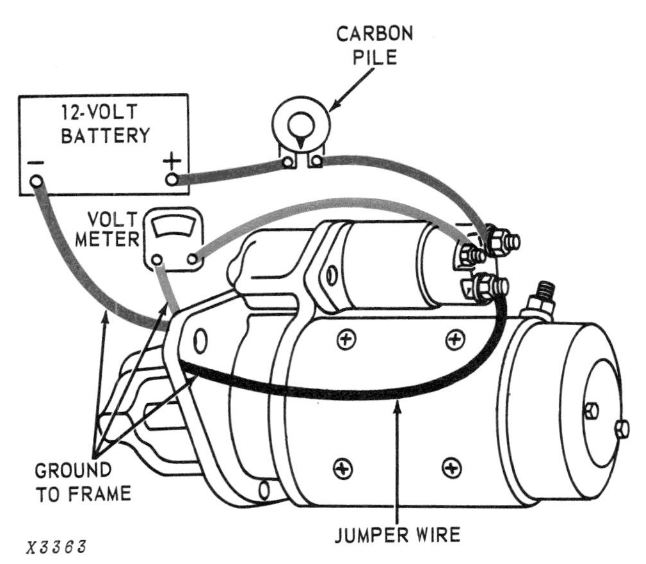

Fig. 69—Solenoid Pull-In Winding Test

TESTING SOLENOID RETURN

Make connections as illustrated in Fig. 70. Close the switch and pull drive out until pinion contacts the pinion stop. When released, the drive should return without hesitation.

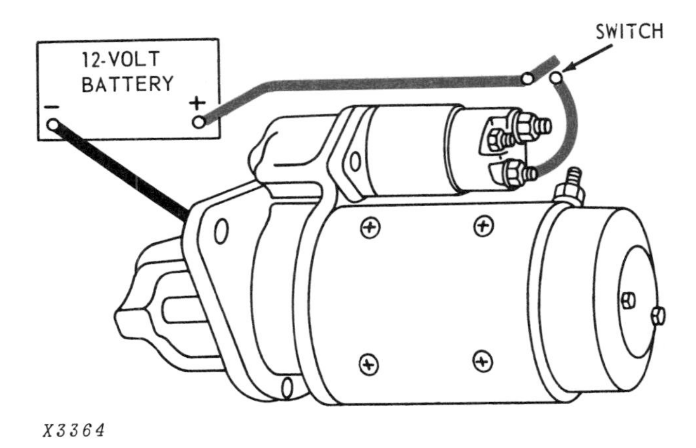

Fig. 70—Solenoid Return Test

EVALUATING ALTERNATE TEST

Fails to Pull In (at 8 Volts)
Defective pull-in winding

Fails to Remain Out
Defective hold-in winding

Fails to Return
Defective pull-in winding

TEST YOURSELF

QUESTIONS

1. (Fill in the blanks.) "In a basic starting circuit, the __________ supplies the energy to the __________ which drives the engine flywheel. The circuit is activated by a __________."

2. What are the four main types of starting motor circuits?

3. True or false? "The coaxial-mounted solenoid switch is on the outside of the motor housing."

4. What is the basic test of the starting motor's internal condition?

5. Replace motor brushes worn to (¾—½—¼—⅓) their original length.

6. When cleaning the commutator, which material should be used? a. Emery cloth. b. No. 00 sandpaper. c. A smooth mill file.

7. (Fill in the blanks.) "To prevent __________, never operate the starting motor for more than __________ at a time without pausing for a few minutes."

(Answers on page 6 at end of text.)

 Litho in U.S.A.

IGNITION CIRCUITS / CHAPTER 8

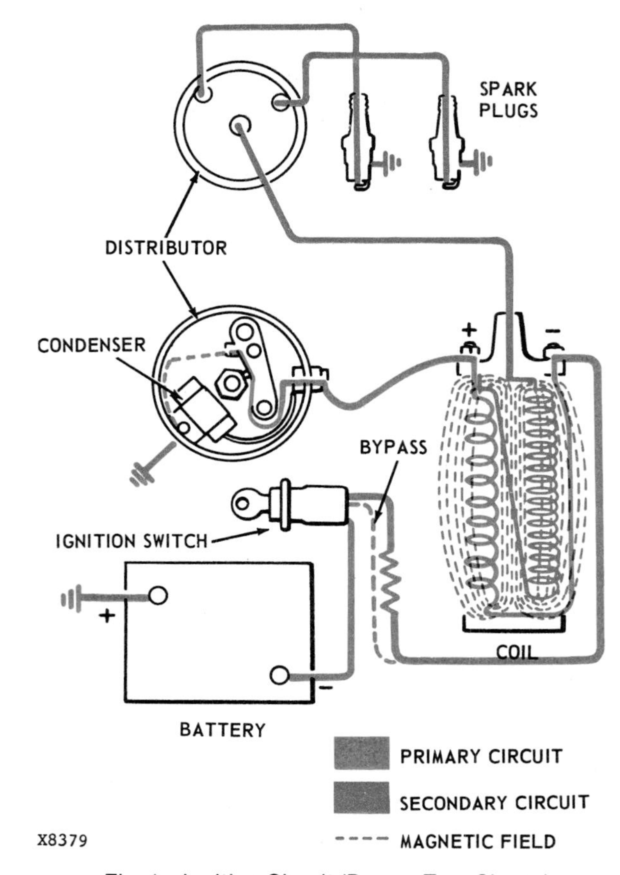

Fig. 1—Ignition Circuit (Bypass-Type Shown)

INTRODUCTION

The ignition circuit creates the spark which ignites fuel and powers the gasoline or LP-Gas engine.

To do this, the ignition circuit must:

1) Step up low- to high-voltage surges

2) Time these surges to the engine

The ignition circuit (Fig. 1) has these parts:

- **Ignition Coil**
- **Condenser**
- **Distributor**
- **Spark Plugs**

The COIL transforms the low voltage from the battery to a high voltage for producing a spark.

The CONDENSER collapses the magnetic field in the coil to produce a high voltage. In doing this it also protects the distributor points against arcing.

The DISTRIBUTOR does three things: 1) Opens and closes the primary circuit, causing the coil to produce high voltage surges, 2) Times these surges to engine rotation, and 3) Directs each high-voltage surge to the proper spark plug.

The SPARK PLUGS ignite the fuel-air mixture within each cylinder of the engine.

The **battery** of the **charging circuit** is the initial power source for the voltage in the ignition circuit, while the **ignition switch** turns on the circuit when it cranks the engine.

HOW THE IGNITION CIRCUIT WORKS

The ignition circuit must take low voltage from the battery and create high voltage to fire the engine. It must do this very accurately and very rapidly—100 or more times per second.

Let's see how the circuit does this complex job.

The ignition circuit has two separate circuits:

- **Primary—low-voltage circuit**
- **Secondary—high-voltage circuit**

The PRIMARY CIRCUIT is the path for low-voltage current from the power source. Fig. 1 shows this circuit in red. It includes these parts:

1) Ignition Switch

2) Coil Primary Winding

3) Distributor Contact Points

4) Condenser

The SECONDARY CIRCUIT is the high-voltage path for current stepped up by the coil. Fig. 1 shows this circuit in blue. It includes these parts:

1) Coil Secondary Winding

2) Distributor Rotor

3) Distributor Cap

4) Spark Plugs

Now let's take these circuits and see how they work.

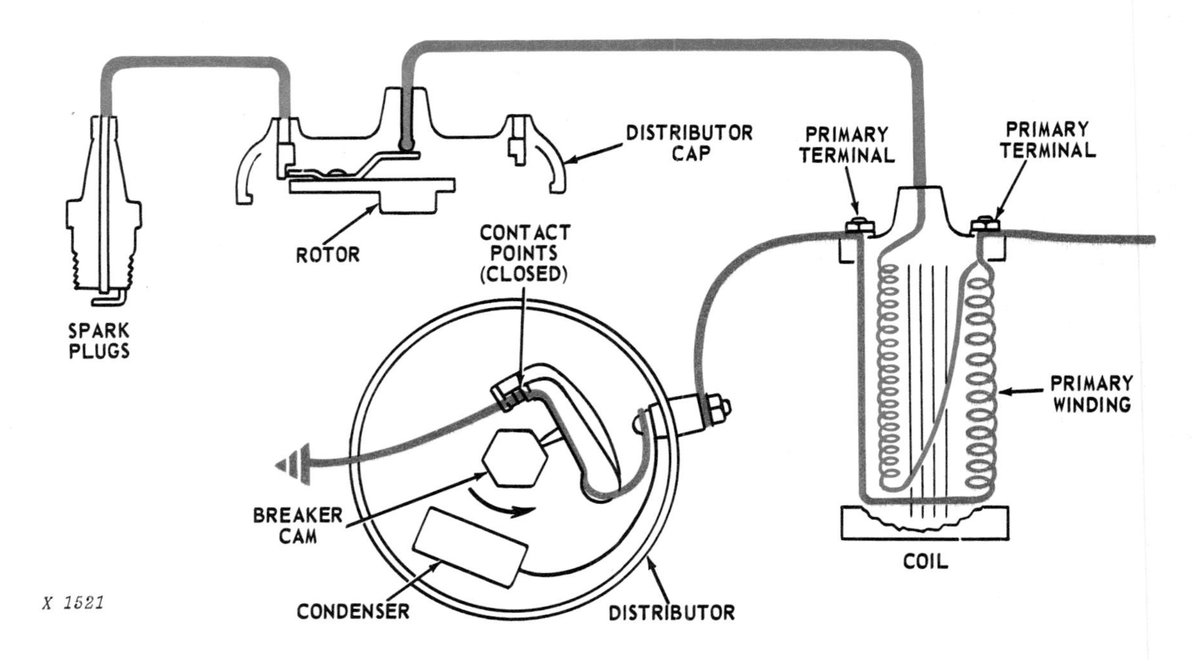

Fig. 2—Operation Before The Distributor Points Open

To simplify, let's divide the operation into two parts—before the distributor points open and after they open.

OPERATION BEFORE THE DISTRIBUTOR POINTS OPEN

Before the engine is started, the distributor points are closed (Fig. 2).

But when the ignition switch is turned on, current flows from the battery into the *primary* windings of the coil as shown in red.

This current creates a magnetic field around the winding.

From the primary winding, the current—at low voltage—simply travels through the closed distributor points and back to ground.

OPERATION AFTER THE DISTRIBUTOR POINTS OPEN

As the engine rotates in starting, it drives the distributor shaft and the breaker cam.

When the breaker cam opens the distributor points, the *second* phase of ignition begins (Fig. 3).

As the points open, the flow of primary current is stopped instantly. Stopping this flow of current allows the magnetic field built up around the windings to collapse instantly. This field collapse creates an induced voltage which causes current to flow in both the primary and secondary windings.

The surge of induced voltage in the primary winding is absorbed by the condenser as explained on page 6—1.

The magnetic field collapsing around the secondary winding induces voltage which causes current to flow. Because the secondary winding is composed of many more turns of much finer wire than the primary winding, a much higher voltage is induced—from 4,000 to 20,000 volts.

This surge of high voltage "pushes" current through the secondary winding and the high-tension terminal into the distributor cap as shown in red in Fig. 3.

The rotor inside the distributor cap turns to a spark plug terminal and directs the voltage "surge" to the correct plug through insulated cables.

At the spark plug, current flows down the center electrode, jumps the gap, and creates the spark.

NOTE: In a "bypass" ignition system (see Fig. 1), there are two primary leads from switch to coil. When the switch is turned to start, full battery voltage flows through the dotted red line, resulting in a hotter spark for first ignition. When the ignition switch is released, primary current flows through the solid line and resistor to the coil. In a 12-volt system, the resistor reduces voltage by

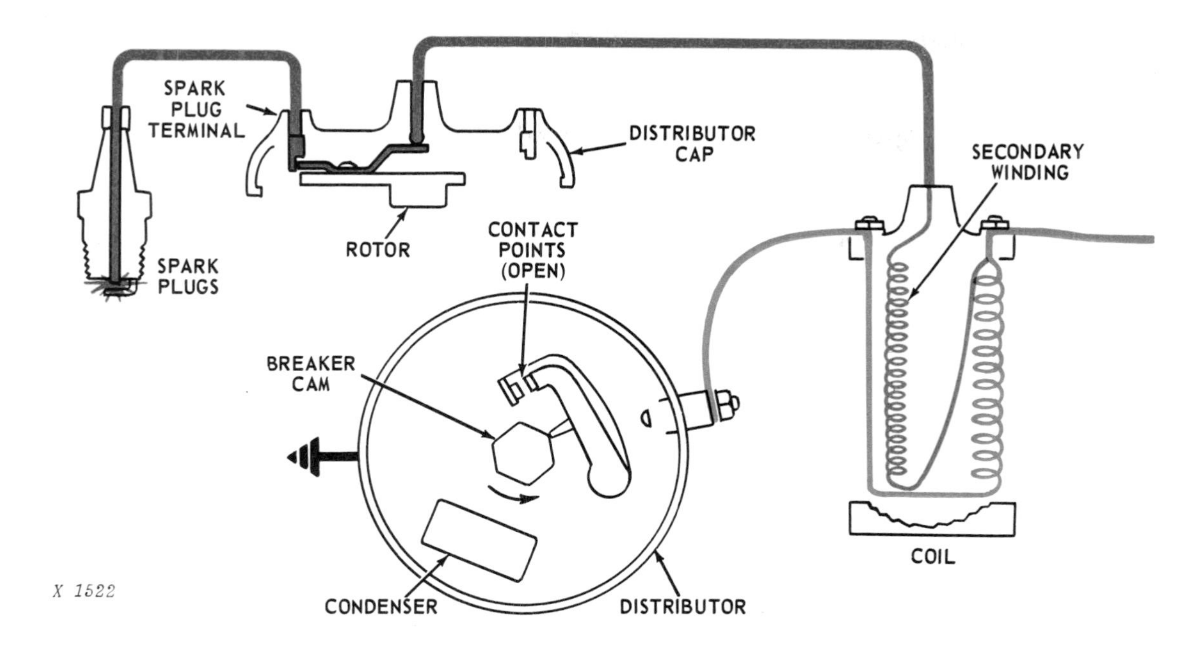

Fig. 3—Operation After The Distributor Points Open

half and allows use of a 6-volt coil. This gives longer life to the distributor points, condenser, and coil because of less heat.

Now that we know the basic operation of the ignition circuit, let's see how each component works.

IGNITION COIL

You have already seen the three main parts of the ignition coil, the **primary** and **secondary windings,** and the **high-tension terminal.** Now let's see how these parts of the coil induce the high voltage surge.

The center of the coil is a soft iron **core** (Fig. 4). The **secondary winding** of fine wire is wrapped around this core. One end of the secondary winding is connected to the **high-tension terminal,** the other end to the primary winding.

The **primary winding** of heavy wire is wrapped around the secondary winding. The two ends of the primary winding are attached to the primary terminals in the coil cap. One of these terminals is connected to the power source; the other is connected to the distributor points.

A shell of laminated material is placed around the windings and core as shown. The core, windings, and shell are then encased in a metal container. The container is filled with either oil or insulating material and hermetically sealed with the coil cap.

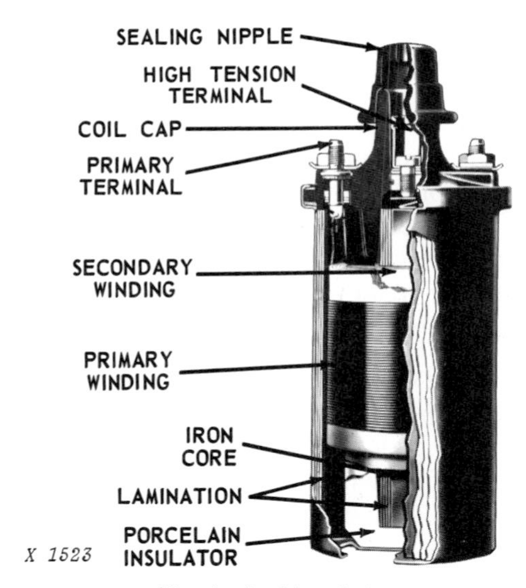

Fig. 4—Ignition Coil

The cap is made of a molded insulating material with the two primary terminals and one high-tension terminal molded into it.

OPERATION OF COIL

The ignition coil is a pulse transformer that steps up the low voltage from the battery as we explained above.

Let's discuss some other factors of coil operation.

The magnetic field around the primary winding does not reach its full potential at once. A moment of time is needed to build up the field strength.

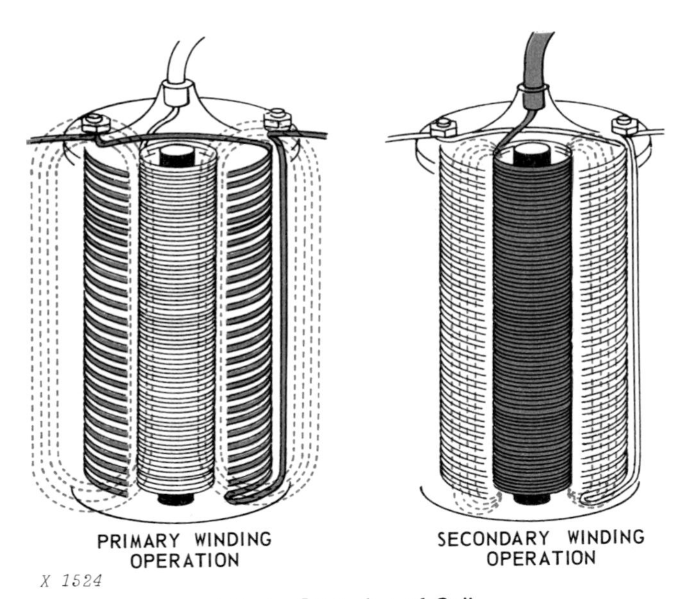

Fig. 5—Operation of Coil

This is because the field build-up induces a momentary counter-voltage in the primary winding. The counter-voltage opposes current flow and must be overcome by circuit voltage so that current flow and field strength can increase.

Only a small fraction of a second is involved, but this build-up time is important: At high speeds, the distributor points are closed for a very short time. If no allowances are made, the current flow and the magnetic field will never reach their full potential. And since the field is not strong enough, it will never induce a high voltage surge when it collapses.

Another factor to consider is the voltages induced in the winding. We have already mentioned that primary winding voltage is much less than secondary winding voltage. However, the primary winding voltage does have some effect. It is absorbed by the condensor, and thus collapses the magnetic field faster.

Voltage in the secondary winding may go as high as 20,000 volts. However, it will only go as high as needed to cause the current to jump the spark plug gap.

Jumping the gap creates another reaction which helps in ignition operation. By jumping the gap, the secondary circuit is completed. Then a magnetic field is created around the secondary winding. This field partly stops the field collapse around the primary winding. The voltage induced by the primary winding field in the secondary winding helps sustain the spark for a short period of time.

TESTING THE COIL

Many electrical system problems can be caused by more than one failing component. In the ignition circuit, it is even harder to find a failure because faulty parts are not the only causes that must be considered.

If the ignition coil is faulty, expect some of the following problems:

1. *The engine will not start.*

2. *The engine is hard to start.*

3. *The engine misfires on a warm humid day.*

4. *The engine suddenly stops.*

Of course, some of these problems can be caused by other failures than in the coil.

Coil Tests

1) Make a *visual* check of the ignition circuit to find broken leads, broken or loose connections, or possible cracks or broken components.

2) Make a complete *electrical* check of the circuit to isolate the faulty component.

Polarity of the Coil

Wrong polarity of the coil is not a serious problem, but can cause damage over a long period of time. A coil that is wrongly connected to the power source and the distributor will require an extra 4000 to 8000 volts to create the spark.

The wrong coil polarity makes the center electrode of the spark plugs have the wrong polarity. This can cause misfiring as the voltage required to jump the spark gap increases.

Fig. 6 illustrates the proper coil connections.

On negative-ground systems, the **negative** primary terminal is connected to the distributor.

On *positive-ground systems,* the **positive** primary terminal is connected to the distributor.

Most coils have the polarity signs imprinted in the cap by each terminal.

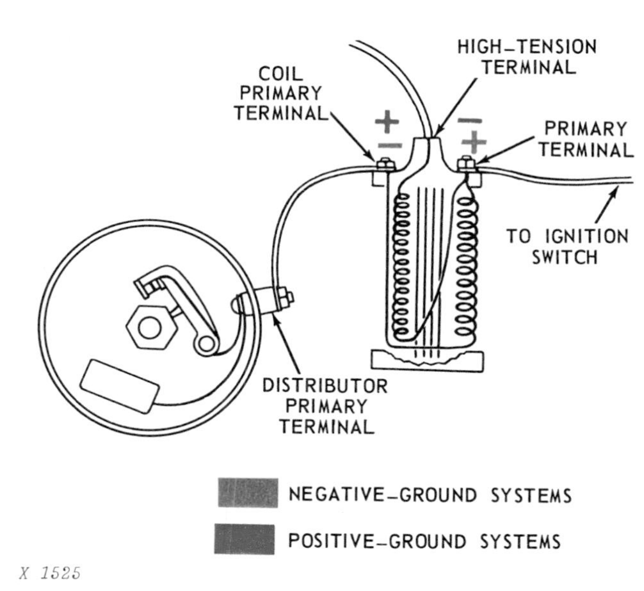

Fig. 6—Correct Polarity of the Coil

Electrical Tests on the Coil

Most well-equipped shops have an electrical service tester that can be used to test many of the ignition circuit components. When testing the coil with one of these testers, follow the recommended test procedures.

Basically, the ignition coil is tested for its voltage strength. Two methods are generally used:

1) The spark gap tester method

2) The meter tester method

TESTING COIL WITH SPARK GAP TESTER

The spark gap tester is connected into the ignition circuit, usually off the high-tension terminal of the coil. When the distributor points open, a spark jumps a gap in the tester. This gap is adjustable. In this way, the distance the spark is able to jump determines the strength of the coil voltage.

When using this tester, you must have a good coil for a comparison. Both the good coil and the coil being tested must be of the same temperature and use identical test leads.

With the spark gap tester, many variables enter into the test results. These include altitude, temperature, and atmospheric conditions.

Because the test procedures differ between the different models of this tester, follow the manufacturer's recommendations.

TESTING COIL WITH METER-TYPE COIL TESTER

This type of tester may be part of an electrical servicer or just a single unit (Fig. 7). It may also be capable of testing the condenser for shorted, open, or grounded coil windings.

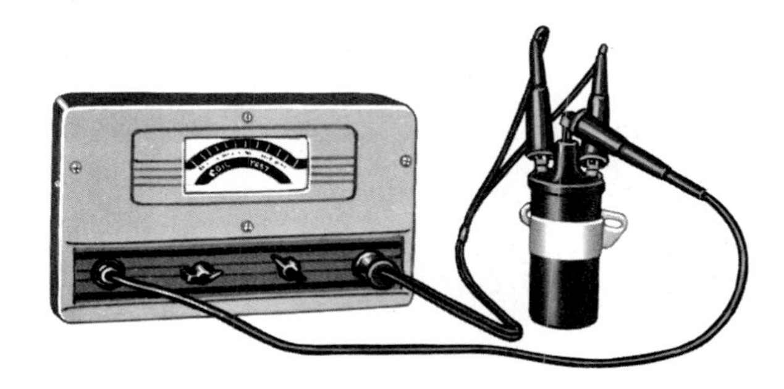

Fig. 7—Meter-Type Coil Tester

The tester uses a graded meter to show the condition of the coil. The meter may have two scales —one scale marked "Bad—Fair—Good"; the other scale marked in numbers from one to ten. The "ten" side of one scale usually corresponds with the "good" range in the other scale.

For each model of meter tester, follow the tester's instructions.

NOTE: Of the two types of testers, the meter-type coil tester is more accurate because it is not affected by other variables.

Testing Coil for Grounded Windings

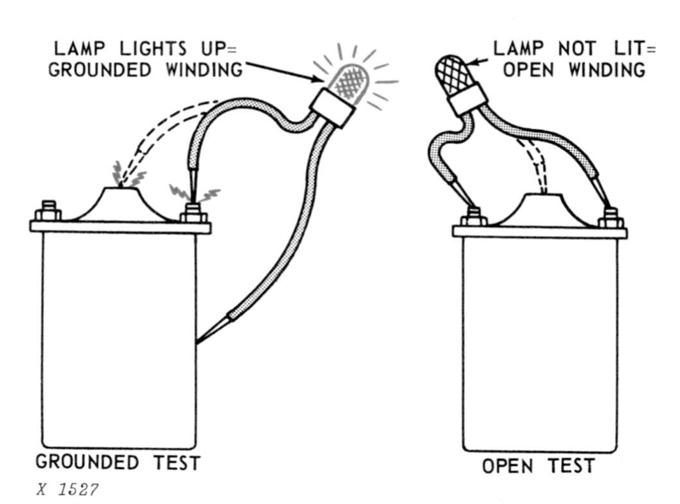

Fig. 8—Testing Coil for Grounded or Open Windings

To test the windings, a test lamp and probes are used (Fig. 8).

To find a grounded primary or secondary winding, place one of the probes on a clean part of the coil case. Place the other probe on the primary terminal or the high-tension terminal as shown in Fig. 8.

If the lamp lights or a spark appears on contact, the windings are grounded.

On some insulated or two-wire systems, this test doesn't apply to the secondary winding. However, the test can always be used on the primary windings.

Testing Coil for Open Windings

To test for an open *primary* winding, place the test lamp probes on the two primary terminals (Fig. 8).

If the lamp does not light, the winding is open.

To test for an open *secondary* winding, place one probe on the high-tension terminal and the other probe on the primary terminal. Rub the probes over the terminals.

No sparking indicates an open winding. (If the windings are okay, a light sparking should occur but the lamp will not light).

SERVICING THE COIL

The only service required on the coil is to keep the terminals and connections clean and tight. The coil itself should also be kept reasonably clean.

Rubber nipples on the high-voltage terminal help in preventing "tracing" or leakage of current across the exposed surfaces.

There is no repair on the coil. If it is cracked or has bad wiring, it must be replaced. A crack in the coil cap breaks the hermetic seal and allows moisture to enter the coil.

IGNITION CONDENSER

The *condenser* prevents arcing at the distributor points when they begin to open.

Excess current flows into the condenser as the points separate.

The coil is controlled by the opening and closing of the distributor points. However, the points cannot do this job alone because:

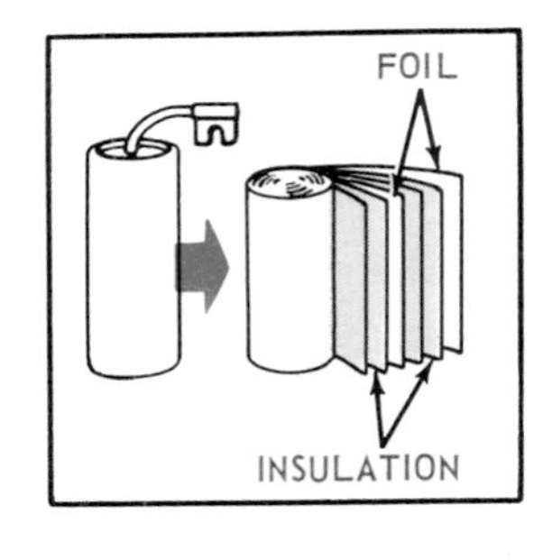

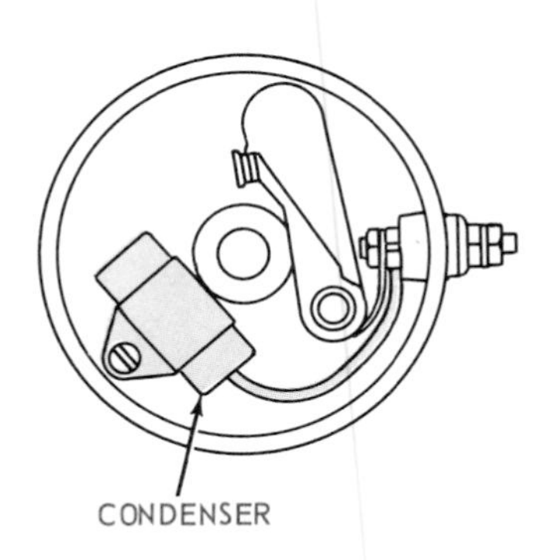

X 1528

Fig. 9—Condenser Installed On Distributor

1. The points open and close mechanically—a fairly slow action.

2. The points open only a short distance.

3. Voltage within the coil can become very high.

Without a condenser, what happens is this:

1. Induced voltage in the coil primary gets too high and pushes current across the gap—burning the points.

2. Current flow does not stop quick enough, and the field collapses too slowly. So the secondary voltage is too low and no high-voltage surge is produced to fire the spark plug.

The condenser helps eliminate these problems by both absorbing the "arc" current, thus helping to induce a higher voltage for the spark plugs.

CONSTRUCTION OF THE CONDENSER

The condenser (Fig. 9) is mounted in the distributor and is connected across the points.

The condenser is made up of two foil plates which are insulated from each other with a special paper. The plates and the paper are wrapped around an arbor to form a winding. Each plate has a wiring lead—one connected to the condenser case and other connected to the main wiring lead. The whole assembly may be encased in metal or epoxy.

OPERATION OF THE CONDENSER

The condenser provides a place for current to flow as the distributor points separate. This current flowing into the condenser prevents an arc between the points.

(1) Steady, straight current flow through coil primary, points closed.

(2) As points open, induced voltage causes current in the primary to flow into the condenser, creating a voltage difference between the insulated foil sheets.

(3) High charge on condenser foil sheet forces current back through coil primary, collapsing the magnetic field faster and creating a hotter spark.

(4) Drained condenser foil sheets now have lower voltage charge than adjacent grounded sheets, so current flow again reverses as shown until all coil energy is used up.

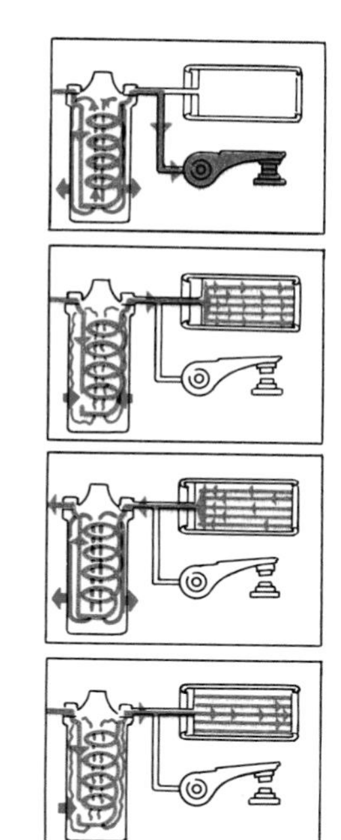

X 1529

Fig. 10—Operation of the Condenser

Fig. 10 explains the operation of the condenser.

The condenser holds only a limited amount of current. It fills up or is "charged" very quickly. This stops the current flow in the primary winding. Then the field around the winding collapses, inducing voltage in both the primary and secondary windings. The greatest voltage, of course, is induced in the secondary winding.

The voltage induced in the primary winding causes a current to flow in the primary winding. This current also helps to charge the condenser, which creates still more opposition to current flow, producing a further collapse of the field and high voltage.

The voltage keeps on rising to try to force the current to flow. But by this time the points are far enough apart so that the voltage never "pushes" current across the gap.

All arcing is not eliminated, however. In spite of the condenser, arcing does occur during the first millionth of a second that the points separate. But it is a small spark much like the one produced when a wiring connector is pulled apart.

TROUBLESHOOTING THE CONDENSER

Listed below are the problems that can be caused by a faulty condenser.

Problem	Cause
Engine cranks, but won't start	*Faulty condenser*
Engine runs, but misses on all cylinders	*Faulty condenser*
Pitted distributor points	*Wrong-capacity condenser*
Burned distributor points	*High resistance in the condenser*

Remember that some of these problems may be caused by other faulty components. Only a complete circuit check will finally pinpoint the cause of the problem.

 Litho in U.S.A.

REPAIRING THE CONDENSER

The condenser is not repairable. If it fails, replace it.

IGNITION DISTRIBUTOR

Fig. 11—Ignition Distributor

The distributor does three jobs:

- **Opens and closes the primary circuit**
- **Times the high-voltage surges**
- **Delivers current to the spark plugs**

Fig. 12 shows the parts of the distributor. We can put all the distributor parts into three groups:

Primary Circuit Operation	Timing	Delivery
• *drive shaft*	• *drive shaft*	• *drive shaft*
• *breaker cam*	• *breaker cam*	• *rotor*
• *breaker plate*	• *centrifugal advance*	• *cap*
• *contact points*	• *contact points*	

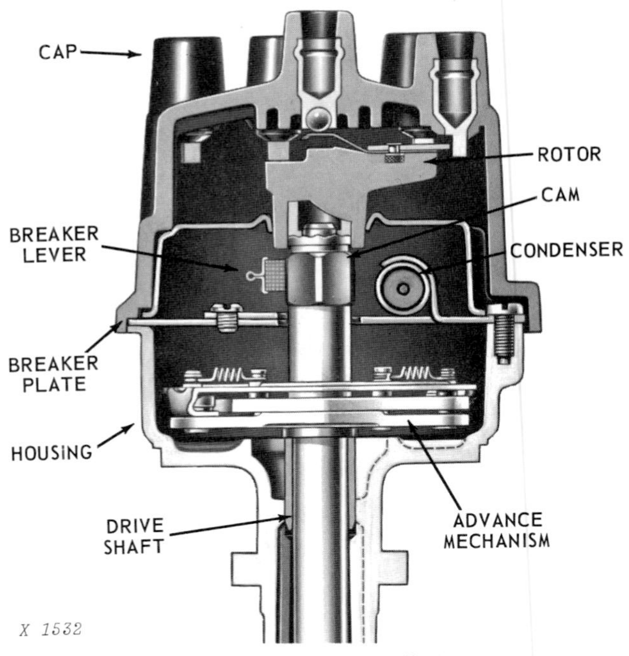

Fig. 12—Cutaway of Distributor

Let's see what each of these distributor parts do and then we'll see how they operate.

Drive Shaft—is driven at one-half engine speed by the engine camshaft. It drives the centrifugal advance mechanism, the breaker cam, and the rotor.

Breaker Cam—is slip-mounted on the drive shaft and pinned to the centrifugal advance. As you can see in Fig. 12, the cam has lobes or corners—one for each engine cylinder. As the cam rotates, each lobe pushes against the contact point breaker lever, opening the contact points.

 Litho in U.S.A.

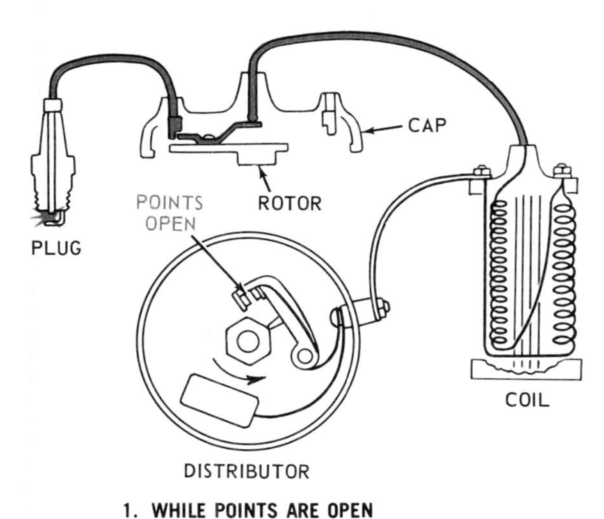

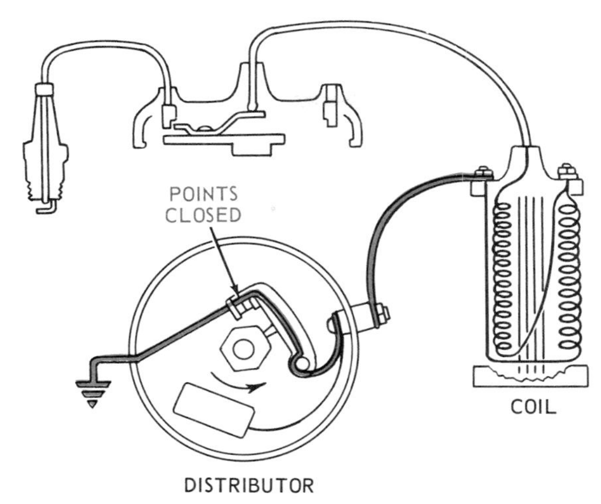

X 1533

Fig. 13—Basic Operation of the Distributor

Breaker Plate—is a mounting for the contact points and condenser. It also has a terminal which connects the points and condenser into the primary circuit.

Contact Points Assembly—two contact points, a breaker lever and a breaker lever spring. All three are mounted on a base which is attached to the breaker plate. The two points are usually made of tungsten. One is fixed to the base. The other is attached to the breaker lever and aligned with the first. The breaker lever is mounted on a pivot pin to the assembly base. The lever is made of metal with a nylon or bakelite rubbing block which contacts the breaker cam lobes. The breaker lever spring is attached to the breaker lever. The spring holds the lever to the cam after each cam lobe passes the rubbing block.

Rotor—is mounted on the upper part of the breaker cam. A flat side of the rotor hub fits on a flat side of the cam. In this way the rotor will fit in only one position. On the top of the rotor, a spring metal piece is in contact with the center terminal of the distributor cap. A rigid piece completes the circuit to each spark plug terminal in the cap as the rotor turns. The rotor itself is molded of a plastic material which makes it a good insulator.

Distributor Cap—is also molded of a plastic material. Brass or copper contact inserts are embedded in the cap. These contacts are equally spaced around the cap and lead to the spark plug terminals in the top of the cap. A carbon button in the center of the cap contacts the rotor and leads to the center high-voltage terminal in the top of the cap. The cap is notched into the housing to prevent a wrong installation.

Later we will cover another feature of the distributor—the advance mechanism.

OPERATION OF THE DISTRIBUTOR

As the drive shaft turns, the breaker cam lobe pushes the breaker lever rubbing block. This action opens the contact points, stopping the current flow in the coil primary circuit. See at left in Fig. 13.

The collapsing field and the resulting high-voltage surge in the coil's secondary winding forces current into the center terminal of the distributor cap as shown in red. The distributor rotor picks up this current and delivers it to the proper spark plug to fire the engine.

Meanwhile, the distributor cam lobe has moved away from the rubbing block and spring tension brings the points back into contact (see at right in Fig. 13). The primary circuit is again complete and current flows until the next lobe opens the points. The cycle then repeats itself.

In this way a spark is created as each lobe of the cam opens the contact points. The entire cycle for each spark takes place at a very high speed.

 Litho in U.S.A.

CENTRIFUGAL ADVANCE MECHANISM

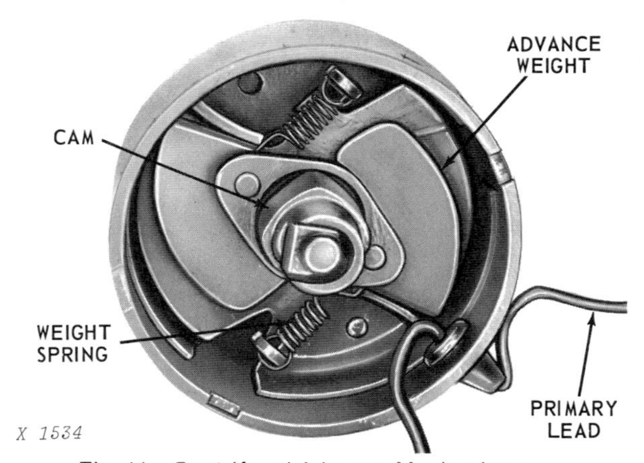

Fig. 14—Centrifugal Advance Mechanisms

Basically, an advance mechanism is a device which times the spark to occur at a certain time as determined by engine speed. Why is a spark advance necessary?

The distributor must deliver the spark to the engine when it is most effective. This is determined by the position of the piston and the time required to ignite the fuel-air mixture in the cylinder.

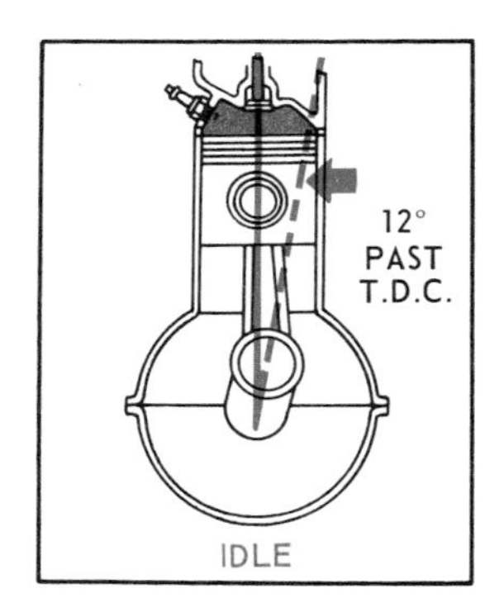

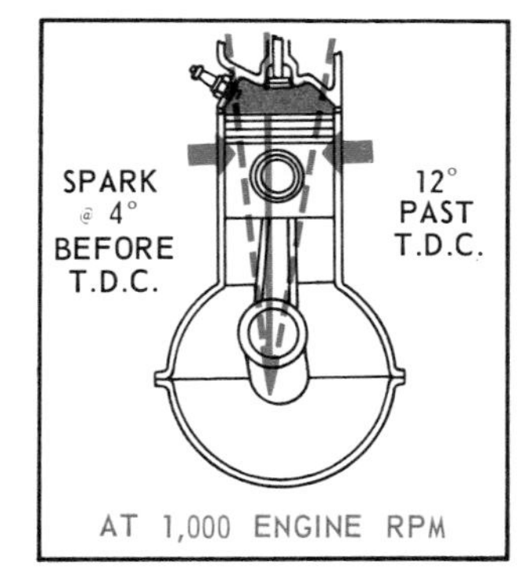

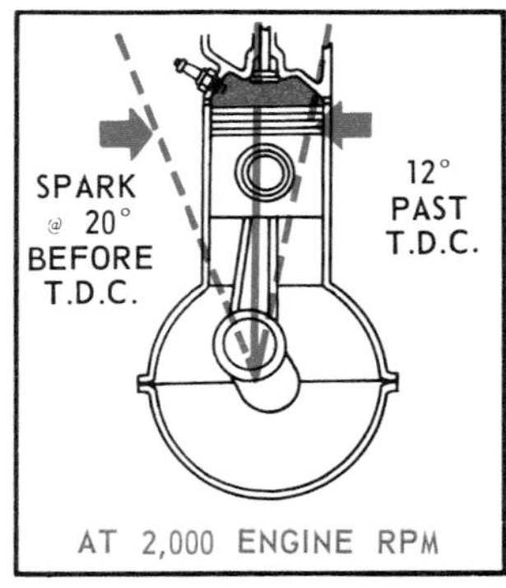

Fig. 15—Spark Timing of the Engine Combustion

Let's assume that the engine piston must be at 12 degrees past top dead center to get the full force from combustion. See Fig. 15. The 12 degrees past top dead center stays the same regardless of the engine speed.

Let's also assume that it takes 0.002 second to reach the full force of combustion, and that at 1000 revolutions-per-minute (rpm) the piston would travel 16 degrees during this time. Therefore, the spark must come at 4 degrees *before* top dead center.

Now if we double the engine speed to 2000 rpm, the distance the piston would travel is also doubled to 32 degrees. The spark must then occur at *20* degrees *before* top dead center to get maximum power.

What the **advance mechanism** does is to adjust the distributor timing to allow for speed changes so that the spark will occur at the right time.

The most popular advance mechanism is the **centrifugal advance.**This device has two weights, a weight base, and two springs.

The weight base is part of the distributor drive shaft. The springs are connected to the base, while the weights are placed on the base. The distributor breaker cam has two pins which connect it to the springs and weights. The pins also set in slots in the base.

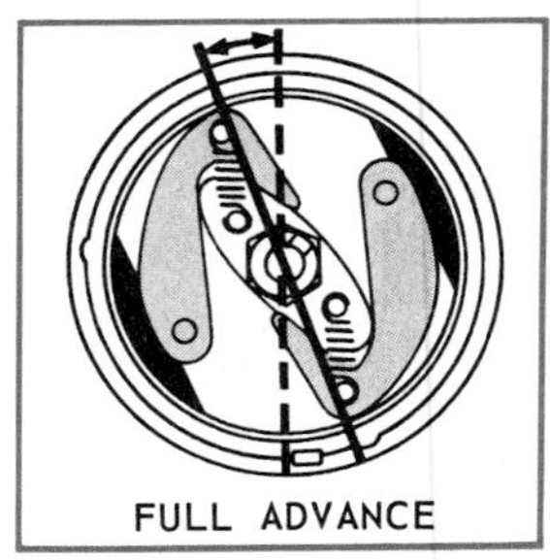

Fig. 16—Operation of Centrifugal Advance

At idle speeds the breaker cam is "pinned" to the base and rotates with the drive shaft. The cam lobes then open the points at a preset time—such as 4 degrees before top dead center.

As the engine speeds up, centrifugal force throws the weights out against spring tension (Fig. 16). This turns the breaker cam so the cam lobes are now striking the breaker lever earlier. Therefore, the contact points open ahead of time.

The higher the speed, the further the weights are thrown out. And the further the weights are thrown, the more they turn the breaker cam and the more the spark is advanced.

When the engine slows down, the springs return the breaker cam and weights to their original position.

VACUUM ADVANCE MECHANISM

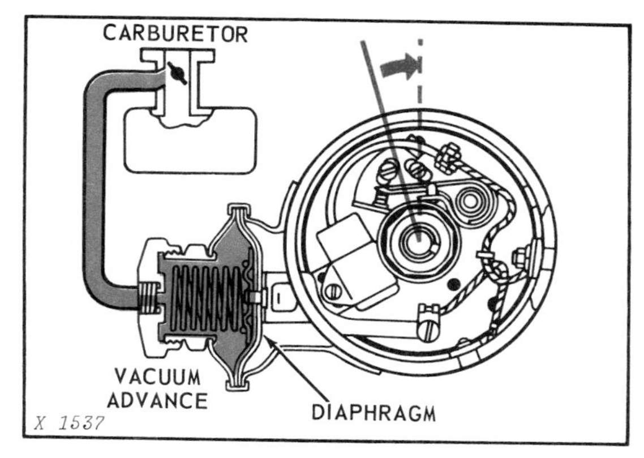

Fig. 17—Vacuum Advance Mechanism

For greater fuel economy, an extra advance mechanism is used on some distributors. This is the **vacuum advance** (Fig. 17).

A vacuum can develop in the engine intake manifold, allowing less fuel and air into the cylinder. Since a lean fuel-air mixture will burn slower, ignition must take place sooner in the cycle than even the centrifugal advance can provide. So a vacuum advance is used to advance the spark still further.

The vacuum advance uses an air-tight diaphragm connected to an opening in the carburetor by a vacuum passage (Fig. 17). The diaphragm is connected by linkage to the distributor housing or the breaker plate.

When a vacuum at the intake manifold draws air from the diaphragm chamber, it causes the diaphragm to rotate the distributor breaker plate in the opposite direction of drive shaft rotation. This moves the breaker lever to contact the breaker cam lobes sooner and thus advances the spark.

EXTRA-DUTY FEATURES OF DISTRIBUTORS

There are extra design features that can be added to a distributor. They may provide a needed service or can increase the reliability. The following is a list of these features:

- **Screw-type distributor cap terminals**
- **Distributor caps and rotors made of mica**
- **Built-in distributor shaft lubrication**
- **Flexible shock-absorbing drive gears**
- **Breaker cam lubricator**
- **Tungsten-tipped rotor**
- **Heavy-duty contact points**
- **Dust sealing cover on breaker compartment**
- **Elastic terminal nipples**

TESTING AND ADJUSTING THE DISTRIBUTOR

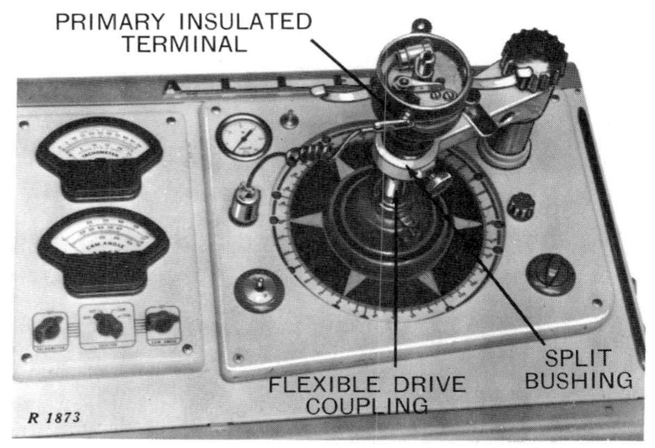

Fig. 18—Distributor Tester

Electrical tests on the distributor are usually made with a tester such as the one shown in Fig. 18. This unit is called a syncrograph and can test most of the distributor operations.

Mechanical tests such as a spring tension and point opening distance are also performed on the distributor. These tests usually call for the use of mechanical test instruments such as a spring gauge, feeler gauge, or micrometer.

The most important test you can make is a good *visual* check. Often many physical failures such as cracks, loose terminal connections, burned points, etc. will be discovered during this check.

Failures of Distributors

Because the distributor plays an important part in the operation of both the primary and secondary circuits, any failure of the distributor will directly affect the ignition circuit and engine operation.

 Litho in U.S.A.

Checking Distributor Caps

A visual check is usually the only "test" performed on the cap.

First, wipe the distributor cap clean. Never clean the cap with a degreasing solution.

Check the cap for the following:

- *Carbon paths*
- *Cracks and chips*
- *Eroded spark plug contacts*
- *Worn center terminal carbon button*

Let's discuss these failures of the cap.

CARBON PATHS

Carbon paths are burnt areas on the distributor cap. Since carbon is a conductor, these burnt areas provide an easy path for current leakage between terminals or between the terminals and ground.

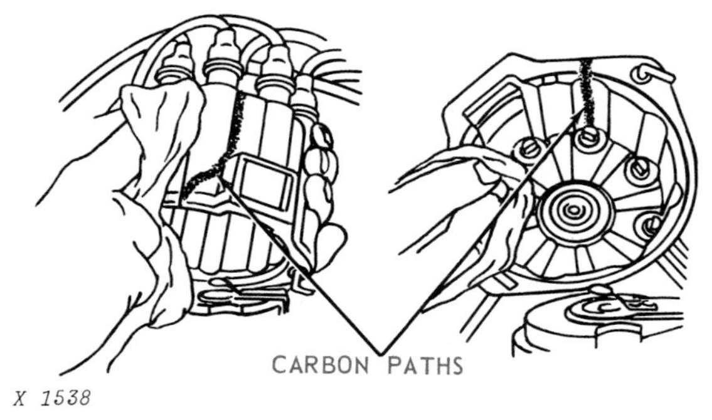

Fig. 19—Carbon Paths on Distributor Caps

Carbon paths can be formed on the outside of the cap by an accumulation of dirt (Fig. 19). Dirt will retain moisture and thus conduct a current. Current leakage, usually from the center terminal to a retaining clip, will create a carbon path. Once the path is formed, it will continue to conduct current.

Carbon paths inside the cap are usually found between the spark plug terminals or between the terminals and a grounded surface. They are caused by moisture and arcing in the cap.

A distributor cap with a carbon path must be replaced. This is because the path will continue to conduct current, especially in damp weather. Eventually, it will burn through the cap.

CRACKS AND CHIPS

Cracks and chips allow moisture and dirt into the cap and breaker lever area. Carbon paths as well as burnt points, breaker cam lobe wear, and arcing can be caused by this. Caps with cracks or chips must also be discarded.

TERMINAL EROSION

Spark plug insert terminals that are eroded will widen the gap between the rotor and the inserts. As the gap increases, secondary voltage requirements also increase, resulting in poor engine performance, overloading, and ultimate failure of the coil. These gaps also promote arcing and resulting carbon path. A cap with eroded inserts must also be replaced.

Servicing The Distributor Cap

The distributor cap should be kept clean. The wiring leads should fit tight in the terminals. The use of elastic nipples over the center terminal and spark plug terminals can reduce some of the problems of dirt and moisture.

Distributor Rotor Checks

The rotor should be checked for cracks or chips and for erosion on the tip of the rigid contact.

Check the rotor spring contact for proper tension against the distributor cap carbon button. Too little tension can create a gap between the contact and button. Too much tension can cause excessive wear. Be careful when bending the spring to adjust the pressure.

Always replace a chipped or cracked rotor.

Distributor Breaker Cam Check

Check the cam lobes periodically for excessive wear and lack of lubrication. Worn lobes will not open the points enough to create a strong spark.

The lobes can be checked visually. However, the syncrograph is the best way to check for cam wear.

The point openings as caused by the breaker cam should be evenly spaced within plus or minus one degree. For example, a four-lobe cam should open the points every 90 degrees, a six-lobe cam every 60 degrees, etc.

Wear on the cam is usually caused by abrasive material such as dirt in the breaker plate area. Replace an excessively worn cam.

Centrifugal Advance Checks

The centrifugal advance should turn freely, check this by turning the breaker cam in the direction of rotation and then releasing it. The advance springs should return the cam to its original position without sticking.

Use a distributor tester to check the centrifugal advance operation.

1) Set the distributor into the tester and drive it at various specified speeds.

2) At each test speed, note the amount of advance. It should be within plus or minus one degree of specifications.

3) When the highest specified speed is reached, speed up for a moment, then return to zero.

4) While returning the speed to zero, check each test speed again. The centrifugal advance should be the same.

5) If the advance is excessive during both acceleration and deceleration, the advance springs are too weak or the wrong springs are installed.

6) If the advance is too little during both acceleration and deceleration, the springs are too tight.

7) If the advance is slow during acceleration but fast during deceleration, the advance weights are sticking.

8) In any case, new springs may be needed or the weights may need cleaning and a light oiling to free them. On some distributors, spring tension can be adjusted by bending the outer spring support.

Distributor Drive Shaft Checks

If you suspect the drive shaft is worn or out-of-line, check for:

- *A worn carbon button in the distributor cap*
- *One side of the breaker cam excessively worn*
- *Engine misfiring*

Check the shaft alignment on a syncrograph tester (Fig. 20). While the distributor is driven by the tester at a specified speed, the distributor case is physically shaken.

If the test shows variations in the firing positions, the shaft or bushings are worn.

Fig. 20—Checking The Distributor Drive Shaft

Testing and Adjustment of Distributor Contact Points

The contact points are the key to good timing.

Check the contact points assembly for these things:

- *Pitted contact points*
- *Burned points*
- *Worn rubbing block*
- *Alignment of points*
- *Worn breaker lever pivot post*
- *Tension of breaker lever spring*
- *Gap of contact points*
- *Angle of cam (dwell)*

PITTED CONTACT POINTS

A visual inspection will usually reveal pitted or rough contact points. However, points in this condition are not necessarily worn.

During normal operation, material may transfer from one point to another. The transferred material will appear rough and gray in color. However, the points will have a *greater* contact area than when they were new. For this reason, don't replace the points until the build-up is over 0.020 inch (0.5 mm).

Litho in U.S.A.

Sometimes pitted points can be reconditioned by removing them and honing each surface smooth and flat. You need not remove all the pits. To clean the points, use a few drops of lighter fluid on a strip of lint-free cloth. Then pull a dry strip through the points to remove the residue. BE CAREFUL when using lighter fluid. **Do not use emery cloth or sandpaper to clean the points.**

During a tune-up, however, replace pitted points to assure good operation.

BURNED CONTACT POINTS

Burned points are usually caused by either high voltage, oil deposits, foreign material, a defective condenser or improper point adjustment.

High voltage cause a high current flow through the points. This heats up the points and rapidly burns them.

Oil or crankcase vapors can seep up through the distributor and deposit on the points, causing them to burn. Look for a smudgy oil line beneath the points.

If the point opening gap is too small, the points will be closed for too long. Then even an average current flow through the points will be too high, burning the points.

WORN RUBBING BLOCK

Rubbing block wear is not a usual cause of trouble. Normally the contact points will wear out before the rubbing block wears enough to affect operation.

In some extreme cases, abrasive foreign material has caused rubbing block wear.

ALIGNMENT OF POINTS

To see if the contact points are aligned, use a distributor tester. With the distributor operating at 1000 rpm, you should be able to see a slight arc between the points.

If the points are aligned, the arc will appear in the *center* of the points when seen from above and from the side.

WORN BREAKER PLATE PIVOT POST

If this pivot post is loose or worn, replace the breaker plate. A worn or loose post will cause the rubbing block and breaker lever to ride erratically on the breaker cam. (On some distributors, the pivot post is part of the contact point assembly. Replace this assembly if the post is loose or worn.)

CHECKING TENSION OF BREAKER LEVER SPRING

Spring tension can be tested with the distributor on or off the distributor tester.

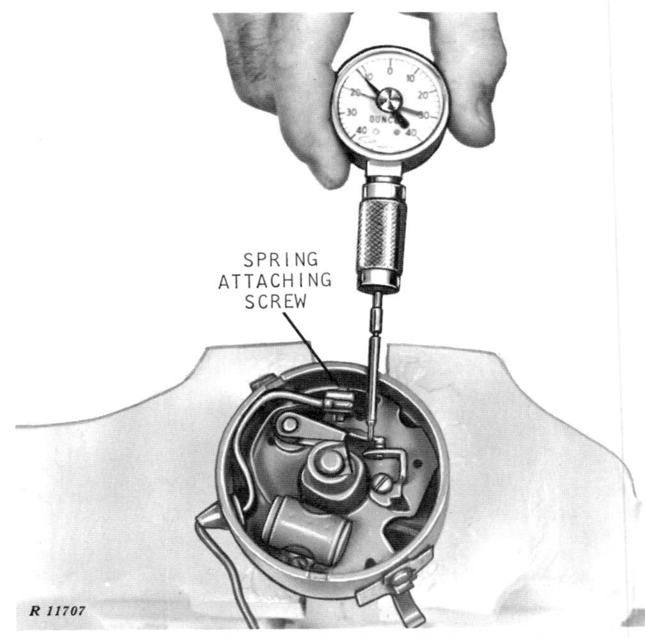

Fig. 21—Checking Breaker Lever Spring Tension

Use a tension tester placed on the movable point and pull it at right angles to the breaker lever (Fig. 21). The instant the points separate, note the reading on the tester.

Since the distributor tester can "show" the instant the points separate, using it will aid in this test.

Proper breaker spring tension is important. Excessive tension can cause the rubbing block, breaker cam and contact point to wear. Too little tension can allow the points to bounce at high speeds, causing arcing, burned points, and engine misfire.

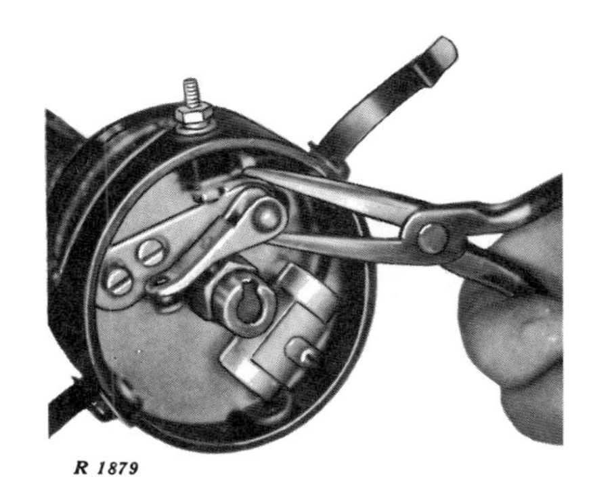

Fig. 22—Adjusting Tension of Breaker Lever Spring

Adjust the breaker spring by bending it toward the breaker lever (to decrease tension) or away from the lever (to increase tension). See Fig. 22. Some springs can also be adjusted by sliding the spring in or out under the attaching screw.

CHECKING CONTACT POINT GAP

The distance the points open and the length of time that they stay open are important in distributor operation.

Too wide a point gap will cause the points to open sooner and stay open longer. This will limit build-up time in the coil and the coil will not produce enough voltage for a spark at high speed.

Too narrow a point gap will allow the points to be closed longer, causing the rubbing block to wear and the points to burn.

X 1539 Fig. 23—Checking Contact Points Gap

A feeler gauge can be used to set the gap on new points. However, old points that are serviceable but are also pitted should not be set with a feeler gauge. The gauge will only measure the *high spots* on the points and so will not give a true picture of the gap.

Therefore, to check old contact points, use a dial indicator (Fig. 23).

CHECKING CAM ANGLE (DWELL)

The *cam angle* is the number of degrees that the breaker cam rotates from the time the points close until they open again (Fig. 24). As the cam angle increases, the point gap decreases, and vice versa.

Too little cam angle can cause engine to misfire at high speed.

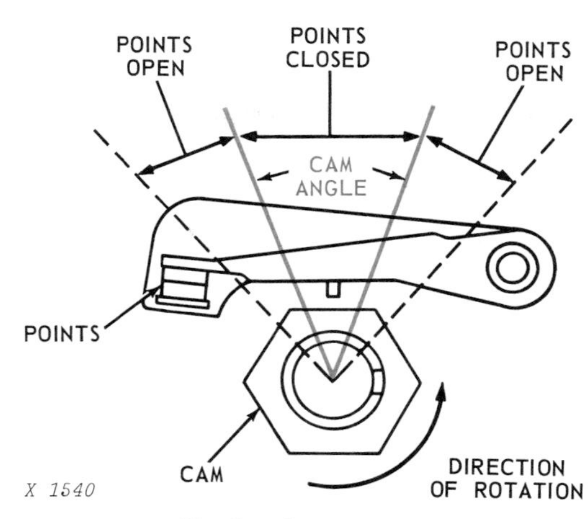

Fig. 24—Cam Angle (Dwell)

Too much cam angle allows the points to close for too long, causing burned points.

Use the dwell meter on the distributor tester to check the cam angle. Then adjust the points to the specified dwell angle.

If the cam angle reading on the meter varies more than two degrees, look for a worn drive shaft or bushings.

If the cam angle and point gap cannot both be set to specifications at the same time, check for these problems:

- *Improper spring tension*
- *Wrong contact point assembly*
- *Worn breaker cam*
- *Points not following cam at high speeds*
- *Bent drive shaft*

SERVICING THE DISTRIBUTOR

This section will give you a general idea of the services which the distributor may require.

In all disassembly, repair, testing, and servicing, follow the machine Technical Manual or the manufacturer's recommendations.

1. Disassemble the distributor only as far as necessary to make repairs or tests.

2. When removing the distributor from the engine, set the No. 1 cylinder at top dead center at the end of the compression stroke. Mark the rotor position on the distributor housing and drive gear. This will allow you to turn the crankshaft while the distributor is off the engine.

3. If necessary, clean the contact points by honing or cleaning them. (See earlier under "Pitted Contact Points".) **Do not use emery cloth or sandpaper to clean the points.** Abrasive material will embed in the points, causing them to arc and burn.

4. Some breaker cams are self-lubricated. Others require a light coating of cam lubricant on the hubs.

5. Some distributors require a few drops of light engine oil on the pivot post at certain intervals.

6. **Avoid too much lubrication of the distributor.** Excess oil may get on the contact points, causing them to burn.

TIMING THE DISTRIBUTOR

After service, the distributor must be timed or "geared" to the engine.

Good timing assures that a spark will occur when it will do the most good during the compression stroke of the piston.

The centrifugal advance has already been set to insure a spark at various speeds. Now we must time the basic spark without the advance mechanism.

There is no hard and fast rule on when this spark should occur. It depends upon the engine design and specifications. It might have to occur before top dead center, at top dead center, or even after top dead center.

The best method of timing the distributor is with a timing light (Fig. 25).

1. Connect the timing light to the No. 1 spark plug wire.

2. Locate the engine timing marks, usually on the flywheel or crankshaft pulley. Normally this mark turns beside a stationary mark on the engine block as shown.

3. Run the engine at its rated speed.

4. Hold the timing light over the timing marks. The light should flash at the instant the two timing marks are aligned. (The flash tells when the No. 1 cylinder fires.)

5. Adjust the distributor timing by loosening the distributor mount and turning the distributor slightly until the spark or flash occurs at the right moment.

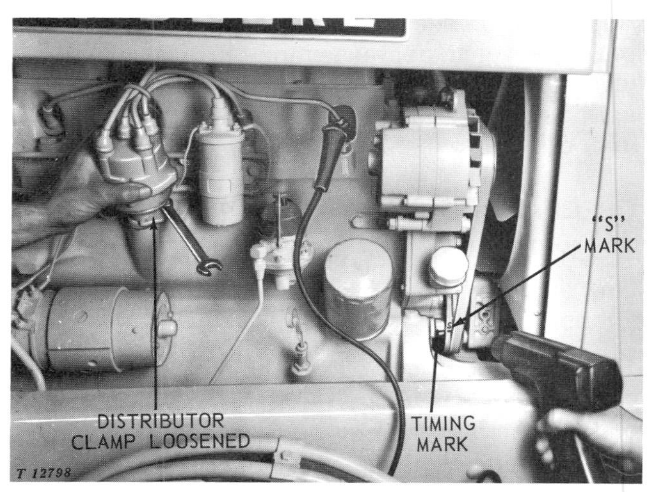

Fig. 25—Timing the Distributor with a Timing Light

6. Tighten the distributor mounting and recheck the timing.

Importance of Timing

Improper timing can affect engine performance in several ways.

For example, if the spark occurs a little late, a distinct loss of power will result. In this case, the piston has traveled part way on its downward power stroke without the aid of the combustion force. Part of the combustion power is then lost and the engine may also overheat.

However, a late spark is not as bad as an advance spark. When the spark occurs too soon, it ignites the fuel-air mixture, resulting in a combustion force. But the combustion power then *opposes* the *upward* travel of the piston, causing a knocking or "pinging" sound. This sound means that great force is being placed on the piston, pins, rings, and connecting rods as it tries to overcome the power of combustions.

If you have an advance spark while using a low-octane fuel, another problem occurs. Normally, the spark ignites the fuel around the spark plug and the resulting flame travels across the combustion chamber. With low-octane fuel, part of the fuel is ignited by the advance spark and part of it is ignited by compression heat on the *other side* of the chamber. The two resulting combustion forces meet head-on and a sharp "ping" is heard. This force can be strong enough to punch a hole through the piston head. With low-octane fuels, the timing should be retarded to allow for the slower burning of the fuel.

IMPORTANT: Improper timing can cause loss of power and damage to engine parts.

SPARK PLUGS

The spark plug ignites the fuel-air mixture in the engine cylinder.

We have learned that there is no current flow in an open circuit. In most cases this is true. However, if the opening in the circuit is small and a strong voltage is present, the circuit can still be completed. In this case, the strong voltage is able to force the current to jump the opening or gap, thus completing the circuit. This is the principle of the spark plug.

OPERATION OF SPARK PLUGS

The plug has two conductors or **electrodes.** One is connected by wire to the distributor cap and the other is connected to ground. The other end of each electrode is separated by a small opening or **gap** from the other.

The high-voltage surge induced in the coil causes a current to flow from the coil to the distributor cap and through a cable to one of the spark plug electrodes. This current then jumps the gap to the other electrode and on to ground. By jumping the gap, the current has completed the circuit and continues to flow.

Completing the circuit is of secondary importance, however. The important fact is that when the current jumps the gap, a *spark* is created. This is the final goal of the ignition circuit.

After all the ignition action, the current jumping the gap to create the spark seems rather simple. However, creating a controlled spark is a different story and this story is told in the construction of the spark plug.

CONSTRUCTION OF SPARK PLUGS

Although the spark plug has no moving parts, each of its parts is designed for a specific purpose. For this reason, many types of spark plugs are available.

Fig. 26 shows the parts of a typical spark plug. Basically, all spark plugs have the same parts; only the design is different. Let's have a look at the parts of the spark plug.

Outer Shell

Each spark plug has a steel outer shell. The top of the shell is hex-shaped for tightening the plug when installing it. The lower part of the shell is threaded and screws into the cylinder head. The grounded electrode extends out from the lower part of the shell.

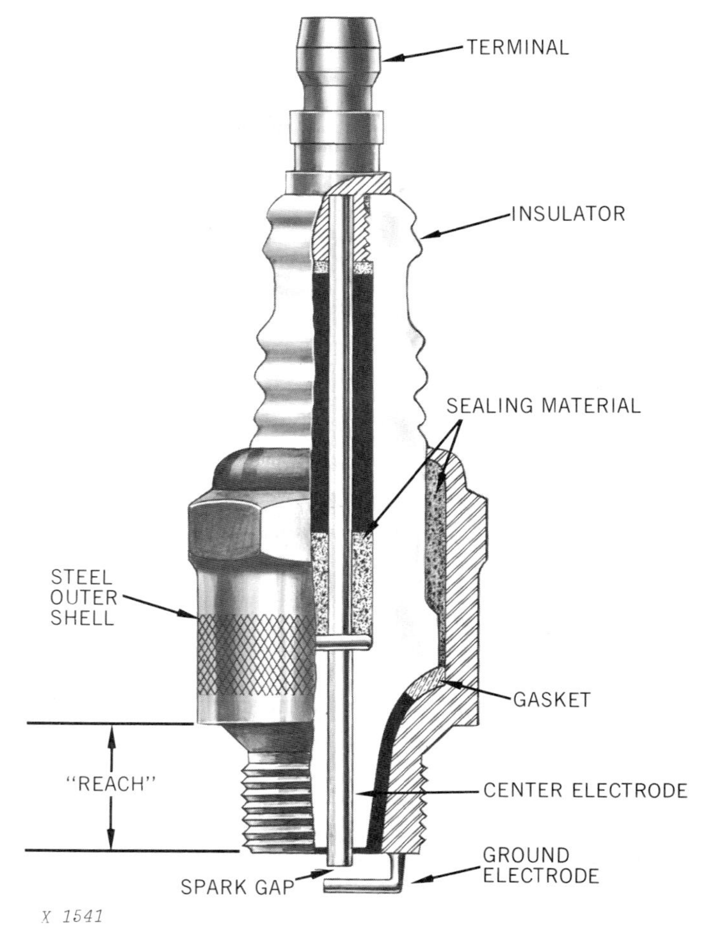

Fig. 26—Spark Plug Construction

A gasket slips over the threaded portion of the plug and rests against the flange at the bottom of the upper part of the shell. The gasket serves two purposes—it seals the plug against compression loss and provides a path for the transfer of heat to the cooling system.

A gasket is not always used. Some spark plugs use a tapered seat instead of a flat flange and this replaces the gasket.

The distance from the flange to the end of the plug threads is called the **reach** (Fig. 26). The reach of a spark plug is very important in plug selection. A plug with too long a reach will extend too far into the combustion area. Not only will the plug run hotter, but it also can be damaged when hit by a piston or valve. A plug with too short a reach will run cold and cause misfiring due to fouled electrodes.

The thread diameter may vary according to the size of the spark plug hole in the cylinder head.

The engine Technical Manual can give you the exact spark plug specifications for each engine.

Spark Plug Insulator

The insulated core or *insulator* is mounted in the outer shell. The outer end carries the spark plug terminal. The core is usually made of such insulating materials as white ceramic or porcelain.

The insulator is held in position and shielded from the outer shell by an inside gasket and sealing compound (Fig. 26).

Besides holding the center electrode, the insulator is a shield for the electrode so that current will flow only through the electrode. It must also withstand extreme heat and cooling and vibrations.

The exposed upper portion of the insulator must be kept clean to prevent current from leaking out. Many plugs have ribbed insulators to discourage this dirt build-up.

Spark Plug Electrodes

The *electrodes* are usually made of a metal alloy to withstand constant burning and erosion.

The *center electrode* extends through the insulator. One end is connected to a stud screwed into the top of the insulator. The other end extends out the nose or cone of the insulator. The electrode is held in position in the insulator by a sealing compound.

The *grounded electrode* is attached to the outer shell. It has a slight bend so that the end is directly beneath the end of the center electrode.

The gap between the two electrodes is the prime factor in plug operation. *This gap must be set to exact engine specifications.*

If the gap is too narrow, the spark will be weak and fouling and misfiring is the result.

Too wide a gap may work okay at low speeds, but at high speeds or loads it will strain the coil, resulting in misfire.

The surfaces of the two electrodes at the sparking point or gap should be parallel and have squared corners. This gives the current a better "jump" across the gap.

HEAT RANGE OF PLUGS

The heat range of a spark plug is as important as the gap setting. In fact, the various heat ranges of different spark plugs are used to classify them.

The term **heat range** refers to the plug's ability to transfer the heat at the firing tip to the cooling systems of the engine. This is determined by the **distance the heat must travel.**

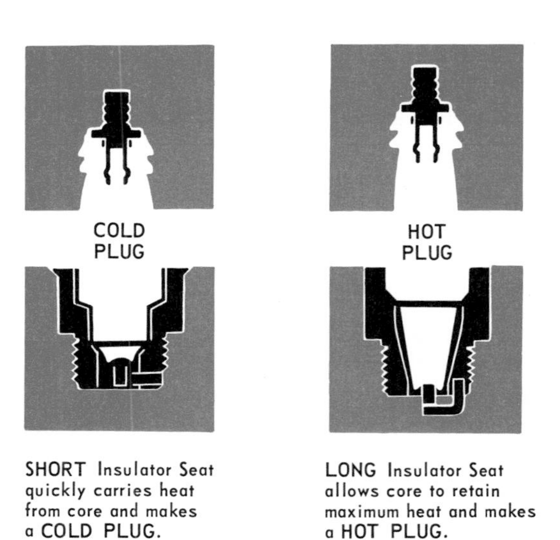

Fig. 27—Heat Range Of Spark Plugs

As you can see in Fig. 27, the end of a plug insulator that has a long nose or cone is further from the cooling system. Therefore, heat at the end will travel further. This type of plug will then run *hot.*

The end of a short insulator cone is closer to the cooling system and heat will transfer faster. A plug with a short insulator cone will then operate *cooler.*

Engine design and operating conditions will decide which type of spark plug—hot or cold—should be used. Generally, an engine which operates at fast speeds or heavy loads, and thus hotter, will require a **cold** plug so that the heat will transfer faster. On the other hand, a **hot** plug will be used in an engine that operates at low or idle speeds most of the time. Hot plugs will burn off the deposits that occur in this type of operation. For normal engine operations, a plug that falls somewhere between *hot* and *cold* is used (Fig. 28).

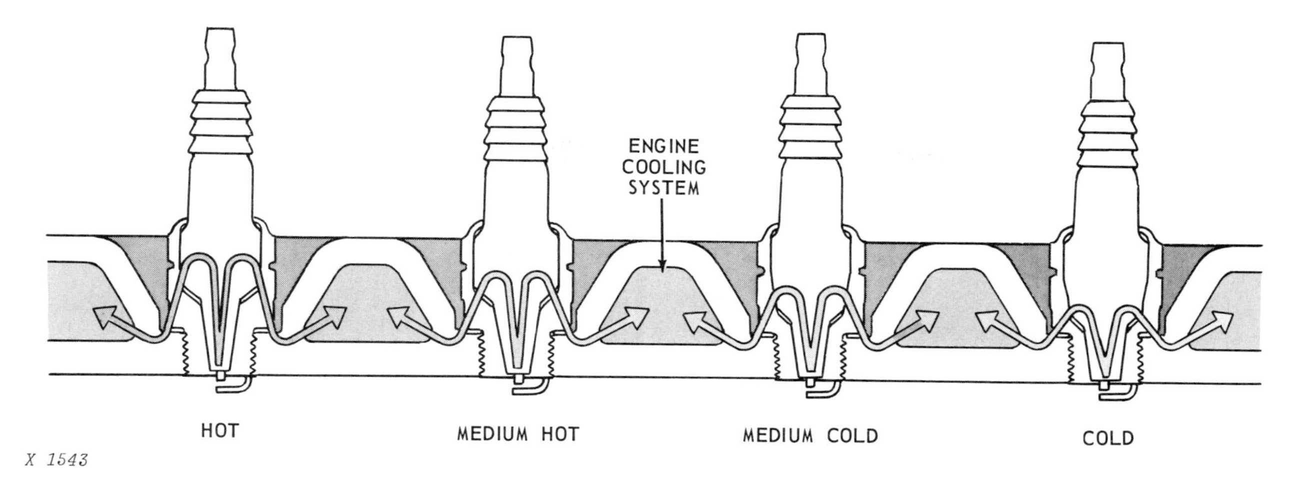

Fig. 28—Spark Plugs of Various Heat Ranges

SPECIAL TYPES OF SPARK PLUGS

As we have said, there are many variations on the standard spark plug. In addition, there are several kinds of spark plugs that are unique in design and that suit a special application. We will briefly describe some of these plugs for you.

Resistor-Type Plugs

Resistor-type plugs have a resistor between the terminal and center electrodes. These plugs are used to avoid radio and television static generated via the ignition circuit.

Auxiliary-Gap Plugs

This plug gives extra protection against fouling in cooler operations. They are useful when deposits can short out the current, preventing a spark. The gap is placed *with-in* the center electrode and holds in the voltage that could be drained off by the shorting deposits.

Surface-Gap Plugs

This plug eliminates the grounded electrode. The spark fires across the insulator tip between the center electrode and the shell. This also gives fouling protection. However, it is used only in special high-voltage systems.

Insulator-Tip Plugs

These plugs have the insulator tip extended further into the combustion area. By doing this, the tip is said to be cooled by the incoming fuel-air mixture and cleaned of deposits by the hot exhaust gases.

REMOVING AND INSTALLING SPARK PLUGS

Observe these practices when removing and installing plugs:

1. Pull the wire from the plug by grasping the terminal, **not** by pulling on the wire.

2. After loosening the plug but before removing it, always clean the area around the spark plug by blowing, wiping, or brushing. (Be sure to protect your eyes.) This will prevent dirt from falling into the cylinder after removal.

3. Use a deep-well socket to remove the spark plugs. Also remove the gaskets with the plugs (if used).

4. If the plugs are to be reused, be sure to note which cylinder each came from. The condition of the spark plug can tell you a lot about the operation of a particular cylinder.

5. When replacing spark plugs, it is best to **replace all of them** at the same time.

6. When installing spark plugs with gaskets, be sure the gaskets are in place. The gaskets act as a seal to prevent loss of compression around the plug. Without a gasket, the reach of the plug would also change, affecting plug operation.

7. Remember another point on gaskets: Most manufacturers recommend installing new gaskets with both new and reconditioned spark plugs. This is due to the crush of a used gasket which might prevent adequate sealing. However, always remove the old gasket from the spark plug. **Never use both the old and the new gaskets.** Here again, spark plug reach is affected.

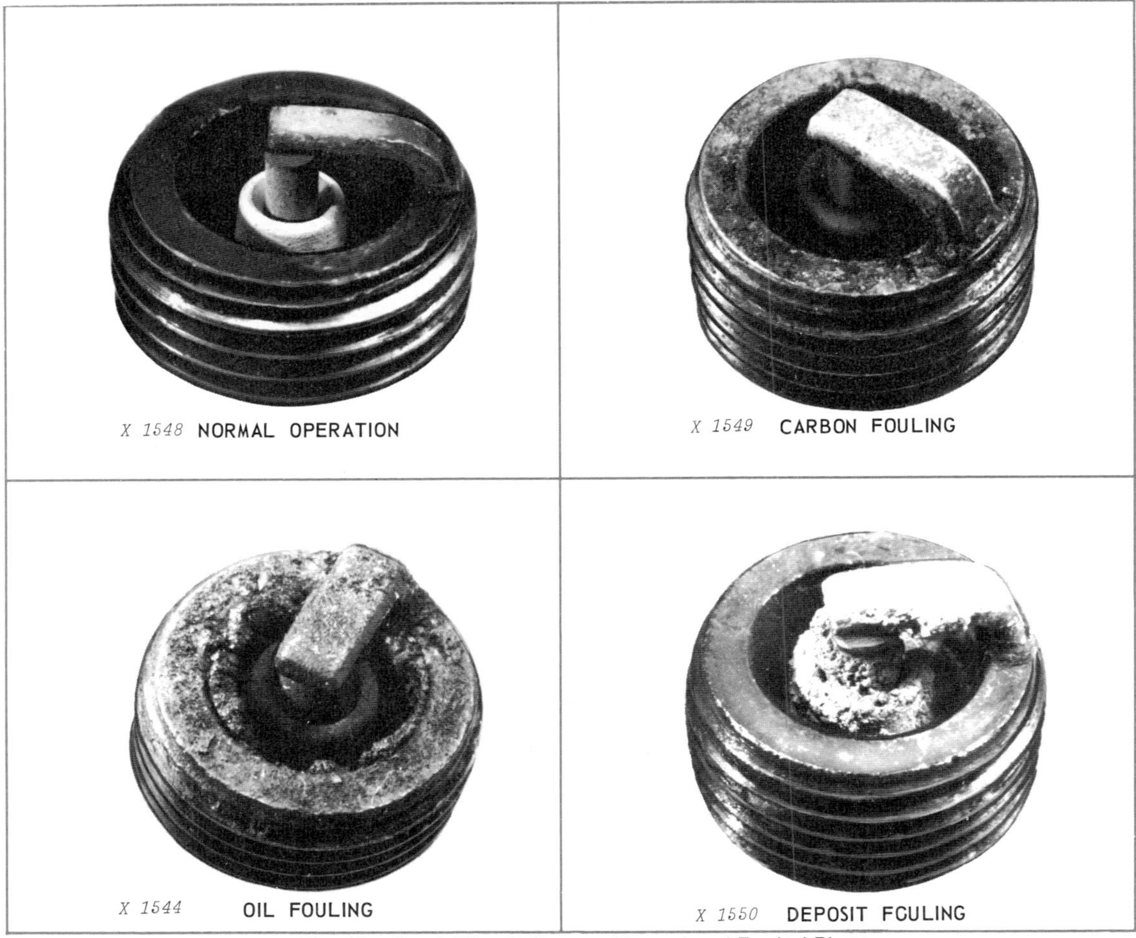

Fig. 29—Spark Plug Failures—Normal and Fouled Plugs

8. Tighten the spark plugs as specified and connect the spark plug wires to the proper plugs. If a torque wrench is not available, tighten the plug until you feel it seat, then turn it ½ to ¾ turn more. (With steel gaskets, tighten ¼-turn after seating).

FAILURES OF SPARK PLUGS

A visual inspection will usually tell the condition of the spark plugs. However, for an electrical check, use a spark plug analyzer and follow its test instructions.

Spark Plug Failures—General

The first sign of failing plugs is when the *engine misfires.*

Faulty plugs are not the only cause of engine misfire nor are the plugs usually the original cause. But when the engine misfires, check the condition of the plugs.

The two main failures of spark plugs are:

- **Fouled plugs (dirty deposits on tips)**
- **Eroded plugs (badly burned tips)**

FOULED PLUGS around the electrodes and tip can result from *lack of heat* to burn off the deposits at the firing point.

ERODED PLUGS are caused by *too much heat* at the firing point.

These plug failures are caused by many things as we'll see now.

Remember that the spark plug can tell you many things about the condition of the engine.

Spark Plug Failures—Fouled Plugs

Pictures of fouled plugs are shown in Fig. 29.

For comparison, a plug in *normal operation* is shown at the top left in Fig. 29. A normal plug will have brown to greyish-tan deposits and a slight wear on the electrodes. This indicates good adjustments of the engine.

A spark plug having this appearance can be cleaned, regapped, and reinstalled.

An *oil-fouled plug* is shown at the bottom left in Fig. 29. Wet, oily deposits with a slight electrode wear can mean that oil is getting into the combustion area. It can be caused by overfilling the engine crankcase, excessive clearance on the valve stem guides, or broken or poorly seated piston rings. Plugs in this condition can usually be degreased, cleaned, and reinstalled. If the oily deposits are more than shown, the plug should be replaced.

A *carbon-fouled plug* is shown at top right in Fig. 29. This plug has dry, fluffy, black deposits. These can be caused by too "rich" a fuel-air mixture or a clogged air cleaner, or by reduced voltage from faulty components in the circuit. After the cause has been corrected, this type of plug can usually be cleaned, regapped, and reinstalled.

A *deposit-fouled plug* is pictured at bottom right in Fig. 29. Red, brown, yellow, and white powdery deposits are usually the by-products of combustion and come from fuel and lubricating oil additives. These powdery deposits are usually not harmful. However, they can cause intermittent misfire at high speeds or heavy loads. If the insulator is heavily coated, replace the plug.

Hard, sandy deposits (not shown) can be caused by a faulty air cleaner or an engine operating in extremely dusty conditions. Replace these plugs.

Plug fouling can be caused by a "cold" plug being used in an engine that operates at low or idle speeds most of the time. Heat is dissipated so quickly that it does not have time to burn the excessive deposits from the firing point. So the deposits build up and the engine misfires.

Spark Plug Failures—Erosion and Breakage

Erosion can be caused by the use of a "hot" spark plug in an engine operating at fast speeds or heavy loads. The extreme heat is not dissipated to the cooling system fast enough, resulting in burned or blistered insulator tips and badly eroded electrodes.

Pre-ignition is another cause of plug erosion (see at center in Fig. 30) Hot spots in the combustion chamber including hot carbon deposits, hot spots in the piston head or cylinder walls, extreme hot valve edges, or hot electrodes can cause preignition. (Pre-ignition is the firing of the fuel-air mixture by a hot spot before the spark occurs.) Pre-ignition can cause blistered insulators and eroded electrodes as shown.

A lean fuel-air mixture can also cause erosion.

Wrong coil polarity can cause the grounded electrode to erode or "dish out" as shown at the right in Fig. 30.

Broken insulator tips and broken electrodes can be caused by three things:

1. Broken insulator tips might be caused by heat shock (see at left in Fig. 30). This is when there is a rapid increase in tip temperature under severe operating conditions. It can usually be prevented by allowing an adequate warm-up period at low speeds and by having the correct timing and the proper octane fuel.

X 1545 HEAT SHOCK FAILURE

X 1551 EROSION FROM PRE-IGNITION

X 1552 GROUNDED ELECTRODE "DISHED"-WRONG COIL POLARITY

Fig. 30—Spark Plug Failures—Erosion and Breakage

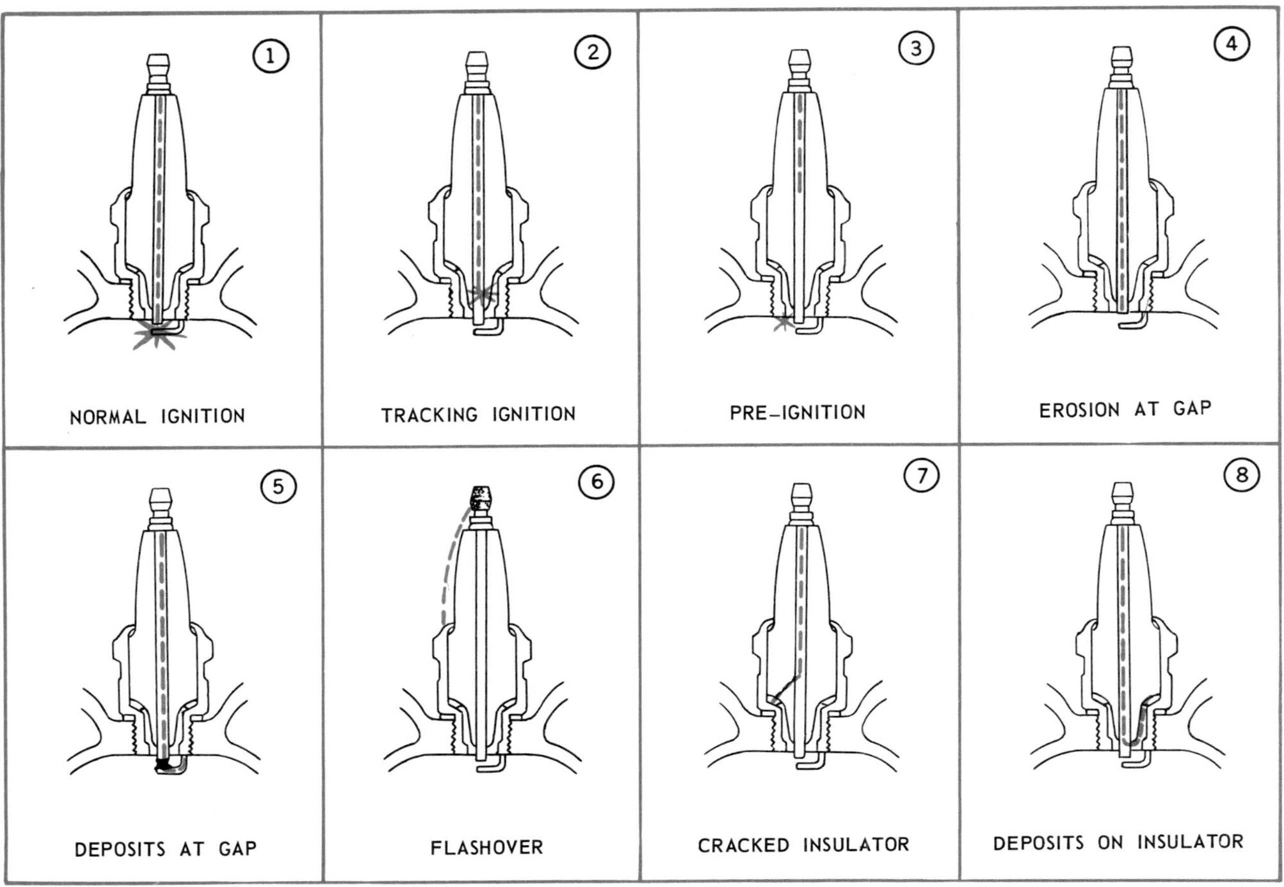

Fig. 31—Spark Plug Misfires

2. The tip can be broken by mishandling during gap adjustment, particularly if an attempt is made to bend the center electrode.

3. The tip and the electrodes can be broken if they are hit by the piston or valves. Using a spark plug with the correct reach will solve this problem.

Replace spark plugs with electrode and tip erosion or breakage.

Causes of Spark Plug Failures

In nearly every case of spark plug failure, the cause must be corrected before new or reconditioned spark plugs are installed. Otherwise, servicing will again be needed in a short time.

Fig. 31 illustrates what takes place in a faulty spark plug during ignition. The following numbered statements tell what causes the misfirings.

1. *Normal ignition* occurs when a spark of adequate energy is delivered at the correct instant across the electrode gap as shown.

2. *Tracking ignition* occurs when the spark, jumping from one deposit "island" to another, ignites the fuel charge at some point along the insulator nose. The effect is to retard ignition timing.

3. *Pre-ignition* occurs when some surface in the combustion chamber becomes hot enough to fire the fuel-air mixture before the spark occurs.

4. *Bad erosion* of the plug electrodes may prevent voltage from jumping the gap.

5. *Deposits* may have bridged the gap so that the coil voltage is drained away without a spark occurring.

6. *Flashover* is caused by moisture or dirt or by a worn out terminal boot. This allows voltage to short across the outside of the insulator.

7. *Cracks* in the plug insulator may allow high voltage to short circuit to the ground.

8. *Deposits* formed on the *insulator* surface may drain away voltage.

SPARK PLUG SERVICING

Spark plug service has three steps:

- **Inspection**
- **Cleaning**
- **Gap adjustment**

Most engine Technical Manuals recommend a specific time interval for ignition services. Follow this timetable very closely.

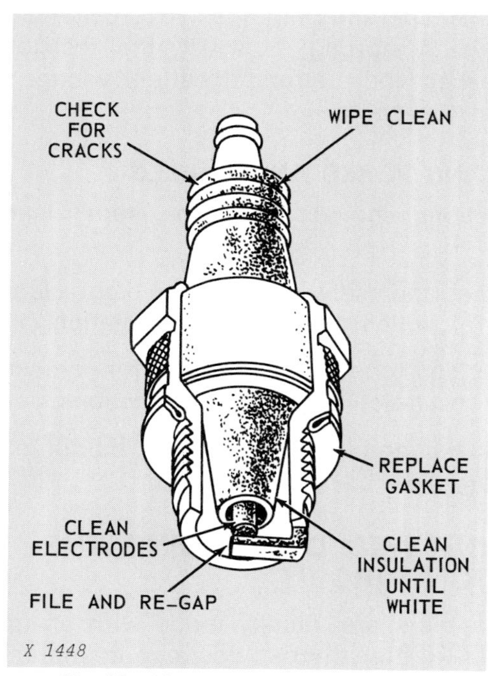

Fig. 32—Maintenance of Spark Plugs

Inspection of Spark Plugs

When inspecting the plugs, look for the normal or abnormal wear we have just described. Then decide whether to recondition or replace the plugs.

Remember, if one plug is replaced, all the plugs should be replaced to get the full advantage of new plug performance and economy. This is not applicable, of course, if unusual conditions cause premature failure to just one in a fairly new set of plugs.

Normally, replace the whole set of plugs after long intervals of use. Normal wear can double the voltage requirements of a spark plug even in a short period of time.

Cleaning of Spark Plugs

There are two types of spark plug cleaning machines that do an acceptable job of removing deposits from the insulator and electrodes. One unit uses an abrasive compound and *air blast* to clean the insulator and electrodes. The other machine uses a strong *liquid cleaner* to do the same job.

Make a careful inspection, however, after using an abrasive blast machine to insure that all abrasive particles have also been removed. Engine damage could result if they were left on the plugs.

NOTE: Do not use a power wire brush to clean the plugs. Most makers of plugs do not recommend this.

Badly fouled plugs should be replaced. It is doubtful sand blasting or liquid cleaning will remove all the deposits from such plugs.

Clean the threads with a wire hand brush or a powered soft wire brush wheel.

Wet oily plugs may require cleaning with a petroleum solvent before abrasive cleaning.

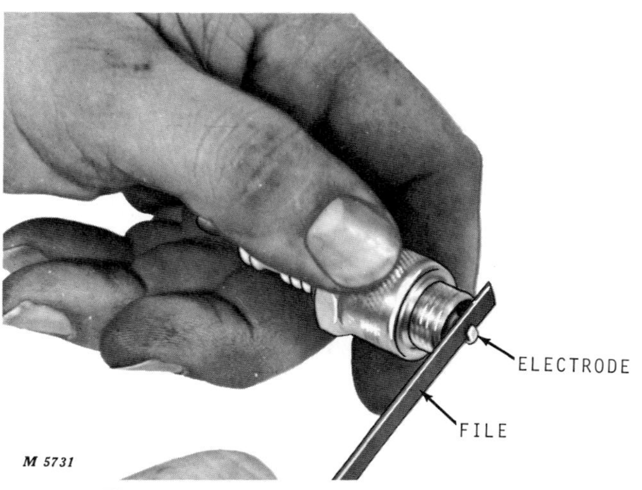

Fig. 33—Filing The Spark Plug Electrodes

Before gapping the electrodes, file them with a small point file to flatten their surface at the firing point and square up the edges (Fig. 33). Squaring the surface edges can reduce voltage requirements, even more than if the plugs are only cleaned.

Adjusting Spark Plug Gaps

Whether the plug is new or used, always check the gap before you install the plug.

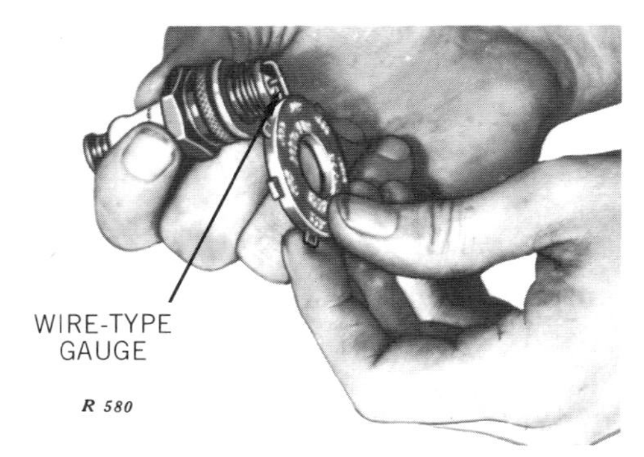

Fig. 34—Checking The Spark Plug Gap

Use a *wire* spark plug gauge to check the gap (Fig. 34).

The wire gauge will give a true gap reading even if the electrode surface is not flat.

If the gap is far too wide, the plug may need replacing.

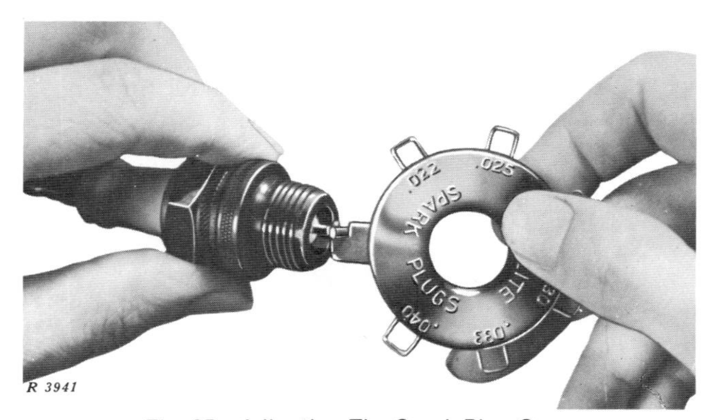

Fig. 35—Adjusting The Spark Plug Gap

Use a bending tool to adjust the gap to specifications (Fig. 35). (Often the bending tool is part of the wire gauge as shown.)

After adjusting, check to be sure the electrode surfaces are parallel.

Never bend the center electrode. Doing so may crack or break the insulator tip.

IGNITION WIRING

The type of wiring used in the ignition circuit is important as well as its condition.

Light wiring is used in the **primary** circuit because of the relatively low voltage.

Heavy wire or cables are used in the **secondary** circuit (coil to distributor cap, distributor cap to spark plugs) because they are subject to high voltage. High-voltage wires usually have metal terminal ends: female and male on the spark plug cables, male on the coil-to-distributor cap cable.

Some ignition circuits have an ignition **resistor** in the form of a special *resistance wire.* When the engine is running, the resistor wire is in series with the coil and reduces the voltage at the coil to normal operating voltage. To provide a better spark when starting, however, the resistor wire is bypassed and a stronger battery voltage is applied at the cable.

SERVICING WIRES AND CABLES

Connections should always be kept clean and tight.

The wires themselves should be kept clean and free of oil which could rot the insulation.

Periodically, check the wiring for cracked or broken insulation and loose connections.

When installing high-voltage wires, *do not let them touch a ground or each other.*

TESTING AND DIAGNOSING THE IGNITION CIRCUIT

Ignition tests are usually made with an ignition analyzer—either the oscilloscope or meter types (see Chapter 2).

Many of these tests can also be made with all components in place on the machine.

Follow the test procedures given with the analyzer, and consult the engine Technical Manual for the expected readings.

TROUBLE SHOOTING CHART FOR IGNITION PROBLEMS

Engine operating problems can be caused by many things, both mechanical and electrical. The chart below lists some of the engine problems and how they can be caused by a faulty ignition circuit.

Problems	Cause
1. Lack of power	a. Incorrect timing b. Pitted distributor points
2. Hard starting	a. Weak spark
3. Engine overheats	a. Advance mechanism sticking
4. Engine knocks	a. Incorrect timing
5. Engine backfires	a. Advance mechanism sticking
6. Engine pre-ignition	a. Faulty spark plugs
7. Engine misfires	a. Dirty spark plugs b. Faulty cables c. Incorrect distributor point gap
8. Engine uses too much fuel	a. Fouled spark plugs b. Incorrect timing
9. Engine runs irregular	a. Faulty ignition
10. Slow acceleration	a. Advance mechanism sticking b. Defective coil or condenser c. Faulty distributor points
11. Engine will not start	a. Faulty coil b. Faulty condenser c. Faulty distributor points d. Coil high tension wire out of socket e. Cracked distributor rotor f. Faulty spark plugs g. Incorrect timing h. Spark plug cables installed incorrectly
12. Engine starts but will not continue to run	a. Faulty coil b. Faulty condenser c. Faulty distributor points d. Faulty bypass resistor
13. Poor ignition of fuel	a. Incorrect spark plug gap b. Dirty spark plugs c. Faulty cables or wiring d. Incorrect timing e. Faulty distributor points f. Faulty condenser g. Defective coil h. Cracked distributor cap or rotor

CHECKING OUT IGNITION PROBLEMS

Once the cause of the problem has been isolated to the ignition circuit, check each component to find the faulty one.

We have already described the tests that are made on the components. Usually, they can be made with the unit on or off the machine. To refresh your memory, we will now give you a brief list of items to check.

Visual Inspection

Check for:

1. Loose or broken connections and cables

2. Cracked coil cap

3. Cracked or broken distributor cap and rotor

4. Carbon paths on and in distributor cap

5. Burned, pitted, or worn distributor points

6. Worn distributor breaker cam and breaker lever rubbing block wear

Electrical Tests

Check these items:

1. Coil for grounded, shorted, or open windings and coil polarity

2. Condenser for leakage, series resistance, and capacity

3. Cam angle (dwell) of distributor

4. Timing of distributor

5. Spark plug strength. (Check this by removing the coil high-tension cable from the distributor cap

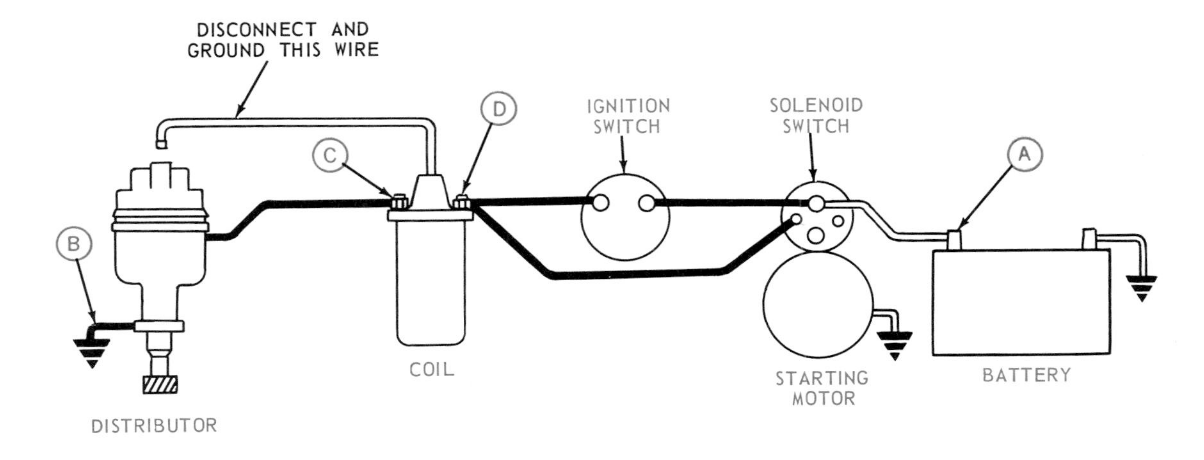

Fig. 36—Ignition Circuit Test Points (For a Typical 12-Volt Negative-Ground Circuit)

and holding it (with insulated pliers) 1/4 inch (6 mm) from the engine while cranking. If a good spark appears, the trouble is probably in the distributor cap or rotor, or the spark plugs or cables.)

6. Ignition circuit for high resistance, open circuits, and grounds. See Fig. 36 and the following instructions for details on this test.

GENERAL TEST OF IGNITION CIRCUIT

1. Use a voltmeter with a range of 0.1 to 20 volts.

2. Make all tests with lights and accessories off.

3. An example of the test results on a typical 12-volt negative-ground electrical system is given in the chart below.

SUMMARY: SERVICING THE IGNITION CIRCUIT

By performing periodic servicing of the ignition circuit, making all adjustments, and operating the engine within the specified limits, the ignition circuit should give long, dependable service.

IGNITION CIRCUIT TEST RESULTS

Voltmeter Connected to Points (Fig. 36)	*Ignition Switch Position*	*Distributor Breaker Points Position*	*Voltmeter Reading If Circuit In Good Condition*
A—D	Cranking	....	1 volt (max.)
B—D	Cranking	....	10 volts (approx.)
B—D	On	Open	Battery voltage
B—D	On	Closed	4.8 volts (approx.)
B—C	On	Closed	0.2 volt (max.)

TEST YOURSELF

QUESTIONS

1. What is the final goal of the ignition circuit?

2. Name the four basic parts of an ignition circuit.

3. Match the items below:

a. Primary circuit	1. High voltage
b. Secondary circuit	2. Low voltage

4. What does an advance mechanism do for a distributor?

5. True or false? "Use emery cloth or fine sandpaper to clean the distributor points."

6. What is the difference between fouled and eroded spark plugs? Which is caused by too much heat? Which by too little heat?

(Answers on page 20 at the end of this book.)

Litho in U.S.A.

 Litho in U.S.A.

ELECTRONIC IGNITION AND FUEL INJECTION / CHAPTER 9

INTRODUCTION

Many modern agriculture and industrial machines have two major electronic systems:

- **Electronic Ignition**
- **Electronic Fuel Injection**

Let's discuss both of these systems in detail.

ELECTRONIC IGNITION

An electronic ignition system does not use breaker points and condenser, which are the main elements of a conventional ignition system distributor (see Chapter 8). Instead of breaker points and condenser, electronic ignition systems (Fig. 1) contain a reluctor and sensor in the distributor and an ignition control unit. The control unit causes the flow of primary current to stop thus inducing high voltage through the secondary winding.

All other ignition circuit components are common to both electronic and conventional systems.

Important: Electronic and conventional ignition coils are similar in construction, but must **not** be interchanged. Electronic ignition coils generally deliver much higher voltage. Use of a coil not designed for a particular ignition system may damage other components of the system. See Chapter 8 for coil construction and operation.

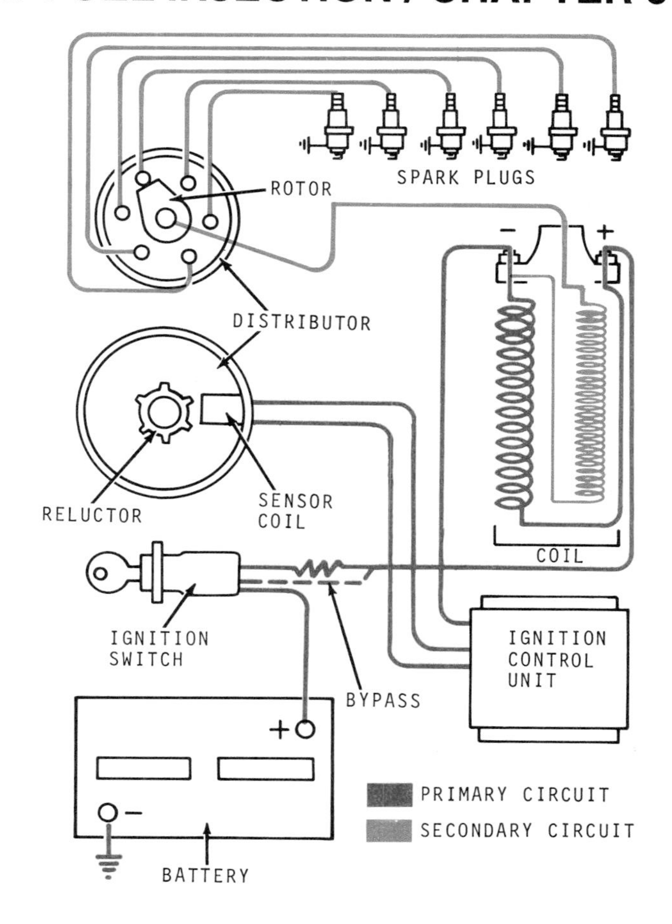

Fig. 1 — Electronic Ignition System

TYPES OF ELECTRONIC IGNITION SYSTEMS

Most electronic ignition systems are constructed and operate like the one just described. One basic difference is the style of the reluctor-sensor assembly (Fig. 2). The self-integrated electronic ignition distributor (D, Fig. 2) has a considerably different appearance and will be discussed in more detail later.

ADVANTAGES OF ELECTRONIC IGNITION

The advantages of electronic ignition systems are:

- No problems caused by malfunctioning distributor points and condenser because there are no points and condenser.
- Quicker starts in all kinds of weather. This places less drain on the battery.
- Electronic ignition produces a hotter spark of longer duration, which will ignite marginal air-fuel mixtures under adverse weather conditions. A conventional ignition system in good condition delivers about 25,000 volts. An electronic ignition system in good condition delivers 35,000 or more volts.
- The hotter spark of longer duration provides more complete combustion of the fuel mixture, which improves fuel efficiency and lessens exhaust emissions.
- The hotter spark of longer duration helps increase spark plug life. The spark plugs can burn off detrimental deposits that settle on electrodes and cause fouling. Electronic ignition and the use of unleaded gasoline have combined to extend spark plug life to over 20,000 miles (32 160 km). Vehicles without electronic ignition and not using unleaded gasoline have an expectant spark plug life of about 10,000 miles (16 080 km) (Fig. 3).

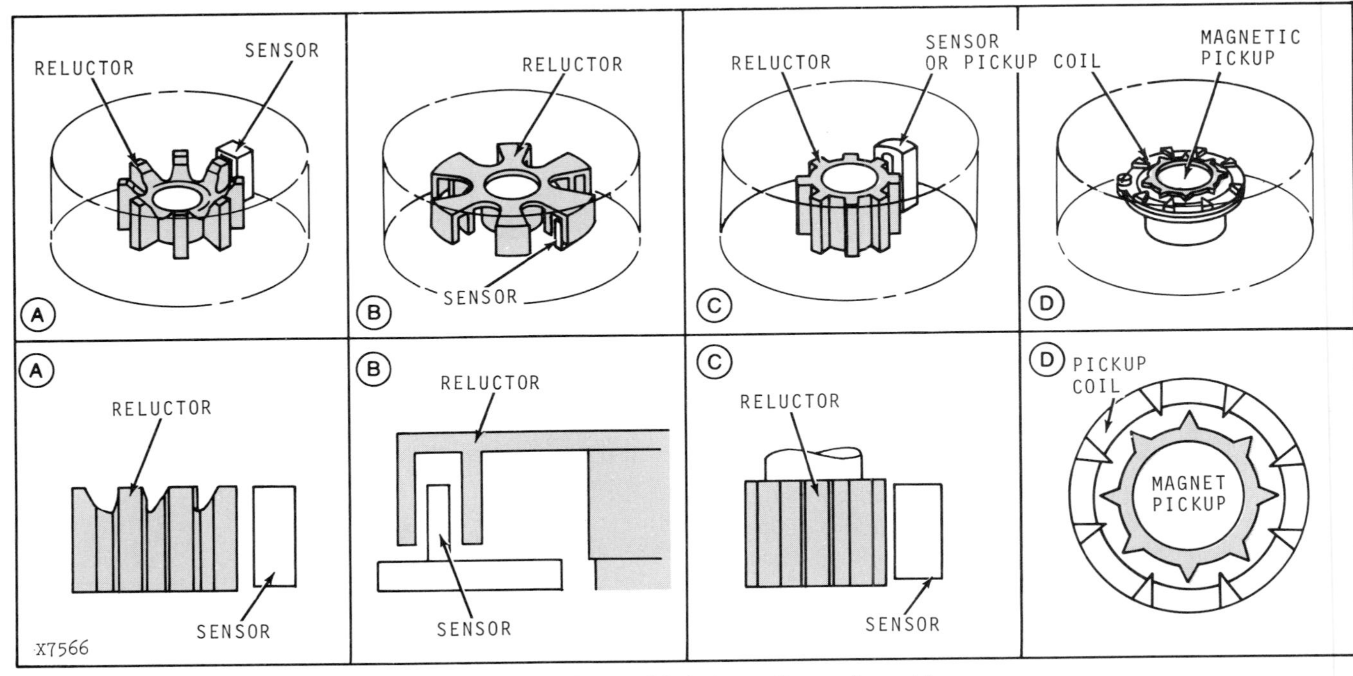

Fig. 2 — Various Styles of Reluctor — Sensor Assemblies

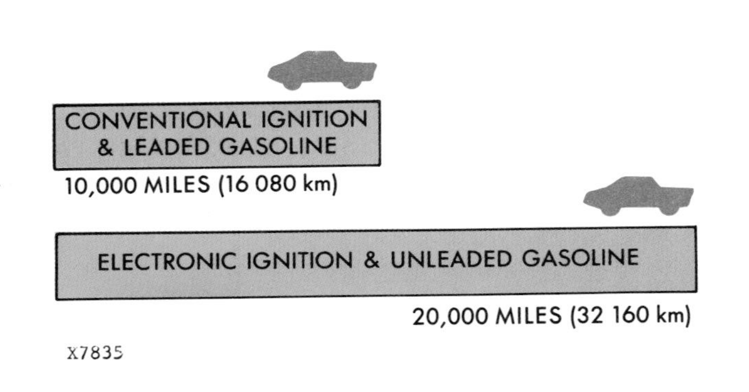

Fig. 3 — Spark Plug Life Comparison

HOW ELECTRONIC IGNITION WORKS

The **ignition control unit** contains solid-state components, permanently sealed in a material that resists vibration and environmental conditions. The unit has built-in reverse polarity and transient voltage protection.

The ignition control unit cannot be repaired. If a malfunction occurs, the unit must be replaced.

When the ignition key is turned on, an oscillator in the ignition control unit excites the sensor in the distributor. The sensor (a small coil of fine wire) develops an electromagnetic field that is sensitive to the presence of metal (Fig. 4).

As the engine camshaft rotates, it drives the distributor shaft which causes the reluctor to rotate. As the leading edge of a metal tooth of the rotating reluctor enters the electromagnetic field of the sensor, the strength of the oscillator in the sensor is reduced. This acts as a signal to a demodulator in the ignition coil unit. The demodulator controls a power switch transistor, which is in series with the ignition coil primary circuit.

The demodulator causes the power switch transistor to switch off the coil primary circuit, inducing high voltage in the ignition coil's secondary winding (Fig. 5). This high voltage is distributed to the spark plugs through the distributor cap, rotor, and spark plug cables.

This distributor does not require periodic maintenance other than to clean the distributor cap and check the cap and rotor terminals for damage. The reluctor and sensor cannot be repaired. If testing reveals a malfunction, they must be replaced. The electronically-timed dwell of the circuit, which is the period of time that the circuit is closed, cannot be adjusted on most electronic ignition systems.

TROUBLESHOOTING ELECTRONIC IGNITION MALFUNCTIONS

Troubleshoot the electronic ignition system by performing the following tests. For other ignition system troubleshooting see Chapter 8.

Test the ignition system when the engine fails to start or when there is no spark at the spark plugs. An ignition malfunction may also cause the engine to not start on the first attempt and stall when running. In these situations, the engine will eventually start or restart if it stalls.

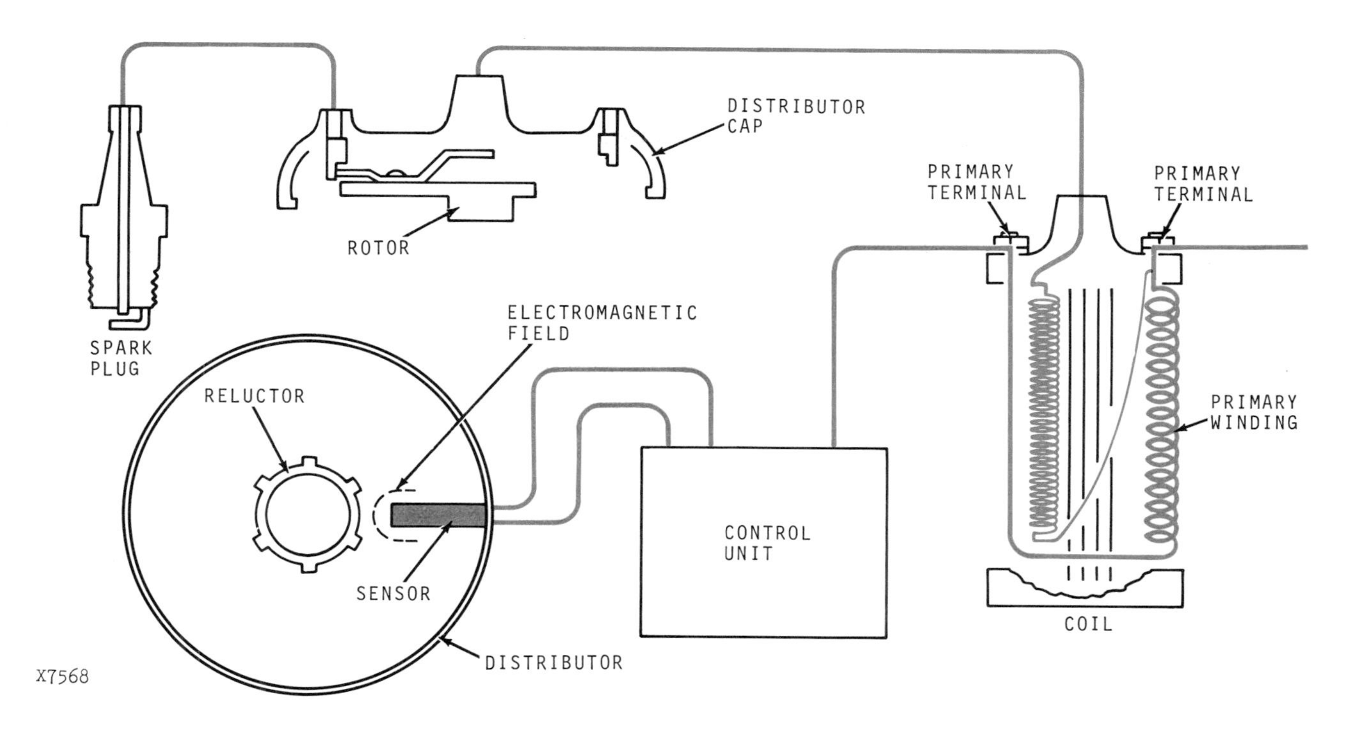

Fig. 4 — Operation Before Reluctor Passes Through Electromagnetic Field

Note: Ignition related problems, such as backfiring and poor starting, can occur if the distributor cap and rotor are damaged, and if the spark plug cables and distributor-to-ignition coil cable are defective. These problems are common to all ignition systems and are discussed in Chapter 8.

IDENTIFY GENERAL LOCATION OF TROUBLE

When an ignition problem is suspected, first test the battery. Be certain to check the battery terminals and make sure the cables are attached securely and not corroded.

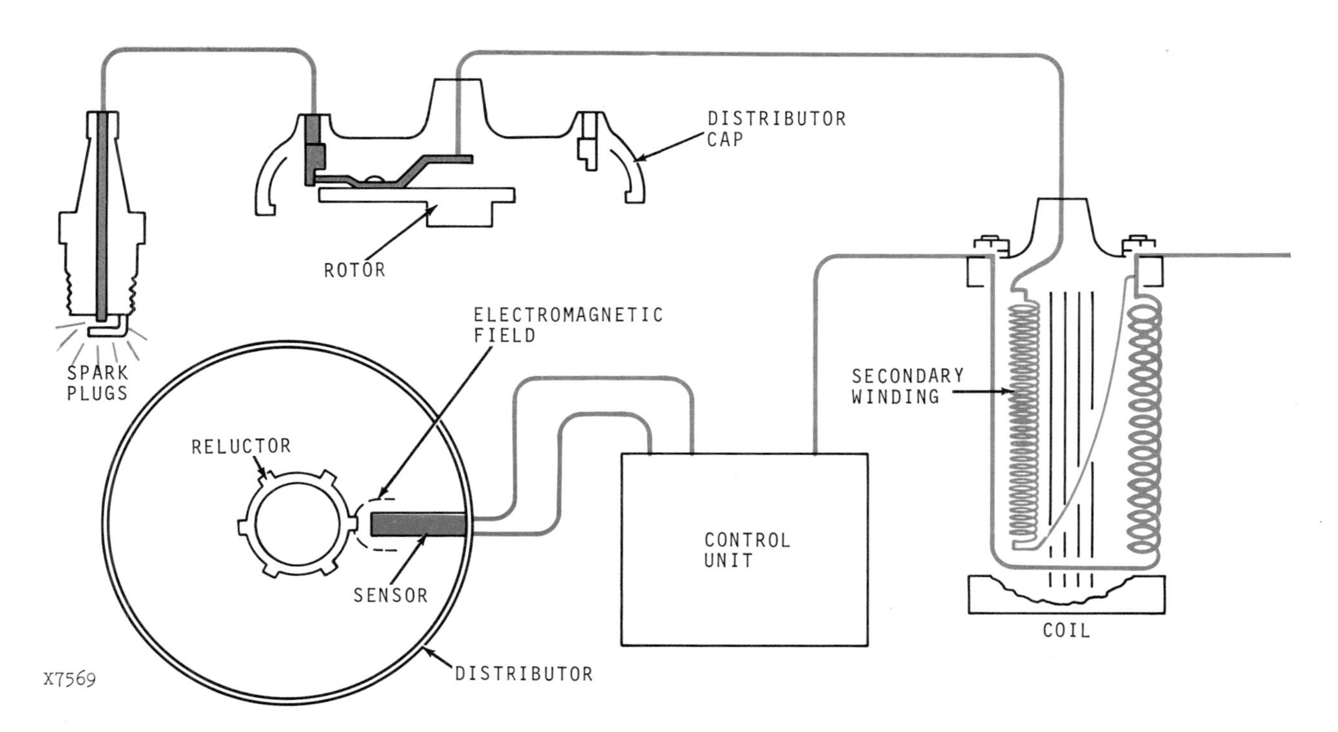

Fig. 5 — High Voltage Induced in Secondary Winding

To avoid shock turn the ignition switch OFF.

1. Then remove the high voltage cable from the center tower of the distributor.
2. Using insulated pliers, hold the end of the cable about 1/4 inch (6 mm) from ground and crank the engine (Fig. 6). If a spark arcs the gap, a problem exists in the distributor cap, rotor, spark plug wires, or spark plugs. If a spark does not arc, the problem exists in the ignition control unit, coil, ignition switch, sensor, or a wire.

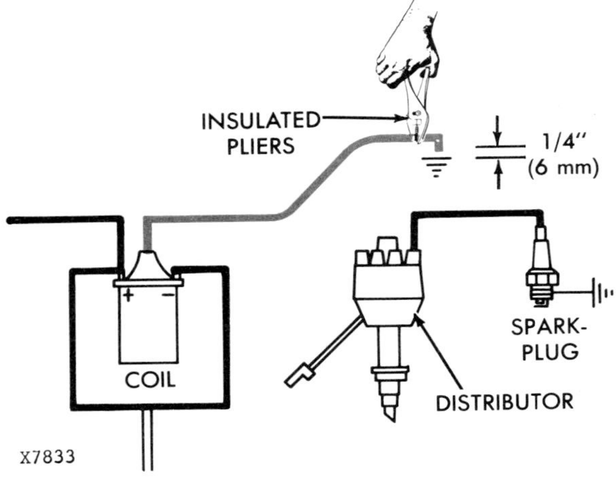

Fig. 6 — Use Insulated Pliers

Troubleshooting electronic ignition systems requires using a test light, voltmeter, ohmmeter, jumper wire, and possibly a tester switch if recommended by the manufacturer.

TESTING PRIMARY CIRCUIT—BATTERY TO COIL

One type electronic ignition system uses a dual ballast resistor that is contained in a unit outside the ignition control unit (Fig. 7). To test the primary circuit wires:

1. Unplug the wiring harness connector from the ignition control unit to reveal five cavities or pins (Fig. 8).

2. Turn the ignition switch to ON and connect the negative lead of the voltmeter to ground.

3. Connect the positive lead of the voltmeter to the wiring harness connector of the No. 1 cavity (Fig. 9). The voltage reading should be within one volt of battery voltage.

Important: All accessories must be off.

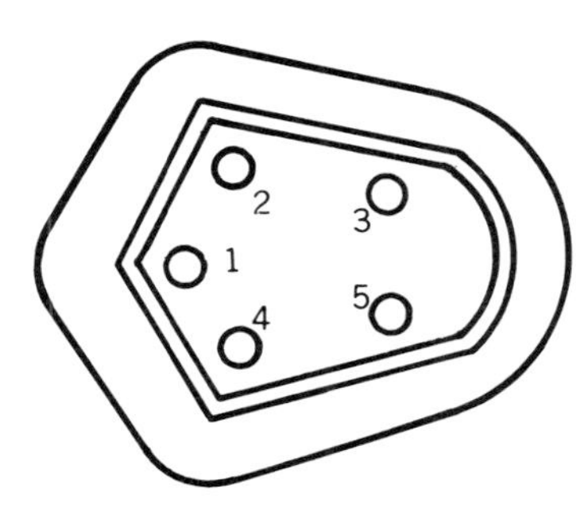

Fig. 8 — Five Cavities of Wiring Harness Connector

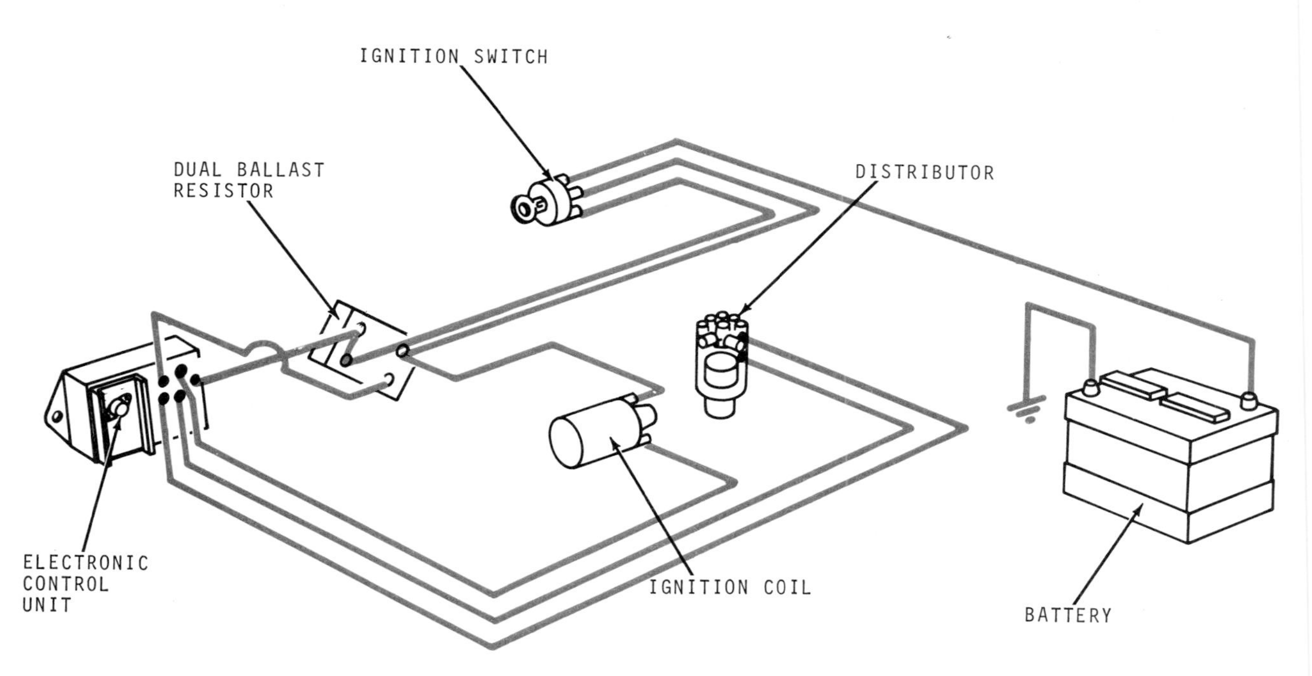

Fig. 7 — System with External Dual Ballast Resistor

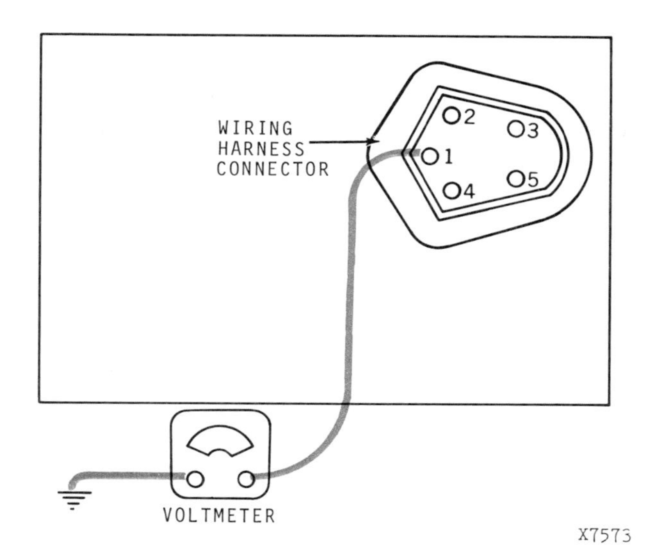

Fig. 9 — Testing No. 1 Cavity

4. Check No. 2 and No. 3 cavities in the same way. If there is more than a one volt difference in any reading, inspect the circuit for that particular wiring harness cavity to be sure all connections are clean and tight.

For ignition systems without the external dual ballast resistor, connect the voltmeter to ground and to the ignition coil positive terminal (Fig. 10). Turn the ignition switch to ON. If the voltmeter does not record battery voltage, check for high resistance between the battery and ignition coil or a defective ignition switch.

TESTING THE IGNITION CONTROL UNIT, COIL, AND SENSOR

There are three methods of testing the ignition control unit depending on the type of electronic ignition system being tested. If malfunctions occur in any system, make sure wires are not broken and connections are tight and free of corrosion.

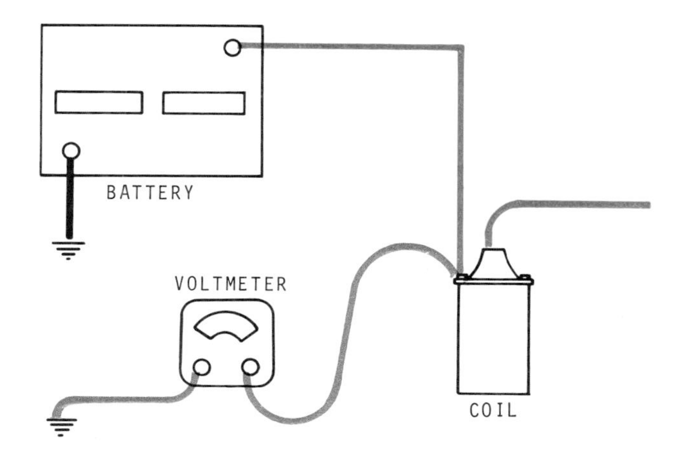

Fig. 10 — Connect Voltmeter Between Coil Positive Terminal And Ground

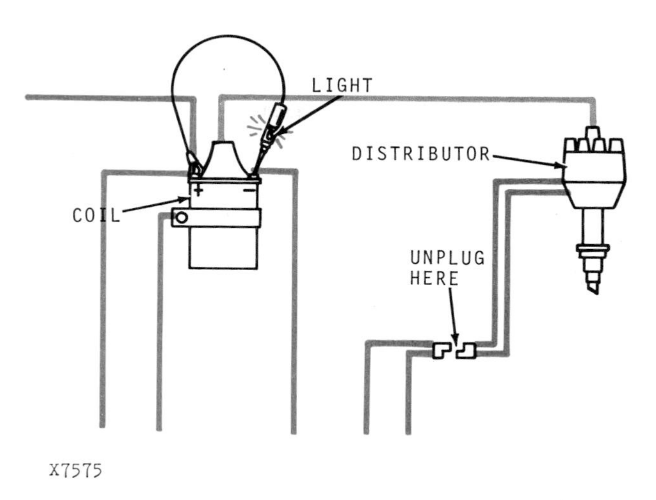

Fig. 11 — Test Light Across Coil Terminals

Testing with a Test Light

When testing the ignition control unit of the system with a reluctor-sensor configuration as in B, Fig. 2:

1. Disconnect the distributor lead connector.

2. Connect a test light across the terminals of the ignition coil (Fig. 11) and turn the ignition switch to the ON position. If the bulb does not light, replace the ignition control unit.

If the bulb lights with the ignition switch in the ON position, place a jumper wire across the ignition control unit lead terminals (Fig. 12). If the bulb remains lit, replace the ignition control unit.

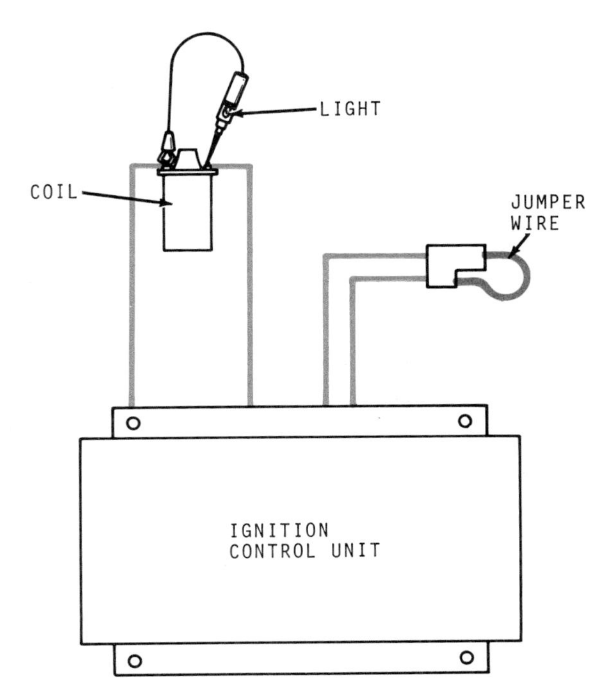

Fig. 12 — Place Jumper Wire Across Ignition Control Unit Lead Terminals

If the bulb goes out when you place the jumper across the ignition control unit's lead terminals, do the following:

1. Disconnect one end of the jumper wire at one of the lead terminals.

2. Remove the high voltage cable from the center tower of the distributor.

3. Using insulated pliers, hold the end of the cable 1/4 inch (6 mm) from ground (Fig. 13).

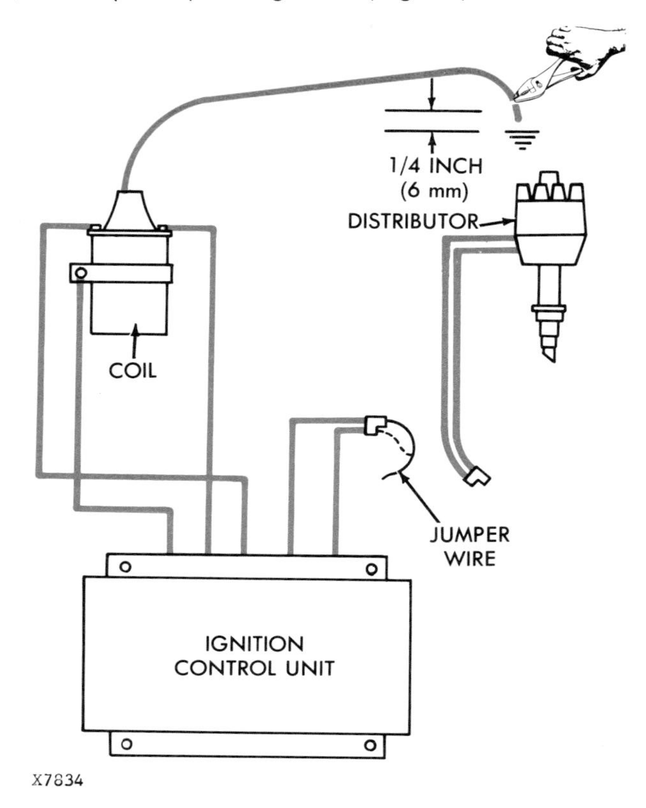

Fig. 13 — Hold End of Wire One-fourth Inch From Ground

4. Have someone crank the engine while you again short out the ignition control unit's lead terminals by reconnecting the jumper wire.

If a spark jumps the gap between the end of the high voltage cable and ground, the sensor in the distributor is defective and must be replaced (see next page). If no spark occurs, replace the ignition coil.

Testing with an Ohmmeter

For the electronic ignition system with an external dual ballast resistor, unplug the wiring harness connector from the ignition control unit and turn OFF the ignition. Connect an ohmmeter to wiring harness connector cavities No. 4 and No. 5 (Fig. 14). The ohmmeter resistance reading should be between 350 and 550 ohms.

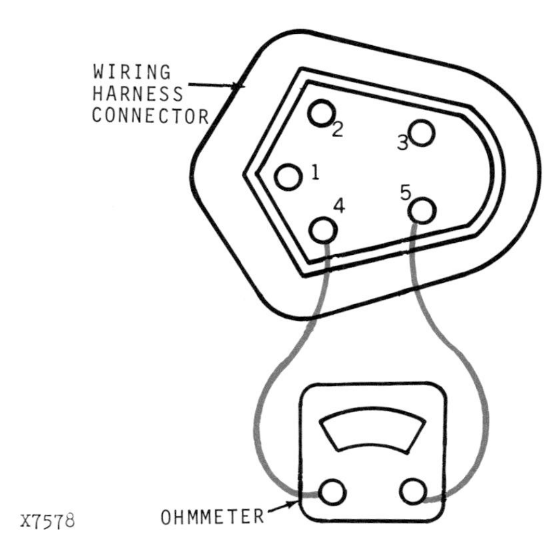

Fig. 14 — Ohmmeter Between Cavities No. 4 and 5

If the reading is less than 350 or more than 550 ohms, disconnect the distributor lead connector coming from the distributor and check resistance across the lead terminals (Fig. 15). If the reading is not between 350 and 550 ohms, replace the sensor in the distributor.

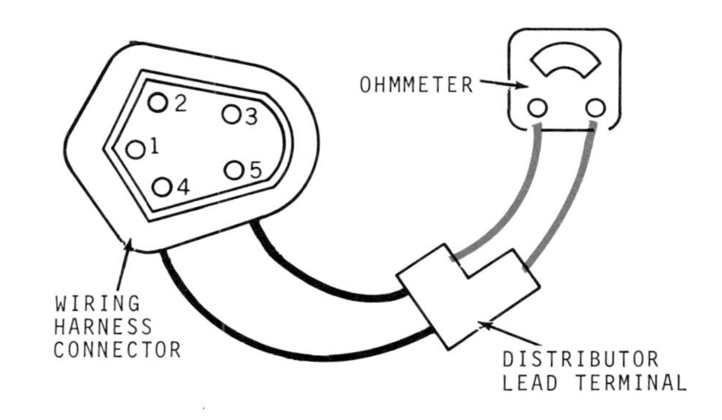

Fig. 15 — Connect Ohmmeter Across Distributor Lead Terminal

If the ohmmeter reading at the distributor-lead connector is between 350 and 550 ohms, inspect the wiring harness from the distributor lead connector to the electronic control unit by connecting the ohmmeter to ground and to either distributor lead terminal pin of the dsitributor harness (Fig. 16). The ohmmeter should show high resistance indicating an open circuit. If the ohmmeter shows low resistance, replace the sensor assembly in the distributor.

Connect the ohmmeter to ground and to the wiring harness connector of the No. 5 cavity (Fig. 17). The ohmmeter should show continuity. If not, tighten the bolts holding the electronic control unit to the firewall.

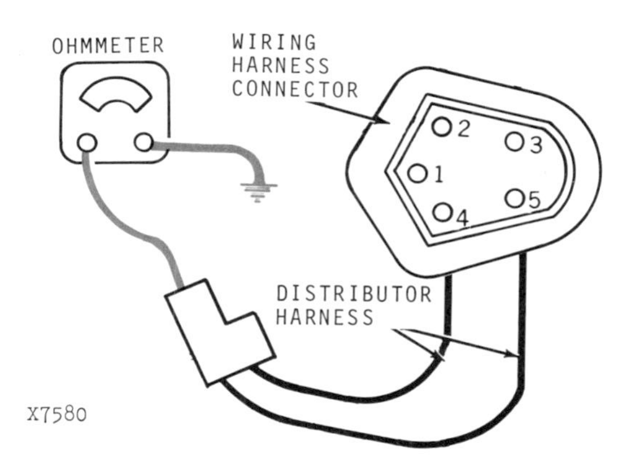

Fig. 16 — Check Resistence

Retest. If continuity does not exist, replace the electronic control unit.

Testing with a Voltmeter

The electronic ignition system with the reluctor-sensor assembly shown in A, Fig. 2 can be tested by using a voltmeter at the distributor lead connection. To perform this test:

1. Disconnect the distributor lead connector.

2. Connect a voltmeter between the two parallel blades (Fig. 18).

3. Set the voltmeter scale at 2.5 volts and crank the engine. The voltmeter needle should fluctuate.

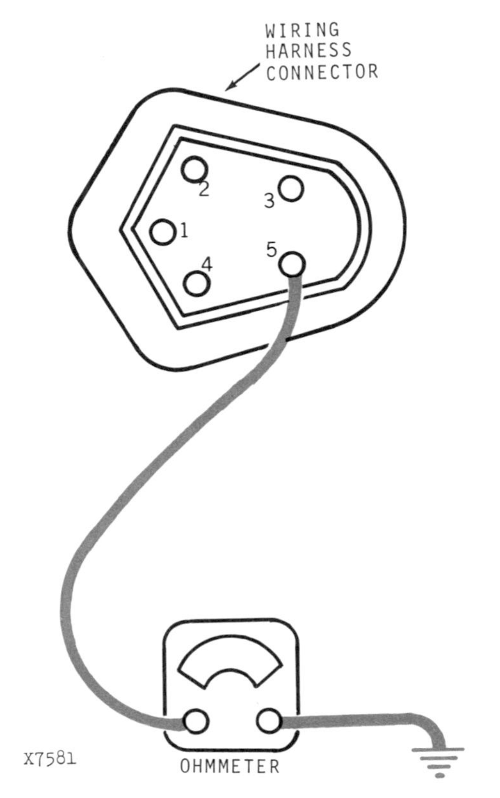

Fig. 17 — Check No. 5 Cavity

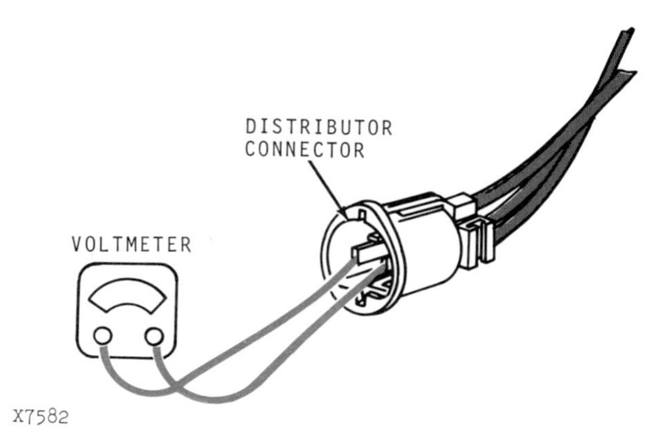

Fig. 18 — Voltmeter Between Parallel Blades

If the voltmeter needle does not fluctuate:

1. Inspect the distributor for a defective cap or rotor, loose reluctor, broken reluctor tooth, or misaligned keeper pin.

2. See that the reluctor rotates when the engine is cranked.

3. If it doesn't rotate, replace the sleeve and advance plate assembly.

4. If this examination does not reveal the reason for the voltmeter needle not fluctuating, install a new sensor assembly in the distributor. Use the exploded view shown in Fig. 19 to reassemble the unit.

Note: When disassembling the distributor, make reference marks on the reluctor and some adjacent part of the distributor. The reluctor must be assembled in the same position as it was removed to assure proper timing.

If the voltmeter needle fluctuates, perform the spark test with the high voltage cable as described on pages 9-3 and 9-4. During the test:

1. If a spark arcs the gap, check the spark plug wires, distributor cap and rotor for damage.

2. If spark does not appear, install a new ignition control unit and perform another spark test.

3. If spark occurs, the old control unit is bad.

4. If still no spark appears, replace the coil.

5. If the problem still exists, test each wiring circuit as outlined in the manufacturers service manual.

REPLACING THE SENSOR

To remove a defective sensor of the type shown in B, Fig. 2 from the distributor:

1. Remove the distributor cap, rotor, and dust shield.

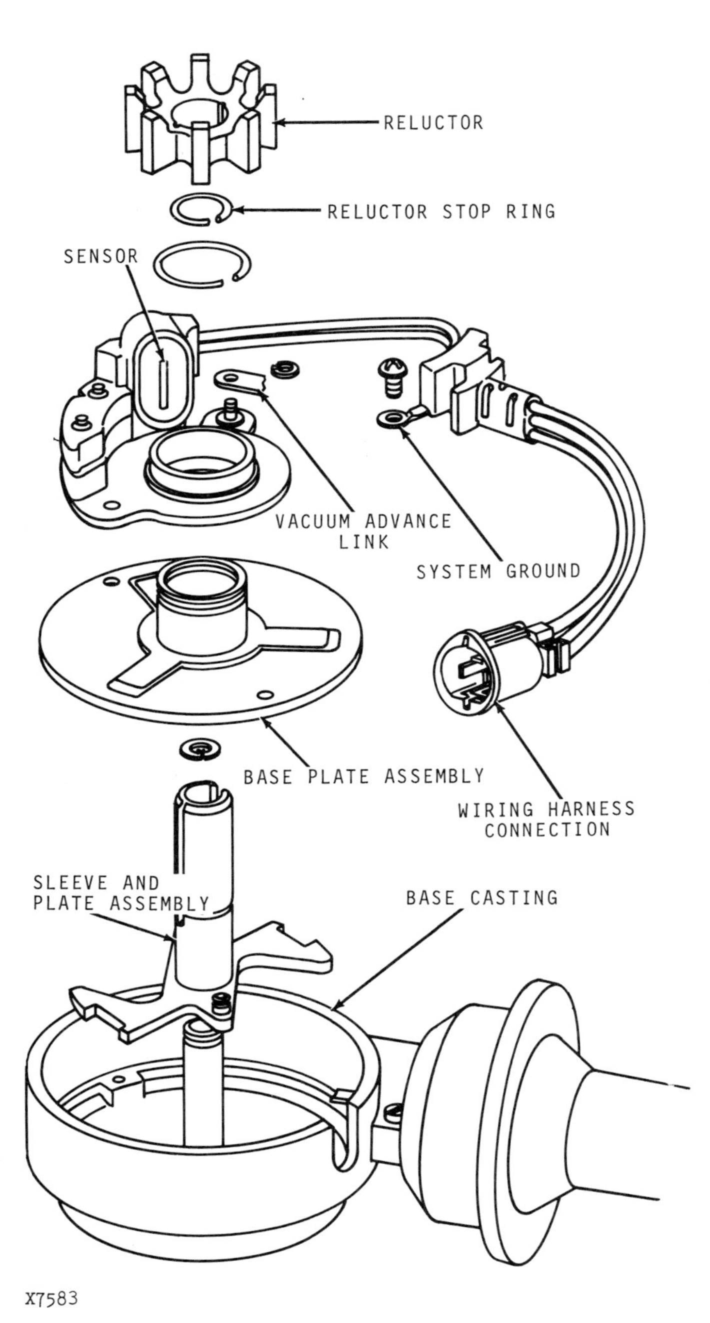

Fig. 19 — Exploded View of Distributor

2. Using a small gear puller, remove the reluctor by placing a spacer (thick flat washer or nut) between the turn-down of the gear puller and the distributor's center shaft (Fig. 20).

Important: Do not press directly on the shaft. Make sure the jaws of the puller grip the inner shoulder of the reluctor.

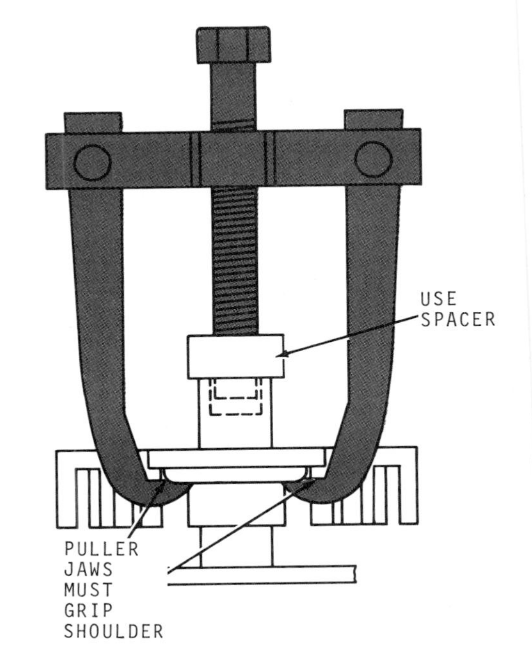

Fig. 20 — Removing Reluctor

3. Remove the sensor locking screw that holds the sensor (Fig. 21).

4. Lift the sensor lead grommet from the distributor, and pull the sensor leads out of the slot around the sensor spring pivot pin. Lift and release the sensor spring, and slide the sensor off the vacuum chamber bracket.

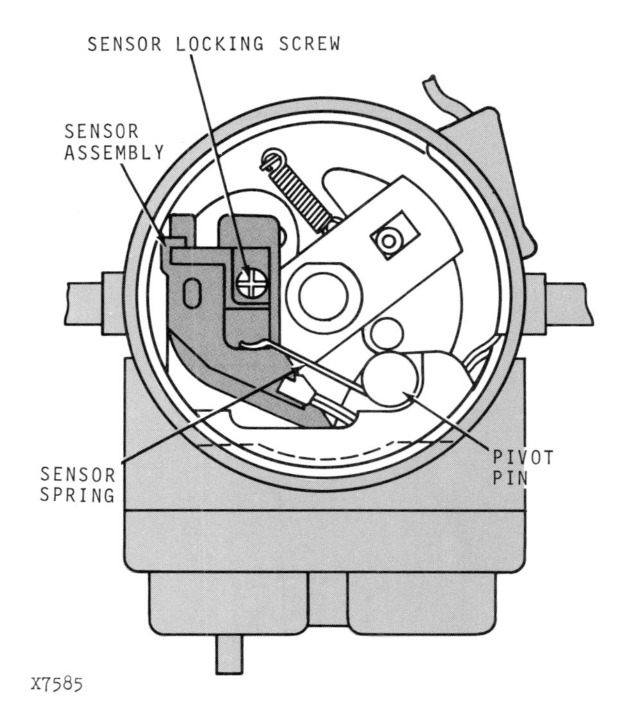

Fig. 21 — Sensor Locking Screw

To install a new sensor:

1. Place the sensor on the vacuum chamber bracket.

2. Put the sensor spring on the sensor.

3. Route the sensor leads around the spring pivot pin.

4. Install the sensor lead grommet in the distributor, and position the leads so they won't be caught by the reluctor.

5. Move the sensor sideways against the flat of the center shaft yoke until you can seat the sensor gauge. The sensor gauge is included in the replacement sensor kit. Tighten the sensor locking screw.

6. The position of the sensor in some systems is set using a sensor gauge. If the sensor is positioned correctly, the sensor gauge may be removed and installed without having to move it sideways.

7. With the sensor in correct position, place the reluctor on the yoke of the distributor shaft, so the sensor core is in the center of the reluctor legs (Fig. 22).

Important: The reluctor legs should not touch the sensor core. If so, the legs may break or wear and result in hard starting.

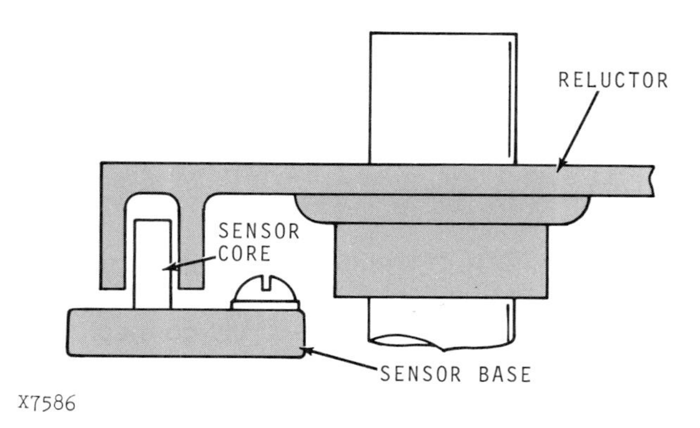

Fig. 22—Correct Position of Reluctor

8. Press the reluctor down on the shaft. Measure clearance below the reluctor as specified by the manufacturer.

9. Apply two drops of a light motor oil to the felt pad on top of the distributor shaft yoke. Install dust shield, rotor, and distributor cap.

10. Set ignition timing to specification.

In the ignition system with the external dual ballast resistor, it is necessary to remove the upper and lower plates to remove the sensor. In these disbributors:

1. Remove the distributor cap and rotor.

2. Remove the screws and lockwashers holding the vacuum control to the distributor housing. Disconnect the vacuum control arm and remove the vacuum control.

3. Remove the reluctor keeper pin. Place two screwdrivers under the reluctor on opposite sides of the distributor shaft. Pry evenly with the screwdrivers until the reluctor comes off the shaft (Fig. 23). Be careful to not damage reluctor teeth.

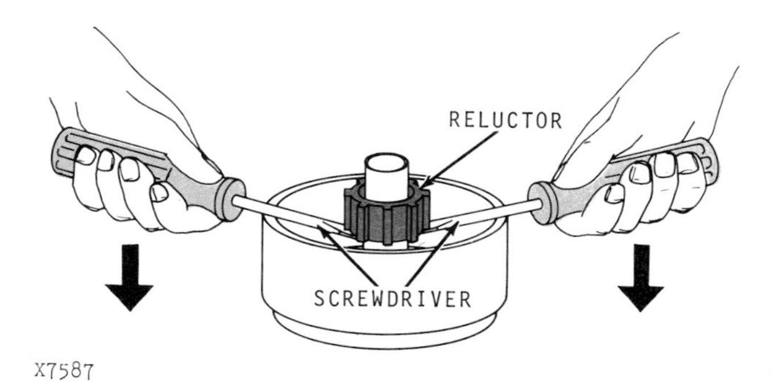

Fig. 23 — Removing Reluctor With Screwdrivers

4. Remove the screws and lockwashers holding the lower plate to the distributor housing. Lift off the lower plate, upper plate, and sensor assembly.

5. Install a new sensor assembly with the upper and lower plates.

6. Install the reluctor and keeper pin. Press the reluctor down firmly.

7. Lubricate the felt pad on top of the reluctor sleeve with a drop of light engine oil.

8. Set the air gap between the sensor and reluctor by loosening the sensor locking screw. Use a nonmagnetic gauge to adjust the sensor until the specified gap is obtained. Tighten the screw.

SELF-INTEGRATED ELECTRONIC IGNITION

The self-integrated ignition system is unique because all components (including the ignition coil) are inside the distributor (Fig. 24). The distributor is larger than a conventional unit and contains terminal studs on the distributor cap where spark plug cables are connected.

The ignition coil of self-integrated system is smaller than a conventional ignition coil. It has more primary and secondary windings than a conventional coil and is constructed like a true transformer; that is, the windings are surrounded by a laminated iron core. In a conventional ignition coil, the iron core is surrounded by the windings.

Other parts of this distributor are a rotor, a magnetic pickup assembly (containing a permanent magnet, a

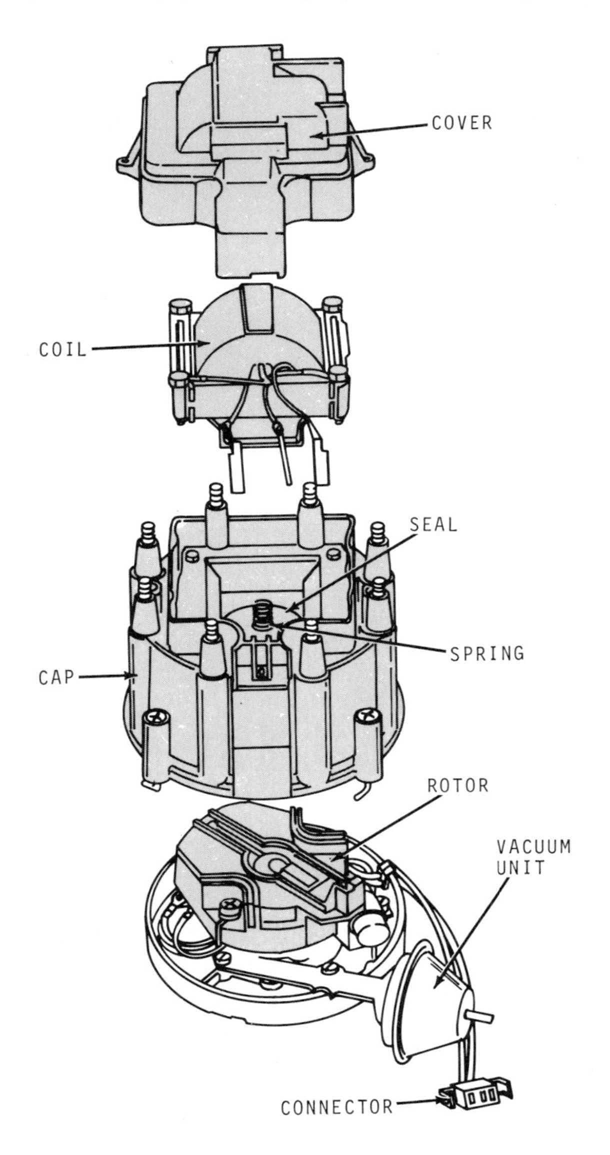

Fig. 24 — Self-integrated Distributor

pole piece with internal teeth, and a pickup coil), and an electronic module.

There is also a capacitor inside the distributor that is used for radio-noise suppression. It has no bearing on ignition.

HOW THE SELF-INTEGRATED SYSTEM WORKS

The teeth of the timer core of the movable magnetic pickup assembly on the distributor shaft rotate inside the pole piece and line up with the teeth of the pole piece (Fig. 25). This induces a voltage in the pickup

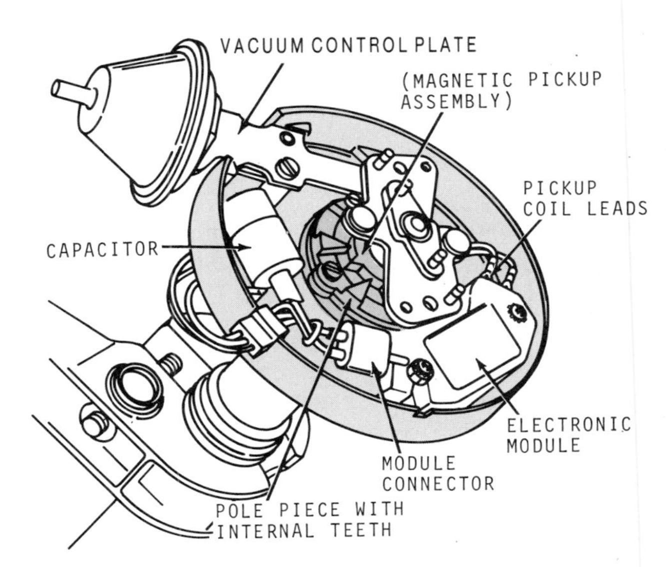

Fig. 25 — Distributor Components

coil. The voltage signals the electronic module to open the ignition coil primary circuit. This interruption of primary current causes high voltage to be induced in the ignition coil secondary winding. High voltage is directed through the rotor and spark plug cables to the spark plugs.

TROUBLESHOOTING THE SELF-INTEGRATED SYSTEM

Test this system using a voltmeter and an ohmmeter. Other test equipment may be recommended by the manufacturer.

IMPORTANT: Before proceeding, make sure the connector on the side of the distributor is secure and all spark plug cables are securely connected.

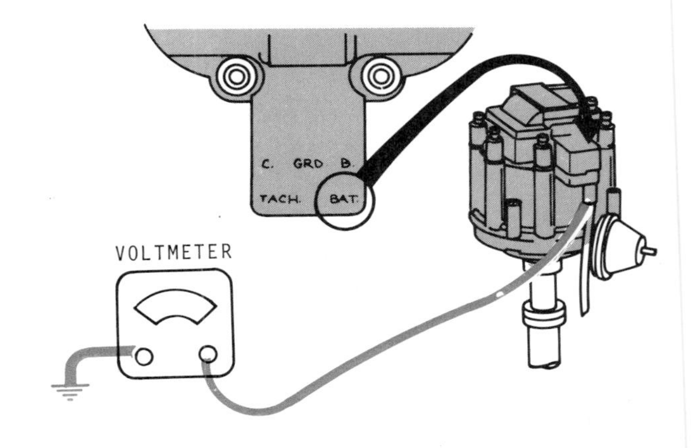

Fig. 26 — Voltmeter Connected to BAT Terminal

 Litho in U.S.A.

- If the engine will not start:

1. Connect a voltmeter between ground and the BAT (battery) terminal lead of the distributor (Fig. 26). Turn ON the ignition switch.

2. If the voltage reading is zero, look for an open circuit between the BAT (battery) terminal and the battery.

3. If the voltmeter records battery voltage, using insulated pliers remove a spark plug cable and hold it one-quarter inch from ground. Use insulated pliers. Crank the engine. If spark occurs, look for worn spark plugs or a fuel system problem. If there is no spark, test distributor components (see below).

- If the engine runs rough:

1. Check to see that fuel is reaching the carburetor.

2. Check vacuum hoses and connections for leaks.

3. With the engine running, check all spark plug cables by looking and listening. If a cable is weak, sparks will be heard or can be seen jumping to ground.

4. Check the ignition timing and centrifugal advance settings with a stroboscopic timing light.

5. Check all spark plugs for correct gap and fouling.

If the problem isn't solved, test distributor components.

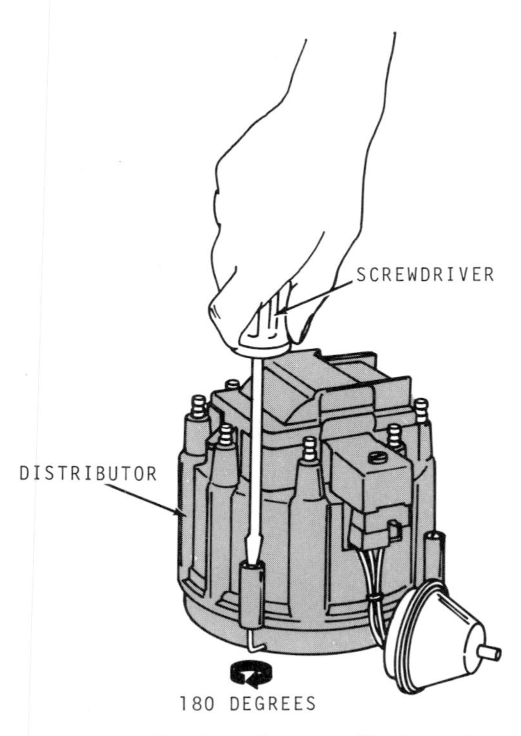

Fig. 27 — Removing Distributor Cap

To test distributor components:

1. Depress each fastener holding the distributor cap and turn 180 degrees to remove the cap (Fig. 27).

2. Replace any distributor cap, ignition coil, or rotor when it is burned or corroded.

3. Connect an ohmmeter between terminals 1 and 2 in the distributor cap (Fig. 28). The meter should show little or no resistance. If it does not, replace the ignition coil (see next page).

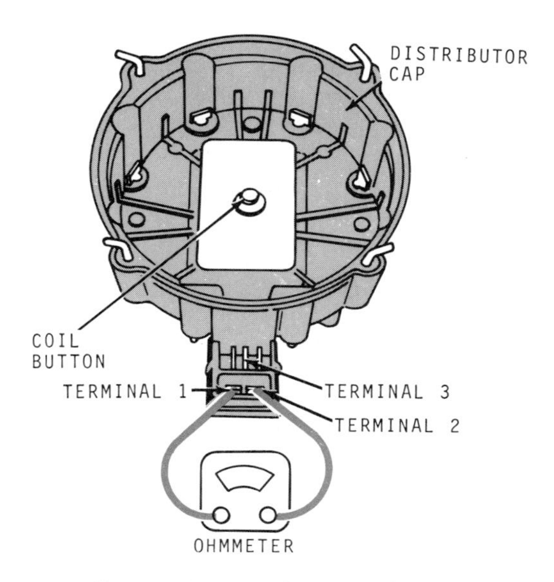

Fig. 28 — Inside of Distributor Cap

4. Connect an ohmmeter lead to the ignition coil button inside the distributor cap (Fig. 29). Connect the other lead first to the No. 2 terminal and, then, to the No. 3 terminal. Set the ohmmeter scale on its highest setting. If both readings show high resistance, replace the ignition coil.

5. Connect a vacuum source to the vacuum advance. The vacuum control plate should move as vacuum is applied. If not, replace the vacuum advance.

6. Disconnect the pickup coil leads (Fig. 30) from the electronic module and connect an ohmmeter between lead No. 1 and ground. Set the ohmmeter on the middle scale and operate the vacuum advance, using the vacuum source, to the full range of travel of the vacuum control plate. The ohmmeter should show very little resistance as this is done. If not, replace the pickup coil (see next page).

7. Connect the ohmmeter to pickup coil leads 1 and 2, and apply vacuum. If the meter reads less than 500 ohms or more than 1,500 ohms while the vacuum is applied, replace the pickup coil.

 Litho in U.S.A.

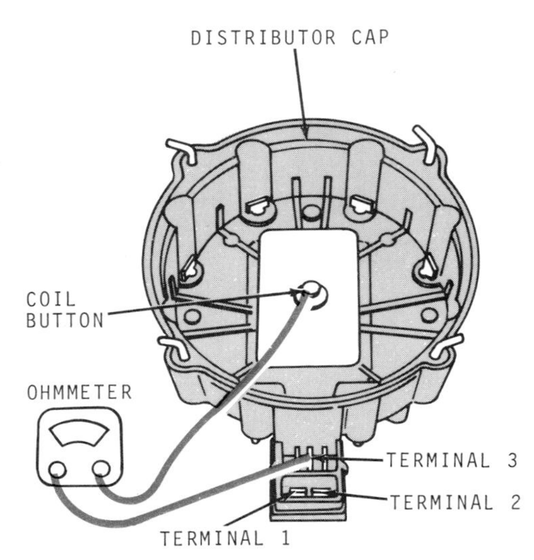

Fig. 29 — Testing Distributor With Ohmmeter

8. If tests to this point fail to reveal the cause of the problem, all that remains is to replace the electronic module. This may be done without removing the distributor from the engine by removing the two screws holding the module (Fig. 30).

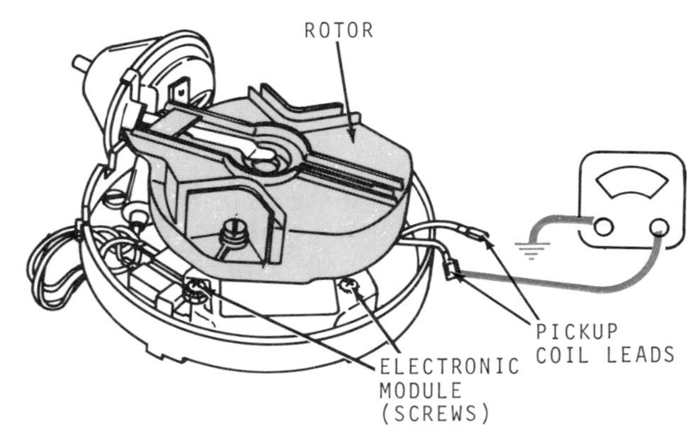

Fig 30 — Pickup Coil Leads and Module Screws

REPLACING THE IGNITION COIL

Remove the three screws holding the coil to the distributor cap (Fig. 31). Remove the ground wires from the coil, push the coil leads from the underside of the connectors, and remove the coil.

Important: A number of different ignition coils are made for this self-integrated system. They work the same way, but the correct one for the particular set-up must be used.

REPLACING THE PICKUP COIL

Remove the distributor and mount it in a workbench vise. Drive the pin from the distributor shaft gear (Fig. 32). Remove the rotor shaft assembly from the

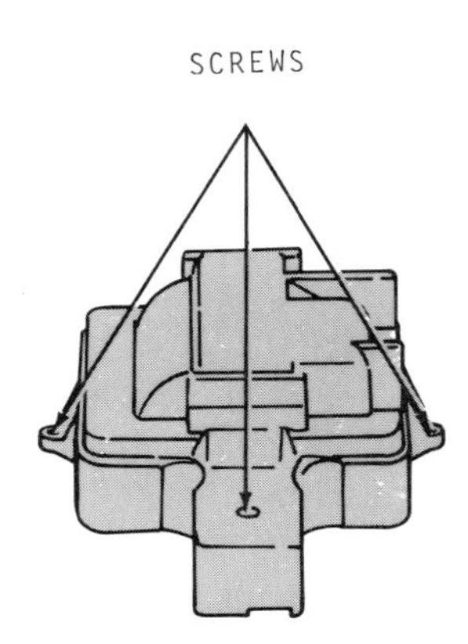

Fig. 31 — Screws Holding Coil to Distributor Cap

housing and take off the thin "C" clip. This releases the pickup coil.

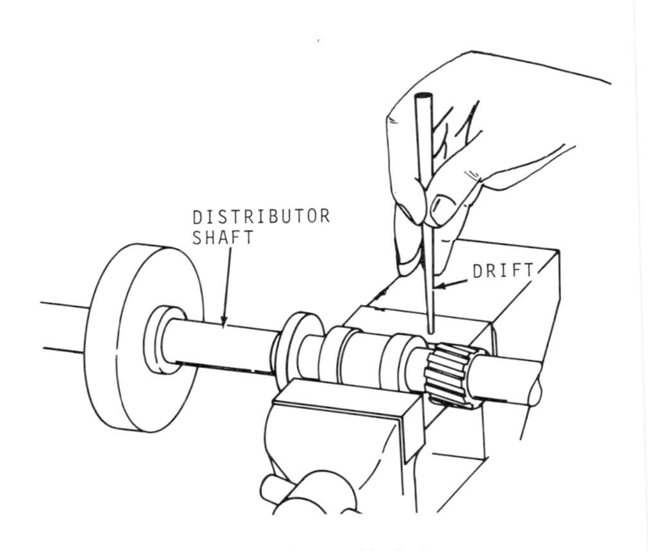

Fig. 32—Distributor Shaft Gear Pin

Install a new pickup coil. Replace the "C" clip and rotor shaft assembly. Place the keeper pin back in the distributor shaft gear and reinstall the distributor.

SUMMARY: ELECTRONIC IGNITION

Electronic ignition systems operate similar to conventional ignition systems except that the spark is electronically timed. Electronic ignition systems do not have points or a condenser so they require little maintenance. They also produce a hotter spark of longer duration for:

- *quicker starts under adverse weather conditions*
- *more complete combustion*
- *and longer spark plug life*

ELECTRONIC FUEL INJECTION

INTRODUCTION

Conventional fuel injection systems include an in-line multiple plunger injection pump equipped with a mechanical fly weight governor and aneroid control. An engine-driven gear on the pump cam-shaft drives the pump at one-half engine speed.

Some of these pumps are distributor-type pumps. In these pumps, a fuel metering valve is controlled by the governor.

A governor-operated control rack is connected to the control sleeves and plungers to regulate the quantity of fuel delivered to the engine.

The hydraulically-actuated aneroid controls rack track travel and, therefore, fuel delivery. The aneroid control is in effect until the manifold pressure is high enough to overcome the aneroid diaphragm spring pressure.

Newer tractors utilize electronic fuel injection systems (Fig. 33). These systems usually do not use an aneroid. They rely on electronic governing to control fuel quantity. These systems consist of a controller, auxiliary speed sensor, a transient voltage protection module (TVP) and an injection pump/actuator assem-bly.

The injection pump/actuator assembly includes the injection pump, actuator solenoid, rack position sensor, primary speed sensor, fuel shutoff solenoid and fuel temperature sensor.

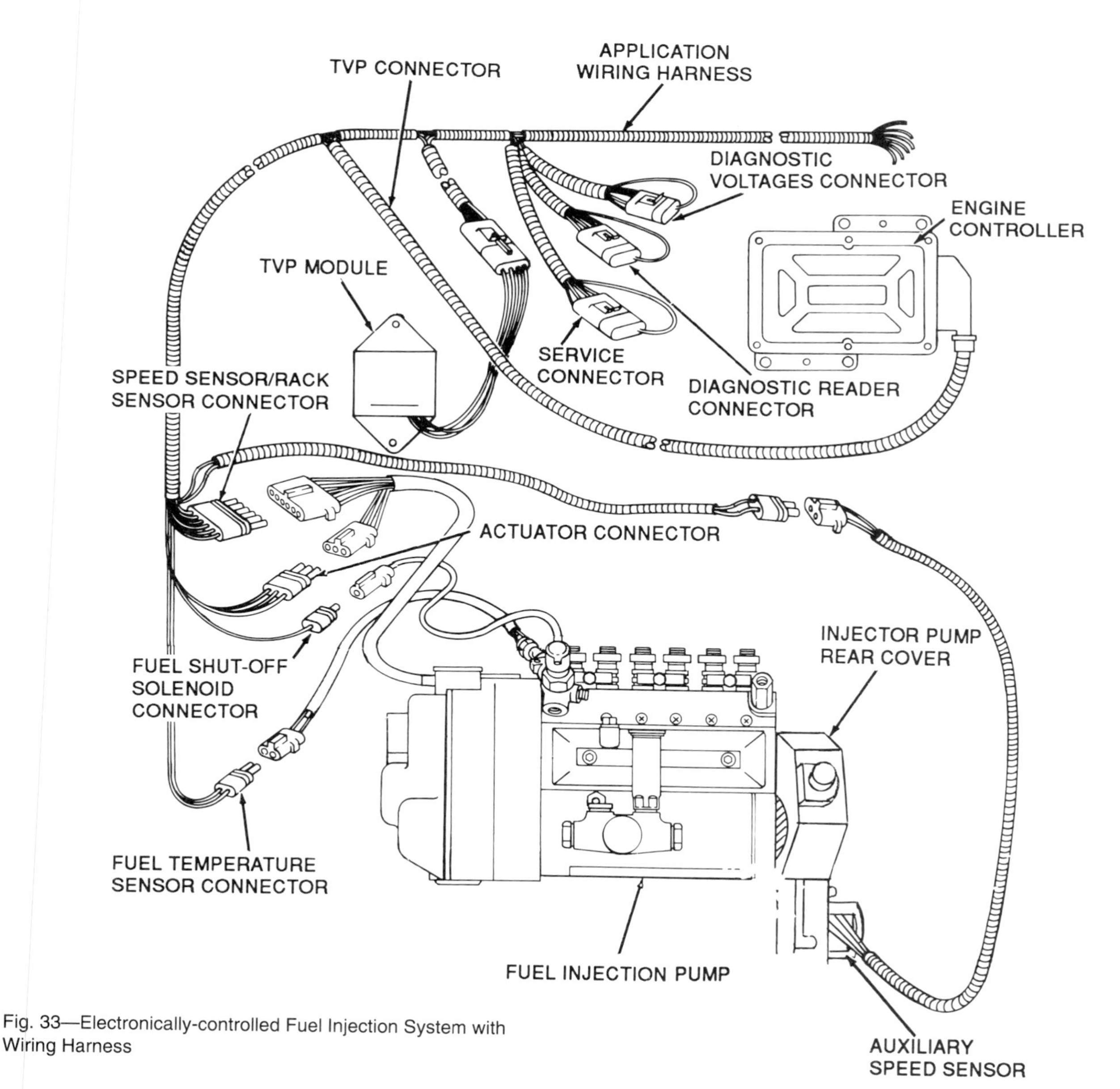

Fig. 33—Electronically-controlled Fuel Injection System with Wiring Harness

Litho in U.S.A.

HOW ELECTRONIC FUEL INJECTION WORKS

In order to understand how electronic fuel injection works, it is first necessary to comprehend **frequency, cycles,** and **waveforms.**

Alternating current and voltage reverse at regular intervals. This forms a curve which can be plotted or graphically represented. This curve is known as a waveform. When current or voltage reaches the top of this curve, it reverses and forms a curve in the opposite direction (Fig. 34). When it again reaches its original starting point, it has completed a cycle.

The frequency of an alternating current or voltage is the number of these cycles completed every second. Electricity travels at such speed that it completes many cycles per second. Therefore, frequency is often measured in milliseconds (one thousandths of a second).

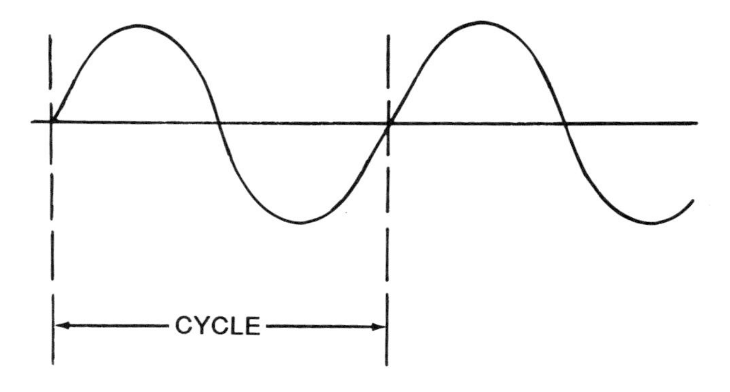

Fig. 34—Frequency Cycles Create a Waveform

In summary, what is actually plotted in a waveform are the instantaneous values of current or voltage at any time. Waveforms represent both the position and the time elapsed while electricity rotates through these positions.

Now let's discuss each of the system components shown in Fig. 33.

ENGINE CONTROL UNIT (ECU)

The engine controller (Fig. 35) is a self-contained module with mounting brackets which contains electronic circuitry and a computer program which performs governor and diagnostic functions. The controller is remotely located from the engine in a protected environment and connects to the engine application through a wiring harness.

The engine controller controls the fuel delivery as a function of engine speed, throttle command and computer programmed delivery values. The controller also

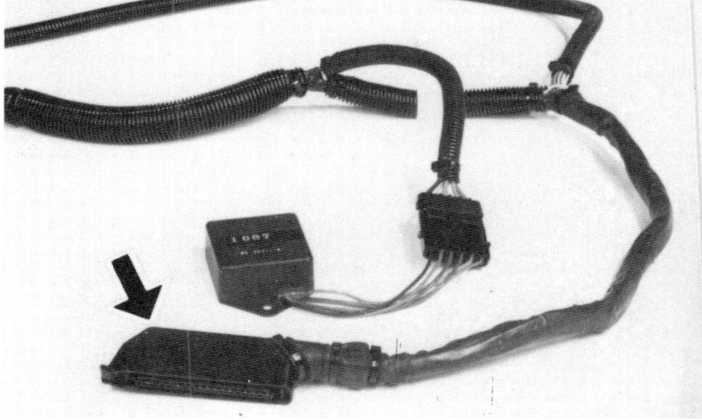

Fig. 35—Engine Controller with Wiring Harness and Connector

controls the fuel limiting for torque curves and the governing speed control. The controller can control fuel delivery to limit smoke without using additional sensors. For more stringent requirements, manifold air density can be monitored by the controller to limit smoke.

INJECTION PUMP

The electronically-controlled in-line injection system uses the same basic hydraulic pumping mechanism used in mechanically governed in-line pumps.

The mechanical governor mechanism is replaced with the actuator assembly which includes the actuator solenoid to move the control rack, rack position sensor, primary speed sensor, and a toothed speed wheel (Fig. 36).

The throttle lever mechanism used on the mechanical pumps is removed and its function is implemented by a throttle position sensor input to the engine control unit (ECU). Fuel is transferred from fuel tanks with a fuel supply pump. The injection pump fuel inlet connection is located at the rear of the pump on the fuel inlet assembly, which includes the fuel shut-off solenoid and the fuel temperature sensor.

 Litho in U.S.A.

Fig. 36—The Fuel Injection Pump and Actuator Assembly

ACTUATOR SOLENOID

The control rack is spring loaded to the fuel shutoff position "zero rack." As increasing current is supplied to the actuator solenoid from the controller, the rack is driven toward full rack position. The engine controller has the capability of controlling the current to the solenoid in order to position the solenoid and control rack anywhere between zero rack and full rack.

On certain engines it is necessary to check this solenoid if the engine will not start and all other systems are operating correctly. To test the solenoid, close the switch (Fig. 37) and observe the ammeter. Current draw should be about 2.5 amps at 12 volts. High current draw indicates shorted solenoid windings. Low or no current draw indicates a high resistance connection, either internal or external, or open windings in the solenoid.

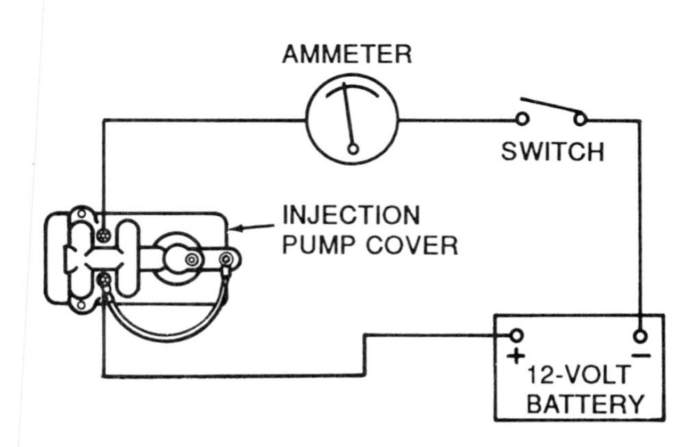

Fig. 37—Testing the Injection Pump Solenoid

RACK POSITION SENSOR

The rack position sensor within the actuator housing supplies rack position information to the controller so that a specific rack position can be controlled.

The sensor includes an electronic module mounted in the actuator housing which provides a voltage to the controller indicating the position of the rack and is used to control rack position for all operating conditions. If this critical sensor were to fail, the controller would be forced to shut down the engine due to loss of control.

The rack position sensor can only be serviced by an authorized shop because of the recalibration which is required if the actuator housing is removed from the pump.

The rack puller can also be tested if the engine will not start and all other systems are operating properly. Check the Technical Manual for the test to use for the appropriate engine application.

PRIMARY SPEED SENSOR

The primary speed sensor is also located within the actuator housing. It is a magnetic pickup which generates voltage pulses to the controller as the teeth on the speed wheel pass by the tip of the sensor. If this sensor were to fail completely, the controller would use the signal coming from the auxiliary speed sensor to get engine speed information.

Because the primary speed sensor is located inside the actuator housing, it can only be serviced by an authorized repair shop.

FUEL SHUT-OFF SOLENOID

The fuel shut-off solenoid will shut off fuel to the injection pump when the key switch is turned off or when the engine controller detects a rack position error.

Testing the Fuel Shut-off Solenoid

The solenoid is a rotary-type electro-mechanical assembly controlled by the key switch (Fig. 38). An adjustable control rod connects the solenoid arm with the injection pump shut-off lever.

Turn the key switch on. The solenoid should energize and the pump shut-off lever will move to "run" position.

Turn the key switch off. The solenoid should de-energize and the return spring on the solenoid arm should move the shut-off lever to "stop" position.

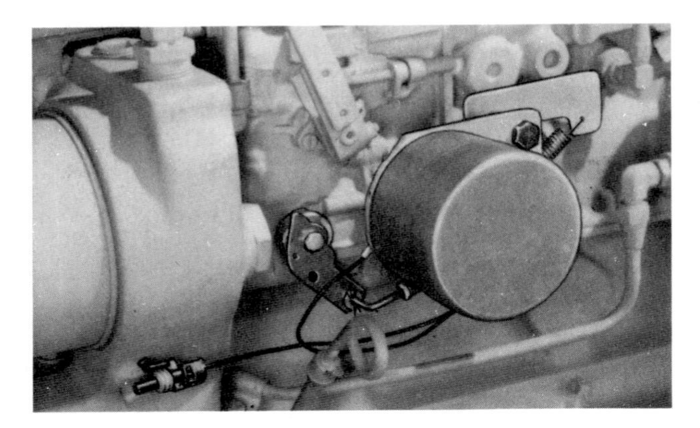

Fig. 38—Testing the Electric Shut-off Solenoid

An improperly adjusted fuel shut-off control rod may result in the engine not being stopped when the key switch is turned off.

Adjust the control rod only when the engine is stopped. The control rod can be lengthened or shortened as required to have the shut-off lever against, or nearly against, the stop on the injection pump.

MAGNETIC SAFETY SWITCH

A magnetic safety switch is installed on engines equipped with an instrument panel. This switch activates the injection pump shut-off solenoid or rack puller if engine oil pressure goes below specification or if coolant temperature rises above specification.

Testing the Magnetic Safety Switch

Disconnect the magnetic safety switch wiring. Tag all wiring for reassembly.

Make connections as shown in Fig. 39. Close the switch and observe ammeter reading. Switch coil should draw 1.6 to 1.8 amperes at 12 volts. If the switch coil is not within specifications, the magnetic switch must be replaced.

FUEL TEMPERATURE SENSOR

The fuel temperature sensor (Fig. 40) is located at the fuel inlet of the pump. It is used to determine the optimum fuel delivery for starting and, depending on the application, is used to maintain constant power over a predetermined temperature range.

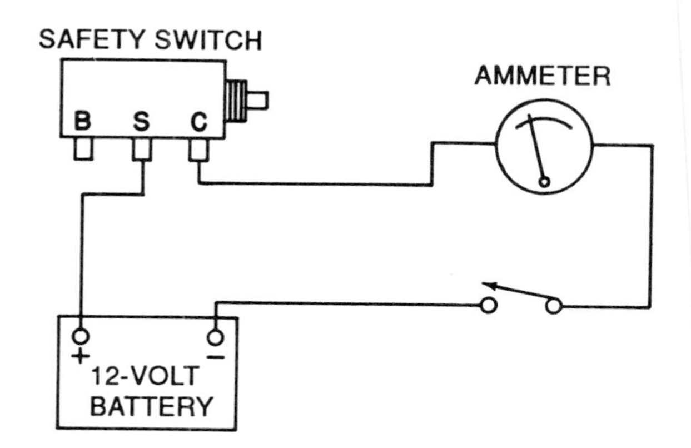

Fig. 39—Testing the Magnetic Safety Switch

If this sensor were to fail, a low temperature would be assumed by the engine controller. In warm weather this might result in a slight drop in maximum torque and smokier starts. The fuel temperature sensor has its own 2-pin connector and is a serviceable part. No recalibration is required when this sensor is replaced.

Fig. 40—Fuel Temperature Sensor

AUXILIARY SPEED SENSOR

The engine speed sensor (magnetic pickup) (Fig. 41) serves as a back-up speed sensor in the event of complete failure of the primary speed sensor.

 Litho in U.S.A.

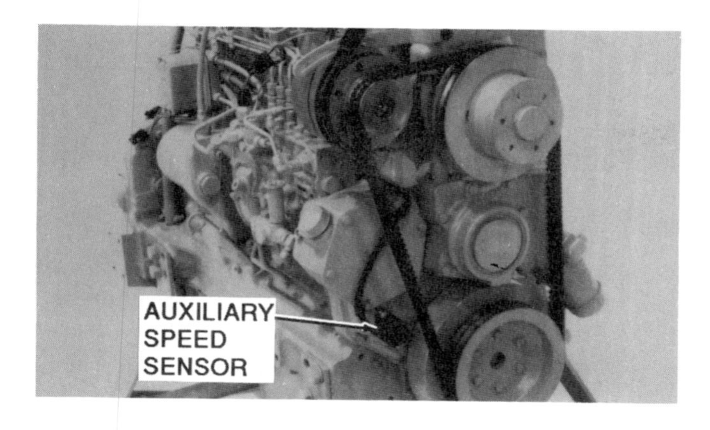

Fig. 41—Auxiliary Speed Sensor

TRANSIENT VOLTAGE PROTECTION MODULE

The main function of the transient voltage protection (TVP) module is to limit high energy voltage transients (from the charging system) to a maximum of 40 volts to protect the electronic circuitry in the engine controller (Fig. 42).

Fig. 42—Transient Voltage Protection Module

HOW THE ELECTRONICALLY-CONTROLLED FUEL INJECTION SYSTEM WORKS

1. When the key switch is turned to the "ON" position, the engine controller receives power and the fuel shut-off solenoid is powered (opening the valve).

2. When the key switch is turned to the "START" position, the controller powers the actuator solenoid moving the rack to starting fuel positions based on fuel temperature and engine speed.

This starting mode is triggered either by a controller input which senses the key switch "START" position, or by an engine speed greater than 60 rpm. Starting fuel quantity is not affected by throttle position.

3. Once the engine has started, fuel delivery is controlled by the engine controller based on various inputs (primarily throttle and engine speed).

4. The engine controller controls rack position by adjusting the current level to the actuator solenoid until the rack position signal from the injection pump matches the commanded signal.

When no fuel is needed, the controller turns off current to the actuator solenoid. If a problem occurs where the controller cannot control rack position, the fuel shut-off solenoid is turned off in addition to the actuator solenoid being turned off.

Starting Control

The engine controller uses engine speed and initial fuel temperature to control rack position during starting. This permits use of excess fuel and retarding for cold temperatures but less fuel and no retard for hot starts. Thus, cold starting is improved and black smoke can be greatly reduced on hot starts.

The throttle input is ignored by the controller until the starting routing is completed.

Governor Modes

The governor mode is preset at the factory. Based on application the engine controller can provide either all-speed governing or minimum-maximum governing. When using all-speed governing, the controller regulates the engine speed based on the throttle command and selected speed regulation (droop).

When using the min-max governor, the controller provides the same minimum (slow idle) and maximum (fast idle) speed governing as with the all-speed governor. However, in between the minimum and maximum speeds, the throttle and engine speed inputs are used by the controller to select a fuel quantity. Thus, the throttle commands fuel quantity rather than engine speed in the min-max governor mode.

The percent of droop can be programmed at the factory to provide three switch selectable combinations of droop, rated speed, and fast idle, including zero percent (isochronous).

Fig. 43 shows these combinations of engine speed: LI is low (slow) idle, Nr is normal rated and FI is fast idle.

The engine application determines which droops will be programmed into the controller and whether or not they will be switch selectable. If no input signal is present, the normal droop/fast idle combination is selected. The engine controller provides isochronous governing at the slow idle speed regardless of the speed regulation selected for governing over the rest of the operating range.

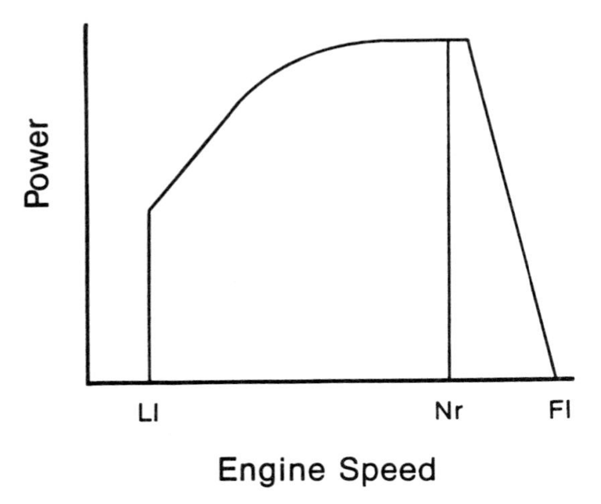

Fig. 43—Combinations of Engine Speed

The slow idle, fast idle, and breakaway speeds are programmed at the factory and are values which are precisely repeatable, eliminating system-to-system variations.

During normal operation which is the governor mode, the diagnostic codes output signal transmits a data signal like the one in Fig. 44.

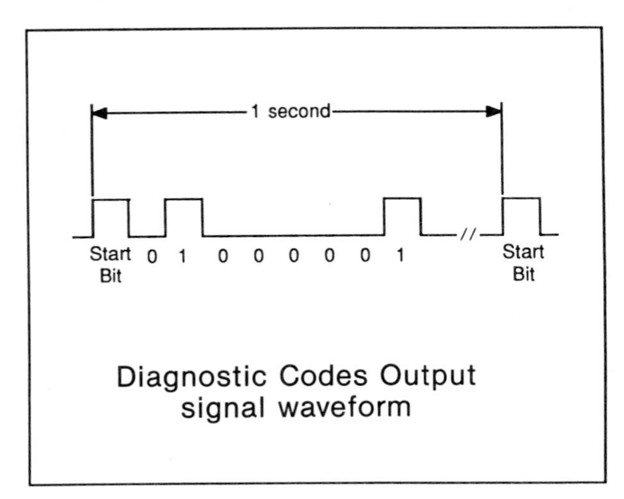

Fig. 44—Diagnostic Codes Data Signal

The signal consists of a stream of high and low pulses which represent code numbers from 0 to 255, only some of which are valid for each application. Each code represents a specific fault condition which could be caused by more than one problem.

Maximum Fuel Quantity Control

The engine controller limits the maximum fuel delivery as a function of engine speed. This function is set at the factory to obtain the proper fuel limit curve (power curve) shape, which is highly repeatable.

An optional feature allows the power curve to be switched to any one of three factory programmed power curves (Fig. 45). Switching between power curves can be done while the engine is running. The three curves may be identical, thus disabling the input. The normal power curve is selected if no input signal is present.

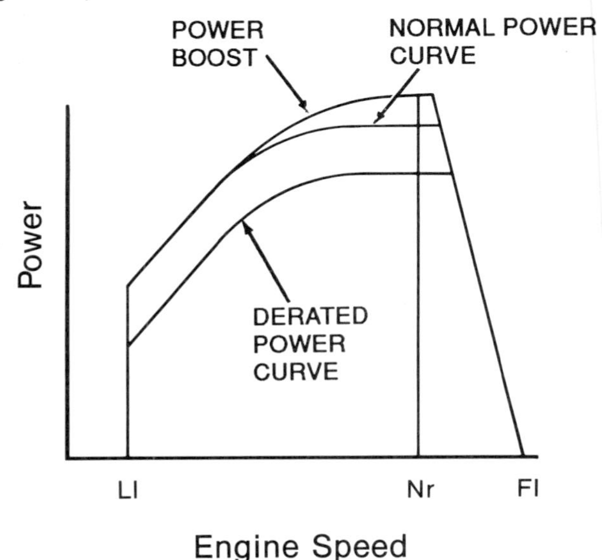

Fig. 45—Fuel Delivery as a Function of Engine Speed

A derated power curve maybe selectable by shorting the input signal to ground. A temperature switch is one method used to select the derated power curve.

The third curve is typically used for a "power boost" mode since operation using this curve can be limited by timers within the engine controller. Power boost mode maybe selected by connecting the input to ground through a 2000 ohm resistor if the engine is designed for it.

If the power boost timers are enabled, the controller will limit power boost operation to the amount of time in the "ON" time. If the "ON" time is reached, the controller will automatically switch back to the normal power curve. The "OFF" time must then be reached before the controller will allow selection of the power boost mode again. The switch must be switched off and back on to reselect the power boost mode.

Use of the optional power curves is dependent on engine application.

Smoke Control

Smoke control is based on mathematical equations which use known engine characteristics. The engine controller implements smoke control based on the instantaneous values and the rates of change of throttle, load and speed. This "math model" smoke control is used for most applications since it provides good smoke control without requiring intake manifold sensors.

Fuel Temperature Compensation

The engine controller monitors the injection pump inlet fuel temperature with the fuel temperature sensor located at the fuel shut-off valve manifold. The controller can provide nearly constant fuel delivery (horsepower) by compensating for changes in fuel density over any desired temperature range. The fuel temperature compensation characteristic is dependent on engine application and is programmed at the factory.

Self-diagnosis and Back-up Features

The engine controller self-diagnoses as many system faults as is practical. These include determining if any of the sensor input voltages are too high or too low, if the engine speed signals are valid, and if the control rack is responding properly.

The controller also monitors its own operation for problems. In most cases, the controller will output a diagnostic code to indicate the specific problem which has been detected (Fig. 46). The controller can also flash a fault lamp.

Fig. 46—Tachometer Display of Diagnostic Codes

The controller will automatically switch to another mode of operation as a backup whenever possible, or will shut down the engine if control cannot be guaranteed. In some cases, little or no degradation in performance is noticed. For example, the engine will continue to operate normally using the auxiliary speed sensor in the event of a primary speed sensor failure.

If the operator is able to keep the engine running during a fault condition, no damage to the engine should result. The problem should be fixed at the earliest convenience.

Fuel Flow/Throttle Output Signal

The engine controller sends a pulse-width-modulated signal which indicates the percentage of full load/rated speed fuel delivery and percentage of full throttle. This signal is primarily intended for use by a transmission controller, but is also useful for monitoring performance if the engine application has compatible electronics.

The illustration in Fig. 47 shows the waveform of the signal. "Pulse-width-modulated" means that the frequency of the signal remains the same, and the width of the pulse is changed to indicate a change in value. The width of the synchronizing pulse remains constant and indicates the beginning of the signal. This signal is repeated every 20 milliseconds.

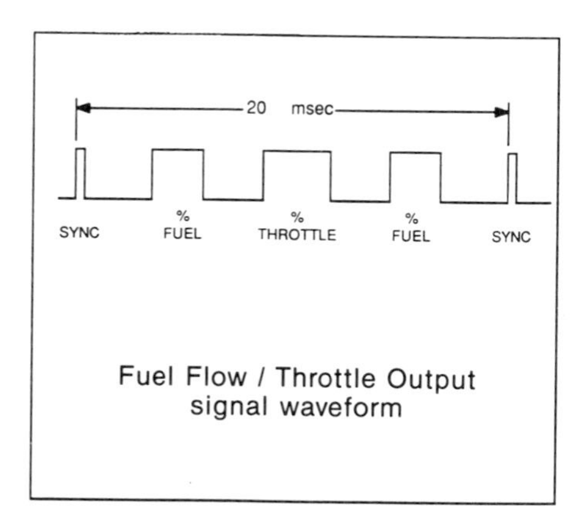

Fig. 47—Pulse-width-modulated Signal Waveform

An electronic governor tester is capable of reading this signal and displaying "percent fuel" and "percent throttle." It is also capable of reading and displaying diagnostic codes (Fig. 48).

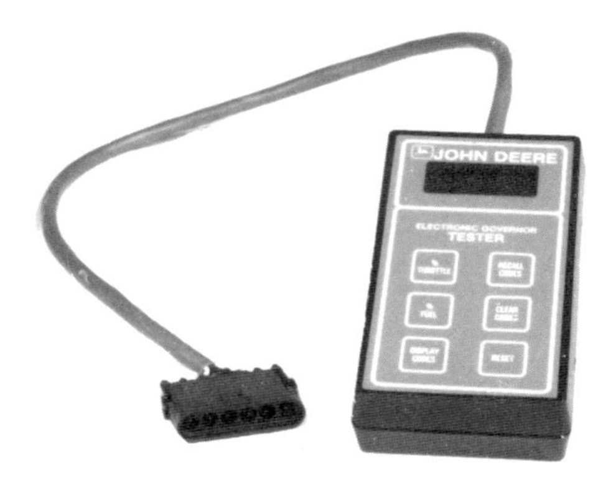

Fig. 48—An Electronically-Controlled Fuel Injection Diagnostic Reader

Auxiliary Speed Output Signal

The engine controller receives the auxiliary speed input from the sensor and then transmits the auxiliary speed output signal for use by other electronic modules such as the tachometer (Fig. 46) and other controllers which are discussed in Chapter 10. The engine controller and only the engine controller receives the signal from the auxiliary speed sensor at the front of the engine.

The illustration in Fig. 49 shows the waveform of the auxiliary speed output signal. The signal is the same frequency as the auxiliary speed sensor. This may be 10 or 23 pulses per engine revolution, depending on the application.

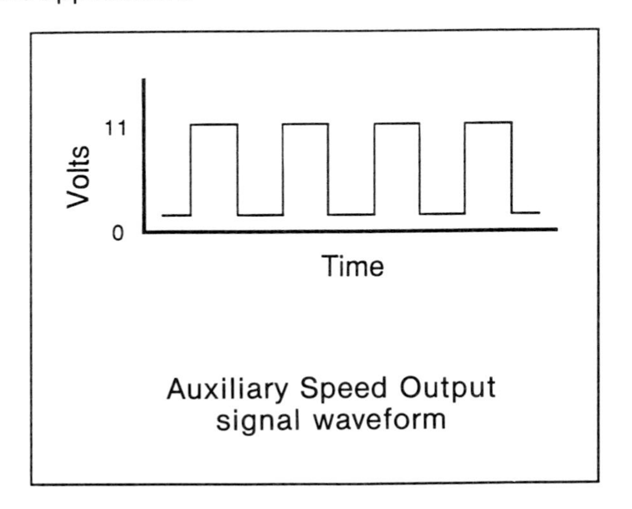

Fig. 49—Auxiliary Speed Output Data Signal

Throttle Options

There are three throttle options available for use with the electronically-controlled fuel injection system.

The selection of a throttle option is dependent on the application and is programmed into the engine controller at the factory. These options are:

1. An analog throttle (continuously variable voltage input) which is normally implemented using a potentiometer.

2. A 3-state throttle which uses a simple switching arrangement to select one of three fixed engine speeds.

3. A pulse-width-modulated (PWM) throttle which can be used alone or with an analog throttle.

Analog Throttle

The analog throttle is commonly used with either all-speed or min-max governing. The engine controller converts the potentiometer (Fig. 50) voltage signal into a percent of full throttle command and controls engine speed or fuel quantity accordingly.

The resistance between the outside terminals of the potentiometer can be measured using a digital ohmmeter.

Fig. 50—Speed Control Potentiometer

3-State Throttle

The 3-state throttle is typically used with all-speed governing and when a maximum of three fixed speeds are desired. Typical applications are generator sets where one or two fixed speeds are desired, and combines which use three fixed speeds.

Pulse-width-modulated Throttle

The pulse-width-modulated (PWM) throttle is a signal received from another electronic module (usually a transmission controller) which uses a pulse width to indicate the desired percent of full throttle.

The PWM throttle input has priority over the analog throttle. This means that if the engine controller starts receiving a PWM throttle signal, the controller will stop using the analog throttle and will start using the PWM throttle to determine the throttle command. If the PWM throttle signal is turned off or is disconnected, the engine controller will start using the analog throttle again.

The illustration in Fig. 51 shows the PWM throttle signal waveform. "Pulse-width-modulated" means that the frequency of the signal remains the same, and the width of the pulse is changed to indicate a change in value. This signal is repeated every 10 milli-seconds.

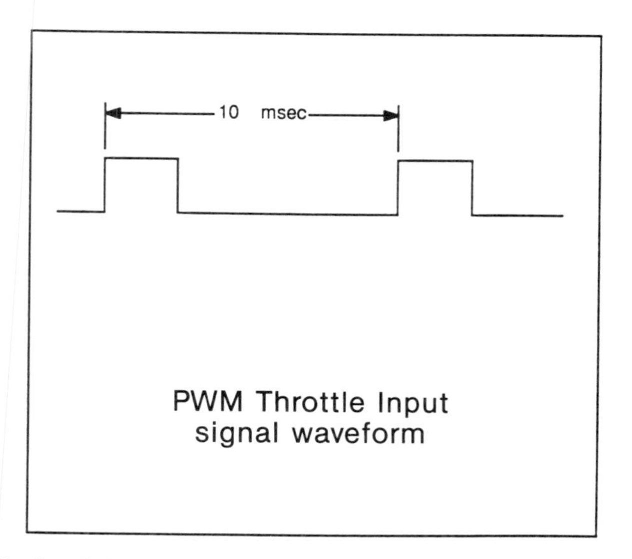

Fig. 51—Pulse-width-modulated Throttle Data Signal

SUMMARY: ELECTRONIC FUEL INJECTION

The electronic fuel injection system consists of an engine controller, injection pump/actuator assembly, auxiliary speed sensor and transient voltage protection module.

The central component of the electronic fuel injection system is the engine controller which contains electronic circuitry and a computer program which performs governor and diagnostic functions. It controls fuel delivery as a function of engine speed and throttle command. It also controls the fuel limiting for torque curves and the governing speed control. Aneroids are eliminated and for most applications the controller can control fuei delivery to limit smoke without using additional sensors.

The electronically-controlled in-line injection system uses the same basic hydraulic pumping mechanism used in mechanically-governed in-line pumps except that the mechanical governor mechanism is replaced with the actuator assembly which includes an actuator solenoid to move the control rack, a rack position sensor, a primary speed sensor and a toothed speed wheel.

The auxiliary speed sensor serves as a back-up speed sensor in the event of complete failure of the primary speed sensor. The transient voltage protection module limits high energy voltage transients from the charging system to protect the electronic circuitry in the engine controller.

The advantages of the electronic fuel injection system include:

- *decreased smoke*
- *maximum fuel quantity control*
- *fuel temperature compensation which delivers nearly constant fuel delivery (HP)*
- *self-diagnostic and back-up features*

TEST YOURSELF

QUESTIONS

1. What three of the following parts are found in electronic ignition systems and not in conventional ignition systems?
   a. ignition coil
   b. sensor
   c. distributor
   d. spark plug wires
   e. reluctor
   f. electronic control unit (or module)

2. What are three advantages of electronic ignition?

3. True or false? The reluctor and sensor may be repaired if they malfunction.

4. Why must the reluctor and sensor not contact each other?

5. How is a self-integrated electronic ignition system different from other types of electronic ignition systems?

6. The number of cycles electricity completes in one second is known as ____________________.

7. Engine control units rely on ____________________ governing to regulate fuel.

8. When the controller regulates the engine speed based on the throttle command and selected speed regulation, this is called ____________________ governing.

9. Electronic fuel injection systems use a "math model" to control ____________________.

(Answers on page 20 at the end of this book.)

LIGHTING AND ACCESSORY CIRCUITS / CHAPTER 10

INTRODUCTION

This chapter covers all the extra equipment which completes the electrical system on modern machines.

We have covered the basic system in the preceding chapters on batteries and charging, starting, and ignition circuits.

Now let's fill in the remainder of the system with the lighting and accessory circuits.

This chapter will cover the following acessories in the order shown:

- **Lighting Circuits (including Lamps, Circuit Breakers, and Fuses)**
- **Wiring Harnesses**
- **Electromagnetic Clutches**
- **Gauges**
- **Meters**
- **Horns and Buzzers**
- **Electric Motors**
- **Convenience Outlets**
- **Cigarette Lighters**
- **Flame Rods**
- **Glow Plugs (for Diesel Engines)**

We will describe each accessory and give a brief story on testing and maintaining it in its circuit.

LIGHTING CIRCUITS

Lights are used on most all farm and industrial machines. They are required by the local government for night driving or towing in most areas. To avoid trouble, know your local regulations.

On modern tractors, the lights are often two or four headlights and one taillight.

Fig. 1—Lighting on a Modern Tractor

In addition, a flashing warning lamp is required in some states. This lamp warns other traffic that a slow-moving vehicle is on the road.

Most motorists on public roads are not farmers. In fact, the percentage of people who have farming backgrounds or those who even know farmers is becoming smaller each year. So, only a small percentage of the motoring public is likely to give much thought to the unique nature of a farm machine when they see it on the highway.

A motorist topping a hill at 55 mph and seeing a tractor 400 feet ahead of him travelling at 15 mph will have a closing speed of 40 mph. The motorist has less than seven seconds to recognize the speed of the tractor, react, and slow down.

Situations similar to the one above lead to thousands of slow-moving vehicle accidents each year. The motorist aren't able to stop in time. Many accidents are fatal.

Every vehicle intended to travel 25 mph or less is considered a slow-moving vehicle and should be identified with a Slow Moving Vehicle (SMV) Emblem visible from the rear (Fig. 2).

The triangular SMV emblem is the universal symbol to tell everyone the vehicle travels 25 mph or less. The emblem surface should be kept clean and in good repair for both day and night identification. When the reflective red border or fluorescent orange center lose their brilliance, the emblem should be

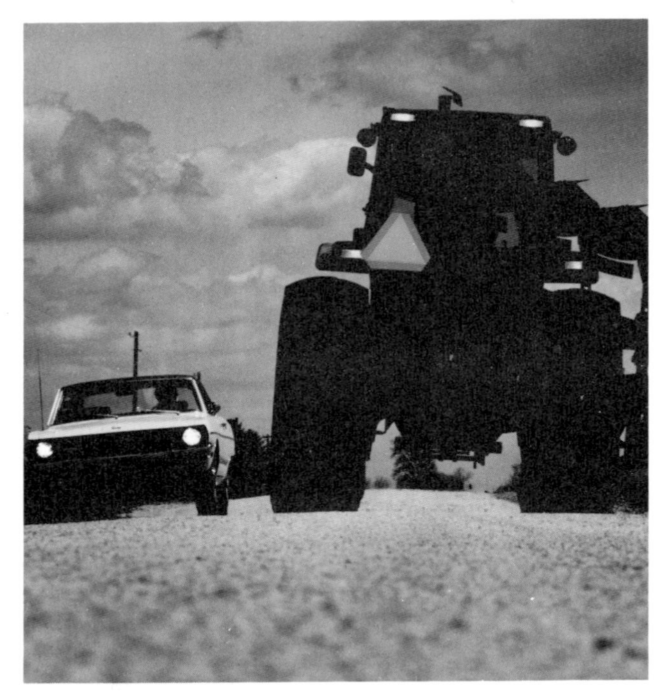

Fig. 2—An SMV Emblem Combined with Proper Machine Lighting Help Prevent Slow-moving Vehicle Accidents

replaced. The emblem should always be mounted securely with a point upward to clearly identify it is a universal symbol as intended.

In addition, lights and reflectors should be kept in good working order for farm machines traveling on public roads. Operate flashing lights both day and night so you can be recognized as an SMV.

On many farm implements, special lighting is used for night work in the field. On a self-propelled combine, for example, a special flood light is used to light up the area where the grain is being cut and fed into the combine. Another flood light shows the operator when the grain tank is full of grain.

LAYOUT OF THE LIGHTING CIRCUIT

The lighting circuit is normally a part of the complete electrical system. It operates on power from the battery with help from the charging circuit.

The parts of a typical lighting circuit are shown in Fig. 3. They are the various **lamps** which are operated by a **light switch** connected by the **wiring harness.** Other accessories shown are an electrical outlet socket and indicator lamps. The circuit is protected from a current overload by the **circuit breakers** and a **fuse.**

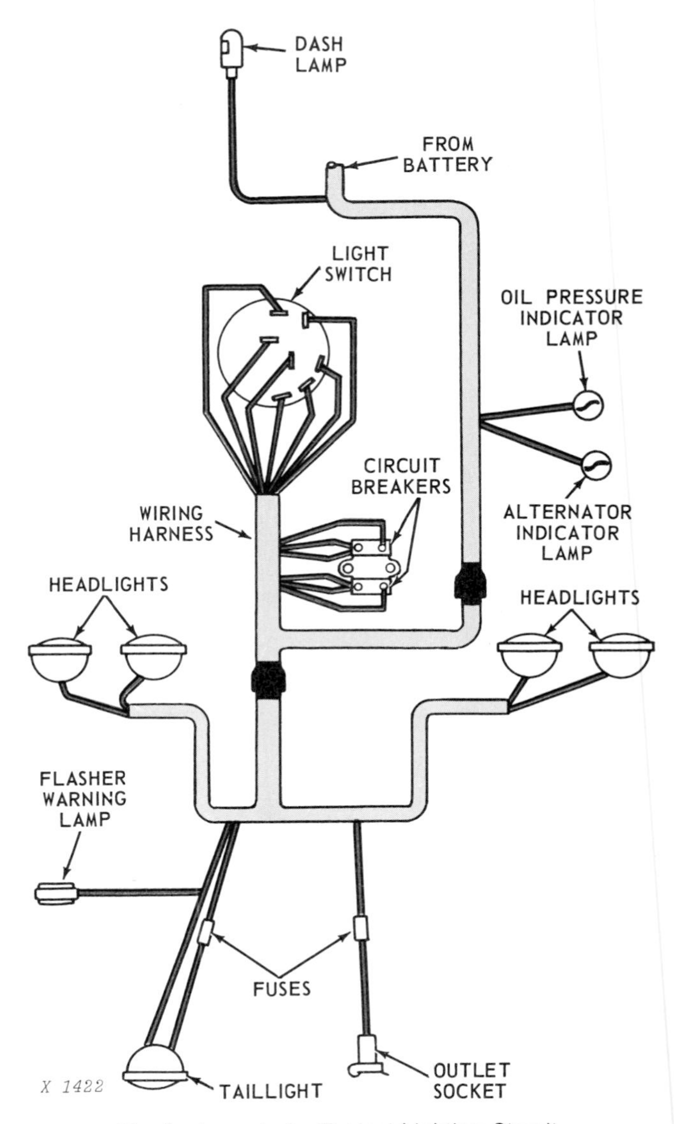

Fig. 3—Layout of a Typical Lighting Circuit

On most machines, the ignition switch must be turned on before the lights can be operated by the light switch. Also some machines have indicator lamps which light up when the ignition switch is turned on and then go out when the engine is running, showing that the generator, oil pressure, etc. are normal.

On machines with two or four batteries, a "split-load" lighting circuit is sometimes used. This divides the load equally between both batteries when the lights are turned on. (For details on balancing batteries in "split-load" systems, see Chapter 5.)

LAMPS

Lamps used for lighting of modern machines are:

- **Sealed-beam units (for headlights and flood lamps)**
- **Light bulbs (for other lights)**

SEALED-BEAM UNITS

Headlights and flood lamps are normally sealed-beam units.

The headlights throw out a high-intensity beam and should always be dimmed when approaching a vehicle at night.

Caution should be exercised when replacing halogen bulbs. Halogen bulbs contain gas under pressure which could cause them to shatter. Turn off the light switch and allow the bulb to cool. Wear eye protection. Handle the bulb by its base. Do not drop or scratch the bulb.

Adjusting Headlights

When headlights are used on public roads, be sure they are adjusted so that the glare of the lights will not shine into the eyes of approaching drivers.

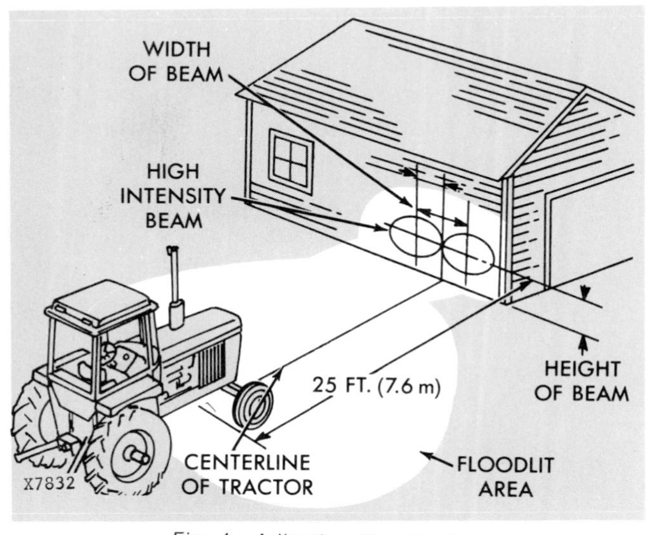

Fig. 4—Adjusting Headlights

Fig. 4 shows how to adjust the headlights on a typical tractor.

Position the tractor as shown on level ground and directly facing a wall 25 feet (7.6 m) away.

Turn the headlights on "bright" and check the height of the beams on the wall. The centers of the light beams should be 1-4 feet (3-12 m) high, depending upon the height of the tractor headlights. Keep the intense part of the beams *at least* 5 inches (127 mm) below the center of the light from which it comes.

The headlights should also be parallel to the tractor centerline. Look down the center of the tractor hood and see if the centers of the beams are at equal distances from the centerline (about four feet apart or two feet off center on most tractors).

On tractors with dual headlights, also check the outer flood lamps on "dim" to see that they shine downward and outward to illuminate the desired area as shown in Fig. 4.

To adjust the lights, loosen the lamp mounting bracket and rotate them as desired. Then retighten the mounting bracket. **Remember: Only the lamp mounting bracket is adjustable—not the sealed-beam unit itself.**

On other machines with headlights mounted higher than those on a tractor, always adjust the lamps downward far enough to avoid glare to oncoming traffic.

LIGHT BULBS

Miscellaneous lights such as taillights, dash lamps, and indicator lamps are normally equipped with single- or double-contact bulbs.

When replacing bulbs, make sure the replacement is of the same style and part number as the old one. Otherwise, the system can be damaged or the bulb may burn out rapidly.

Special light bulbs for special lighting effects may be required for some taillights, flasher lamps, or indicator lamps.

Indicator lamps are often used in place of gauges for generator, oil pressure, and temperature checks.

If the indicator lamp lights up or glows while the engine is running, this shows a failure in the system being monitored. For example, an oil pressure light may glow when the engine is low on oil.

When an indicator lamp comes on, be sure to stop the engine at once and check out the possible causes. But also remember that a burnt-out indicator lamp never lights up, even while the engine is being damaged.

Litho in U.S.A.

FAILURES OF LAMPS

Failure of lamps may be caused by a defective unit, a broken wire, a disconnected wire, a corroded connection, a switch failure or an open circuit breaker.

To check a light bulb, visual inspection is best. A good bulb will be clear while a burnt-out bulb may be dark. Where the bulb glass is not dark, the filament wire will be visible. If it is broken, the bulb is defective.

To check a lighting circuit, test the faulty part for voltage drop. For details on making voltmeter tests, see Chapter 2.

If the voltage is correct but the lamp is dim, look for a poor wiring connection between the lamp and the main wiring harness.

If there is no voltage across the lamp, look for a broken wire, poor contact in the light switch, or a disconnected wiring connector. Replace or repair any defective parts.

If the lights all go out suddenly, check for a tripped circuit breaker or a blown fuse. For details, see "Circuit Breakers" and "Fuses" which follow in this chapter.

FLASHING LAMPS

Fig. 5—Flashing Lamp

When a flashing warning lamp is prohibited by local regulations, disconnect the flasher unit and the short wire connected to the flasher. (See the operator's manual.) If you want to burn the warning lamp continually, connect the wire from the lamp directly into the lighting circuit wiring harness.

A circuit breaker protects an electrical circuit from current overload. The breaker acts as a switch and opens when the current passing

CIRCUIT BREAKERS

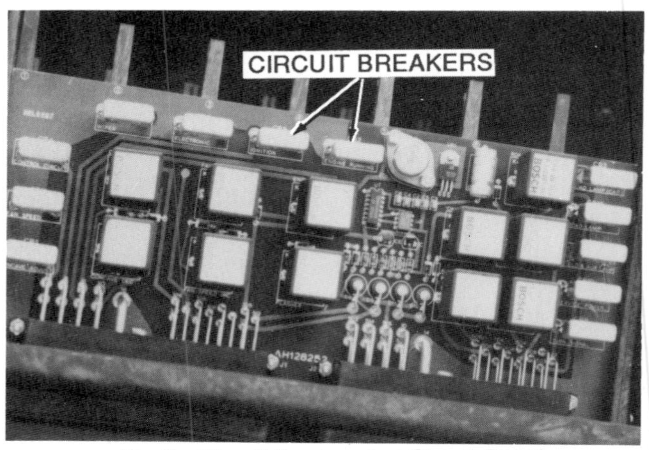

Fig. 6—Circuit Breakers in a Circuit Board

through the circuit exceeds the rated level. The breaker may close automatically after a period of time, after the switch is shut off.

The breaker may also be closed manually. Both automatic and manual reset circuit breakers are used in electrical systems.

In general, circuit breakers are used when heavy loads may be instantly placed on the circuit.

To check for a defective circuit breaker, use an ohmmeter or a voltmeter. If an ohmmeter is used, a good circuit breaker will show zero resistance.

A defective breaker will show full resistance. If a voltmeter is used, and the circuit breaker is defective, the battery side will show voltage and the load side will not.

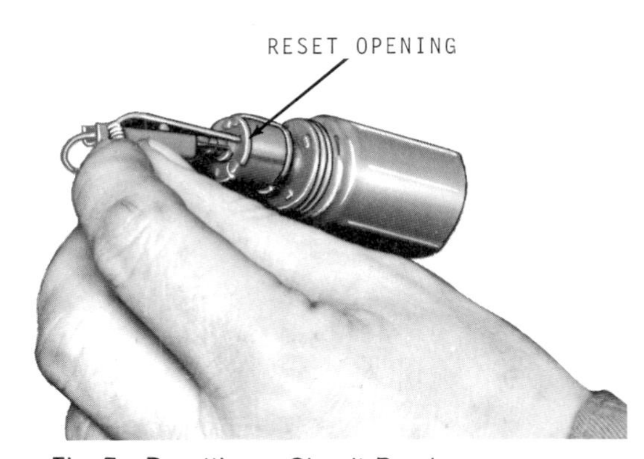

Fig. 7—Resetting a Circuit Breaker

Some circuit breakers are equipped with a reset opening. When the breaker trips, it may be reset by inserting a small wire in the opening and pushing against the spring tension (Fig. 7).

On modern machines, if a circuit breaker does not reset, replace it. If it does reset and trips again, determine the cause and correct it.

These circuit breakers are plug-in type devices and should be replaced if defective. Circuit breakers are made with a tight fit for good contact. Pull them straight out. Do not attempt to pry them out with a screwdriver or similar tool. Each circuit breaker is marked, by circuit, on the power distribution circuit board (Fig. 6).

Circuit boards can also be replaced. Make sure all lights, switches and the key switch are off before replacing a circuit board. To eliminate damage from static electricity, keep the new replacement board in its protective packaging until you are ready to install it. Hold it by its edges when installing. Circuit boards are very fragile and expensive and should be handled with care.

Transfer any circuit breakers or relays to the new circuit board. Recalibrate the switches on the new board as the old ones are calibrated.

FUSES

Fuses protect an electrical circuit from current overload. When too much current passes through the circuit, the fuse blows, breaking the flow of current and preventing damage to the circuit.

A fuse consists of a fine wire or thin metal strip enclosed in glass or fire-resistant material. Most fuses used for modern farm and industrial equipment are small replaceable types. They have a contact on each end and are held in place in a fuseholder or load center panel. The holder is usually a two-piece tube which can be easily separated for installing or replacing fuses (Fig. 8). The load center panel is a plastic or fiber board which contains many fuses. The panel keeps the fuses in one place and aids in fuse replacement.

 **CAUTION: Make sure the key switch is off when checking or changing load center fuses.**

Blown fuses are usually caused by:

1. A short circuit in the electrical circuit caused by defective wiring or a defective component (lights, motor, etc.).

2. An overload in the circuit caused by a surge of electricity passing through the circuit.

3. Poor contacts in the electrical circuit or the components.

4. Overheating in the circuit caused by overloads or poor contacts.

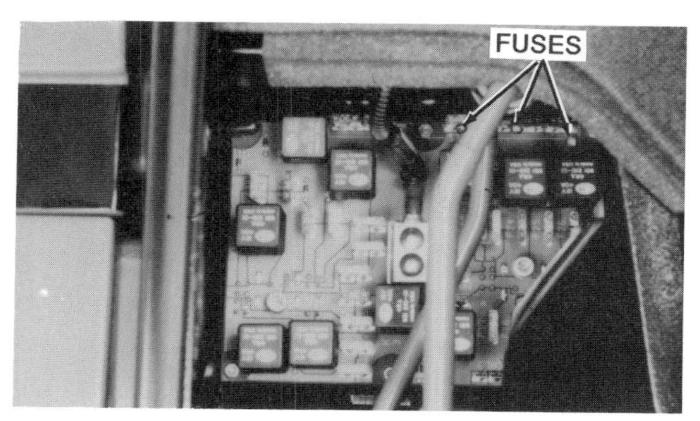

Fig. 8—Fuses in an Electrical Load Center Panel

5. Use of incorrect size fuses in the circuit.

6. Fuse located too near a hot area such as an engine or a heater.

7. Vibrations near the fuse, causing the contacts to come loose.

Types of Fuses

Two types of fuses are widely used in machine electrical circuits:

- **Quick-blowing fuses**
- **Slow-blowing fuses**

A QUICK-BLOWING fuse blows instantly whenever a too-heavy load occurs.

A SLOW-BLOWING fuse will allow an overload for a short period before it blows.

How to Tell What Caused a Fuse to Blow

By looking at a blown fuse, you can often tell what failure in the circuit caused it to blow.

Fig. 9 shows a general check for causes of blown-out fuses—whether from overload or from a short circuit.

QUICK-BLOWING FUSES:

If an *overload*—glass case will be clear because fuse link overheats and simply melts away.

If a *short circuit*—glass will be dark, stained by the fuse link which suddenly "burns up".

SLOW-BLOWING FUSES:

If an *overload*—fuse link will break at solder, which has melted.

If a *short circuit*—fuse link will break at small wires because of sudden heat.

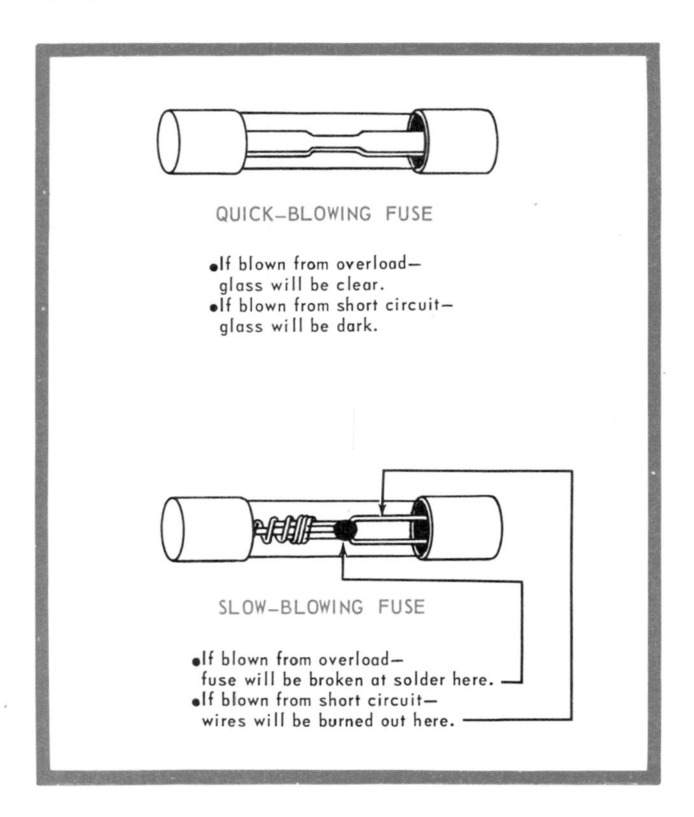

Fig. 9—How to Tell What Caused a Fuse to Blow

Remember that a fuse is meant to protect against overloads. It will also protect against short circuits, but these are abnormal and should be repaired at once.

If a quick-blowing fuse tends to blow out frequently due to small overloads, replace it with a slow-blowing type of the same size.

If the fuse still keeps on blowing, the circuit is overloaded and is not meant to handle the loads being placed on it.

IMPORTANT: Always replace a blown fuse with one of the same size.

Never use a fuse with a higher amperage rating as it may result in serious damage to the circuit it protects.

FUSIBLE LINK

A *fusible link* is another way to protect a wiring harness from damage by inserting a short length of smaller gauge wire in the circuit. If the circuit is accidentally grounded, the heavy overload melts the fusible link first and prevents the wiring harness from being burnt.

The fusible link is usually a 5-inch (127 mm) length of wire that is four wire gauges smaller than the wire gauge in the circuit that is being protected.

Always position the fusible link so that it has a sagging loop. Then when the wire melts, the melted ends at the bottom of the loop will fall apart and prevent further melting of the insulation. Do not position the link vertically because the heat would travel up the wire and might cause the wire insulation to burn.

See the machine Technical Manual for fusible link locations and specifications.

FUSES AND CIRCUIT BREAKERS— THE APPLICATIONS

Fuses are less expensive but are not reusable. They are used mainly in circuits where "blow outs" from heavy loads are not common.

Circuit breakers are more expensive but can be reset without replacement. So they are used in heavy-duty circuits when safety or other factors justify their use.

WIRING HARNESSES

A wiring harness is the trunk and branches which feed the electrical circuit. Wiring leads from one part of the circuit enter the trunk or sheath, joining other wires, and then emerge at another point in the circuit (Fig. 10). The harness sheath is normally made of rubber, cloth, electrical tape, or plastic tubing.

Be careful when installing a wire harness. Disconnect the battery negative cable first. The harness must not interfere with moving parts of the machine. Also make certain that the clips which hold the harness do not pinch through the harness and cut the wires. This can cause a short in the circuit. Make sure the harness is routed away from hot parts of the equipment and away from sharp objects.

Individual wires in a harness may be replaced by cutting off the defective wire at each end of the harness. Discard the removed ends of the wire. Run the new wire around the harness; do not try to thread the wire through the harness. Place the new wire in clips with the harness or attach to the harness with electrical tape. Avoid any sharp bends when installing the harness.

The proper gauge or size of an electrical wire depends upon:

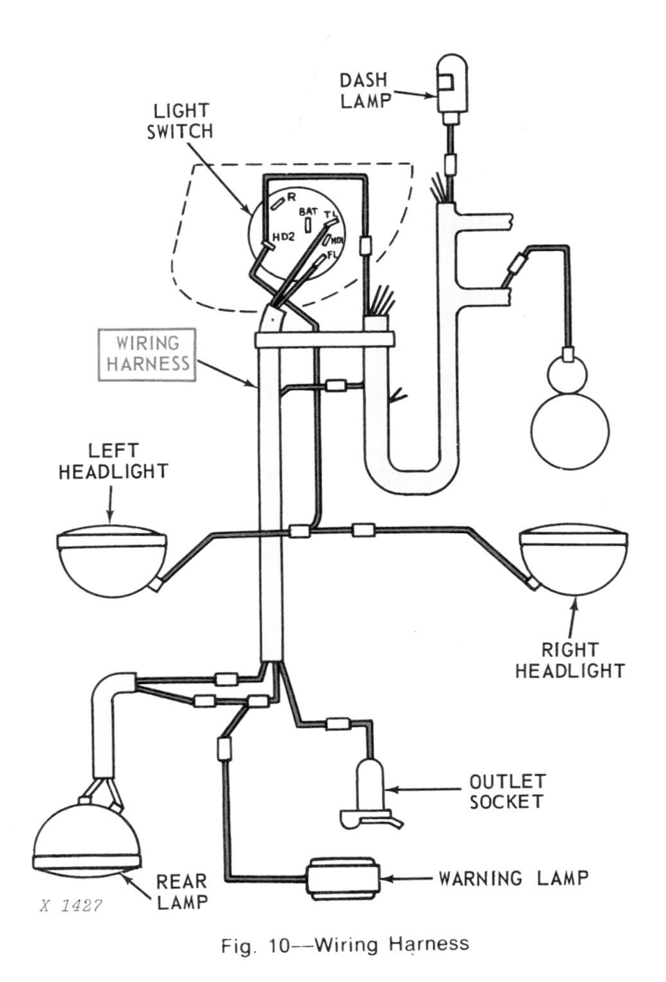

Fig. 10—Wiring Harness

1. *Total **length** of the wire in circuit.*
2. *Total **amperes** that the wire will carry.*

But when replacing a defective wire in a circuit, remember:

Always use the same gauge of wire for replacement. Never use an undersized wire as it will not carry the required load and will overheat.

TESTING AND DIAGNOSIS OF WIRING

In Chapter 2, we said that a wiring circuit may fail in three ways: 1) *open or break,* 2) *ground,* 3) *short.*

The chart below tells you how to test for each of these three failures.

The right-hand column gives the test results you can expect if the wiring has failed.

WIRING TEST CHART

Type of Failure	*Test Unit and Expected Results If Wiring Failed*
Open (broken wire)	Ohmmeter — Infinite resistance at other end of wire. Infinite to adjacent wire. Infinite to ground. Voltmeter—Zero volts at other end of wire.
Ground (bare wire touching frame)	Ohmmeter — Zero resistance to ground. Infinite to adjacent wire. May or may not be infinite to other end of wire. Voltmeter—Instead of testing, normally look for blown fuse or tripped circuit breaker.
Short (rubbing of two bare wires)	Ohmmeter — Zero resistance to adjacent wire. Infinite to ground. Zero to other end of wire. Voltmeter—Voltage will be read on both wires.

ELECTROMAGNETIC CLUTCHES

An electromagnetic clutch is an electric magnet device which stops the operation of one part of a machine while the other part of the unit keeps on operating.

One use of an electromagnetic clutch is on a grain combine where it is desirable to stop the cutting operation and yet continue the separation and cleaning operations (see Fig. 11).

The clutch consists of a field coil assembly, rotor unit, face plate, condenser, and operating switch.

The basic theory of electromagnetism is covered in Chapter 1. The clutch works as follows:

When the switch is actuated, current through the field coil inside the rotor assembly sets up a magnetic field which draws the face plate against the clutch facing on the rotor assembly. (The face plate is free to slide on the drive studs of the mechanism.) Power is then transmitted from the

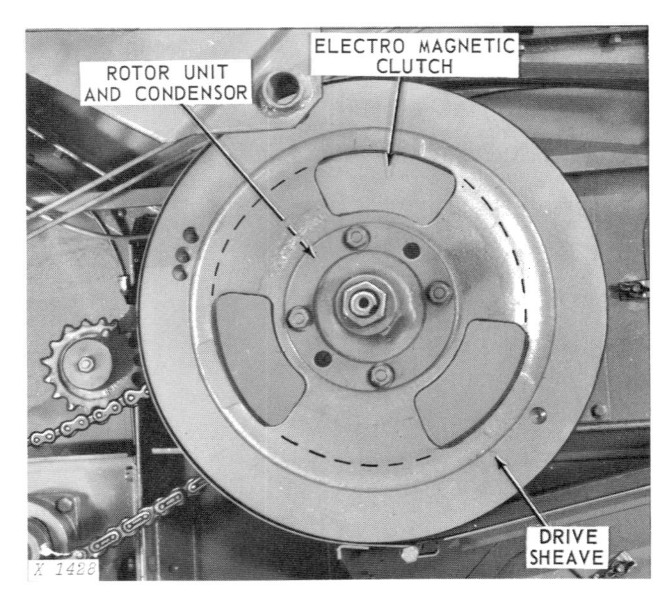

Fig. 11—Electromagnetic Clutch on a Grain Combine

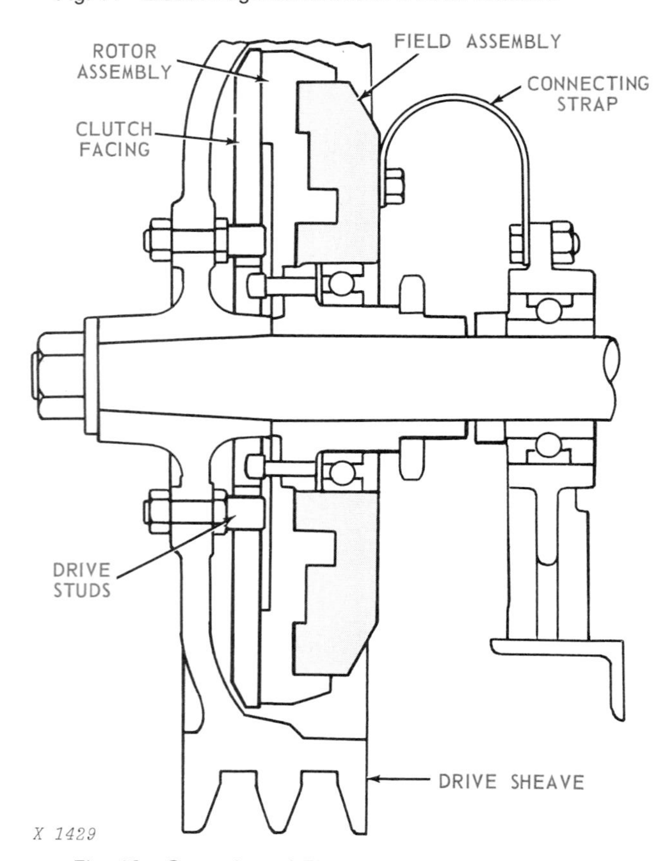

Fig. 12—Operation of Electromagnetic Clutch

drive pulley through the face plate, rotor, and hub to the drive mechanism.

When the switch is operated again, the magnetic field is collapsed. The face plate is freed from the rotor and the transmission of power to the mechanism is instantly stopped.

TESTING AND DIAGNOSIS

If the clutch suddenly loses power and fails to function, it is probably due to a failure in the electrical circuit.

First check all electrical connections, wires, switch, and circuit breaker in the switch. If these components are okay, then check the voltage to the field and the amperage of the field.

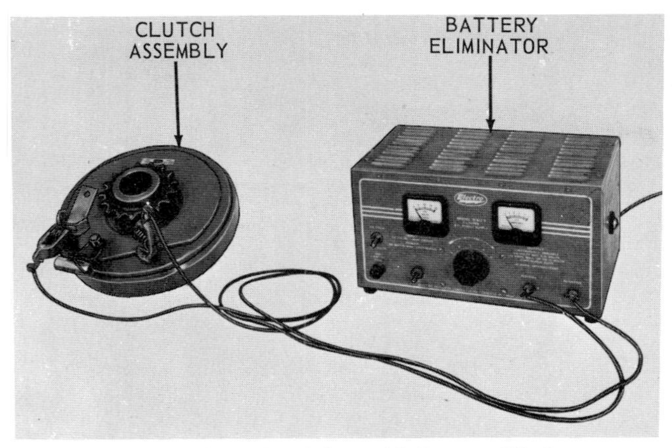

Fig. 13—Checking Field Coil with Battery Eliminator

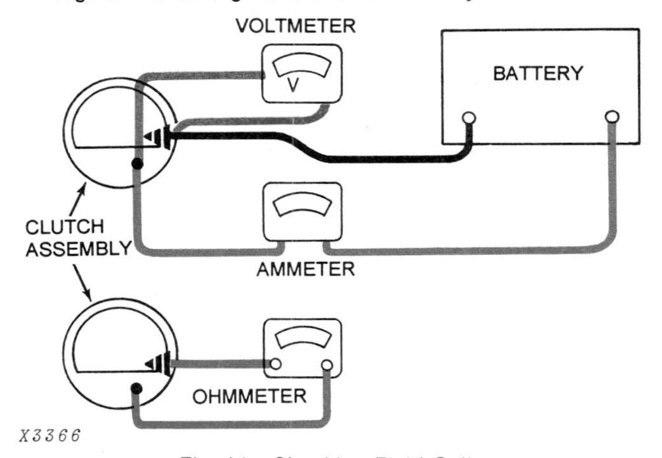

Fig. 14—Checking Field Coil

To check the voltage to the field, the amperage of the field or the resistance of the field, use a battery eliminator or a storage battery and a volt-ohmmeter with a 0 to 15 volt scale and a 10 amp scale. (See Figs. 13 and 14.) Check the individual specifications on each clutch for the correct voltage input, amperage and resistance.

The condenser on an electromagnetic clutch is used to absorb the surge of high voltage when the clutch is disengaged. This prevents arcing and burning out of the electric clutch switch.

If the condenser seems defective, the best check is to install a new condenser and see if this remedies the problem. If not, check the rest of the circuit for defects.

If diodes are used in the clutch, they may be checked with a diode tester or an ohmmeter. When using an ohmmeter, be sure it has 1½ volts or less to avoid damaging the diodes.

A **good diode** *will have infinite resistance in one direction and zero resistance in the other direction.*

A **defective diode** *will show zero resistance or infinite resistance in both directions.*

GAUGES

Fig. 15—Gauges on the Dash of a Modern Machine

Gauges are used on modern machines to keep the operator informed on the various functions of the machine systems. Examples of these gauges are: fuel gauge, water temperature gauge, and oil pressure gauge. (See Fig. 15.)

The fuel gauge is controlled by a sending unit located in the fuel tank and the water temperature gauge and oil pressure gauge are controlled by sending units located in the engine radiator and cylinder block.

The sending units are all variable resistance types and operate the gauges in the following manner: The higher the water temperature, oil pressure, or level of fuel, the lower (or higher) the resistance in the sending unit. This change in resistance causes more (or less) current to pass through the connecting wire to the gauge coil, which in turn causes a new reading on the gauge.

NOTE: Resistance in gauges is normally affected as follows: 1) Fuel and pressure gauges: higher level=higher resistance in gauge coils. 2) Temperature gauges: higher temperature=lower resistance in gauge coil.

TESTING AND DIAGNOSIS OF GAUGES

Fig. 16 shows the circuits for three common gauges.

Diagnose the failures of gauges as given below.

If a gauge does not register, the cause could be:

1. Lack of current to the gauge.

2. Poor ground connection.

3. Connecting wire grounded to implement.

4. A defective sending unit or gauge.

If a gauge consistently registers too high, the cause could be:

1. Poor connection between gauge and connecting wire.

2. Broken connecting wire.

3. Poor ground at sending unit.

4. Failure of gauge or sender, usually the sender.

To test a gauge, use a commercial gauge tester. Follow the manufacturer's instructions closely.

If a commercial gauge tester is not available, substitute a new gauge and make sure it is satisfactory. A new sending unit may also be installed to check or to replace a defective unit.

Indicator lamps often use small light bulbs which "glow" to tell of a failure.
If these lamps do not glow when starting the engine, first check for a defective bulb. Then, if necessary, check out the other causes given above.

METERS

AMMETERS

The ammeter is an instrument for measuring the strength of an electric current in terms of amperes. (See Fig. 15.)

Normally the ammeter is connected directly to the regulator to measure the flow of current through the electrical system.

A typical low-cost ammeter consists of a moving magnet with attached needle placed close to a conductor between the ammeter terminals. Current flow through the conductor creates a magnetic field that deflects the moving magnet and causes the needle to deflect away from zero on the meter scale.

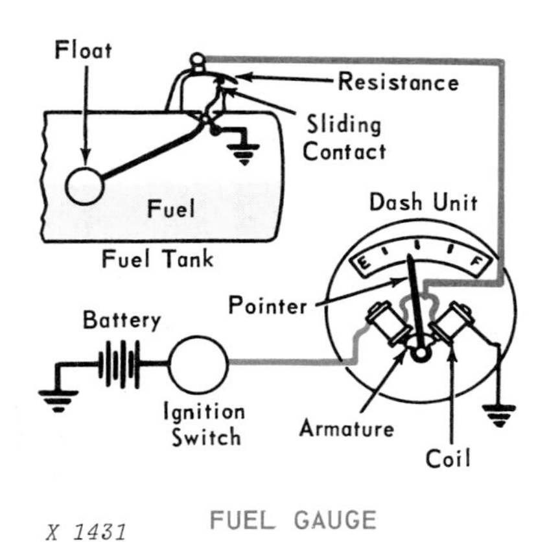

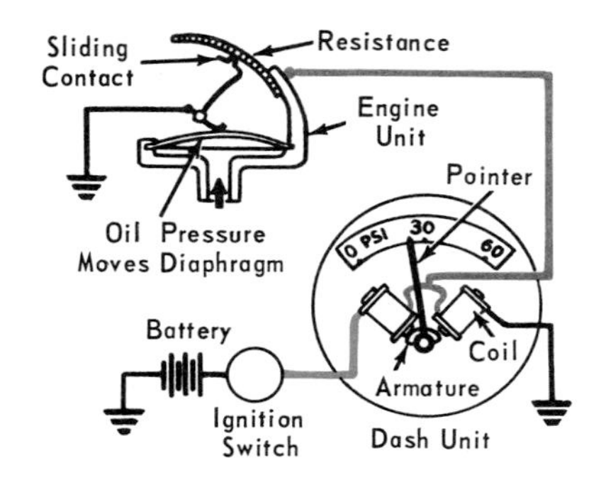

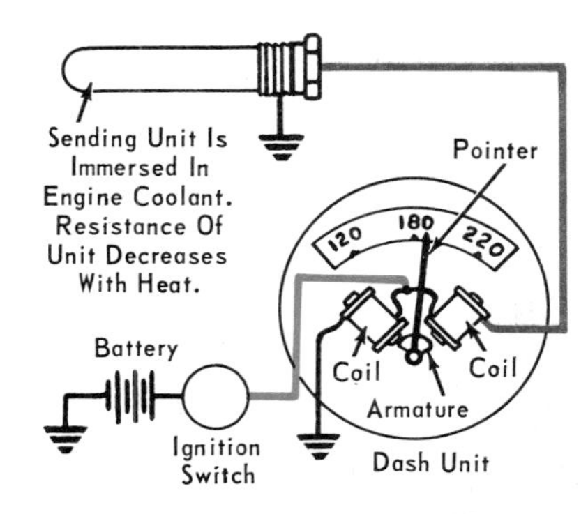

Fig. 16—Three Gauges and Their Circuits

If an ammeter does not register correctly, replace it with a new ammeter. Do not attempt to repair a defective ammeter.

VOLTMETERS

Voltmeters are used to indicate the voltage of the electrical input in a circuit.

A typical low-cost voltmeter consists of a needle attached to a moving iron vane placed inside two stationary coils. The coil winding is parallel with the needle pivot and one winding tends to keep the needle at zero while the other tends to move the needle to the full-scale deflection.

Voltmeters are connected in parallel with the voltage to be measured. Since the voltmeter has a high resistance, adding this component to the circuit will change the total circuit current very little, and the voltage reading obtained shows the true voltage present without the meter in the circuit.

If a voltmeter does not function properly, a new voltmeter must be installed. A defective voltmeter cannot be repaired.

ELECTRIC HOUR METER

Electric hour meters are used to show the operating time of a machine while the engine is operating at its rated speed.

The hour meter records the time in hours and only operates when the engine is running.

Do not attempt to repair a defective hour meter or its sending unit. When they fail, new components must be installed.

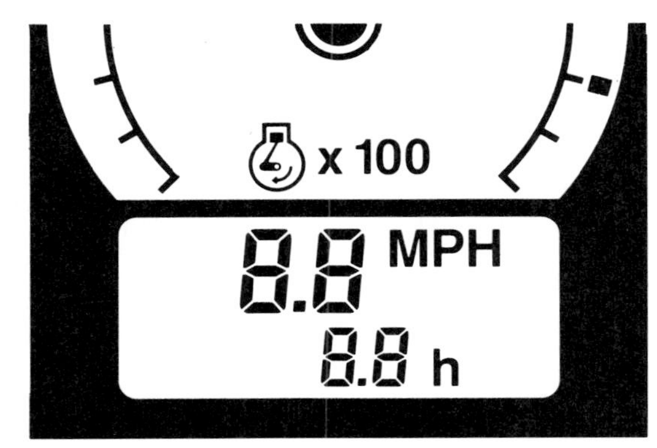

Fig. 17—Electric Hour Meter

MOISTURE METER

Fig. 18—Moisture Meter

The moisture meter is a portable electronic device used to measure the moisture content of grain (Fig. 18). In grain drying, the moisture content is the percentage (by weight) of water in the grain.

Moisture in the grain is measured by penetrating the sample with high frequency radio waves. The effect the grain has on the radio waves varies with the amount of moisture. This is measured electronically by observing the change required to rebalance the meter when the grain is added. Because the waves penetrate the grain, the surface condition of the grain or mixed wet and dry grain will not affect the accuracy of the meter.

The moisture meter is accurately calibrated at the factory and should provide accurate measurements for years of normal use. To insure accuracy and long life, avoid exposing the meter to rain and excessive dampness. When out of the carrying case, keep the meter covered with the plastic cover except when in use.

The electronic circuit cannot be repaired by a radio or television repairman or any other local electronic repair shop without destroying the factory calibration. Never remove the case from the meter. If the moisture meter fails, return it to the factory where it can be repaired and accurately calibrated.

Some parts of the meter can be replaced in the field. In case of breakage or loss, the plastic index hand, knobs, scales, grain cup, small weight, and drawer can be replaced.

HORNS AND BUZZERS

Horns and buzzers are used on machines as signaling and warning devices.

Fig. 19—Horn

For example, on some grain combines, two different horns are used. One horn is used as a "call horn" to call the truck to the combine for unloading the grain tank when it is full of grain. The other horn is a signal device located inside the combine separator to warn the operator when the separator is overloading.

On farm and industrial tractors, horns are used to warn pedestrians and other vehicles that a moving machine is in the area. Horns are also used to warn the operator of some equipment when the engine is heating up.

Horns and buzzers are the same in design and operation. The only difference is in the sound. This is achieved by the use of different air columns. In a horn the air column is formed into a compact seashell form. This shape produces the maximum volume from the sound generated by the diaphragm.

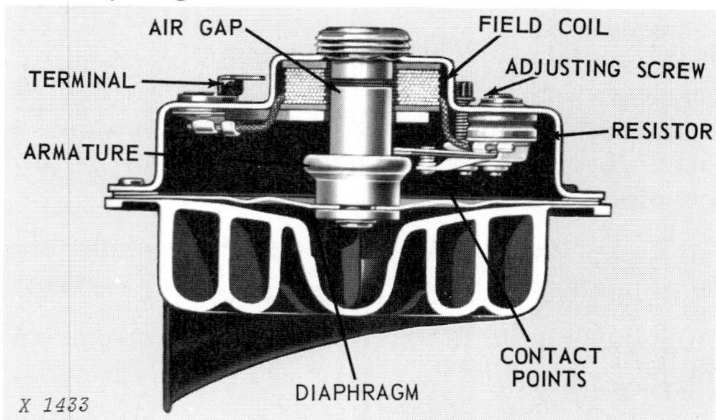

Fig. 20—Cutaway View of Horn

Two types of horns or buzzers may be used: The **type S air-tone horn** is shown in Fig. 20. The **type C air-tone horn** is similar to a type S horn except that it is smaller in size and does not have a resistor.

The horn or buzzer has a vibrating power unit to actuate a diaphragm which produces the sound as a warning signal.

When energized, current flows through the field coil to the contact points, then to the ground. The field coil magnetic field pulls the armature into the field coils, moving the diaphragm and opening the contact points. Opening the points de-energizes the field coil and the spring force of the diaphragm pulls the armature from the field coils, closing the points. This cycle is repeated 360 times per second on a high-note horn. Some horns are capable of sounding three notes.

The adjusting screw (Fig. 20) controls the time of point opening, which in turn controls current draw and frequency.

On a type S horn, the resistor connected across the contact points reduces the arcing of the points.

TESTS AND ADJUSTMENTS

IMPORTANT: Do not turn the horn adjusting screw more than ¼-turn. To do so may damage the horn.

Before adjusting the horn, check the available

voltage when the current draw is 10 amps. Adjusting the horn to compensate for excessive voltage drop will shorten the service life of the horn.

If the horn fails to operate with normal voltage, energize the horn and tap it lightly. If the horn now operates, the horn contacts were probably held open by a foreign particle. The horn should continue its normal operation when it is re-energized.

If the contacts are open as a result of wear, it will be necessary to tap the horn again after it is re-energized. A rough current adjustment can be made by turning the adjusting screw ¼-turn counterclockwise.

To make further tests, connect an ammeter and a voltmeter to the horn.

Apply voltage to the horn and check the readings as follows:

1. No current indicates a broken lead or open circuit due to overheating. Overheated horns have the usual odor of burned insulation and should be replaced. On type C horns, no current flow also indicates open contacts. Turn the adjusting screw counterclockwise and recheck the horn.

2. On type S horns, a reading of approximately 2.5 amps indicates open contact points and that current is flowing through the horn resistor only. Turn the adjusting screw counterclockwise and recheck the horn.

3. A reading of approximately 20 to 25 amps indicates that the contact points are not opening. Turn the adjusting screw clockwise. One-quarter turn clockwise is usually enough unless the horn has been tampered with.

4. When the horn is operating, adjust the current draw to the horn specifications in the machine Technical Manual. Turn the adjusting screw clockwise to decrease current. Turn the adjusting screw only 1/10 turn at a time.

ELECTRIC MOTORS

Small DC electrical motors are used to perform auxiliary functions on some machines. For example, motors are used to operate air conditioner blower, ventilating and heating fans, and windshield wipers. (See Fig. 21.)

Small electric motors operate on the same principle as a starting motor. Refer to Chapter 5 for complete theory and operation.

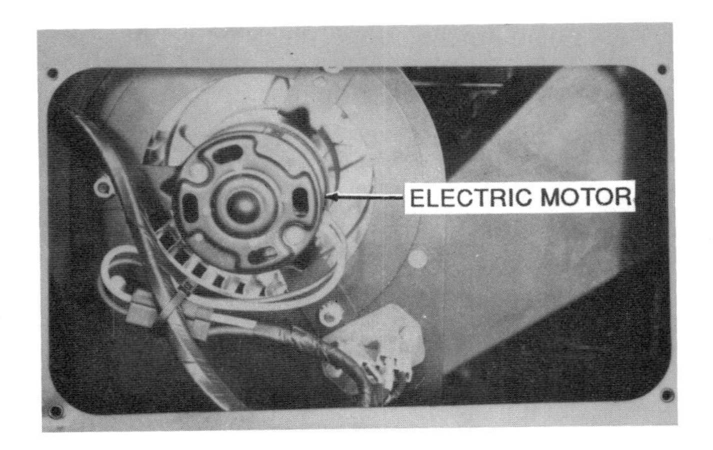

Fig. 21—Electrical Motor Used to Operate Air Conditioner Blower Fan

If the proper equipment is not available, take defective motors to a good electrical shop for servicing.

CIGARETTE LIGHTERS

The cigarette lighter contains a heating element which contacts the electrical circuit when the lighter is pushed in. This causes the element to heat up and glow. The element will remain heated long enough to permit lighting a cigarette or cigar.

The lighter employs a circuit breaker to cut off the electrical current to the lighter element when it has heated to its peak.

Failure of the cigarette lighter may be caused by a broken wire, disconnected wire, burned-out element, defective lighter shell, or a tripped circuit breaker.

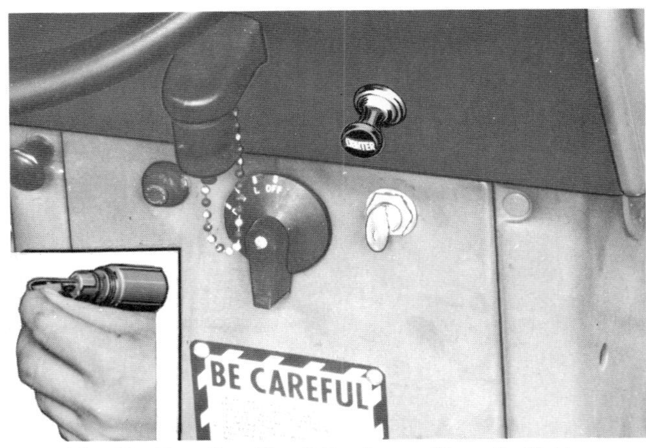

Fig. 22—Resetting Circuit Breaker on Cigarette Lighter

Some circuit breakers on lighters are equipped with a reset opening (Fig. 22). If the lighter fails to operate, the circuit breaker may be open and must be reset.

To do this, insert a small wire in the small hole in the end of the lighter as shown and push in against the spring tension.

If a lighter has a burned-out element or a defective shell, these parts must be replaced.

CONVENIENCE OUTLETS

Some machines are equipped with electrical outlets for use with trailers or implements (Fig. 23). Always use auxiliary light on a towed implement when the tractor rear signals and other lights are obscured.

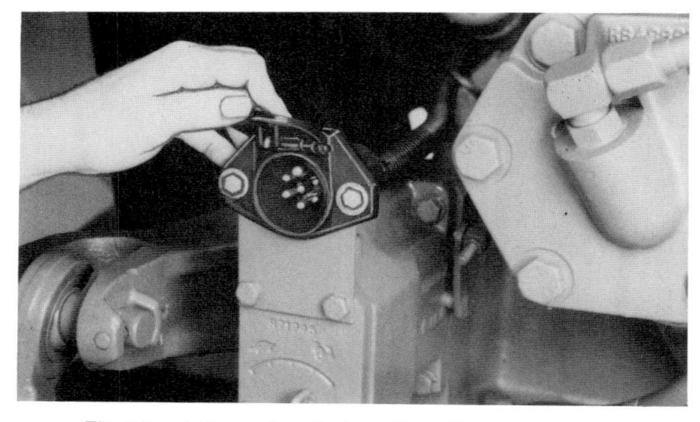

Fig. 23— A Seven-terminal Auxiliary Electrical Outlet

In addition, some machines have a 12-volt electrical outlet used for connecting auxiliary electrical equipment, such as seed monitors or implement control boxes (Fig. 24). If the outlet is already used, you may use two unused terminals of the load center for an auxiliary power source: the IGN and ACC terminals. These terminals are controlled by the IGN and ACC key switch positions respectively.

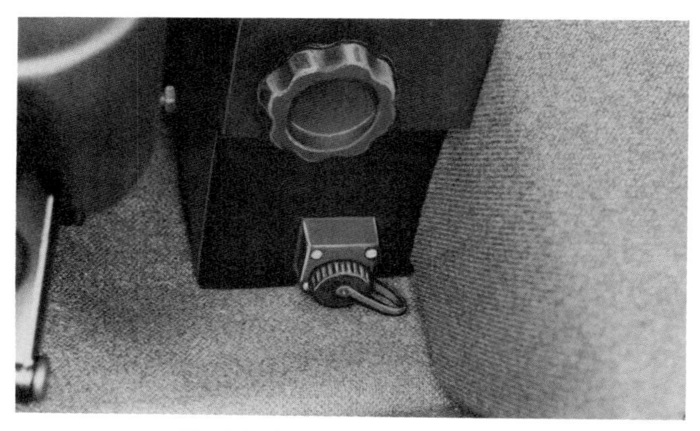

Fig. 24—Accessory Electrical Outlet

FLAME RODS

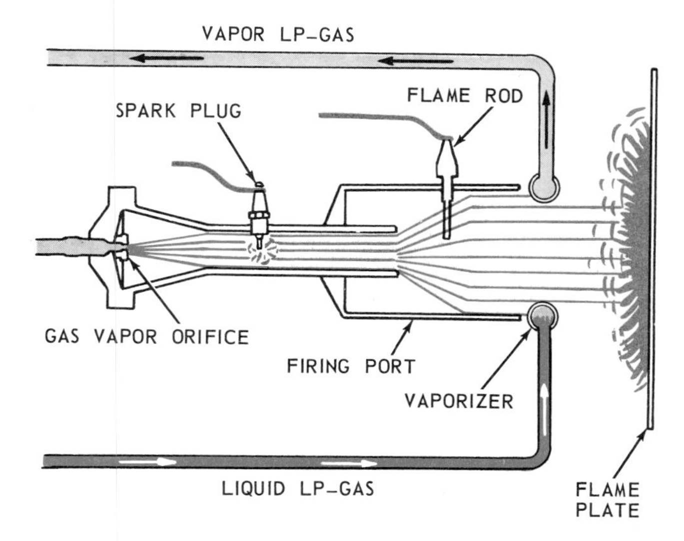

Fig. 25—Flame Rod in Operation

A flame rod is a device used in a gas-fired crop dryer to sense the presence of flame in the dryer. The flame rod is located in the firing port (Fig. 25) and works as follows:

AC electrical current is fed from the Protectorely to a flame rod which is in direct contact with the normal burner flame. The flame completes the electrical circuit between the rod and the ground (the flame plate and firing port). However, because of the relatively small size of the flame rod compared with the ground, the current is rectified, or converted to DC. This small current flow indicates "flame" and signals the Protectorelay to hold the solenoid valves open. But if the flame loses contact with the rod, the circuit is broken and the solenoid valves are de-energized within 2 to 4 seconds. Thus the supply of gas to the burner is stopped as soon as the flame goes out.

Short circuits or high resistance grounds cannot simulate the presence of a flame. A component failure within the system has the same effect as a loss of flame.

These are truly "safety shutdowns" as there is no attempt at automatic restarting.

The burner must be restarted manually after the cause of the shutdown is corrected.

For service of flame rods, see the crop dryer operator's manual. Also see the proper manuals for operation of the complete dryer and its LP-Gas and electrical systems.

Litho in U.S.A.

GLOW PLUGS

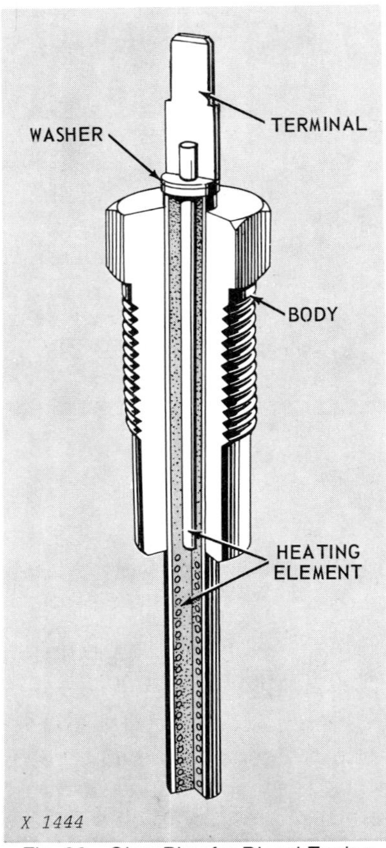

Fig. 26—Glow Plug for Diesel Engine

Some diesel engines have electrical pre-heating devices called glow plugs (Fig. 26). They are used as a starting aid when starting a cold engine and are mounted in the precombustion chamber of each cylinder.

TESTING GLOW PLUGS

Glow plugs are tested most accurately with the use of an ammeter.

1. Remove the wiring leads from all the glow plugs.

2. Connect the ammeter in series between the glow plug terminal and the lead wire.

3. A glow plug should be replaced if the amperage is either too low or too high. (See the machine specifications.)

Check each glow plug by this method. If all glow plugs check out to be too low, but the readings on them are equal, the starter switch or wiring harness is probably at fault.

TEST YOURSELF

QUESTIONS

1. What is the difference between a circuit breaker and a fuse? Which is used where heavy loads are instantly placed on the circuits? Which is normally replaced after it activates?

2. True or false? "Horns and buzzers are basically the same in design and operation and only differ in the sound they make."

3. True or false? On an agriculture tractor, lights are adjusted by the lamp brackets and not by the sealed-beam unit itself.

4. Glow plus are preheating devices used as a _____ ____________ aid for a cold engine.

(Answers on page 20 at the end of this book.)

 Litho in U.S.A.

CONNECTORS / CHAPTER 11

INTRODUCTION

With the increasing use of electrical equipment and electronic monitoring systems on modern agricultural machines and implements, more and more connectors of all types, sizes, and shapes are being used. And with this more widespread use comes the need for improved skills in maintaining and repairing the connectors and their associated wiring, pins, and sockets.

Another contributing factor to increased requirement for maintenance and repair is the harsh environment in which these connectors must operate. Connectors on various types of agricultural equipment have been designed to operate in, but can be damaged by extremes of heat and cold, dirt, dust, moisture, and even chemicals of both dry and wet types.

CURRENT FLOW IN CONNECTORS

The purpose of a connector is to pass current from one set of wires to another. To do this, connectors have what is known as mating halves which are shown in Fig. 1. One mating half houses the male pin contacts and the other contains the female socket contacts. When the connector halves are mated, the pins fit inside the sockets and make contact, thus enabling the current flow to continue. Sounds simple.

But pins and sockets, when mated, have resistance. And as we learned in Chapter 1, resistance impedes current flow so we strive for as little as possible by eliminating or at least minimizing the causes.

What are some causes of resistance in connectors?

ASPERITY

The pins and sockets used in connectors are merely extensions of the wires they are connected to. And as we mentioned, their purpose is to pass or conduct the current in a continuing manner from one wire to the other. To do this, the surfaces of the pins must contact the surfaces of the socket.

When we look at and feel a pin contact, its surface seems very smooth. But the surface actually contains microscopic peaks and valleys — a condition known as asperity. The inside of the socket has this same condition. Because of this, when the pin and socket are mated, only about one percent of the actual surfaces contact each other. Fig. 2 illustrates this condition in a highly exaggerated form. When the surfaces of the pin and socket meet, electrons (again see Chapter 1) must converge to pass from one conducting surface to the other (Fig. 3) rather than being able to pass at any point (Fig. 4). This convergence causes the electrical path length to increase, thereby increasing the effective resistance at the contact interface. This effect, called constrictive resistance, contributes to the total contact resistance.

PLATING MATERIAL

In contact pins and sockets, we strive for minimum resistance. This is directly dependent on surface finish, contact pressure, and the kind of metal used. Tin is soft enough to allow film-wiping but it has high resistivity. Conversely, copper has low resistivity but is hard. So in striving for minimum resistance and thus reducing the asperity condition discussed above, low-resistance copper contacts are often plated with tin.

What is film-wiping? As we just mentioned, tin is soft enough for this condition to occur. When a pin contact and socket that are plated with tin are mated or plugged together, the tin on the two mating surfaces has a tendency to "wipe" together and actually smooth out some of the peaks and valleys of the asperity condition, thus reducing resistance. Copper alone is too hard to permit this. Gold and silver also make good plating materials. Neither material will oxidize and both are good conductors. In these two respects they are much superior to tin but they are also much more expensive, especially gold.

Contaminants

Contaminants of all types are another major contributing factor to resistance in connectors. As we mentioned, connectors used on agricultural equipment operate in rather harsh environments. A tractor, for example, will operate in dusty fields pulling a planter that could be loaded with dry or liquid insecticides and herbicides. Then this same tractor could be used in the winter to remove heavy wet snow. These contaminants along with oxide films and oils all contribute to increased resistance on the pins and sockets used in connectors.

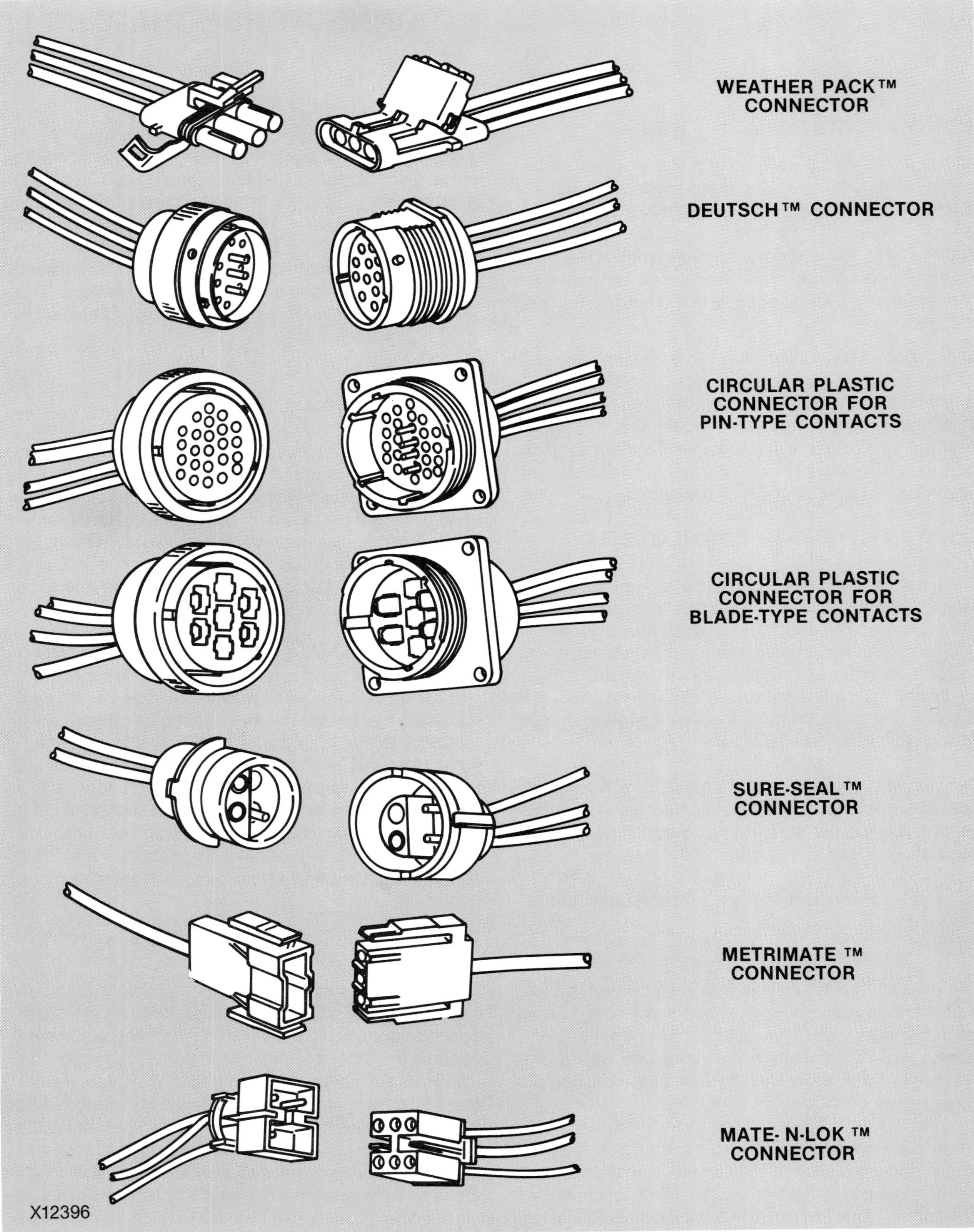

Fig. 1 — Common Types of Connectors

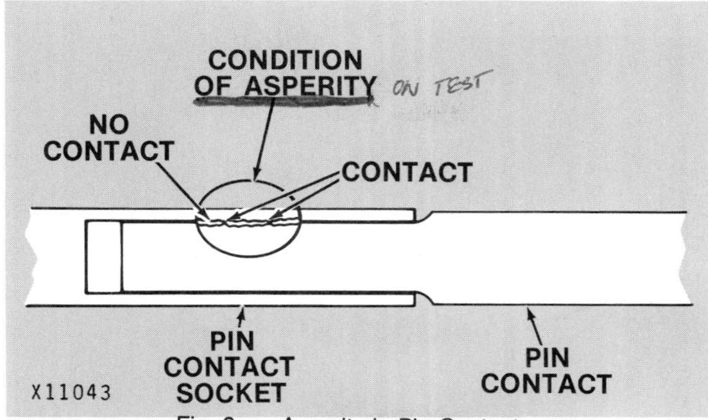

Fig. 2 — Asperity in Pin Contacts

ELECTRICAL CONNECTOR HANDLING

Electrical connectors must not be forcibly mated or unmated. All are designed to be mated easily. If you have to use tools, you may be doing something wrong. Prying or forcing connectors may cause permanent damage to the locking mechanism, contacts, or both.

When working on connectors, make sure you are working on the correct terminal. Remember that male and female halves are mirror images of each other. Look for the terminal number on the connector body. The connection of improper electrical circuits can cause unusual electrical symptoms.

When an electrical connector is repaired, it is important that the proper terminals are used. In some of these connectors, different terminals are used to carry different currents. If contacts of different materials are mated, corrosion may develop that could affect performance.

When removing a terminal from a connector, it is very important to use the correct extraction tool and gently remove the terminal. The connector body can be damaged if terminals are just "jerked" out of it. The damage caused will prevent the new terminal from staying in the connector and will result in replacement of the connector body.

Fig. 3 — Electrons Converging

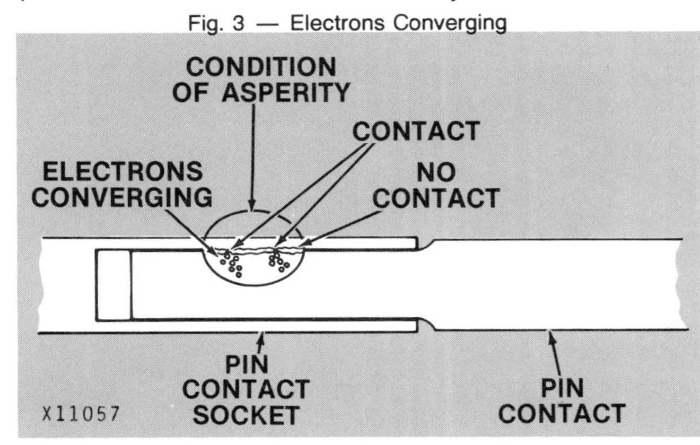

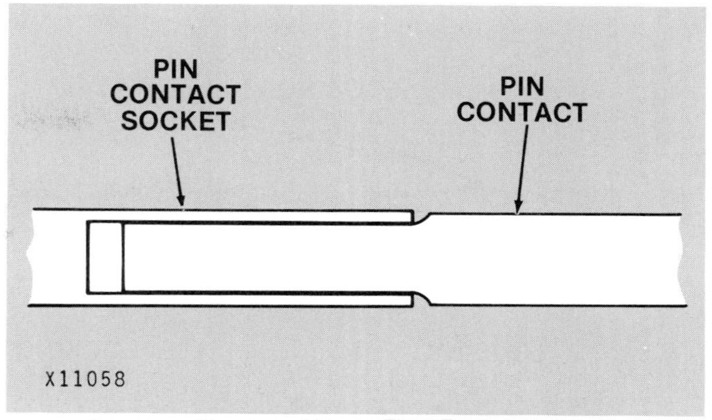

Fig. 4 — Electrons Able to Flow From Any Point

COMMON TYPES OF CONNECTORS

In the remainder of this chapter, we will discuss seven common types of connectors used in electrical systems of agricultural equipment. These connectors, listed below, are shown in Fig. 1. We will also discuss repair procedures and service tools required for the repair.

- Circular Plastic Connector (CPC)™ with Pin-Type Contacts
- Circular Plastic Connector (CPC)™ with Blade-Type Contacts
- Sure-Seal™ Connector
- Metrimate™ Connector
- Mate-N-Lok™ Connector
- Deutsch™ Connector
- Weather Pack™ Connector

All seven connectors are considerably different from one another because they are used in different applications, but they do have three characteristics in common:

1. The two mating halves have a locking mechanism so that they will not inadvertently disconnect.

2. The mating halves have either a mating guide mechanism or pin contact arrangement so that the two halves cannot be wrongly connected.

3. All connectors are designed so that an individual pin contact can be removed and replaced should it become bent or broken.

CIRCULAR PLASTIC CONNECTOR — PIN-TYPE CONTACTS

The circular plastic connector (CPC) shown in Fig. 5 will accommodate 37 pin contacts. However, this same design of CPC is available with many different pin

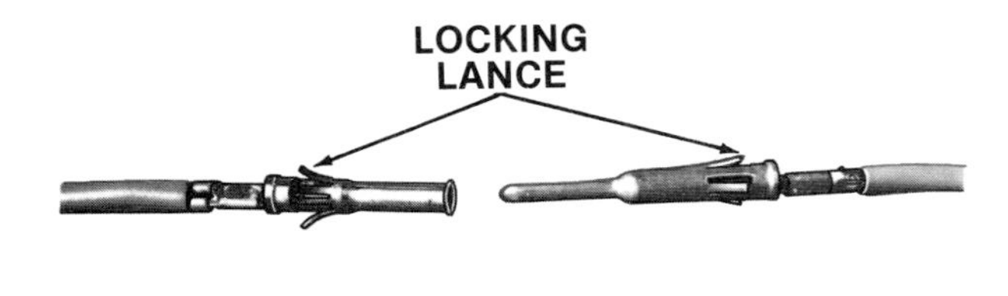

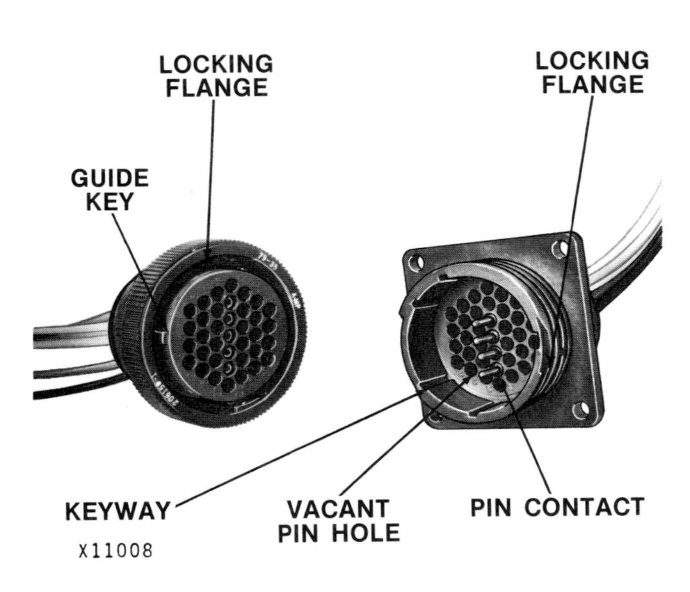

Fig. 5 — Circular Plastic Connector for Pin-Type Contacts

contact configurations. Fig. 6 shows a connector which will accommodate 16 pin contacts. Not all pin cavities in the connector have to be used as can be seen in Fig. 5. If you do not need all pin cavities, just make sure that the corresponding cavity in each mating half is used because the connector halves will only mate one way.

Fig. 6 — Connector Body, Pins, and Sockets

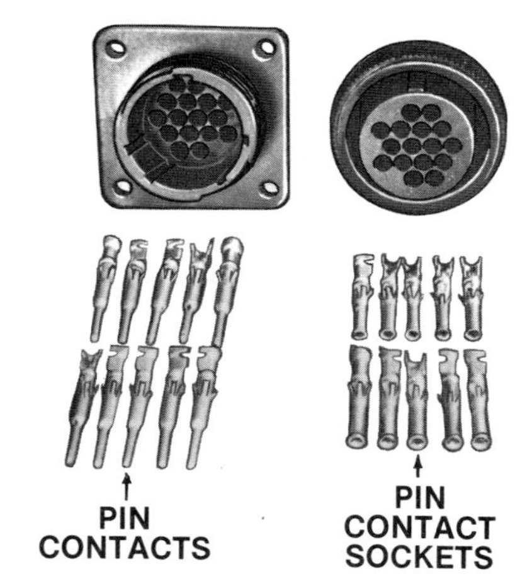

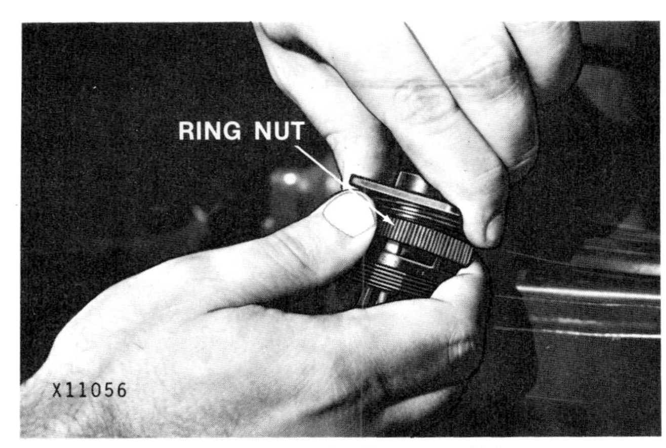

Fig. 7 — Mating the Connector Halves

To mate the connector halves, align the guide keys with the keyways (Fig. 5), push the two halves together, and then turn the knurled connector half clockwise to engage the locking flange (Fig. 7). When turning the connector, you will feel an initial resistance. Turn past this point until you feel a solid resistance. You will then have a secure connection with practically no danger of it coming apart. This connector also has provisions for mounting the female half (which actually contains the male pin contacts) onto a panel, mounting flange, or bulkhead.

Installing Pin Contacts

No special tools are required to install the pin contacts into the connector halves. Both female and male pin contacts are simply pushed through from the rear of the connector (Fig. 8). Just make sure that the pins are pushed far enough into the connector to engage the locking lances on the pin (Fig. 5). Fig. 9 shows a cutaway of the locking lance inside the connector cavity. Also, make sure that the proper pin contact is installed in the proper connector body.

Fig. 8 — Inserting Pins or Sockets

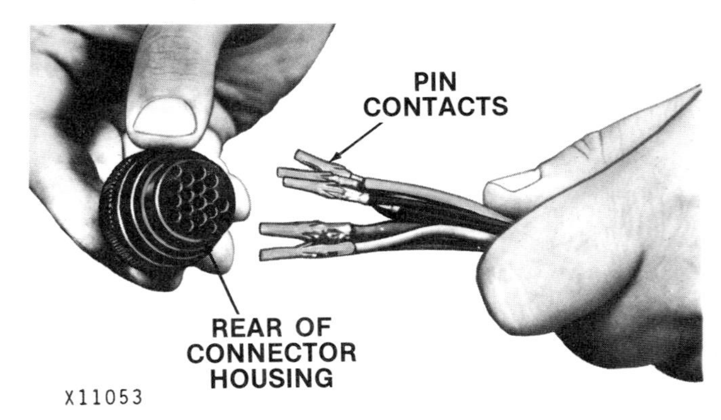

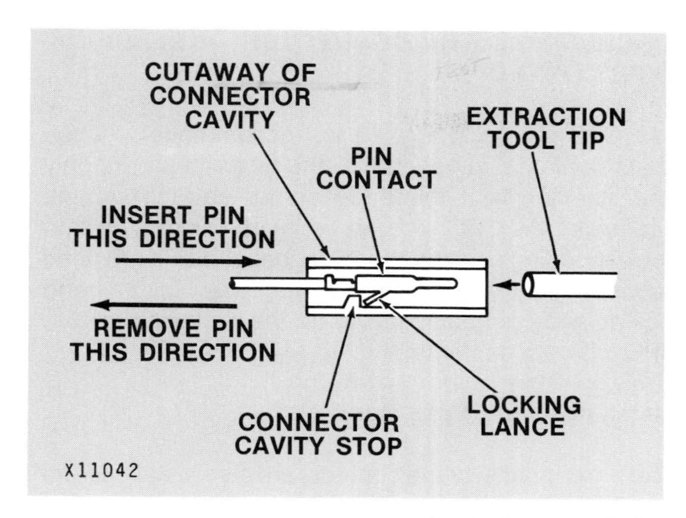

Fig. 9 — Pin Contact Locking Mechanism in Connector Body

Extracting Pin Contacts

It is not necessary to replace an entire connector body unless it is broken, cracked, or chipped. But quite often it is necessary to replace pin contacts because they become corroded, broken, or bent. Basically, all you have to do to remove a pin contact is depress the locking lance (Fig. 5) and pull the wire out from the rear. But you need a special tool to do this--one that will fit over and around the pin contact between the pin and cavity of the connector. As the tool is inserted over the pin, it depresses the locking lance so the pin can be removed (Fig. 9). So the size of the tool is critical. It must be large enough to fit over the pin, yet small enough to fit inside the connector cavity. The following steps explain how to remove a pin contact.

Fig. 10—Extractor Tool for Pin Contacts in Circular, Metrimate ™, and Mate-N-Lok™ Connectors

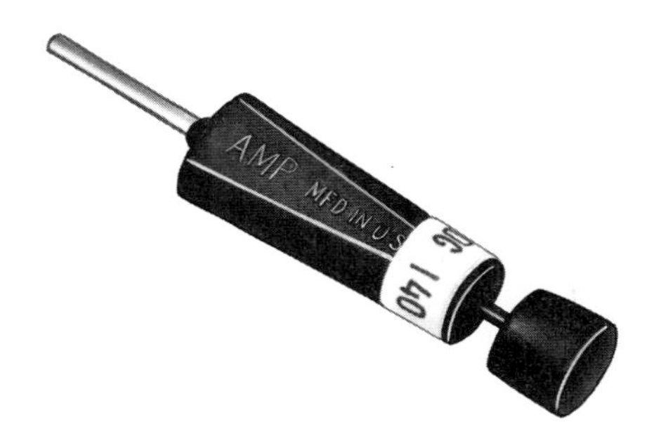

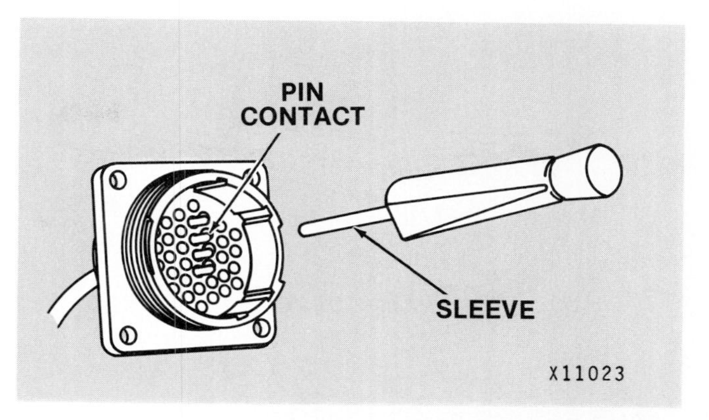

Fig. 11 — Removing Contact from Circular Plastic Connector Housing (Not Inserted)

1. Select the proper tool (Fig. 10). Align the sleeve of the tool (Fig. 11) with the contact to be removed.

2. Push wire in slightly.

3. While holding the handle, insert the sleeve of the tool into the cavity until it bottoms (Fig. 12). Allow the push rod to back out during insertion (Fig. 13).

4. Rotate the handle of the tool in either direction to assure the release of the contact locking lances.

5. Keep the sleeve firmly bottomed in the cavity and depress the push rod button (Fig. 14) to eject the contact.

NOTE: If you intend to reinstall the pin contact, bend back the locking lance with a knife blade as shown in Fig. 15.

Fig. 12 — Inserting the Tool

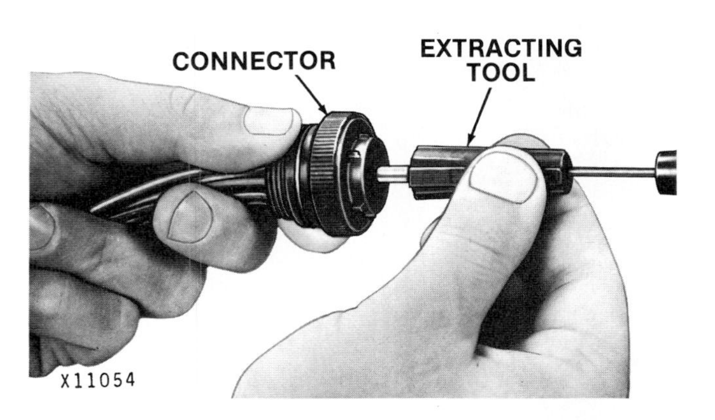

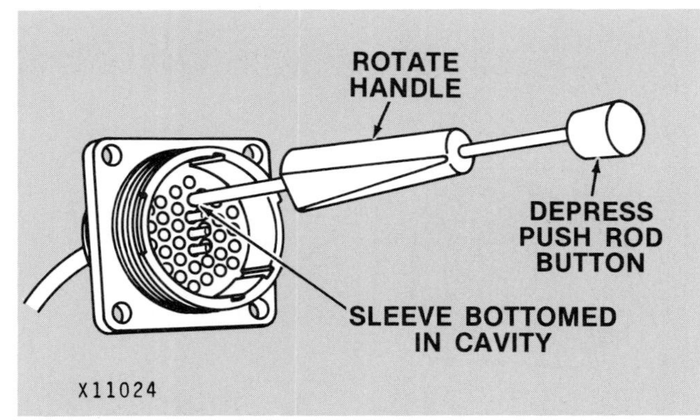

Fig. 13 — Removing Contact from Circular Plastic Connector Housing (Inserted)

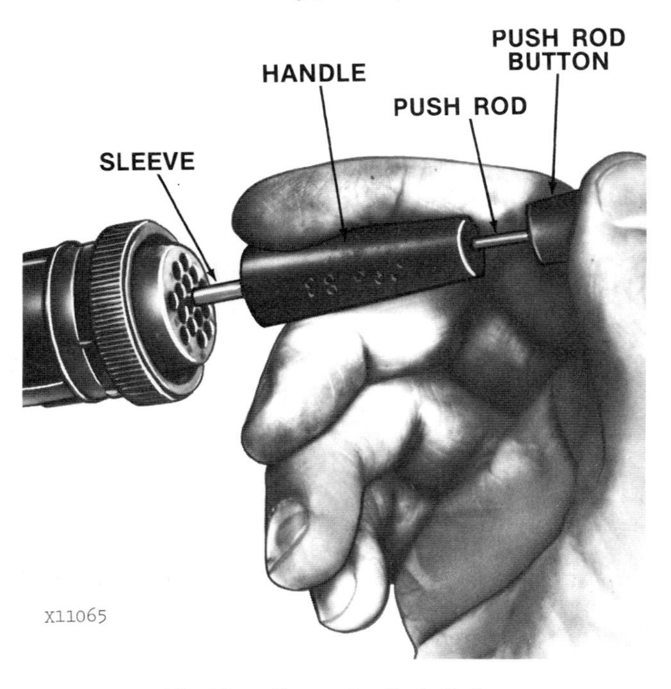

Fig. 14 — Depressing Push Rod

Fig. 15 — Resetting Locking Lance

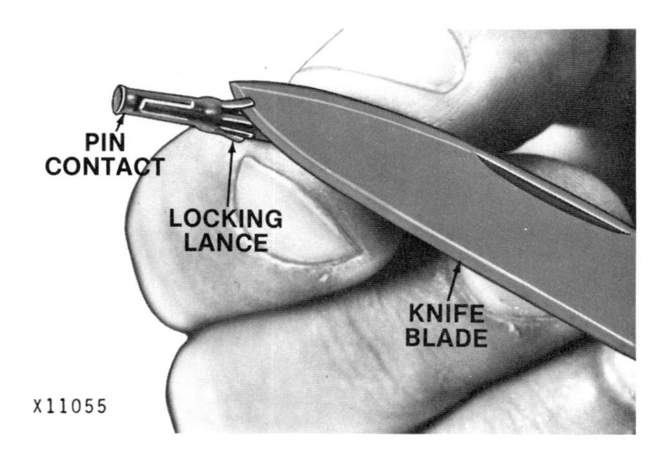

CIRCULAR PLASTIC CONNECTOR — BLADE-TYPE CONTACTS

The CPC shown in Fig. 16 will accommodate seven blade contacts, although the one pictured shows only four installed with the three center cavities vacant. Again, like the CPC with pin contacts, make sure that corresponding cavities in each connector half are used because they will only mate one way. Mating and locking the connector halves is the same as for the CPC with pin contacts.

Installing Blade-Type Contacts

Like pin contacts, blade contacts are also installed from the rear of the connector halves by simply pushing them in. Make sure that the male contacts are installed in the female connector half and vice versa just like for the CPC with pin-type contacts.

Fig. 16 — Circular Plastic Connector for Blade-Type Contacts

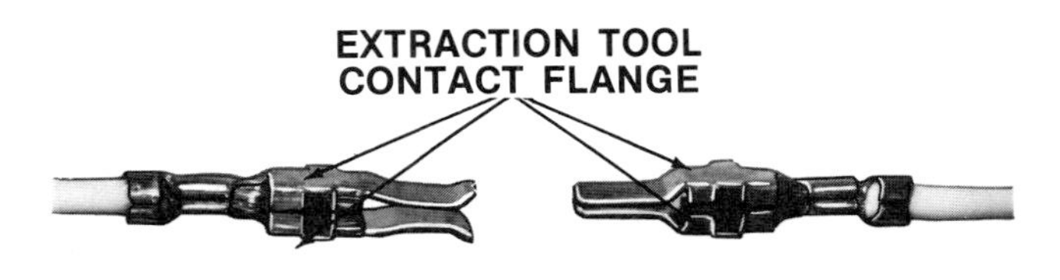

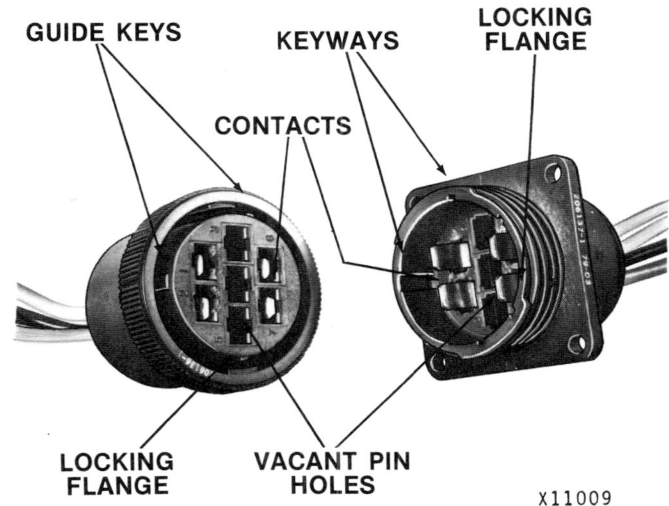

X11014

Fig. 17 — Extractor Tool for Blade-Type Contacts

Extracting Blade-Type Contacts

1. Select the appropriate tool (Fig. 17).

2. Push the tips of the tool into the offsets in the cavity of the housing until they bottom. It may be necessary to wiggle the tool during insertion to get it to bottom completely. But if it is necessary, be sure to wiggle the tool only in the direction of the edges of the tool tips as shown by the arrows in Fig. 18. You may bend the tool tips if you work the tool up and down against the flat surfaces of the tool tips.

Fig. 18 — Removing Blade-Type Contacts from Connector Housing

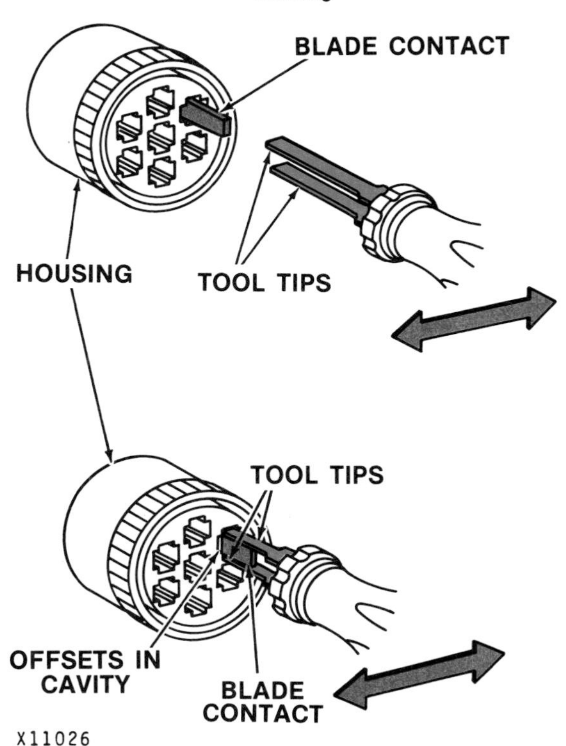

X11026

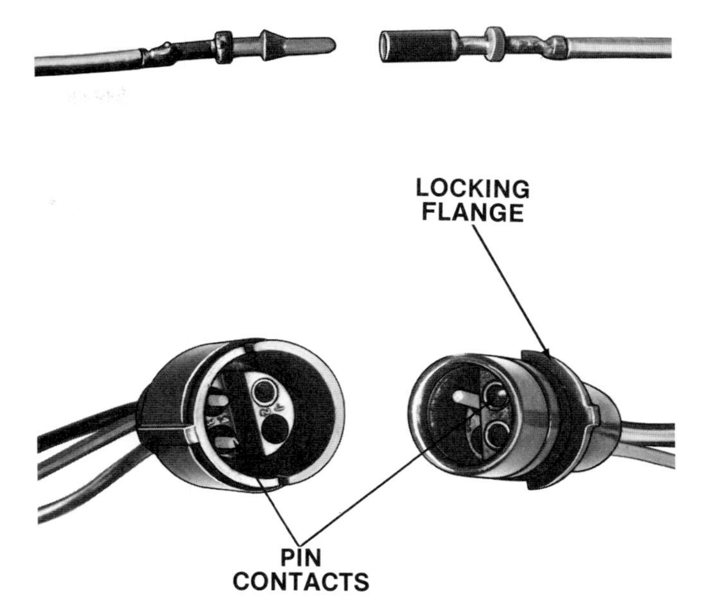

X11010

Fig. 19 — Sure-Seal™ Connector and Pin Contacts

3. Carefully, pull the wire lead to remove the contact.

SURE-SEAL™ CONNECTOR

The Sure-Seal™ connector and pin contacts are shown in Fig. 19. These connector housings also have provisions for accurate mating between the two halves, but instead of using guide keys and keyways, the connector bodies are molded such that they will not mate incorrectly. Unlike the other connectors discussed in this chapter, the Sure-Seal requires two special tools for installing the pin contacts (Fig. 20).

Fig. 20 — Insertion Tool and Holding Plate for Installing Pin Contacts in Sure-Seal™ Connector

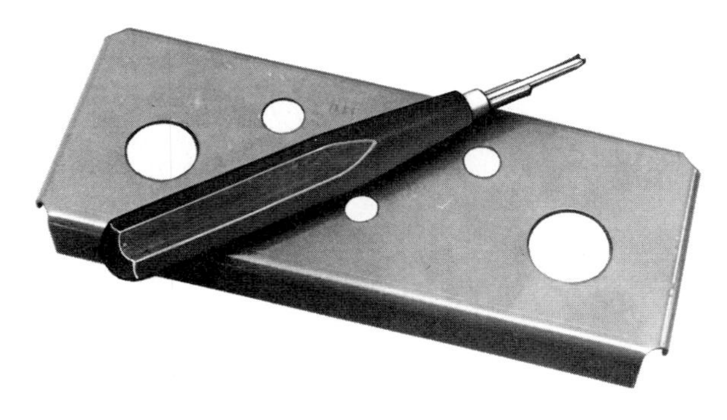

X11015

Litho in U.S.A.

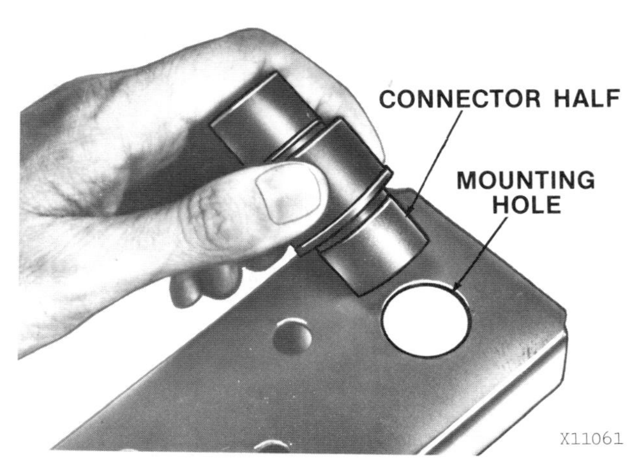

Fig. 21 — Sure-Seal™ Connector and Holding Plate

Installing Sure-Seal™ Pin Contacts

1. Determine the number of the hole in which the pin contact is to be installed.

2. Mate this connector half to the other connector half and place in the proper size hole of the holding plate (Fig. 21).

3. Start the contact into the connector (Fig. 22) and position the wire inside the tip of the contact insertion tool (Fig. 23) so that the tip of the tool butts against the contact shoulder.

4. Holding the insertion tool firmly, insert the contact into the cavity to the depth of the insertion tool shoulder (Fig. 24). Remove the tool and pull the wire slightly to insure that the contact is properly seated.

Fig. 22 — Positioning Pin Contact

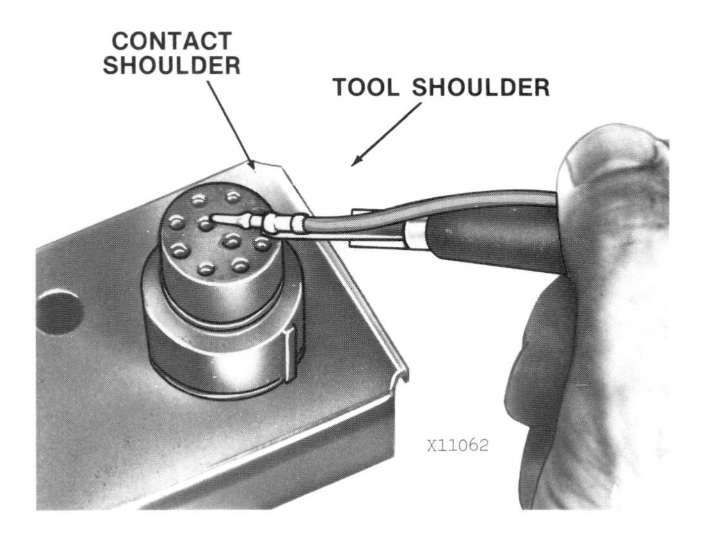

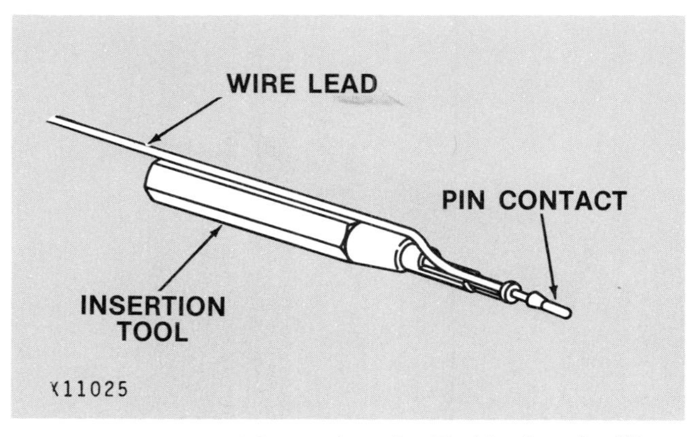

Fig. 23 — Use of Contact Insertion Tool for Sure-Seal™ Connector

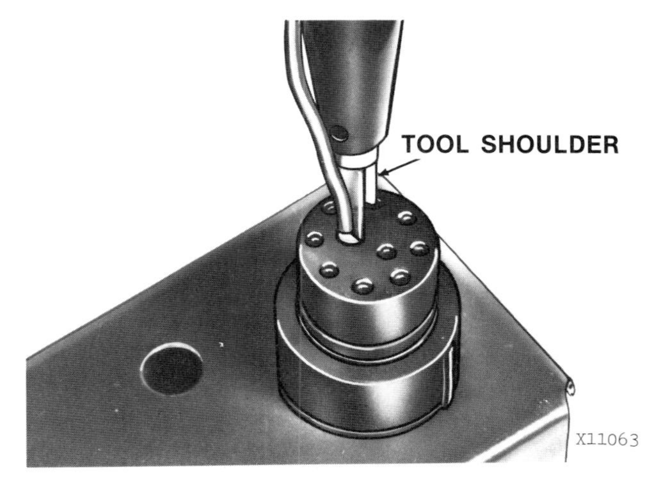

Fig. 24 — Inserting Pin into Connector

5. Install cavity plugs (Fig. 25 and 26) in all unused holes in the connector housings.

NOTE: Plugs must be installed from the wire side.

Fig. 25 — Connector Cavity Plug

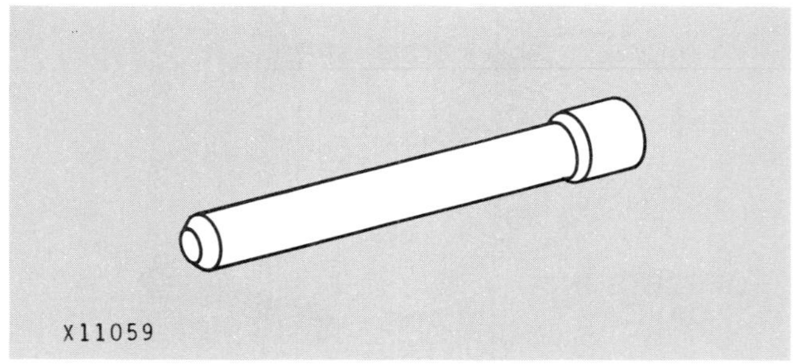

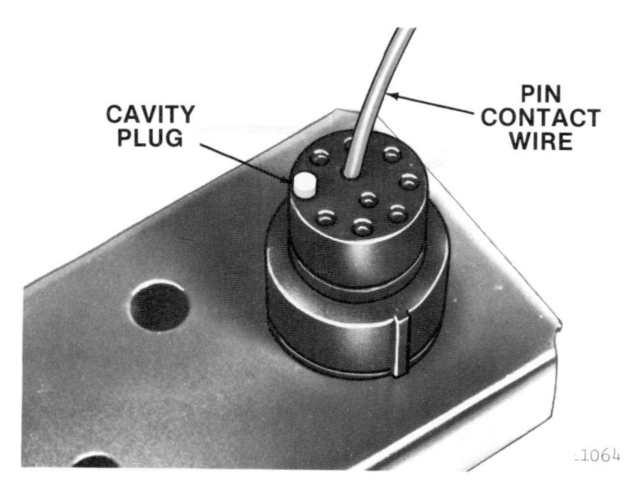

Fig. 26 — Connector with Cavity Plug Installed

Removing Sure-Seal™ Pin Contacts

There is no special tool for removing the pin contacts from the Sure-Seal™ connector housing. If a contact is damaged, first try pulling it out by grasping the wire. If the wire comes loose, use the tip of a small screwdriver or any other suitable tool to press the contact from the connector housing.

Fig. 27 — Metrimate™ Connector and Pin Contacts

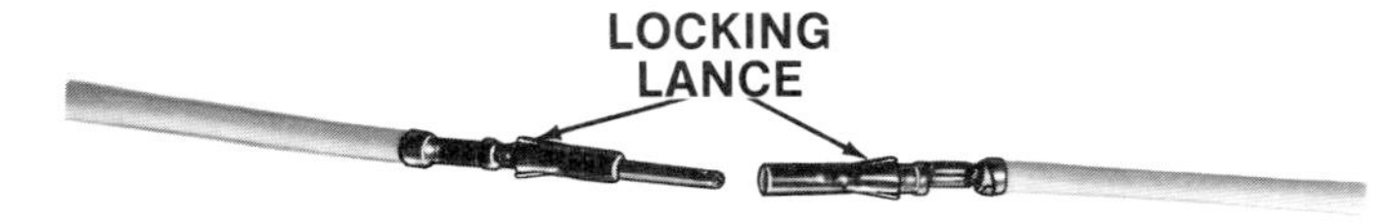

METRIMATE™ CONNECTOR

The Metrimate™ connector and pin contacts are shown in Fig. 27. Like the other connectors, it too has a safeguard so that it can be mated only one way. The locking mechanism consists of clips and to separate the connector halves, the clips must be depressed while pulling them apart. The procedures for removing and installing pin contacts are the same as for the circular plastic connector for pin-type contacts discussed earlier.

MATE-N-LOK™ CONNECTOR

The Mate-N-Lok™ connector shown in Fig. 28 has provisions for six pin contacts although in this picture, only three are installed. Like the other connectors, the female part of the contact or socket is installed in the male half of the connector body and the male part or pin, in the female half. The locking mechanism engages as the connector halves are mated. To disengage, the locking clips on the male connector half must be depressed as the two connector halves are pulled apart.

Fig. 28—Mate-N-Lok™ Connector and Pin Contacts

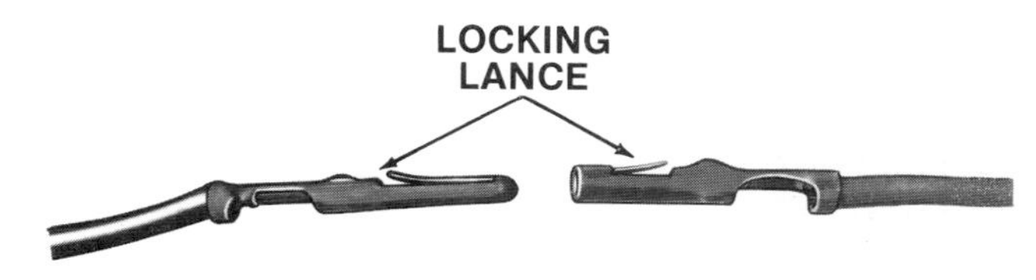

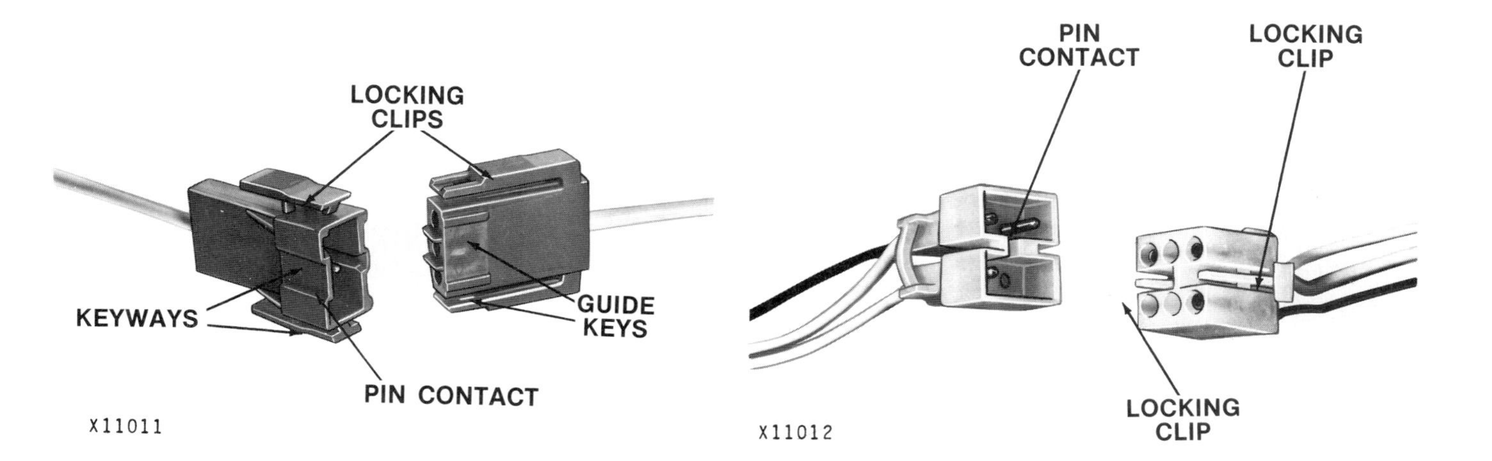

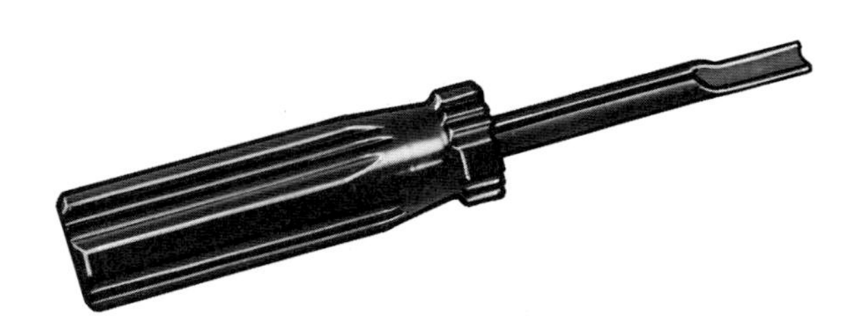

Fig. 29—Extractor Tool for Pin Contacts in Mate-N-Lok™ Connector

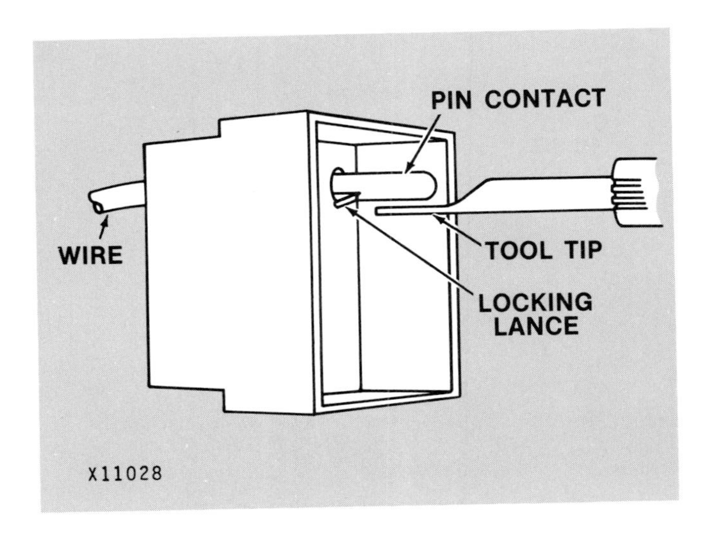

Fig. 31—Removing Pin Contact from Mate-N-Lok™ Connector Housing

Installing Mate-N-Lok™ Pin Contacts

These pin contacts are easily installed, and like the circular plastic connector, are simply pushed in from the rear of the connector housing. Just make sure that the locking lance is engaged so the contact will not pull out.

Extracting Mate-N-Lok™ Pin Contacts

Removing or extracting the pin contacts is also a simple procedure which involves nothing more than depressing the locking lance and pulling the contact from the connector housing.

Fig. 30—Removing Pin Contact Socket from Mate-N-Lok™ Connector

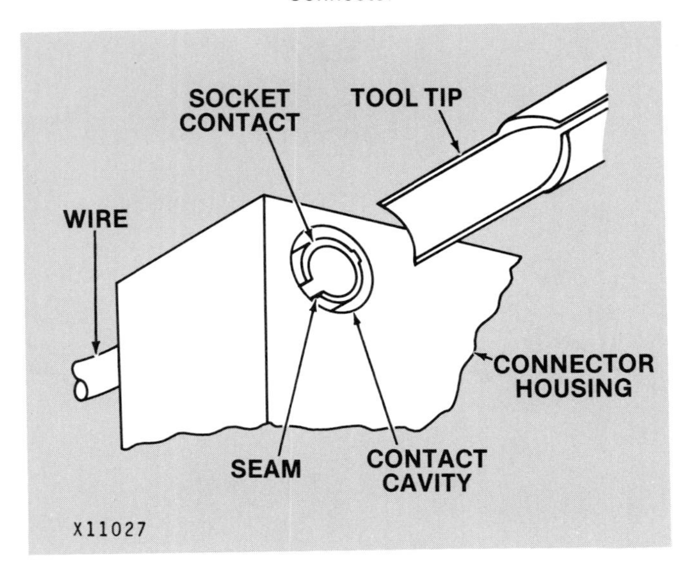

To remove the female pin contacts or sockets:

1. Push the socket into the cavity of the connector as far as it will go. Align the tool (shown in Fig. 29) so that the tip is positioned opposite the socket seam (Fig. 30).

2. Bottom the tool in the cavity. Pull the wire to remove the contact.

To remove the male pin contacts or pins:

1. Push the pin into the cavity as far as it will go. Place the tool tip (Fig. 29) against the locking lance (Fig. 31).

2. Depress the locking lance with the tool tip and pull the wire to remove the pin contact.

Not all Mate-N-Lok™ connectors use the same size pin contacts. Other tools that can be used to remove the contacts are shown in Fig. 32 and Fig. 33.

Fig. 32—Extractor Tool for Pin Contacts in Circular, Metrimate™, and Mate-N-Lok™ Connector

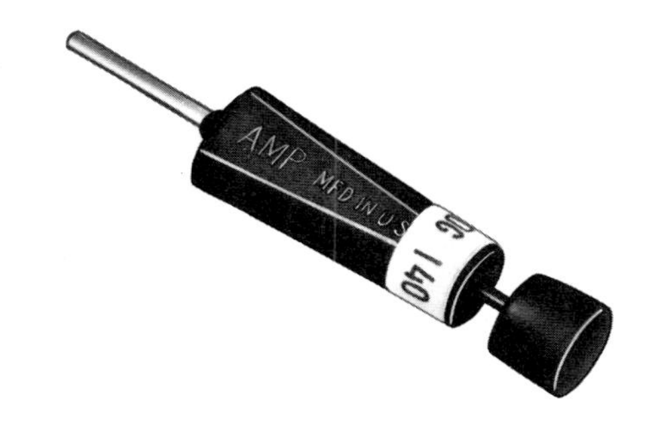

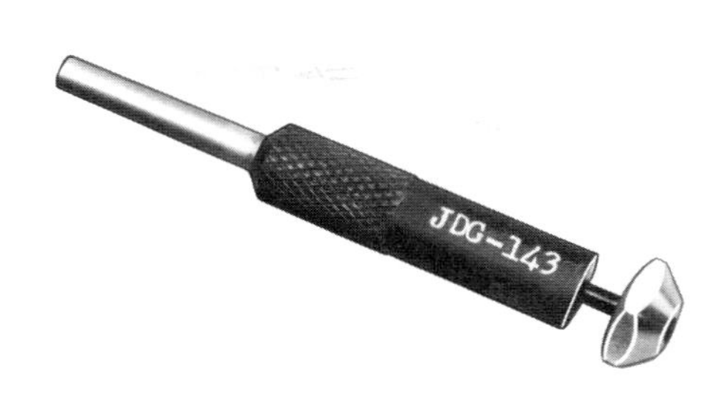

Fig. 33 — Extractor Tool for Large Pin Contacts in Mate-N-Lok™ Connector

DEUTSCH™ CONNECTOR

The Deutsch™ connector (Fig. 34) is circular like the circular plastic connectors discussed earlier but is made from metal rather than plastic. It also has soft rubber around the cavity holes to seal out moisture, dust, or any other type of contaminant. This connector is available in different sizes to accommodate varying numbers of pin contacts, and some models will take two different sizes of pins.

Extracting Pin Contacts

Special extracting tools are available in different sizes to remove the pin contacts from the Deutsch™ connectors. The size tool you use depends on the gauge of the wire lead. But note that this tool is used from the **wire** side of the connector rather than the pin side like all the others discussed. The following steps explain how to remove the pin contacts.

Fig. 34 — Deutsch™ Connector and Pin Contacts

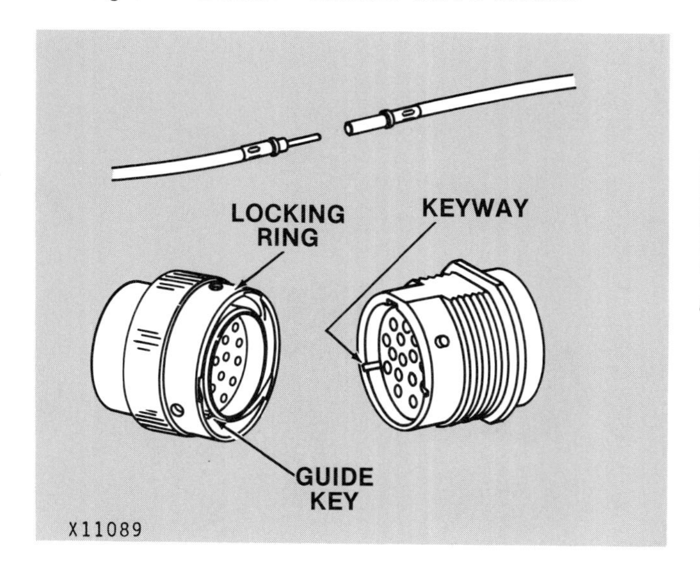

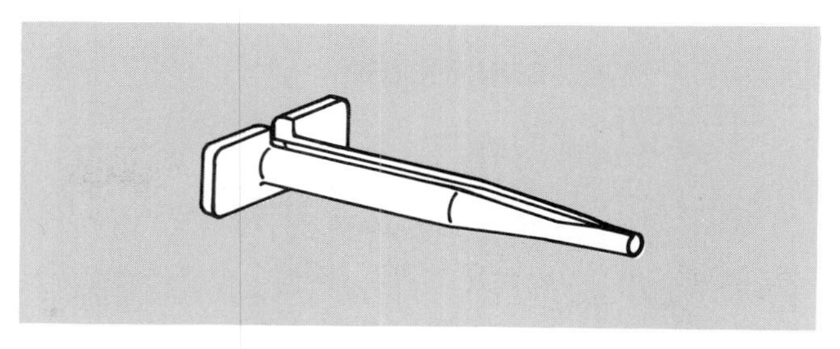

Fig. 35 — Extraction Tool for Deutsch™ Connector Pins

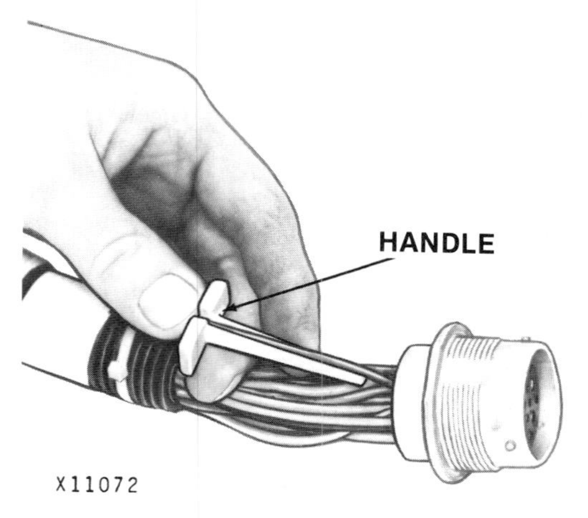

Fig. 36 — Starting Extractor Tool Onto Wire

1. Select correct size tool (Fig. 35) for the size of the wire to be removed.

2. Insert the wire into the tool starting at the handle end (Fig. 36). Slide the tool rearward along the wire (away from the connector) allowing the wire to slide into the tool until the tool tip snaps onto wire (Fig. 37). You may have to spread the tool tips slightly with your fingernail, but be very careful not to break the tip.

Fig. 37 — Wire Inside Extractor Tool

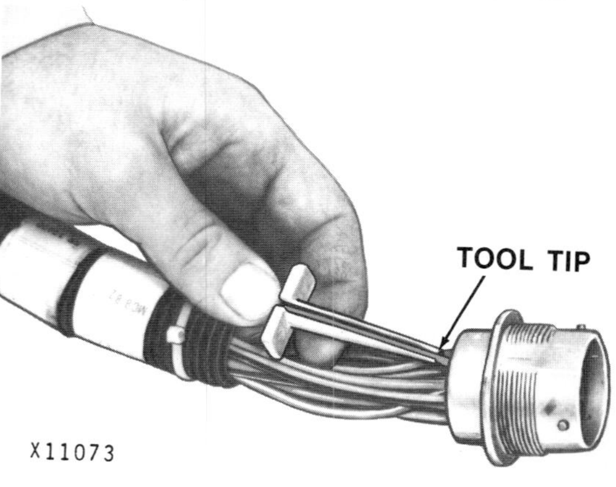

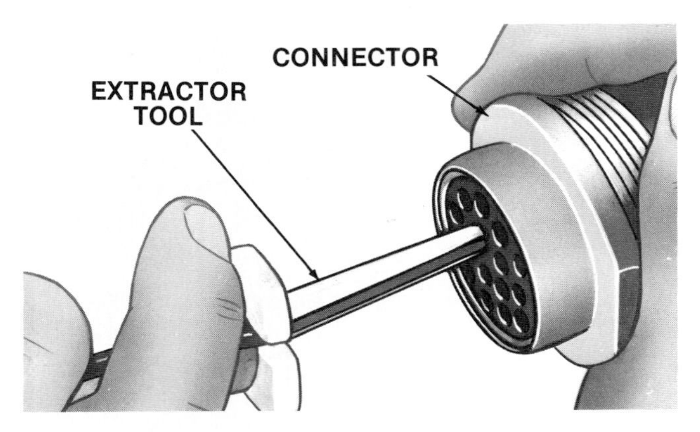

Fig. 38 — Inserting Extractor Tool Into Connector

3. Slide the tool along the wire toward the connector and into the pin cavity until the tool bottoms inside the connector (Fig. 38).

IMPORTANT: Do not twist the tool when inserting into the connector. The twisting action may break the tool tips.

4. Pull the wire and the extraction tool from the connector body (Fig. 39).

Installing Pin Contacts

Like the other connectors, the pin contacts are installed from the rear of the connector (Fig. 40).

1. Push pin contact straight into connector body until you feel a positive stop. You will also hear a soft click as the locking mechanism inside the connector body locks into place.

2. Pull on the wire slightly to make sure the pin is locked into place.

Fig. 39 — Removing Wire, Contact, and Tool from Connector

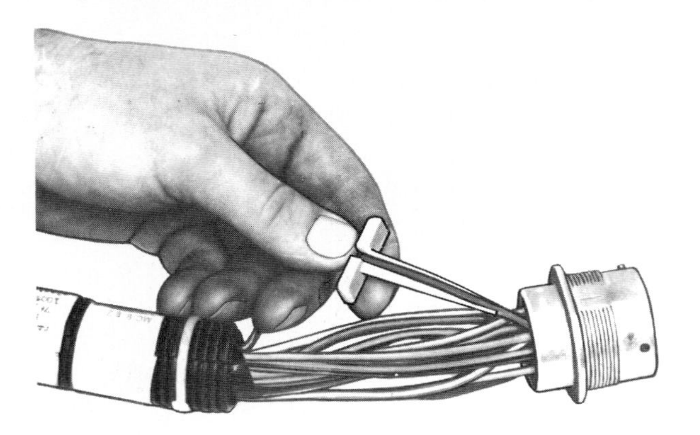

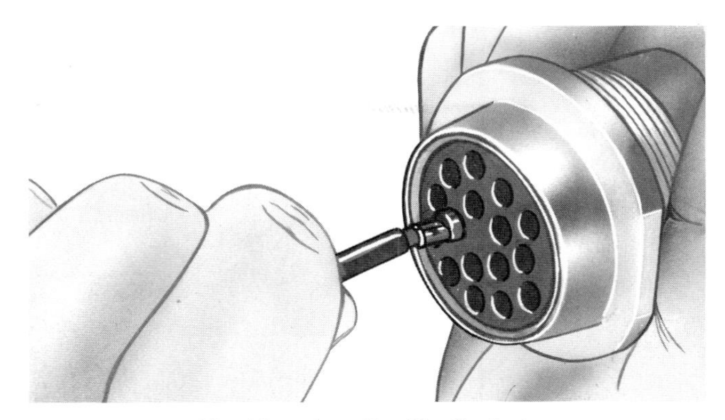

Fig. 40 — Inserting Pin Contact

WEATHER PACK™ CONNECTOR

The Weather Pack™ connector has a molded self-lubricating silicone seal that comes assembled to the male connector half (Fig. 41). When the two connector halves are mated, the seal creates an effective environmental seal between the connector halves. This connector, like the others, is equipped with a locking mechanism between the two halves. To keep moisture and other contaminants from entering the connector at the wire leads, cable seals (Fig. 42) are used on each wire lead. In vacant cavity holes, cavity plugs (also in Fig. 42) are installed.

Extracting Pin Contacts

The Weather Pack™ connector also requires a special tool to remove the pin contacts. The tool is similar to the one used for the circular plastic connector but does not have a plunger. However, the tool is hollow, even the handle, so that a nail or wire can be inserted all the

Fig. 41 — Weather Pack™ Connector and Pin Contacts

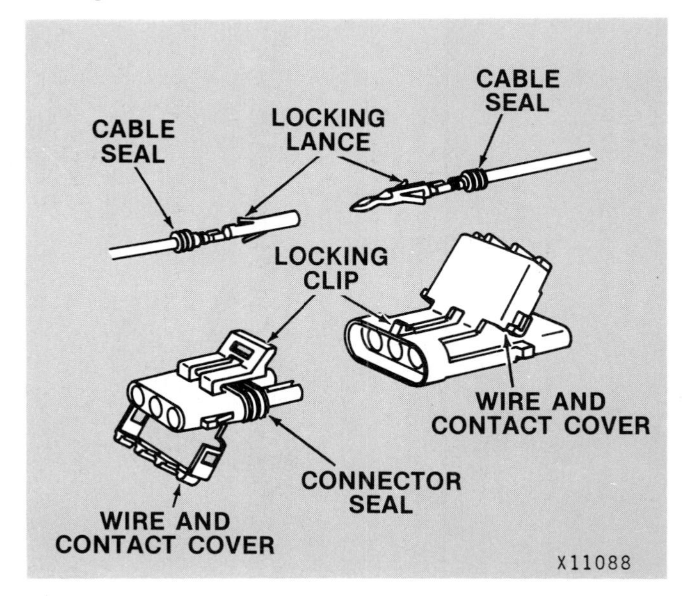

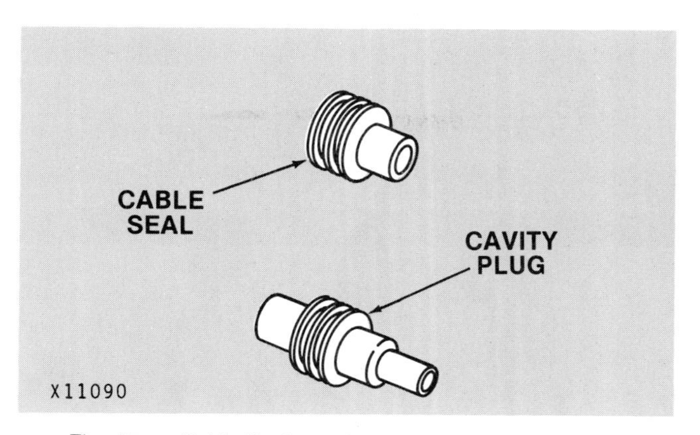

Fig. 42 — Cable Seal and Cavity Plug for Weather Pack™ Connector

way through the tool to push the pin contact out should it be necessary.

The procedure for removing the contact pins follows:

1. Open the wire cover protector (Fig. 43).

2. Insert the extractor tool over the pin contact in the connector body (Fig. 44). Make sure the tool bottoms in the connector so that the locking lances on the contact will be depressed.

3. Hold the extractor tool fully seated (Fig. 45) and pull wire and pin contact from the connector body.

NOTE: If the pin contact cannot be removed, insert a wire or nail through the tool handle and push the pin out.

Installing Pin Contacts

1. Push pin contact straight into connector body from the rear (Fig. 46) until you feel a positive stop. Make sure the cable seal is flush with the connector body.

Fig. 43 — Weather Pack™ Connector with Wire Cover Open

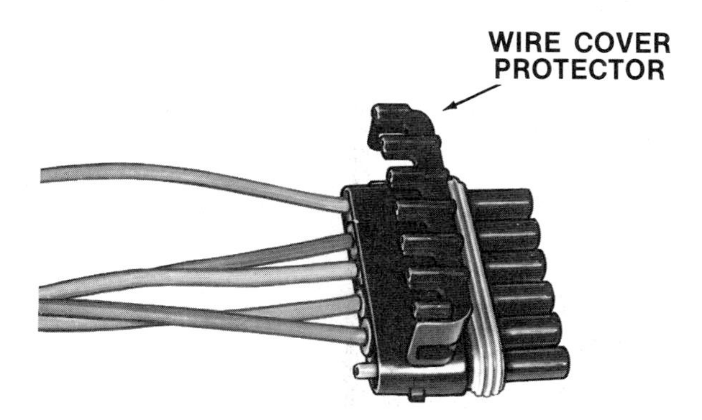

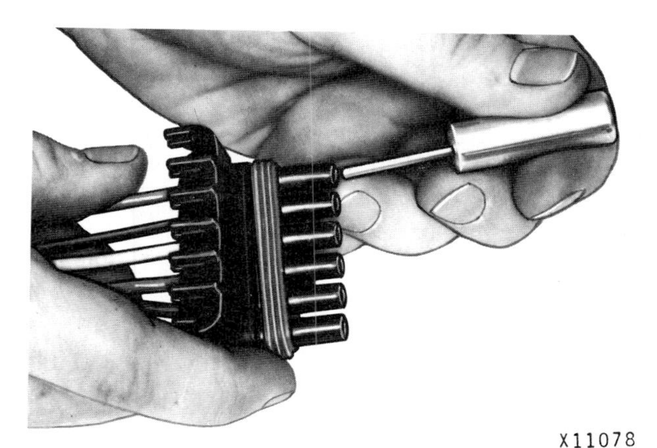

Fig. 44 — Inserting Extractor Tool

2. Pull on the wire slightly to make sure the pin contact is locked into place.

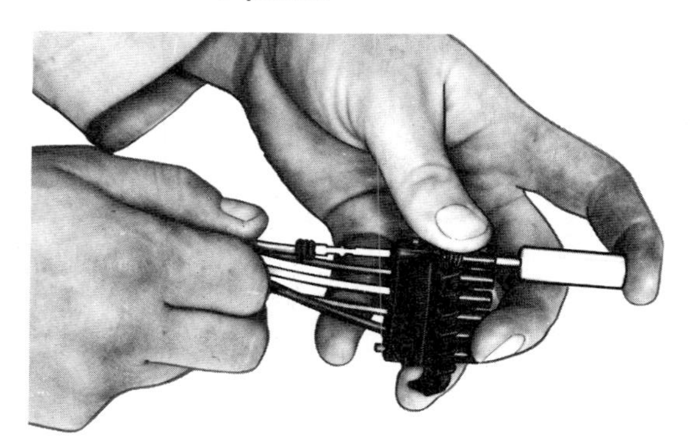

Fig. 45 — Removing Pin Contact

Fig. 46 — Installing Pin Contacts

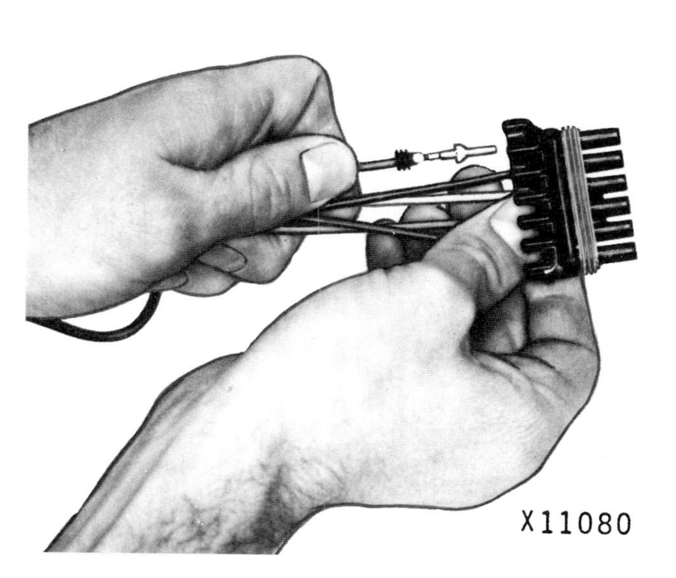

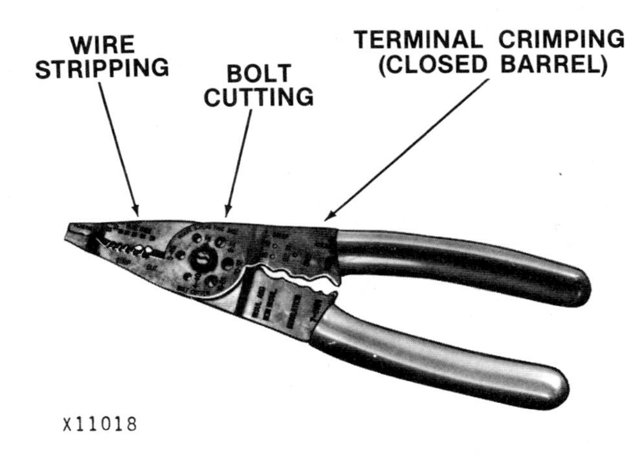

Fig. 47 — Universal Electrician's Plier

STRIPPING, CRIMPING, AND BOLT CUTTING

In the remaining part of this chapter, we will discuss wire stripping, crimping terminals to wires, soldering terminal connections, and bolt cutting in relation to using universal electrician's pliers (Fig. 47) and the terminal applicator (Fig. 48). These tools, especially the terminal applicator, are fairly expensive but are practically indispensable when it comes to repairing electrical wiring and connectors. And you do need both tools. Even though they do have two common functions, only the terminal applicator will crimp open-barrel terminals and only the electrician's plier will strip wires. Two other types of wire strippers and crimpers are shown in Fig. 49.

Fig. 48 — Terminal Applicator

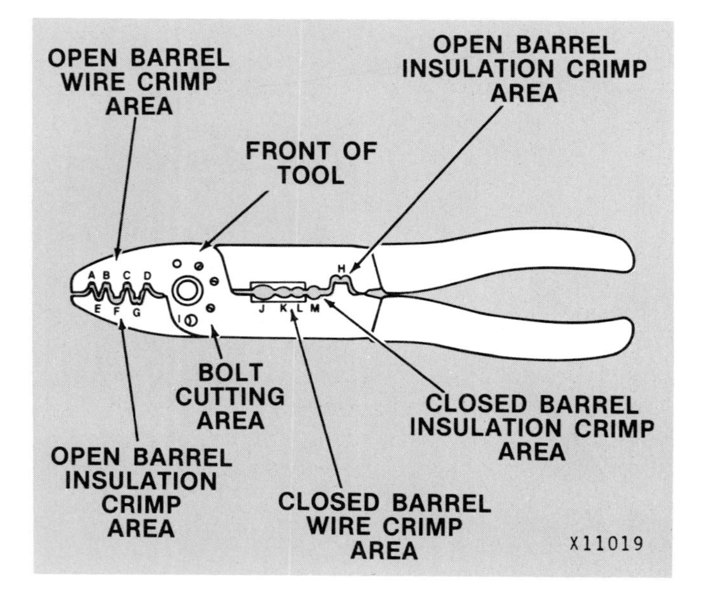

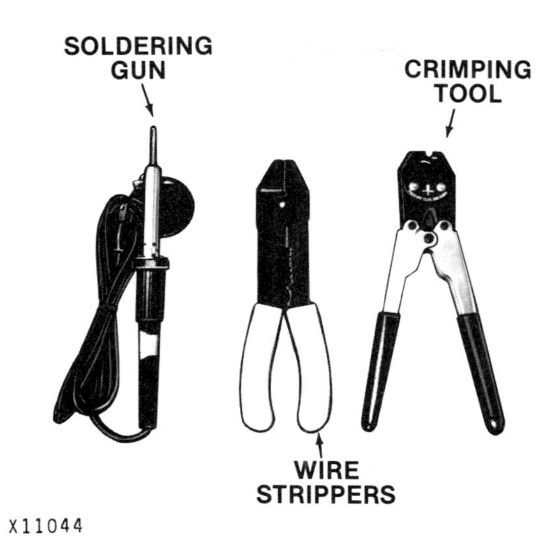

Fig. 49 — Special Tools

WIRE STRIPPING

The electrician's plier (Fig. 47) makes wire stripping a simple and easy task. But to use them properly, you have to know the gauge of the wire you intend to strip because the plier has several sizes of stripping holes. If you use a stripping hole that is too small, you will not only cut through the insulation but also through part of the wire strands. If the hole is too large, you will not cut through the insulation and then not be able to strip it off.

As we said, wire stripping is a simple task, but like anything else, the more practice you have, the better you will become at doing it. Generally, about all you have to do is match the gauge of the wire to the proper stripping hole on the plier (Fig. 50), insert the wire, close

Fig. 50 — Stripping Wires

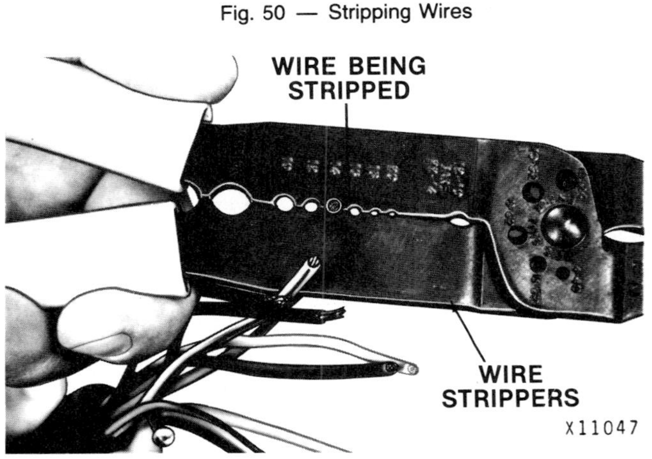

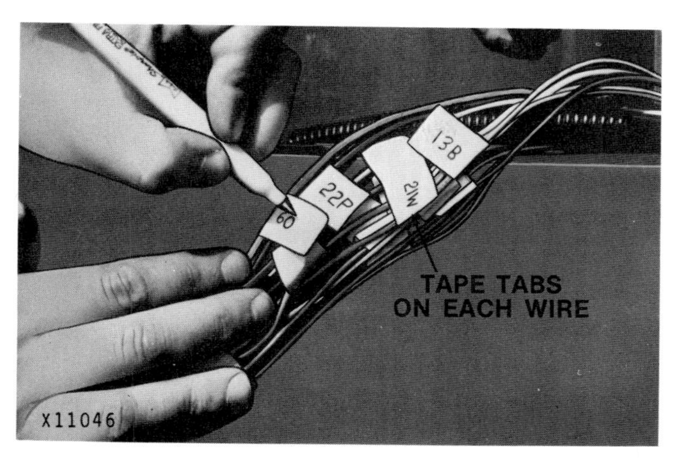

Fig. 51 — Marking Wires with Circuit Code Numbers

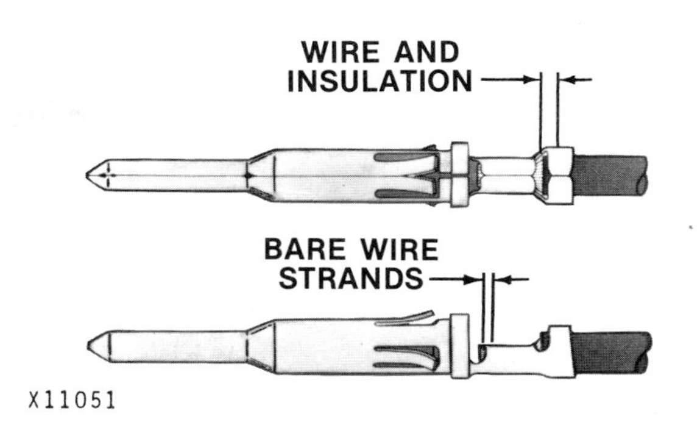

Fig. 52 — Relationship Between Stripped Wire and Pin Contact

the plier, and pull the insulation off. Sometimes, the most difficult task is determining the gauge of the wire. If the gauge is not printed directly on the insulation, you will have to refer to the wiring diagram for whatever piece of electrical equipment or machine you are working on. Also, if you are replacing several pins in a connector, you should label the wire according to the wiring diagram (Fig. 51). Do not depend solely on color. It may have faded or have grease or dirt on it.

IMPORTANT: The wires in any one connector may be of different gauges. More than likely, they are. Do not assume they are all the same gauge.

How much insulation should be stripped from the wire? That depends entirely on the terminal, pin, or contact you will be using. If you will notice in Fig. 52, the pin has two points where it is crimped. One point is around the bare wire strands; the other, around the end of the insulation. So the amount of insulation stripped from the wire is critical. You will simply have to measure this amount on the particular terminal you are using.

After the insulation is stripped, **do not** twist the exposed bare wire strands as you may be accustomed to doing when replacing a plug on a lamp or toaster. If you twist the strands, they will overlap each other and not run parallel with the terminal. This may cause you to cut some of the strands with the crimp wings.

To review the wire stripping process, we can put it into four distinct steps:

1. Determine the amount of insulation to be stripped depending on the terminal or pin to be used. Again, refer to Fig. 52.

2. Select the stripping hole on the plier that corresponds to the gauge of the wire.

3. Insert the wire to the point determined in step 1 and cut the insulation. Release the plier just slightly, rotate the wire about 180 degrees, and again close the plier to make sure that the insulation is cut all the way around.

4. Pull the wire from the plier to remove the cut insulation.

As we mentioned, with practice, wire stripping will become practically automatic. It would be a good idea to obtain several pieces of scrap wire of different gauges and practice stripping insulation of various lengths.

CRIMPING

In working with electrical connectors, wiring, and terminals, you will be faced with two types of crimping — open-barrel and closed-barrel. Both of these types are illustrated in Fig. 53. The pin contacts shown in Fig. 40, which are installed in connector bodies, are of the open-barrel type. Most terminals are closed-barrel.

Open-Barrel Crimping

Since the open-barrel pin contacts are installed in connector bodies, the crimping process is critical but certainly not difficult, at least not with a little practice. Crimping these pin contacts is actually a two-step procedure whereby you first crimp the bare wire strands in the wire barrel part of the contact and then the tip

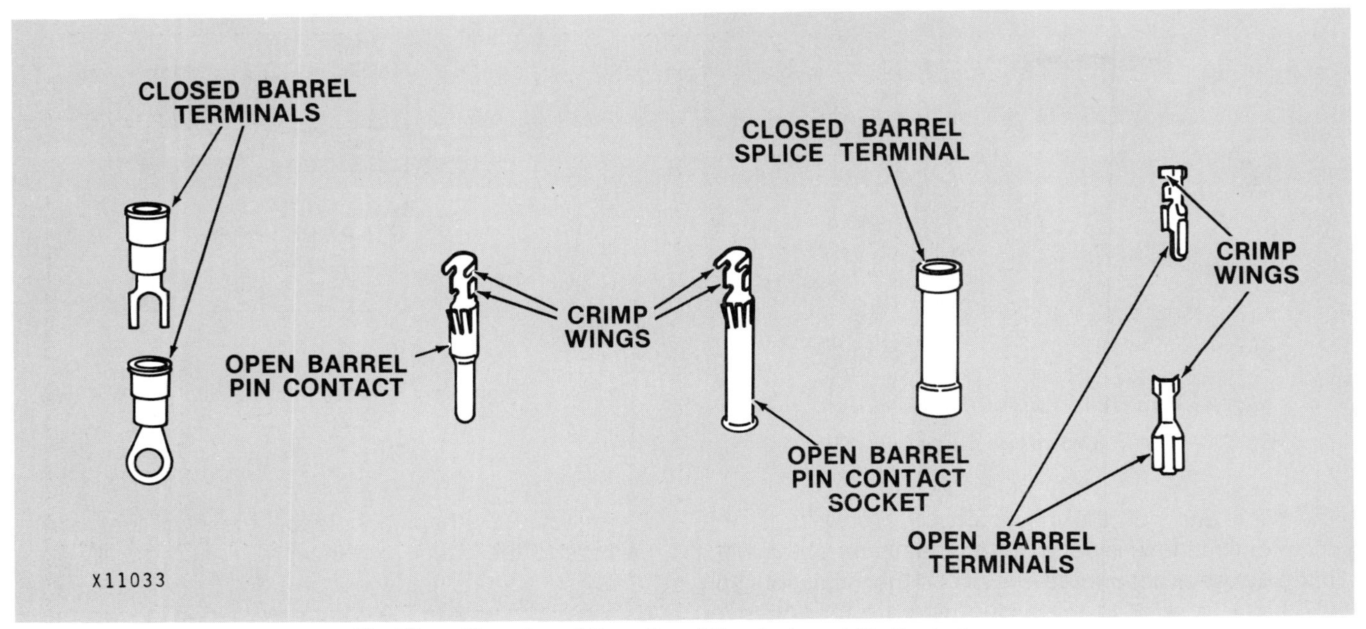

Fig. 53 — Closed- and Open-Barrel Terminals and Pin Contacts

of the insulation in the insulation barrel. The wire barrel part of the contact is toward the center of the contact and has shorter but wider crimp wings. The insulation barrel is on the end and has the long and narrow crimp wings.

To perform the crimping procedure, you will need a crimping tool (Fig. 54). Fig. 55 shows another type in actual use. Follow these steps to obtain a good crimp.

Fig. 54 — Terminal Applicator

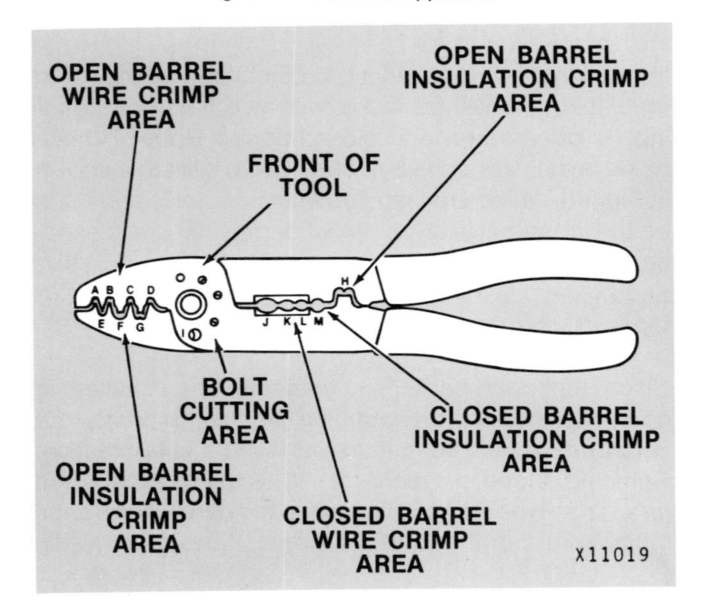

1. Position the wire barrel of the contact in the best suited area (A, B, C, or D) with the wire barrel opening facing the letter as shown in Fig. 56, view A. Close the handles of the tool just enough to grasp the contact and hold it in place.

IMPORTANT: The pin or socket must be positioned in the crimping tool with the crimp wings centered. Be certain that only the exposed wire is crimped. Do not insert the wire too far and crimp the wire barrel of the pin or socket on any insulation. Correct crimping is critical because of the low current flow in many of these circuits.

Fig. 55 — Crimping Pin Contact to Wire

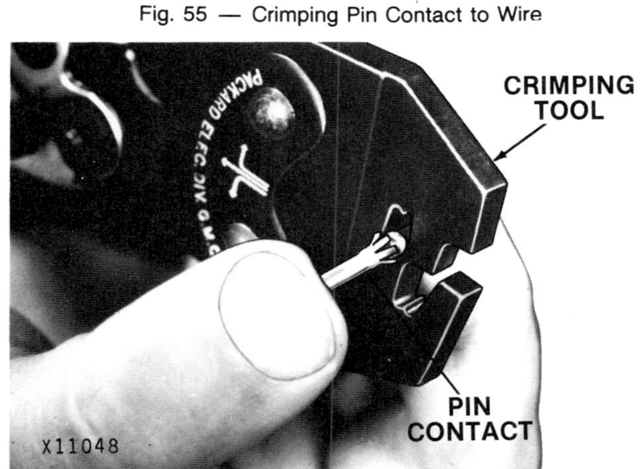

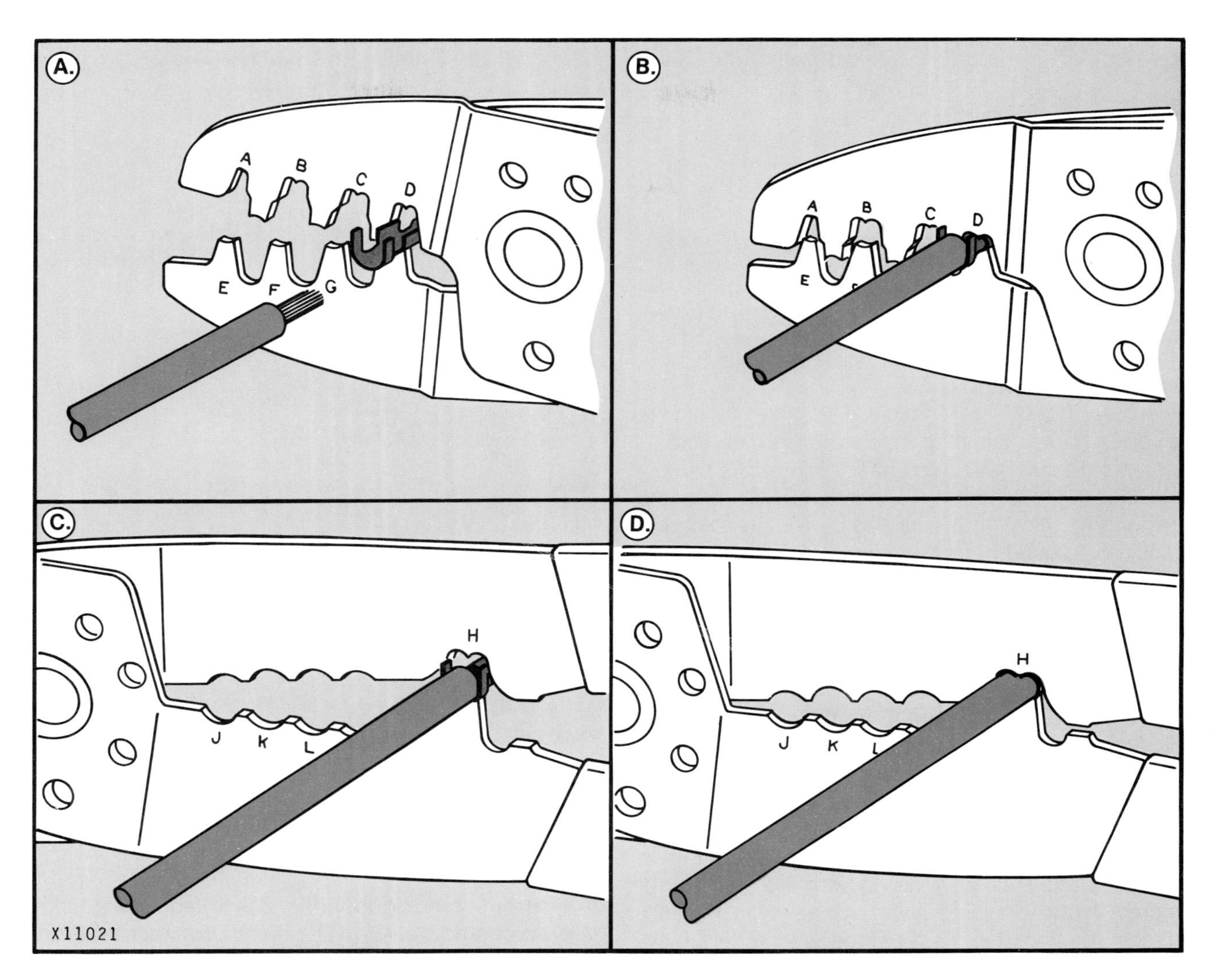

Fig. 56 — Open-Barrel Crimping Using Terminal Applicator Tool

NOTE: Unless you are very steady of hand, it is a good idea to rest the tool on a solid surface (table or workbench top) while grasping the contact and inserting the wire.

Fig. 57 — Pin Socket Before Insulation Crimp

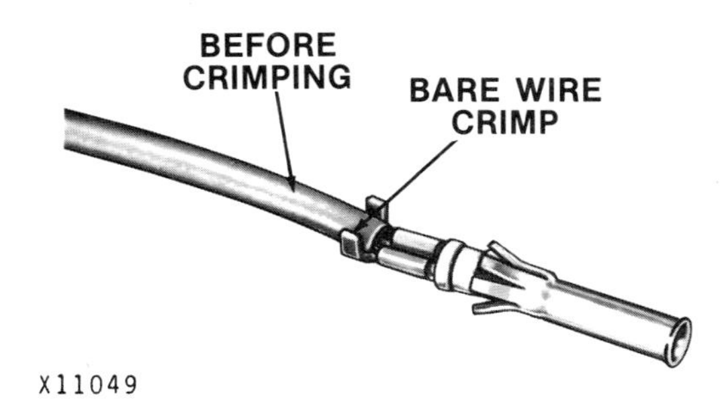

2. Insert a properly stripped wire into the wire barrel (Fig. 56, view B) and position it as shown in Fig. 52. Hold the wire in place and squeeze the tool handles to complete the crimp. At this point the contact and wire should be as shown in Fig. 57.

3. Position the insulation barrel in the best suited crimp area (E, F, G, or H) of the crimping tool (Fig. 56, view C) with the opening of the insulation barrel facing the letter on the tool.

4. Hold the insulation and wire in place and squeeze the tool handles to complete the crimp. At this point, after the completion of both crimps, the contact and wire should look as shown in Fig, 56, view D and removed from the tool in Fig. 58.

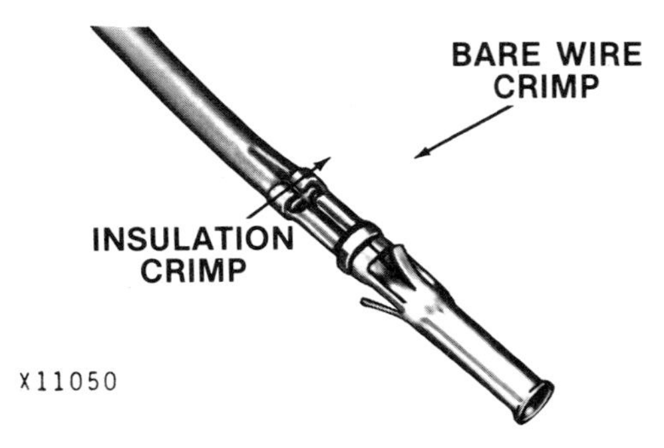

Fig. 58 — Pin Socket After Both Crimps

Closed-Barrel Crimping

Like open-barrel crimping, closed-barrel is also a two-step process. But closed-barrel crimps are usually associated with terminal connections like the two shown in Fig. 53. With the closed-barrel, no soldering is required. Closed-barrel terminals also have a wire barrel part and an insulation barrel part. These two parts of the barrel are clearly evident as Fig. 53 shows. The insulation barrel is on the end of the terminal and is larger in diameter than the wire barrel.

Either the electrician's plier (Fig. 47) or the terminal applicator (Fig. 48) may be used for closed-barrel crimps. Follow these steps to crimp a closed-barrel terminal.

1. Position the wire barrel of the terminal or contact in the best suited area (J, K, or L) with the wire barrel centered in the crimping jaws as shown in Fig. 59. Squeeze the tool handles just enough to hold the terminal in place.

2. Insert a properly stripped wire into the wire barrel as shown in Fig. 60. Hold the wire in place and squeeze the tool handles to complete the crimp.

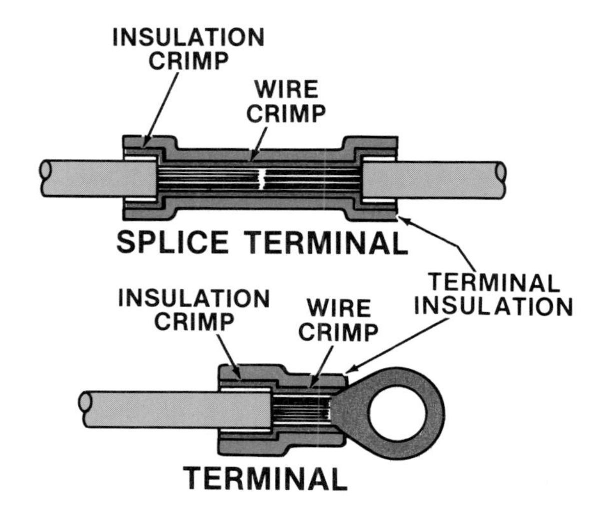

Fig. 60 — Cross Section View of a Closed-Barrel Splice and Terminal

3. Position the insulation barrel in the crimp area marked M so that it is centered in the crimping jaws as shown in Fig. 59.

4. Hold the terminal in place and squeeze the tool handles to complete the crimp.

Crimping Pin Contacts For Deutsch™ Connectors

A different crimping tool (Fig. 61) is required to crimp the pin contacts used in Deutsch™ connectors; therefore, the actual crimping procedure is different.

Follow these steps to crimp pin contacts for Deutsch™ connectors.

1. Strip 1/4 inch (6 mm) of insulation from wire.

Fig. 59 — Closed-Barrel Crimping Using Terminal Applicator Tool

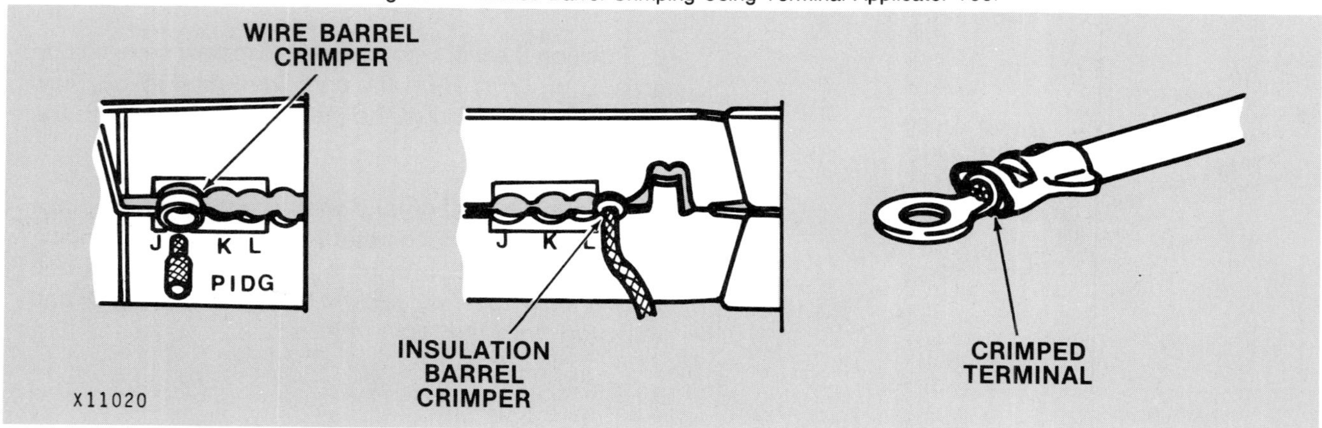

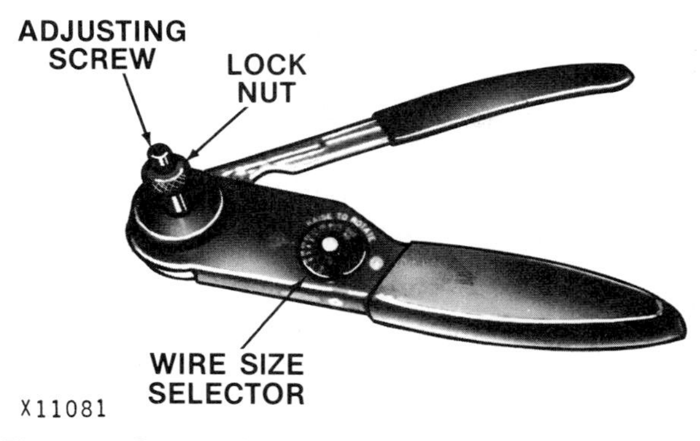

Fig. 61 — Crimping Tool for Deutsch™ Pin Contacts

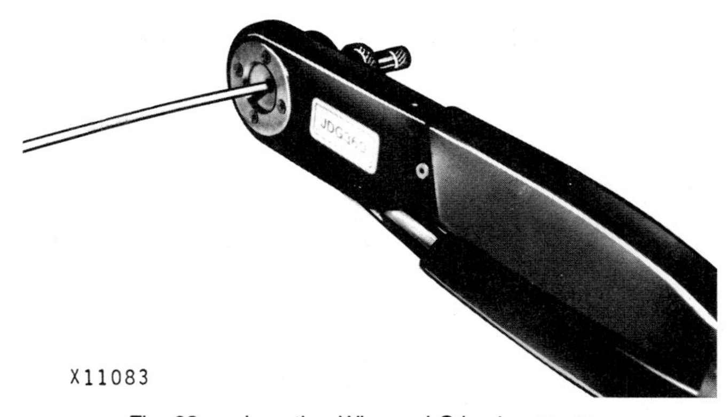

Fig. 63 — Inserting Wire and Crimping Pin Contact

2. Adjust wire size selector on crimping tool (Fig. 61) to correspond with the gauge of the wire.

3. Loosen the lock nut and turn adjusting screw in until it stops.

4. Insert a pin contact and turn adjusting screw until the contact is flush with the cover (Fig. 62). Tighten lock nut.

5. Close handle on crimping tool just enough to hold pin contact snug. While doing this, observe that the contact is centered between the crimping indentors in the tool.

6. Insert wire into the pin contact (Fig. 63) and close handle until it touches the stop.

7. Release handle and remove the contact.

8. Inspect the contact (Fig. 64) to be certain that all wires are crimped in the barrel.

Fig. 62 — Adjusting Crimping Tool

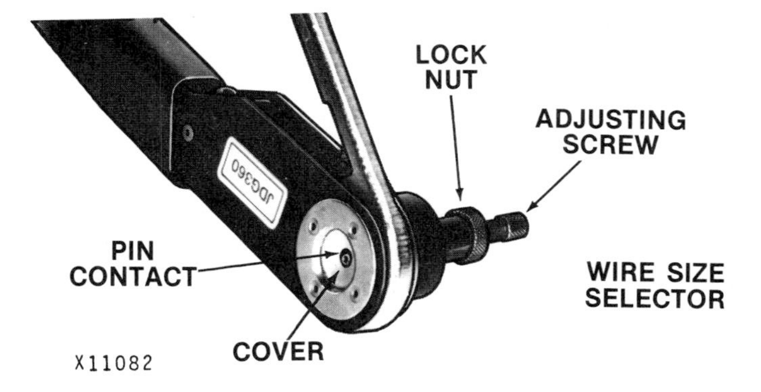

Crimping Pin Contacts For Weather Pack™ Connectors

The procedure for crimping contacts for Weather Pack™ connectors is actually the same as for the others (except Deutsch™). However, we are providing it separately because another part is involved in the crimping — the cable seal (Fig. 65).

Cable seals are color coded for three sizes of wire.

- Green — 18 to 20 gauge wire
- Gray — 14 to 16 gauge wire
- Blue — 10 to 12 gauge wire

The procedure for crimping Weather Pack™ pin contacts follows:

1. Slide the correct size cable seal onto wire. Refer to the color coding above.

2. Strip 1/4 inch (6 mm) of insulation from wire. Align cable seal with edge of the insulation (Fig. 65).

3. Position wire barrel of the contact in the crimping tool (Fig. 66). Close the handles of the tool just enough to grasp the contact and hold it in place. Be sure the crimp wings are centered.

Fig. 64 — Pin Contact After Crimping

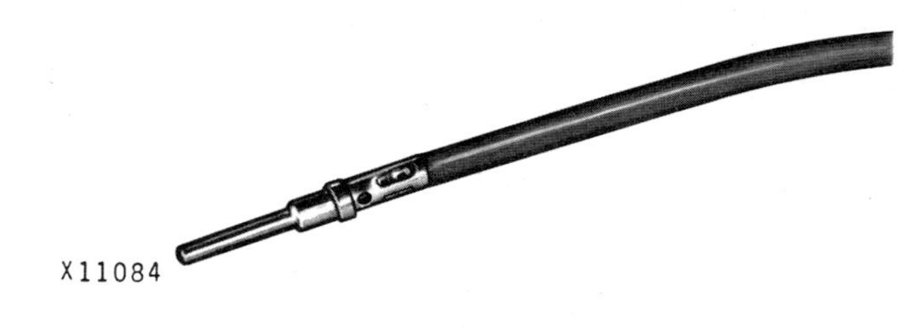

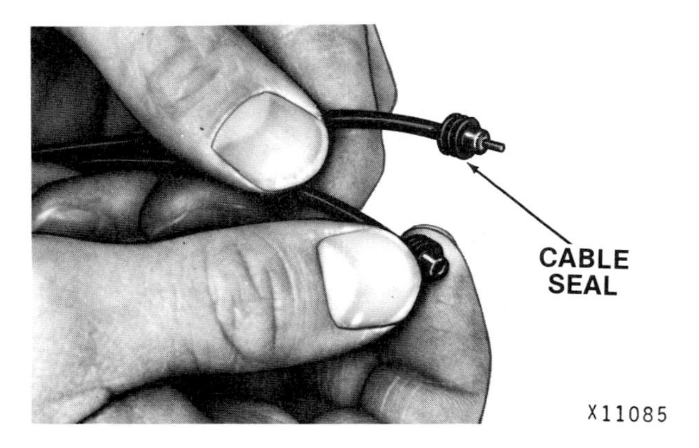

Fig. 65 — Installing Cable Seal

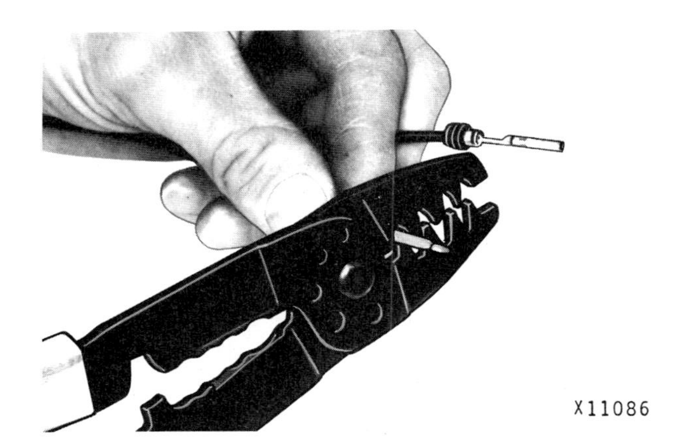

Fig. 66 — Crimping Insulation and Cable Seal

4. Insert a properly stripped wire into the wire barrel part of the contact. Squeeze the tool handles to complete the crimp.

5. Position the insulation barrel in the crimping tool. Make sure the small part of the cable seal is within the crimp wings of the contact as shown in Fig. 67. Close the tool handles to complete the crimp.

Crimping A Splice

To splice two wires, you will need to use a closed-barrel splice. As Fig. 53 shows, the splice, like the terminal, has an insulation crimp area and a wire crimp area. The difference is that the splice has an insulation crimp at each end. The procedure for crimping a splice is the same as for a closed-barrel terminal except that you repeat the procedure for each of the two wires you are splicing. In other words, each splice requires two wire crimps and two insulation crimps. Fig. 60 shows a cross section of a splice with the wires inserted.

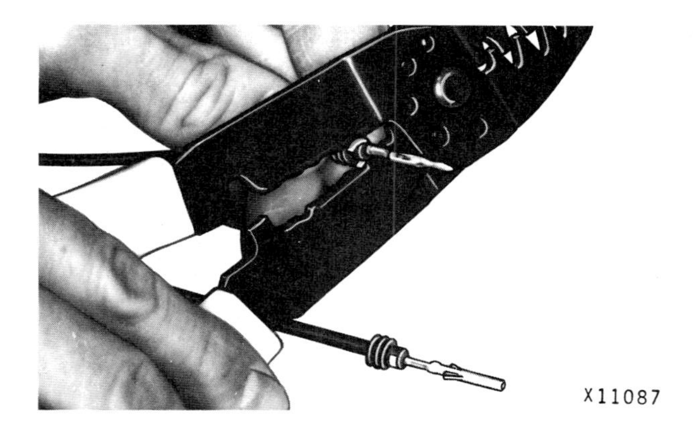

Fig. 67 — Crimping Insulation and Cable Seal

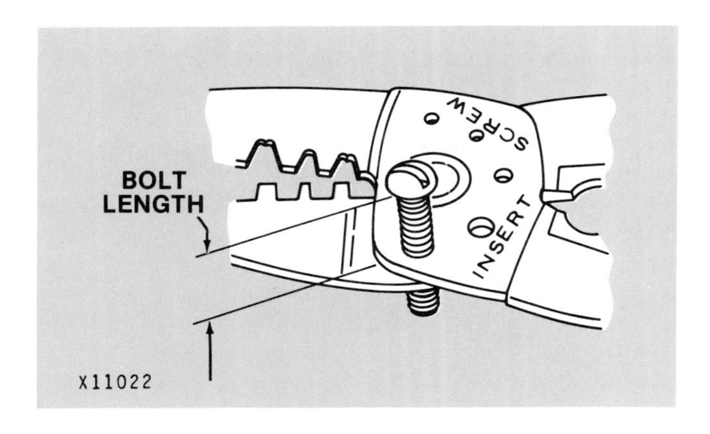

Fig. 68 — Bolt Cutting Using Terminal Applicator Tool

BOLT CUTTING

Either the electrician's plier or the terminal applicator may be used for cutting bolts. The sizes that each will accept are marked on the tool itself. For either tool, follow the same procedure.

1. Select the proper bolt cutting hole according to bolt size and the marking on the tool.

2. Thread the bolt into the side of the tool indicated on the tool itself. Measure the desired length for the cut bolt as indicated in Fig. 68.

3. Squeeze the tool handles to cut the bolt.

4. Remove the bolt from the side of the tool. This will clean any burrs from the bolt threads.

SUMMARY

The electrical connectors discussed in this chapter are some of the more common types that are in use. These connectors all have a locking mechanism to keep the two mating halves from disconnecting, a mating guide mechanism or pin contact arrangement to prevent connecting them wrong, and are designed so that individual contacts can be replaced.

Specialized tools are required for repairing the connectors particularly for removing and installing the contacts. Two universal tools that can be used with all connectors and wiring are the electrician's plier and terminal applicator. The electrician's plier is used primarily for stripping wire and the terminal applicator, for crimping terminals and splices. Both tools can be used as bolt cutters.

Stripping wires and crimping terminals and pin contacts are critical processes in the repair of connectors, but are easily performed if you follow the procedures.

TEST YOURSELF

QUESTIONS

1. What is the purpose of a connector?

2. What is asperity?

3. What three characteristics do the seven connectors discussed in this chapter have in common?

4. True or False? It is not necessary to use all pin contact cavities in the connector bodies.

5. Which connector can accommodate the most contacts?

6. True or False? After stripping, the end of the wire should be twisted tightly before installing the terminal.

7. Why do most crimping operations involve a two-step procedure?

(Answers on page 21 at the end of this book.)

MONITORS AND CONTROLLERS / CHAPTER 12

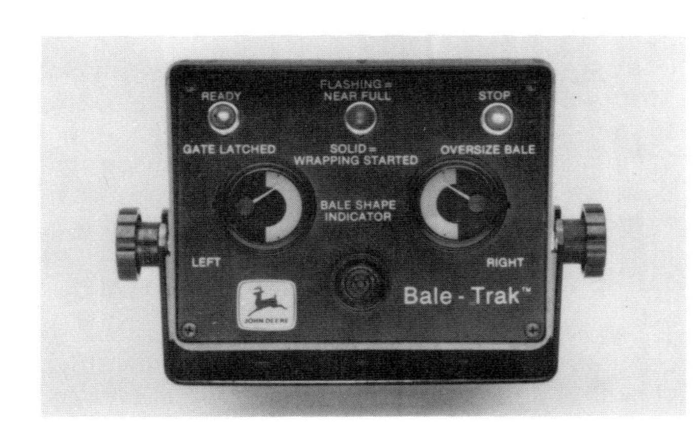

Fig. 1—Monitors Keep Track of What Your Machine and/or Implement is Doing

INTRODUCTION

We will discuss two types of electronic devices (Fig. 2):

- **Monitors** keep track of the performance of a function of the machine and signal the operator when something goes wrong.

- **Controllers** cause a particular function of a machine to operate in a programmed and automatic manner to help the operator obtain greater productivity from the machine.

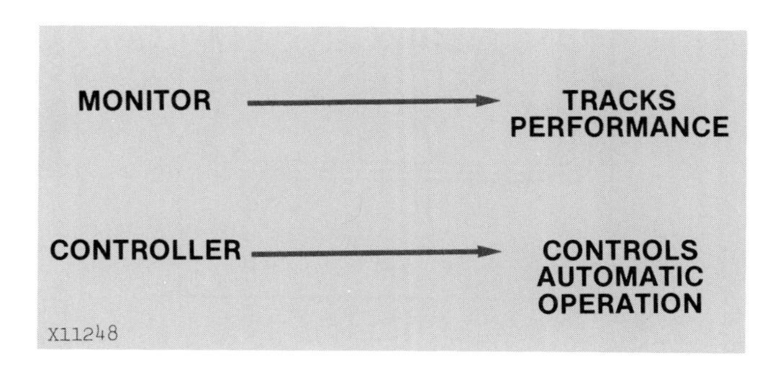

Fig. 2—Monitors and Controllers Perform Different Tasks

AGRICULTURAL AND INDUSTRIAL APPLICATIONS

The applications of electronic systems used in agricultural and industrial equipment and presented in this chapter include:

- Grain loss monitors on combines.
- Tachometer speed monitoring system on combines.
- Low shaft speed monitoring system on combines.
- Planter monitor.
- Seeder monitor.
- Bale size monitoring system on round balers.
- Metal detection system on forage harvesters.

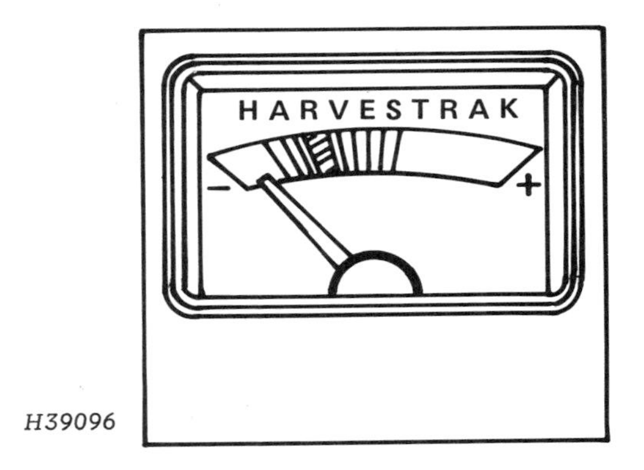

H39096

Fig. 3—Grain Loss Monitor

- Engine, hydraulic, and power train monitors on tractors.
- Automatic blade control system on motor graders.
- Automatic transmission control system on scrapers.

Let's look briefly at each one before we get into a more detailed discussion later in the chapter.

Fig. 4—Tachometer Speed Monitoring System

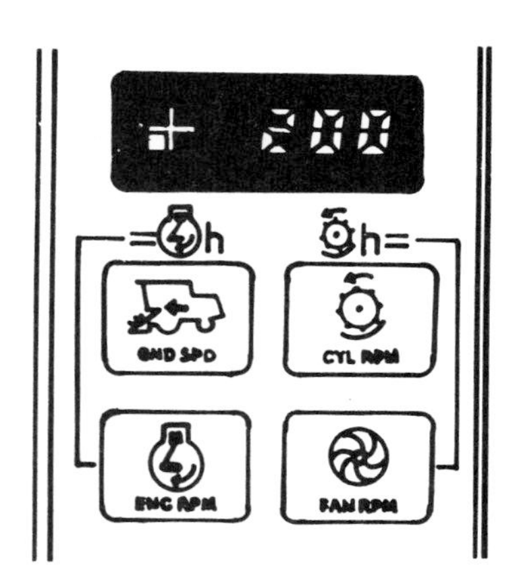

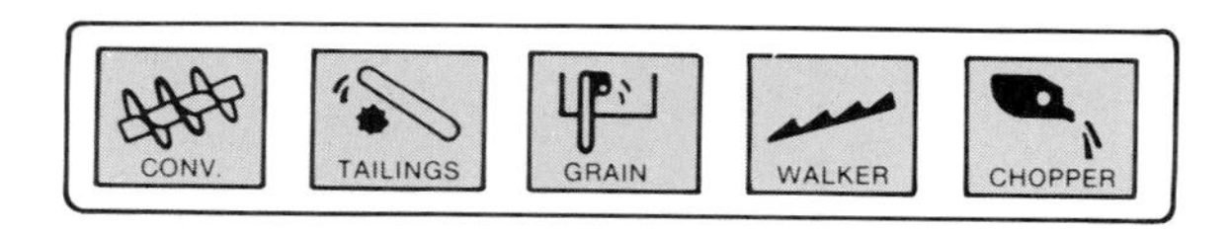

X11251

Fig. 5—Low Shaft Speed Monitoring System

GRAIN LOSS MONITORS

These type monitors (Fig. 3) are used on combines. They monitor the performance of the combine to enable the operator to use maximum combine capacity. The operator adjusts the combine and header to an acceptable loss level and then adjusts the display needle to the green arc. Sensors at the straw walkers and cleaning shoes sense the level of grain loss and relay this information to the monitor inside the cab. Based on the preset loss level, the operator knows whether to increase or decrease ground speed of the combine.

The monitor can also detect plugged or closed straw walkers, sieves and chaffers. A sudden move to the + side of the meter, when field conditions are unchanged, can indicate that one of these components is plugged or closed. It is then necessary to stop the combine and correct the problem.

TACHOMETER SPEED MONITORING SYSTEMS

These systems (Fig. 4) are also used on combines. Sensors located on the engine, cylinder drive, cleaning fan, and transmission monitor rpm and ground speed. This information is than available on a digital display inside the cab.

LOW SHAFT SPEED MONITORING SYSTEMS

The five drive shafts on a combine—conveyor augers, tailings elevator, clean grain elevator, straw walkers, and straw chopper—are monitored by this system. Sensors on each shaft provide a signal to the monitor system which activates a light and buzzer inside the cab when these shafts are operating at less than their designed speeds (Fig. 6).

Fig. 6—Planter Monitor

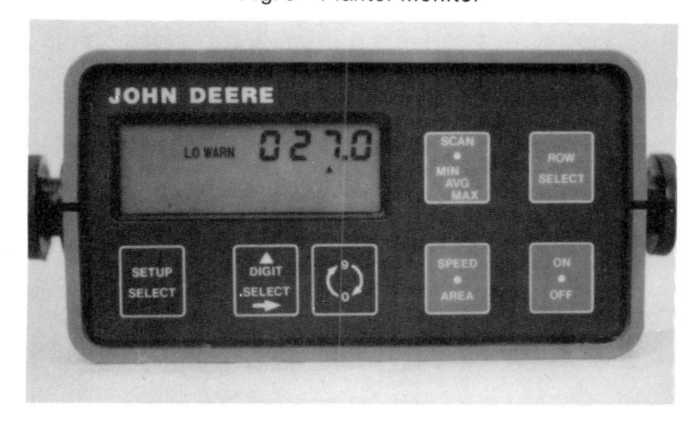

Fig. 7—Seeder Monitor

PLANTER MONITORS

Used on multi-row planters, these monitors (Fig. 6) keep track of the performance of each planting unit on the planter and signal the operator when this performance level is not being met. Some highly sophisticated monitors keep track of number of rows planted, number of acres, and size of each field. Some newer monitors are equipped with a tractor-mounted radar sensor to indicate ground speed of plant population.

SEEDER MONITORS

These monitors (Fig. 7) are used on central metering seeders. Like planter monitors keep track of each planter unit, seeder monitors monitor each seed tube for normal flow of seed. They may also keep track of levels of seed and fertilizer in the bin sections of the seeder and measure the area that has been seeded.

BALE SIZE MONITORING SYSTEM

Bale size monitors (Fig. 8) are used on some round balers. As the hay bale is being formed; they monitor its size and shape so that the operator can achieve evenly shaped and uniformly sized bales.

Fig. 8—Bale Size Monitor

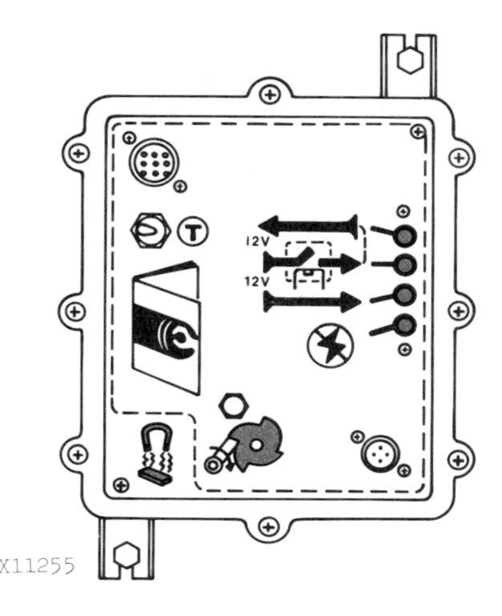

Fig. 9—Metal Detection System

METAL DETECTION SYSTEM

These systems (Fig. 9) are used on some forage harvesters to detect and prevent material containing iron from entering the cutterhead area of the forage harvester. This reduces possibility of damage to the cutterhead from large pieces of scrap metal and ingestion of small bits of wire by livestock. The metal detector is a sensing and controlling device rather than a monitor since it will stop rotation of the feedroll automatically without intervention by the operator if metal is detected.

TRACTOR AND COMBINE MONITORS

Some monitoring systems on tractors and combines (Fig. 10) monitor all critical functions of the machine and warn the operator by gauges, beeping horns, and lights when something is wrong. Functions monitored include voltage level, fuel, oil pressure, coolant temperature, transmission control pressure, transmission lube pressure, and clutch temperature.

Fig. 10—Tractor Monitor

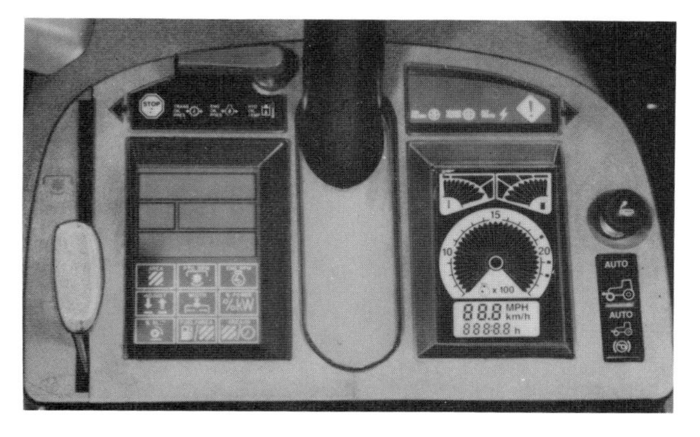

Fig. 11—Blade Control System

BLADE CONTROL SYSTEMS

Automatic blade controllers (Fig. 11) are used on some motor graders. Their purpose is to sense changes in grade and slope of the ground and then signal the blade lift cylinders so they can adjust the blade accordingly without operator intervention.

AUTOMATIC TRANSMISSION CONTROL SYSTEMS

Automatic transmission controllers (Fig. 12) are used on scrapers. These controllers automatically shift the transmission up or down depending on ground speed and whether the elevator is in the load or transport mode.

PRINCIPLES OF OPERATION

Essentially, the principles of operation for all the systems are the same. They all have sensors of some type that “sense” a function such as the speed of a wheel or rotating staff, temperature of a coolant, or number or amount of something contacting them or passing through them (Fig. 13). The sensors then transmit this information, in the form of electrical signals, to a control console or monitor. The monitor contains the necessary electronic circuitry to process this information, which in turn will display data allowing the operator to react.

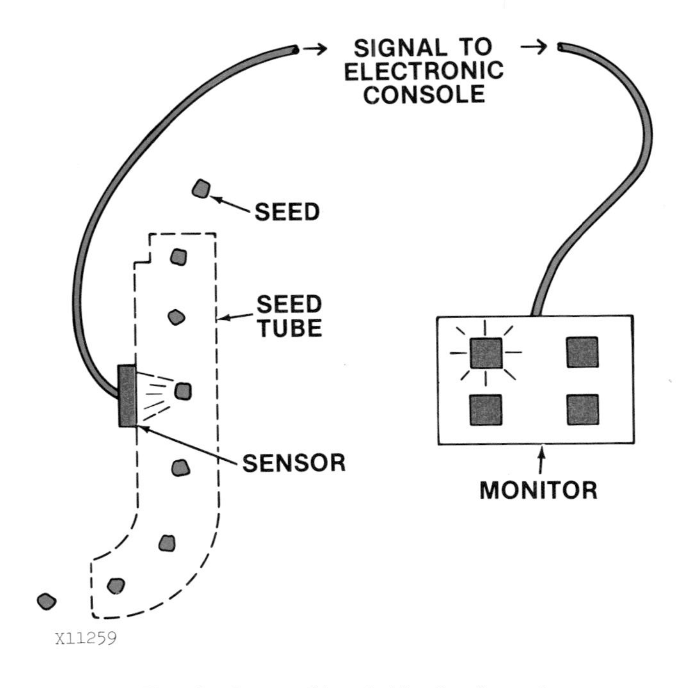

Fig. 13—Sensor Signals Monitor Console

Fig. 12—Automatic Transmission Control System

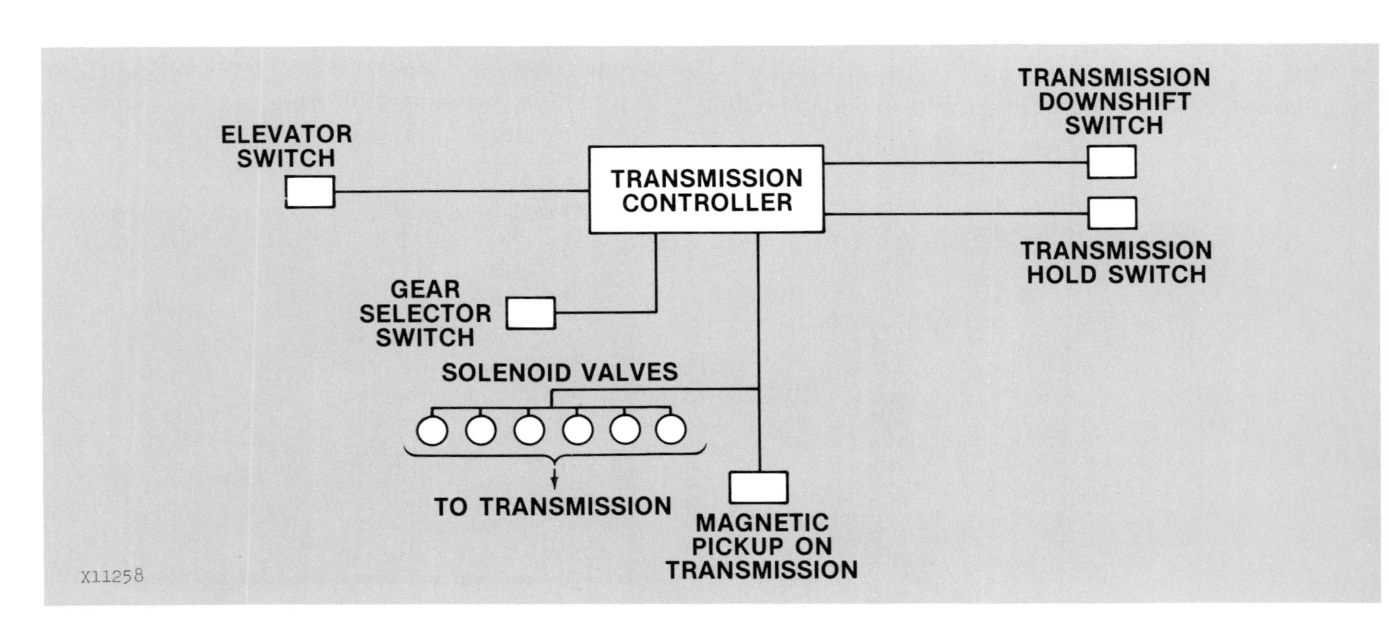

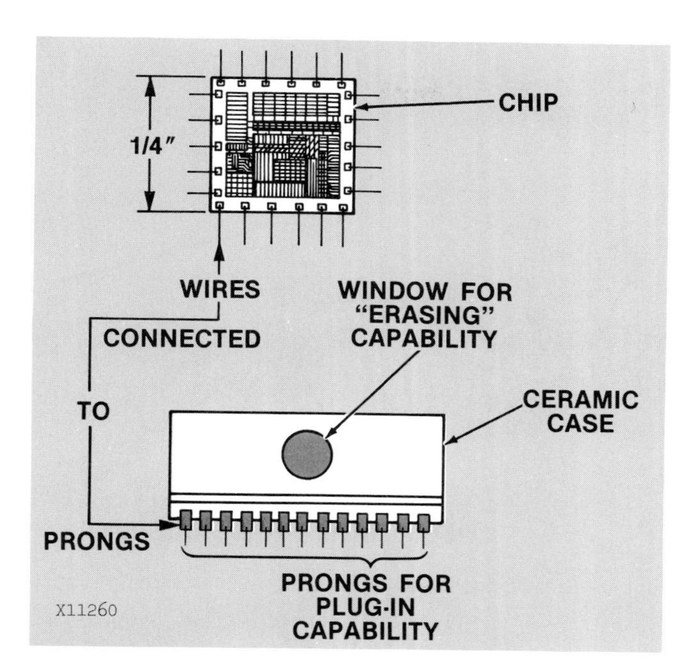

Fig. 14—Computer on a Chip is Sealed in a Ceramic Case

For example, if the planter monitor indicates that no seed, or very few seeds are going into the ground from planter unit number 6, the operator would stop the tractor and investigate. Maybe the problem would be nothing more than a clogged seed meter. Or it could be a broken drive chain on that particular planter unit. In any event a problem exists that has to be corrected or that row simply will not be planted.

GENERAL MAINTENANCE, REPAIR, AND DIAGNOSTICS

Before beginning any maintenance on a monitor or control system, familiarize yourself with the seven basic steps of a good diagnostic and testing procedure as explained in Chapter 14.

Also, when diagnosing electrical systems, first do the following:

1. Verify that the system is being used properly.

2. Use the self-diagnostics if the machine is so equipped.

3. Visually inspect the electrical system:

— Look for bare wires that could ground a component or short across to another component.

— Look for missing or worn conduit.

— Look for loose or broken connectors and wires.

4. Inspect batteries for:

Corroded terminals
Loose terminals or battery posts
Dirty condition
Damp condition
Cracked case
Proper electrolyte level

5. Check alternator belt tension.

6. After the machine has been shut down for five minutes, inspect it for overheated parts. They will often smell like burned insulation. Put your hand on the alternator. Heat in this area when the unit has not been operated for some time is a sure clue to charging circuit problems.

7. If your visual inspection does not indicate the possible malfunction, but your inspection does indicate that the machine can be run, turn the key switch to the IGN position. Try out the accessory circuits, indicator lights, gauge lights, etc. Do these components work properly? Look for sparks or smoke which might indicate shorts.

8. Start the machine. Check all gauges for good operation and check to see if the system is charging or discharging. Look for anything unusual.

9. Check for electrical circuit malfunctions:

High-resistance circuit
Open circuit
Grounded circuit
Shorted circuit

10. Check fuses and sensors.

11. Use wiring diagrams and schematics to trace the circuit in question and diagnose the appropriate electrical subsystems:

Power circuit
Start circuit
Charging circuit
Display module and logic module circuit
Indicator circuit
MFWD circuit
Start aid circuit
Fuel shut-off circuit
Reverse alarm circuit
Dome light circuit
Wiper/washer circuit
Blower circuit
Drive and work light circuit
Park brake/clutch disconnect circuit
Horn circuit
Turn signal, flasher and brake light circuit
Beacon circuit

Litho in U.S.A.

Return-to-dig circuit
Gauge and hourmeter circuit
Accessory valve side shift circuit

12. Check the monitor and controller console.

As a general statement, the monitor or control console is probably one of the most reliable and trouble-free components on the entire tractor, implement, combine, or whatever other piece of equipment you may be operating. The reason for this is the integrated circuits or chips (FIg. 14) now used in most electronic devices. The chip itself is ultrareliable because it has virtually no components that can come loose or break. We are not saying that the control console will never fail—any component can fail at some point in time. What we are saying is that if the monitoring system does malfunction, do not just assume the control console is at fault and remove it and send it back to the factory. Check all other components first (Fig. 15).

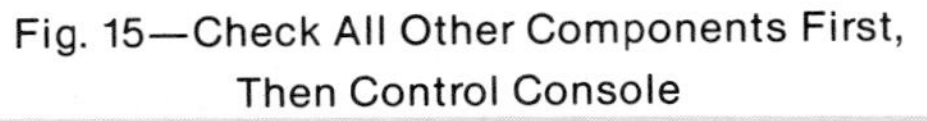
Fig. 15—Check All Other Components First, Then Control Console

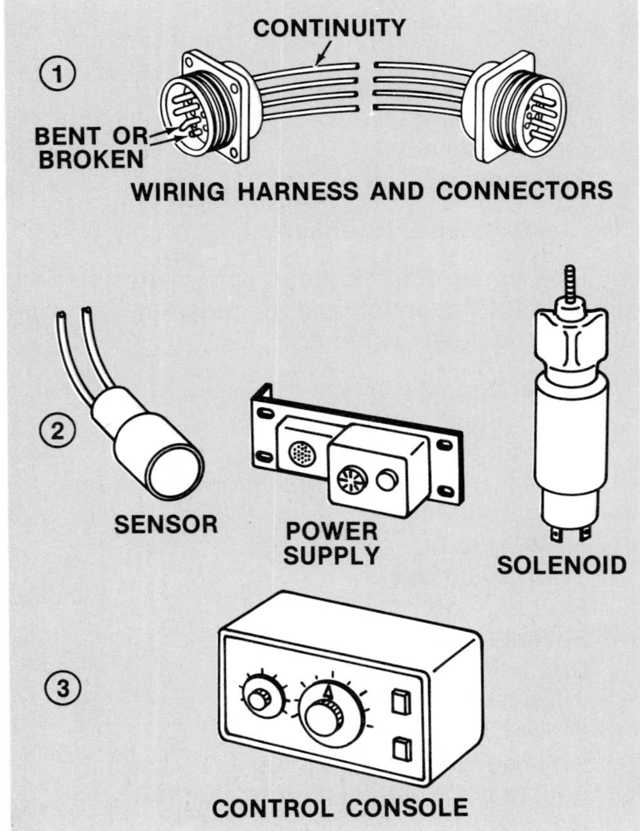

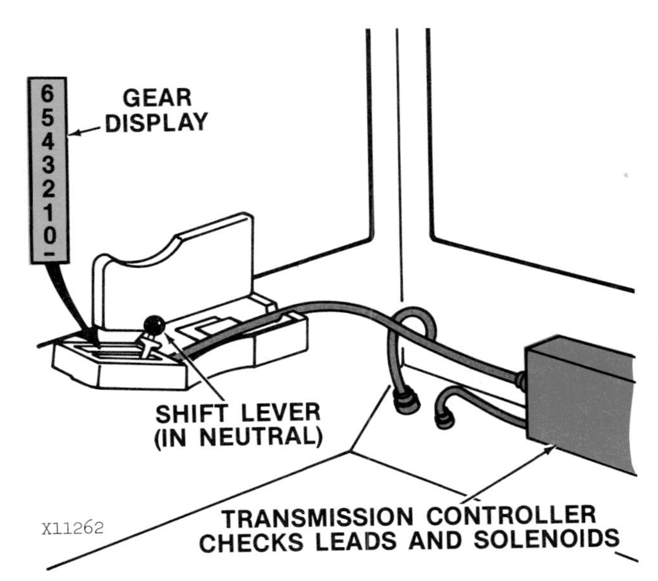

Fig. 16—Self-Diagnostics of Automatic Transmission Controller

SELF-DIAGNOSTICS

The first thing you should do if the system malfunctions is to run it through its self-check procedure, provided, of course, that the system has one. Some of them do, such as the low shaft speed monitoring system on combines, engine control system on tractors, metal detector on forage harvesters, and automatic transmission control system on scrapers. Let it tell you itself where it "hurts," so to speak! Most self-check procedures will be found in the applicable operator's manual for the particular machine or implement. And the manual will probably recommend that you run the self-check daily just prior to putting the machine to use. These procedures usually only take a minute or two.

For example, the automatic transmission controller on a scraper (Fig. 16) automatically checks all leads and the solenoids that control the actual shifting of the transmission each time the machine is started. The gear display indicates by flashing lights if there is a bad solenoid or short in the wires.

A tractor monitor/control system self-diagnostic system operates like this:

1. The system will first perform a lamp check and horn check. If the system voltage exceeds 16 volts, the unit cancels this check. If the warning system fails the check, test circuits before continuing.

2. Test tachometer modes:

Normal Operating Mode
Diagnostic Mode

Internal Data Mode
Troubleshooting Mode
Clear Codes Mode
Calibration Mode

Diagnostic code numbers are preceded by letters which indicate their source:

SLF — Tachometer
HCU — Hitch Control Unit
ECU — Engine Control Unit

Code numbers refer to that area of the above which is having the difficulty.

3. Using the calibration mode, enter the applicable calibration number in the correct position. Check calibration and operation of any of the following:

Tire rolling radius
Tachometer digital speed read out
Engine oil pressure sensor and tel-light circuit
Engine coolant temperature sensor and gauge circuit
Hydraulic oil temperature sensor and tel-light circuit
Axle/PTO lube pressure sensor and tel-light circuit
Transmission control pressure sensor and tel-light circuit
Transmission control pressure sensor and tel-light circuit
Hydraulic and transmission oil filter restrictor sensors and tel-light circuit
Air filter restrictor sensor and tel-light circuit
Fuel level sensor and gauge circuit
Transmission magnetic output speed sensor and circuit
Rated engine speed or rated PTO speed
Ground speed
Hourmeter
Turn signals and warning lamps
PTO overspeed indicator
Performance monitor, radar sensor and associated circuits

WIRING HARNESSES

Check continuity of all wires and connectors (Fig. 17). A control console cannot process an electrical signal that it never receives! You will need some type of volt-ohm meter (Chapter 2) to check that there is a complete current path along the entire length of the wires.

A good general procedure for checking wire continuity is presented in the following four steps.

1. Disconnect all connectors on the wiring harness.

2. Using a volt-ohm meter, check the continuity from each switch or sensor to the connector. Refer to the applicable technical manual for wire colors and connector pin numbers.

3. If continuity is present, the wires are good.

4. If continuity is not present and a broken wire is suspected, connect a volt-ohm meter to wire end and connector end of harness. Starting at one end, flex harness while watching the volt-ohm meter and work along entire length of harness. A reading on the volt-ohm meter will indicate that the wire is broken in that approximate area.

Do not forget the connectors. Because of the low voltages used for most electronic devices, contamination and corrosion of connector pins and sockets can be real problems. But as the use of gold-plated contacts and sealed connectors become more widespread, corrosion is not as serious a problem as it used to be. See Chapter 11 for servicing connectors.

VALVES, POTENTIOMETERS, AND SOLENOID VALVES

Some automatic control systems will use associated valves, potentiometers, and solenoid valves. These components should also be checked.

Fig. 17—Check Continuity of Wiring and Connectors

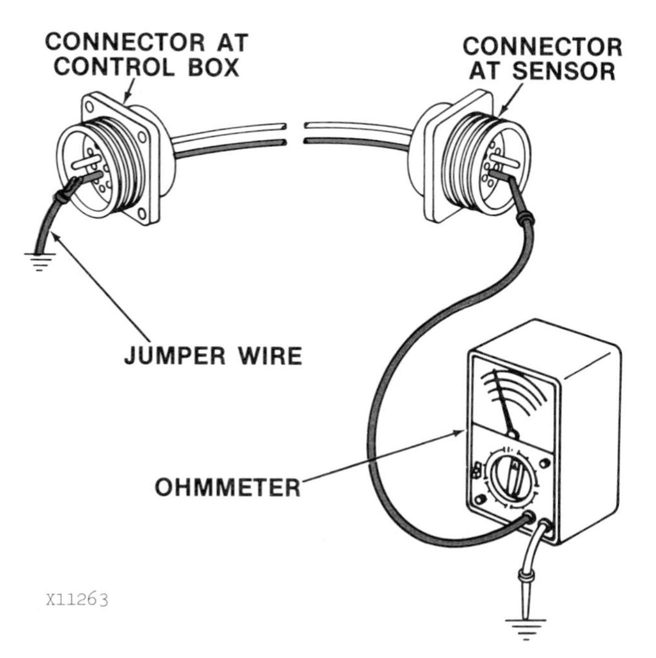

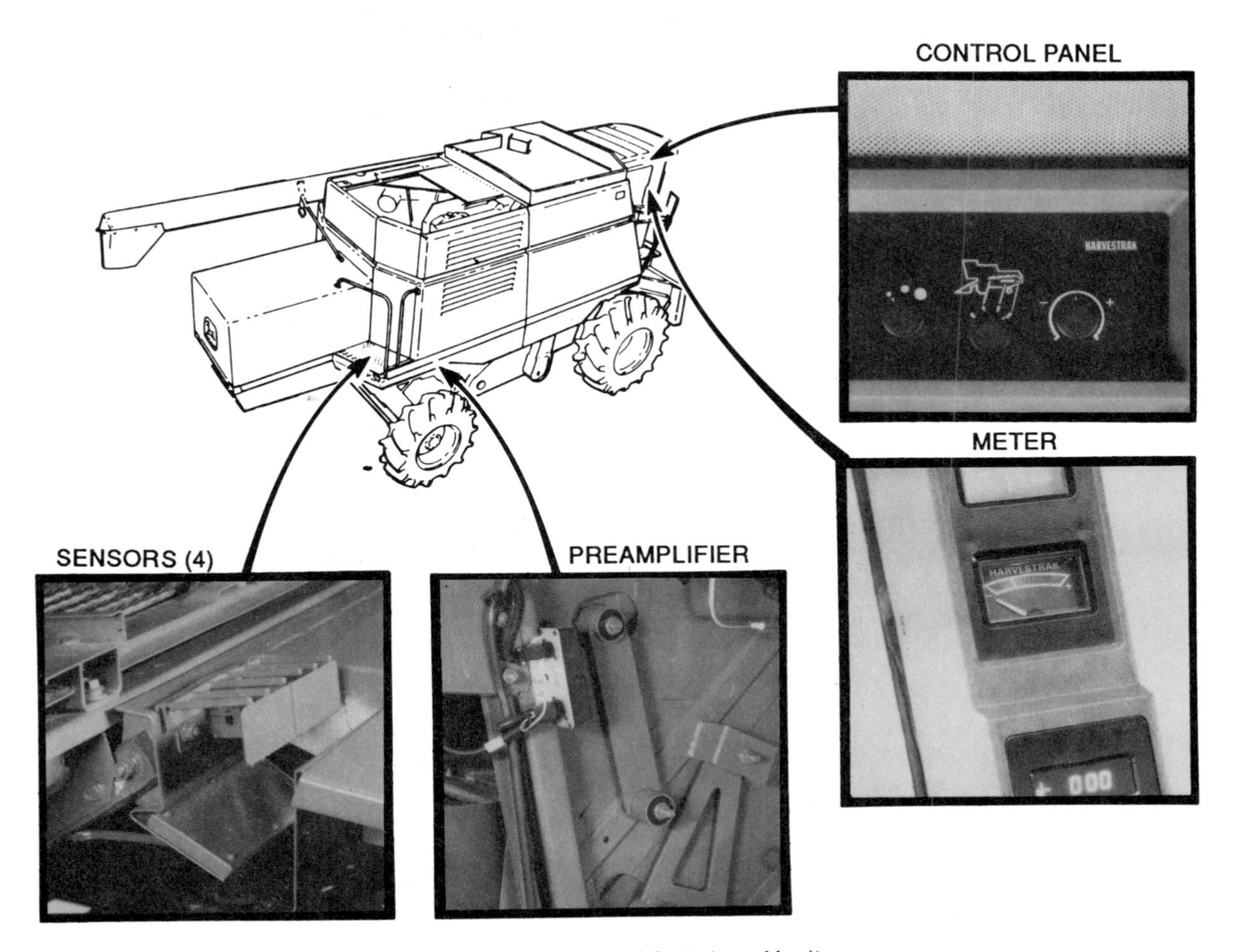

Fig. 18—Components of Grain Loss Monitor

CONTROL CONSOLE OR MONITOR CONSOLE

If all other components of the system check out to be working correctly, then the control console can and should be checked at this point. Some systems have specially designed test sets that are used for this purpose.

Now, let's look at specific monitoring systems.

COMBINES

Combines have three major monitoring systems:

- Grain loss monitors
- Tachometer speed monitoring system
- Low shaft speed monitoring system

Let's look at each of these three systems and see how they operate and how they are maintained.

GRAIN LOSS MONITORS

Grain loss monitors (Fig. 18) measure grain loss by measuring a representative sample of losses over the cleaning shoes and straw walkers. The monitor system consists of an ignition switch and electric clutch switch, a control panel, preamplifier circuit box, two cleaning shoe sensors, and two straw walker sensors. These components are connected with a wiring harness. Any change in the loss rate is indicated by the meter on the control panel (Fig. 19).

The monitor continuously monitors combine performance to enable the operator to use maximum combine capacity. After the operator has adjusted the combine and header to a loss level that is acceptable, the monitor can then be set to this level indicating whether to increase or decrease ground speed.

 Litho in U.S.A.

Fig. 19—Control Panel for Grain Loss Monitor

The four sensors (Fig. 18) detect impacts from grain coming off the cleaning shoe and straw walkers and transmit these impacts as electrical impulses to the preamplifier. The preamplifier eliminates or "filters out" those signals that result from impacts on the sensors from straw stems, chaff, or cobs.

The sensors detect grain by a principle similar to a microphone. When grain strikes the sensor board, it causes the board to vibrate. This vibration is then transmitted to an electric element inside the sensor causing the element to distort. The distortion causes the electric element to send out a voltage signal whose amplitude and frequency are characteristic of the material striking the sensor board. The difference in voltage signals is how the preamplifier tells whether the sensor board was struck by a kernel of grain or a piece of cob.

Next, the control panel (Fig. 19) inside the cab receives the impulses from the preamplifier and displays the grain loss on the meter (Fig. 21).

Fig. 20—Function of Grain Loss Monitor

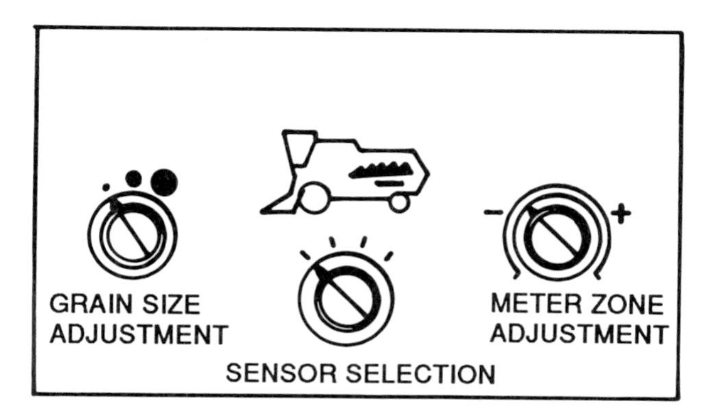

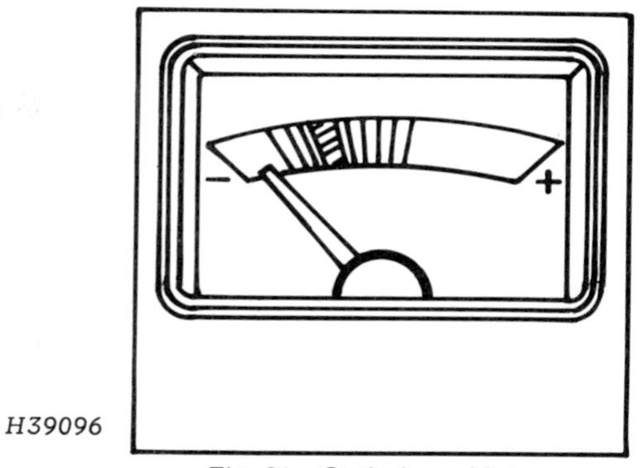

H39096

Fig. 21—Grain Loss Meter

Operational Check

A very simple and fast operational check can be performed on the grain loss monitor by performing the nine steps listed below. An advantage to this check is that each of the four sensors and the individual leads of the wiring harness from the sensors are checked individually.

1. Turn ignition switch of the combine to ON but do not start engine.

2. Turn on the header drive switch.

3. Turn the grain size knob (Fig. 21) to the small grain position.

4. Turn sensor selection knob to the cleaning shoe symbol.

5. Turn the meter zone adjustment knob to the +.

6. Tap rapidly and lightly on the left cleaning shoe sensor (Fig. 22) with a hard object such as a small screwdriver while another person watches the meter needle. The needle should move into the green scale. Repeat for the right cleaning shoe sensor.

7. Turn sensor selector knob to straw walker symbol B. Check both straw walker sensors in the same manner.

8. Turn light switch to the fourth position (fieldlights). Check that the light for the meter (Fig. 21) is lit.

9. Turn off key switch, separator and header.

The operational check can lead to further diagnostic testing. For example, if tapping the left cleaning shoe sensor (Fig. 22) causes the meter to read correctly but tapping the right one causes no reading at all, probably one of two things is wrong. Either the wiring harness lead to that sensor is bad or the sensor itself has failed. No needle movement at all from tapping all of the sensors would probably indi-

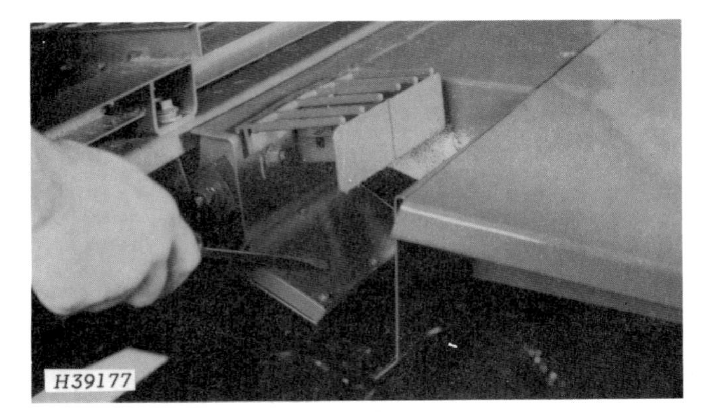

Fig. 22—Tapping on Sensor Board

cate no power, bad harness connection at the preamplifier, a malfunctioning preamplifier, bad meter, etc. The sensors would probably be all right since it is not likely that all four would fail at the same time.

Repair of Grain Loss Monitor

The control panel, preamplifier, and sensors are all sealed units and cannot be repaired. If found defective, they must be replaced. Environmental factors have little or no effect on the sensors. However, they should be kept clean because excessive dirt (caked mud) could cause them to lose sensitivity.

DIGITAL TACHOMETER

The tachometer in many modern combines displays four different functions on a digital display (Fig. 23: ground speed and rpm functions of the cylinder, cleaning fan, and engine). Engine and separator hours are also available for display.

The tachometer is accurate within ± 10 rpm in displaying all three rpm functions and is accurate within ± 1/2 km/h (1/10 mph) in displaying ground speed. If ground speed drops below 3/4 km/h (1/2 mph) the digital display shows 0.0.

Fig. 23—Digital Tachometer

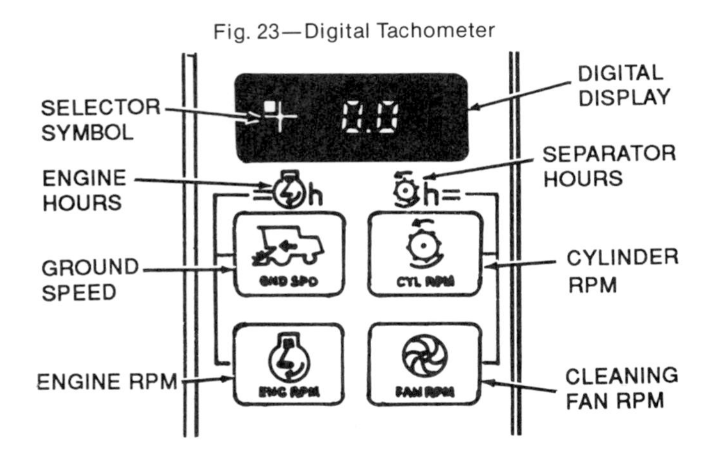

The engine and separator hour meters are accurate to ± one hour. Only full hours are shown, but the computer memory is updated in quarter hours.

The digital display dims by 50% when the headlights are turned on and can be dimmed even more with the dimmer switch on the light panel.

DIGITAL TACHOMETER ENGINE CONTROL DIAGNOSTIC CODES

On some combines, the tachometer may display a number of diagnostic codes relating to engine operation.

The diagnostic codes will appear as —E on the tachometer. If any of these codes appear, you can continue to operate the combine, but write the code on a piece of paper and call your dealer at your earliest convenience. However, certain codes may appear which indicate that the combine engine might die or may not start.

After making a written note of the code, press the desired speed symbol to return the tachometer to normal operation. The diagnostic code will not appear again until the engine is turned off, restarted and the problem occurs again.

TACHOMETER OPERATION

When the key is turned on, the tachometer (Fig. 23) first shows ground speed.

Select the function to be shown and press that symbol. The selector symbol square will move to show the function displayed.

The engine hourmeter shows whole hours when ground speed and engine rpm symbols are touched at the same time.

The separator hourmeter shows whole hours when cylinder rpm and cleaning fan rpm symbols are touched at the same time.

The tachometer also sends an alarm to the light and buzzer in the overhead panel for low engine speed and low cylinder speed.

The engine alarm turns on when rpm drops below a certain level.

The cylinder speed alarm turns on when speed drops below 80 percent of setting and engine is at full rpm with header drive engaged.

SENSOR LOCATIONS

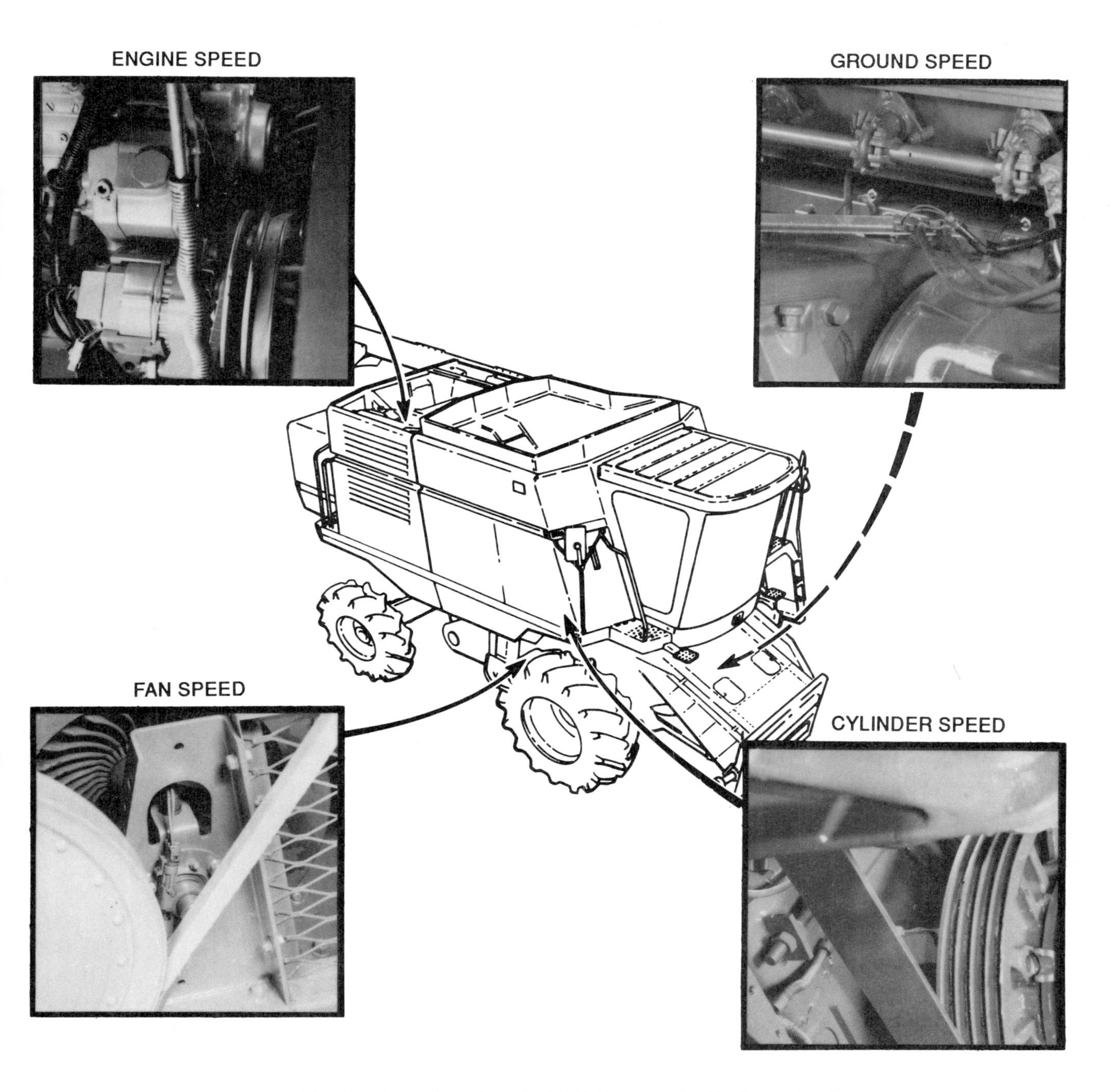

Fig. 24—Sensor Locations for Digital Tachometer Monitoring System

Litho in U.S.A.

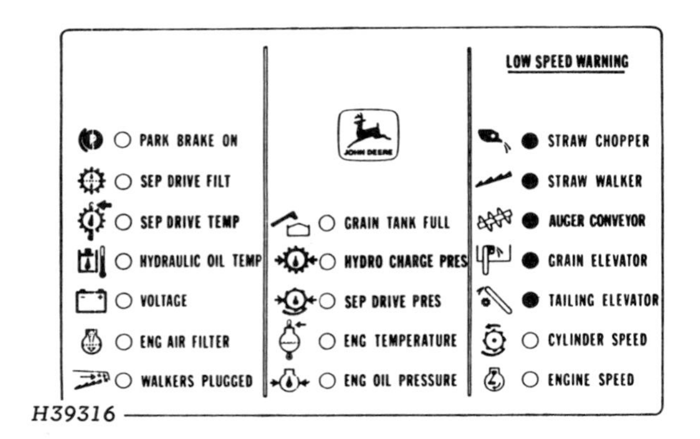

Fig. 25—Low Shaft Speed Monitor

LOW SHAFT SPEED MONITOR

The low shaft speed monitor (Fig. 25) shows with lights and a buzzer when the straw chopper, straw walkers, conveyor augers, grain elevator and tailings elevator are operating at less than 70 percent of their designed speed.

The electric clutch and low shaft speed monitor are both on the same circuit. To check operation of the monitor, turn the key to "ON" (do not start engine) and engage the separator and header drive switches. All lights must turn on and the buzzer must sound. If they do not, see TROUBLESHOOTING.

If all the lights come on and the buzzer sounds, start the engine, engage the separator and header drive switches, and run the engine at fast idle. If the lights continue to stay on, disengage the separator and header switches and idle engine. Shut off the engine, remove the key, and see TROUBLESHOOTING.

The five sensors (Fig. 26) are not adjustable. If the combine is not equipped with a straw chopper, be certain the sensor leads are connected to ground plugs to prevent a false alarm.

The five sensors are:
A. Straw chopper sensor
B. Straw walker sensor
C. Tailings elevator sensor
D. Clean grain elevator sensor
E. Conveyor augers

LOW SHAFT SPEED MONITOR TROUBLESHOOTING

CAUTION: Turn off the engine and remove the key before working on combine.

A yellow light comes on and the buzzer sounds whenever any of these five functions are running below proper speed. The separator and header drive switches must be on to turn on this system.

When the buzzer sounds, check for:

Plugged straw chopper, straw walkers or augers, or a broken belt. Then check for:

Broken or disconnected sensor wires.

Damaged sensor.

Loose sensor actuator on shaft collar.

If the light and buzzer still stay on after checking those items, short both sensor wires to ground. If the light and buzzer then go out, replace that sensor.

If, after replacing the sensor, the light and buzzer still stay on, see your dealer.

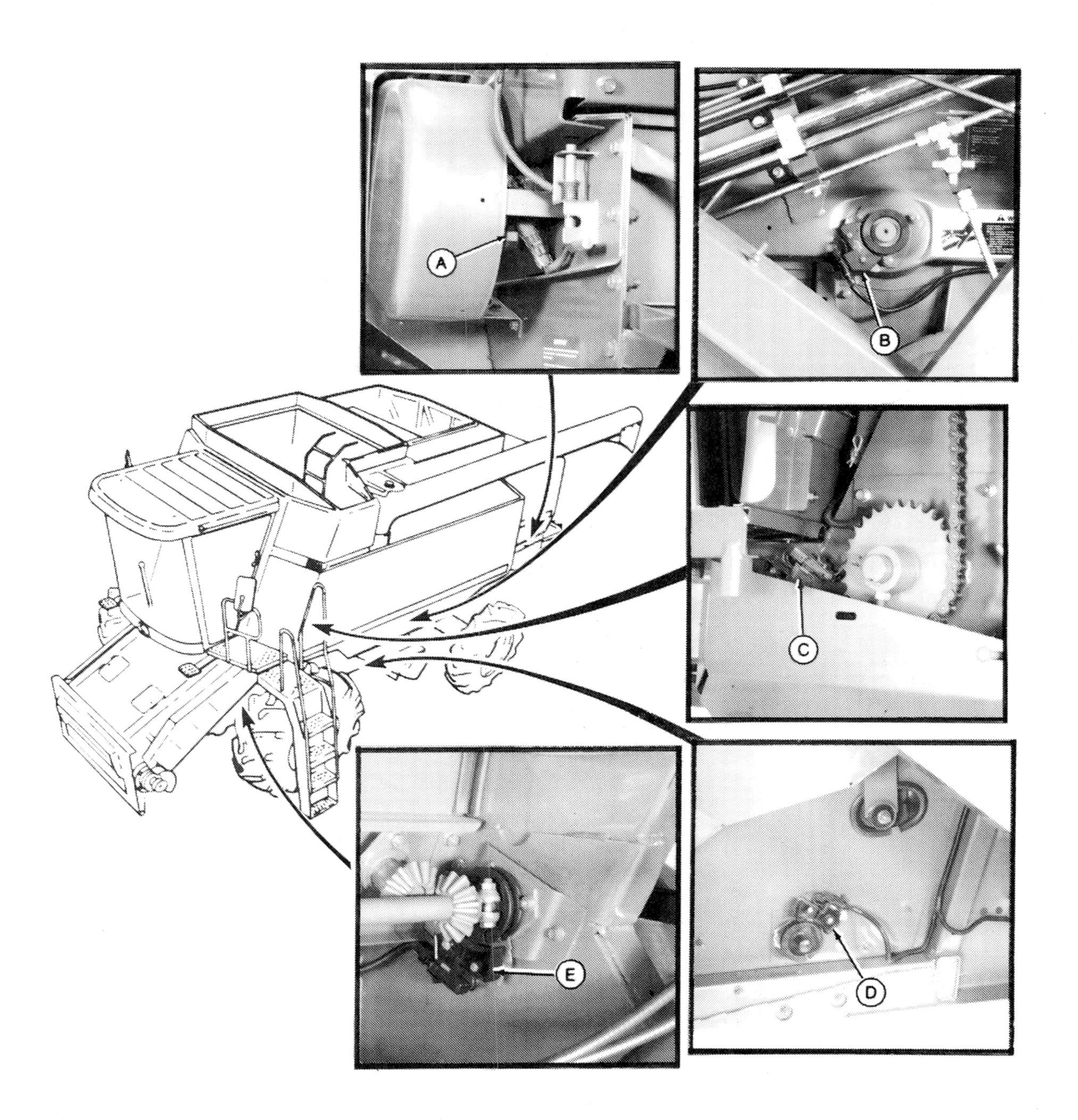

Fig. 26—Sensor Locations for Low Shaft Speed Monitoring System

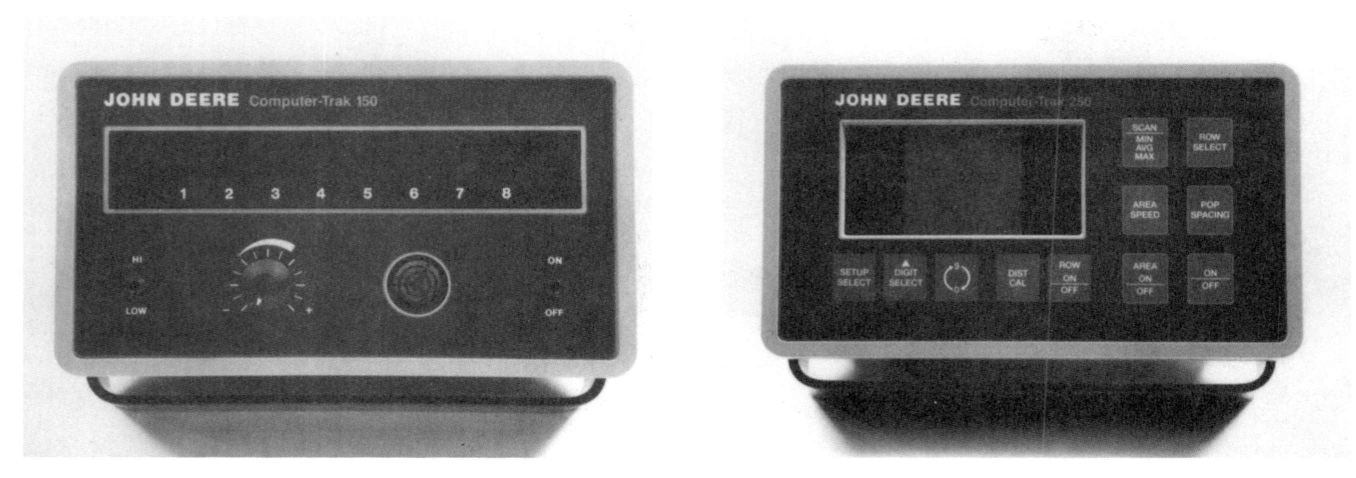

Fig. 27—Planter Monitors

PLANTERS AND SEEDERS

PLANTER MONITORS

Planter monitors (Fig. 27) let the operator know if the planter is actually doing its job. The monitor constantly checks each planting unit on the planter (Fig. 28) so that the operator knows whether or not any of the units are plugged. With today's highly sophisticated planters, some planting up to 24 rows at a time, the farmer has to know that the correct amount of seed at the proper spacing is going into the ground.

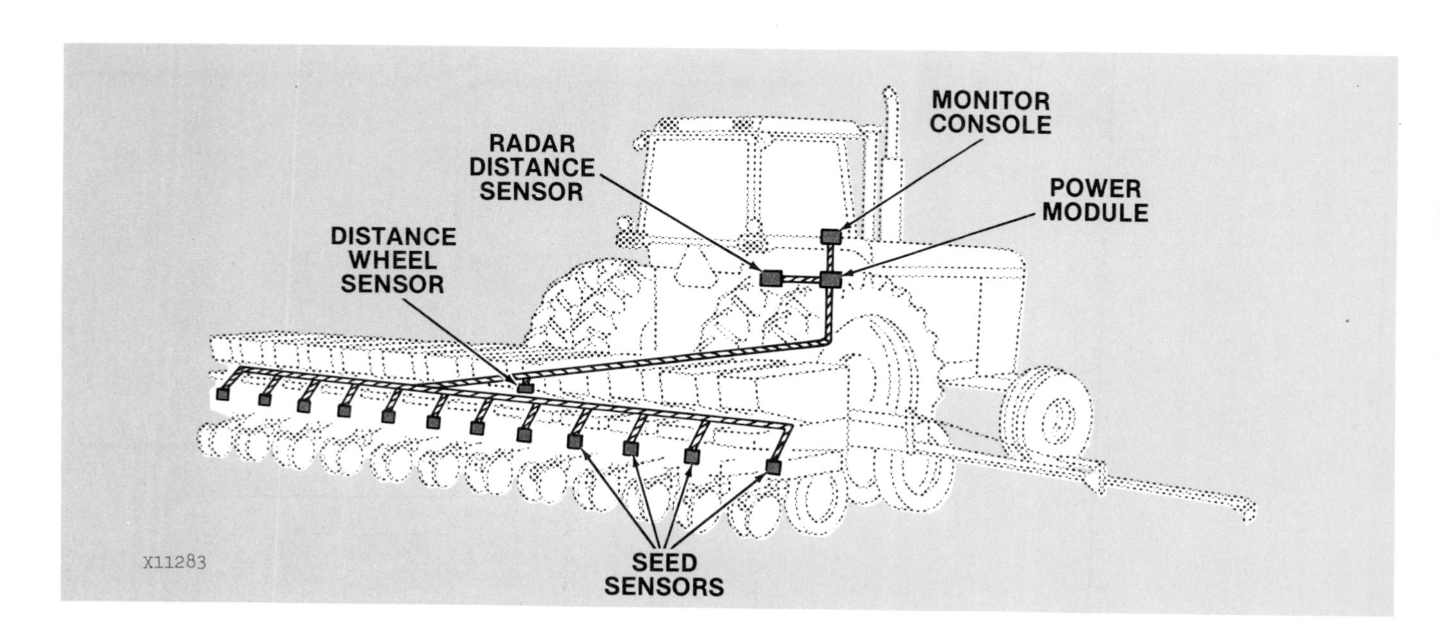

Fig. 28—Components of Planter Monitor

The monitor system (Fig. 28) consists of a sensor located at each planter unit, a monitor console and power module inside the tractor cab, and associated wiring harnesses. The actual sensing device is located inside the seed tube (Fig. 29). It is a photo-electric cell that senses the presence of each seed as it falls through the seed tube and relays this information to the monitor inside the tractor. Without a monitor, a planter unit could be clogged and the farmer could miss planting an entire row on each pass and not know it until he stopped to fill the seed hoppers.

Fig. 29—Seed Tube Sensor

On monitors which give a digital display of corn population per acre, distance information is provided to the console from either a sensor located on a distance measuring wheel or from a radar sensor (Fig. 30).

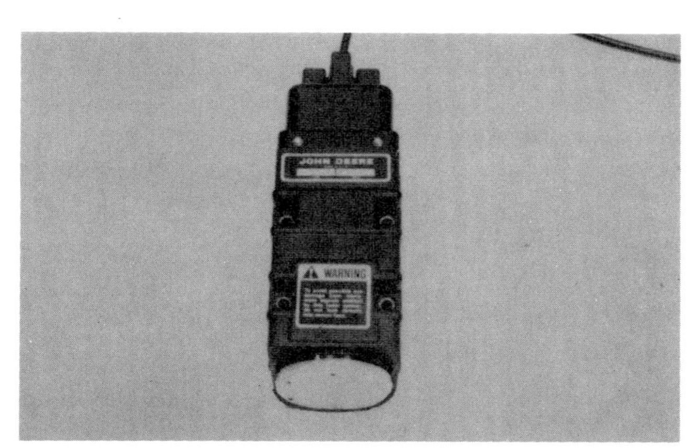

Fig. 30—Planter Radar Sensor

The monitors do not tell the planter what to do, rather, they monitor what the planter is doing. The farmer knows the calibration setting of the planter and tractor combination. The operator programs the monitor to provide a warning when the seed population in each planting unit drops below or rises above the calibration setting of the planter.

Some monitors are equipped with indicator lights (Fig. 31) and a rate control knob to program the monitor to the calibration setting of the planter. Other monitors (Fig. 32) are much more sophisticated, containing digital readouts and additional features that can be programmed into the monitor. For example, one monitor contains a sophisticated computer program which uses the information programmed into the monitor by the operator—row width, number of rows, high and low warning limits,

Fig. 31—Planter Monitor with Indicator Lights

and distance calibration—to compute and display speed, area planted, individual row, and average population as well as to alert the operator when population has gone outside the programmed limits. There are also monitors that will even keep track of different field sizes and then total them when planting is complete.

Troubleshooting Planter Monitors

CONSOLE OPERATES INTERMITTENTLY.

Monitor is not receiving full 12 volts.

ONE ROW SHOWS FAILURE.

Seed tube is clogged.

Plateless planters: Unit roller chain is off sprocket.

Wrong seed plate is installed.

Unit is not planting.

Photocells in shank are dusty.

Plateless planters: Fingers are malfunctioning.

Sensor or planter harness is shorted.

Litho in U.S.A.

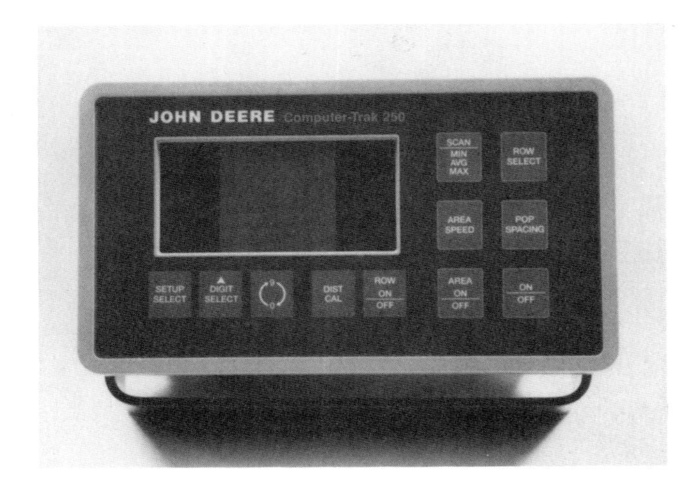

Fig. 32—Programmable Planter Monitor with Digital Displays

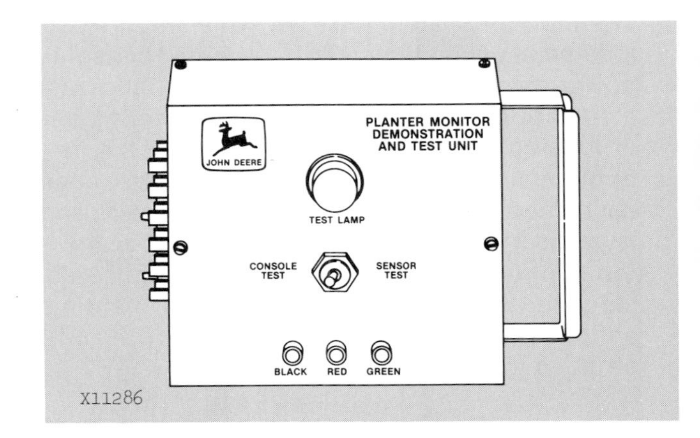

Fig. 33—Standard Test Set for Planter Monitoring Systems

FAILURE IS SHOWN FOR HALF OF PLANTER UNITS.

Drill shaft pin is sheared.

ALL ROWS FAIL.

Drive chain in center of planter is off sprockets.

Sensor or planter harness is shorted.

FAILURE IS SHOWN INTERMITTENTLY.

Dampness is in coupler or connectors.

Electrical interference coming from gasoline engine.

Interference coming from alternator or generator.

Sensor or planter harness is shorted.

Testing Planter Monitors

Since planter monitors use highly sophisticated seed and distance detection systems, specialized test sets (Fig. 33 and Fig. 34) have been developed for testing all components of the monitor systems.

The test unit actually simulates seed and distance signals to allow the technician to determine whether the components of the main system are performing satisfactorily.

Testing is divided into three separate phases:

- Connecting to power source.
- Preliminary setup.
- Functional tests.

CONNECTING TO POWER SOURCE

If possible, perform the test with console connected to the tractor on which it is used. If not, attach the power module battery leads to a well-charged 12-volt battery or to a 120V AC-12V, DC converter (Fig. 35).

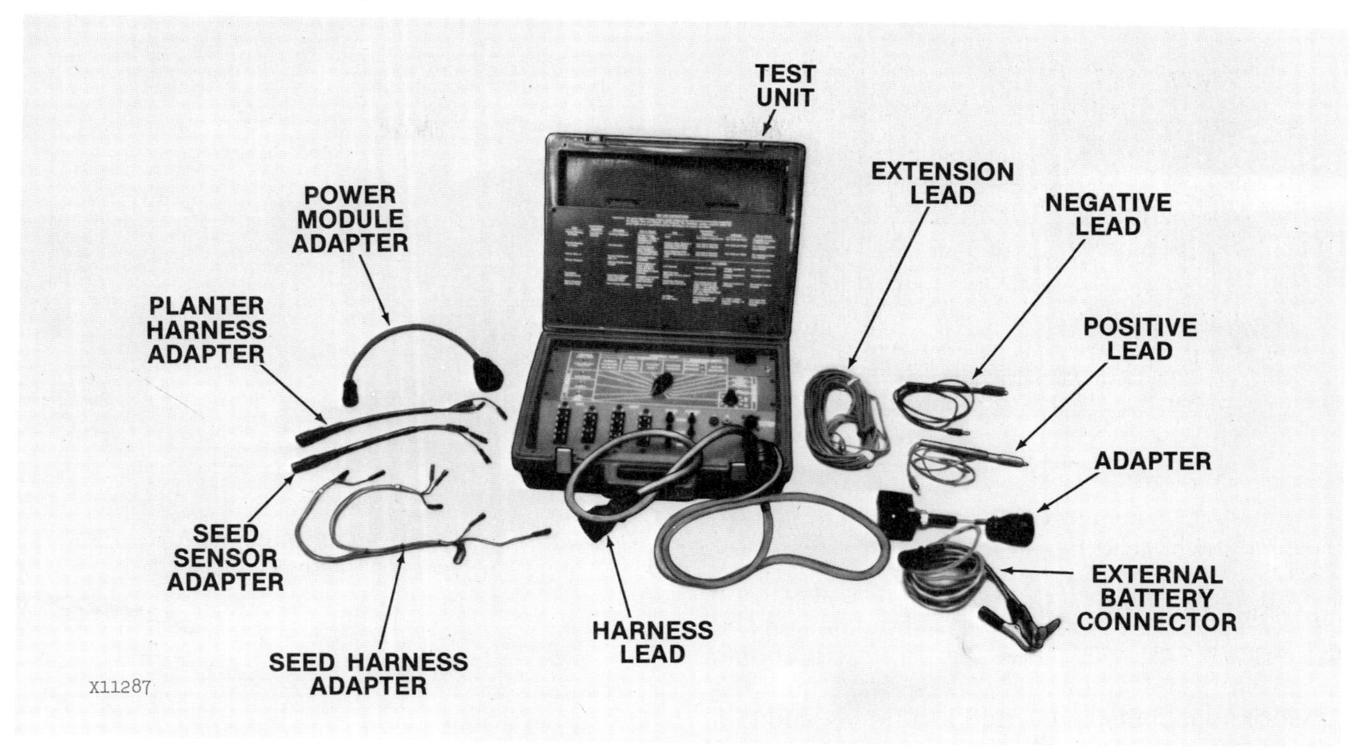

Fig. 34—Deluxe Test Set Showing Attachments

Make certain that the positive (+) console power lead goes to the positive terminal and that the negative (−) lead goes to the negative terminal. Also, be certain the power module has two good 3 amp fuses.

IMPORTANT: NEVER USE A BATTERY CHARGER FOR A POWER SUPPLY. IF THE BATTERY YOU INTEND TO USE IS CONNECTED TO A CHARGER, DISCONNECT THE CHARGER DURING THE TEST.

PRELIMINARY SETUP

Make sure the selector switch on the test set is turned off.

1. Connect lead A from the monitor console to the fuse side of the power module (Fig. 36).

2. Connect the TO CONSOLE lead B of the test set to the back of the power module (Fig. 36).

Fig. 35—Hookup of Power Module

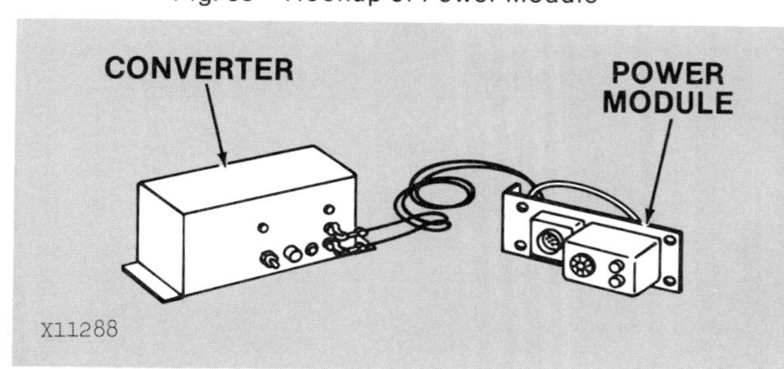

Fig. 36—Test Setup for Planter Monitor

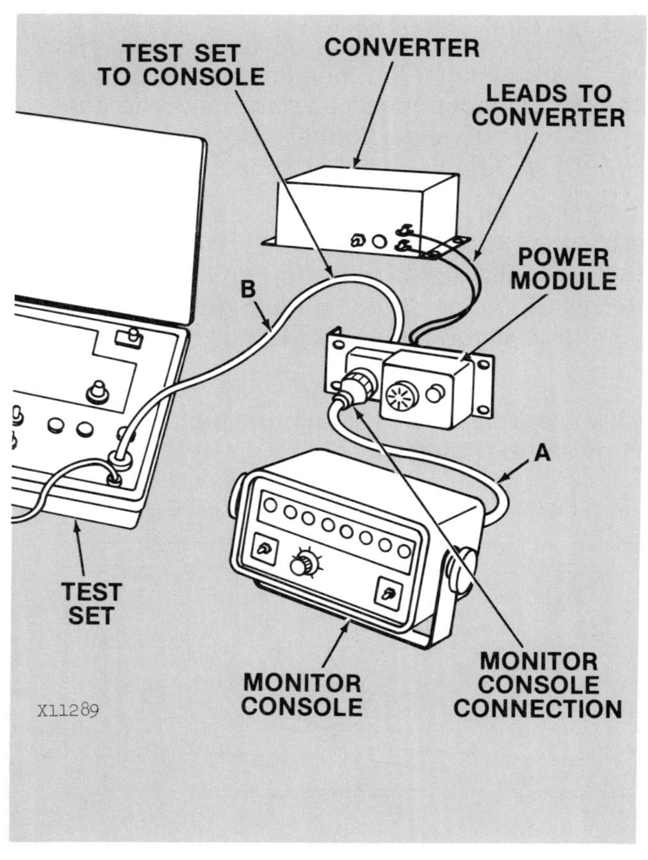

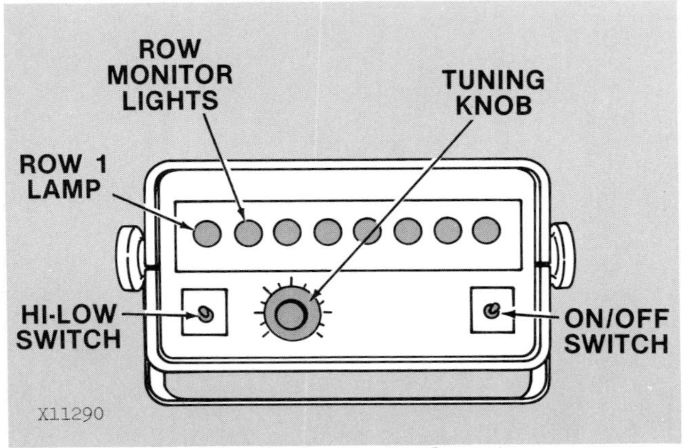

Fig. 37—Planter Monitor Console

3. Turn HI-LOW switch on the monitor console to low and power switch on (Fig. 37).

4. Turn tuning knob (Fig. 37) as far as it will go clockwise. All row monitor lights should be on and the PROCEED WITH TEST light (Fig. 38) should come on. If not, check the fuses, make sure the monitor console is connected to 12 volts, and that the leads are not damaged. If the light still does not come on, either the monitor console or power supply is defective.

FUNCTIONAL TESTS

You are now ready for the functional tests. However, if at any point the monitor console or power module does not perform as described, the power module must be checked first before you proceed any further. Test the monitor console for:

- High rate. All lamps should be on when test set is in NORMAL PLANTING (HI RATE) position (Fig. 39), monitor is set to HI, and tuning knob is turned clockwise. Lamps should all dim out at about the same time as tuning knob is turned counterclockwise.

- Low rate. This test is the same as high rate except for the settings on the test set (Fig. 40) and monitor.

Fig. 38—Proceed With Test Light Must Be On

X11291

X11292

Fig. 39—High Rate Test

- Row 1 failure. Row 1 lamp (Fig. 37) should light and alarm must sound when selector switch on the test set is turned to FAILURE ROW 1 (Fig. 41).

- Planter stopped. All lamps must light and alarm sound briefly when selector switch is turned to PLANTER STOPPED (Fig. 42).

Fig. 40—Low Rate Test

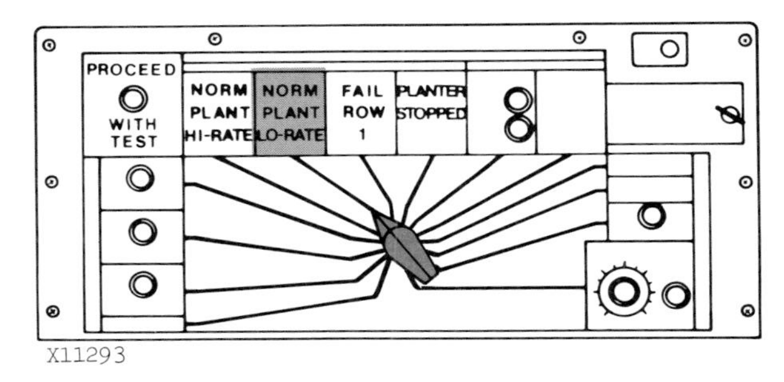

Fig. 41—Failure Row 1 Test

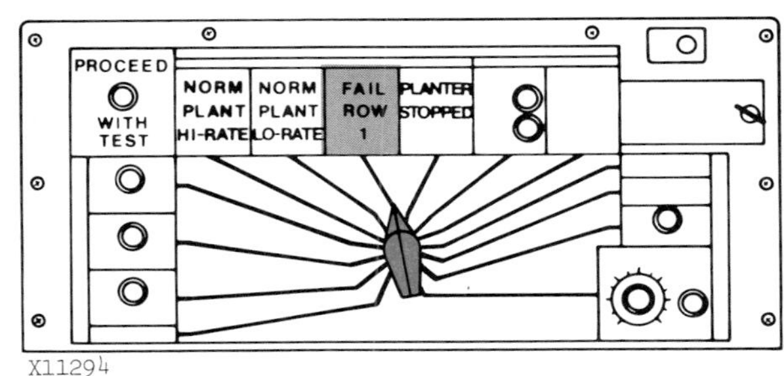

Fig. 42—Planter Stopped Test

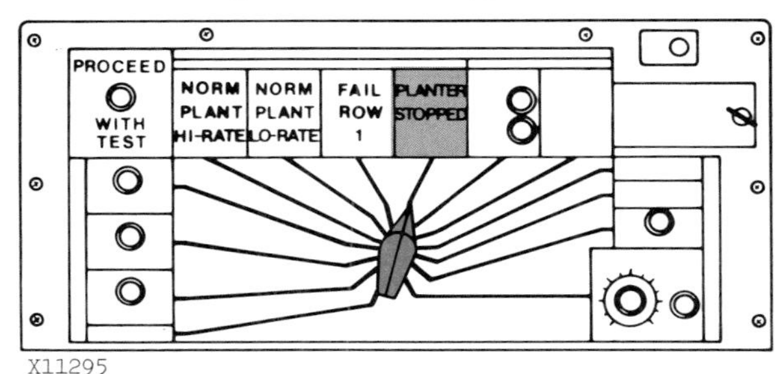

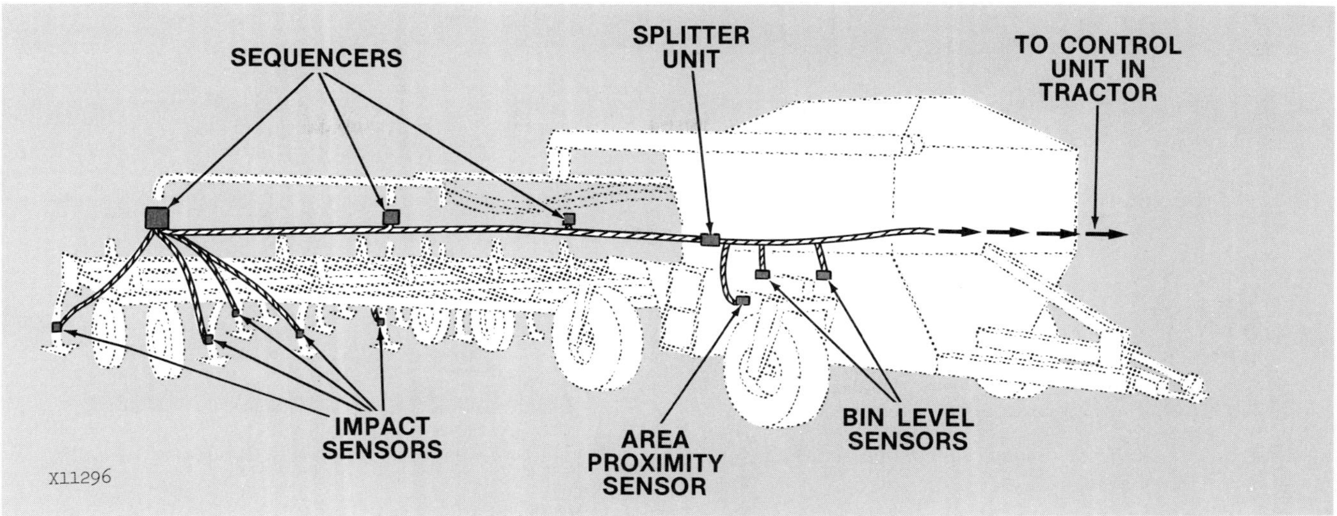

Fig. 43—Components of Seeder Monitor

If a problem occurred during any one of the four tests above, your first task is to determine which unit is defective—the power module or monitor console. Substitute another power module or console and repeat the tests. If the problem was eliminated, the unit which was replaced is defective.

Fig. 44—Block Diagram of Seeder Monitor

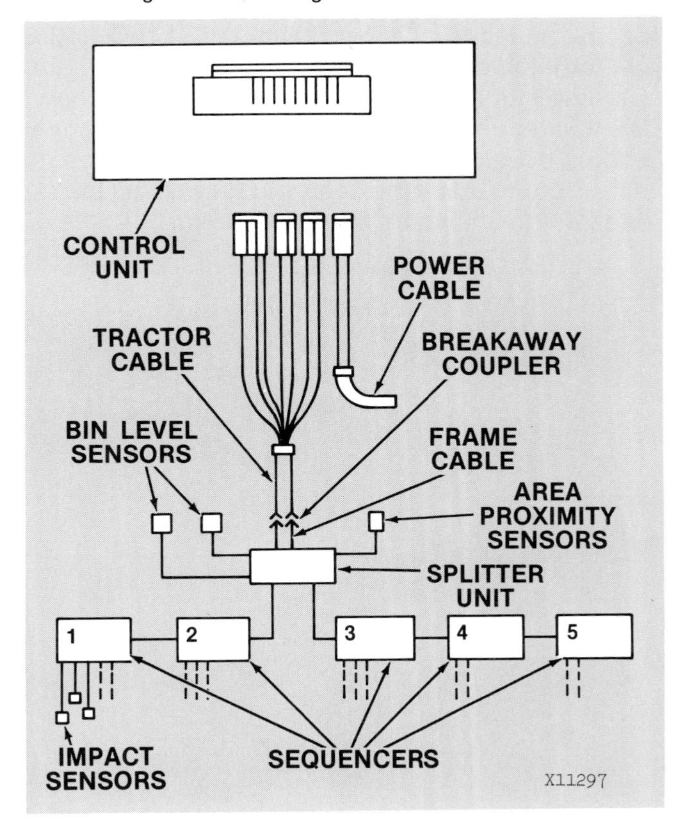

SEEDER MONITOR

The monitor (Fig. 43) provides the operator with information on the functions of the seeder. Each seed tube is monitored to assure normal flow of seed, the seed bin sections in the seeder are monitored to prevent running out of seed or fertilizer, and the area seeded can be measured and displayed.

The seed flow monitoring system consists of six basic components in addition to the wiring harnesses:

- Impact sensors
- Sequencers
- Splitter unit
- Bin level sensors
- Area proximity sensor
- Control unit

Fig. 44 is a block diagram of the system.

Impact Sensors

The impact sensors (Fig. 45) consist of a stainless steel pin with a vibration sensing crystal unit mounted and sealed in a plastic sensor assembly. The sensor is mounted on the seed boot with the steel pin inserted directly into the flow of seed (Fig. 46). Each time a seed or fertilizer particle strikes the steel sensor pin, the sensing crystal detects the vibration and sends a signal pulse to the sequencer. These sensors should be checked periodically because they can become encrusted and lose sensitivity.

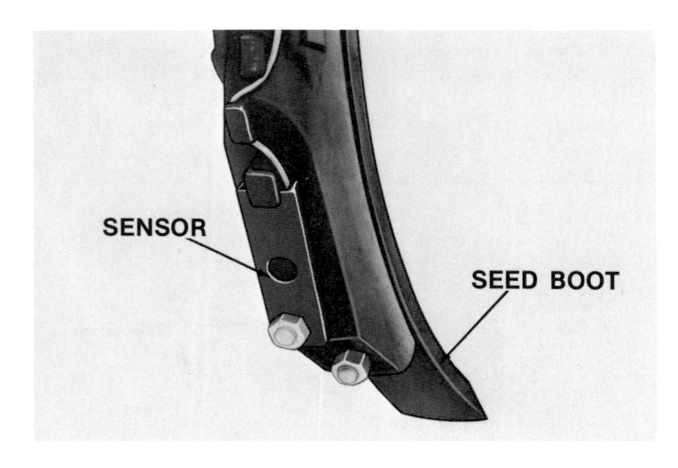

Fig. 45—Impact Sensor on Seed Boot

Sequencers

Groups of individual impact sensors are connected to one sequencer (Fig. 47). The sequencer samples each impact sensor in order, starting at sensor number one and proceeding to the last one (Fig. 48). Each sensor is sampled for a fixed period of time which is adjustable from about one half a second to seven seconds. The sequencer indicates which sensor is being sampled at any point in time. When a signal is received from the impact sensor being sampled, the sequencer progresses to the next sensor in the sequence. However, if at the end of the sample time, no signal has been received from the sensor being sampled, the sequencer halts its scan at that sensor and signals the splitter unit that a problem exists.

Splitter Unit

The splitter unit (Fig. 49) receives signals from the sequencers, bin level sensors, and area proximity sensor. After processing the signals from the bin level sensor and the area proximity sensor, the input from all the sources (including sequencers) is transmitted to the control unit mounted inside the tractor cab (Fig. 50).

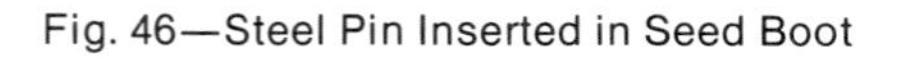

Fig. 46—Steel Pin Inserted in Seed Boot

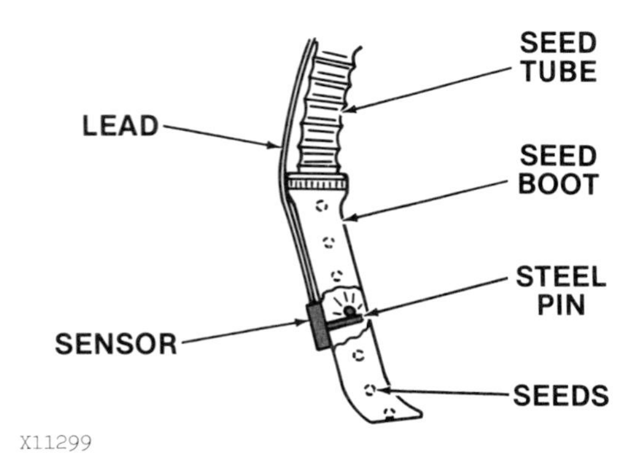

Fig. 47—Sequencer Showing Connection of Impact Sensors

Bin Level Sensors

The bin level sensor (Fig. 51) is an opto-electronic device that projects a beam of light of invisible frequency across a gap near the bottom of the seed bin section. As long as material is present to block the passage of the light beam, no signal is created. However, if the material level drops below the light beam, the bin level sensor detects the completion of the light circuit and transmits a low bin level warning signal to the splitter unit. These sensors are located inside the seed bins. The optical lens should be kept clean because excessive contamination could block the beams just as a high level of seed in the bin would block the beam.

Fig. 48—Sequencer Samples Each Sensor in Order

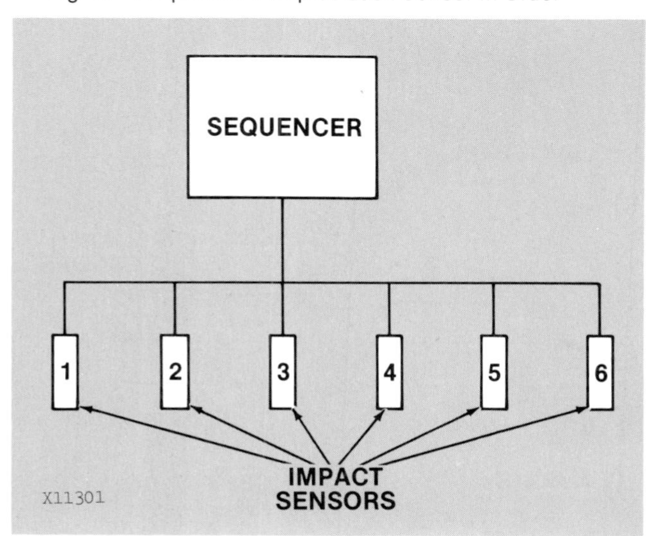

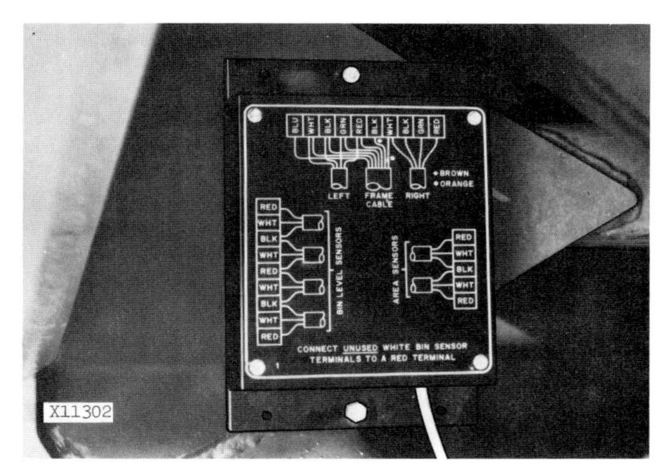

Fig. 49—Splitter Unit

Area Proximity Sensor

The feed roll transmits the seed from the bin to the seed tubes. The area proximity sensor (Fig. 52) signals each revolution of the feed roll on the seeder by electronically detecting the passage of a target cap screw on the feed roll shaft. As the target cap screw passes the sensor tip, a small LED (light-emitting diode) on the sensor body blinks, and a signal is transmitted to the splitter unit.

Control Unit

The control unit (Fig. 53) receives input from the splitter unit and signals the operator with lights and alarm of problem conditions at either the low bin level sensors or the impact sensors on the seed boots. The control unit also displays the area count (calculated from revolutions of the feed roll), adjusts the time interval for the impact sensors, and controls the power to the other components of the monitor.

Fig. 50—Splitter Unit Receives and Processes Signals and Sends to Control Unit

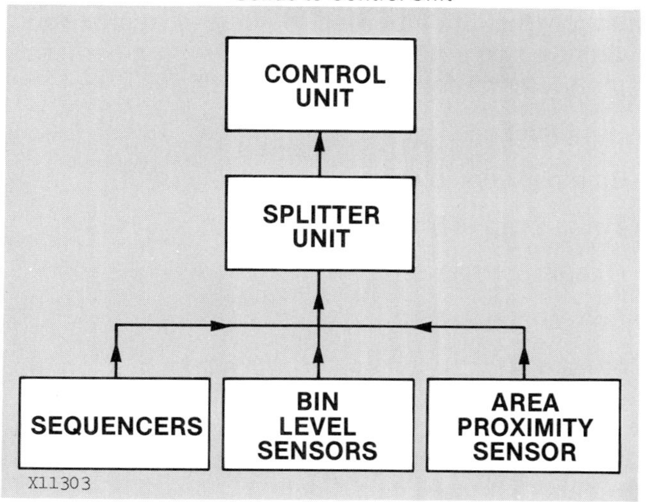

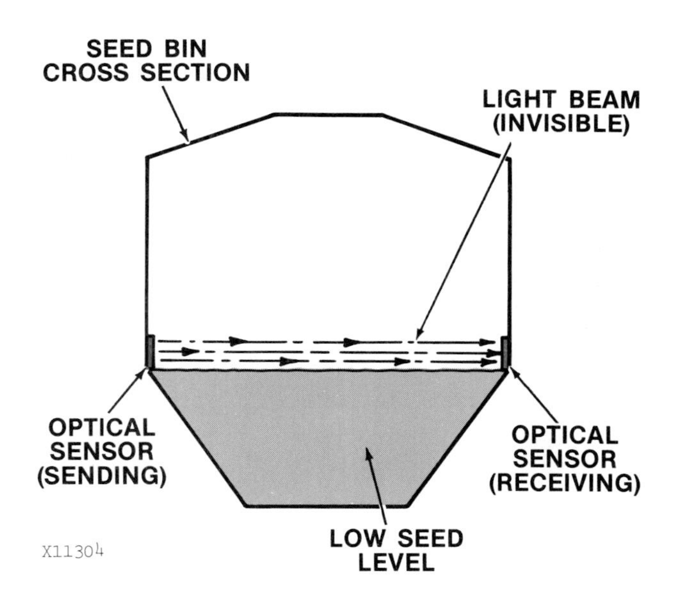

Fig. 51—Optical Bin Level Sensor Operation

Seeder Monitor Troubleshooting Chart

CONTROL UNIT LIGHTS ARE OFF EVEN THOUGH POWER SWITCH IS ON AND MACHINE IS NOT SEEDING.

Tractor key switch is off.

Power cable is broken or shorted.

Power cable is improperly connected.

Tractor cable is shorted.

One or more sequencers are shorted.

Splitter unit is shorted.

One or more sensors are faulty.

Fig. 52—Area Proximity Sensor

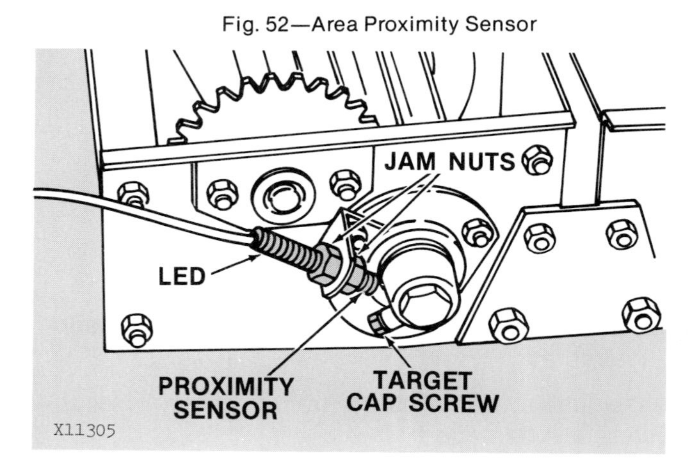

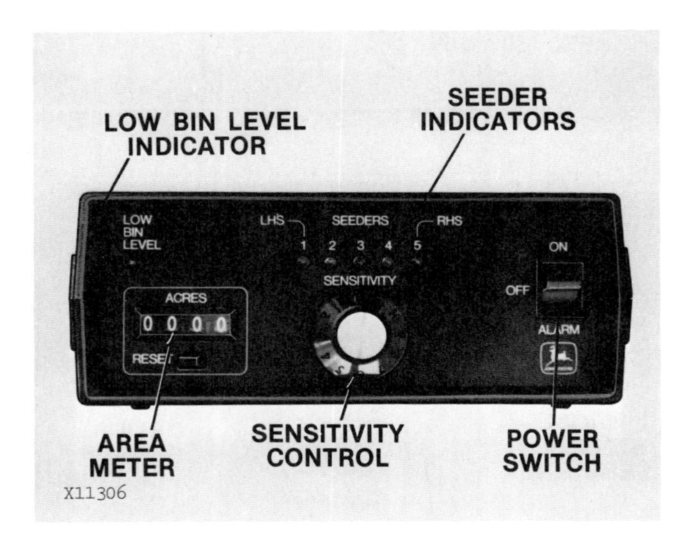

Fig. 53—Control Unit

SEEDER LIGHTS ON ONLY ONE SIDE OF MONITOR FLASH WHILE SEEDER IS NOT SEEDING.

Cables are faulty to that side of monitor.

Sequencers are faulty on that side.

Control unit is defective.

TRACTOR KEY SWITCH IS ON BUT NO SEEDER LIGHTS COME ON AFTER POWER SWITCH IS TURNED ON.

Breakaway connector is not making connections.

Power cable is faulty.

Control unit is defective.

Cable between control unit and splitter unit is defective.

LOW BIN LEVEL LIGHT FLASHES; BOTH BINS ARE FULL.

Bin sensors are defective.

Tractor or frame cable is damaged.

Connectors are damaged.

Cables are improperly connected.

Splitter unit is defective.

Control unit is defective.

LOW BIN LEVEL INDICATOR ON CONTROL UNIT DOES NOT LIGHT WHEN BIN IS EMPTY.

Light path between two halves of bin level sensor is obstructed.

Sensor cable, frame cable, tractor cable, or connectors are damaged.

Cables are improperly installed.

Sensor is defective.

Splitter unit is defective.

ONE OR MORE SEEDER LIGHTS FLASH WHILE SEEDING.

Seed tubes are plugged or not attached.

Main seed tube is plugged or leaking.

Sensitivity setting is too low.

Applicable sequencer is not receiving power.

Applicable sequencer is defective.

Sensor unit wire is broken.

Sensor pin is blocked or is contaminated.

Sensor is defective.

ALL SEEDER LIGHTS FLASH WHILE SEEDING.

Seed bins are empty. (Bin level light should be flashing.)

Fan and metering system is not functioning properly.

Sensitivity setting is too low.

Control unit is defective (sensitivity rheostat).

ROUND BALERS

BALE SIZE MONITORING SYSTEM

The bale size monitoring system (Fig. 54) found on some round balers monitors both the size and shape of round hay bales. The system consists of a monitor (Fig. 55) mounted inside the tractor, four switches, two potentiometers, and wiring harness. It monitors the following functions:

- Gate latched
- Bale near completion
- Twine wrap starting
- Oversized bale
- Bale shape

The **gate latch switches** (Fig. 56) are located on the right and left sides of the baler behind the bale tension spring. When the gate is latched, the switch contact will be depressed and the green light will indicate that the gate is latched.

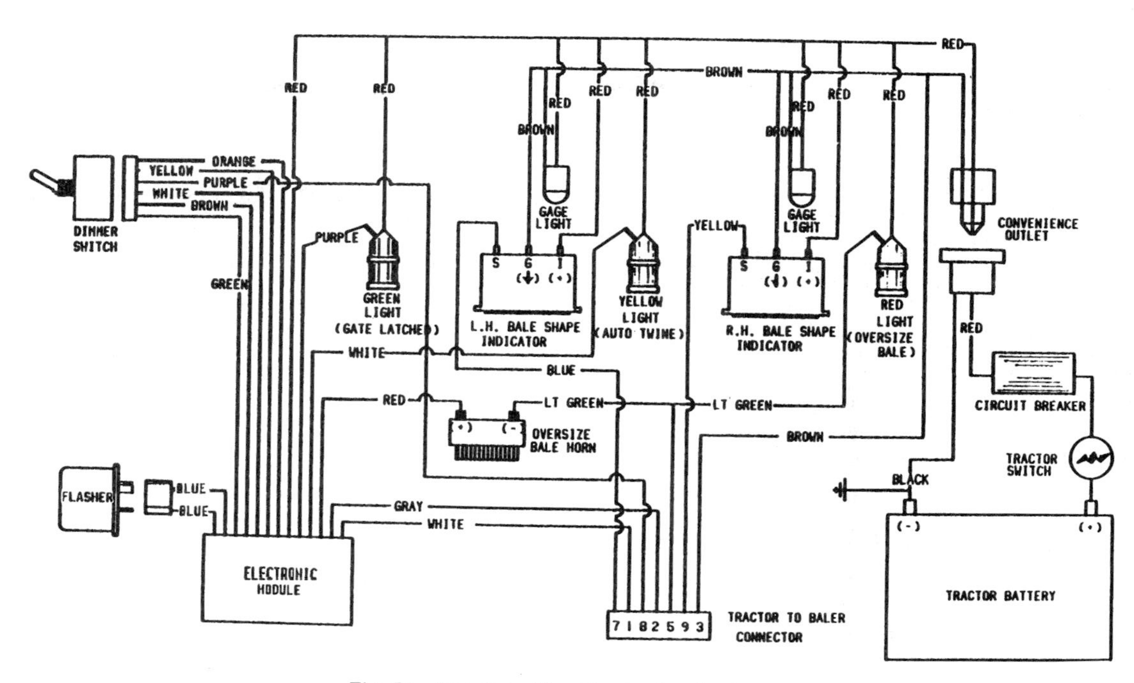

Fig. 54—The Bale Size Monitoring System

The **bale size switch** (Fig. 57) is activated when the bale size adjustment knob reaches the end of the slot in the linkage. It depresses the switch contact and causes the yellow light on the bale monitor to flash. This flashing light tells the operator that the bale is approaching the preset size and that the pointer indicators should be evened and preparations made to stop.

The **twine arm switch** (Fig. 58) is depressed during most of the baling cycle. When the twine wrapping cycle begins, the twine arm moves and activates the switch. This changes the flashing yellow light activated by the bale size switch to a solid yellow light. At this point forward motion of the tractor should be stopped.

If the tractor is not stopped and the bale is permitted to grow to the maximum possible diameter, the **oversize bale arm** will pivot and depress the contact on the microswitch (Fig. 59). The red light on the monitor will light and the alarm will sound to alert the operator that baling must be stopped immediately.

The **shape** of the bales is monitored by a potentiometer mounted on each outside belt on the rear of the baler (Fig. 60). Each potentiometer uses a bell crank which pivots to monitor the slack in the belt. Signals are sent to the left and right bale shape indicators (Fig. 55) which tell the operator how evenly hay is entering the baler. The pointers on the two indicators should be kept the same as near as possible so as to obtain an evenly shaped bale.

Fig. 55—Bale Monitor

Fig. 56—Gate Latch Switch

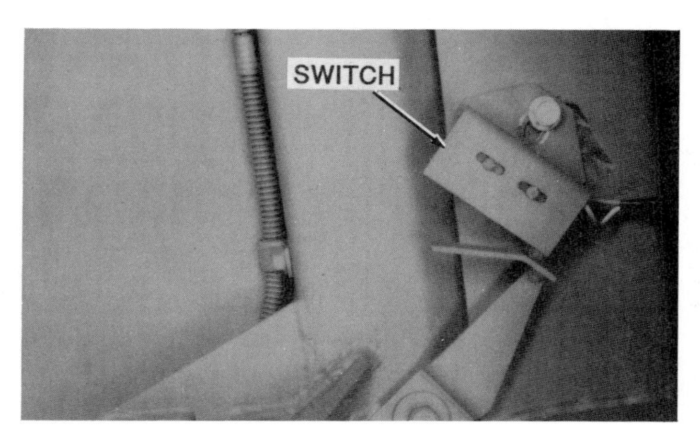

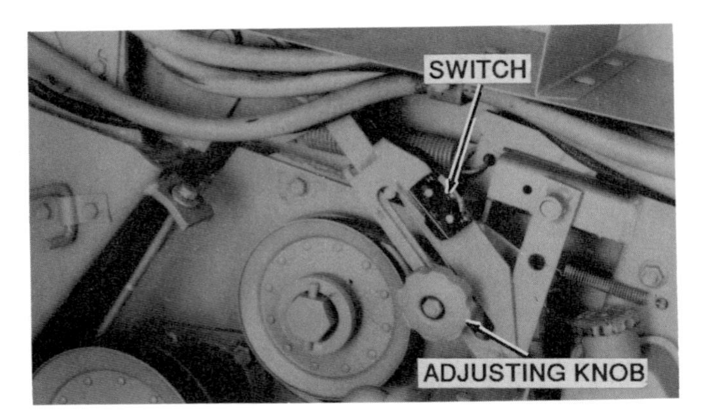

Fig. 57—Bale Size Switch

Troubleshooting

RED LIGHT COMES ON, SOLID YELLOW LIGHT DID NOT COME ON AND TWINE ARM DID NOT CYCLE.

Twine trip bell crank arm is out of adjustment.

Twine trip rod clevis is out of adjustment.

Red light switch is not adjusted properly.

SOLID YELLOW LIGHT ON, TWINE ARM IS IN HOME POSITION.

Switch is not adjusted properly.

Switch is defective.

White wire from twine arm switch is shorted to baler or tractor frame.

NO FLASHING YELLOW LIGHT, YELLOW LIGHT COMES ON SOLID AND TWINE ARM GOES THROUGH ITS NORMAL CYCLE.

Switch is not adjusted properly.

Extra light bulb inside of monitor box is burned out.

Flasher is defective or connection loose.

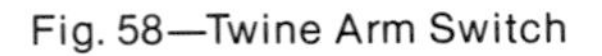

Fig. 58—Twine Arm Switch

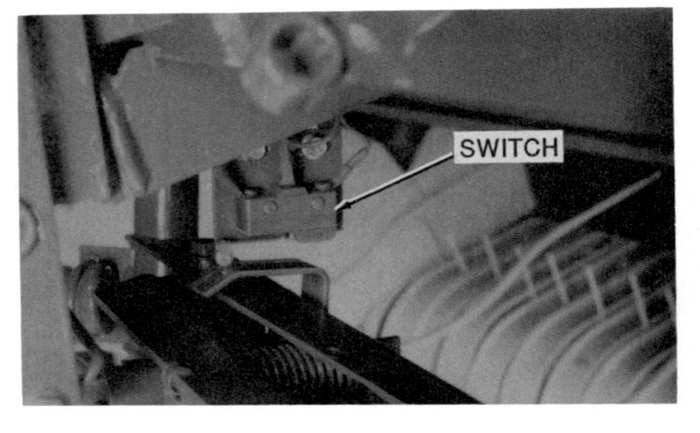

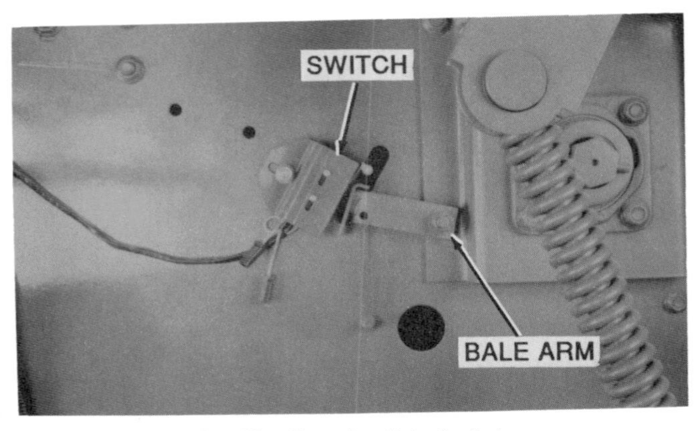

Fig. 59—Oversize Bale Switch

Switch is defective.

Voltage is too low.

GREEN LIGHT DOES NOT COME ON WHEN GATE IS CLOSED.

Gate lockout lever is engaged.

Gate switch is not adjusted properly.

Bulb or switch is defective.

Wire is broken or not connected properly.

GREEN LIGHT GOES OUT WHILE BALING.

Gate latch switch is not adjusted properly.

Hydraulic system has air in it.

Gate hydraulic cylinder has an internal leak.

GAUGES READ LOW OR UNEVEN WITH TIGHT WELL-SHAPED BALE.

Gauge sending units are not adjusted properly.

Gauge or sending unit is defective.

Fig. 60—Bale Shape Potentiometer

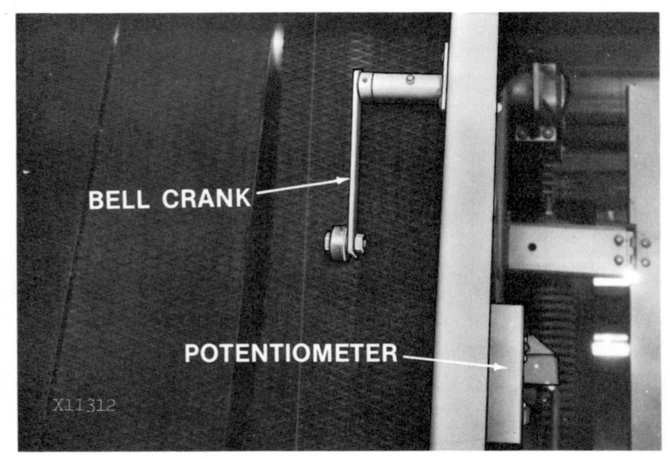

GATE IS NOT LATCHED. GREEN LIGHT IS ON.

Switch is defective.

BALE SHAPE GAUGES WILL NOT FUNCTION. LIGHTS ARE ALL RIGHT.

Polarity is reversed on hookup.

Checks and Adjustments

Checks and adjustments consist of:

- Checking wire continuity.
- Adjusting switches and potentiometers.
- Checking switches, potentiometers, and gauges.

CHECKING WIRE CONTINUITY

The first thing that should be checked in this or any other electrical system if it is not working properly is continuity of the wires and wiring harness. Because if a switch is not receiving current, it certainly will not work. Perform the following steps to check wire continuity. Also, refer to the applicable technical manual.

1. Disconnect the baler wiring harness at the connector on the tractor.

2. Using a volt-ohm meter, check the continuity from each switch to the connector. Refer to the technical manual for wire colors and connector pin numbers.

3. If continuity is present from switch to connector, check continuity from connector to monitor inside the tractor cab.

4. If continuity is present from the connector to monitor, connect the two halves of the wiring harness and check continuity from switches to the monitor inside the tractor cab.

5. If continuity is still present, this indicates that all wiring and the connector are good. If continuity is not present, the connector is at fault. Chapter 9 discusses servicing and repair of connectors.

6. If a broken wire is suspected, connect a volt-ohm meter to wire end and connector end of harness. Starting at one end, flex harness while watching the volt-ohm meter and work along entire length of harness. A reading on the volt-ohm meter will indicate that the wire is broken in that approximate area.

ADJUSTMENTS

The switches and potentiometers on the round baler can be adjusted. For example, you know that the gate is closed but the monitor in the tractor cab is not telling you that it is. Or the monitor is indicating that the bales are correctly shaped when in fact they are not. More than likely the problem is that a potentiometer or switch needs adjustment. The monitor itself is probably in good working order.

Adjusting the switches and potentiometers is a mechanical operation and usually involves loosening the mounting hardware of the switch, repositioning the switch, and tightening the screws and nuts.

As an example, Fig. 61 shows a gate latch switch. When the gate that releases the finished bale is closed, the switch arm should be contacting the switch body so that the switch roller is centered on the short leg of the ramp. But if the mounting cap screw had come loose allowing the switch bracket to pivot, the switch arm would not contact the switch and the green light would not indicate that the gate was closed when in fact it was. See applicable technical manuals for proper adjustment procedures and dimensions.

CHECKING SWITCHES, POTENTIOMETERS, AND GAUGES

A volt-ohm meter can be used to check the switches and potentiometers. A known good potentiometer and a battery can test the two gauges on the monitor. Use the following procedures for these electrical checks.

Fig. 61—Adjusting Gate Latch Switch

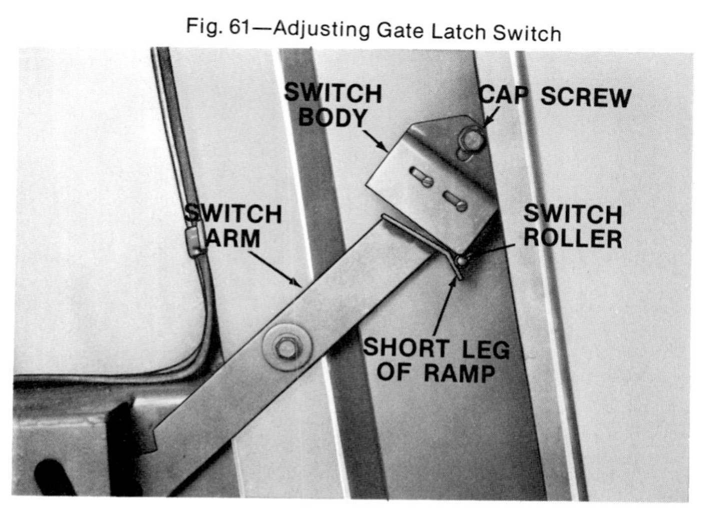

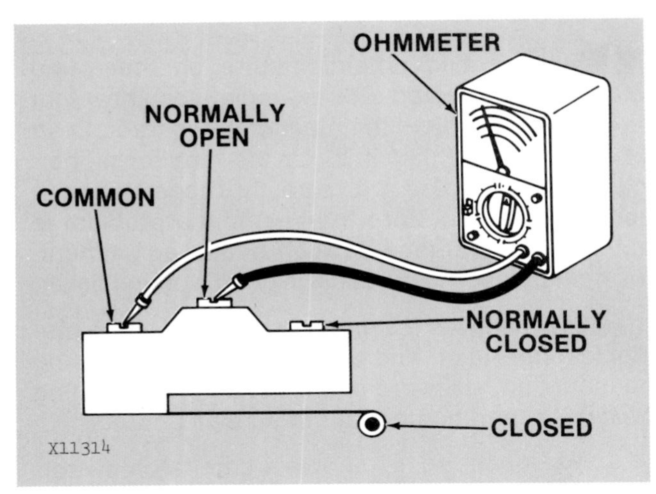

Fig. 62—Checking Switch in Closed Position

CHECKING SWITCHES

1. Unplug baler-to-tractor harness connection.

2. Using a volt-ohm meter, check across COMMON and NORMALLY OPEN (Fig. 62) with switch contact in closed position. If the volt-ohm meter does not register, replace switch. If it does register, go to step 3.

3. Check across COMMON and NORMALLY CLOSED (Fig. 63) with contact switch in open position. If volt-ohm meter does not register, replace switch.

CHECKING POTENTIOMETERS

1. Attach volt-ohm meter to potentiometer.

2. Move the arm of the potentiometer (Fig. 64) slowly taking about five seconds from stop to stop.

Fig. 63—Checking Switch in Open Position

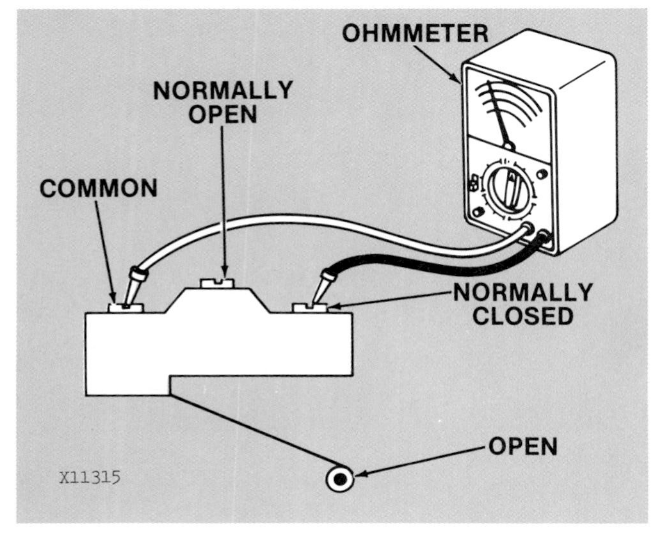

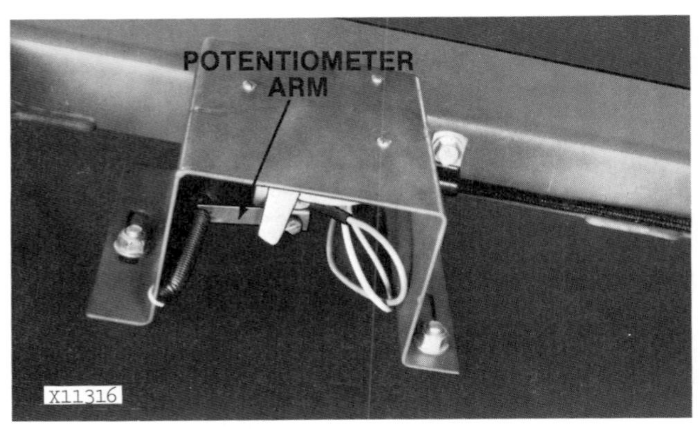

Fig. 64—Checking Potentiometer

Needle movement on the volt-ohm meter should correspond to bell crank arm (Fig. 65). There should be no dead spots in the needle movement. Check the technical manual for proper reading on volt-ohm meter.

3. Replace potentiometer if there are dead spots in needle movement or if resistance is not within approximate range.

CHECKING BALE SHAPE INDICATOR GAUGE

1. Connect a known good potentiometer, 12-volt battery, and the gauge as shown in Fig. 66.

2. Move potentiometer arm slowly taking about five seconds from stop to stop. Needle on the gauge should correspond to movement of the potentiometer arm (Fig. 67). If it does not, replace the gauge.

FORAGE HARVESTERS

METAL DETECTION SYSTEM

The metal detection system (Fig. 68) is designed to prevent material containing iron from entering the cutterhead area of the forage harvester. This reduces the possibility of cutterhead damage and hardware disease in livestock. The system consists of a metal detector sensor located inside the lower front feedroll, a metal detector module, a relay mod-

Fig. 65—Bell Crank Arm and Volt-Ohm Meter Move With Potentiometer

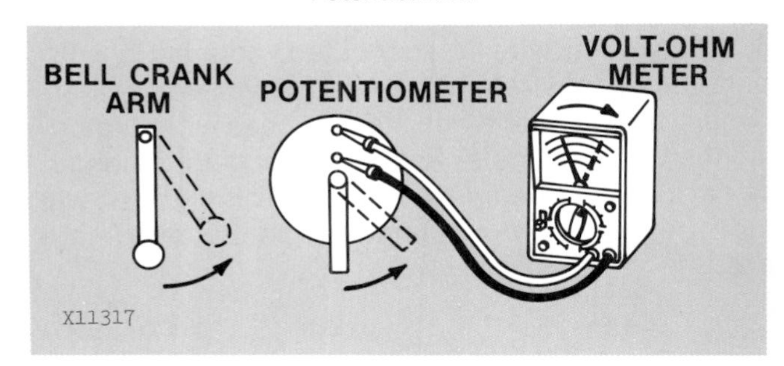

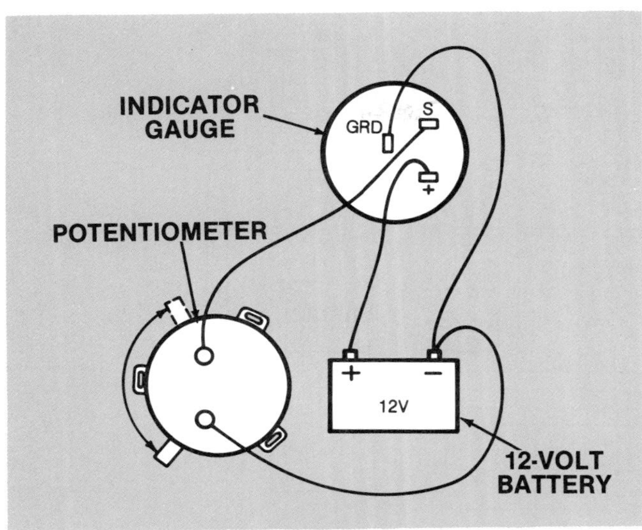

Fig. 66—Checking Bale Shape Indicator Gauge

ule, a solenoid and stopping mechanism, an electric clutch, warning lights and alarm, and wiring harnesses. An electrical diagram is shown in Fig. 69.

The sensor (Fig. 70) is made from coil-wrapped magnets encased in an aluminum channel and then mounted on a stationary shaft inside the lower front feedroll (Fig. 71). The feedroll itself is made from nonmagnetic stainless steel. When a piece of metal is drawn into the feedroll, the metal distorts the magnetic field of the sensor which then creates a voltage signal. This signal goes directly to the

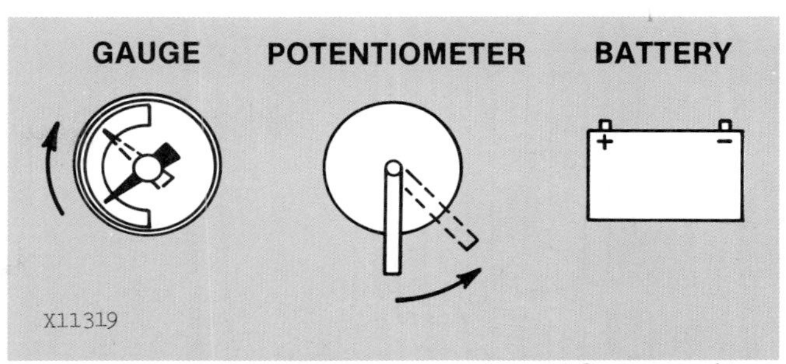

Fig. 67—Gauge Needle Movement Should Correspond to Potentiometer Arm Movement

metal detector module (Fig. 72) which stops the flow of electricity to the solenoid deactivating it (Fig. 73).

When the solenoid is deactivated, it drops the stopping pawls (Fig. 74) into the ratchet plates to stop the feedrolls (Fig. 75). At that time the electric clutch is deactivated and the red warning light and alarm are activated. The operator then shuts down the forage harvester, removes the metal from the feedrolls, resets the metal detector, and resumes operation.

Testing the Metal Detector

The metal detection system has what is known as on-board diagnostics—it will test itself. The diagnostics are performed using the lights and test buttons on the metal detector module and relay module.

Fig. 68—Components of Metal Detection System

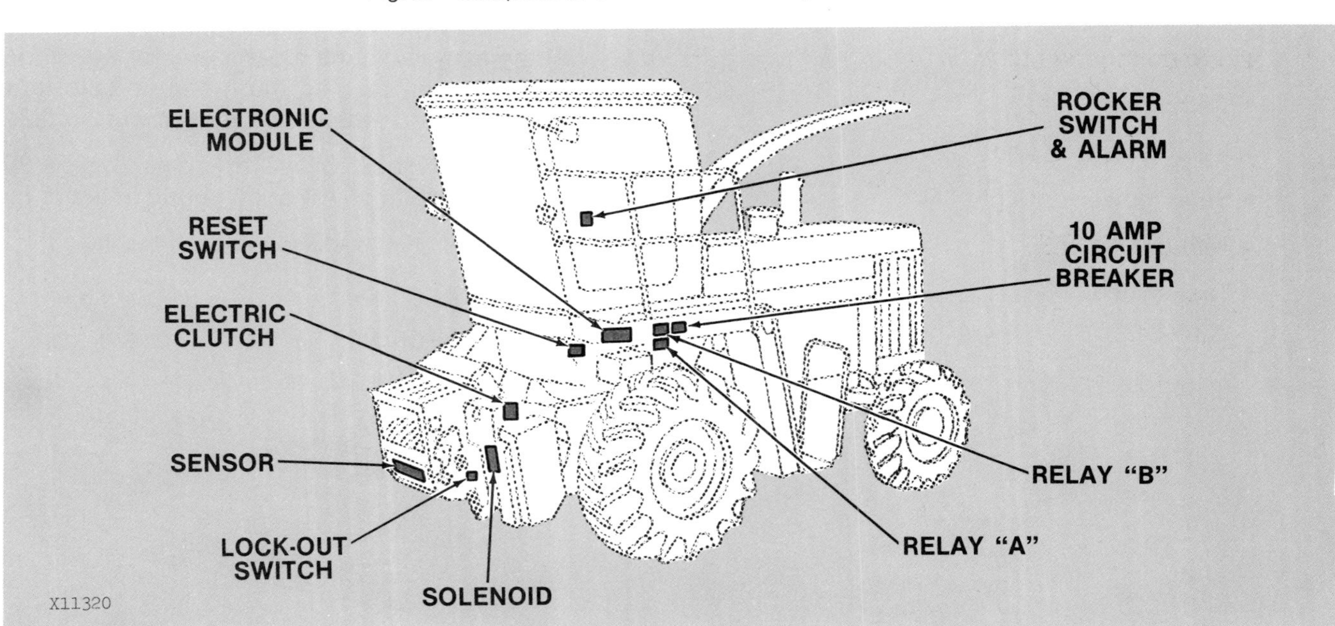

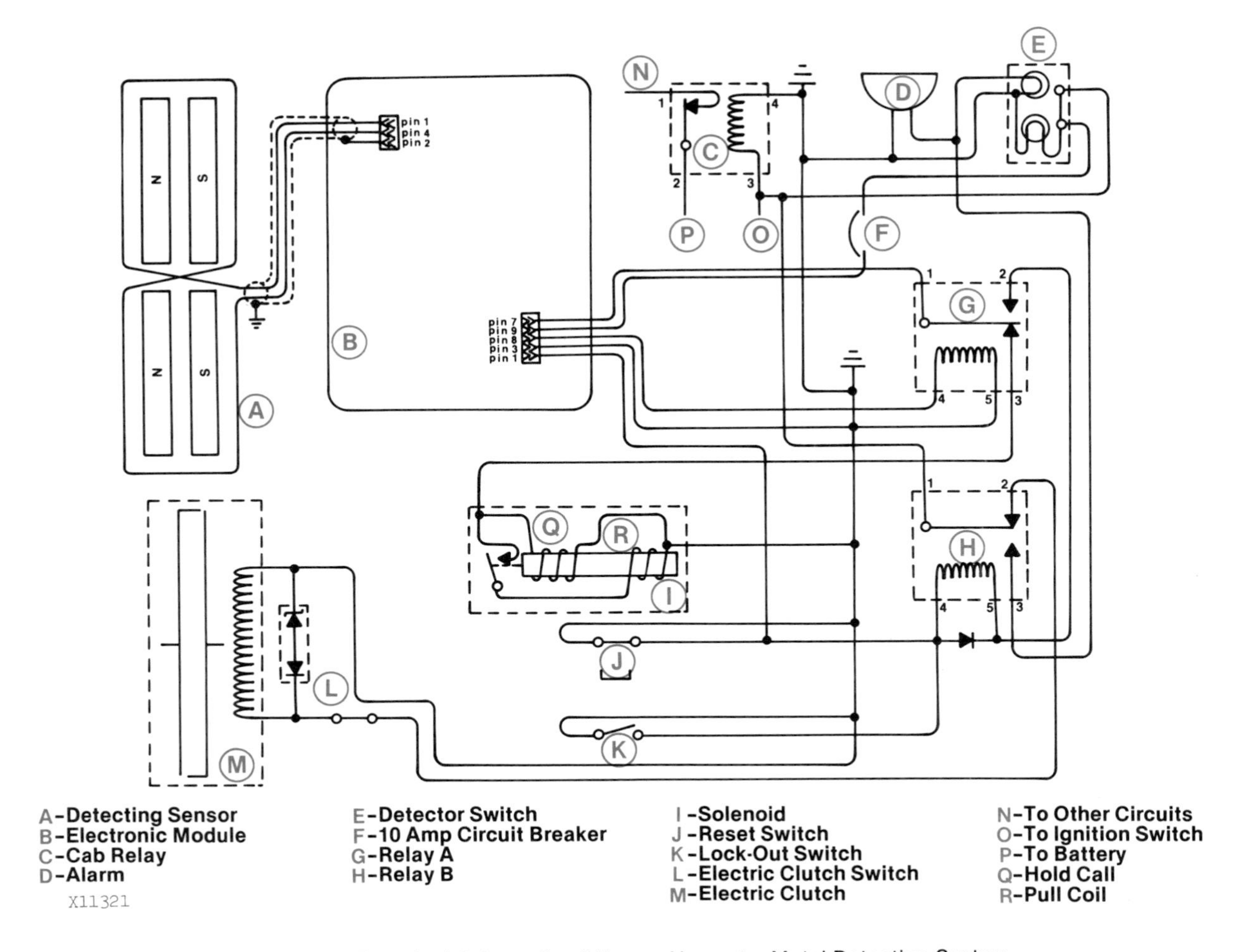

Fig. 69—Electrical Schematic of Forage Harvester Metal Detection System

The metal detector module has four power lights and two test buttons (Fig. 72).

- Low voltage light "A"
- Main power light "B"
- Relay power-in light "C"
- Relay power-out light "D"
- Test button "E"
- Test switch "F"

Fig. 70—Sensor in Metal Detector

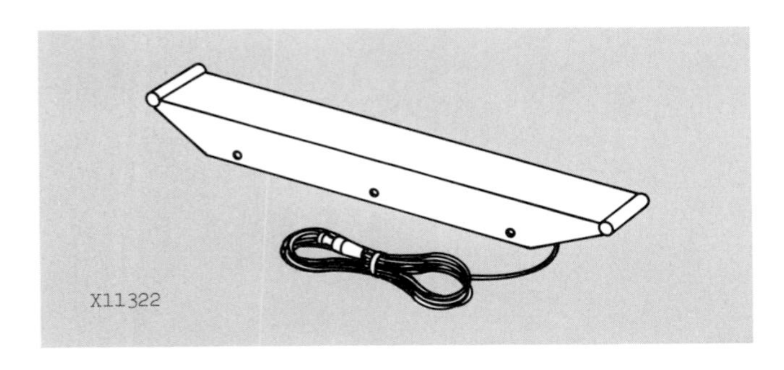

When the power to the metal detector module falls below 10 volts, low voltage light "A" will come on. At the same time the metal detector will "trip." The light will stay on until the metal detector is reset. This light also indicates a power interruption.

Main power light "B" will be on as long as there is at least 10 volts power to the metal detector module

Fig. 71—Sensor Is Inside Feedroll

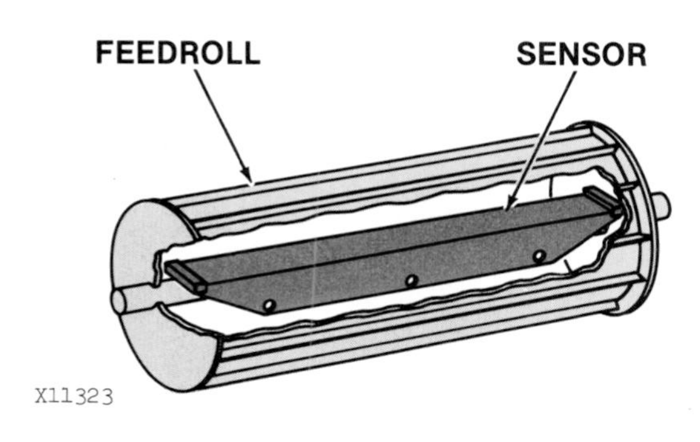

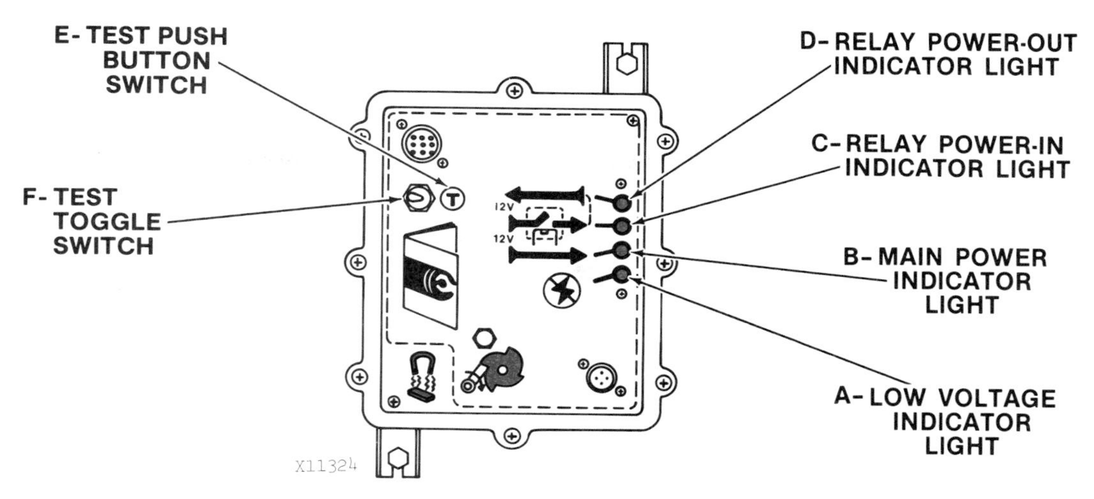

Fig. 72—Metal Detector Module in Metal Detection System

and will be on during normal operation. Both the low voltage light "A" and main power light "B" monitor power at the same point. The power supply is protected by a 10-amp circuit breaker in the relay module.

Relay power-in light "C" indicates that there is power to the relay in the metal detector module. The light is on during normal operation.

Relay power-out light "D" indicates that there is power to the coil in the metal detector module relay. The light is on during normal operation.

Test push button "E" (Fig. 72) induces a voltage into the metal detector module. When this button is pushed, power relay lights "C" and "D" should go out and the system should trip causing the stopping pawls (Fig. 75) to fall into the ratchet plates.

Test toggle switch "F" (Fig. 72) checks the reset coil assembly. When it is pushed, power is sent to the relay coil in the metal detector module bypassing all the logic circuits and amplifier. At this time the relay power-out light "D" should be on.

The relay module has one test button and four indicator lights (Fig. 76).

- Indicator light "A"
- Clutch shift light "B"
- Solenoid light "C"
- Reset light "D"
- Test push button "E"

Fig. 73—Solenoid on Metal Detector

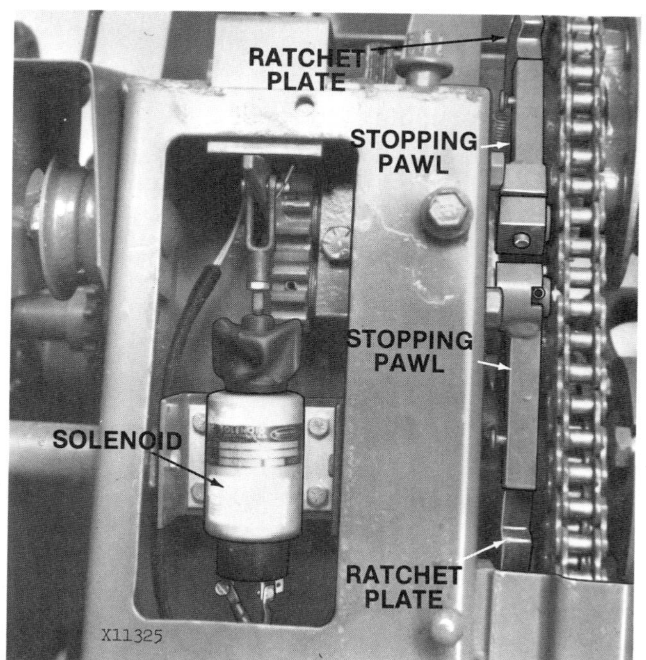

Fig. 74—Stopping Pawls

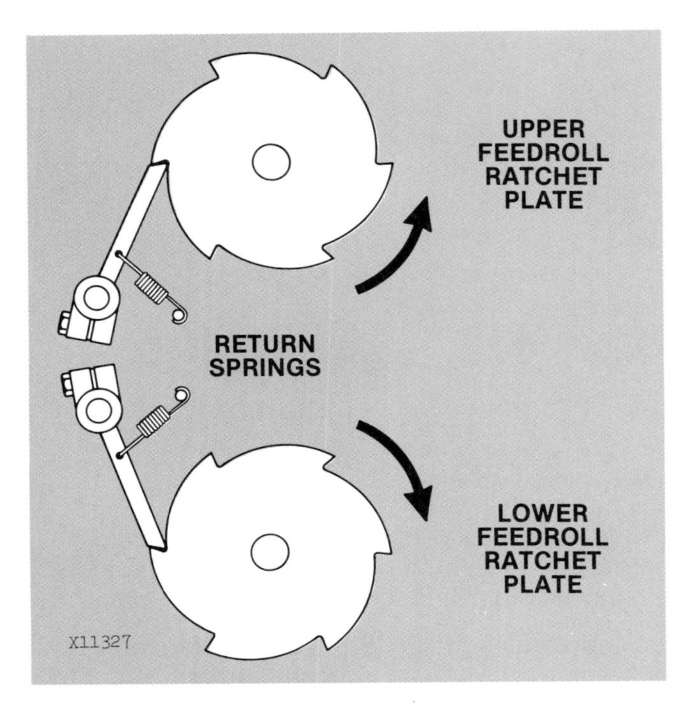

Fig. 75—Stopping Pawls Engaging Ratchets

Test light "A" is not used. Light "B" is the clutch shift light. It should be on when the feedrolls are being shifted to forward. It also indicates that the electric clutch has power. This light should always be on during normal operation for self-propelled machines only. Light "C" is the solenoid light. It should be on during normal operation and will go off when the system "trips." Light "D" is the reset light and is on during normal operation. Test push button "E" is located above the diagnostics lights. When this button is pushed, approximately 0.5 volt is sent to the sensor causing the system to "trip" just as if a piece of metal had been detected during normal operation of the forage harvester. The trip switch will test the sensor and both the metal detector module and relay module.

Metal Detection System Troubleshooting Chart

DETECTOR WILL NOT RESET.

Reset switch is not working properly.

Lockout switch is closed.

Loose connections or defective wiring.

Detector is resetting, but solenoid is not releasing stopping pawls.

Reset switch, lockout switch, or electronic module has failed.

STOPPING PAWLS DO NOT DISENGAGE.

Pawls are trapped by ratchet plate.

Pawls do not move freely.

Voltage at solenoid is low.

Pawl clearance not adjusted.

Detector not resetting.

10-amp circuit breaker, solenoid, or relay A has failed.

DETECTOR TRIPS IMMEDIATELY AFTER SETTING.

Plunger of solenoid unable to retract completely.

Pull coil not switched to hold. High amperage opens circuit breaker.

Failure of detecting sensor electronic module, solenoid, or circuit breaker.

Metal is lodged in, or wrapped around feed roll.

METAL DETECTOR DOES NOT TRIP.

Metal does not contain iron or improper test procedure.

Fig. 76—Relay Module

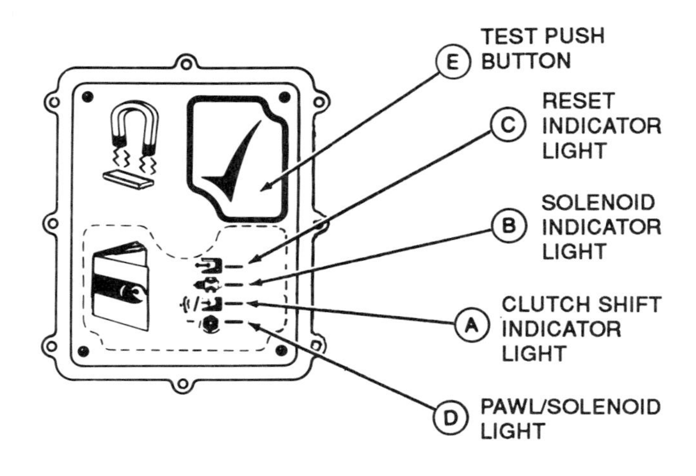

Detecting sensor harness damaged.

Poor electronic module connections.

Detecting sensor not installed correctly.

Failure of detecting sensor, electronic module, or relay A.

NUMEROUS FALSE TRIPS.

Loose connections or intermittent short at pawl solenoid.

Detected metal is too small to find easily.

Metal is lodged in, or wrapped around feed roll.

Detector sensor harness is damaged.

Loose connections or damaged wire in power supply circuit.

Stopping pawls improperly adjusted.

Upper or lower front feed roll assembly damaged, or worn bearings.

Repair welds using ferrous metal have been made on front feed rolls.

Detecting sensor is picking up a strong outside signal.

A feed roll shaft or one of the rear rolls is magnetized.

Failure of detecting sensor, electronic module, circuit breaker, or rocker switch.

GREEN PILOT LIGHT DOES NOT COME ON.

Ignition or detector switch not in "ON" position.

Loose connections at detector switch.

Battery voltage low, poor connections.

Faulty detector switch.

ALARM AND RED LIGHT DO NOT COME ON.

Feed roll shift lever is in reverse position.

Reset switch not positioned correctly.

Fig. 77—Components of Tractor Monitoring System

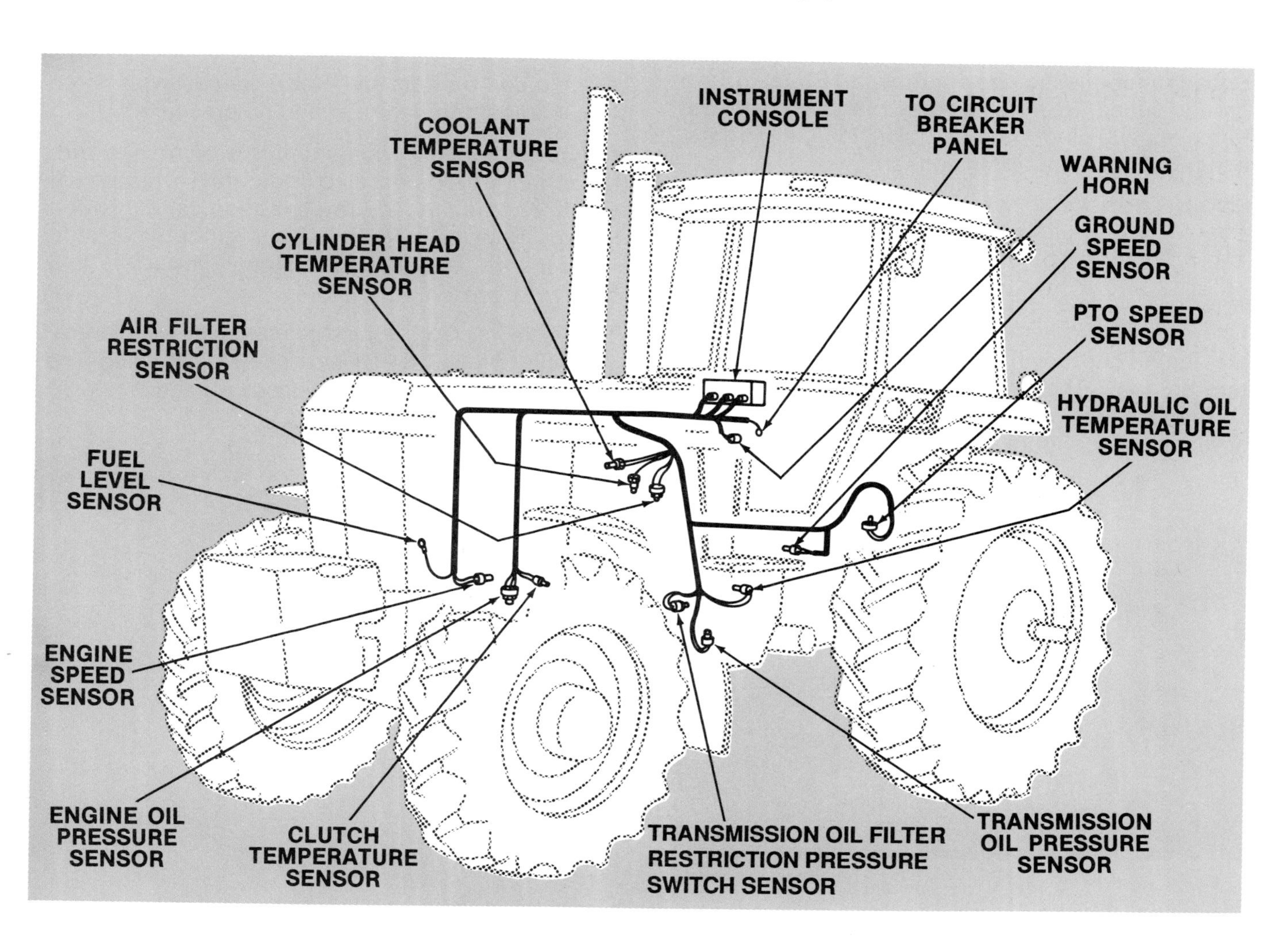

Failure of relay A, relay B, circuit breaker, alarm, detector switch or reset switch.

ALARM AND RED LIGHT REMAIN ON WITH FEED ROLLS REVERSED.

Pawls locked out. Lock-out bail not in position under spring clip retainer.

Feed roll shift lever is not between reset switch and magnet in reverse position.

Failure of reset switch, lockout switch, or relay B.

ELECTRIC CLUTCH DOES NOT ENGAGE.

Loose connections, damaged wiring.

Retaining pin for electric clutch field is loose.

Failure of relay B.

TRACTORS

GENERAL INFORMATION

Tractor monitoring systems (Fig. 77) monitor two main functions:

- Engine, hydraulic, and power trains systems.
- Speed functions.

Speed functions include engine rpm, PTO rpm, and ground speed. To perform these functions, the system uses two electronic modules, indicator lights, and numerous electronic sensors.

ENGINE ANALYSIS

One electronic module (Fig. 78) analyzes the condition of the engine and power train and indicates where problems exist. The unit also indicates whether the problem is critical requiring immediate engine shutdown or is noncritical requiring attention soon. The unit consists of four gauges—voltmeter, fuel, oil pressure, and coolant—and seven indicator lights which will be explained later.

Fig. 78—Electronic Modules for Tractor Monitor

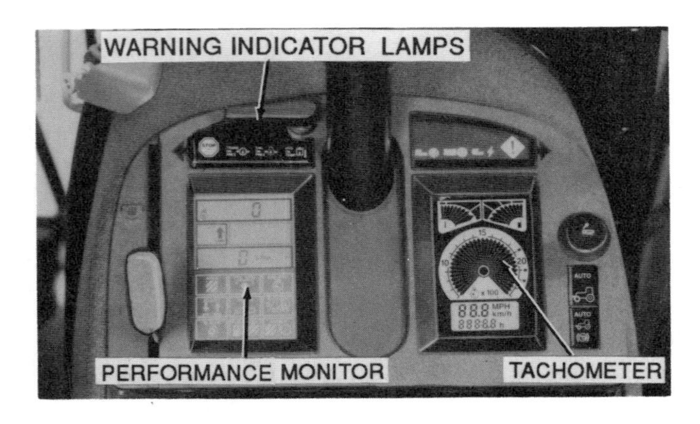

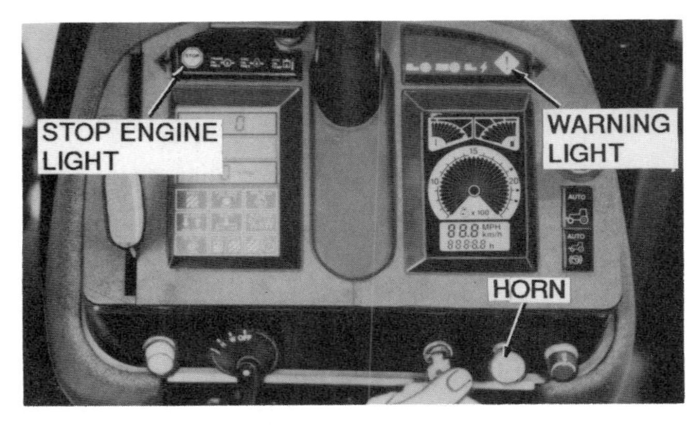

Fig. 79—Warning Lights and Horn

Critical and noncritical problems are indicated by a red stop engine light or amber warning light and a horn (Fig. 79). A critical problem is indicated by a steady horn sound, a flashing stop engine light, and a lighted indicator light or gauge needle in the warning or red scale of the gauge. A noncritical problem is indicated by a steady service message light, an indicator light or gauge needle in warning or red scale, and on some systems, by a beeping horn.

Gauges

Most older tractors use four gauges: voltmeter, fuel, coolant temperature, and oil pressure.

Newer tractors use indicator lights combined with a tachometer module to indicate voltage, fuel level, engine speed, rated speed, and ground speed. They also include an hourmeter, a stop engine light, a service alert signal, and a warning horn or buzzer.

The voltmeter (Fig. 80) indicates tractor system voltage when the key switch is on and while running. The bands on the meter indicate various voltages.

Fig. 80—Voltmeter

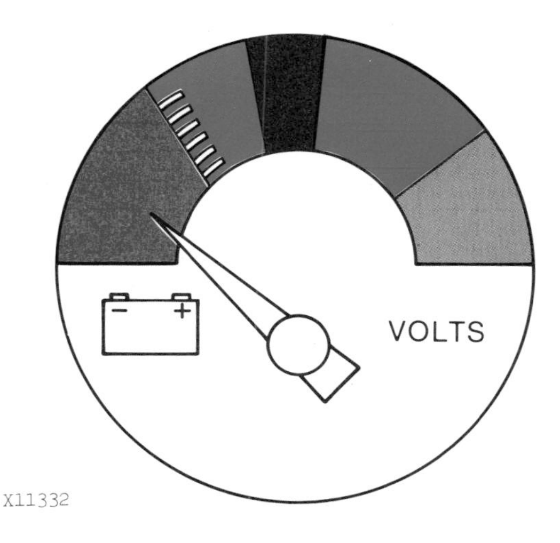

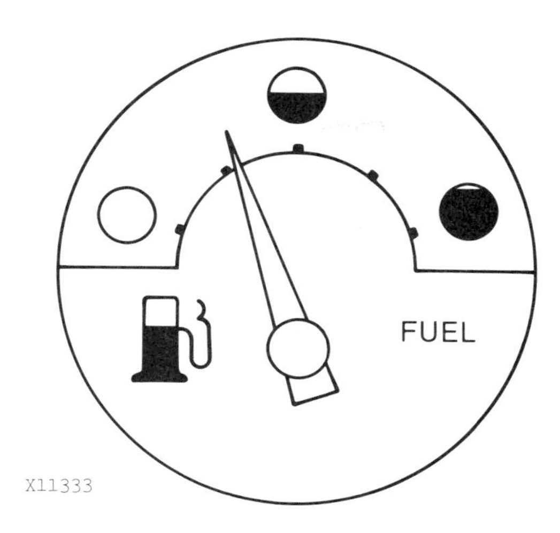

Fig. 81—Fuel Gauge

In the **fuel gauge** circuit (Fig. 81), a float and potentiometer mechanism causes fuel level sensor resistance to increase as level rises. So as the tank is filled with fuel, resistance at the sensor increases causing the needle in the gauge to move toward the full mark.

Coolant temperature (Fig. 82) is measured by a variable resistance temperature sensor in the cylinder head. Increasing temperature increases sensor resistance values.

Engine oil pressure (Fig. 83) is sensed by a pressure sensor that indicates decreasing voltage with decreasing pressure. When the pressure drops below the predetermined pressure for the particular engine rpm, the gauge needle moves to the red band, engine oil pressure indicator light comes on, stop engine light flashes, and the horn sounds. Low or no engine oil pressure is a critical problem.

Fig. 82— Water Temperature Gauge

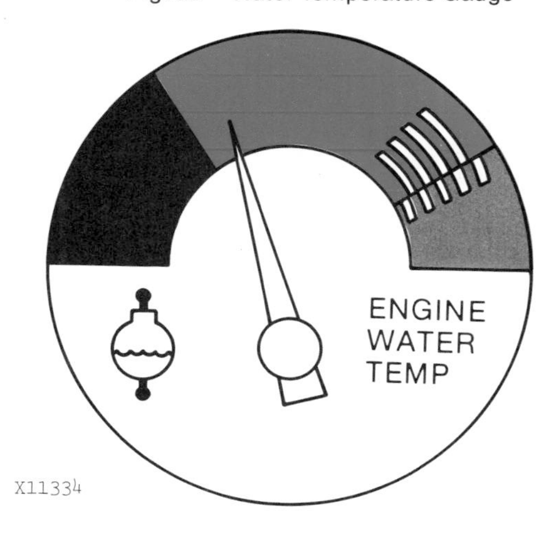

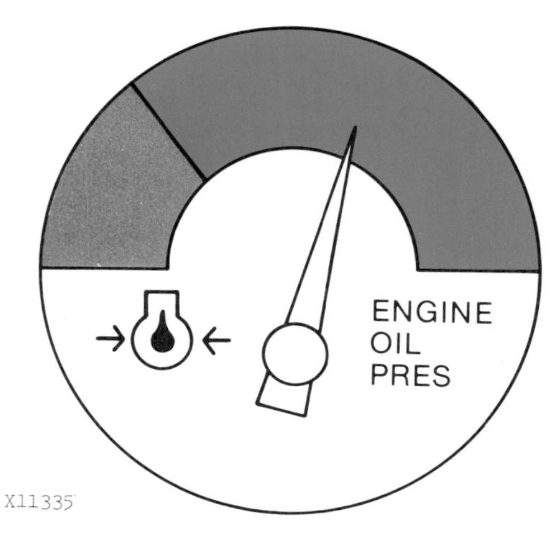

Fig. 83—Oil Pressure Gauge

Indicator Lights

The indicator lights (Fig. 84) give priority to a critical problem over a noncritical problem. Two or more lights may signal a noncritical warning at the same time. However, if any one of the functions become critical, all concritical warnings are cancelled and a critical problem is signaled by the appropriate indicator light, stop engine light, service alert signal, horn, and gauges, if equipped.

On tractor start-up, the module conducts a lamp test (all lamps light) for five seconds and a horn sound for two seconds.

The seven indicator lights alert the operator of critical and noncritical problems pertaining to:

- Transmission control pressure.
- Engine oil pressure.
- Voltage level.
- Hydraulic oil temperature.

Fig. 84—Indicator Lights

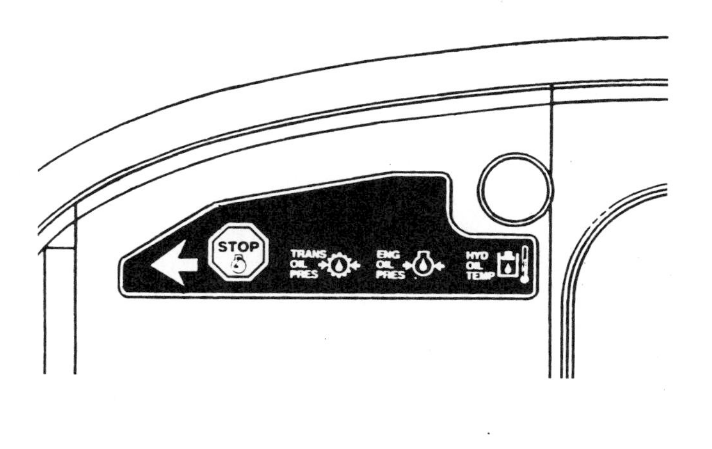

- Air filter.
- Transmission oil filter.
- Transmission lube pressure (on early model 4-wheel drive).

Common Troubleshooting

ALL INDICATOR LIGHTS COME ON AND STAY ON.

The 37-pin connector lost ground at pin 4.

INDICATOR LIGHTS, STOP ENGINE, AND SERVICE ALERT LIGHTS WILL NOT ACTIVATE.

The 37-pin connector lost ground at pin 22.

FILTER INDICATOR LIGHT, STOP ENGINE LAMP, AND HORN ARE ON WHEN KEY SWITCH IS ON AND ENGINE IS OFF.

Transmission pressure sensor and transmission filter sensor F2 wires are switched.

FILTER INDICATOR LIGHT AND SERVICE ALERT LIGHT ARE ON WHEN KEY SWITCH IS ON AND ENGINE IS OFF.

Transmission pressure sensor and transmission filter sensor F1 wires are switched.

TRANSMISSION PRESSURE INDICATOR LIGHT, STOP ENGINE LIGHT, AND HORN COME ON.

The 37-pin connector lost ground at pin 17.

SERVICE ALERT LAMP ONLY COMES ON.

Alternator diode is faulty.

CALIBRATIONS AND CHECKS

Procedures follow for:

- Calibration of voltmeter.
- Low voltage warning check.
- High voltage warning check.
- Speed monitoring.

Calibration of Voltmeter

1. Remove the voltmeter.

2. Place two of the module hold-down screws in holes #1 and #3 on the back of the voltmeter as shown in Fig. 85.

3. Connect a jumper lead from #3 of voltmeter to ground.

4. Connect green and black leads of universal gauge tester between # of voltmeter and #1 of voltmeter socket.

5. Connect digital multimeter between #1 of voltmeter and ground.

6. With key switch on and engine at 1,000 rpm, adjust 50-ohm potentiometer on the universal gauge tester so as to position voltmeter needle on each of

Fig. 85—Voltmeter Calibration Test Setup

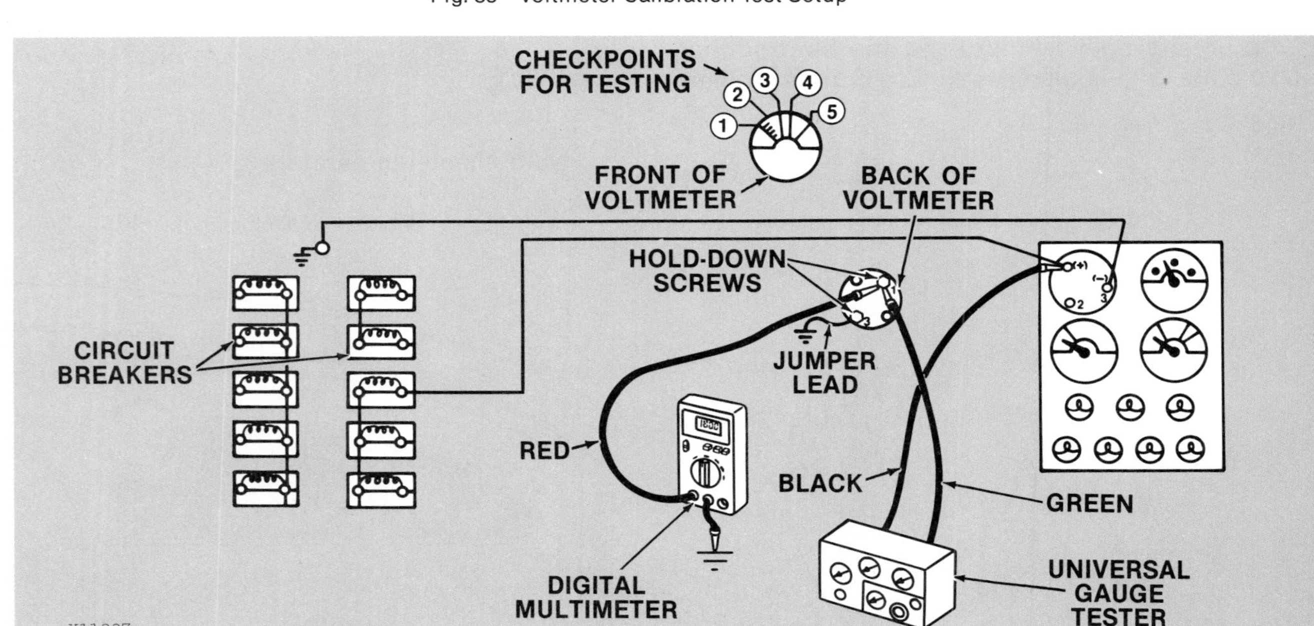

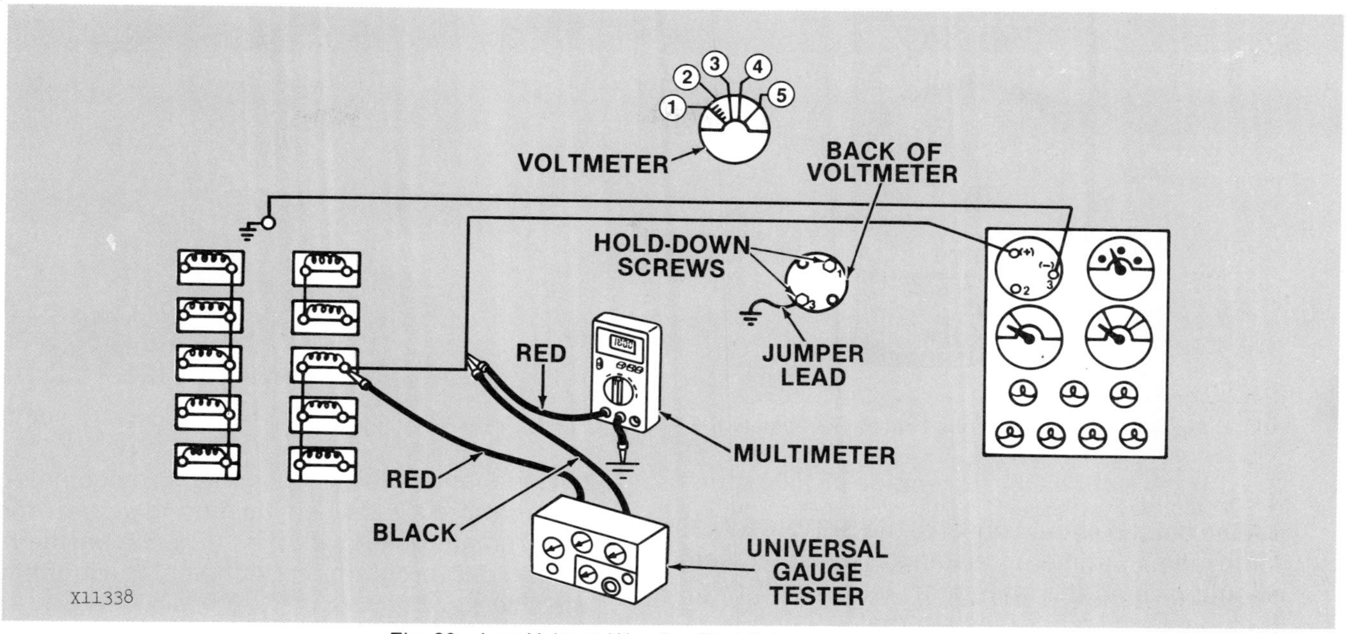

Fig. 86—Low Voltage Warning Test Setup

the five checkpoints one at a time as shown at the top of Fig. 85.

7. Read volts on multimeter and compare the readings as follows:

Checkpoint	Volts
1	11.2 + 0.35
2	11.8 + 0.35
3	12.6 + 0.35
4	13.2 + 0.35
5	15.5 + 0.35

Low Voltage Warning Check

1. Connect universal gauge tester in series with the lead at third right-hand circuit breaker (Fig. 86). Connect red lead to circuit breaker as shown.

2. Connect digital multimeter as shown in Fig. 86.

3. Turn key switch ON. After automatic lamp test cycle is completed, hold test button on gauge tester down and adjust the 25-ohm potentiometer so that the SERVICE ALERT lamp comes on. The digital multimeter should show 10.5 to 11.5 volts. The tractor voltmeter needle should be at checkpoint 1 (Fig. 86).

High Voltage Warning Check

1. Connect red lead of digital multimeter to starter solenoid battery terminal and ground the other lead (Fig. 87).

Fig. 87—High Voltage Warning Test Setup

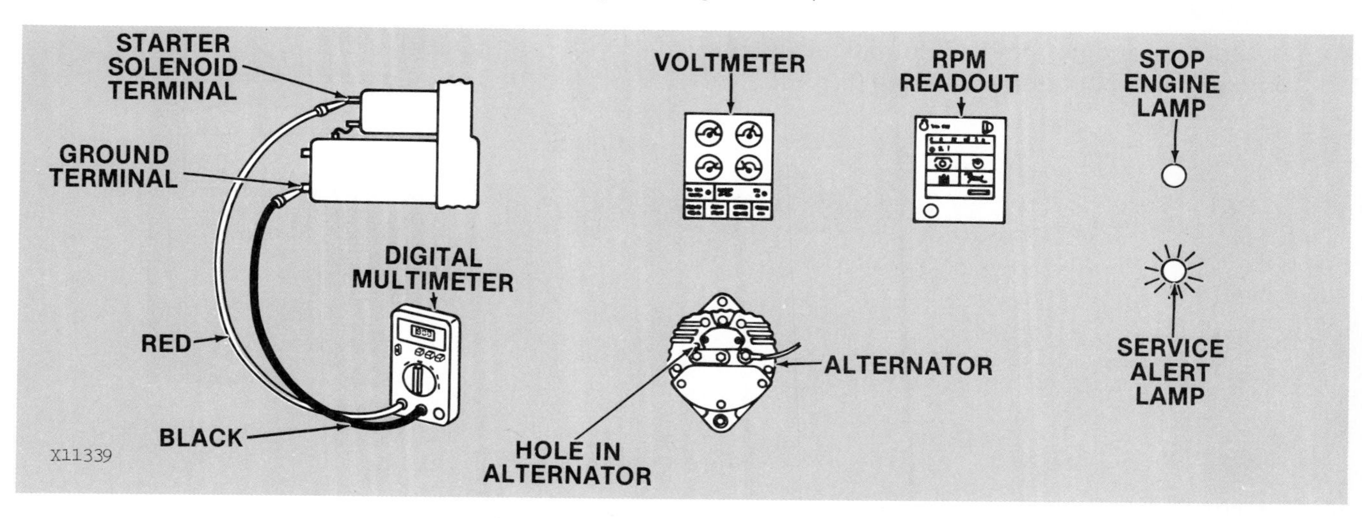

Fig. 88—Speed Monitoring Module (Tachometer)

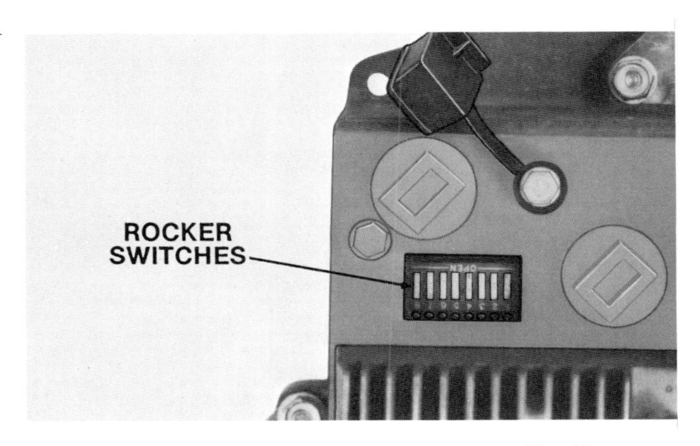

Fig. 89—Programming Tachometer to Tire Size

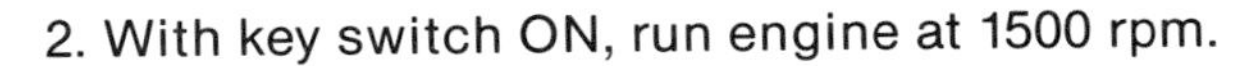

2. With key switch ON, run engine at 1500 rpm.

3. One person should observe the SERVICE ALERT lamp and multimeter reading. The other person should ground the alternator field circuit through the hole in the rear of the alternator frame.

CAUTION: Make sure you know how to perform this procedure. If not, check the applicable technical manual.

IMPORTANT: Do not ground voltage regulator longer than necessary to observe the warning lamp or reach a maximum of 18 volts.

4. The SERVICE ALERT lamp should light above 16 volts. The tractor voltmeter needle should be in the right-hand amber band (top of Fig. 86).

Speed Monitoring

As we mentioned earlier, the system also monitors functions pertaining to speed. A six-function digital tachometer (Fig. 88) displays engine rpm, rated speed, and ground speed of the tractor. It also includes an hourmeter, an engine coolant temperature gauge, and a fuel gauge.

At start-up, engine rpm is automatically displayed both in miles per hour and in rpm (Fig. 88).

The tachometer was designed to accommodate different tire sizes when measuring ground speed. Eight small rocker switches (Fig. 89) on the back are used to program the tachometer according to tire size.

Another module (Fig. 90), the performance monitor, contains touch switches which indicate engine speed and the distance traveled in feet or meters. It has an adjust touch switch which indicates the position of an implement and also displays various values (and increases them in increments) during preset.

To preset a desired distance, touch the distance switch repeatedly until the digit to be changed flashes. Pressing the adjust switch as the digit is flashing causes the digit to increase. Then, press the distance switch again for the next digit.

Fig. 90—Performance Monitor

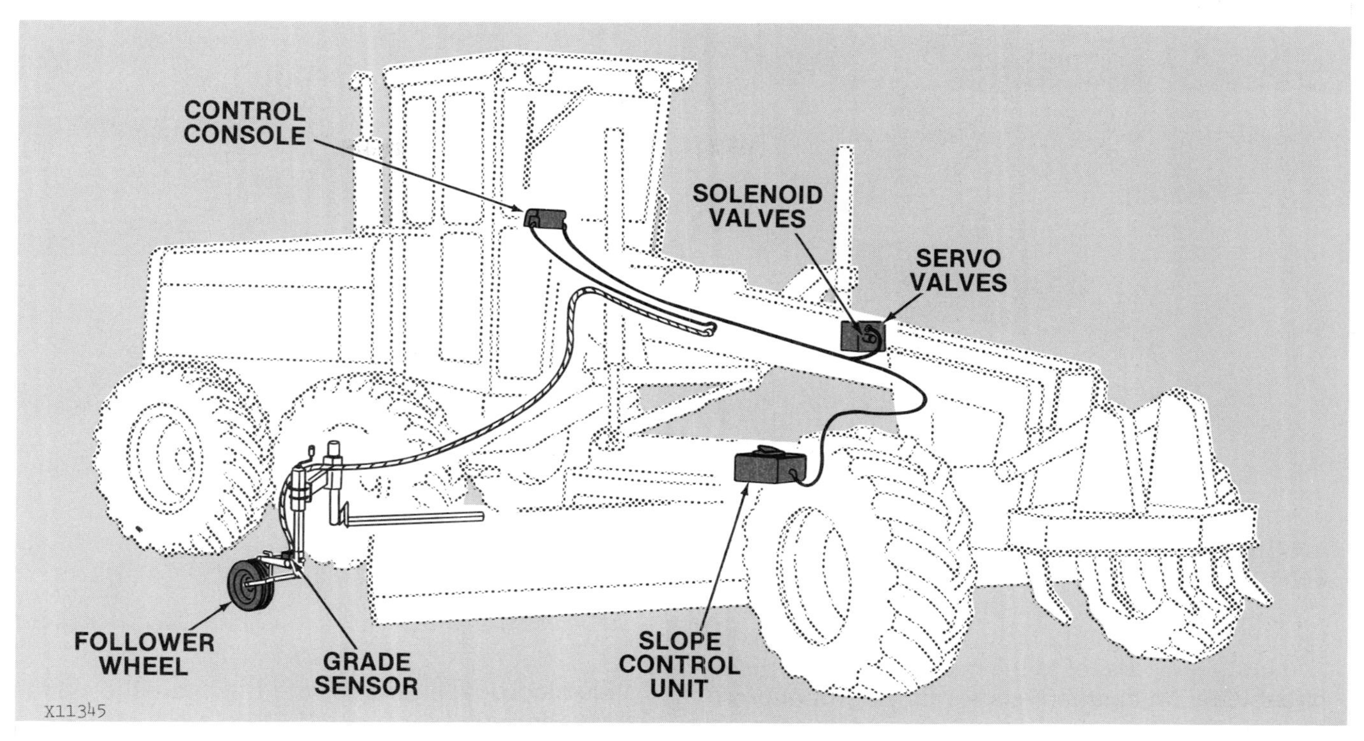

Fig. 91—Components of Automatic Blade Control System

MOTOR GRADERS

AUTOMATIC BLADE CONTROL SYSTEM

The purpose of the automatic blade control system in a motor grader is to sense changes in grade and slope of the ground so that the blade lift cylinders can adjust the blade accordingly (Fig. 91).

The system consists of:

- Slope control unit.
- Control console.
- Grade sensor.
- Solenoid valves.
- Servovalves.
- Wiring harnesses.

The **slope control assembly** (Fig. 92) is mounted on the draft frame of the grader. The slope sensor (Fig. 93), mounted inside the slope control assembly,

Fig. 92—Slope Control Unit

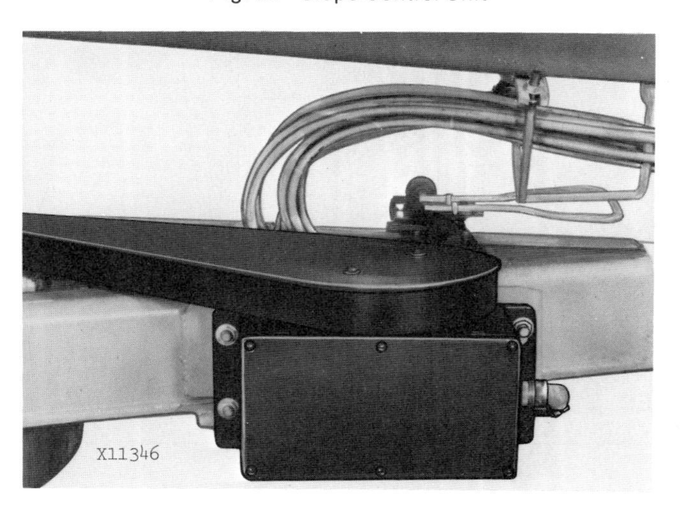

Fig. 93—Slope Sensor

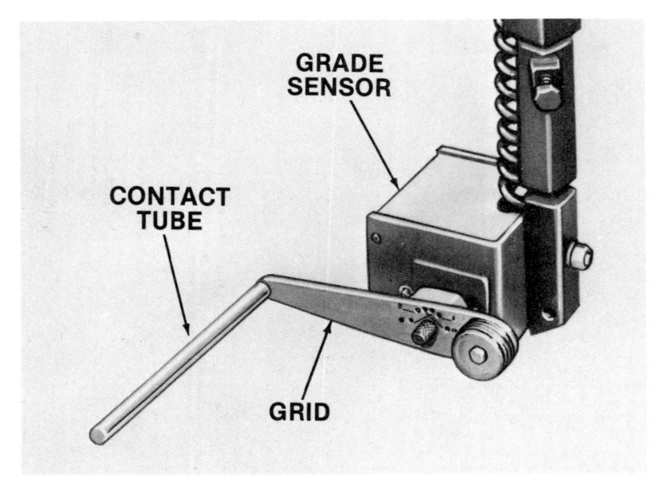

Fig. 94—Grade Sensor

electromagnetically measures the deviation of the sensor and its mount from a gravity reference. That is why it is so important that the motor grader be leveled first and then the slope control unit leveled and installed parallel to the main frame. See the applicable technical manuals for these procedures.

The **slope sensor** (Fig. 93) consists of a pendulum and a blade angle switch. The pendulum is suspended from a shaft mounted on ball bearings so that it can rotate with respect to the gravity reference.

The **grade sensor** assembly (Fig. 94) attaches to the blade of the motor grader. It senses changes in grade by means of the contact tube. This tube can either follow a preleveled stringline attached to stakes or a wheel can be attached (Fig. 95) to follow a curb, previously cut surface, or pavement.

The function of the **servovalves** (Fig. 96) is to provide electrical interface with the hydraulic system. The servovalves, of which there are two—one for each lift cylinder, are mounted on the hydraulic manifold. The manifold contains integral check and shutoff valves which isolate the automatic blade control system from the manual blade control hydraulics.

Fig. 96—Servovalve

The **control console** (Fig. 97) receives signals from the sensors and in turn causes the servovalves to actuate the lift cylinders.

Operation

The relationships between the servovalves, sen-

Fig. 95—Grade Sensor With Wheel Attached

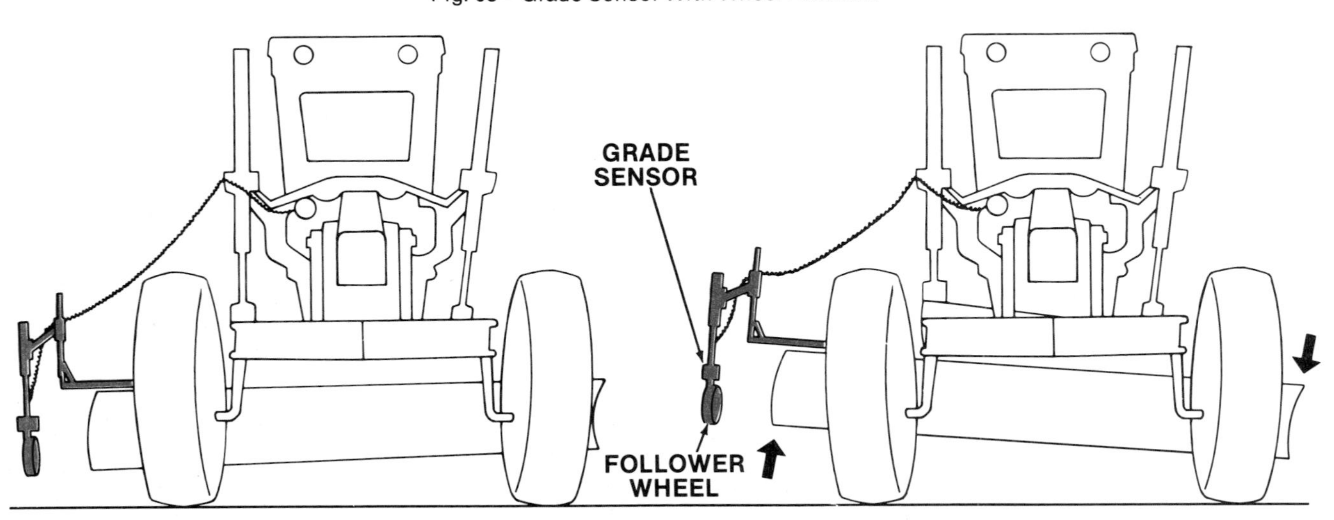

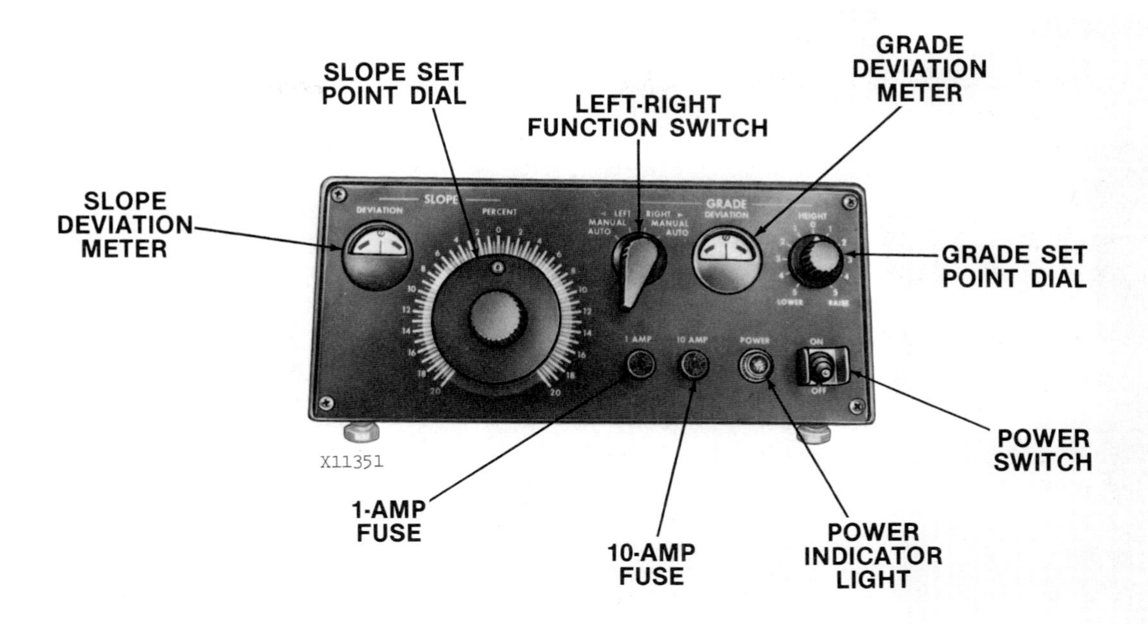

Fig. 97—Control Console

sors, and amplifier or control console can best be described by the block diagram of Fig. 98. As the sensor actuator moves because of a mechanical motion (1) such as the grid arm raising or lowering, a sensor electrical signal (2) is transmitted to the control console. The control console then sends an electrical signal (3) that powers the servovalve. The servovalve then allows hydraulic flow (4) to the blade lift cylinder which positions the lift cylinder (5) in the proper direction. The lift cylinder then positions the blade (6). As the blade is positioned, the sensor actuator returns to its normal position and is again ready to sense a change in grade or slope.

Fig. 98—How a Blade Control System Works

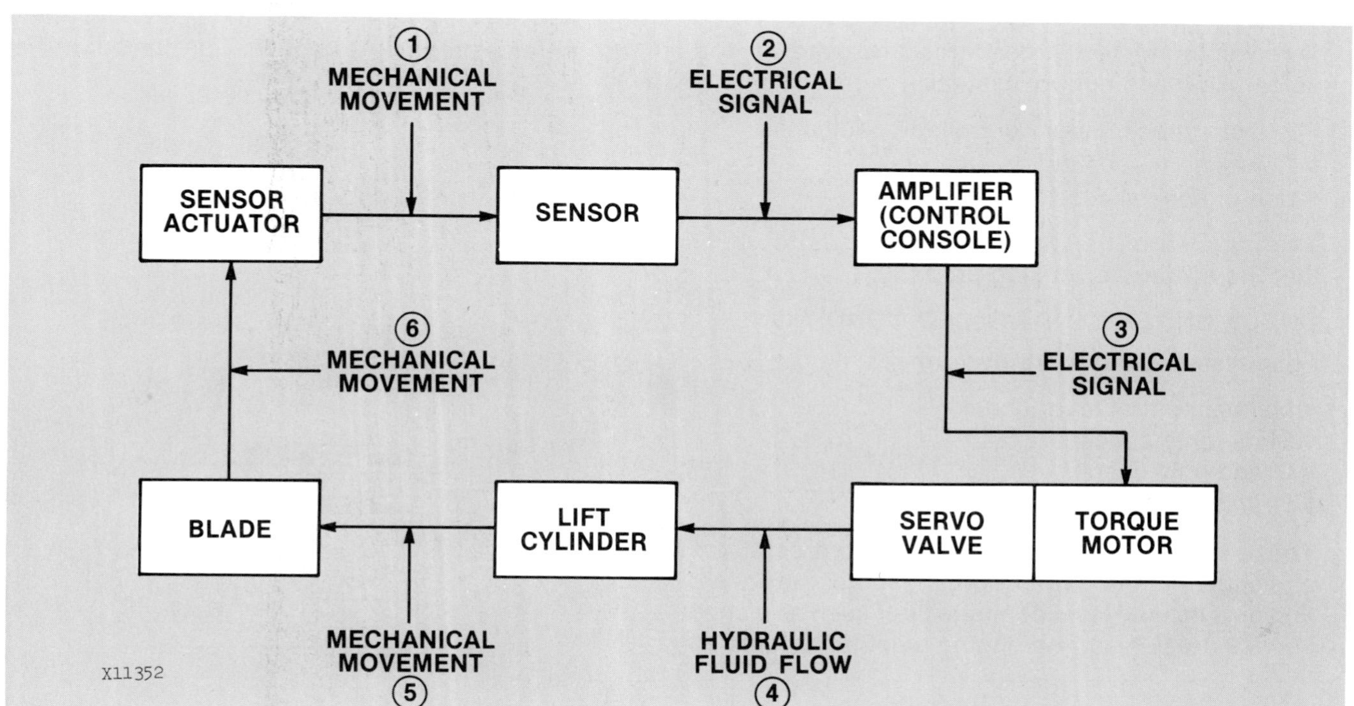

Troubleshooting

AUTOMATIC BLADE CONTROL SYSTEM OPERATES, BUT WITH A LOSS OF ACCURACY.

Tire pressure on motor grader is uneven.

Blade is not centered under draft frame.

Grade sensor bracket is not tight in its mount.

Grade sensor is too far from the blade.

Rear tires are running in loose material.

Servovalves are not centered.

Wear points are not tight (circle support shoes, moldboard side shift, blade lift, blade tilt, and center shift linkages).

LEFT OR RIGHT CYLINDER LOSES ACTION WITH AUTOMATIC BLADE CONTROL SYSTEM ON.

Hydraulic system has a malfunction.

Servovalves are defective.

Automatic blade control system has an electrical malfunction.

AUTOMATIC BLADE CONTROL SYSTEM IS COMPLETELY INOPERATIVE EVEN THOUGH POWER IS ON.

One-amp fuse in control console is blown.

SLOPE METER CONSISTENTLY READS OFF CENTER IN SAME DIRECTION.

Servovalve is out of adjustment.

System Testing

Three different test procedures are used on the automatic blade control system:

- System check under controlled conditions.
- Component substitution.
- Use of specialized test set.

The purpose of these tests or checks is to verify that the system is working properly.

CHECK UNDER CONTROLLED CONDITIONS

Four system checks are performed:

- Manual control lever check.
- Slope only check.
- Grade only check.
- Grade and slope full system check.

These checks are normal operational procedures and should be performed with the motor grader sitting on a flat and level concrete floor such as a shop service area. Also, the engine must be running so that the operation of the blade lift cylinders, blade pitch, and circle can be observed in relation to the particular settings on the control console.

COMPONENT SUBSTITUTIONS

The second test procedure is used when the first one indicates that something is not performing as it should. This procedure involves changing electronic boards and meters in the control console in an attempt to isolate a failed component.

The slope amplifier and grade amplifier are identical printed circuit boards and can be interchanged to verify a defective board. The boards are easily interchanged by removing the two screws in the corners of the boards and gently pulling and rocking the boards up and out of the console. Fig. 99 shows one board removed from the console.

Check **slope amplifier board** in the following manner:

1. With engine off turn key switch to ACC and control console to ON.

2. Place left/right function switch (Fig. 103) in manual left position.

3. Rotate slope set point dial left to 2 degrees slope (Fig. 101). Slope deviation meter should deflect to the left. If it does not, the slope amplifier circuit board inside the control console is probably defective.

4. Interchange the two amplifier boards and repeat the procedure (Fig. 99).

5. If the slope deviation meter still does not deflect, the meter is probably defective. This can be verified

Fig. 99—Changing Amplifier Boards

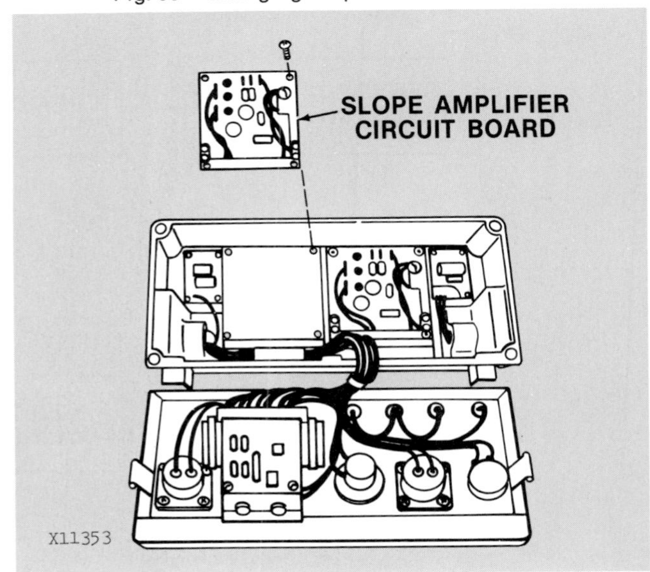

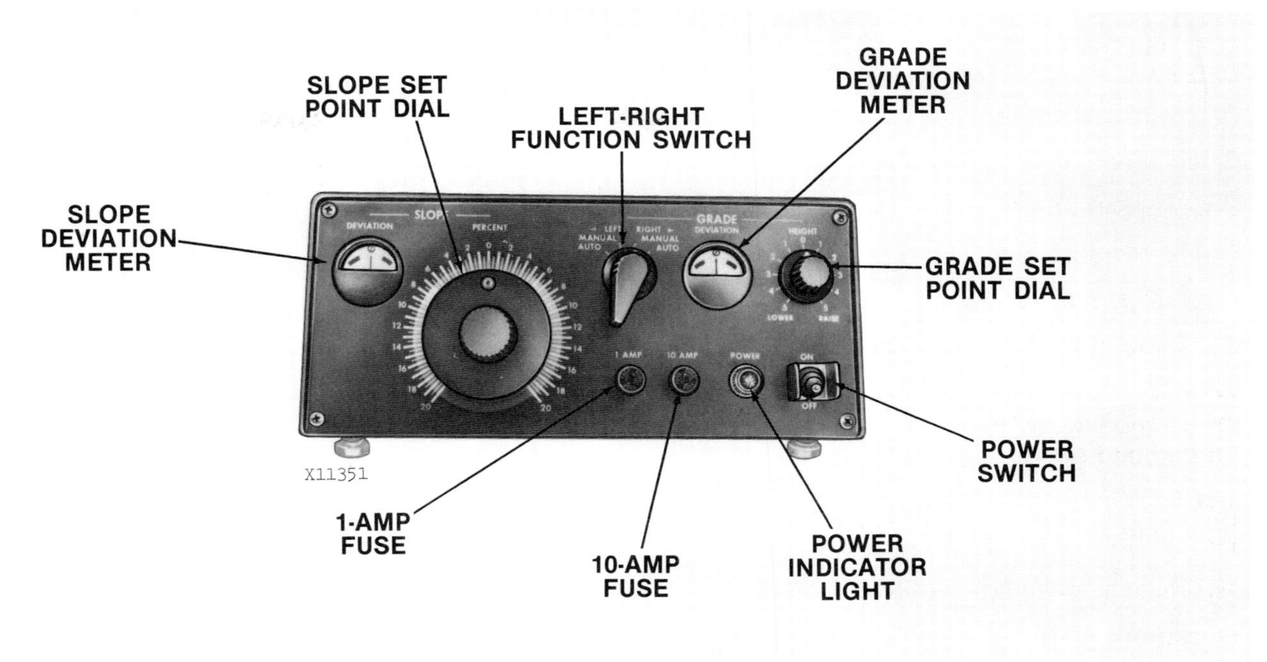

Fig. 100—Control Console

by interchanging the slope and grade deviation meters (Fig. 102) by disconnecting the wires and removing the attaching screws.

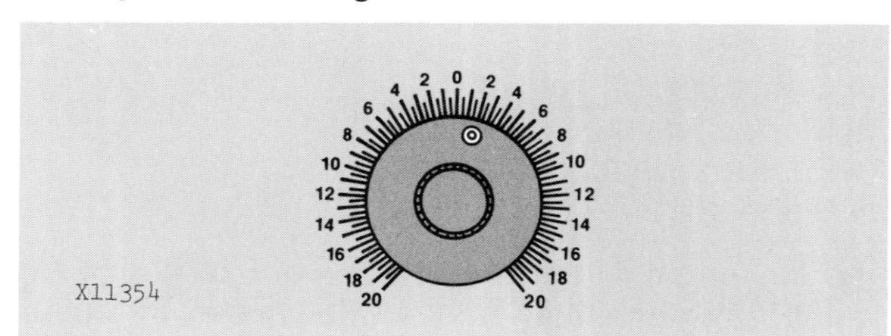

Fig 101—Rotate Dial to Two Degrees of Slope

Check the **grade amplifier board** in the following manner:

1. Disconnect grade sensor cable at the middle of the main frame of the motor grader.

Fig. 102—Changing Deviation Meters

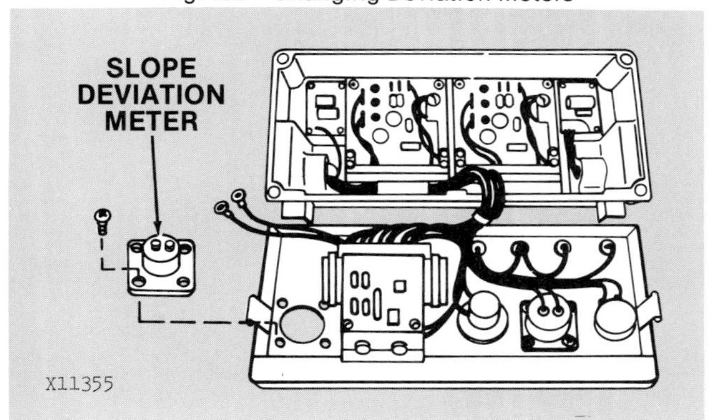

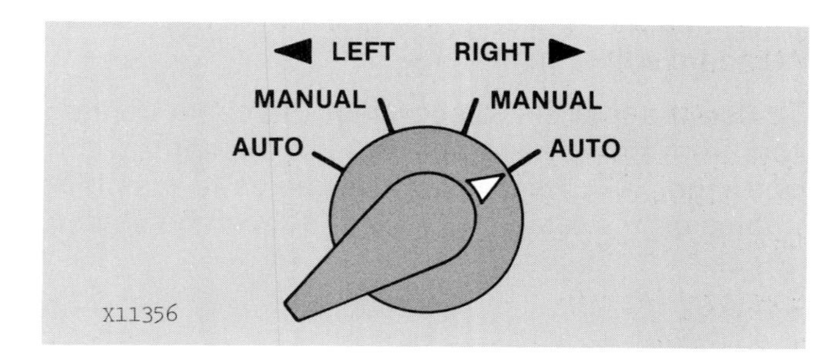

Fig. 103—Set Function Switch to "Auto Right" Position

2. Set left/right function switch to "Auto Right" position (Fig. 103).

3. Rotate grade set point dial (Fig. 104) up and down from 0 position.

Fig. 104—Rotate Set Point Dial Up and Down

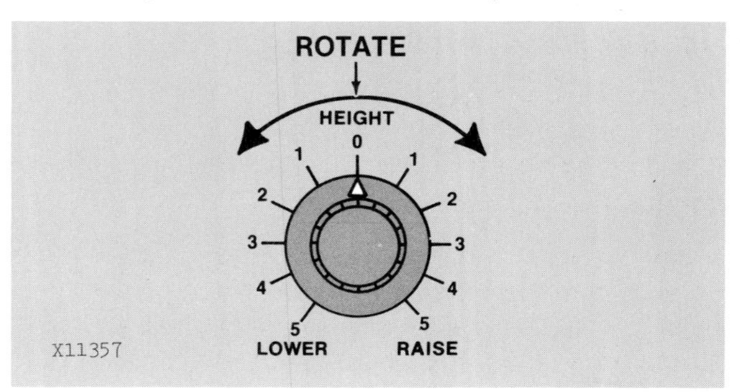

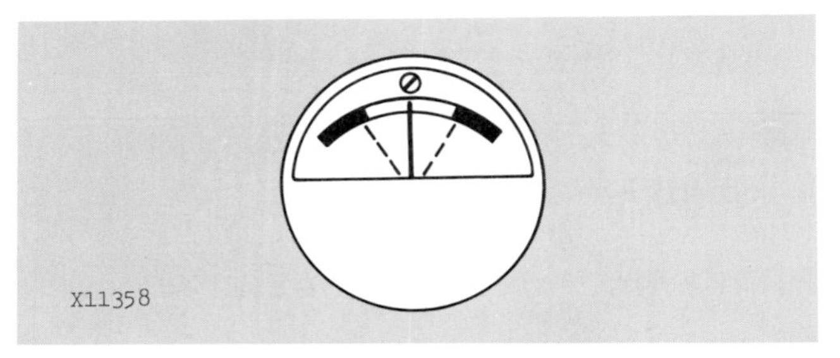

Fig. 105—Grade Deviation Meter Should Deflect

4. The grade deviation meter should deflect to the left and right (Fig. 105). If it does not, the grade amplifier board is probably defective.

5. Interchange the two amplifier boards and repeat the procedure.

6. If the grade deviation meter still does not deflect, the meter is probably defective. Verify by interchanging the slope and grade deviation meters.

USING SPECIAL TEST SET

The third test procedure uses a special test set (Fig. 106) that provides a means of quickly performing resistance continuity and ground defect checks on sensors, valves, slope control unit, and interconnecting wiring. The test set is used with a standard volt-ohm meter.

To use the test set, disconnect the cable connectors from the left and right side of the control console (Fig. 107). Reconnect the cables to matching cables on the test set. The test set can be used to check:

- Power.
- Solenoid valves.

Fig. 106—Test Set for Automatic Blade Control System

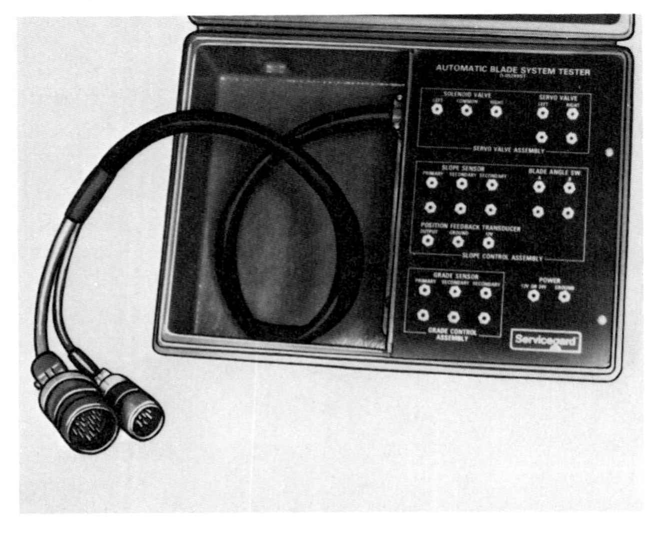

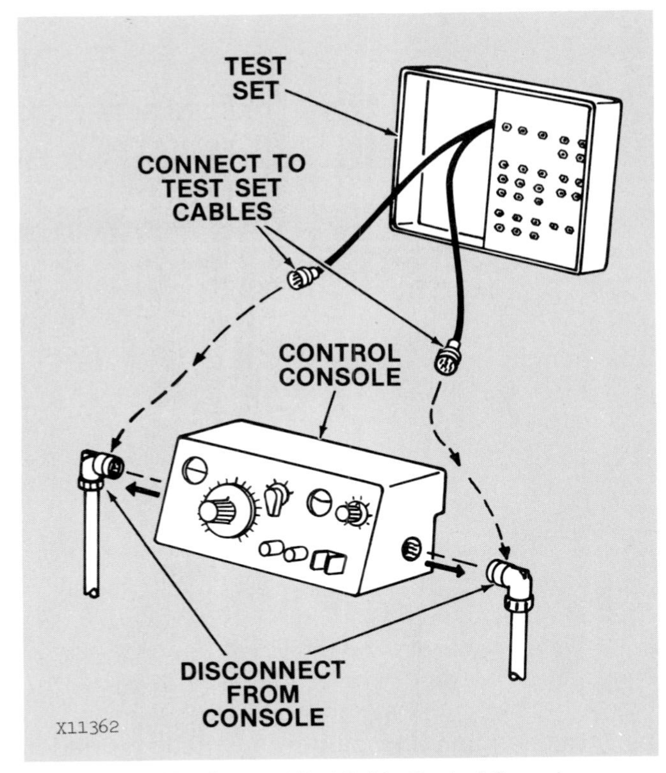

Fig. 107—Connect Test Set to Control Console

- Servovalves.
- Grade control sensor.
- Slope sensor.
- Position feedback transducer.
- Blade angle switch.

When running these tests, be sure to check the applicable technical manuals for the exact specifications pertaining to voltage and resistance readings.

Fig. 108—Test Points for Power Check

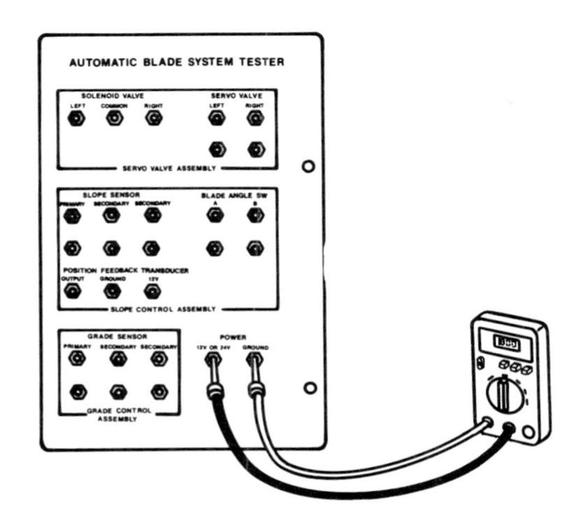

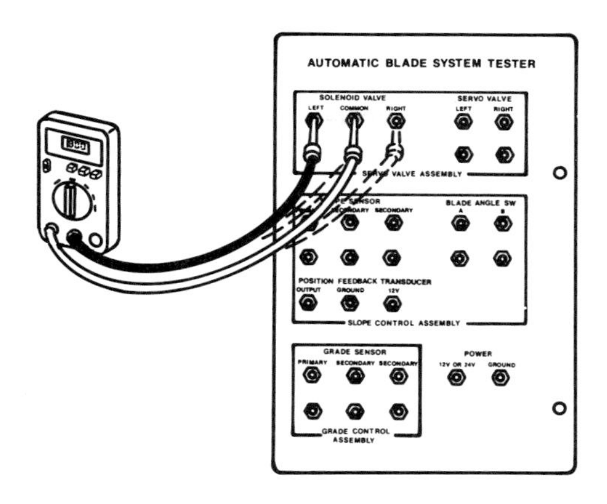

Fig. 109—Test Points for Solenoid Valves

Power

1. Turn key switch to ACC.

2. Using the volt-ohm meter, check for proper voltage (see technical manual) on the test panel between the points indicated in Fig. 108.

Solenoid Valves

1. Turn key switch to OFF.

2. Check resistance between the left test points (Fig. 109) and then the right test points. The reading should be approximately 25 ohms.

3. If not, disconnect wiring to the solenoids and make the same test.

4. If the test results were not correct at the solenoid valve, replace the solenoid.

5. If they were correct at the solenoid but not at the test panel, the harness is probably defective.

Fig. 110—Test Points for Servovalves

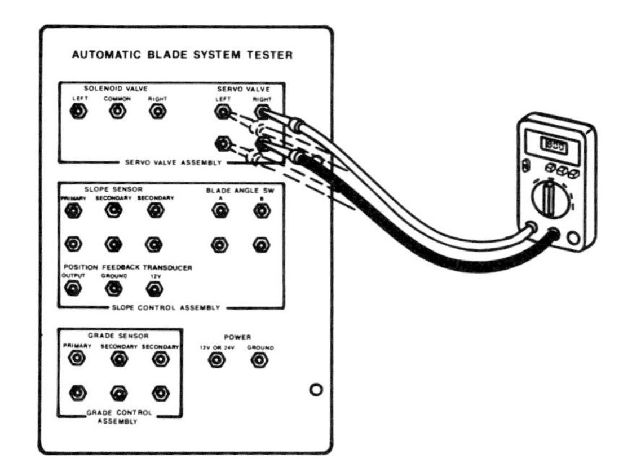

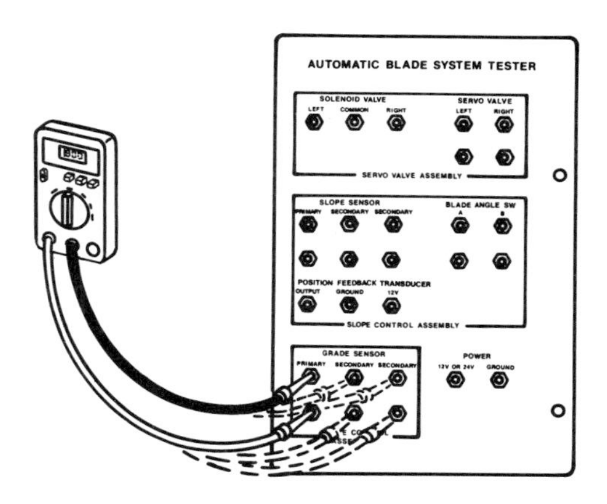

Fig. 111—Test Points for Grade Sensor

Servovalves

1. Check for resistance (see technical manual) between the two left test points (Fig. 110) and then the two right test points.

2. If the readings at the test panel are not correct, disconnect the wires at the servovalves and test at the valves.

3. If the reading at the servovalves is not correct, replace them.

4. If the reading is correct, but not at the test panel, then the harness is probably defective.

Grade Control Sensor

1. Check for resistance (see technical manual) between the primary test points (Fig. 111) and between the two secondary test points.

2. If results of any of the test points were not correct, disconnect the grade sensor cord at the mainframe and check the cord and sensor.

3. If the readings still are not correct, disconnect the cord from the sensor and check the sensor.

4. If the sensor reads incorrect, replace it.

5. If sensor reads correct but incorrect at the cord, replace the cord.

Slope Sensor

1. Check for resistance (see technical manual) between the primary test points (Fig. 112) and between the two secondary test points.

2. If any of the checks were incorrect, disconnect the connector at the slope control unit and make the same checks at the unit itself. Pins A and B are

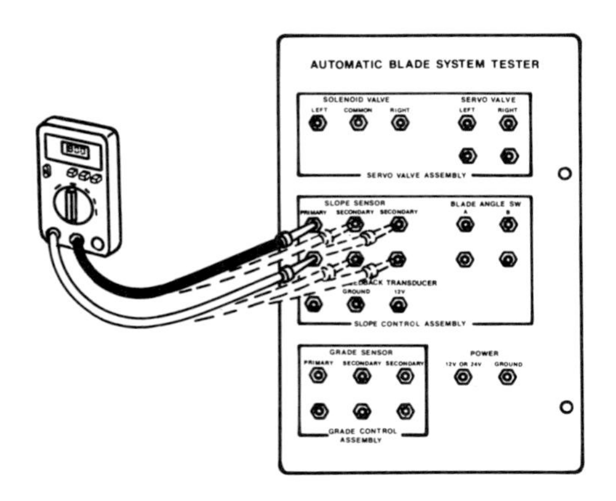

Fig. 112—Test Points for Slope Sensor

the primary, pins C and D one secondary, and E and F the other secondary.

3. If readings at slope control unit are incorrect, the pendulum (Fig. 113) or the wiring from the pendulum to the slope control connector is defective.

4. Test wiring with an ohmmeter with the wiring disconnected at the pendulum. If the wiring is good, replace the pendulum.

Position Feedback Transducer

1. Turn on key switch.

2. Check for voltage (see technical manual) between output and ground (Fig. 114).

3. If no voltage is present, either the wiring harness or feedback transducer (Fig. 113) is defective.

4. Turn off key switch.

5. Disconnect the connector at the slope control unit.

Fig. 113—Slope Sensor

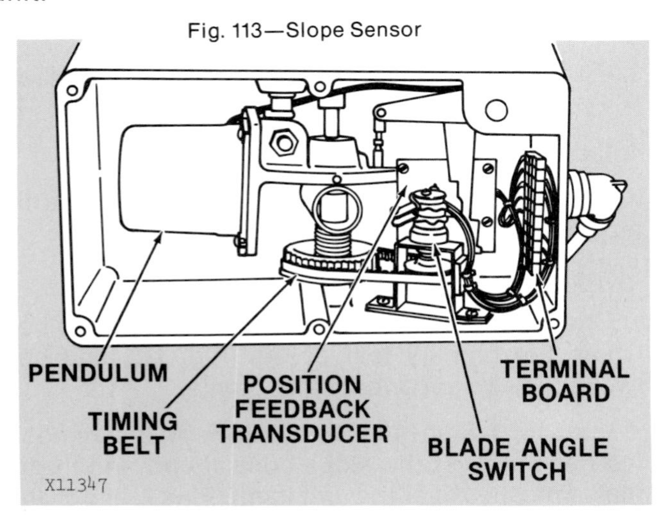

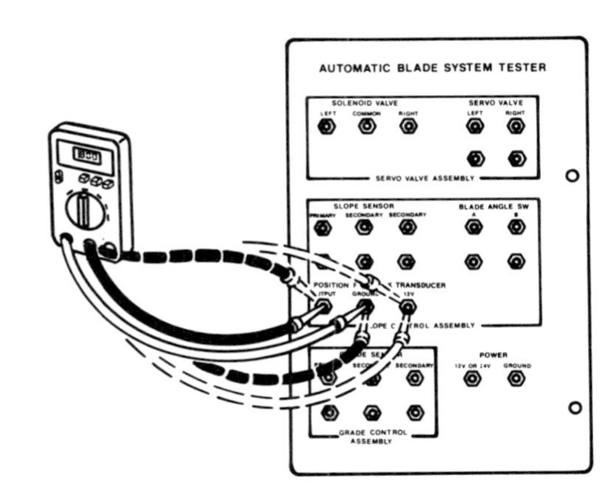

Fig. 114—Test Points for Position Feedback Transducer

6. Check resistance between output and ground, 12-volt and ground, and output and 12-volt as shown in Fig. 114. Resistance should be high (infinity). If there is a resistance reading, there is a short in the harness.

7. If a high reading was obtained, place jumper wires in terminals V, M, and L of the wiring harness at the slope control unit end.

8. Tie the three wires together making sure a good electrical connection is made.

9. Again check resistance between output and ground, 12-volt and ground, and output and 12-volt. There should be very little resistance (see technical manual). If any pair of terminals does not read near zero, the wiring harness is open.

10. If the harness checks out and the initial voltage check between output and ground did not read the correct voltage (see technical manual), replace the position feedback transducer.

Fig. 115—Test Points for Blade Angle Switch

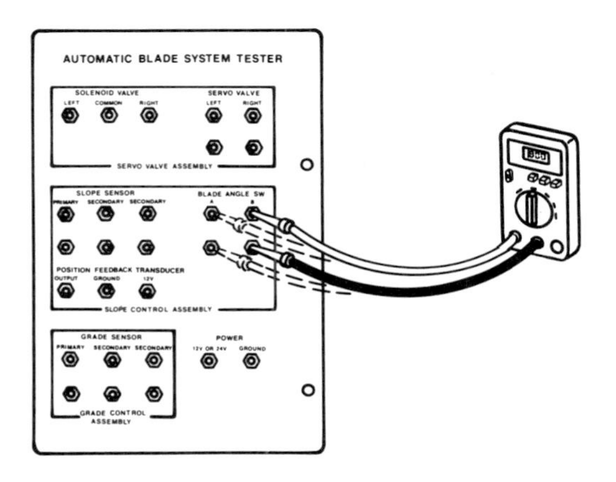

Blade Angle Switch

NOTE: Check the technical manual for correct readings in steps 2, 3, 4, 5, and 6.

1. Start the motor grader and position the blade perpendicular to the main frame.

2. Check resistance between the two terminals of switch A (Fig. 115) and switch B.

3. Circle the blade 15 degrees either direction. Repeat step 2.

4. Circle the blade to 30 degrees. Resistance should be about 25 ohms.

5. Circle the blade to 45 degrees. Resistance should increase to about 45 ohms.

6. Run the blade back to perpendicular. The resistance should return to zero.

7. Circle the blade the other direction and repeat steps 3 through 6. Readings should be the same.

8. If only one or two resistances are incorrect, the blade angle switch (Fig. 113) is defective.

9. If there was no resistance reading on one or both sets of terminals in any degree of blade angle, the switch or wiring harness could be at fault.

10. Repeat steps 2 through 7 but connect the ohmmeter to the connector pins on the slope control unit. Use pins G and K for switch A and H and J for switch B.

11. If the readings are incorrect, the blade angle switch is defective. If the readings are correct at the slope control unit and incorrect at the test panel, the wiring harness is defective.

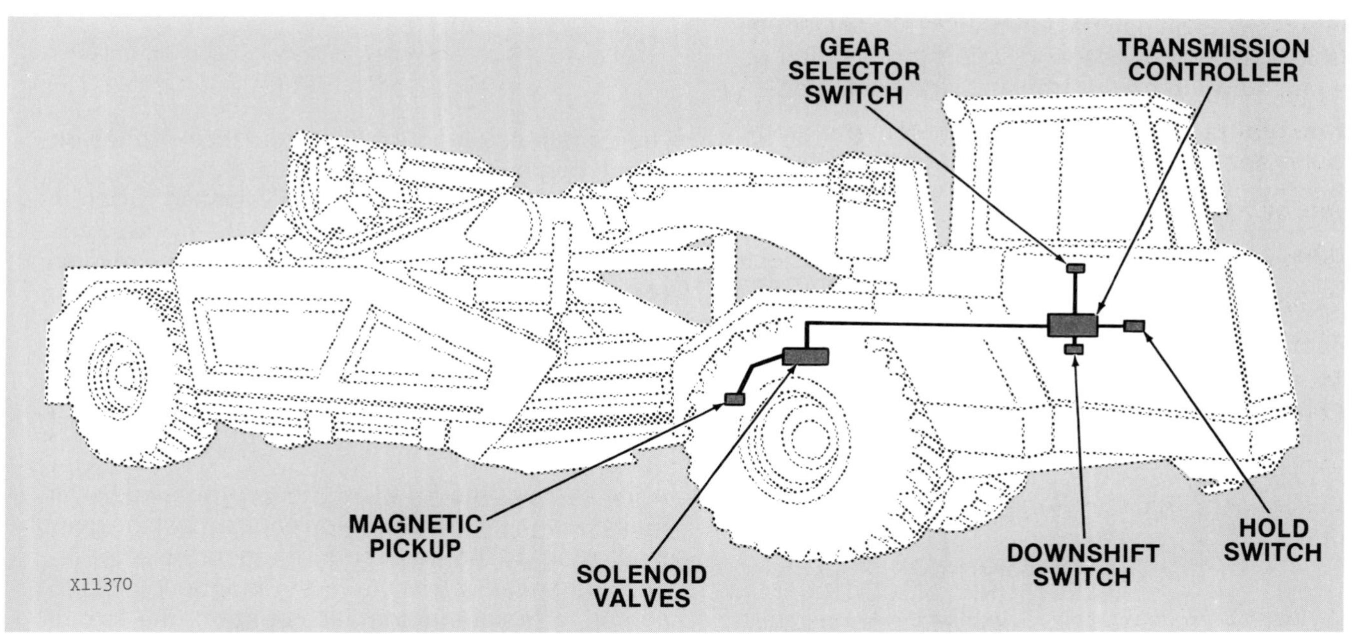

Fig. 116—Components of Automatic Transmission Control System

SCRAPERS

AUTOMATIC TRANSMISSION CONTROL SYSTEM

The automatic transmission control system (Fig. 116) consists of:

- Transmission controller.
- Gear selector switch.
- Magnetic pickup.
- Six solenoid valves.
- Hold switch.
- Downshift switch.
- Elevator switch.

A neutral start switch and reverse warning alarm switch are contained in the circuitry of the transmission controller. The elevator circuit on the scraper is also connected to the transmission controller since the position of the elevator determines the gear that the controller selects.

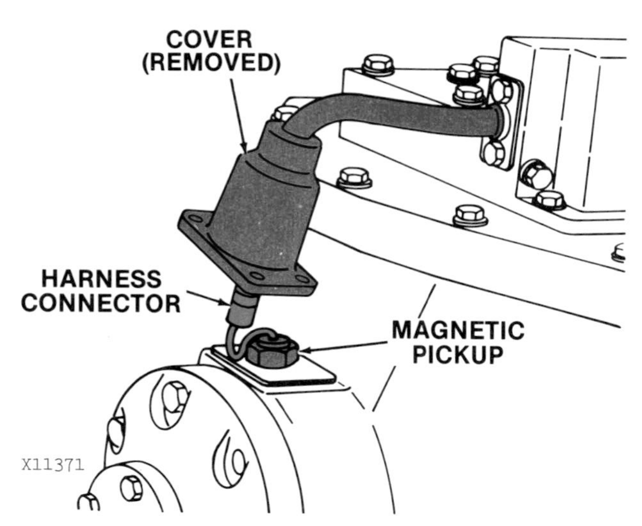

Fig. 117—Magnetic Pickup on Transmission Housing

The transmission can be operated in either manual or automatic mode, depending on the position of the mode switch on the transmission controller.

When the transmission is in the automatic mode, ground speed of the scraper determines whether the transmission shifts up or down. As the scraper slows, the transmission shifts down, and as speed increases, the transmission shifts to a higher gear just as your car does. However, the transmission will not shift any higher than the gear the gear selector lever is in, just as your car will not shift into high if you have the lever in second. A magnetic pickup (Fig. 117), located on the transmission housing, senses ground speed.

Fig. 118—Transmission Controller

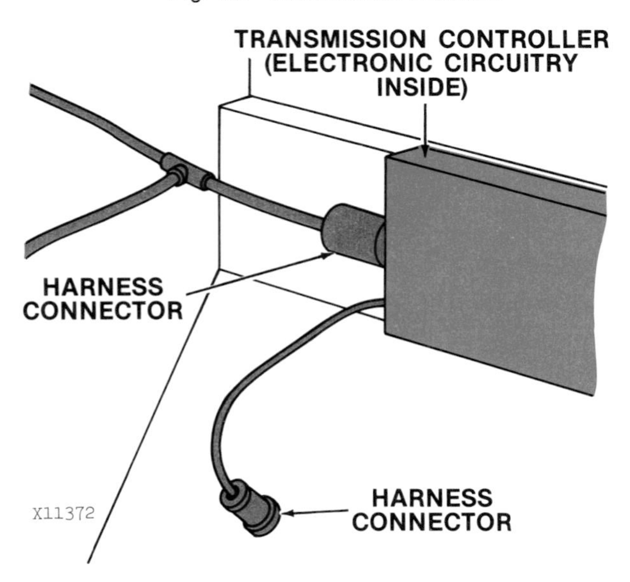

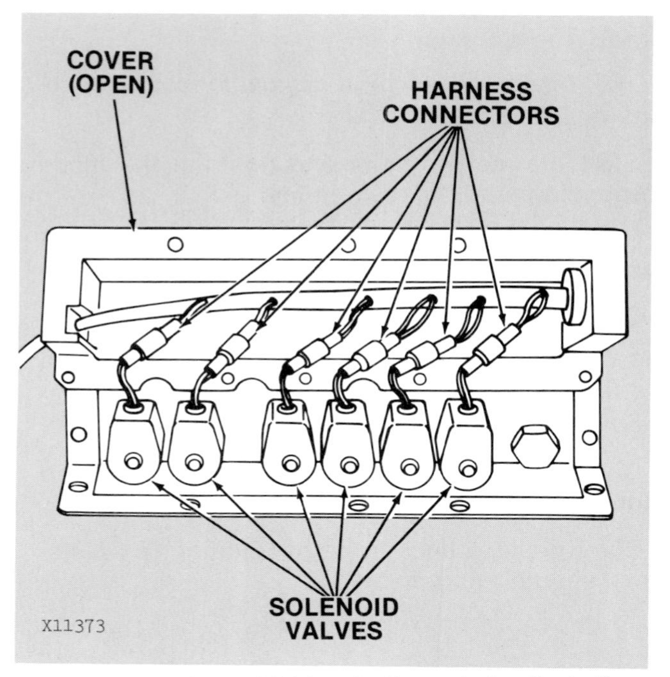

Fig. 119—Solenoid Valves for Transmission Controller

The transmission controller (Fig. 118) switches different combinations of six solenoid valves (Fig. 119) on and off. These solenoid valves in turn control two clutch and four brake packs in the transmission to obtain six forward speeds, one reverse, and torque converter lockup. Fig. 120 gives solenoid and brake and clutch pack engagement for each gear. Solenoid number 3 is on in all gears. When this solenoid is off, the transmission is in neutral. Solenoid number 4 controls torque converter lockup and number 6 is on only in reverse.

In the manual mode, even though the signal still goes through the transmission control box, the gear is selected by the position of the gear selector lever. The speed signal (from the magnetic pickup) is used to provide automatic control of the torque converter lockup clutch.

Fig. 120—Transmission Gear Engagement Chart

GEAR	SOLENOIDS	CLUTCH AND/OR BRAKE PACKS
Reverse	3, 5, 6	B2, B3
Neutral		B4
1	2, 3, 5	B1, C2
2	3, 5	B2, C2
3	2, 3	B1, B4
4	3	B2, B4
5	1, 2, 3, 5	C1, C2
6	1, 2, 3	C1, B4
TC (Lockup)	4	Converter Clutch

X11374

Fig. 121—Transmission Will Start in 3rd and Shift Up to 5th with Elevator Switch in Off or Reverse

In the automatic mode, the gear the transmission starts in is determined by the position of the elevator switch and gear selector lever. When the elevator switch is off or in reverse (elevator in transport mode) a certain portion of the control box circuitry is grounded through the elevator switch. As long as the gear selector is in third gear or higher (Fig. 121), the transmission will start in third gear and shift up as high as the position of the gear selector if ground speed is adequate, and down to third as ground speed decreases. If the gear selector is in first or second, the transmission will start in the gear selected. The torque converter will lock and unlock automatically in gears four, five, and six. In the transport mode the torque converter will lock and unlock in third gear only if the gear selector is in third.

When the elevator switch is in forward (loading mode), the ground connection, through the elevator switch, is removed which causes the transmission to start in first gear (Fig. 122). The transmission will shift up as high as the position of the gear selector if ground speed is adequate and back down to first as ground speed decreases. The torque converter will not lock up in the loading mode.

A hold switch (Fig. 123), located above the hold pedal, is incorporated into the automatic mode. This switch is normally open. When the hold pedal

Fig. 122—Transmission Will Start in 1st and Shift Up to 5th with Elevator Switch in Forward

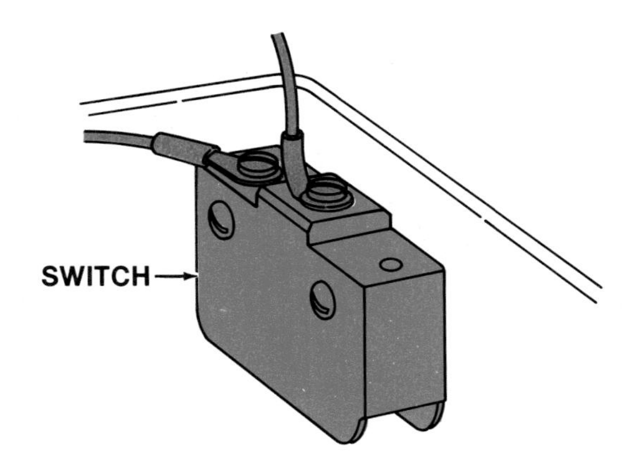

Fig. 123—Transmission Hold Switch

is depressed causing the switch to close, the transmission gear and torque converter mode are "locked in" into their existing gear and mode, respectively, for as long as the hold pedal is depressed and there is no loss of speed signal. Releasing the hold pedal returns the system to normal automatic operation.

A downshift switch (Fig. 124), located next to the fuel injection pump, is activated by the throttle linkage. It is normally closed. At about 1200 rpm, the switch is activated. When the foot throttle is released, the switch is closed and the transmission

Fig. 124—Transmission Downshift Switch

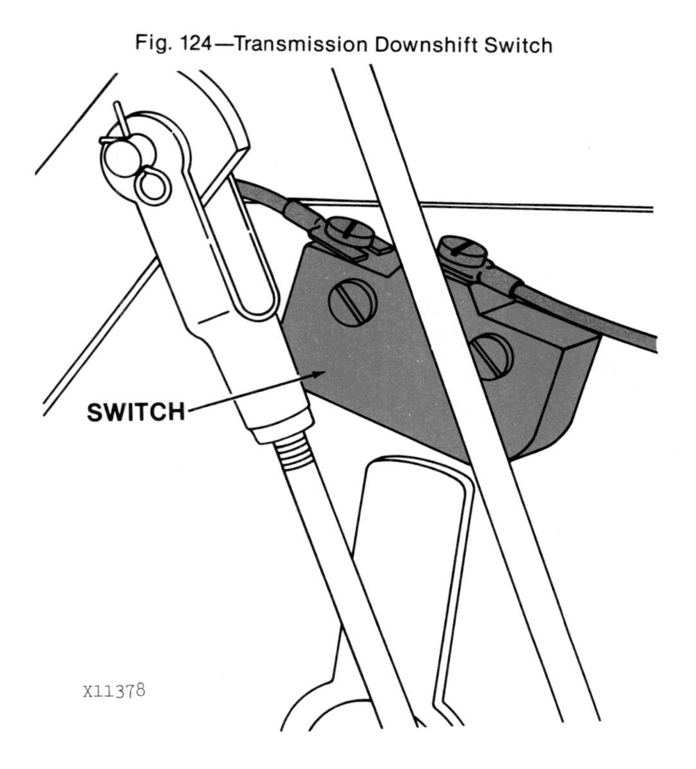

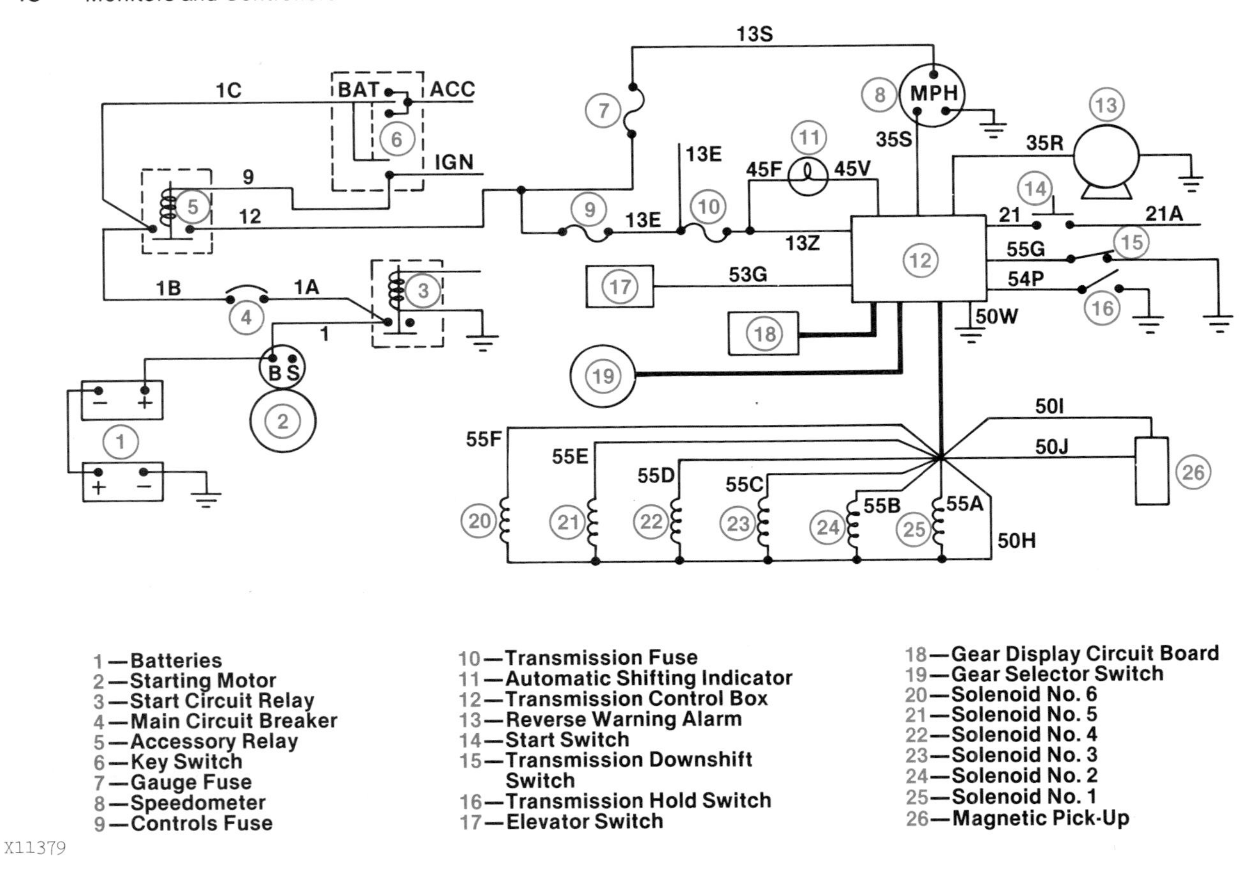

Fig. 125—Automatic Transmission Control Circuit

downshifts from 6th to 5th to 4th with the torque converter locked up and then down to 3rd with the torque converter unlocked when the scraper is in the transport mode. Fig. 125 shows the circuitry of the automatic transmission control system.

Testing the Transmission Control Circuit

Two different diagnostic procedures are used to check the scraper transmission control circuit. The first is a self-diagnostic system built into the transmission control box. The second uses a special scraper transmission/elevator diagnostic test set.

The second test procedure is used only when the first test procedure does not reveal the problem in the transmission control circuit.

NOTE: The first test procedure will only check the solenoids and their associated wiring and not other components of the transmission control circuit.

SELF-DIAGNOSTIC TEST

The first diagnostic procedure is initiated every time the scraper is started. The control box does not check itself but automatically checks all solenoids and leads for short or open circuits.

Place the gear selector lever in neutral. Check the gear display for a neutral (0) indication.

If the gear display is flashing a sequence of gears, the control box has diagnosed one or more of the solenoid valve circuits as defective. The flashing sequence shows the remaining usable gears.

NOTE: The transmission gears are represented by the numbers 1 through 6 (1 for 1st gear, etc.). Reverse is represented by a negative (–) and neutral by a zero (0).

TEST	RESULTS*	CAUSE
Scraper battery	Between 18 - 30 volts.	Battery, cable, or fuse is defective.
Hold switch (key off)	Meter should indicate open circuit.	Switch or cable is shorted.
Hold switch (key on)	Meter should indicate less than 10 ohms.	Switch or cable is open.
Elevator switch (key off, switch in F)	Meter should indicate open circuit.	Switch or cable is shorted.
Elevator switch (key off, switch in N or R)	Meter should indicate less than 10 ohms.	Switch or cable is open.
Downshift switch (key off, full throttle)	Meter should indicate an open circuit.	Switch or cable is shorted.
Downshift switch (key off, throttle released)	Meter should indicate less than 10 ohms.	Switch or cable is open.
Gear selector (repeat for each gear)	Meter should indicate less than 10 ohms.	Switch or cable is open.
Solenoids (repeat for each)	Meter should indicate more than 5 but less than 12 ohms.	Less than 5, valve or cable is shorted. More than 12, they are open.
Gear display (repeat for each display)	Meter should indicate more than 25 but less than 250 ohms.	Less than 25, display or cable is shorted. More than 250, they are open.
Gear selector switch (in N)	Meter should indicate an open circuit.	Switch is defective.
Gear selector switch (in all other gears)	Meter should indicate less than 10 ohms.	Switch is defective.
Magnetic pickup (key on, engine off)	Meter should indicate more than 1 but less than 6 volts.	Pickup or cable is shorted if less than 1; open if more than 6 volts.
Speedometer output (key off)	Meter should indicate more than 1,000 ohms but less than 10,000.	Speedometer or cable is shorted if less than 1,000; open if more than 10,000 ohms.
Starter relay (key off, push starter button)	Meter should indicate more than 10 ohms but less than 20.	Starter relay or cable is shorted if less than 10; open if more than 20 ohms.
Reverse warning horn (key on, engine off) (Press button on test set.)	Horn should sound.	Cable or horn is defective if horn does not sound.

*Volts and ohms listed in this column are typical. Check the applicable technical manual for the exact specifications for your machine.

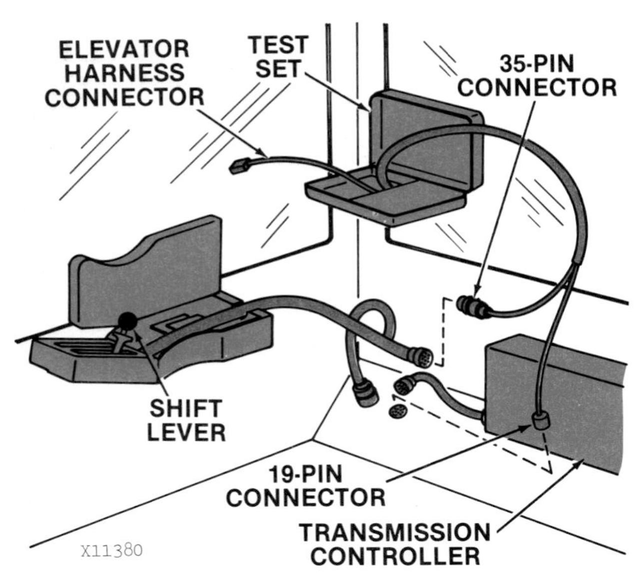

Fig. 126—Connecting Test Set to Transmission Controller

USES FOR DIAGNOSTIC TEST SET

When the self-diagnostic test fails to reveal the problem, then a special diagnostic test set must be used. Fig. 126 shows how the test set is connected to the transmission controller.

The test set will test the following components:

- Scraper battery.
- Transmission hold switch.
- Elevator switch.
- Downshift switch.
- Gear selector.
- Transmission solenoids.
- Gear display.
- Gear selector switch.
- Magnetic pickup.
- Speedometer output.
- Starter relay.
- Reverse warning horn.

The table on page 12—49 lists the tests, the results that should be expected from the tests, and the probable cause if the test results are not met.

SUMMARY

As we said earlier, no attempt was made to present every application of electronics and microprocessors in agricultural and industrial equipment. Our primary aim was to show you how vastly different and varied the uses are and to present the fundamentals of service on these systems.

One final note: Do not be intimidated by the electronic systems on a machine. They were installed on the machine for one main purpose—to improve productivity of the machine.

TEST YOURSELF

QUESTIONS

1. What is the difference between a monitor and a controller?

2. True or false. The main purpose of a sensor is to sense a function and transmit this information to a gauge or console.

3. What three components do practically all monitoring systems have in common?

4. The sensor in a metal detector is ____________ (optical, magnetic).

5. True or false. The sensor in a planter monitoring system is of the magnetic type.

6. True or false. Buildups of chaff and dirt have little or no effect on opto-electrical sensors.

7. The __________________monitor indicates with lights and a buzzer when a combine straw chopper, straw walker, conveyor auger, grain elevator, or tailings elevator are operating at less than ________ percent of their designed speed.

(Answers on page 21 at the end of this book.)

GENERAL MAINTENANCE / CHAPTER 13

INTRODUCTION

Maintaining the electrical system is not difficult—many of the components require little or no maintenance. However, there are parts that are extremely important, and some of these are often neglected. Regular maintenance takes little time and effort, and can prevent many problems.

Other chapters in this manual give you detailed descriptions of electrical components, as well as diagnosis and testing procedures. This chapter will be confined to regular maintenance—the care that will keep the system operating properly and insure a long service life.

Before proceeding, review the safety information in Chapter 1. Closely follow safety precautions when working with batteries and other electrical devices.

STORAGE BATTERIES

The battery is the heart of the electrical system (Chapter 5). It is perishable, but with the proper care you can keep it operating for the longest possible time.

Here are the most common causes of poor battery performance or failure:

- **Low electrolyte level**
- **Low specific gravity**
- **External damage, dirt, and corrosion**
- **Faulty connections**

Let's take a closer look at these.

CHECKING THE ELECTROLYTE LEVEL

Fig. 1—Proper Level of Battery Electrolyte

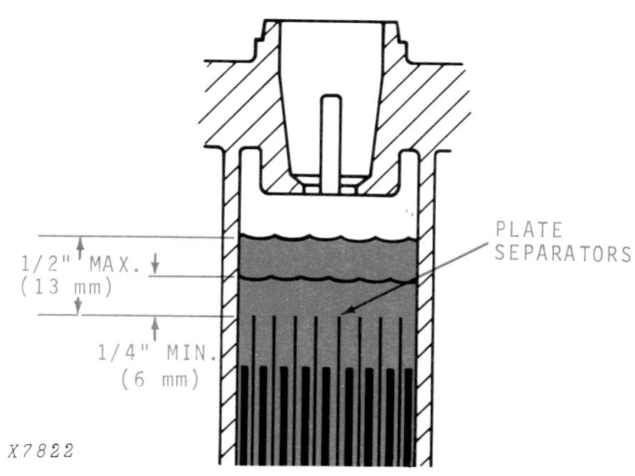

Once a week, or after 50 hours of operation, make sure the electrolyte level is above the battery plates (Fig. 1). If the level is low, add water to each cell until the level rises to the bottom of the split ring in the vent well.

IMPORTANT: If the battery electrolyte level is allowed to stay low, the charging circuit will not work properly. Also, the battery plates will dry, and permanent damage may result.

Don't add water you wouldn't drink! **If distilled water is available, use it.** But remember that even using hard water is better than allowing the electrolyte level to get low.

Don't overfill—you'll lose electrolyte. Overfilling also can cause the electrolyte to splatter, causing acid burns. Always follow proper safety precautions when working with electrolyte.

CHECKING THE SPECIFIC GRAVITY

Fig. 2—Checking the Specific Gravity

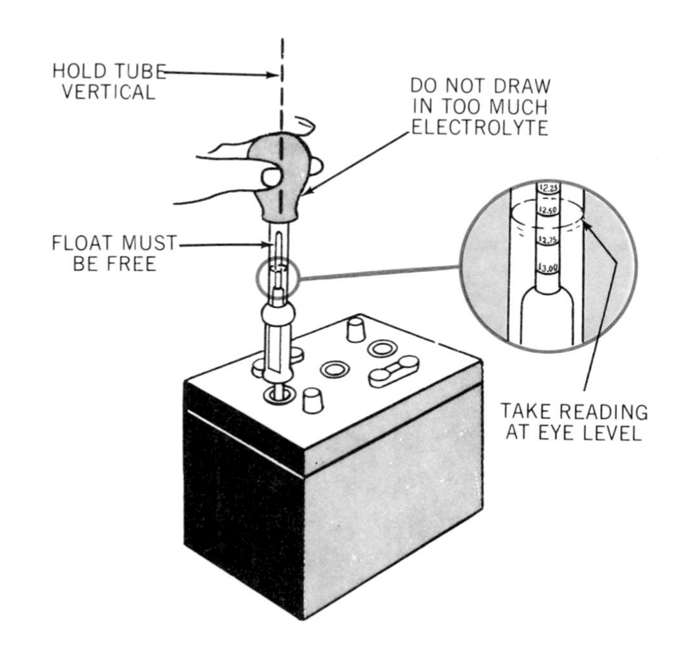

The specific gravity reading tells you the battery charge. Check periodically using an accurate hydrometer (Fig. 2). Readings should be from **1.225 to 1.280,** corrected per the electrolyte temperature as described in Chapter 3.

Never allow the specific gravity to fall below 1.225. Keep the battery at full charge to prevent sulfation and get longer service life.

On the other hand, don't overcharge the battery. A sign of this is when the battery uses too much water.

PREVENTING DAMAGE TO BATTERIES

When you service the battery, always check for external damage (Fig. 3). A broken case or cracked cell cover will result in battery failure. Battery hold-down clamps must be tight enough to hold the battery securely, yet not so tight as to warp the case.

KEEPING THE BATTERY CLEAN

If acid film, corrosion, and dirt are present on the top of the battery, current will flow between the terminals, causing slow discharge. Use diluted ammonia or baking soda and water solution to clean the battery, flush with clear water. But keep the vent plugs tight so that none of the solution gets into the cells.

When disconnecting the battery cables for cleaning, disconnect the ground strap first to prevent arcing. After cleaning apply a coat of petroleum jelly to terminals and cable clamps to prevent further corrosion.

Fig. 3—Maintenance Problems with Batteries

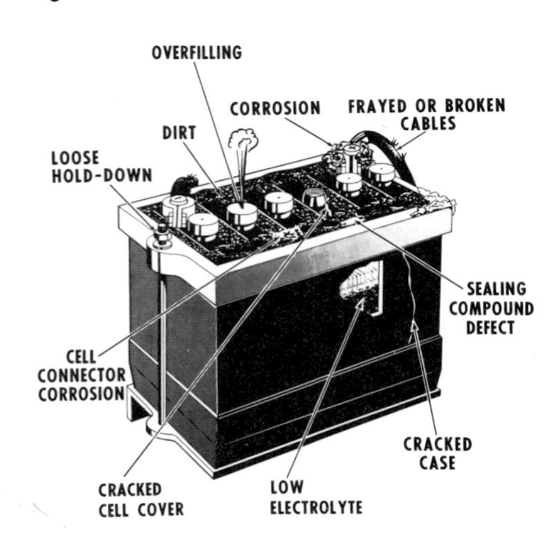

CHECKING THE BATTERY CONNECTIONS

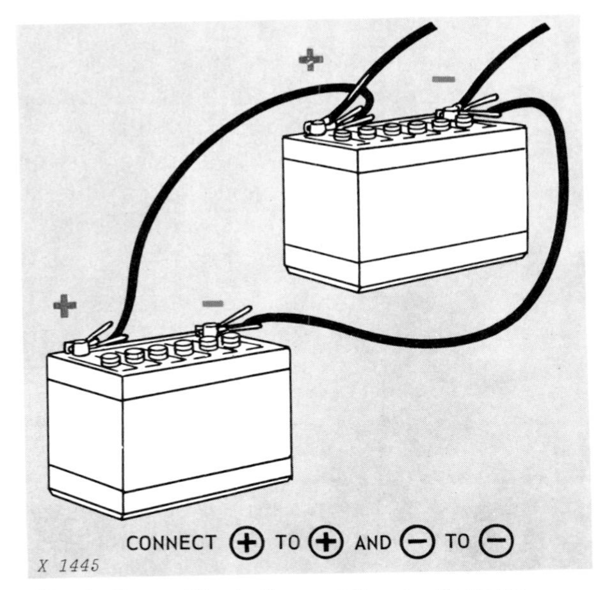

Fig. 4—Proper Way to Connect Booster Batteries

Poor battery connections will impede the flow of current. Clamps should hold cables firmly to battery terminals. If necessary, remove some metal from the clamp jaws to insure a snug fit when tightened.

When connecting booster batteries, be sure to connect positive to positive poles, and negative to negative poles (Fig. 4).

LOW-MAINTENANCE BATTERIES

Visually inspect low-maintenance batteries and their cables as you would any other battery. Do not add water to these batteries. If the battery is of a type which allows viewing of the electrolyte levels in the cells or is equipped with hydrometers which indicate electrolyte levels and the electrolyte levels are found to be low, the battery should be replaced.

Low-maintenance batteries can be checked with a voltmeter. If voltage is below 12.4 volts, charge the battery as described in Chapter 5. The battery charger should include a charge duration control. If the battery fails to charge the first time, attempt to charge it again. If the battery fails to accept a second charge, replace it.

If voltage is 12.4 volts or above, load test the battery (see fixed load testing in Chapter 5). If voltage is less than the minimum specified in Chapter 5 replace the battery. If voltage is at or above minimum specified, return the battery to service.

DC CHARGING CIRCUITS

GENERATOR

Fig. 5—Generator Oil Cups

For detailed description, testing, and servicing of the DC generator, refer to Chapter 6 of this manual.

Aside from periodic lubrication and keeping the drive belt in proper adjustment, the generator requires no regular maintenance.

Every 200 hours, add SAE 10W or SAE 5W-20 oil to both front and rear generator oil cups (Fig. 5). Only 8 to 10 drops are required. **Do not over lubricate!** This can cause more damage to electrical components than a lack of lubrication.

Don't forget to check the generator belt tension at this time. The amount of flex at a certain tension is specified by the manufacturer.

Also note the condition of the belt. Excessive wear or glazing may mean replacement is necessary.

Remember that a V-belt shoud ride on the **sides** of the pulley grove—not on the bottom. If the belt bottoms in the pulley, it is either too tight or excessively worn.

If the generator brushes are visible, examine the commutator and brushes for dirt, oil, wear, and other defects.

REGULATOR

For complete instructions covering the generator regulator, refer to Chapter 6.

The regulator requires no periodic maintenance.

On a **DC** charging circuit, however, the **generator must be polarized** after the generator starting motor, or battery has been disconnected for any reason.

Before starting the engine, momentarily touch a jumper lead to both the "GEN" and "BAT" terminals of the regulator. This will establish correct polarity. Failure to do this may damage the regulator or the wiring.

AC CHARGING CIRCUITS

AC charging circuits are covered in detail in Chapter 6.

ALTERNATOR

Most alternators need no regular maintenance. Bearing lubricant is sealed in at the factory—no further lubrication is required.

Check the fan or alternator belt for proper tension (Fig. 6).

Also note whether belt is worn or glazed, or bottoming in the pulley, and replace belt if necessary.

Loose or slipping belts can cause the battery to run down.

Fig. 6—Adjusting Fan or Alternator Belt Tension

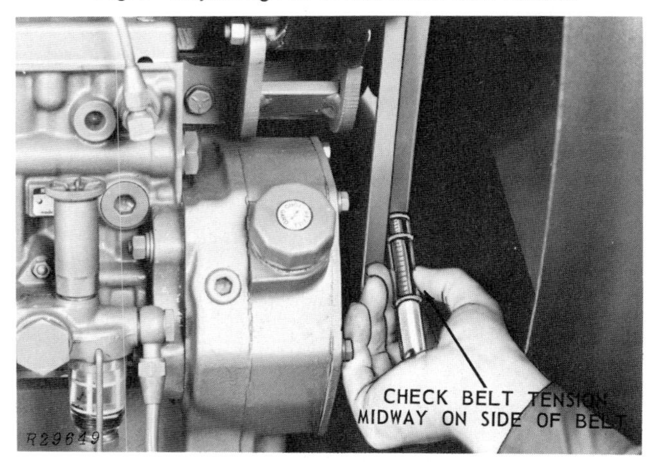

REGULATOR

The regulator normally requires no maintenance. For service and testing, see Chapter 6.

SAFETY RULES FOR AC CHARGING CIRCUITS

To prevent damage to the AC or alternator charging circuit, remember:

1. Disconnect the battery ground straps when working on the alternator or regulator. Damage could occur if terminals are accidentally grounded.

2. NEVER POLARIZE THE ALTERNATOR. Never ground an alternator or regulator terminal, or connect a jumper wire to any terminal (unless recommended for a test).

3. If the alternator or regulator wiring is disconnected, connect it properly before connecting batteries.

4. Never disconnect or connect any alternator or regulator wiring with batteries connected or with alternator operating.

5. Always connect batteries or boosters in the correct polarity. Most AC electrical systems are NEGATIVE-grounded.

6. Don't use a battery charger as a booster to start the engine.

7. Never disconnect the batteries when the engine is running and the alternator is charging.

STARTING CIRCUITS

Periodically check for corroded or loose terminals on the starting motor. Most starting motors require no other regular maintenance. (Some models may have armature shaft wicks that need regular oiling.) See Chapter 7 for complete information.

IGNITION CIRCUITS

Ignition circuits are covered in detail in Chapter 8.

DISTRIBUTOR

Every 500 hours, check the condition of the distributor contact points. Rough or pitted contacts must be replaced when metal transfer is excessive.

Clean the points with a few strokes of a clean contact file. Don't try to remove all roughness—just scale or dirt. Never use emery cloth or sandpaper —particles will embed in the points and cause them to burn.

If the points are burned, find the cause before operating the machine. Contact point burning could be caused by one or more of the following:

- **Too-high voltage**
- **Presence of oil or other foreign material**
- **Defective condenser**
- **Improper point adjustment**

Check the distributor point gap periodically (Fig. 7). For the correct gap, see the machine Technical or Operator's Manual.

Also check the timing of the distributor. If it seems erratic, adjust as given in Chapter 8.

Fig. 7—Adjusting Distributor Point Gap

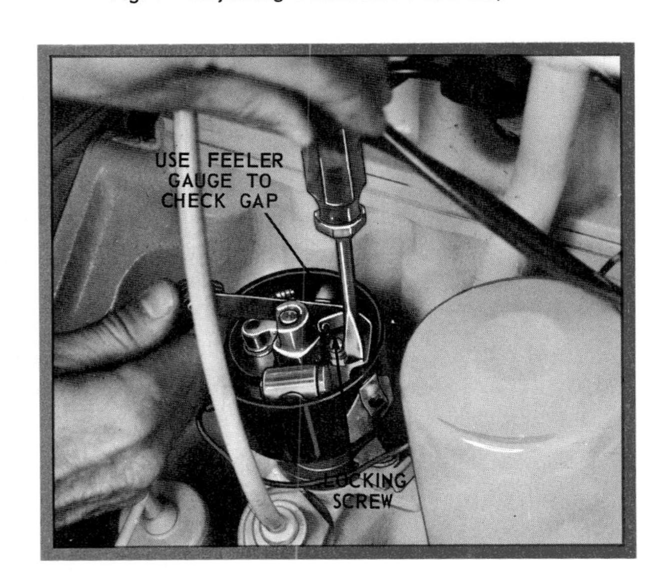

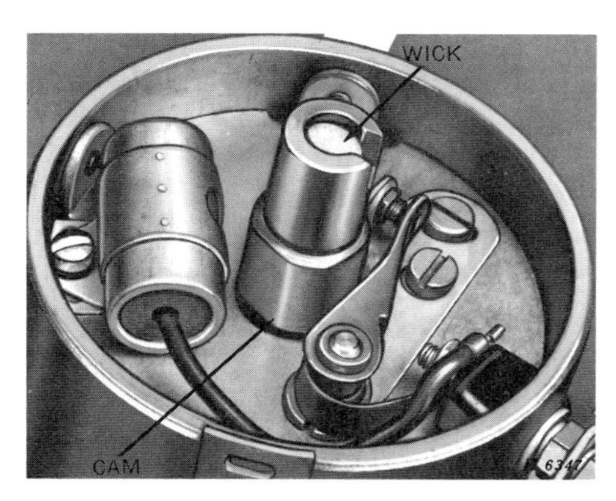

Fig. 8—Lubricating the Distributor

Inspect the distributor cap. If you find defects, replace the cap. Some of these defects are: cracks, chips, carbon tracks, worn or corroded tower inserts, or worn or damaged rotor button. If the cap is still serviceable, clean it before installing.

Periodically lubricate the distributor cam, using only a TRACE of cam lubricant or high-temperature grease. **Too much lubricant will ruin the distributor points.**

Some distributors also have an oil wick in the top of the cam (Fig. 8). In this case, place 4 or 5 drops of SAE 30 oil on the wick.

IGNITION COIL

Ignition coils don't require service—but check the coil for damage or carbon tracks.

SPARK PLUGS

For the complete story on spark plugs, refer to Chapter 8, which covers description, installation and testing.

Periodically check the external condition of the spark plugs (Fig. 9).

Look for carbon tracks or dirt which might contribute to poor performance.

If the ceramic insulator is cracked, the plug must be replaced.

Fig. 9—Spark Plug Maintenance

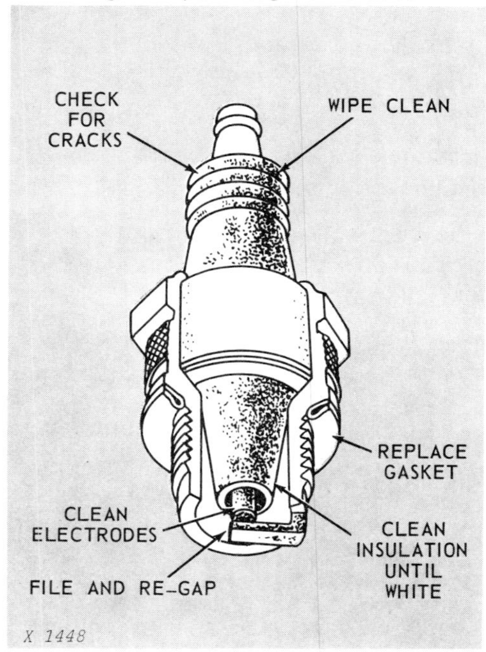

When replacing plugs, remember:

- **Use the right plug.**
- **Gap the plug as specified.**
- **Install properly, using a new gasket.**
- **Tighten the plug to the specified torque.**

WIRING

When servicing electrical components, take the time to check all wiring.

Loose or corroded connections can contribute to poor performance, as can cracked or damaged insulation.

These conditions cause excessive voltage drops in the system because of high resistance.

Unless the connections are tight and wiring is in good condition, batteries may be undercharged and the spark plugs will misfire.

Also check the protective boots on the secondary ignition circuit. These can become cracked or oil-soaked, causing low ignition voltage. Carbon tracks inside the boots also permit current leakage. Replace faulty boots.

TEST YOURSELF

QUESTIONS

1. (Fill in the blanks.) When connecting booster batteries, connect positive terminals to terminals.

2. True or false? "After disconnecting any part of the AC charging system, always polarize the alternator."

3. What difficulties can be caused by faulty wiring or poor connections?

4. Why is it necessary to remove acid film and dirt from the battery top? Why remove corrosion?

(Answers on page 21 at the end of this book.)

DIAGNOSIS AND TESTING OF SYSTEMS / CHAPTER 14

INTRODUCTION

"We've had some faulty regulators lately, so that must be your problem. We'll replace the regulator!"

"I don't really see what the owner's maintenance program has to do with troubleshooting!"

"I never talk to the operator. Operators sometimes know more about electrical systems than I do."

Here are three statements you might hear from "experienced" troubleshooters in a shop. Then again, maybe they aren't troubleshooters.

Regulator problems on the last job do not mean that the regulator is bad in every situation. What if the alternator had failed or a wiring lead was broken? The customer cannot afford the time and expense of replacing parts when unsure if the part is bad or requires replacement. That first quotation above wasn't made by a troubleshooter—it was made by a hit-or-miss parts replacer.

Is maintenance a key to troubleshooting? It certainly is. Problems are soon caused if the electrolyte level in a battery is never checked, or if the alternator drive belt tension is never checked. The second quotation above couldn't have been made by a troubleshooter, because the effects of poor maintenance are being overlooked.

What was being done when the equipment broke down? Is the problem erratic, or does it happen every time? What was done after the breakdown? These are a few of the questions that can be answered by talking to the operator. These answers help the troubleshooter to recreate the problem and accurately diagnose the cause. The third quotation was not made by a troubleshooter either. In fact, the person who made this statement knows little about the system being worked on.

Then what is a good troubleshooter?

Before answering this, let's realize that the electrical and electronic systems of today require diagnosis and testing at a minimum of cost. Equipment breakdown means a loss of time and money for the operator-owner, and we cannot expect the owner to absorb the extra cost due to poor troubleshooting.

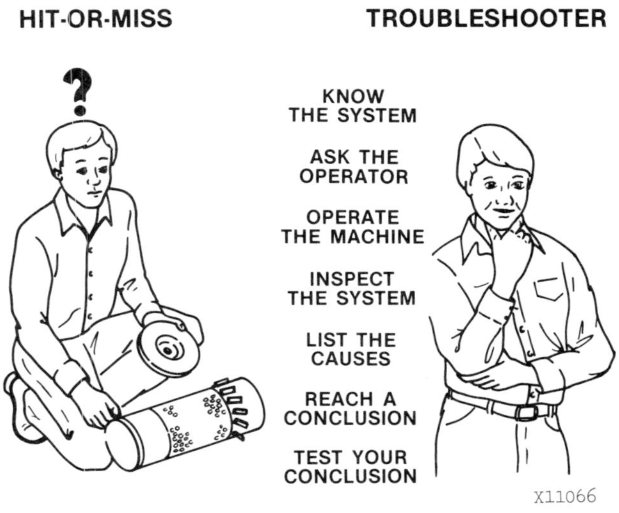

HIT-OR-MISS
TROUBLESHOOTER
KNOW THE SYSTEM
ASK THE OPERATOR
OPERATE THE MACHINE
INSPECT THE SYSTEM
LIST THE CAUSES
REACH A CONCLUSION
TEST YOUR CONCLUSION
X11066

Fig. 1—Which Would You Rather Be?

So a good troubleshooter starts out by using *common sense*, getting all the facts and examining them until the trouble has been pin-pointed. Then checks out the diagnosis by testing it. And only then does the process of replacing parts begin.

With the complex systems of today, diagnosis and testing is the only way.

SEVEN BASIC STEPS

A good program of diagnosis and testing has seven basic steps:

1. Know the System

2. Ask the Operator

3. Inspect the System

4. Operate the Machine (If Possible)

5. List the Possible Causes

6. Reach a Conclusion

7. Test Your Conclusion

We now have a formula for a troubleshooter:

TROUBLESHOOTER = COMMON SENSE + SEVEN BASIC STEPS.

Now let's see what these seven basic steps mean.

 Litho in U.S.A.

Fig. 2—Know the System

1. KNOW THE SYSTEM

In other words, do your homework. Find out all you can about the electrical and electronic systems of the machine. Is it a positive or negative-grounded system? Is the starting system 12 or 24 volts? Answers to these and many other questions are in the machine Technical Manual. Study this manual, especially the diagrams of the system. Schematics are an important tool—you should know how to read them. The glossary at the end of this book explains the various terms and symbols used in electrical diagrams. Study them and be able to recognize them at a glance.

Be familiar with the key specifications for the system given at the end of each section or group in the machine Technical Manual.

Keep up with the latest service bulletins. Read them and then file. The problem on your latest machine may be in this month's bulletin, giving the cause and remedy.

You can be prepared for any problem by knowing the system.

2. ASK THE OPERATOR

A good reporter gets the full story from a witness —the operator.

What work was the machine doing when the trouble was noticed? Is the trouble erratic or consistent? What did the operator do after the breakdown? Was an attempt made to fix the problem?

These are just a few of the many questions a good troubleshooter will ask the operator. Often a passing comment by the operator will provide the key to the

Fig. 3—Ask the Operator

problem. For example, the operator may have tried to adjust or repair some of the electrical system components. Perhaps the operator attempted to "jump start" the machine and damaged the alternator or reversed the polarity of the system.

Ask about how the machine is used and when it was serviced. Many problems can be traced to poor periodic maintenance programs or abuse to the machine.

Fig. 4—Checklist for Inspecting the Electrical System

3. INSPECT THE SYSTEM

Carefully inspect the electrical and electronic components for "tips" on the malfunction. Check to see if the machine can be operated without further damage to the system.

Always check these items before turning on switches or running the machine:

- *Look for bare wires that could cause grounds or shorts and dangerous sparks. Shorted wires can damage the charging system.*
- *Look for loose or broken wires. In the charging system, they can damage the regulator.*
- *Inspect all connections, especially battery connecting points. Acid film and dirt on the battery may cause current flow between the battery terminals, resulting in current leakage. Check the battery ground strap for proper connection.*
- *Check the battery electrolyte level. Continued loss of electrolyte indicates overcharging.*
- *Check the generator or alternator drive belt tension.*
- *Inspect for overheated parts after the machine has been stopped for awhile. They will often smell like burnt insulation. Put your hand on the alternator or regulator. Heat in these parts when the machine has not been operated for some time is a sure tip-off to charging circuit problems.*

In general, look for anything unusual.

Many electrical failures cannot be detected even if the machine is started. Therefore, a systematic and complete inspection of the electrical and electronic systems is necessary.

Many times the problem can be detected without turning on the switch or starting the machine.

While inspecting the electrical and electronic systems, make a note of all trouble signs.

4. OPERATE THE MACHINE (IF POSSIBLE)

If your inspection shows that the machine can be run, first turn the starting switch to the accessory position. Try out the accessory circuits—lights, cigarette lighter, etc. How do each of these components work? Look for sparks or smoke which might indicate shorts.

Turn the starting switch to the "on" position. The indicator lights should glow. (Some lights will also glow when the switch is in the "start" position.)

Now start the machine. Check all gauges for good operation and see if the system is charging or discharging.

Fig. 5—Operate the Machine (If Possible)

Fig. 6—List the Possible Causes

5. LIST THE POSSIBLE CAUSES

Now we are ready to make a list of the possible causes. What were the signs you found while inspecting the machine? What is the most likely cause? Are there other possibilities? Remember that one failure often leads to another.

6. REACH A CONCLUSION

Look over your list of possible causes and decide which are most likely and which are easiest to verify. Use the troubleshooting charts at the end of this chapter as a guide.

Fig. 7 — Reach a Conclusion

Fig. 8—Test Your Conclusion

7. TEST YOUR CONCLUSION

Before you repair the system, test your conclusions to see if they are correct. Many of the items can probably be verified without further testing. Maybe you can isolate the problem to a particular circuit, but not to an individual component. This is where test instruments will help you further isolate the trouble spot.

The next part of this chapter will tell you how to test the system and locate troubles.

But first let's repeat the seven rules for good troubleshooting:

1. **Know the System**
2. **Ask the Operator**
3. **Inspect the System**
4. **Operate the Machine (If Possible)**
5. **List the Possible Causes**
6. **Reach a Conclusion**
7. **Test Your Conclusion**

TESTING THE SYSTEM

The use of many types of electrical test equipment is most effective after the failure has been isolated to a particular circuit in the system. Other types of test equipment are specialized and are designed to isolate problems in specific systems. See Chapter 12.

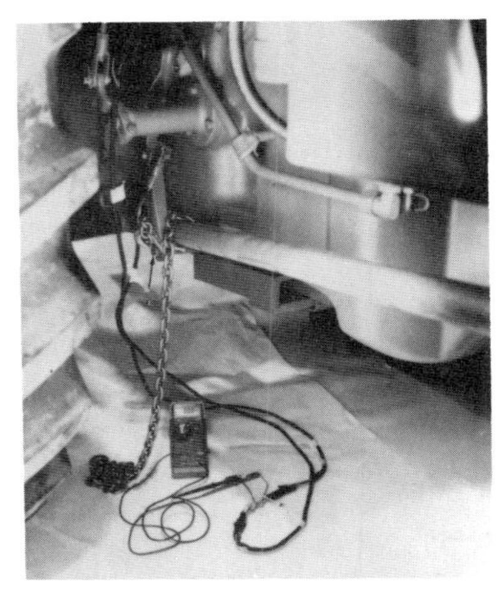

Fig. 9—Electrical Testing

Chapter 2 lists the most common types of test equipment required to test circuits and components. Instruction manuals are also furnished with these special tools. Study this information and take all precautions. Improper use of test equipment can damage the electrical components.

Other chapters of this manual have sections on testing of components and circuits:

Chapter 5—Storage Batteries

Chapter 6—Charging Circuits

Chapter 7—Starting Circuits

Chapter 8—Ignition Circuits

Chapter 9—Electronic Ignition Systems

Chapter 10—Lighting and Accessory Circuits

Chapter 11—Connectors

Chapter 12—Monitors and Controllers

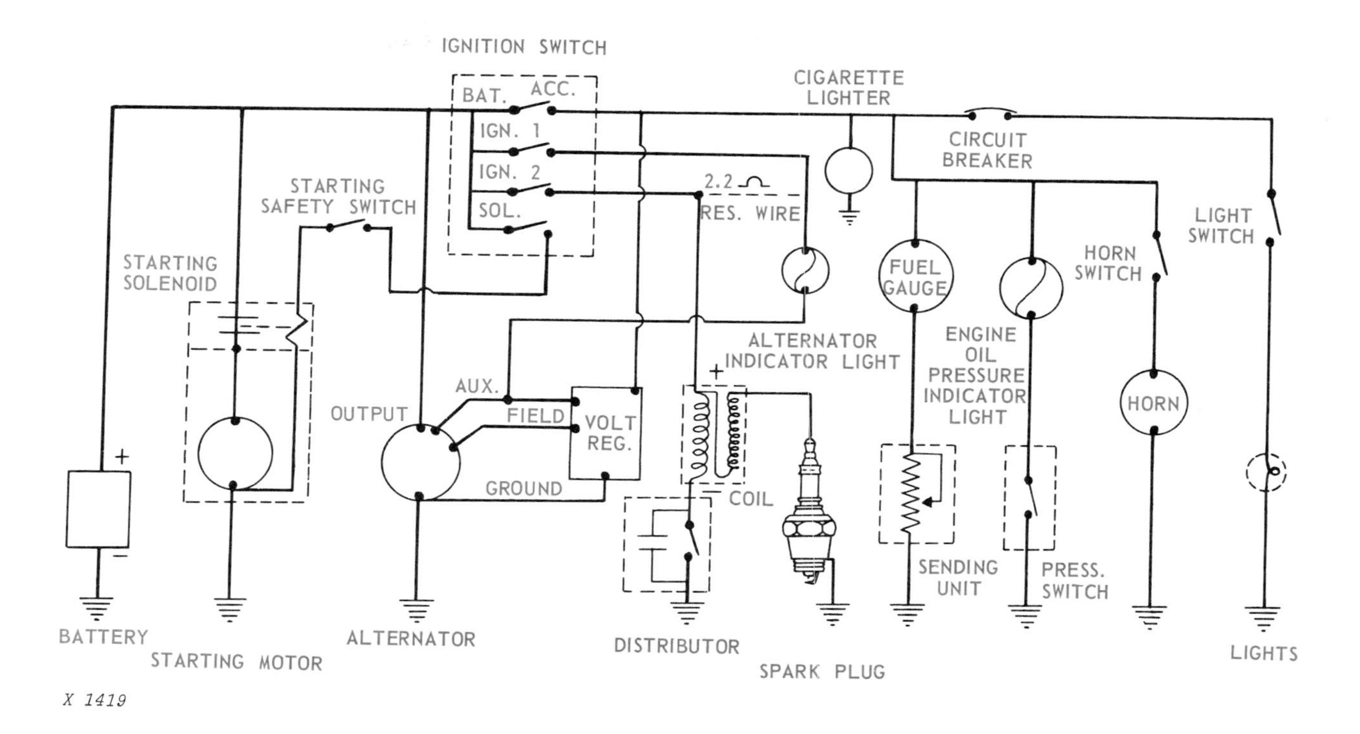

Fig. 10—Layout of "Problem" Electrical System

REMEMBER THE BATTERY!

The battery is a prime factor in each circuit of any electrical or electronic system. However, it is often overlooked while troubleshooting the system.

Before you begin most circuit tests, remember to do the following.

- **Check battery electrolyte level.**
- **Look for corroded terminals.**
- **Check for acid film and dirt on top of battery.**
- **Check battery polarity.**
- **Test the charge of the battery.**

For all battery tests, see Chapter 5.

TESTING PROCEDURES

Most problems with electrical systems require the same basic reasoning. Let's see how a good troubleshooter approaches an electrical problem and isolates the cause.

A sample problem follows. A schematic of the system is shown in Fig. 10.

Problem

The troubleshooter, when first arriving on the scene, located the operator, asked what had happened and asked if the operator had noticed anything unusual about the operation of the machine.

"The battery was 'run-down' yesterday morning after the machine had been idle overnight," said the operator. "So we started it with a slave battery. The alternator indicator light was okay—it stayed off during operation. We started the unit several times during the day without a slave battery. But this morning we tried to start it and it was discharged again."

After talking to the operator, the troubleshooter felt that a visual inspection would be helpful since the unit had not yet been started.

In checking the battery, the troubleshooter noticed that the electrolyte level was only slightly low. The battery terminals were not corroded and all connections were tight. Further inspection of the wiring did not provide any clues either.

The charging circuit was then checked.

The alternator was warm to the touch—not extremely hot or smoking, just moderately warm. This could possibly be the cause of the problem. A warm alternator in a case like this could mean that current is flowing in the alternator field circuit.

Since there didn't seem to be anything unusual that would hinder starting, the troubleshooter made an operational check of the unit. During this check the warm alternator was kept in mind.

When the starting switch was turned on and the dash gauges checked, the engine oil pressure light was on, but it was very dim. This could be expected, since battery voltage was low. However, the alternator indicator light was out. This could be another clue to the malfunction. If the alternator light was out, current might be flowing in the charging circuit.

The troubleshooter asked the operator whether or not the light came on at all while starting. The operator hadn't noticed. But, even if it didn't come on, the machine still operated.

At this point the troubleshooter felt that a tentative conclusion could be made. One clue was that a shorted isolation diode in the alternator would discharge the battery even if the machine was idle. Also, there were two exact clues to a shorted diode: *1) the warm alternator, and 2) the indicator light that would not light.*

To test this conclusion, the troubleshooter turned the switch off and, using a voltmeter, found that there was battery voltage at the alternator auxiliary terminal. (See Fig. 10.) This condition is not normal and is usually caused by a shorted isolation diode.

To clearly pinpoint the isolation diode as the problem, the troubleshooter asked the operator to start the machine.

Using a slave battery, the voltage drop across the isolation diode was checked. During normal operation there is a 1-volt drop across this diode. However, the test indicated no voltage drop at all.

This confirmed the troubleshooter's conclusion and pinpointed the trouble. The isolation diode was replaced and the voltage rechecked. The reading was normal, so the problem was corrected.

After correcting the problem, a second inspection was made of the charging circuit to be sure that other parts were not damaged.

The battery was then checked before the unit was started and returned to normal operation.

SUMMARY

Did you notice how the troubleshooter used the seven basic troubleshooting steps? Let's review the sequence of the story and look for these steps.

The first thing the troubleshooter did was to **Ask the Operator.** The machine operation and the symptoms of the malfunction were kept in mind. The troubleshooter did not assume the operator was correct, but tactfully double checked the operator's comments. Checking the indicator light was a good example of this.

The troubleshooter then **Inspected the System,** for it was during this inspection that the first clues to the problem were obtained. To show the importance of inspecting the system, notice that the troubleshooter was able to reach a tentative conclusion without even starting the machine.

While inspecting the system, the troubleshooter checked on whether it was possible to **Operate the Machine** without further damage. Although the unit was not actually started, the troubleshooter knew that it could be successfully and safely started.

There are many ways of **Listing Possible Causes.** This troubleshooter made a mental list of clues to the malfunction.

We should also point out that the troubleshooter **Knew the System.** When a clue became apparent, the cause and effect were recognized as well as how one symptom was related to another. Because of knowledge of the system, the troubleshooter was able to reach a tentative conclusion from these clues.

The last of our seven basic steps is **Testing Your Conclusion.** The troubleshooter did not assume that the first test had pinpointed the cause. An extra test was made to confirm that there was only one trouble spot and that it was definitely the cause of the trouble. Only after positive proof of malfunction was the isolation diode replaced.

We can see from the example that the **seven basic steps** fall easily into place during troubleshooting. Proper troubleshooting can save the machine owner time and money. The time spent in replacing many parts suspected of being faulty can be better utilized in diagnosis through proper diagnosis, the trouble spot can be quickly identified. Then time can be spent in correcting the problem without wasting time and money.

TROUBLESHOOTING CHARTS

Use the charts on the following pages to help in listing all the possible causes of trouble when you begin diagnosis and testing of the electrical system.

Once you have located the cause, check the item in the chart again for the possible remedy.

When you have pinpointed the problem to one circuit or component, refer to the other chapters of this manual for more details.

The Technical Manual for each machine supplements all these charts by giving more detailed and specific causes and remedies.

BATTERIES

LOW BATTERY OUTPUT

High Resistance in Circuit.
Check for resistance with voltmeter. See machine Technical Manual for maximum permissible voltage drop.

Low Electrolyte Level.
Add distilled water to proper level.

Low Specific Gravity.
See "Low Battery Charge."

Defective Battery Cell.
Replace battery.

Cracked or Broken Battery Case.
Replace battery.

Low Battery Capacity.
Always replace battery with one of adequate capacity.

BATTERY USES TOO MUCH WATER

Cracked Battery Case.
Replace battery.

Overcharged Battery.
See "High Charging Circuit Voltage." (A shorted cell in a 24-volt system may cause other batteries to be overcharged.)

LOW BATTERY CHARGE

Excessive Loads from Added Accessories.
Remove excessive loads.

Excessive Engine Idling.
Idle engine only when necessary.

Lights or Acessories Left On.
Be sure electrical switches are off before leaving machine.

Continuous Drain on Battery.
Check for leakage on dirty battery top. Disconnect battery ground and connect voltmeter between the ground battery terminal and a good ground. (On a 24-volt split-load system, also place voltmeter in the series connection between batteries to test for defective circuits that are not grounded. Disconnect circuits until ground or shorted component is located.)

Faulty Charging Operation.
See "Low Charging Circuit Voltage" or "Low Charging Circuit Output."

CHARGING CIRCUITS

LOW CHARGING CIRCUIT VOLTAGE

High Resistance in Charging Circuit Connections.
Check voltage drop to locate resistance. Be sure to use pin connector at battery to locate resistance between battery post and battery cable.

Defective Wiring.
Check voltage drop in wire to locate broken wire strands or undersized replacement wire.

Low Amperage Output of Generator or Alternator.
See "Low Charging Circuit Output."

Poor Regulator Ground.
Clean regulator ground connection.

Dirty Voltage Regulator Contact Points.
Clean points or replace regulator. (Broken resistors will cause excessive pitting of contact points.)

Regulator Out of Adjustment.
Adjust to specifications.

Dirty Cutout Relay Points.
Clean points or replace regulator.

Defective Regulator.
Replace regulator.

Open-Circuited Isolation Diode in Alternator.
Replace isolation diode assembly.

LOW CHARGING CIRCUIT OUTPUT

Slipping Drive Belts.
Adjust belt tension.

Excessively Worn or Sticking Brushes.
Repair or replace. Check commutator or slip rings.

Dirty or Out-of-Round Commutator or Slip Rings.
Clean commutator or slip rings.

Grounded Field Circuit (24-Volt, Split-Load System).
Repair or replace defective component.

NOTE: This defect places continuous drain on battery.

Dirty Current Regulator Contact Points.
Clean points or replace regulator.

Defective Diodes in Alternator.
Replace diode or diode plate assembly.

Defective Electrical Windings in Generator or Alternator.
Repair or replace windings. (If alternator stator windings are defective, be sure regulator will control alternator voltage.)

HIGH CHARGING CIRCUIT VOLTAGE

Grounded Generator Field Wire or Field Terminal.
(Does not apply to 24-volt generator on split-load system.)

Repair or replace wire or terminal.

High Resistance at Alternator-Regulator Connections.
Clean and tighten connections.

Defective Alternator Regulator.
Replace regulator.

Faulty Regulator.
Check and clean voltage regulator points is necessary. Adjust voltage regulator. Replace regulator if it does not respond to adjustment.

EXCESSIVE GENERATOR OUTPUT (DC CIRCUITS)

Grounded Generator Field Wire or Field Terminal.
(Does not apply to 24-volt generator on split-load system.)

Repair or replace wire or terminal.

Faulty Current Regulator in Generator Regulator.
Check and clean current regulator points if necessary. Adjust current regulator. Replace regulator if it does not respond to adjustment.

NOISY GENERATOR OR ALTERNATOR

Defective or Badly Worn Drive Belt.
Replace belt. Adjust to proper tension.

Generator Brushes Not Seated.
Seat brushes with No. 00 sandpaper or brush seating stone.

Generator Commutator Needs Reconditioning.
Turn and undercut commutator.

Worn or Defective Bearings.
Replace bearings.

Loose Mounting or Loose Drive Pulley.
Tighten mounting and pulley.

Misaligned Drive Belt or Pulley.
Check pulley condition. Align pulley.

Alternator Rectifier Shorted or Open.
Replace diode or diode plate assembly.

BUZZING IN ALTERNATOR-REGULATOR FIELD RELAY (AC CIRCUITS)

Open Negative Diode in Alternator Causing Reduced Voltage to Relay Winding.
Replace rectifier assembly.

SLUGGISH STARTING MOTOR OPERATION

Low Battery Charge.
Charge battery and check specific gravity. If battery does not respond to charging, install a new one.

High Resistance in Circuit.
Clean and tighten all connections. Repair or replace faulty wiring.

Defective Starting Motor.
Service and repair starting motor where necessary.

Starting Motor Bearings Dry.
Lubricate bearings with oil of proper viscosity.

Excessive Engine Drag Due to Tight Bearings.
Recheck engine overhaul procedures. If problem still exists after break-in period, test and service starting circuit.

Extremely Cold Weather.
Warm up battery befort starting the engine.

Too-High Engine Oil Viscosity.
Drain oil and replace with lower viscosity oil as recommended.

STARTING MOTOR WILL NOT OPERATE

Low Battery Charge.
Charge battery and check specific gravity. If battery does not respond to charging, install a new battery.

Starter Safety Switch Open.
Put shift lever in neutral or park position.

Improperly Adjusted or Defective Starter Safety Switch.
Adjust or replace switch.

Defective Starting Switch.
Replace switch.

High Resistance in Starting Circuit or Defective Wiring.
Clean and tighten all connections and replace faulty wiring.

Faulty Solenoid Switch on Starting Motor.
Repair or replace switch.

Faulty Starting Motor.
Service and repair motor.

STARTING MOTOR SOLENOID SWITCH FLUTTERS

Low Battery Charge.
Charge battery and check specific gravity. If battery does not respond to charging, install a new battery.

High Resistance in Circuit.
Clean and tighten all connections and replace faulty wiring.

Open Circuit in Starter Solenoid Hold-in Winding Circuit.
Repair or replace solenoid or wires.

MISFIRING OF ENGINE

Improper Spark Plugs Heat Range.
Replace with "hotter" or "colder" plugs as required.

Worn Spark Plug Electrodes or Dirty Spark Plugs.
Clean, file and regap plugs. Replace if necessary.

Defective Spark Plug.
Replace Spark Plug.

Incorrect Distributor Timing.
Reset timing.

Insufficient High-Tension Voltage Available to Spark Plug.
See "Low Available Voltage at Spark Plug," below.

Cold Starting Conditions.
Warm battery before starting engine.

LOW AVAILABLE VOLTAGE AT SPARK PLUG

Worn or Improperly Spaced Distributor Points.
Reset point gap.

Dirty, Burned, or Pitted Points.
Clean or replace points.

Defective Condenser (Leakage, Shorts, High Resistance, or Wrong Capacity).
Replace condenser.

Dirt or Moisture in Distributor Cap.
Clean distributor cap.

Cracked Rotor or Distributor Cap.
Replace cap or rotor.

Too-Wide Distributor Rotor Gap.
Replace rotor. Replace cap if deep path is worn on contacts.

Paint Covering Ignition Cables.
Remove paint. Replace defective wiring if necessary.

Faulty Spark Plug Cables.

Defective Ignition Coil.

Loose Coil Tower—High Tension Cable or Primary Coil Leads.
Clean and tighten all connections. Replace faulty wiring.

Cracked Coil Cap or Carbon Tracks on Cap.
Clean and tighten all connections. Replace faulty parts.

Defective Wiring.
Check all connections. Replace faulty wiring.

Low Battery Charge.
Charge battery and check specific gravity. If battery does not respond to charging, install a new battery.

High Resistance in Resistor Wire to Coil.
Replace resistor wire.

BROKEN OR CHIPPED DISTRIBUTOR CAP

Cocked Distributor Cap.
Replace distributor cap and install correctly.

HIGH-RESISTANCE CORROSION ON DISTRIBUTOR CAP TERMINALS

Arcing Inside the Cap Tower Due to Exposed Leads.
Push all leads snugly into cap towers.

FLUTTER OR "FANNING" OF IGNITION TIMING MARKS WHEN TIMING

Excessive Wear in Bushing or Distributor Drive Train.
If flutter exceeds 3 degrees, replace worn parts.

BUILD-UP OF MATERIAL ON DISTRIBUTOR POINTS

Over-Capacity Condenser: *Causes build-up of metal on stationary contact with negative-grounded battery, or on movable contact with positive-grounded battery.*

Replace with condenser of proper capacity.

Under-Capacity Condenser: *Causes build-up of metal on movable contact with negative-grounded battery, or on stationary contact with positive-grounded battery.*

Replace with condenser of proper capacity.

EXCESSIVE WEAR ON DISTRIBUTOR CONTACT POINTS

Improper Method of Cleaning Points.
Use contact file, crocus cloth, or lint-free tape to clean points.

EXCESSIVELY CONTAMINATED DISTRIBUTOR POINTS

Improper Cam Lubricant or Too Much Lubricant.
Use proper amount of specified lubricant.

Excessive Lubrication of Breaker Plates or Distributor Advance Mechanism.
Use proper amount of lubricant.

Clogged Oil Filler Breather Cap.
Clean breather cap.

Clogged Filters in Crankcase Vent Pipe.
Clean filters.

Worn Drive Shaft Bushing Letting Oil Fumes Enter Distributor.
Replace bushing.

ARCING AND BURNING OF DISTRIBUTOR CONTACT POINTS

Loose Lead or High Internal Resistance in Condenser.
Tighten lead or replace condenser.

Improper Method of Cleaning Distributor Points.
Use proper contact file and polish with stone or crocus cloth. Use lint-free tape to clean parts.

Points Improperly Adjusted.
Readjust points.

Oil or Other Foreign Material on Contacts.
Clean and remove all foreign material. Correct the cause (see "Excessively Contaminated Distributor Points").

High Voltage Due to Faulty or Improperly Adjusted Voltage Regulator.
Adjust voltage regulator. Replace unit if faulty.

Shorted By-Pass Resistor.
Replace resistor.

Wrong Capacity Condenser.
Replace condenser.

DAMAGE TO METAL TERMINALS IN DISTRIBUTOR CAP DUE TO ARCING

Ignition Cable Leads Loose.
Clean terminals. Push leads snugly into cap towers. Replace cap if erosion is severe.

DIM LIGHTS

High Resistance in Circuit or Poor Ground on Lights.
Clean and tighten all connections. Replace faulty wiring.

Low Battery Charge.
Charge battery and check specific gravity. If battery does not respond to charging, install a new battery.

Defective Light Switch or Starting Switch.
Replace switch.

Defective Battery Ground Wire (24-Volt Split-Load System).
Repair or replace ground wire.

LIGHTS BURN OUT PREMATURELY

Excess Voltage Due to Faulty Regulator or Ground Wire on Alternator or Generator "F" Terminal.
Adjust or replace regulator. Replace faulty wiring.

Defective Battery Ground Wire (24-Volt Split-Load System).
Repair or replace ground wire.

GENERATOR OR ALTERNATOR INDICATOR LAMP GLOWS DIMLY OR INTERMITTENTLY

Excessive Resistance in Battery Lead to Generator Regulator.
Clean and tighten all connections. Replace faulty wiring.

Excessive Internal Resistance in Generator Regulator.
Clean and adjust cutout relay contact points.

Excessive Internal Resistance in Alternator Regulator.
Replace regulator.

Defective Generator or Alternator.
Service or replace unit.

OIL PRESSURE INDICATOR LAMP WILL NOT LIGHT

Burnt-Out Bulb.

Open Circuit or Excessive Resistance in Wiring.
Clean and tighten all connections. Replace faulty wiring.

Defective Lamp Body.

Faulty Oil Pressure Switch.
Replace switch.

OIL PRESSURE LAMP REMAINS ON WITH STARTING SWITCH OFF

Defective Lamp Body.

Grounded Wire to Oil Pressure Switch.
Repair or replace wiring.

Faulty Oil Pressure Switch.
Replace switch.

TEST YOURSELF

QUESTIONS

1. Give the seven basic steps for good trouble shooting.

2. During which of the seven steps should you begin replacing parts?

3. What electrical component should be checked out even before testing most circuits?

(Answers on page 21 at the end of this book.)

 Litho in U.S.A.

DEFINITIONS OF TERMS

A

ACTUATOR SOLENOID—The solenoid in the actuator housing on the back of the injection pump which moves the control rack as commanded by the engine controller.

ALTERNATOR—A device which converts mechanical energy into electrical energy. For details, see Chapter 6 on "Charging Circuits."

ALTERNATING CURRENT (AC)—A flow of electrons which reverses its direction of flow at regular intervals in a conductor.

AMBIENT TEMPERATURE—The temperature of the surrounding medium, such as gas, air or liquid, which comes into contact with a particular component.

AMMETER—An instrument for measuring the flow of electrical current in amperes. Ammeters are always connected in series with the circuit to be tested.

AMPERE—A unit of measure for the flow of current in a circuit. One ampere is the amount of current flow provided when one volt of electrical pressure is applied against one ohm of resistance. The ampere is used to measure electricity much as "gallons per minute" is used to measure water flow.

AMPERE-HOUR—A unit of measure for battery capacity. It is obtained by multiplying the current (in amperes) by the time (in hours) during which current flows. For example, a battery which provides 5 amperes for 20 hours is said to deliver 100 ampere-hours.

AMPLIFIER—A device of electronic components used to increase power, voltage, or current of a signal.

AMPLITUDE—A term used to describe the maximum value of a pulse or wave. It is the crest value measured from zero.

ANALOG IC—Integrated circuits composed to produce, amplify, or respond to variable voltages. They include many kinds of amplifiers that involve analog-to-digital conversions and vice versa, timers, and inverters. They are known as Operational Amplifier Circuits or OP-Amps.

ANALOG GAUGE—A display device utilizing a varying current to cause a mechanical change in the position of its needle.

ARMATURE—The movable part of a generator or motor. It is made up of conductors which rotate through a magnetic field to provide voltage or force by electromagnetic induction. The pivoted points in generator regulators are also called armatures.

ARTIFICIAL MAGNETS—A magnet which has been magnetized by artificial means. It is also called, according to shape, a bar magnet or a horseshoe magnet.

ATOM—A particle which is the smallest unit of a chemical element. It is made up mainly of electrons (minus charges) in orbit around protons (positive charges).

AUXILIARY SPEED SENSOR—The engine speed sensor located on the engine timing gear cover. It serves as a back-up to the primary engine speed sensor.

B

BENDIX DRIVE—One type flywheel engaging device for a starting motor. It is said to be mechanical because it engages by inertia. See Chapter 7 for details.

BREAK—See "Open."

BRUSH—A device which rubs against a rotating slip ring or commutator to provide a passage for electric current to a stationary conductor.

C

CALIBRATION—The determination or rectification of the graduations used on a testing instrument.

CAPACITOR—A device which stores electrical energy. Commonly used for filtering out voltage spikes.

CHARGE—To restore the active materials in a storage battery by the passage of direct current through the battery cells in a direction opposite that of the discharging current.

CIRCUIT—A continuous, unbroken path along a conductor through which electrical current can flow from a source, through various units, and back to the source.

CIRCUIT BREAKER—A device used to protect an electrical circuit from overloads. The breaker acts as a switch and opens when the current passing through the circuit exceeds the rated level. Some circuit breakers must be reset manually; others may reset automatically after a period of time or when the current drops to the rated capacity. Also see "Fuse."

COIL—A number of turns of wire wound on an iron core. When current flows through the wire, the assembly becomes an electromagnet.

COIL (IGNITION)—An electrical device with two coil windings, whereby, through electromagnetic induction of one coil to another, the input voltage is stepped up to produce a high voltage ignition spark.

COLD RATING—The cranking load capacity of a battery at low temperatures.

COMMUTATOR—A device for assuring a one-direction (DC) flow of current from a generator.

CONDENSER—An automotive term which describes a capacitor.

CONDUCTOR—A substance or body through which an electrical current can be transmitted. Elements which are good conductors have less than four electrons in the outer rings of their atoms. Also see "insulator."

CONTROL CIRCUIT—The circuit of a control device or system which carries the electrical signals directing the performance of the controller, but does not carry the main power circuit.

CONTROLLER—An electronic device consisting of necessary circuitry, sensors, console, and wiring harnesses to automatically control a function of a machine.

CONVENTIONAL THEORY—The theory which states that the direction of current flow is from positive to negative in a circuit. This manual follows this theory regarding the direction of current flow.

COVALENT BONDING—The joining of electrons in the outer ring of one silicon atom with the electrons of other silicon atoms so that the atoms share electrons in their outer rings. This forms silicon crystal which is then "doped" by adding other materials and which creates a very good insulator.

CURRENT—Movement of electricity along a conductor. Current is measured in amperes.

CURRENT FLOW—The flow or movement of electrons from atom to atom in a conductor.

CYCLE—The change in an alternating electrical sine wave from zero to a positive peak to zero to a negative peak and back to zero.

CYCLING—The process by which a battery is discharged and recharged.

D

DIAGNOSTIC CODE—A number which represents a problem detected by the engine controller. Diagnostic codes are transmitted for use by on-board displays or a diagnostic reader so the operator or technician is aware there is a problem and in what part of the fuel injection system the problem can be found.

DIFFERENTIATOR CIRCUIT—A circuit that consists of resistors and capacitors designed to change a DC input to an AC output. It is used to make narrow pulse generators and to trigger digital logic circuits. When used in integrated circuits it is known as an inverter.

DIGITAL IC—Integrated circuits that produce logic voltage signals or pulses that have only two levels of output that are either ON or OFF (yes or no). Some component output examples are: Diagnostic Codes Output, Pulse-Width-Modulated (PWM) Throttle Output, Auxiliary Speed Output, and Fuel Flow/Throttle Output.

DIODE—An electrical device that will allow current to pass through itself in one direction only. Also see "Zener diode."

DIRECT CURRENT (DC)—A steady flow of electrons moving steadily and continually in the same direction along a conductor from a point of high potential to one of lower potential. It is produced by a battery, generator, or rectifier.

DISCHARGE—To remove electrical energy from a charged body such as a capacitor or battery.

DISTRIBUTOR (IGNITION)—A device which directs the high voltage of the ignition coil to the engine spark plugs.

DISTRIBUTOR LEAD CONNECTOR—A connection plug in the wires that lead from the sensor in the distributor to the electronic control unit.

DYER DRIVE—One type of flywheel engaging mechanism in a starting motor. See Chapter 7.

E

ELECTRICAL FIELD—The region around a charged body in which the charge has an effect.

ELECTRICITY—The flow of electrons from atom to atom in a conductor.

ELECTROCHEMICAL—The relationship of electricity to chemical changes and with the conversions of chemical and electrical energy. A battery is an electrochemical device.

ELECTRO-HYDRAULIC VALVE—A hydraulic valve actuated by a solenoid through variable voltage applied to the solenoid coil.

ELECTROLYTE—Any substance which, in solution, is dissociated into ions and is thus made capable of conducting an electrical current. The sulfuric acid-water solution in a storage battery is an electrolyte.

ELECTROMAGNET—A core of magnetic material, generally soft iron, surrounded by a coil of wire through which electrical current is passed to magnetize the core.

ELECTROMAGNETIC CLUTCH—An electromagnetic device which stops the operation of one part of a machine while other parts of the unit keep on operating.

ELECTROMAGNETIC FIELD—The magnetic field about a conductor created by the flow of electrical current through it.

ELECTROMAGNETIC INDUCTION—The process by which voltage is induced in a conductor by varying the magnetic field so that lines of force cut across the conductor.

ELECTRON—A tiny particle which rotates around the nucleus of an atom. It has a negative charge of electricity.

ELECTRON THEORY—The theory which explains the nature of electricity and the exchange of "free" electrons between atoms of a conductor. It is also used as one theory to explain direction of current flow in a circuit.

ELECTRONICS—The control of electrons (electricity) and the study of their behavior and effects. This control is accomplished by devices that resist, carry, select, steer, switch, store, manipulate, and exploit the electron.

ELECTRONIC CONTROL UNIT (ECU)—General term for any electronic controller. See "controller."

ELECTRONIC GOVERNOR—The computer program within the engine controller which determines the commanded fuel delivery based on throttle command, engine speed, and fuel temperature. It replaces the function of a mechanical governor.

ELECTRONIC IGNITION SYSTEM—A system in which the timing of the ignition spark is controlled electronically. Electronic ignition systems have no points or condenser, but instead have a reluctor, sensor, and electronic control unit.

ELEMENT—(1) Any substance that normally cannot be separated into different substances. (2) The completed assembly of a battery consisting of negative plates, positive plates, and separators mounted in a cell compartment.

ENGINE CONTROLLER—The electronic module which controls fuel delivery, diagnostic outputs, back-up operation, and communications with other electronic modules.

F

FIELD EFFECT TRANSISTOR (FET)—A transistor which uses voltage to control the flow of current. Connections are the source (input), drain (output) and gate (control).

FIXED RESISTOR—A resistor which has only one resistance value.

FREQUENCY—The number of pulse or wave cycles that are completed in one second. Frequency is measured in Hertz, as in 60Hz (hertz) per second.

FUNDAMENTAL LAW OF MAGNETISM—The fundamental law of magnetism is that unlike poles attract each other, and like poles repel each other.

FUSE—A replaceable safety device for an electrical circuit. A fuse consists of a fine wire or a thin metal strip encased in glass or some fire-resistant material. When an overload occurs in the circuit, the wire or metal strip melts, breaking the circuit. Also see "Circuit Breaker."

G

GATE—A logic circuit device which makes a YES or NO (one or zero) decision (output) based on two or more inputs.

GENERATOR—A device which converts mechanical energy into electrical energy. See Chapter 6 for details.

 Litho in U.S.A.

GRID—A wire mesh to which the active materials of a storage battery are attached.

GROUND—A ground occurs when any part of a wiring circuit unintentionally touches a metallic part of the machine frame.

GROUNDED CIRCUIT—A connection of any electrical unit to the frame, engine, or any part of the tractor or machine, completing the electrical circuit to its source.

GROWLER—A device for testing the armature of a generator or motor.

H

HYDROMETER—An instrument for measuring specific gravity. A hydrometer is used to test the specific gravity of the electrolyte in a battery.

I

IGNITION CONTROL UNIT—The module that contains the transistors and resistors that controls the electronic ignition.

INDUCTANCE—The property of an electric circuit by which an electromotive force (voltage) is induced in it by a variation of current either in the circuit itself or in a neighboring circuit.

INDUCTOR—A coil of wire wrapped around an iron core.

INSULATED GATE FIELD EFFECT TRANSISTOR (IGFET)—A diffused transistor which has an insulated gate and almost infinite gate-channel resistance.

INSULATOR—A substance or body that resists the flow of electrical current through it. Also see "Conductor."

INTEGRATED CIRCUIT(IC)—An electronic circuit which utilizes resistors, capacitors, diodes, and transistors to perform various types of operations. The two major types are Analog and Digital Integrated Circuits. Also see "Analog IC" and "Digital IC."

INTEGRATOR CIRCUIT—A circuit that consists of resistors and capacitors and functions as a filter which can pass signals only below a certain frequency.

INVERTER—A device with only one input and one output; it inverts or reverses any input.

ION—An atom having either a shortage or excess of electrons.

ISOLATION DIODE—A diode placed between the battery and the alternator. It blocks any current flow from the battery back through the alternator regulator when the alternator is not operating.

L

LIGHT EMITTING DIODE (LED)—A solid-state display device that emits infrared light when a forward-biased current flows through it.

LINES OF FORCE—Invisible lines which conveniently illustrate the characteristics of a magnetic field and magnetic flux about a magnet.

LIQUID CRYSTAL DISPLAY (LCD)—A display device utilizing a special crystal fluid to allow segmented displays.

M

MAGNET—A body which has the property of attracting iron or other magnets. Its molecules are aligned. See Chapter 1.

MAGNETIC FIELD—That area near a magnet in which its property of magnetism can be detected. It is shown by magnetic lines of force.

MAGNETIC FLUX—The flow of magnetism about a magnet exhibited by magnetic lines of force in a magnetic field.

MAGNETIC INDUCTION—The process of introducing magnetism into a bar of iron or other magnetic material.

MAGNETIC LINES OF FORCE—Invisible lines which conveniently illustrate the characteristics of a magnetic field and magnetic flux about a magnet.

MAGNETIC MATERIAL—Any material to whose molecules the property of magnetism can be imparted.

MAGNETIC NORTH—The direction sought by the north pole end of a magnet, such as a magnetic needle, in a horizontal position. It is near the geographic north pole of the Earth.

MAGNETIC PICKUP ASSEMBLY—The assembly in a self-integrated electronic ignition system that contains a permanent magnet, a pole piece with internal teeth, and a pickup coil. These parts, when properly aligned, cause the primary circuit to switch off and induce high voltage in the secondary windings.

MAGNETIC SOUTH—The opposite direction from magnetic north towards which the south pole end of a magnet, such as a magnetic needle, is attracted when in a horizontal position. It is near the geographic south pole of the Earth.

MAGNETIC SWITCH—A solenoid which performs a simple function, such as closing or opening switch contacts.

MAGNETISM—The property inherent in the molecules of certain substances, such as iron, to become magnetized, thus making the substance into a magnet.

MICROPROCESSOR—An integrated circuit combing logic, amplification and memory functions.

MILLIAMPERE—1/1,000,000 ampere.

MOLECULE—A unit of matter which is the smallest portion of an element or compound that retains chemical identity with the substance in mass. It is made up of one or more atoms.

MONITOR—An electronic device that through sensors, gauges, console, and wiring harnesses keeps constant watch on functions of machines and alerts the operator of possible trouble.

MOTOR—A device which converts electric energy into mechanical energy. See Chapter 7 for details.

MULTIMETER—A testing device that can be set to read ohms (resistance), voltage (force), or amperes (current) of a circuit.

MUTUAL INDUCTION—Occurs when changing current in one coil induces voltage in a second coil.

N

NATURAL MAGNET—A magnet which occurs in nature, such as a lodestone. Its property of magnetism has been imparted by the magnetic effects of the Earth.

NEGATIVE—Designating or pertaining to a kind of electricity. Specifically, an atom that gains negative electrons is negatively charged.

NEUTRON—An uncharged elementary particle. Present in all atomic nuclei except the hydrogen nucleus.

NON-MAGNETIC MATERIAL—A material whose molecules cannot be magnetized.

NORMALLY OPEN and NORMALLY CLOSED—These terms refer to the position taken by the contacts in a magnetically operated switching device, such as a relay, when the operating magnet is deenergized.

O

OHM—The standard unit for measuring resistance to flow of an electrical current. Every electrical conductor offers resistance to the flow of current, just as a tube through which water flows offers resistance to the current of water.

One ohm is the amount of resistance that limits current flow to one ampere in a circuit with one volt of electrical pressure.

OHMMETER—An instrument for measuring the resistance in ohms of an electrical circuit.

OHM'S LAW—Ohm's Law states that when an electric current is flowing through a conductor, such as a wire, the intensity of the current (in amperes) equals the electromotive force (volts) driving it, divided by the resistance of the conductor. The flow is in proportion to the electromotive force, or voltage, as long as the resistance remains the same.

OPEN OR OPEN CIRCUIT—An open or open circuit occurs when a circuit is broken, such as by a broken wire or open switch, interrupting the flow of current through the circuit. It is analogous to a closed valve in a water system.

OPERATIONAL AMPLIFIER—A high-voltage gain, low-power, linear amplifying circuit device used to add, subtract, average, etc.

OVERRUNNING CLUTCH—One type of flywheel-engaging member in a starting motor. See Chapter 7.

P

PARALLEL CIRCUIT—A circuit in which the circuit components are arranged in branches so that there is a separate path to each unit along which electrical current can flow.

PERMANENT MAGNET—A magnet which retains its property of magnetism for an indefinite period.

PIEZO-ELECTRIC DEVICE—A device made of crystalline materials, such as quartz, which bend or distort when force or pressure is exerted on them. This pressure forces the electrons to move.

Litho in U.S.A.

PLATE—A solid substance from which electrons flow. Batteries have positive plates and negative plates.

POLARITY—A collective term applied to the positive (+) and negative (−) ends of a magnet or electrical mechanism such as a coil or battery.

POLE—One or two points of a magnet at which its magnetic attraction is concentrated.

POLE SHOES — Iron blocks fastened to the inside of a generator or motor housing around which the field or stator coils are wound. The pole shoes may be permanent or electro-magnets.

POSITIVE—Designating or pertaining to a kind of electricity. Specifically, an atom which loses negative electrons and is positively charged.

POTENTIOMETER—A variable resistor used as a voltage divider.

POWER SWITCH TRANSISTOR—The part responsible for switching off the primary circuit that causes high voltage induction in the secondary winding in an electronic ignition system.

PRIMARY SPEED SENSOR—An engine speed sensor located inside the actuator housing on the back of the injection pump.

PRINCIPLE OF TURNING FORCE— Explains how magnetic force acts on a current-carrying conductor to create movement of an armature, such as in an electric motor. See Chapter 1.

PRINTED CIRCUIT BOARD—A device used to hold integrated circuit components in place and provide current paths from component to component. Copper pathways are etched into the board with acid.

PROTON—A particle which, together with the neutron constitutes the nucleus of an atom. It exhibits a positive charge of electricity.

PULSE—A signal that is produced by a sudden ON and OFF of direct current (DC) within a circuit.

PULSE-WIDTH-MODULATED (PWM)—A digital electronic signal which consists of a pulse generated at a fixed frequency. The information transmitted by the signal is contained in the width of the pulse. The width of the pulse is changed (modulated) to indicate a corresponding change in the information being transmitted, such as throttle command.

R

RECTIFIER—A device (such as a vacuum tube, commutator, or diode) that converts alternating current into direct current.

REGULATOR—A device which controls the flow of current or voltage in a circuit to a certain desired level.

RELAY—An electrical coil switch that uses a small current to control a much larger current.

RELUCTANCE—The resistance that a magnetic circuit offers to lines of force in a magnetic field.

RELUCTOR—A metal cylinder, with teeth or legs, mounted on the distributor shaft in an electronic ignition system. The reluctor rotates with the distributor shaft and passes through the electromagnetic field of the sensor.

RESISTANCE—The opposing or retarding force offered by a circuit or component of a circuit to the passage of electrical current through it. Resistance is measured in ohms.

RESISTOR—A device usually made of wire or carbon which presents a resistance to current flow.

RHEOSTAT—A resistor used for regulating a current by means of variable resistance; rheostats allow only one current path.

RIGHT-HAND RULE—A method used to determine the direction a magnetic field rotates about a conductor, or to find the north pole of a magnetic field in a coil. See Chapter 1.

ROTOR—The rotating part of an electrical machine such as a generator, motor, or alternator.

S

SELF-INDUCTION—Voltage which occurs in a coil when there is a change of current.

SEMICONDUCTOR—An element which has four electrons in the outer ring of its atoms. Silicon and germanium are examples. These elements are neither good conductors nor good insulators. Semiconductors are used to make diodes, transistors, and integrated circuits.

SENDING UNIT—A device, usually located in some part of an engine, to transmit information to a gauge on an instrument panel.

SENSOR—A small coil of fine wire in the distributor on electronic ignition systems. The sensor develops an electromagnetic field that is sensitive to the presence of metal. In monitors and controllers, they sense operations of machines and relay the information to a console.

SEPARATOR—Any of several substances used to keep one substance from another. In batteries a separator separates the positive plates from the negative plates.

SERIES CIRCUIT—A circuit in which the parts are connected end to end, positive pole to negative pole, so that only one path is provided for current flow.

SERIES—PARALLEL CIRCUIT—A circuit in which some of the circuit components are connected in series and others are connected in parallel.

SHORT (OR SHORT CIRCUIT)—This occurs when one part of a circuit comes in contact with another part of the same circuit, diverting the flow of current from its desired path.

SHUNT—A conductor joining two points in a circuit so as to form a parallel circuit through which a portion of the current may pass.

SLIP RING—In a generator, motor, or alternator, one of two or more continuous conducting rings from which brushes take, or deliver to, current.

SOLENOID—A tubular coil used for producing a magnetic field. A solenoid usually performs some type of mechanical work.

SOLID-STATE CIRCUITS—Electronic (integrated) circuits which utilize semiconductor devices such as transistors, diodes and silicon controlled rectifiers.

SPARK PLUGS—Devices which ignite the fuel by a spark in a spark-ignition engine.

SPECIFIC GRAVITY—The ratio of a weight of any volume of a substance to the weight of an equal volume of some substance taken as a standard, usually water for solids and liquids. When a battery electrolyte is tested the result is the specific gravity of the electrolyte.

SPRAG CLUTCH DRIVE—A type of flywheel engaging device for a starting motor. See Chapter 7.

STARTER MOTOR—A device that converts electrical energy from the battery into mechanical energy that turns an engine over for starting.

STATOR—The stationary part of an alternator in which another part (the rotor) revolves.

STORAGE BATTERY—A group of electrochemical cells connected together to generate electrical energy. It stores the energy in a chemical form. See Chapter 5.

SULFATION—The formation of hard crystals of lead sulfate on battery plates. The battery is then "sulfated."

SWITCH—A device which opens or closes electrical pathways in an electrical circuit.

SYNCHROGRAPH—An all-purpose distributor tester.

T

TACHOMETER—An instrument for measuring rotary speed; usually revolutions per minute.

TEMPORARY MAGNET—A magnet which loses its property of magnetism quickly unless forces act to re-magnetize it.

THERMISTOR—A temperature-compensated resistor. The degree of its resistance varies with the temperature. In some regulators, it controls a Zener diode so that a higher system voltage is produced in cold weather, when needed.

TRANSFORMER—A device made of two coil windings that transfers voltage from one coil to the next through electromagnetic induction. Depending upon the number of windings per coil, a transformer can be designed to step-up or step-down its output voltage from its input voltage. Transformers can only function with alternating current (AC).

TRANSIENT VOLTAGE PROTECTION MODULE (TVP)—A device which protects the engine controller electronics against high energy voltage transients such as alternator load dumps.

TRANSISTOR—A device constructed of semi-conductors that is used in circuits to control a larger current by using a smaller current for operation. Its function is the same as a relay.

TRIMMER RESISTOR—A resistor used in applications where only a small resistance change is needed.

Litho in U.S.A.

V

VACUUM FLORESCENT DISPLAY (VDC)—An anode-controlled display which emits its own light. It works like a television tube, directing streams of electrons to strike phosphorescent segments.

VARIABLE RESISTOR—A resistor that can be adjusted to different ranges of value.

VISCOSITY—The internal resistance of a fluid, caused by molecular attraction, which makes it resist a tendency to flow.

VOLT—A unit of electrical pressure (or electromotive force) which causes current to flow in a circuit. One volt is the amount of pressure required to cause one ampere of current to flow against one ohm of resistance.

VOLTAGE—That force which is generated to cause current to flow in an electrical circuit. It is also referred to as electromotive force or electrical potential. Voltage is measured in volts.

VOLTAGE REGULATOR—A device that controls the strength of a magnetic field produced by a generator or alternator. It prevents the battery from being over- or undercharged during high- or low-speed operation of the generator or alternator.

VOLTMETER—An instrument for measuring the force in volts of an electrical current. This is the difference of potential (voltage) between different points in an electrical circuit. Voltmeters are connected across (parallel to) the points where voltage is to be measured.

W

WATT—A unit of measure for indicating the electrical power applied in a circuit. It is obtained by multiplying the current (in amperes) by the electrical pressure (in volts) which cause it to flow. That is: watts = amperes × volts.

WATT-HOUR—A unit of electrical energy. It indicates the amount of work done in an hour by a circuit at a steady rate of one watt. That is, watt-hours = ampere-hours × volts.

WAVE—A signal that is produced by varying a continuous flow of current within a circuit. Waveforms can be created by either AC or DC current.

WAVEFORM—A graphical representation of electrical cycles which shows the amount of variation in amplitude over some period of time.

WINDING—The coiling of a wire about itself or about some object. Often identified as a series winding, a shunt winding, etc.

WIRING HARNESS—The trunk and branches which feed an electrical circuit. Wires from one part of the circuit enter the trunk, joining other wires, and then emerge at another point in the circuit.

Z

ZENER DIODE (Reverse Bias Direction Diode)—A semiconductor device that will conduct current in the reverse direction when the voltage becomes higher than a predetermined voltage.

SUGGESTED READINGS

Sensors and Circuits, Sensors, Transducers, and Supporting Circuits for Electronic Instrumentation, Measurement and Control; Joseph J. Carr, Prentice Hall, Englewood, New Jersey 07632, 1993

Basic Electronics; Sixth Edition; Bernard Grob; Mcgraw-Hill

Instrumentation and Measurement for Environmental Sciences; Second Edition; Bailey W. Mitchell, Editor; American Society of Agricultural Engineers, 2950 Niles Road, St. Joseph, Michigan 49085, 1983

Your Guide to the Electronic Control of Fluid Power; National Fluid Power Association; 3333 N. Mayfair Road, Milwaukee, Wisconsin 53222-3219, 1992

Agricultural Sensors; Glen E. Vanden Berg; American Society of Agricultural Engineers, 2950 Niles Road, St. Joseph, Michigan 49085, 1988

Agricultural Electronics – 1983 and Beyond; Volume 1, Field Equipment, Irrigation and Drainage; Proceedings of the National Conference on Agricultural Electronics Applications, December 11-13, 1993, Chicago, Illinois; American Society of Agricultural Engineers, 2950 Niles Road, St. Joseph, Michigan 49085, 1983

 Litho in U.S.A.

MEASUREMENT CONVERSION CHART

Metric to English

LENGTH

1 millimeter = 0.03937 inches ....in
1 meter = 3.281 feet..ft
1 meter = 1.094 yard.......................................yd
1 kilometer = 0.621 miles ....mi

AREA

1 meter^2 = 10.76 feet^2 ....ft^2
1 hectare = 2.471 acres ....acre
(1 hectare = 10,000 m^2)

MASS (WEIGHT)

1 kilogram = 2.205 pounds................................lb
1 tonne (1000 kg) = 1.102 short ton................ton

MASS (VOLUME)

1 meter^3 = 35.31 foot^3 ....ft^3
1 meter^3 = 1.308 yard^3 ....yd^3
1 meter^3 = 28.38 bushel..................................bu
1 liter = 0.02838 bushel ....bu
1 liter = 1.057 quart..qt

PRESSURE

1 bar = 14.50 pound/in^2 (psi) ....psi
(1 bar = 10^5 pascal)

STRESS

1 megapascal or
1 newton/millimeter^2 = 145 pound/in^2 (psi) ......psi
(1 N/mm^2 = 1 MPa)

POWER

1 kilowatt = 1.341 horsepower (550 ft-lb/s)......hp
(1 watt = 1 Nm/s)

ENERGY (WORK)

1 joule = 0.0009478 British Thermal Unit .....BTU
(1J = !Ws)

FORCE

1 newton = 0.2248 pounds force.......................lb

TORQUE OR BENDING MOMENT

1 newton meter = 0.7376 pound-foot.............lb-ft

TEMPERATURE

Fahrenheit to Celsius conversion
°F = °C (1.8) + 32 ....°F

English to Metric

LENGTH

1 inch = 25.4 millimeters ....mm
1 foot = 0.3048 meters ....m
1 yard = .9144 meters ....m
1 mile = 1.608 kilometers ....km

AREA

1 foot^2 = 0.0929 meter^2 ....m^2
1 acre = 0.4047 hectare ....ha
(1 hectare = 10,000 m^2)

MASS (WEIGHT)

1 pound = 0.4535 kilograms ....kg
1 ton (2000 lb) = 0.9071 tonnes ....t

MASS (VOLUME)

1 foot^3 = 0.02832 meter^3 ....m^3
1 yard^3 = 0.7646 meter^3 ....m^3
1 bushel = 0.03524 meter^3 ....m^3
1 bushel = 35.24 liters L
1 quart = 0.9464 liters L
1 gallon = 3.785 liters L

PRESSURE

1 pound/in^2 (psi) = 0.06895 bar.......................bar
(1 bar = 10^5 psacal)

STRESS

1 pound/in^2 (psi) = .006895 megapascal ......MPa
or newton/mm^2 ....N/mm^2
(1 N/mm^2 = 1MPa)

POWER

1 horsepower (ft-lb/s) = .7457 kilowatt.............kw
(1 watt = 1 Nm/s)

ENERGY (WORK)

1 British Thermal Unit = 1055 joules ....J
(1J = 1Ws)

FORCE

1 pound = 4.448 newtons ....N

TORQUE OR BENDING MOMENT

1 pound-foot = 1.356 newton-meters ....N-m

TEMPERATURE

Fahrenheit to Celsius conversion
°F = °C (1.8) + 32 ....°C

Litho in U.S.A.

INDEX

A

Page

B

Page

C

D

 Litho in U.S.A.

Litho in U.S.A

Page

S

Page

 Litho in U.S.A.

Page

Litho in U.S.A.

ALPHABETICAL LISTING OF DEVICES AND THEIR LETTERS

Device	Identifying Letter
A	
ABS control unit	A
Air conditioner	E
Air-flow sensor	B
Alarm switch	S
Alarms, general	H
Ammeter	P
Amplifier	N
Analog to digital converter	U
Anti-theft alarm system	A
Assembly	A
Audible alarms	H
Auxiliary-air device	Y
Auxiliary loudspeaker	B
B	
Backup light	E
Backup light switch	S
Battery	G
Battery charger	G
Battery relay	K
Battery switch	S
Bimetallic release device	F
Blower motor	M
Blower switch	S
Brake-fluid warning light	H
Brake light	H
Brake-light switch	S
Brake-lining warning light	H
Break contact switch	S
Breaker points	S
Breakerless regulator	N
C	
Capacitors	C
Car alarm	A
Car antenna	W
Car heater	E
Car heater switch	S
Car intercom, two-way radio	A
Car radio	A
Car telephone	A
Centralized plug-in module	X
Changeover contact	S
Changeover switch	S
Charge warning light	H
Charging unit	G
Choke coil	L
Cigarette lighter	E
Clock	P
Clutch, electromagnetic	Y
Coaxial conductor	W
Coil	L
Cold start valve	Y
Combination instruments	A
Combination switch	S
Combined headlight & ignition switch	S
Common ground conductor	W
Condensers	C
Conductor bundle	W
Connector	X
Consumption meter	B
Contact field rheostat	N
Contact sensor, brake lining	B
Control relay	K
Control unit	A
Converter	G
Cooling-water thermostat	B
Current protection circuit	F
Current transformer	T
D	
Darlington transistor	V
DCconverter	U
DCmotor	M
Diagnostic connector	P
Digital equipment	D
Dimmer switch	S
Diode	V
Door contact switch	S
Door lock, electrical	Y
Drive control	B
Driving switch	S
Dropping resistor	R
E	
Edge connector	X
Electric cutout	F
Electric fuel pump	Y
Electric windshield wiper	Y
Electromagnetic clutch	Y
Electromechanical voltage regulator	N
Electronic voltage regulator	A
Electronic windshield wiper delay control unit	A
Emergency flasher.	K
Emergency warning flasher	K

ALPHABETICAL LISTING OF DEVICES AND THEIR LETTERS

Device	Identifying Letter
Emergency warning flasher switch	S
Engagement solenoid	Y
Engine compartment light	E
Engine compartment light switch	S
Exhaust-gas recirculation valve	Y
Exhaust gas sensor	B
Exhaust sample pickup	B

F

Device	Identifying Letter
Failsafe protection device	F
Fan motor	M
Fanfare horn	B
Field rheostat	N
Filter network	Z
Flame glow plug	R
Fog light	E
Fog light switch	S
Fog warning light	E
Fog warning light switch	S
Frequency converter	U
Fuel consumption meter	P
Fuel gauge	P
Fuel-level sending unit	P
Fuel pump, electrical	Y
Fuse	F
Fusebox, fuse holder	F

G

Device	Identifying Letter
Gear motor, electrical	M
Gear shift, electrical	A
Gearbox control, electrical	Y
Generator	G
Generator regulator	N
Generator—ignition unit	G
Glove compartment light	E
Glove compartment light switch	S
Glow plug	R
Glow plug resistor	R
Governor, electrical	N

H

Device	Identifying Letter
Hall generator	B
Headlight flasher button	S
Headlight switch	S
Headlight vertical aim control	Y
Headlight wash-wipe system motor	M
Headlight wiper motor	M
Headlight, general	E
Heated rear window	E
Heater	E
Heating flange	R
Heating plug	R
Heating resistor	R
Helical heating wire	R
Helical heating wire regulator	R
High beam indicator light	H
High/low beam headlight	E
High tension ignition cable	W
Horn	B
Horn button	S
Hot start solenoid	Y

I

Device	Identifying Letter
Idle cutoff valve	Y
Ignition breaker points	S
Ignition coil	T
Ignition distributor	E
Ignition switch	S
Ignition transformer	T
Ignition trigger unit	B
Ignition distributor connector	X
Ignition/start switch	S
Indicator light	H
Inductive pickup	B
Injection solenoid valve	Y
Instrument panel light	E
Instrument panel light switch	S
Integrated circuit	N
Integrated circuit, analog	N
Integrated circuit, digital	D
Interference suppression capacitor	C
Interference suppression filter	Z
Interference suppression resistor	R
Interference suppression boxes	Z
Interior light	E
Interior light switch	S
Intermittent relay	K
Intermittent wiper relay	K
Intermittent wiper switch	S
Isolating link	S

K

Device	Identifying Letter
Kickdown solenoid valve	Y
Kickdown switch	S

L

Device	Identifying Letter
Lambda probe	B
License plate light	E
Light	E
Lighting, general	E

ALPHABETICAL LISTING OF DEVICES AND THEIR LETTERS

Device	Identifying Letter
Limit switch	S
Line coupler, electrical	X
Line coupler, plug, connector	X
Load sensor	B
Loudspeaker	B
Low oil pressure switch	B
M	
Magneto-magnet generator	G
Main light switch	S
Make contact	S
Measuring instrument	P
Measuring point of instrument	P
Memory	D
Microphone	B
Mini-relay	K
Momentary switch	S
Motor switch	S
Multicontact plug, socket	X
N	
Notching relay	K
NTC resistor	R
O	
Oil-pressure switch	B
Oil-pressure warning light	H
Optocoupler	V
Overvoltage protection device	F
P	
Parking brake switch	S
PC board relay	K
Permanent magnet	Y
Pickup	B
Plug, electrical	X
Plug-in modules	X
Power antenna	W
Power pack	G
Power window drive motor	M
Power window switch	S
Pressure control valve	Y
Pressure sensor	B
Pressure switch	B
Printed circuit	N
PTC resistor	R
Pulse generator	B
Pushbutton	S
Pushbutton switch	S
R	
Rear window defogger	E
Rear window defogger switch	S
Rectifier	V
Reference Mark sensor	B
Relay	K
Relay plate	K
Release device	F
Resistor	R
Reverse polarity interlock	F
Revolving emergency light	H
Rheostat	R
Rotary light switch	S
Rotational speed relay	K
Rotational speed sensor	B
S	
Safety switch	S
Schmitt trigger	U
Selector switch	S
Semiconductor device	V
Sensor	B
Sheathed glow plug element	R
Shielded conductor	W
Shields, electrical	W
Side marker light	E
Small power motor	M
Socket, electrical	X
Solenoid	Y
Solenoid actuation	Y
Solenoid switch	K
Solenoid-operated air valve	Y
Solenoid-operated brake	K
Solenoid-operated cold start valve	Y
Solenoid-operated injection valve	Y
Solenoid-operated valve	Y
Spark plug	E
Spark plug connector	X
Speed sensor	B
Speedometer	P
Start valve	Y
Start-locking relay	K
Start-locking switch	S
Start-repeating relay	K
Starter	M
Starter-generator ignition unit	G
Starter-generator unit	G

 Litho in U.S.A.

ALPHABETICAL LISTING OF DEVICES AND THEIR LETTERS

Device	Identifying Letter
Starting relay	K
Static converter	G
Strip terminal	X
Subassembly, electrical	A
Sunroof motor	M
Sunroof motor switch	S
Super-loud horn	B
Suspension system, electrical	Y
Switch	S
Switch, brake fluid level	S
Switch, hydropneumatic suspension	S
Switching valve, electrical	Y
T	
Tachometer	P
Taillight	E
Tape recorder	D
Taxi alarm unit	A
Temperature gauge	P
Temperature regulator	B
Temperature sensor	B
Temperature switch	B
Tempomat control unit	A
Tempomat, momentary switch	S
Terminal	X
Terminal studs	X
Terminals, electrical	X
Thermo-time relay	K
Thermo-time switch	K
Thermostatic switch	B
Three phase alternator	G
Throttle valve switch	S
Thyristor	V
Time delay relay	K
Time lag relay	K
Tone sequence control device	S
Transducer, general	U
Transformer	T
Trigger box	A
Trigger box, ignition, transistorized	A
Trip recorder	P
Trunk light	E
Trunk light switch	S
Turn signal flasher	K
Turn signal indicator light	H
Turn signal light	H
Turn signal relay	K
Turn signal switch	S
Two-tone horn system	B
U	
Underfloor heater	E
V	
Varactor	V
Variode	V
Vehicle plug, electrical	X
Vehicle socket, electrical	X
Visible alarm device	H
Voltage regulator	N
Voltage stabilizer	N
Voltage transformer	T
Voltmeter	P
W	
Warm-up regulator	Y
Washer-wiper switch	S
Windings	L
Window switch	S
Windshield washer motor	M
Windshield washer switch	S
Wiper motor	M
Wiper switch	S
Wiper-washer switch	S
Wiper speed switch	S
Wirewound resistor	R
Wiring harness	W
Z	
Zener diode	V

Litho in U.S.A.

ANSWERS TO TEST YOURSELF QUESTIONS

ANSWERS TO CHAPTER 1 QUESTIONS

1. First blank—“electrons.” Second blank—“conductor.”

2. “Conductors.”

3. First blank—"repel." Second blank—"attract." Third blank—"ohms."

4. First blank—“amperes.” Second blank—“volts.” Third blank—“ohms.”

5. First blank—"electrons." Second blank—"behavior." Third blank—"effects."

6. "Opposition."

7. First blank—"field." Second blank—"force."

8. True.

9. First blank—"stronger." Second blank—"current."

10. A *coil of* wire wound over a soft *iron core.*

11. A—3; b—1; c—2.

ANSWERS TO CHAPTER 2 QUESTIONS

1. First blank—"voltage." Second blank—"resistor." Third blank—"conductor."

2. "Series, parallel, and series-parallel."

3. Answer to first question—"series circuit." Answer to second question—"parallel circuit."

4. "Voltage."

5. First blank—"on." Second blank—"off." Third blank—"current."

6. "Waves."

7. First blank—"cycles." Second blank—"hertz."

8. First blank—"analog." Second blank—"digital."

9. First blank—"parallel." Second blank—"series."

10. a—3; b—1; c—2.

11. "That all voltage has been cut off within the circuit."

ANSWERS TO CHAPTER 3 QUESTIONS

1. "Silver."

2. "Diodes and transistors."

3. First blank—"flow." Second blank—"stopping."

4. First blank—"one." Second blank—"direction."

5. "Control, selection, or sensing."

6. First blank—"small." Second blank—"larger."

7. "Current."

8. First blank—"resistance." Second blank—"ohms."

9. "Store."

10. "Integrator."

11. "Solenoids.'

12. First blank—"binary." Second blank—"logic."

13. "Gates."

ANSWERS TO CHAPTER 4 QUESTIONS

1. First blank—"bypass." Second blank—"start."

2. "Gasses."

3. "Safety goggles."

4. "Grounded."

5. "Danger," "warning," or "caution."

6. First blank—"current." Second blank—"path."

7. First blank—".006." Second blank—"amperes."

Litho in U.S.A.

ANSWERS TO CHAPTER 5 QUESTIONS

1. False. Specific gravity readings must be adjusted for temperature.

2. Battery electrolyte is made up of *water* and *sulfuric acid.*

3. False. They are activated in the field by the dealer or customer just prior to use.

4. Recharge the battery. Then test it again.

5. Recharge the battery. The difference between cells is less than 0.05 volts, so the battery is probably good. However, two cells are below 1.95, showing a need for recharging.

6. The two methods of charging batteries are *slow* charging and *fast* charging.

7. Slow charging.

8. First blank—"series." Second blank—"20."

9. First blank—"50%." Second blank—"cold cranking ampere."

ANSWERS TO CHAPTER 6 QUESTIONS

1. 1) Recharge the battery, and 2) generate current during operation.

2. False. Both circuits generate an alternating current (AC). The difference is in the way they rectify the AC current to DC current.

3. a—2; b—1; c—3.

4. An *armature* and *magnetic poles* (or field).

5. By moving the armature or conductor through the magnetic field,

6. First blank—"alternator." Second blank—"generator."

7. Rotor assembly, stator assembly, and rectifier assembly.

8. The rectifier assembly.

9. The diodes convert AC to DC current.

10. True. This might damage the rectifier unit.

11. The transistorized model.

12. After.

13. Diode trio.

ANSWERS TO CHAPTER 7 QUESTIONS

1. First blank—"battery." Second blank—"starting motor." Third blank—"switch."

2. Series-wound, compound-wound, parallel-wound, and series-parallel wound.

3. False. The coaxial mounting is enclosed in the motor housing.

4. The no-load test.

5. 1/2.

6. b. No. 00 sandpaper.

7. First blank—"overheating." Second blank—"30 seconds."

ANSWERS TO CHAPTER 8 QUESTIONS

1. The *spark* which ignites the engine fuel.

2. Coil, condenser, distributor, and spark plug.

3. a—2; b—1.

4. Times the sparks to occur at the correct time for the engine speed and piston cycles.

5. False. Remove and hone them or clean with a lint-free cloth and lighter fluid.

6. Fouled plugs have dirty deposits on their tips, while eroded plugs have badly burned tips. Too much heat causes eroded plugs, while lack of heat causes fouled plugs.

Litho in U.S.A.

ANSWERS TO CHAPTER 9 QUESTIONS

1. Reluctor, sensor, and electronic control unit (or module).

2. Any three of the following

   a. No problems with distributor points and condenser

   b. Quicker starts in all kinds of weather

   c. Hotter spark of longer duration will ignite marginal air-fuel mixtures under adverse weather conditions

   d. More complete combustion

   e. Increases spark plug life

3. False. They must be replaced.

4. The reluctor teeth may break or wear and result in hard starting,

5. All components (including the coil) are inside the distributor.

6. Frequency.

7. Electronic.

8. All-speed.

9. Smoke.

ANSWERS TO CHAPTER 10 QUESTIONS

1. *A circuit breaker* is a protective switch which trips when excess current passes through its circuit, and is reset manually or automatically. A *fuse* is protective element which "blows out" when current gets too high. Circuit breakers are used where heavy loads are instantly placed on the current, while fuses are normally replaced after they activate.

2. True.

3. True.

4. "Starting."

ANSWERS TO CHAPTER 11 QUESTIONS

1. Pass current from one set of wires to another.

2. The microscopic roughness or peaks and valleys on the surfaces of pin contacts which causes resistance.

3. Mating half locking mechanism, safeguard against being connected wrong, and design for individual pin contact replacement.

4. True. But vacant cavities should be plugged.

5. Circular plastic connector with pin-type contacts.

6. False. You may cut the wire strands with the crimp wings.

7. Bare wire strands are crimped first then the insulation just above.

ANSWERS TO CHAPTER 12 QUESTIONS

1. A monitor keeps track of a particular function on the machine. A controller causes a function of the machine to operate in a programmed and automatic manner.

2. True.

3. Monitor or control console, sensors, and wiring harness.

4. Magnetic.

5. False. It is an opto-electrical sensor.

6. False. Chaff and dirt could block the light beam causing an erroneous reading.

7. First blank—"low shaft speed." Second blank—"70."

 Litho in U.S.A.

ANSWERS TO CHAPTER 13 QUESTIONS

1. Positive.

2. False. NEVER polarize an alternator. On the other hand, a DC generator-equipped system MUST be polarized.

3. Faulty wiring or poor connections cause high resistance and excessive voltage drop. Batteries will not charge properly, and spark plugs may not fire.

4. Acid film and dirt on the battery will allow current to leak between terminals. Corrosion will impede current flow to the system.

ANSWERS TO CHAPTER 14 QUESTIONS

1. 1) Know the system, 2) Ask the operator, 3) Inspect the system, 4) Operate the machine, 5) List the possible causes, 6) Reach a conclusion, and 7) Test your conclusion.

2. During *none* of these steps. Do all these things *before* you start repairing the system.

3. The *battery.* It is a prime factor in all the circuits and must be in good order or the whole system will be affected.

FOS—20 Litho in U.S.A.